# 32nd International Congress of SIAR of Automotive and Transport Engineering 2022



IOP Conference Series: Materials Science and Engineering
Volume 1311

Timisoara, Romania
26 – 28 October 2022

ISBN: 979-8-3313-0378-5
ISSN: 1757-8981

**Printed from e-media with permission by:**

Curran Associates, Inc.
57 Morehouse Lane
Red Hook, NY 12571

**Some format issues inherent in the e-media version may also appear in this print version.**



Institute of Physics
Dirac House, Temple Back
Bristol BS1 6BE UK

Phone: 44 1 17 929 7481
Fax: 44 1 17 920 0979

techtracking@iop.org

**Additional copies of this publication are available from:**

Curran Associates, Inc.
57 Morehouse Lane
Red Hook, NY 12571 USA
Phone: 845-758-0400
Fax: 845-758-2633
Email: curran@proceedings.com
Web: www.proceedings.com

# TABLE OF CONTENTS

IOP Conf. Series: Materials Science and Engineering **1311** (2024) 011001 doi:10.1088/1757-899X/1311/1/011001

# Foreword

The Congress of Automotive and Transport Engineering is an annual scientific event initiated in 1990 by the Society of Automotive Engineers of Romania (SIAR) and various Romanian universities with an interest in Automotive and Transport Engineering. This Congress is endorsed by EAEC (European Automotive Engineers Cooperation).
The XXXII-nd edition of SIARS's International Congress with the theme "Automotive and Transport Engineering" and the XVII-th European Automotive Congress was organized for the third time by Department of Mechanical Machines, Equipment and Transportation from the Politehnica University Timisoara.

The Congress purpose is to gather members from academia, industry and government affiliated presenting their possibilities for investigations and latest research, to achieve collaborations in the near future in the field of automotive and transport engineering. The Congress hosted 201 participants from Europe, United States of America, Moldova presenting their scientific achievements within plenary sessions and congress sections.

Thus, the auditorium had the opportunity to:

• Gain insight into the cutting edge automotive and transport industry and ideas for future developments;

• Update their skills and knowledge by attending focused technical sessions;

• Network with potential new partners, clients and suppliers;

• View the latest technology products and services in the technical exhibition.

The papers included in this volume were selected by the Scientific Committee amongst the ones presented within the Congress and are covering the latest issues such as advanced powertrain and propulsion, automotive design and testing, terrain vehicles, advanced engineering methods, materials, automotive technology and maintenance, road safety, mobility of things.

**EAEC-MVT 2022 Student Congress** was integrated into the main congress and offered students from around the world the opportunity to participate at an international conference and to present their papers on the main congress topics. This was also the meeting place for the young students (SIAR junior's members) in an International Congress.

We hope that these papers will be of interest to experts from academia and industry around the world and we wish them to create a favourable framework for addressing key issues in the field of Automotive and Road Transport Engineering that concern contemporary society today.

Yours faithfully,
EAEC-MVT 2022 Proceedings Editors

**Acknowledgement**

HONORARY COMMITTEE

Ioan BOLDEA – Romanian Academy

Mihai BOLDIJAR – Bosch România

Nicolae BURNETE – SIAR President

IOP Conf. Series: Materials Science and Engineering **1311** (2024) 011001 doi:10.1088/1757-899X/1311/1/011001

IOP Conf. Series: Materials Science and Engineering **1311** (2024) 011001 doi:10.1088/1757-899X/1311/1/011001

## ORGANIZING COMMITTEE

**Congress Chairmen**

Professor Liviu MIHON - Politehnica University Timisoara

**President of Society of Automotive Engineers of Romania**

Professor Nicolae BURNETE- Technical University of Cluj-Napoca, Romania

**SIAR General Secretary**

Professor Minu MITREA

**Website Coordinator**

Dinu COVACIU - Transilvania University of Braşov, Romania

**Exhibition**

Adrian CIOABLA - Politehnica University Timisoara

Luisa DUNGAN- Politehnica University Timisoara

**Student Contests**

Dorin LELEA

Andrei BORBOREAN

Liviu MIHON

Ramon BALOGH
Corneliu BIRTOK-BĂNEASĂ
Andrei BORBOREAN
Adina BUDIUL
Adrian CIOABLĂ
Adrian CIPLEU
Amalia DASCĂL
Luisa Izabel DUNGAN
Eugen GHITA
Attila GÖNCZI
Sorin HOLOTESCU
Ioana IONEL
Imre KISS
Dorin LELEA
Nicolae LONTIȘ
Nicolae Liviu MIHON
Arina Speranța NEGOIȚESCU
Daniel OSTOIA
CameliaPINCA-BRETOTEAN
Francisc POPESCU
Sorin RAȚIU
Dan STEPAN
Floriana Dana STOIAN
Virgil STOICA
Patrick ȘTEFĂNESCU
Gabriel TRIF-TORDAI
Gabriel URSU-NEAMȚ
Ionel VANDICI
Lucia VÎLCEANU
Ion VETREȘ

IOP Conf. Series: Materials Science and Engineering **1311** (2024) 011001 doi:10.1088/1757-899X/1311/1/011001

**SPONSORS**

The organizing committee would like to thank the following companies for their support:

*Robert Bosch România*

*Beespeed Technical Engineering Center / ZF*

*AVL List GmbH Austria*

*Renault Technologie Roumanie*

*Asociația Specialiștilor Auto din România*

*TechnoVolt*

*Groupama*

*Analytik Jena*

*RMB Timișoara*

*Verbiță Arad*

*Paul Trans Timișoara*

**Peer Review Declaration**

Peer review declaration

The Organiser and/or the Editor(s) are required to declare details about their peer review processes. Therefore, please provide the following information:

- Type of peer review: Single-blind (the identity of reviewers is not known by the authors)

Criteria used by reviewers:

The papers underwent assessment based on their scientific substance, the novelty of the information they presented, language accuracy, and the quality of their presentation. Reviewers were provided with a structured review form encompassing sections on novelty, contextual relevance, manuscript coherence (including communication effectiveness and result clarity), abstract adequacy, language proficiency, and absence of commercial bias. Additionally, reviewers were required to provide constructive comments and suggestions for authors in designated sections. They were further tasked with recommending papers for publication, revision, or rejection. Subsequently, reviewers' feedback, facilitated by the track director, was relayed to the authors, who were then responsible for revising their papers in accordance with the reviewers' input and resubmitting them. Final acceptance (or rejection) decisions were made by the conference track director only after authors had made the necessary revisions.

- Conference submission management system: Open Conference System (https://siarcongress.eu/index.php/mvt/2022 )
- Number of submissions received: 107
- Number of submissions sent for review: 67
- Number of submissions accepted:53
- Acceptance Rate (Number of Submissions Accepted / Number of Submissions Received X 100): 49.53 %
- Average number of reviews per paper: 2
- Total number of reviewers involved: 20
- As additional info on review process the papers were checked for plagiarism with www.turnitinn.com platform (the papers ware uploaded in the platform, verified and presented a report with
- Contact person for queries: Liviu MIHON, Politehnica University of Timisoara, Romania, email: liviu.mihon@upt.ro

IOP Conf. Series: Materials Science and Engineering **1311** (2024) 012001 doi:10.1088/1757-899X/1311/1/012001

# Study of the diesel fuel drop vaporization and combustion at a hydrogen-diesel fuel dual fuelled diesel engine

**A Cernat[1*], C Pana[1], N Negurescu[1] and C Nutu[2]**

[1]Department of Thermotechnics, Engines, Thermal and Frigorific Equipment, University Politehnica of Bucharest, Romania
[2]Department of Automotive, University Politehnica of Bucharest, Romania

*Email: cernatalex@yahoo.com

**Abstract**. The present paper shows the analysis of the results of the theoretical model developed by authors for diesel fuel drops vaporization study, air-diesel fuel-hydrogen mixture forming estimation and diesel fuel drops combustion in the presence of air-hydrogen mixtures set in the cylinder before the start of conventional fuel injection. The model is applied to a single cylinder diesel engine fuelled with diesel fuel and hydrogen, in dual mode by diesel gas method, at the engine speed of 900 rev/min. For a diesel fuel drop the vaporization speed, mixture formation speed, combustion speed, flame radius are presented and analysed. At hydrogen use, the cyclic amount of diesel fuel is reduced and the vaporization duration of the diesel fuel drop is reduced comparative to classic fuelling. At dual fuelling, the combustion of the burning of air-hydrogen mixtures in the vicinity of diesel droplets leads to reduced vaporization time, with 43%, an increased vaporization rate, with 26.9% and an accelerated formation of the mixture between diesel fuel and air-hydrogen mixture. Further, the combustion speed increase with 6.9%, the combustion time is decreased with 6% and the flame radius increase with 6.7%.

## 1. Introduction

The use of alternative fuels can be a viable solution to decrease the levels of the pollutant emissions and of the greenhouse gases from the diesel engines exhaust gases. Compared to other energy solutions of the powertrain proposed for future use, the dual-fuel diesel engine with two fuels is one of the best solutions in terms of efficiency, autonomy, time allocated to restore the amount of energy required for operation. Hydrogen can be a viable alternative fuel for future generation diesel engines from years of 2030-2050 [1], [2] due to its good combustion properties and good perspectives in reduction of the pollutant emissions level and of the carbon dioxide emission level from the exhaust gases. The use of hydrogen as alternative fuel for diesel engines requires special fuelling solutions, applied to assure the normal starting and operation of the diesel engine in dual fuelling mode.

Santoso [3] use different percent of hydrogen to fuel a single cylinder diesel engine. Santoso also use a software to evaluate combustion parameters, considering an homogeneous mixture of air-hydrogen is established into the cylinder at the begin of the calculus. Santoso observes that low hydrogen percents use leads to the decrease of maximum pressure during combustion, which also starts later, being necessary the increase of hydrogen amount in order to reach the same engine power in dual fuelling mode [3], thus, the combustion is accelerated especially for significant amounts of diesel fuel [3]. Ghazal [4] concluded that hydrogen has a higher combustion speed and its use leads to the acceleration of the combustion, the maximum pressure being increased for low speed regimes [4]. The CO and soot emissions were decreased at the increase of hydrogen amount [4] and water is used to decrease the $NO_x$ emission level. Karagoz [5] uses hydrogen in different additions percent's to fuel a diesel engine and observes that CO and smoke emissions starts to decrease their levels once the hydrogen share is raised. Hydrogen use leads to the rise of the maximum values reordered for in-cylinder pressure and for the heat release rate [5], fact that can be correlated with the increase of combustion intensity of air-hydrogen mixture. Ekin [6] uses hydrogen to fuel a diesel-natural gas engine and find that addition of hydrogen leads to the increase of the domain of in flammability limits of air-fuel mixture, the thermal efficiency being increased for small amounts of hydrogen. The CO and soot emission level were decreased by hydrogen use, but the $NO_x$ emission level was higher comparative to classic fuelling because of in-cylinder temperature increase [6]. Das [7] uses different hydrogen flows in the engine inlet to fuel a diesel engine fuelled with biodiesel. Das observes that hydrogen addition leads to the increase of engine thermal efficiency and decrease of energy consumption due to higher hydrogen caloric power and CO, CO2 and HC emission level were decreased [7]. At hydrogen addition the combustion has a shortened duration in correlation with the increase of heat release rate, gases pressure, ignition delay comparative to classic fuelling [7]. Zareei [8] shows that addition of hydrogen, to natural gas, leads to the increase of temperature during the combustion process, in correlation with a raised value of $NO_x$ emission, since the CO and HC emission level were decreased. Zhunqing [9] shows that the increase of hydrogen addition leads to the increase of the maximum pressure during combustion, and the engine performance is correlated with the combustion speed and the caloric value of the in-cylinder mixture. In generally, the presence of hydrogen leads to the acceleration of the combustion process. If hydrogen amount increase the level of the CO emission is increased, but the levels of the HC and NOx were decreased [9]. Kavtaradze [10] shows that flame propagation in air-hydrogen mixtures is extremely times higher comparative to air-diesel fuel mixture, for stoichiometric state (2,3 m/s for $H_2$ versus 0,4 m/s for diesel fuel). Usually, this leads to a shorter combustion duration and the increase of heat release and in-cylinder pressure [10]. Koten [11] analyses the effect of the hydrogen addition on a diesel engine performance and in general the increase of hydrogen cyclic quantity leads to the increase of engine performance. Koten explains these results as effect of hydrogen addition on better mixture formation and higher combustion speed [11]. Koten [11] also observes increase of NOx emission level, decrease of CO and soot emissions levels.

The present paper shows the analysis of the results of the theoretical model developed by authors for diesel fuel drops vaporization study, air-diesel fuel-hydrogen mixture forming estimation and diesel fuel drops combustion in the presence of air-hydrogen mixtures set in the cylinder before the start of conventional fuel injection. The model is applied to a single cylinder diesel engine fuelled with diesel fuel and hydrogen, in dual mode by diesel gas method, at the engine speed of 900 rev/min. For a diesel fuel drop the vaporization speed, mixture formation speed, combustion speed, flame radius are presented and analysed. An analyse of the combustion process for dual fuelling is helpful for understanding the in-cylinder process of the dual fuelled diesel engines and such a study can be completed by aspects regarding vaporization, mixture formation and combustion of diesel fuel drops in presence of hydrogen, in the case of 20% hydrogen energetic amount.

## 2. Research method

The author's concept a thermodynamic bi-zone model to study vaporization, mixture formation and combustion of diesel fuel droplets in air-hydrogen mixture, model which represents a novelty aspect of the research, a fuel drop level detailed combustion analysis being possible.

The thermodynamic model is based on the hypotheses specific to modern concept of spherical model for combustion of the fuel drop [12], according to figure 1.

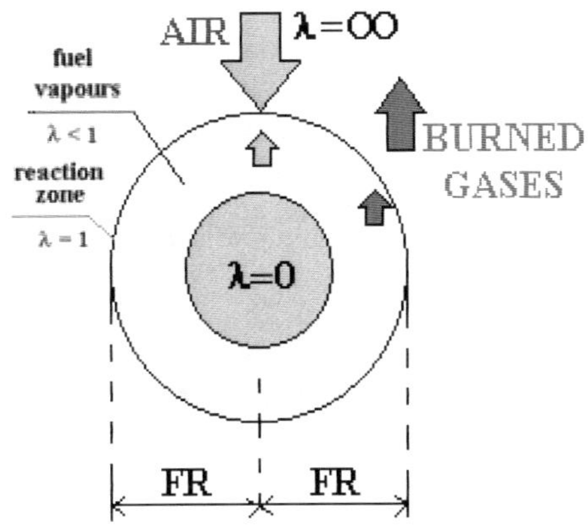


**Figure 1.** Diesel fuel drop spherical model, [12].

Vaporization time of a single fuel drop is evaluated based on the formula:

$$VT = \frac{\rho_f \cdot d_d^2}{8 \cdot \alpha_g \cdot \rho_g \cdot ln(1+B)} \quad (1)$$

were $\rho_f$ is the fuel density, $d_d$ is diameter of the drop, $\alpha_g$ is the thermal diffusivity, $\rho_g$ is density of the gaseous medium, B is the mass transfer number, [13], [14], [15], [16].
The vaporization speed established on the drop surface can be evaluated with the formula:

$$VS = \frac{\alpha_g \cdot \rho_g \cdot ln(1+B)}{0,5 \cdot d_d} \quad (2)$$

The speed of mixture formation influence the combustion speed and can be estimated with the relation which take into consideration the mass quantity of the injected fuel and the partial pressure of the oxygen in the area of mixture forming. At the liquid surface of the drop, were the boiling temperature is achieved, the flow of the vaporized fuel reflects the combustion speed, estimated with a relation similar with the previous one, in which the mass transfer number, B, is take in into consideration the fact that the vaporization process is followed by combustion. Thus, the combustion time can be evaluated for a fuel drop with the relation, [13], [14], [15], [16]:

$$CT = \frac{\rho_f \cdot d_d^2}{8 \cdot \alpha_g \cdot \rho_g \cdot ln(1+B)} \quad (3)$$

The combustion speed of diesel fuel drop is evaluated with a formula similar to the one of vaporization speed determination, taking into account the B mass transfer number evaluated in case of vaporization process is followed by combustion. The gaseous area established around the fuel drop is surrounded by an infinitesimal thickness flame and its position is established with the flame radius:

$$FR = \frac{VS \cdot 0,5 \cdot d_d^2}{\alpha_g \cdot \rho_g \cdot ln(1+c \cdot y_o)} \quad (4)$$

with *c* as stoichiometric mass ratio of fuel-oxygen, $y_o$ oxygen mass fraction in surrounding gas in the infinite medium, [13], [14], [15], [16].

On smooke emission level estimation the soot formation and oxidation speeds are estimated at diesel fuel use and for hydrogen addition. For soot formation speed the following relation is used:

$$SFS = C \cdot m_{gf} \cdot p_{O_2}^{0,5} \cdot exp\left(-\frac{6290}{T}\right) \quad (5)$$

with constant C=0.0025....0.005, $m_{gf}$ mass of gaseous fuel, $p_{O2}$ partial pressure of oxygen and T global temperature, [13], [14], [15], [16]. The soot oxidation speed is estimated by the relation:

$$SOS = 0.000955 \cdot m_s \cdot p_{O_2}^{0,5} \cdot exp\left(-\frac{5000}{T}\right) \quad (6)$$

were $m_s$ is mass of formed soot. The smoke emission evaluate based on soot concentration is given by the difference between the formation and oxidation rates, [13], [14], [15], [16].

## 3. Results

The figure 2 presents the estimated variation of the vaporization time for diesel fuel drop for classic fuelling (xc=0%) and for dual fuelling (xc=20%) In the nearby of the air-hydrogen-diesel fuel vapour mixture combustion the vaporization time of a diesel fuel drop starts to decrease with 43% comparative to classic fuelling.

This aspect is in correlation with the increase of the vaporization speed, figure 3, which increase with 26% at dual fuelling comparative to classic fuelling.

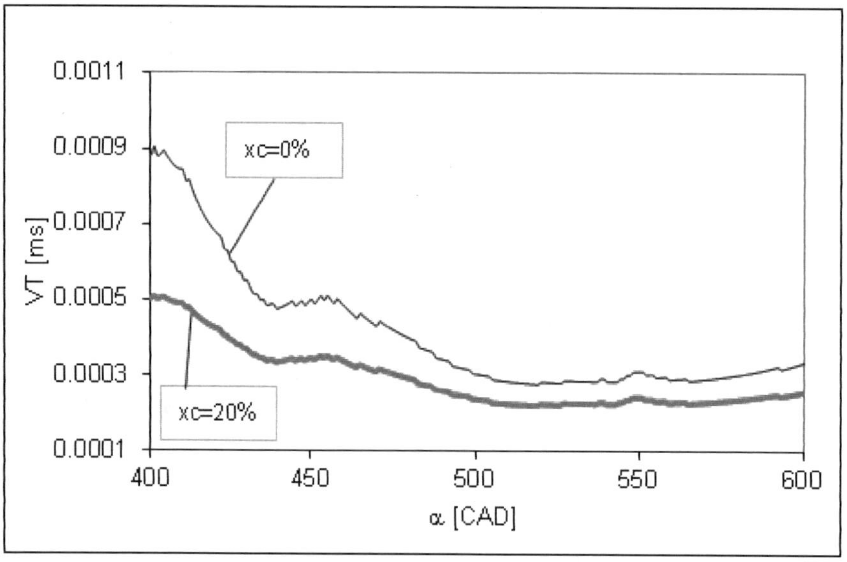


**Figure 2.** Vaporization time for diesel fuel drop.

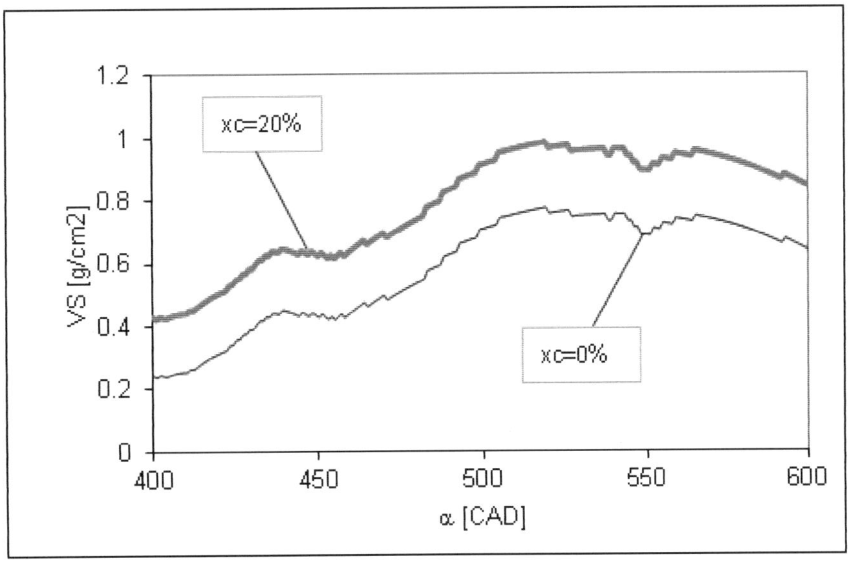


**Figure 3.** Vaporization speed for diesel fuel drop.

The mixture formation is accelerated by the presence of hydrogen. From one hand, the hydrogen addition leads to the reduction of the diesel fuel cyclic quantity, which means less diesel fuel need to vaporize and mixture with air, and on the other hand the accelerated vaporization of diesel fuel drops leads to the increase of mixture formation speed, estimated in the figure 4.

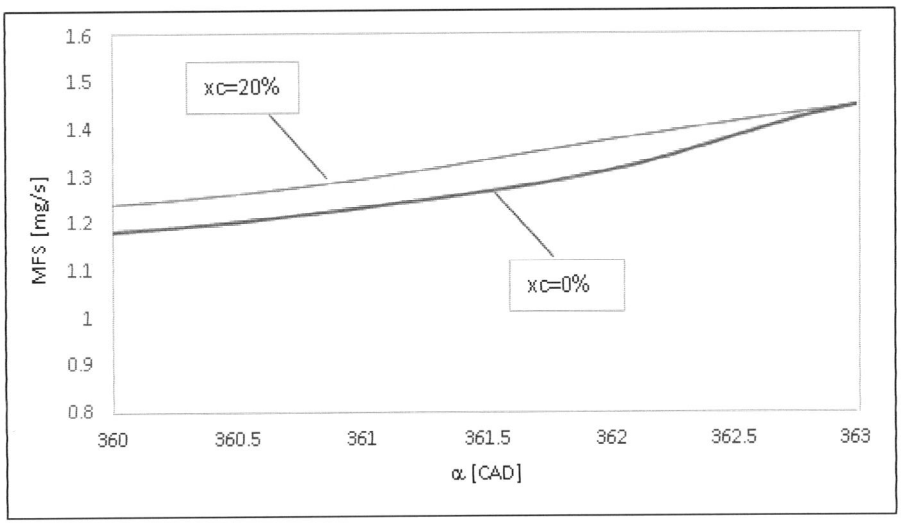


**Figure 4.** Mixture formation speed.

The combustion time of diesel fuel drop is reduced by the hydrogen in-cylinder presence in the way that air-hydrogen mixture combustion accelerates also the combustion of the diesel drops. At 20% hydrogen use the combustion time of diesel fuel drop decrease till 6% comparative to classic fuelling, figure 5.

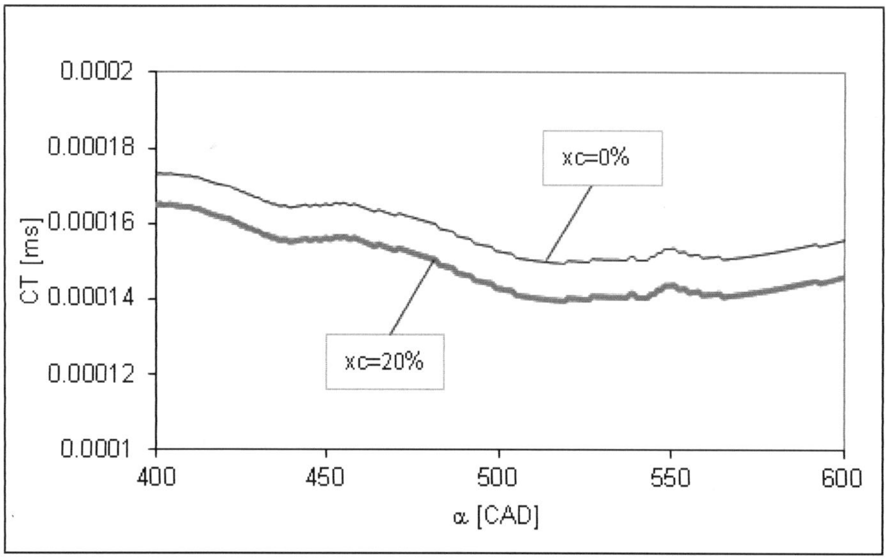


**Figure 5.** Combustion time of diesel fuel drops.

This aspect is in correlation with the increase of the combustion speed of the diesel drop up till 6.9% versus classic fuelling, figure 6. The increase of the diesel fuel drop combustion speed is related with the decrease of combustion time decrease, mixture formation speed increase and acceleration of fuel vaporization at dual fuelling.

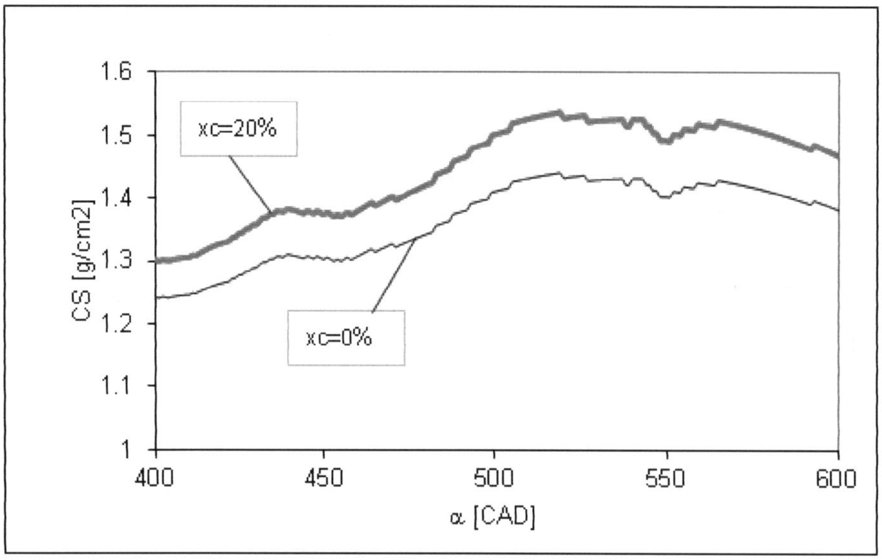


**Figure 6.** Combustion speed of diesel fuel drops.

At hydrogen use, the flame radius developed around diesel fuel drops during combustion process, figure 7, is with 6.7% larger comparative to classic felling, due to hydrogen higher combustion speed and caloric value. Theoretically, the lower heating value of the final mixture at dual fuelling is with 37% higher comparative to diesel fuel use.

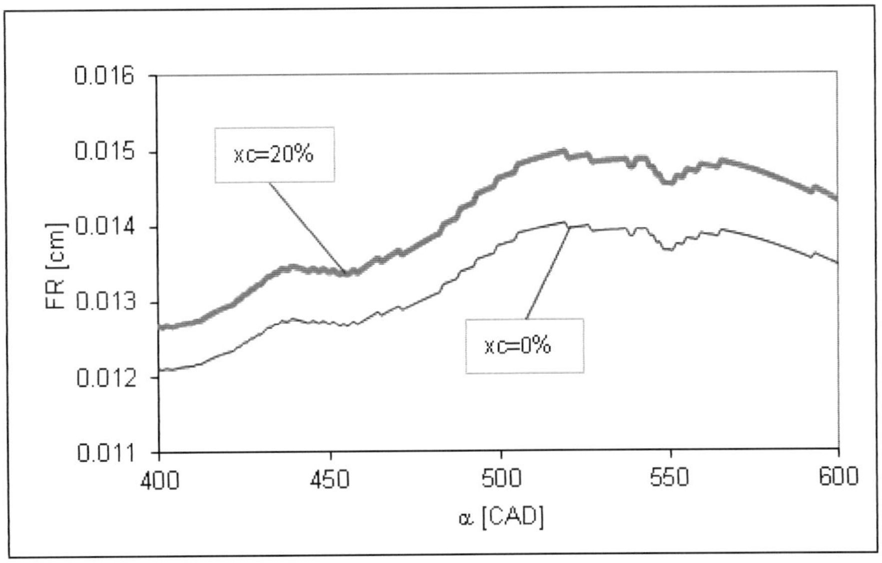


**Figure 7.** Flame radius of diesel fuel drop.

The maximum of CO emission formation speed decrease with 19% at hydrogen use comparative to diesel fuelling due to reduction of the carbon content at the substitution of the diesel fuel by hydrogen, figure 8.

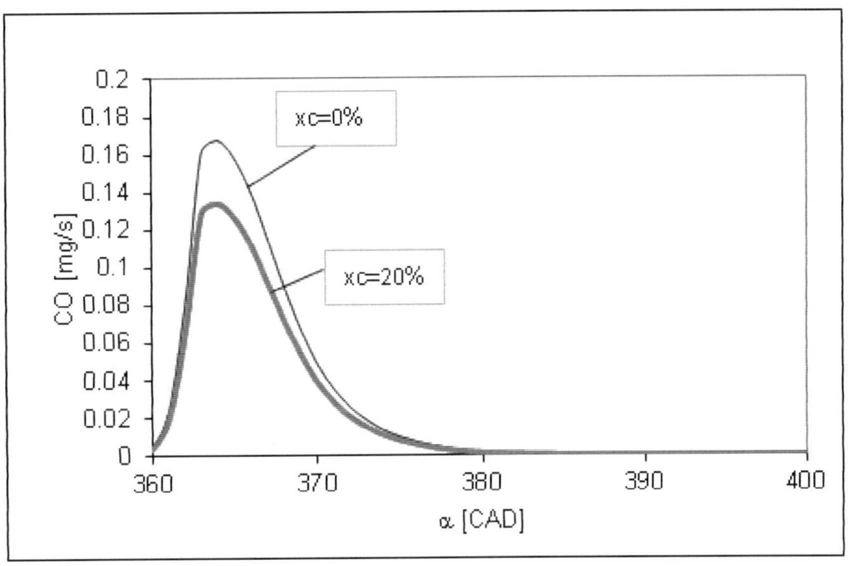


**Figure 8.** Speed of CO formation for xc=0 and 20.

Also the speed of smoke formation is with 8.4% lower at hydrogen use comparative to diesel fuel, figure 9, due to reduction of the carbon content in the final mixture.

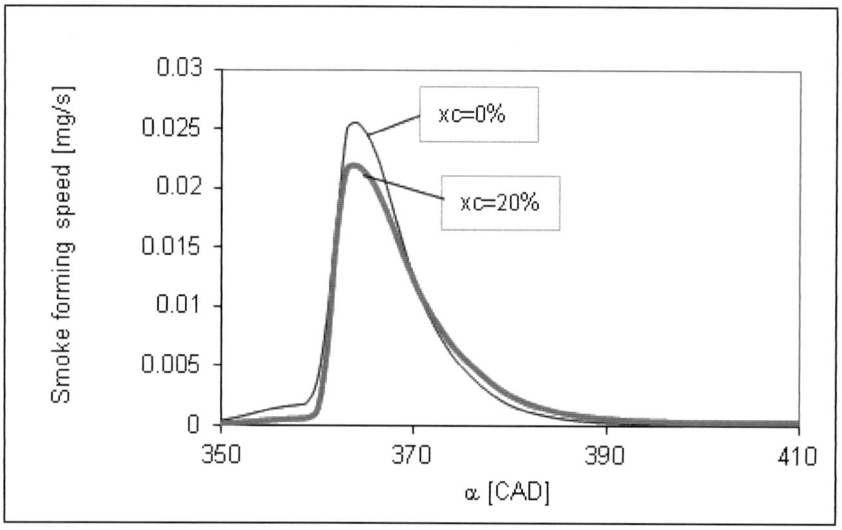


**Figure 9.** Smoke forming speed for xc=0 and 20.

Versus classic fuelling, when the hydrogen is used the smoke oxidation speed increases with 83%, figure 10. For dual fuelling the addition of hydrogen leads to the acceleration of the soot oxidation in conditions of a reduction of the smoke formation, figures 9 and 10.

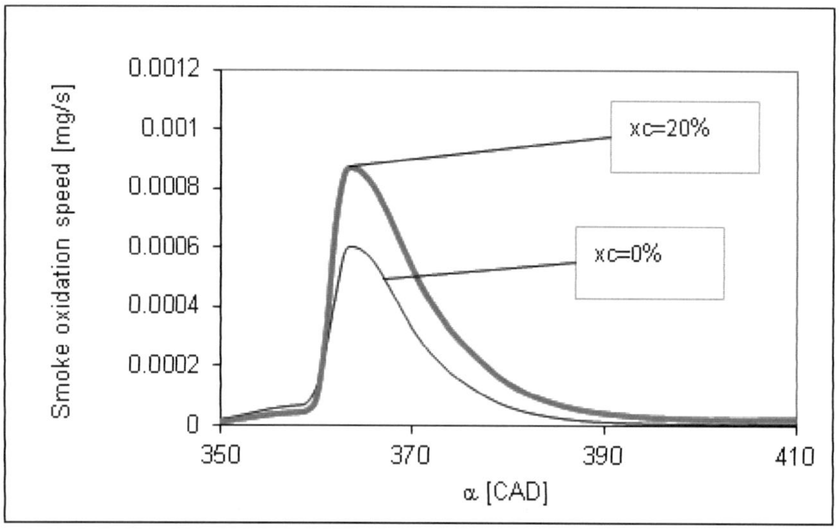


**Figure 10.** Smoke oxidation speed for xc=0 and 20.

This aspects leads to a lower soot quantity in the exhaust gases at hydrogen use, the estimated smoke emission level for classic and dual fuelling being presented in the figure 11.

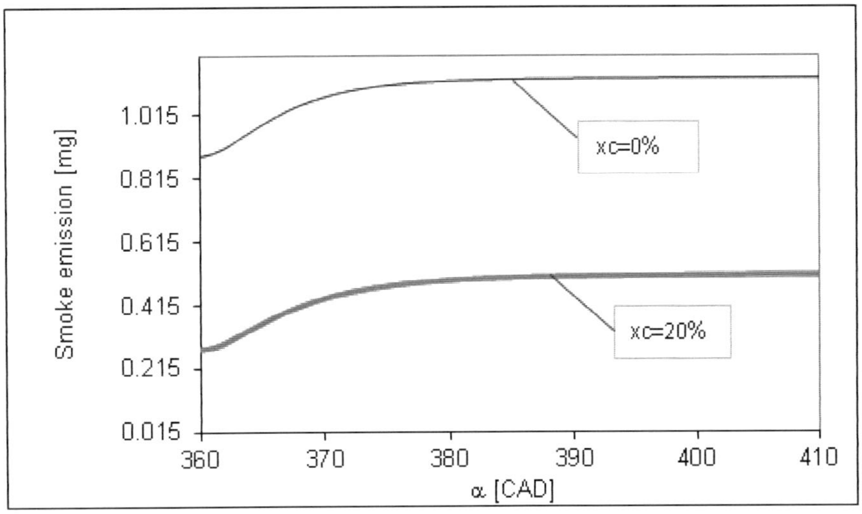


**Figure 11.** Estimated smoke emission level for xc=0 and 20.

For dual fuelling, the use of hydrogen leads to the decrease of smoke emission level with almost 21% versus diesel fuel use. Also, the CO2 emission level is decreased at hydrogen use due to reduction of the in-cylinder carbon content when hydrogen substitutes the diesel fuel, a reduction of 15% being estimated. As estimation the NO emission level may increase with 3.4% at hydrogen use because of increase in maximum of combustion conventional temperature with 31%.

## 4. Conclusions

At hydrogen use, the cyclic amount of diesel fuel is reduced and the vaporization duration of the diesel fuel drop is reduced comparative to classic fuelling. At dual fuelling, the combustion of the burning of air-hydrogen mixtures in the vicinity of diesel droplets leads to reduced vaporization time, with 43%, an increased vaporization rate, with 26.9% and an accelerated formation of the mixture between diesel fuel and air-hydrogen mixture. Further, the combustion speed increase with 6.9%, the combustion time is decreased with 6% and the flame radius increase with 6.7%. The maximum of heat release rate is with 32% higher at dual fuelling comparative to classic fuelling. The levels of the pollutant emissions and of the $CO_2$ are reduced at hydrogen use. The estimated level of the pollutant emissions decrease with 19.7% for CO, with 32.3% for smoke at hydrogen use comparative to classic fuelling. At hydrogen use the speed of soot forming from the diesel fuel drops starts to decrease with 8%, and also the soot oxidation being highly accelerated at combustion of air-hydrogen mixtures. As estimation, a slight increase of 3% in nitrogen oxides may occur at dual fuelling.

**Acknowledgments**

This work was supported by a grant of the Ministry of Research, Innovation and Digitization, CCCDI-UEFISCDI, project number PN-III-P2-2.1-PED-2021-0427, within PNCDI III. The authors address special thanks to AVL GmbH Graz Austria for providing the necessary equipment's.

IOP Conf. Series: Materials Science and Engineering **1311** (2024) 012001 doi:10.1088/1757-899X/1311/1/012001

**References**

[1] https://www.fuelseurope.eu/clean-fuels-for-all/vision-2050/ (accessed on February 11th, 2022)

[2] https://www.sustainablefuels.eu/vision050/#:~:text=By%202050%2C%20climate%20neutral%20fuels,most%20affected%20by%20the%20changes (accessed on February 11th, 2022)

[3] Santoso W B, Bakar R A, Nur A 2013 Combustion Characteristics of Diesel-Hydrogen Dual Fuel Engine at Low Load *Energy Procedia* **32**

[4] Ghazal O H 2019 Combustion analysis of hydrogen-diesel dual fuel enginewith water injection technique *Case Studies in Thermal Engineering* **13**

[5] Karagoz Y, Guler I, Sandalici T, Yuksek L, Dalkilic A S 2016 Effect of the hydrogen enrichment on combustion characteristics, emissions and performanceof a diesel engine *International Journal of Hydrogen Energy* **41** (1) pp 656-665

[6] Ekin F, Ozsoysal O A, Arslan H 2022 The effect of using hydrogen at partial load in a diesel-natural gas dual fuel engine *International Journal of Hydrogen Energy* **47** (42) pp18532-18550

[7] Das S, Kanth S, Das B, Debbarma S 2022 Experimental evaluation of hydrogen enrichment in a dual-fueled CRDI diesel engine *International Journal of Hydrogen Energy* **47** (20) pp11039-11051

[8] Zareei J, Rohani A, Alvarez J R N 2022 The effect of EGR and hydrogen addition to natural gas on performance and exhaust emission in a diesel engine by AVL fire multi-domain simulation, GPR model, and multi-obiective generic algoritm *International Journal of Hydrogen Energy* **47** (50) pp21565-21581

[9] Zhunqing H, Nannan S, Yi J, Xin Z 2022 Experimental research on a blended hydrogen-fuel engine *International Journal of Hydrogen Energy* Avaible online 30 May 2022

[10] Kavtaradze R Z , Natriashvili T, Gladyshev 2019 Hydrogen-Diesel Engine: Problems and Prospects of Improving the Working Process *Conference:WCX SAE World Congress Experience PAPER 2019-01-0541*

[11] Koten H 2018 Hydrogen effects on the diesel engine performance and emissions *International Journal of Hydrogen Energy* **43** (22)

[12] Popa M G, Pana C and Negurescu N 2003 *Diesel Engines. Processes*. vol 1-2 (Bucharest: Matrixrom) p 706

[13] Apostolescu N, Chiriac R 1998 *Combustion process in Internal Combustion Engines, Fuel Economy and Emission Control* (Bucharest: Technica) p 677

[14] Pimsner V, Vasilescu C, Radulescu G 1964 *Energetics of Internal Combustion Turbo-Engines* (Bucharest: RPR Academy) p 600

[15] Merker G, Schwarz C, Teichmann R 2011*Internal Combustion Engines Fundamentals: Running, Simulation, Measuring* (Berlin: Springer) p 500

[16] Haiyan M 2001 *Fundamental Studies of Combustion Phenomena in Dual Fuel Engines* (New South Wales:Thesis) p 256

# A study on the influence of utilizing hydrogen at fuelling spark ignition engines

**R Georgescu[1], C Pană[1], N Negurescu[1], A Cernat[1], C Nuțu[2] and C Sandu[1]**

[1]Thermotechnics, Engines, Thermical and Frigorific Equipment Department, University Politehnica of Bucharest, Bucharest, Romania

[2]Automotive Engineering Department, University Politehnica of Bucharest, Bucharest, Romania

E-mail: raresgeorgescu@yahoo.com

**Abstract.** In the last 30 years, remarkable progress has been made in reducing pollutant emissions and fuel consumption in conventional internal combustion engines. Active and passive methods of reducing exhaust emissions have caused them to decrease significantly over the level of the 1980s. However, the problems of air pollution and the greenhouse effect are far from being solved. International organizations and institutions are pushing for measures to reduce emissions, especially the implementation of "zero-emission" (ZEV) vehicles. Due to its excellent combustion properties and the very low pollutant emissions resulting from its combustion, hydrogen is seen as an alternative fuel of the future. Hydrogen can be used as a fuel in two ways: in internal combustion piston engines or in fuel cells. Research into the use of hydrogen as a fuel for spark ignition engines is divided into two directions: additive fuel or a total gasoline substitution with hydrogen. Compared to the gasoline-powered engine, the fully hydrogen-powered engine runs more economically at low loads but it can create a decrease of performance at high loads due to the high volumes of space needed in the combustion chamber. Because of the very low density, high volumes of hydrogen are difficult to storage, requiring also high pressure, so running only on hydrogen wouldn't be practical. Combining the advantages of the two fuels came the use of gasoline with addition of hydrogen. An important aspect is that emissions are significantly reduced due to the very good combustion properties of hydrogen and the high diffusivity that helps to form a homogeneous mixture. This paper presents aspects of fuelling a spark ignition engine with gasoline and hydrogen. The influence of hydrogen on the cylinder mixture, on the energy and emissions of the engine is analysed. We will focus on: engine power, fuel consumption, the level of pollutant emissions and greenhouse gases, at the studied regimes.

## 1. Introduction

In the past years, environmental pollution has grown seriously and as way to fight climate changes, in the 2015 Paris agreement, 196 countries signed an agreement on the global climate defining a plan to keep the global warming bellow 2 ºC compared to the pre-industrial levels and preferably limit the increase to 1.5 ºC [1]. To stay under the limit, emissions need to be cut by 2030 by around 50% compared with 1990 levels. Concerning Europe, the carbon neutrality goal is set to be reached by 2050 [2]. The main cause of these environmental concerns is the widespread use of fossil fuels to produce energy. Studies indicate that during this century, the fossil fuels will be consumed and the humankind must find alternative solutions to produce energy.

A statistic from 2018 shows that the energy demand of the world was 14,282 million tonnes of oil equivalent (Mtoe) and 81% was produced from fossil fuels [3]. The transport sector consumed 2,890 Mtoe, which is equivalent to 20% of the global energy demand. Furthermore, there is a growing demand for fuel for transport because of the constant increase of vehicles. Thereby, the transport sector constitute a key point and also a challenge of the energy transition because of the ongoing increase of emissions and greenhouse gases. In 2018, the energy demand for passenger cars and light-duty vehicles represent 46% of the energy needed for transport, specifically 1,323 Mtoe [4].

Hydrogen, ammonia and alcoholic fuels are among the most promising renewable fuels. Hydrogen could be one of the most important energy source for the transport sector. Hydrogen can be produced by various processes including: electrochemical processes, thermochemical processes, photochemical processes, photo-electrochemical processes, photocatalytic processes. Nowadays, hydrogen is produced from fossil fuels, water or biomass. Regrettably, fossil fuels are the main resource for large scale production which aggravates the carbon cycle. Because of environmental criteria, its production should be from renewable sources (green hydrogen). Tashie-Lewis B et al [5] concluded that the major limitations of widespread use of hydrogen are concerns about fuel safety and lifecycle costs (investment cost, production cost, supply cost, market price, demand, storage and distribution costs). Li H et al [6] focused on problems regarding hydrogen storage and transportation and presented a detailed review of the research progress on hydrogen leak diffusion characteristics, leak spontaneous combustion mechanisms and material hydrogen damage mechanisms.

Aryal K et al [7] and Wang L et al [8] conducted a study which pointed out the properties that make hydrogen a great alternative fuel for internal combustion engines. Due to its excellent combustion properties (table 1), hydrogen can be used blended with other fuels and also as a single fuel. The large range of flammability and the high speed of flame propagation enables hydrogen to run well on very lean mixtures. Low density which leads to exceptional diffusivity helps hydrogen mix with other fuels and high self-ignition temperature allows a higher compression ratio which leads to better efficiency. Nevertheless, hydrogen`s very low density results in volumetric sizing problems requiring special tanks. Due to its simple chemical formula, the combustion of hydrogen produces only water and NOx in certain conditions.

Shivaprasad K V [9] observed an increase of thermal efficiency when fuelling with a mix of gasoline and hydrogen till a fraction of 20% hydrogen. HC and CO emissions drop with the increase of hydrogen but high $NO_x$ values are observed when operating at rich mixtures.

Studies conducted by [10], [11], [12] and [13] showed that compared with the original gasoline engine, direct injection of hydrogen improved performance on combustion characteristics, but also had a disadvantage of higher levels of $NO_x$ emissions. Direct-injection of hydrogen provides a great advantage in eliminating the risk of intake misfire (backfire) and also doesn`t affect the inlet air quantity. Using moderate EGR rate of 15-20% and under 15-20% addition of hydrogen can produce a significant improvement of combustion and emission performance. A remarkable reduction of $NO_x$ levels was achieved as well by using exhaust gas recirculation (EGR) by [14] or water injection by [15]. Most of the modern diesel engines use also selective catalytic reduction (SCR) to reduce even more the $NO_x$.

**Table 1**. Properties of hydrogen, gasoline and diesel [7]

| Properties | Hydrogen | Gasoline | Diesel |
|---|---|---|---|
| Formula | H2 | $C_nH1.87n$ ($C_4$–$C_{12}$) | $C_nH1.8n$ ($C_8$–$C_{20}$) |
| Density at 16°C and 1.01 bar (kg/m$^3$) | 0.0838 | 721-785 | 833-881 |
| Auto Ignition Temperature (K) | 858 | 533-573 | 530 |
| Flammability limits (vol. % in air) | 4-75 | 1.4-7.6 | 0.7-5 |
| Min. ignition energy (mJ) | 0.02 | 0.24 | - |
| Stoichiometric air fuel ratio on mass | 34.3 | 14.6 | 14.5 |
| Net heating valve (MJ/kg) | 119.93 | 43.9 | 42.5 |
| Limits of flammability (equivalence ratio) | 0.1-7.1 | 0.7-3.8 | - |
| Flame velocity (cm/s) | 265-325 | 37-43 | 30 |
| Diffusivity in air (cm2 /s) | 0.63 | 0.08 | - |
| Quenching gap in NTP air (cm) | 0.064 | 0.2 | - |
| Cetane number | - | 13-17 | 44-55 |
| Octane number | 130 | 92-98 | 30 |

Thawko A and Tartakovsky L [16] study focus on particle formation in a direct injection internal combustion engine ICE indicating that the particles formed during hydrogen combustion are entirely generated from the lubricant coverings of the internal cylinder surfaces. The results reveal that non-premixed hydrogen combustion produces significantly more particles compared to gasoline combustion. This is the result of hydrogen`s much lower flame quenching distance that amplifies the lubricant evaporation and the interaction between the core jet and the lubricant vapour. Also, jet-wall interaction duration increase results in more lubricant vapour into the chamber bulk which leads to more particles.

EL-Kassaby et al [17] carried out an experiment to investigate the effects of hydroxyl gas HHO addition on performance and pollutant emissions of a gasoline fuel 1.3L GLXi Skoda Felicia engine. The results showed that even though the HHO cell provided a maximum productivity of 18L/h, there was a 10% improvement of thermal efficiency, 34% reduction in fuel consumption, a 15% reduction of HC and a 14% improvement of CO. The most interesting result is the 15% reduction of NOx which the authors explain happens because of the introduced HHO gas in the intake manifold that decreases the amount of gasoline which leads to a leaner mixture and to a reduced flame temperature. A hydrogen production device that uses the electrolysis principle was used by Synák F et al [18] to generate HHO gas which was added into the engine`s intake manifold on five different vehicles (Škoda Felicia 1 1.9 D, Škoda Felicia 1 1.3 MPI, Škoda Fabia 2 1.4 MPI, Kia Sportage 1.6 CRDI, Kia Ceed 1.6 CVVT). The study showed a low impact of HHO addition of 2 L/min$^{-1}$ on the emissions and a slightly loss of engine power and torque. Also, because the hydrogen generator is taking electric energy from the vehicle the fuel consumption increased by 20% but at idling speed, the fuel consumption decreased with 10%. Adding H2 in the intake manifold substitutes a part of the combustion air. At a point where the hydrogen content is high such that the charge that enters in the cylinder doesn't have enough oxygen concentration to sustain a complete combustion. Moreover, hydrogen`s lower density highly affects the engine volumetric efficiency. The cleaning effect of HHO was not proven, after functioning a few hours with HHO addition. A similar device that uses electrolysis process was used by Hamaad A et al [19] resulting an increase of power of 11.8 %.

To eliminate the storage challenge of hydrogen, Dinesh M H and Kumar G N [20] have shifted their focus on using a hydrogen-rich carrier fuel like ammonia (NH3) with $H_2$ addition (from 5% to 21%) and studied the effect of compression ratio CR (from 12 to 15) over combustion stability, performance and emissions. Due to hydrogen's excellent burning properties, brake power BP increased by 31.2% and brake thermal efficiency BTE improved by 39%. The peak cylinder pressure, temperature and heat release are modest due to ammonia`s low energy content and rather slow burning speed, which improves to some degree with CR and remarkably with the percentage of hydrogen added. NOx emission increased with 42.34% from 5% hydrogen at a CR12 to 21% hydrogen at CR15.

The study conducted by Li X et al [21] on a 2L direct injection hydrogen engine showed that the start of injection SOI could cause a remarkable influence on cycle variation due to the closing of the intake valve and shortening mixing time causing an increase of cycle variation. The highest effect is at low engine loads/equivalence ratio and at high engine speed. The experimental results indicate that cycle variation increases from 0.67% to 1.02% with the increase of engine speed and decreases from 1.52% to 0.64% with the increase of engine load. When the equivalence ratio increases, the cycle variation decreases at first from 2.52% to 0.35% and then stabilizes.

Wang S et al [22] analysed the combustion and emission of a hydrogen blended gasoline engine at high loads and lean conditions emphasizing the improvement of HC and CO emissions after the hydrogen addition and the stability of combustion with the only disadvantage of $NO_x$ levels that increased. Chitragar P R et al [23] focused their study on combustion and emission on a gasoline engine using pure hydrogen and LPG at idle condition with similar results like other authors regarding the evolution of HC and CO but with a substantial decrees of $NO_x$ when fuelled with hydrogen. The use of hydrogen has improved the cylinder pressure by 18%.

Candelaresi D, Valente A, Iribarren D et al [24] research showed that the environmental life-cycle performance of three different passenger cars fuelled by hydrogen: an internal combustion engine vehicle ($H_2$-ICE), a hybrid electric car (HEV $H_2$-ICE) and a fuel cell electric vehicle (FCEV). Also, two more vehicles that use a mix of hydrogen and natural gas or gasoline were considered. The infrastructure was considered to be the main obstacle and environmental burden but the study shows that hydrogen vehicles are great decarbonisation solutions when fuelled with green hydrogen even though, from a technical point of view, the HEV $H_2$-ICE and $H_2$-ICE suffer from a low driving range because of the on-board hydrogen storage systems.

The purpose of this paper is to study the effects of small amounts of hydrogen on the engine performance and emissions at fuelling an automotive spark ignition engine with 1500 $cm^3$ displacement.

## 2. Experimental test bench and procedures

### *2.1 Experimental setup*

The experimental test bench is presented in the figure 1.

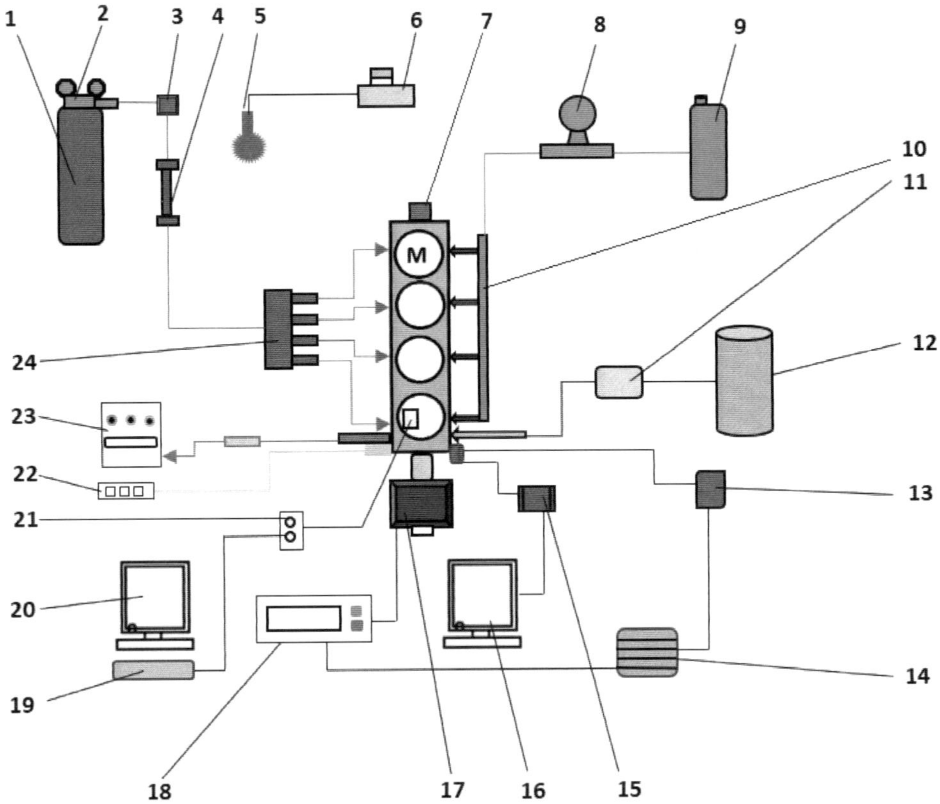


**Figure 1.** Test bench schema.

The main components of the test bench are: 1 – hydrogen storage tank; 2 – pressure regulator; 3 – hydrogen flowmeter; 4 – flame trap; 5 – hydrogen sensor; 6 – acoustic and light warning device; 7 – crankshaft position sensor; 8 – fuel mass flowmeter; 9 – fuel tank; 10 – fuel injectors; 11- airflow meter; 12 – air reservoir; 13 – throttle body actuator; 14 – power unit for throttle body actuator; 15 – hydrogen injectors ECU; 16 – PC for hydrogen ECU; 17 – eddy current dyno; 18 – brake control panel; 19, 20 – PC for data acquisition system; 21 – load amplifier; 22 – display for cooling system, oil and air temperatures; 23 – gas analyser; 24 – hydrogen injectors.

## *2.2 Experimental procedure*

The engine used is a 4 cylinder, 1500 cm$^3$ turbocharged Daewoo A15MF that equipped models like Nubira, Cielo (Nexia) and Espero. Initially, the engine was fuelled only with gasoline at the engine load of $\chi = 55\%$ and speed of 2500 min$^{-1}$. Parameters such as: water and air temperature, intake air temperature, air consumption, fuel consumption, boost pressure and emissions are recorded when fuelling with gasoline. The current tests are aimed at analysing the performance and emission levels of hydrogen enriched gasoline engine.

## 3. Results

### *3.1. CO emission*

In figure 2 are presented the effects of hydrogen addition on CO emission at different mixtures. CO emission diminish as the hydrogen gasoline blend percentage increases. At rich mixtures, because of the high carbon content of the mixture, the CO emission isn`t reduced as much, but at stoichiometric mixture the decrease is notable - 46% less CO. At lean mixtures, the values are rather similar.

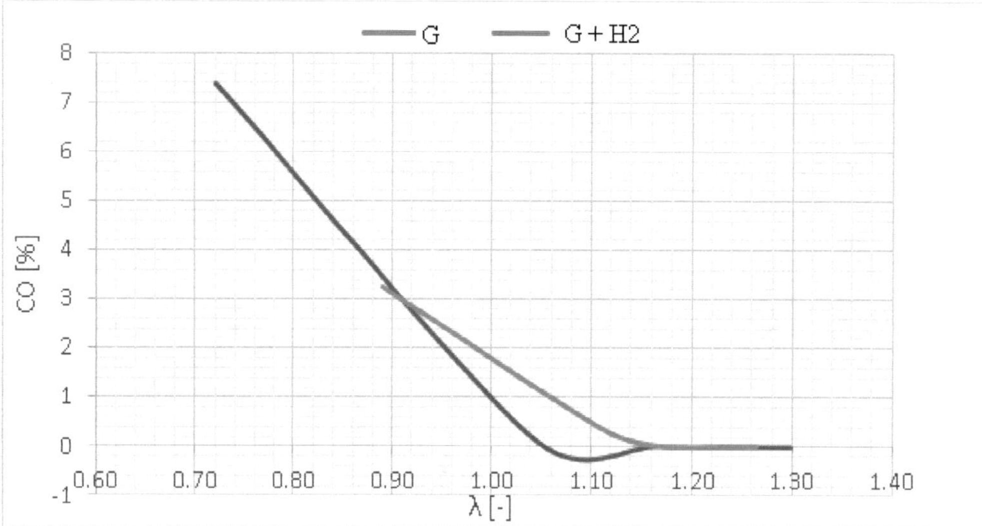


**Figure 2.** The CO emission level versus air-fuel ratio.

Due to the high flame speed of hydrogen and stoichiometric air-to-fuel ratio, the fast combustion of hydrogen rapidly consumes the air, which forms lean-oxygen zones in the cylinder and blocking the formation of CO. As a result of a short combustion and post-combustion period, the necessary time for CO oxidation reaction is significantly reduced resulting in lower CO emission.

### *3.2. $CO_2$ emission*

$CO_2$ emission can be observed in fig. 3 indicating an improvement in using hydrogen gasoline blends. The reduction is also a consequence of complete combustion. The combustion process for the gasoline-hydrogen fuel was improved, especially around stoichiometric mixtures, marking a 16 % reduction of $CO_2$ emissions.

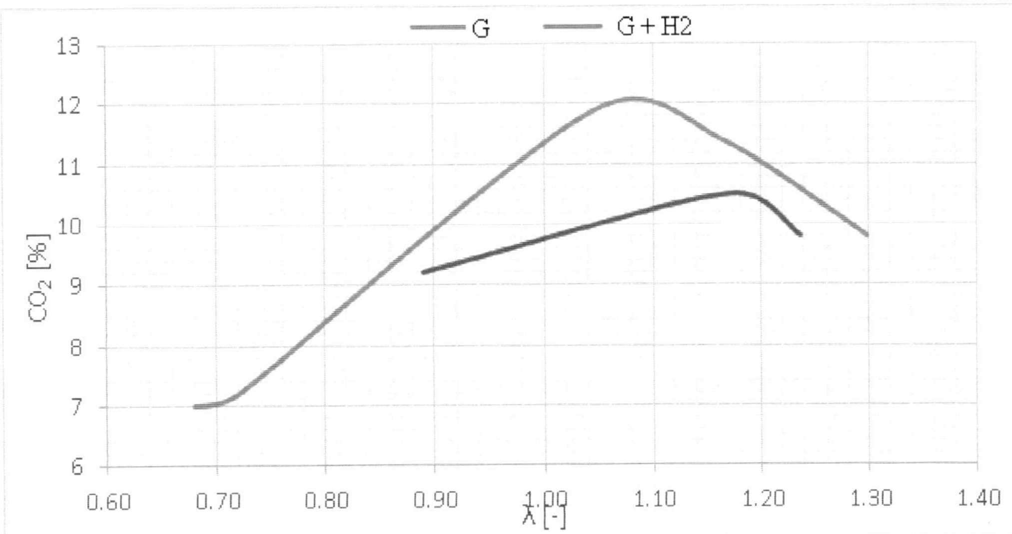


**Figure 3.** The $CO_2$ emission level versus air-fuel ratio.

### *3.3. HC emission*

Gasoline-hydrogen mixtures burn more completely and release less HC emission especially at lean mixtures. The short quenching distance of hydrogen allows the flame to travel near the cylinder wall thus improving combustion. The high diffusivity property of hydrogen is another reason for better combustion because of a more homogenous mixture.

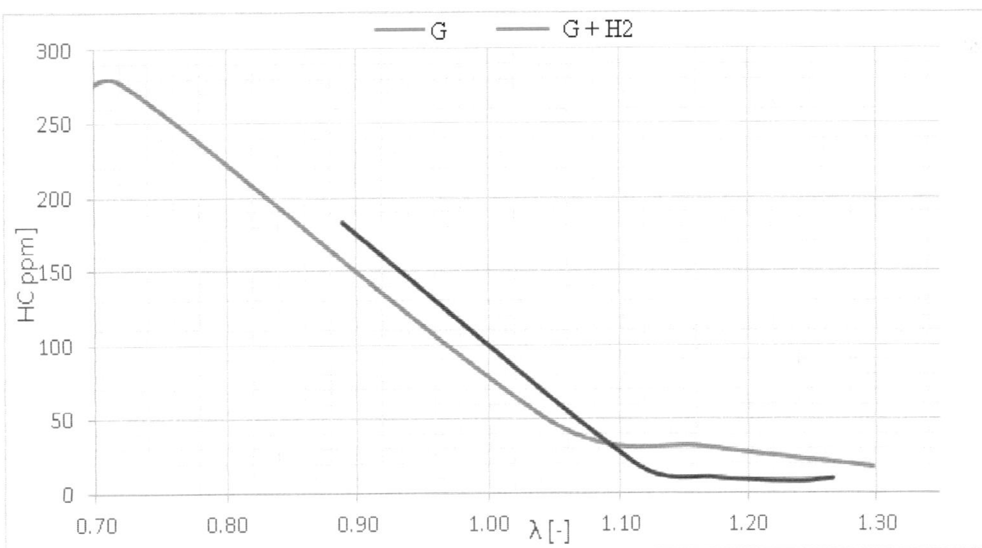


**Figure 4.** The HC emission level versus air-fuel ratio.

At rich mixtures, because of the small percentage of hydrogen in the fuel, an increase was noted. At lean mixtures, the gasoline-hydrogen mix showed a 44% improvement regarding HC emission.

*3.4. $NO_x$ emission*

In fig. 5 is presented the variation of $NO_x$ emission which increases even due to the small percentages of hydrogen addition. The higher hydrogen addition results in a higher combustion temperature which leads to more dissociation and more $NO_x$ generated. The use of lean mixture in the area of λ=1.2…1.3, and 3-7% $H_2$ addition, the $NO_x$ emission decrease with 14.8%.

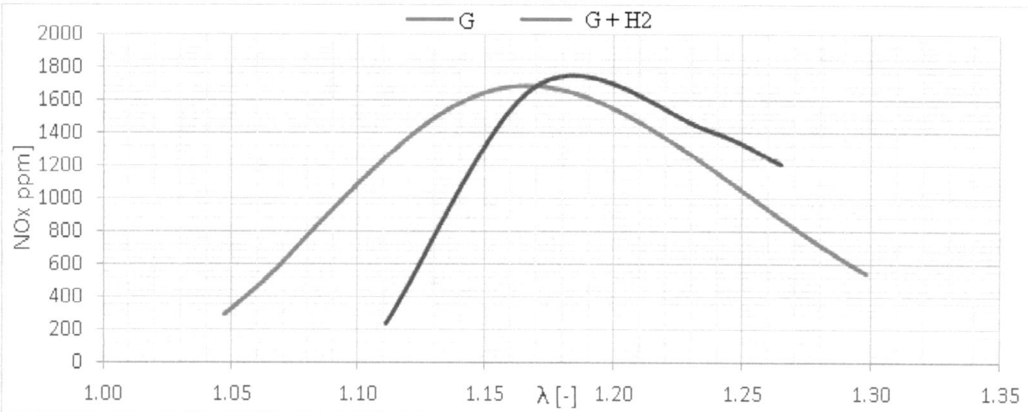


**Figure 5.** The $NO_x$ emission level versus air-fuel ratio.

Spark timing tune or mixture leaning over λ=1.3 can assure further $NO_x$ emission level decrease.

*3.5. Brake thermal efficiency*

Maximum *brake thermal efficiency* (fig. 6), under the test conditions, increases from 27.8% with no hydrogen addition to 29.1% with 6.5% $H_2$ addition. As a consequence of flame speed of hydrogen which is 6-8 times higher than gasoline and the flammability range of hydrogen which is much wider, the gasoline-hydrogen fuel will have a rapidly combustion speed and an extended flammability limit which leads to a more complete combustion.

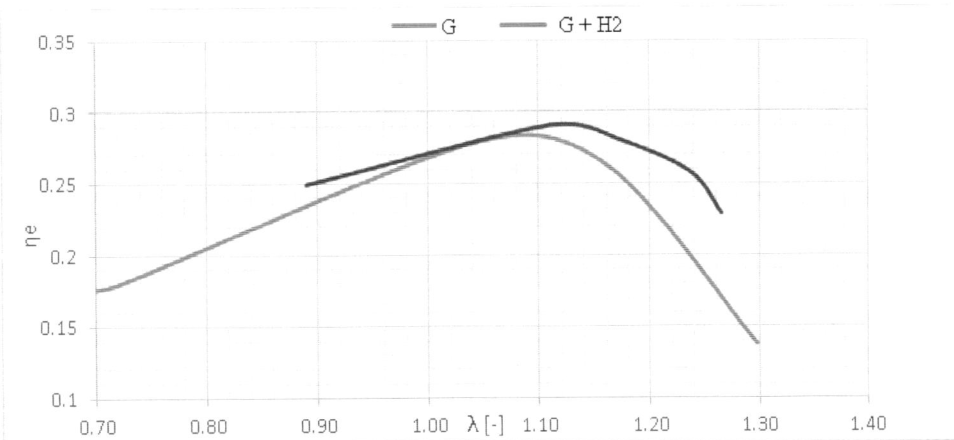


**Figure 6.** Brake thermal efficiency versus air-fuel ratio.

At leaner mixtures, BTE improves with more than 2% because of the excellent combustion properties of hydrogen comparative to gasoline.

### *3.6. Effective power*

Regarding the effective power (fig. 7), the results indicated a small increase of power at rich mixtures. Because of the wide flammability limits of hydrogen, at lean mixtures $\lambda$=1.17, the power improved with 11.8% while at $\lambda$=1.24, the power increased with 28% with a 6.5% hydrogen addition in both tests.

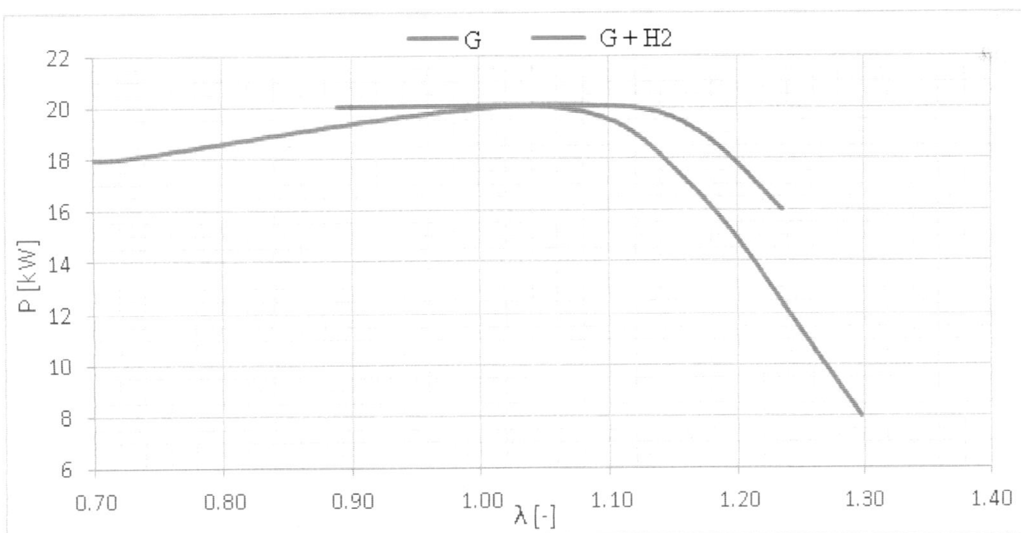


**Figure 7.** The effective power versus air-fuel ratio.

### *3.7. Energetic brake specific fuel consumption*

By using a 5.2% addition of hydrogen, EBSFC was reduced by 10.9% at rich mixtures around λ=0,90. At lean mixtures λ=1.24 even with the effective power increase, there was still a reduction of fuel consumption by 9.6% comparative to gasoline use.

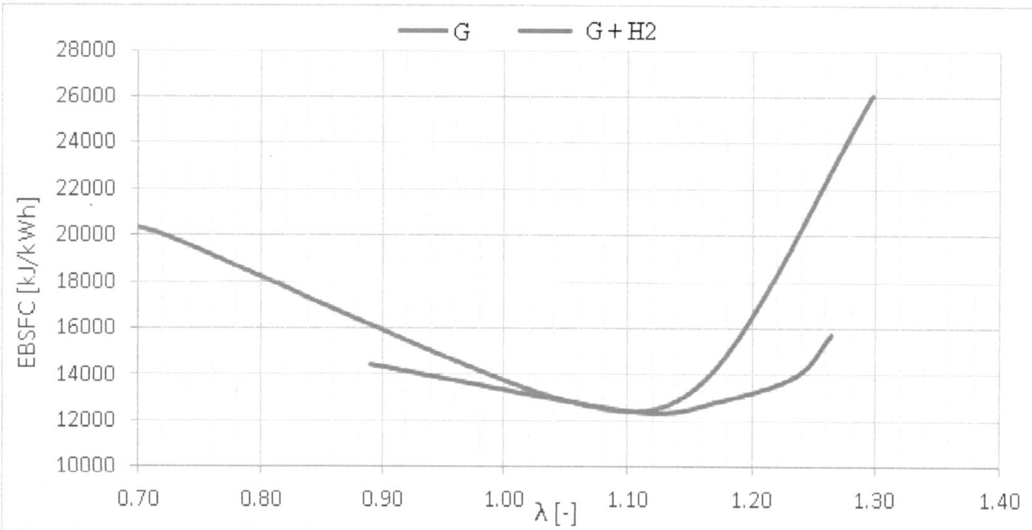


**Figure 8.** The Energetic Brake Specific Fuel Consumption versus air-fuel ratio.

## 4. Conclusions

The experiment was carried out with different percentages (up to 7%) of hydrogen injected in the intake manifold. The main conclusions of the experimental investigations are:

CO, $CO_2$ and HC emissions reduces with the increase of hydrogen addition due to the high flame speed of hydrogen and the short quenching distance which allows the flame to travel near the cylinder wall thus improving combustion. The high diffusivity property of hydrogen is another reason for better combustion because of a more homogenous mixture. At stoichiometric mixture, the level of CO was reduced with 46% and $CO_2$ emission marked a 16% cut. At lean mixtures, the gasoline-hydrogen mix showed a 44% improvement regarding HC emission.

The higher hydrogen addition results in a higher combustion temperature which leads to more dissociation and more $NO_x$ emission causing a 14.8% decrease at lean mixtures.

Brake thermal efficiency under the test conditions, increased from 27.8% with no hydrogen addition to 29.1% with 6.5% $H_2$ addition.

Regarding the effective power, the results indicated a small increase at rich mixtures but at lean mixtures such as λ=1.17, the power improved with 11.8% while at λ=1.24, the power increased with 28% with a 6.5% hydrogen addition in both tests.

By using a 5.2% addition of hydrogen, EBSFC was reduced by 10.9% at rich mixtures around λ=0,90. At lean mixtures λ=1.24, even with the effective power increase, there was still a reduction of fuel consumption by 9.6%.

Due to its excellent burning properties, hydrogen has proved to be an excellent renewable fuel which improves performance and emissions of internal combustion engines.

## 5. References

[1] United Nations Framework Convention on Climate Change. Adoption of the Paris Agreement 2015
[2] European Commision. The European. The European Green Deal-COM 2019
[3] International Energy Agency. Key world energy statistics
[4] 2020 U.S. Energy Information Administration. International energy outlook 2019 – with projections to 2050
[5] Tashie-Lewis B, Nnabuife S 2021 Hydrogen Production, Distribution, Storage and Power Conversion in a Hydrogen Economy - A Technology Review *Chemical Eng. J. Advances* **8** 100172
[6] Li H, Cao X, Liu Y, Shao Y, Nan Z, Teng L, Peng W, Bian J 2022 Safety of hydrogen storage and transportation: An overview on mechanisms, techniques and challenges *Energy reports* 8 6258-6269
[7] Aryal K 2017 Hydrogen Fuelled Internal Combustion Engine: A Review *Int. J. for Research in Applied Science & Engineering Technology* **5** 498-506
[8] Wang L, Hong C, Li X, Yang Z, Guo S, Li Q 2022 Review on blended hydrogen-fuel internal combustion engines: A case study for China *Energy Reports* **8** 480-6498
[9] Shivaprasad K V, Raviteja S, Parashuram C, Kumar G N 2014 Experimental Investigation of the Effect of Hydrogen Addition on Combustion Performance and Emissions Characteristics of a Spark Ignition High Speed Gasoline *Engine Procedia Tech.* **14** 141-148
[10] He L, Jingyuan L, Xiumin Y, Mengliang L, Tian Y 2018 Numerical Study on Combustion and Emission Characteristics of a PFI Gasoline Engine with Hydrogen Direct-Injection *Energy Procedia* **158** 1449-1454
[11] Negurescu N, Pană C, Cernat A 2012 Aspects of using Hydrogen in SI engine *U.P.B. Sci. Bull. Series D, Vol.* **74**, Iss. 1
[12] Rusu E, Pană C, Negurescu N 2013 Experimental study of a spark ignition engine fuelled with gasoline and hydrogen in addition *U.P.B. Sci. Bull. Series D, Vol.* **75**, Iss. 4
[13] Pantile V, Pană C, Negurescu N 2013 Improvement of the performance of a spark ignition engine by using hydrogen *U.P.B. Sci. Bull. Series D, Vol.* **75**, Iss. 3
[14] Tamilarasan P, Loganathan M 2016 Experimental Study on the use of EGR in a Hydrogen-Fuelled SI Engine *Int. J of Sci. & Eng. Research, vol.* **8** 336-342
[15] Falfari S, Bianchi G, Cazzoli G, Forte C, Negro S 2018 Basics on Water Injection Process for Gasoline Engines *Energy Procedia* 148 50-57
[16] Thawko A, Tartakovsky L 2022 The Mechanism of Particle Formation in Non-Premixed Hydrogen Combustion in a Direct-Injection Internal Combustion Engine *Fuel* **327** 125187
[17] EL-Kassaby M, Eldrainy Y, Khidr M, Khidr K 2015 Effect of hydroxy (HHO) gas addition on gasoline engine performance and emissions *Alexandria Eng. J.* **55** 243-251

[18] Synák F, Synák J, Skrúcaný T 2021 Assessing the addition of hydrogen and oxygen into the engine`s intake air on selected vehicle features *Int. J. of Hydrogen. Energy* **46** 31854-31878
[19] Hamaad A, Tawfik M, Khattab S, Newir A 2016 Device for Using Hydrogen Gas as Environmental Friendly Fuel for Automotive Engine (GREEN & ECO $H_2$) *Procedia Env. Sci.* **37** 564-571
[20] Dinesh M H, Kumar G N 2022 Effects of compression and mixing ratio on $NH_3/H_2$ fuelled SI engine performance, combustion stability and emission *Energy Conversion and Management*: **X**
[21] Li X, Sun B, Zhang D, Wang X, Bao L, Luo Q 2022 Experimental study on the cycle variation characteristics of direct injection hydrogen engine *Energy Conversion and Management*: **X 15** 100260
[22] Wang S, Ji C, Zhang B, Zhou X 2014 Analysis on combustion of a hydrogen-blended gasoline engine at high loads and lean conditions *Energy Procedia* **61** 323-326
[23] Chitragar P R, Shivaprasad K V, Nayak V, Bedar P, Kumar GN 2016 An experimental study on combustion and emission analysis of four cylinder 4-stroke gasoline engine using pure hydrogen and LPG at idle condition Injection *Energy Procedia* **90** 525-534
[24] Candelaresi D, Valente A, Iribarren D, Dufour J, Spazzafumo G 2021 Comparative life cycle assessment of hydrogen fuelled passanger cars *Int. J. of Hydrogen Energy* **46** 35961-35973

## Acknowledgments

The authors express their gratitude to AVL List GmbH Graz, Austria, for providing the necessary equipment's. This work was supported by a grant of the Ministry of Research, Innovation and Digitization, CCCDI - UEFISCDI, project number PN-III-P2-2.1-PED-2021-0427, within PNCDI III. This work was supported by a grant of the Ministry of Research, Innovation and Digitization, CNCS - UEFISCDI, project number PCE 5/2022, PN-III-P4-PCE-2021-0777, within PNCDI III.

IOP Conf. Series: Materials Science and Engineering **1311** (2024) 012003 doi:10.1088/1757-899X/1311/1/012003

# A calculation algorithm for the heater of a Stirling engine

**T. Mitran[1*], D. Craciun[2], 3. A. Rus[1], G. Dragomir[1], D. Chira[1]**

[1] Department of Mechanical Engineering and Automotive, University of Oradea, Universitatii 1, 410087 Oradea, Romania

[2] Department of Mechatronics, University of Oradea, Universitatii 1, 410087 Oradea, Romania

[*]Email: tudor_mitran@yahoo.com

**Abstract**. The paper presents a calculation algorithm for the heater of a Stirling engine. In the current context of the energy crisis, Stirling engines can be used to produce mechanical work from renewable energy sources. The heat exchangers are one of the key elements regarding the efficiency of a Stirling engine. So, a calculation algorithm for the heater of a Stirling engine can be used for the dimensioning of this component, according to the requirements for the functioning of the system. To validate this algorithm, a dedicated program for the calculation of the heat exchangers based on the finite element analyse was used.

**Keywords:** Stirling engine, heat exchanger, heat transfer, renewable energy

## 1. Introduction

The theoretical cycle of a Stirling engine is the same, regardless of the engine's configuration, and it's composed by two isothermal and two isochoric processes. This is an external combustion engine. The scheme of the beta configuration is presented in fig. 1.

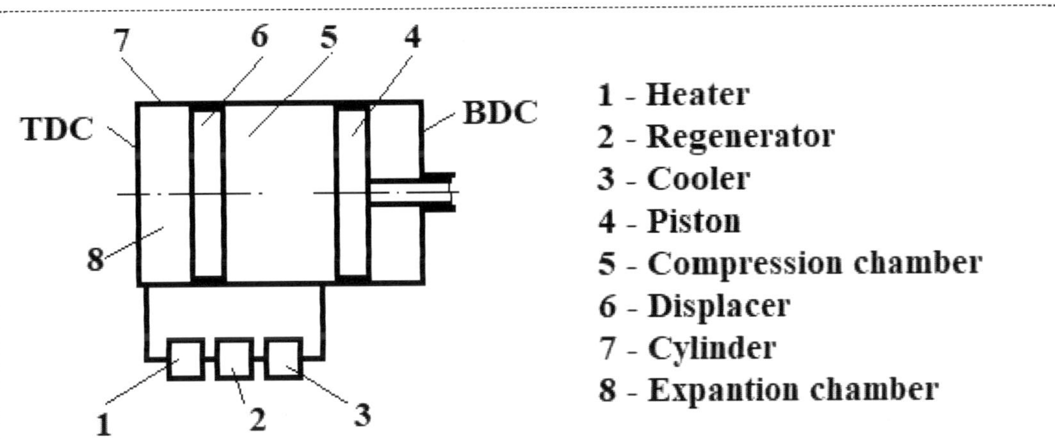


Fig. 1 The beta configuration of a Stirling engine

The characteristic of its functioning is the existence of a warm source (the heater, position 1 in fig. 1) and of a cold source (the cooler, position 3 in fig. 1). Usually, the cold source is the environment. To rise the efficiency of the engine, between the heater and the cooler, a regenerator is placed. The mechanism is put in movement by a starter. The piston (4) moves from the BDC to TDC and compresses the working agent (air). This forces the working agent to move from the compression chamber (5) to the expansion chamber (7). During this process, the working fluid passes through the cooler (3), the regenerator (2) and the heater (1). The particularity of the functioning of a Stirling engine is that the working agent moves in both senses, first from the compression chamber to the expansion chamber, and then in the reverse sense. So, when the working agent moves from the expansion chamber to the compression one (it is supposed that the temperature inside the warm chamber is equal to the one inside the heater), after it passes the heater, it can transfer heat to the regenerator. This has two advantages. The first one is the temperature drop in the regenerator, that means that the heat that must be exchanged in the cooler is smaller. The second one is the fact that, after the compression, the working fluid passes through the cooler (it is supposed that the temperature of the cooler is equal to that of the compression chamber) and after that it passes through the regenerator, and recovers a part of the energy supplied by the gases that have passed through the system in the previous cycle and, in continuation, it is heated in the heater.

One of the Stirling engine's advantages is that it can use as source to achieve the temperature of the warm source any kind of fuel: fossil or renewable.

Some applications of the engines functioning based on the Stirling cycle are the following [1]:

a. Electricity generation

a.1. Nuclear powered Stirling engines in electric powered plants.

The warm source is the cooling agent of the nuclear plant. In this way, it is possible to eliminate the need of the water that is used to lower the temperature of the cooling agent.

a.2. Solar energy

Solar panels can be used to produce the warming agent for the functioning of a Stirling engine. The intensity of the solar radiation can be increased by using a parabolic mirror. This will lead to a better efficiency, comparable to the one provided by non-concentrated photovoltaic cells, and it can be compared to the solution with concentrated photovoltaic panels. A small system to produce electric current by using a Stirling engine by using solar panels was developed at the Electrical Engineering and Computer Sciences University of California at Berkley [2].

a.3. The production of combined heat and power

In the case of a system that can generate heat and power, it can produce mechanical or electrical energy. The waste heat emitted by the engine can be used to supply another application.

a.4. The use of the waste heat coming from thermal power plants.

b. Stirling motor for mechanical output and propulsion

b.1. Automotive engines

The disadvantages of using Stirling engines for the propulsion of the road vehicles are a too low power to wight ratio, a too high cost and a too much time to start.

b.2. Aircraft engines

This type of engine can be used for aircraft propulsion if it is possible to achieve a higher power density and a lower cost. As benefits, they are quieter, more reliable, produce much less vibration, less polluting and can use safer, more less explosive fuels.

b.3. Electric vehicles

Stirling engines, used as part of a hybrid electric drive system can avoid the design challenges or drawbacks of a non-hybrid Stirling car. The Swedish company PRECER has developed a patented Solid Biofuel Technology that uses a Stirling engine to generate electric current [3]. This electric current is used to supply an electric motor for the propulsion of small and middle-size vehicles. The fuel used is a solid one, pellets.

b.4. Marine engines

There is the possibility to use a Stirling engine as a submerged power system. Research in this field was developed by General Motors, that includes the thermal storage for undersea applications. The company United Stirling from Malmo, Sweden, was developing an experimental Stirling engine that uses hydrogen peroxide as an oxidizer in underwater fuelling systems.

So, the Stirling engines can cover a wide range of applications and can function using renewable sources of energy.

The functioning of a Stirling engine depends very much on the heat exchangers. The processes that take place in a heat exchanger are very complex and can be described only by empirical or semi-empirical relations. As mentioned before, another problem is that the working fluid moves in both senses in a Stirling engine heat exchanger.

This paper presents a calculation algorithm for the heater of a Stirling engine in Beta configuration. The calculated values were compared to those obtained by using a dedicated program for the calculation of the heat exchangers, based on the finite element analyse.

## 2. The calculation algorithm for the heater

The heater is a recuperative heat exchanger. The warm and cold fluids are simultaneously passing through the exchanger. The two fluids are separated by a wall. The heat exchange takes place through this wall.

To design a recuperative heat exchanger, it is necessary to go through the following steps:

- the thermal and the hydrodynamic calculation;
- the mechanical calculation;
- the manufacturing project;

Following the thermal calculation, it can be determined the heat exchange surface and how it is arranged in space, corresponding to some thermal imposed data, depending on the working fluids flows and temperatures. The hydraulic calculation aims the correlation of these elements with the admissible pressure losses in the system and with the pumping energy of the working fluids.

The mechanical calculation establishes the constructive solution, taking into consideration: the operating regime, the temperatures and the pressures of the working fluids, the corrosion characteristics of these fluids and the tensions produced by the temperature gradients. Also, it must be taken into consideration the link between the heat exchanger and other equipment.

The heater of a Stirling engine is a heat exchanger with continuous operation without the mixing of the heating agent with the working fluid.

Depending on the relative circulation sense of the fluids in the device, the exchanger can be:

- in concurrently current (fig. 2) – both fluids pass through the exchanger in the same direction and in the same sense;

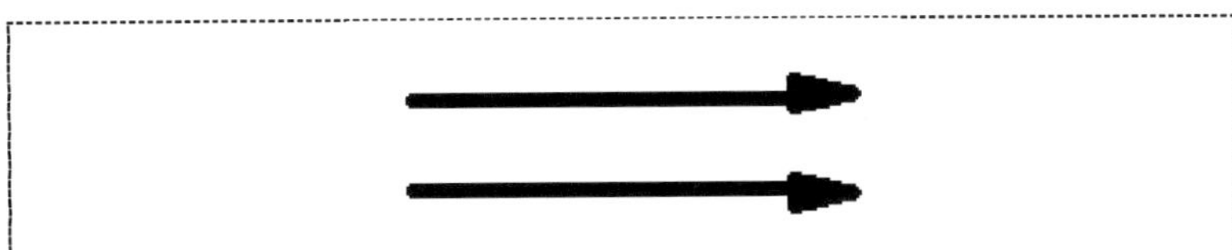

Fig. 2 The concurrently flow

- in counter current (fig. 3) – both fluids pass through the device in the same direction, but in opposite senses;

IOP Conf. Series: Materials Science and Engineering **1311** (2024) 012003 doi:10.1088/1757-899X/1311/1/012003

Fig. 3 The counter current flow

The temperature diagrams for this type of device are presented in fig. 4.

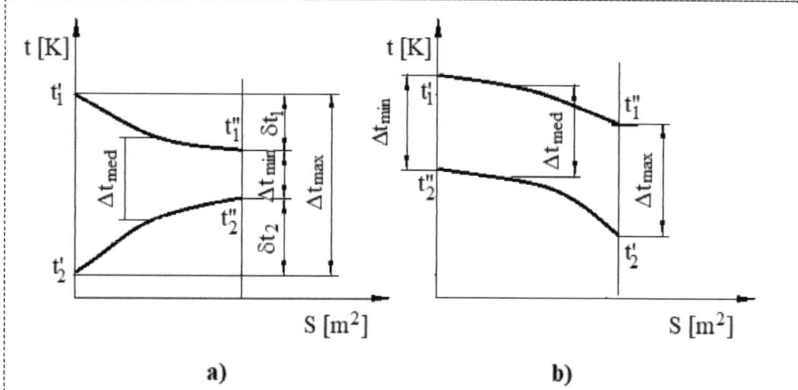


Fig. 4 The temperatures diagrams for the heat exchange devices;
a) in concurrent; b) in counter current;

As mentioned before, in the case of a Stirling engine, the heater is functioning both in counter current during the compression stoke and in concurrently current during the expansion stroke.

The determination of the global heat exchange coefficient for the tubular devices, in a general case, is realized based on the scheme in the fig. 5.

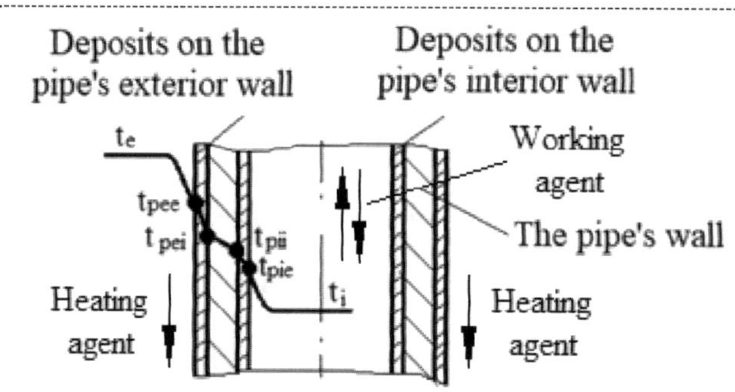


Fig. 5 The heat exchange scheme through a cylindrical wall for a Stirling engine heater

The heating agent is moving outside the tubular pipes, in one sense only. The temperature of the heating agent ($t_e$) is constant, and it's considered to be the temperature of the warm source. During the functioning of a heat exchanger, because of corrosion or of other chemical processes, on the outside or on the inner surfaces of the pipe deposits can appear. These deposits act like thermal insulating layer. This can produce a big impact on the heat transfer and on the functioning of the Stirling engine. In a theoretical approach it is difficult to estimate the effect of these deposits. So, in this theoretical calculus the eventual presence of deposits on the surfaces of the heater's pipes were not taken into consideration. Because of this, the exterior temperature of the pipe's wall is considered to be equal to the one of the heating agent ($t_{pee}=t_{pei}=t_e$), and the interior pipe's wall temperature is considered to be equal to the temperature of the working agent ($t_{pie}=t_{pii}=t_i$).

The calculation of the heater was realized during the cooling stroke, when the heat exchanger is working in counter current.

This calculus aims to determine the dimensions of the heater's elements to realize the transfer of a specific heat during the compression stroke.

A configuration with tubular pipes was adopted for the heater.

Because there is no danger of explosion during the exploitation, the chosen working fluid is the air. There are some initial data that must be used for this calculus, for the heating agent and for the working agent. The specific heat at constant pressure ($c_p$), the specific heat at constant volume ($c_v$), the density ($\rho$), the thermal conductivity ($\lambda$), the dynamic viscosity ($\eta$), the kinematic viscosity ($\nu$) and the Prandtl number (Pr). These parameters were determined based on the data from [4]. All these parameters depend on the temperature. This is not a problem, because in this calculation it is considered that temperature of the heating agent is constant. This leads to a functioning of the engine in steady conditions.

During the cooling stroke, the working agent moves from the cold source to the warm source, through the cooler, the regenerator, and the heater. This movement takes place in opposite direction to the moving of the heating agent. The material chosen for the heater's pipes is cupper.

As heating agent was choose the mono ethylene glycol. For the calculation is necessary the knowing of its following properties:

- the specific heat: $c_{ai}$=2300 J/kgK;
- the density: $\rho_{ai}$=1110 kg/m$^3$;
- the kinematic viscosity: $\nu_{ai}$=1,52467 m$^2$/s;
- the thermal conductivity: $\lambda_{ai}$=0,29 W/mK;

The heat flux in the heater is:

$$\dot{Q}_{inc} = \dot{m}_{ali} \cdot c_{val} \cdot \Delta T_{ali} \text{ [W]} \tag{1}$$

where: - $\dot{m}_{ali}$ [kg/s] – is the mass flow of the working agent through the heater;

- $c_{val}$ [J/kgK] – is the specific heat at constant volume of the working agent;
- $\Delta T_{ali}$ [K] – is the temperature variation of the working agent at the pass through the heater;

For the determination of the mass flow of the working agent it is necessary to know the mass of the working agent that passes through the heater in the transfer stroke from the compression chamber to the expansion chamber. At first, the dimensions of the working cylinder were adopted (fig. 6).

where: - C – is the cooler;

- R – is the regenerator;
- H – is the heater;
- $C_m$ – is the motor cylinder;
- $D_{mc}$ – is the inner diameter of the motor cylinder;
- $h_1$ – is the length of the compression stroke;
- $h_2$ – is the length of the piston ($P_2$) movement until it reaches the piston ($P_1$) in its position at the end of the compression stroke;
- $h_1+h_2$ – is the expansion stroke;

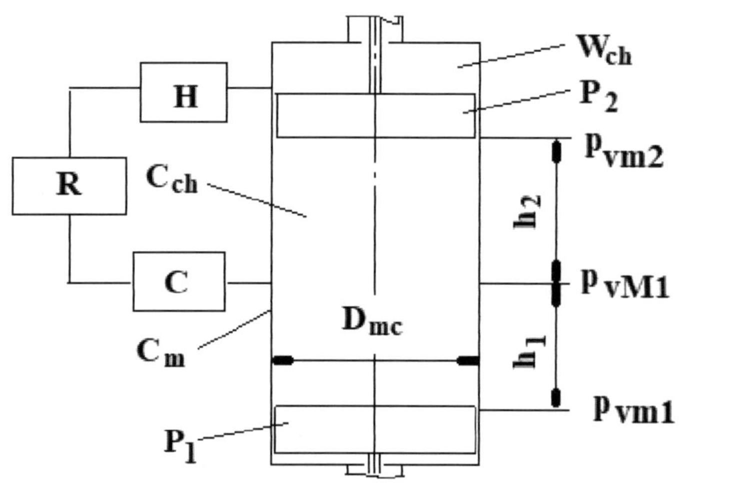


Fig. 6 The dimensions of the working cylinder

At first, the position of the two pistons ($P_1$ and $P_2$) are the ones in the fig. 6 (the piston $P_1$ is in $p_{vm1}$ and the piston $P_2$ is in $p_{vm2}$). When commanded, the piston ($P_1$) moves to the piston ($P_2$) with the compression stroke ($h_1$) and reaches the point $p_{vM1}$. Because the compression chamber ($C_{ch}$) is linked with the expansion chamber ($W_{ch}$), the pressure is the same in the system, and it considered that the working agent do not move through the heat exchangers. After that, the piston ($P_2$) moves, also by command, until it reaches piston ($P_1$) in the point $p_{vM1}$. This is the cooling stroke. The working agent passes through the cooler (C), the regenerator (R) and the heater (H). In this process the working agent's temperature rises from the cold source ($T_c$) to the one of the warm source ($T_c$), equal to the temperature of the warming agent.

The volume of the compression chamber at the end of compression stroke:

$V_c$=0,515·10$^{-3}$ m$^3$.

The pression in the compression chamber at the end of the compression stroke: $p_2$=2 MPa.

The temperature in the compression chamber at the end of the compression stroke: $T_1$=313 K (40 $^0$C).

The working agent mass in the compression chamber at the end of the compression stroke:

$$m_{ali} = \frac{10^6 \cdot p_2 \cdot V_c}{R_{aer} \cdot T_1} \text{ [kg]} \tag{2}$$

It was adopted the value of the piston's ($P_2$) mean displacement speed in the heating stroke $W_{pi}$=0,5 m/s

The duration of the transfer stroke from the compression chamber to the expansion chamber:

$$\tau_i = S_{tr} / W_{pi} \text{ [s]} \tag{3}$$

$S_{tri}$=42 mm – the transfer stroke's length;

$\tau_i$=0,042/0,5=0,084 s;

The mass flow of the working agent:

$$\dot{m}_{ali} = \frac{m_{ali}}{\tau_i} \text{ [kg/s]} \tag{4}$$

The following parameters were adopted:

- the working agent's temperature at the entrance in the heater: $T_{2'}$=343 K (70 $^0$C);

- the working agent's temperature at the exit from the heater: $T_{2''}$ =423 K (150 $^0$C);
- the temperature of the heating agent at the entrance in the heater: $T_{1'}$ =453 K (180 $^0$C);

The mean temperature of the working fluid in the heater:
$T_{mal}=(T_{2'}+T_{2''})/2=(343+423)/2=383$ K (110 $^0$C)

The specific heat at constant volume of the working agent at the mean temperature in the heater: $c_{val}$=712 J/kgK;

The volumetric flow of heating agent can be adjusted by using a pump. The chosen value is: $\dot{V}_{ai} = 0{,}5\ m^3/h$;

The mass flow of working agent in the heater:

$$\dot{m}_{al} = \dot{V}_{al} \cdot \rho_{al}\ \text{[kg/s]} \tag{5}$$

The temperature variation of the heating agent at the pass through the heater:

$$\Delta T_{al} = \frac{\dot{Q}_{inc}}{\dot{m}_{al} \cdot c_{pal}}\ \text{[K]} \tag{6}$$

The temperature of the heating agent at the exit from the heater:
$T_{1''}=T_{1'}-\Delta T=180-22=158$ $^0$C;

The mean temperature in the heater:

$$\Delta T_m = \frac{\Delta T_{max} - \Delta T_{min}}{\ln\left(\frac{\Delta T_{max}}{\Delta T_{min}}\right)}\ \text{[K]} \tag{7}$$

The difference between the temperature of the working agent at the exit from the heater and the temperature of the heating agent at the entrance in the heater:
$T_{1''}-T_{2'}=158-70=88$ $^0$C;

The difference between the temperature of the heating agent at the exit from the heater and the temperature of the working agent at the entrance in the heater:
$T_{1'}-T_{2''}=180-150=30$ $^0$C;
$\Delta T_{max}= T_{1''}-T_{2'}=88$ $^0$C and $\Delta T_{min}= T_{1'}-T_{2''} =30$ $^0$C;

The volumetric flow of the working agent that passes through the heater:

$\dot{V}_{al} = \frac{V_c}{\tau_i} = \frac{0{,}515 \cdot 10^{-3}}{0{,}084} = 0{,}006131\ m^3/s$ (lipseste numarul!!, după renumerotare vazut si in text)!

The economic speed, recommended for the flow of the working agent through the heater is: $W_{ali}$=3 ... 6 m/s [4]. The adopted value is: $W_{ali}$=4 m/s.

The total flow section of the working fluid through the pipes necessary to realize this speed:

$$S_{tt} = \frac{\dot{V}_{al}}{W_{ali}}\ [\text{m}^2] \tag{8}$$

For the heater were adopted cupper pipes with the inner diameter $d_i$=6 mm and the exterior diameter $d_e$=8 mm.

The number of pipes necessary for the realization of the total flow section:

$$n_t = \frac{S_{tt} \cdot 4}{\pi \cdot d_i^2} \tag{9}$$

The adopted number of the pipes is: $n_{ta}$=54.

The total flow section of the working fluid through the heater is recalculated:

$$S_{tt} = \frac{\pi \cdot d_i^2}{4} \cdot n_{ta} \text{ [m}^2\text{]} \tag{10}$$

The flow speed of the working agent through the heater:

$$W_{ali} = \frac{\dot{V}_{ali}}{S_{tt}} \text{ [m/s]} \tag{11}$$

In the heat exchanger the heat transfer is realized through convection between the heating agent and the pipes walls, through conduction in the pipes walls and again through convection from the pipes walls to the working agent (fig. 7).

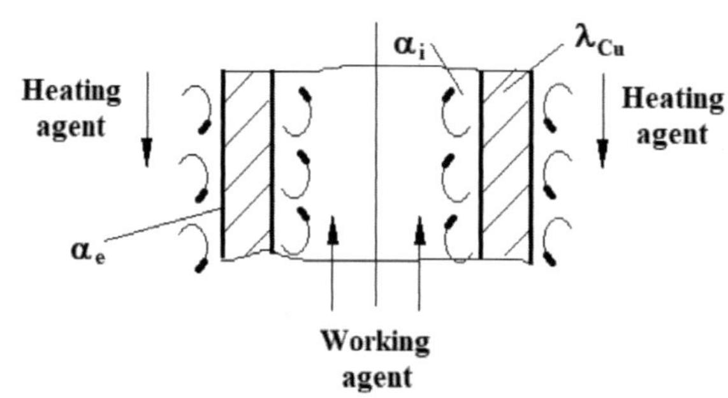


Fig. 7 The scheme of the heat transfer in the heater

For determining the total heat transfer surface necessary to realize the calculated heat flux, it must be known:

- the convection coefficient from the heating agent to the pipe's wall ($\alpha_e$);
- the convection coefficient from the pipe's wall to the working agent ($\alpha_i$);

The chosen variant is with a rhombuses placement of the pipes (fig. 7). The recommended distance between the pipes axels in the case of the mandrel pipes [4] is: t=(1,3 ... 1,5)·$d_e$ [mm] (fig. 8);

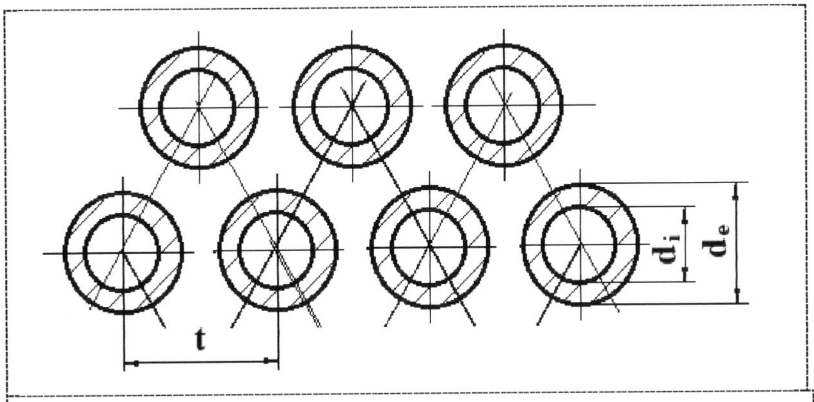


Fig. 8 The arrangement mode of the pipes in the heater

The adopted value is: t=1,5·$d_e$=1,5·8=12 mm;

The inner diameter of the heater mantle was determined based on the scheme of the pipes placement in transversal section: $D_{mti}$=108 mm;

The total flow section of the heating agent through the heater is:

$$S_{cai} = \frac{\pi \cdot D_{mti}^2}{4} - \frac{\pi \cdot d_e^2}{4} \cdot n_{ta} \ [\mathrm{m}^2] \tag{12}$$

The flow speed of the heating agent through the heater is:

$$W_{ai} = \frac{\dot{V}_{ai}}{S_{cai}} \ [\mathrm{m/s}] \tag{13}$$

The equivalent diameter of the flow section of the heating agent through the heater:

$$d_{ech} = \frac{D_{mti}^2 - n_{ta} \cdot d_e^2}{D_{mti} + n_{ta} \cdot d_e} \ [\mathrm{m}] \tag{14}$$

The Reynolds number for the heating agent:

$$\mathrm{Re}_{ai} = \frac{W_{ali} \cdot d_{ech}}{\nu_{ai}} \tag{15}$$

The convection coefficient from the heating agent to the exterior wall of the pipe is:

$$\alpha_e = \frac{Nu_{ai} \cdot \lambda_{ai}}{d_{ech}} \ [\mathrm{W/(m^2/K]} \tag{16}$$

where: - $Nu_{ai}$ – the Nusselt criteria for the heating agent;

For the calculation of working agent Nusselt criteria it is used the Wessman relation [4], in which the calculation is developed for the mean temperature of the heating agent:

$$Nu_{mai} = C \cdot \mathrm{Re}_{ai}^{0,8} \cdot \mathrm{Pr}_m^{1/3} \tag{17}$$

C=0,026·(t/$d_e$)-0,006 for t/$d_e$=1,1 ... 1,5 [4].

$$\Pr_m = \frac{\eta_{ai} \cdot c_{pai}}{\lambda_{ai}} \tag{18}$$

Because the high value of the thermal conductivity of the pipes material (copper) and of the small thickness of the pipe walls (1 mm) the difference between the outer and the inner walls of the copper pipes walls is insignificant.

The convection coefficient from the inner wall of the pipe to the working agent:

$$\alpha_i = \frac{Nu_{mal} \cdot \lambda_{al}}{d_i} \quad [\mathrm{W/(m^2/K]} \tag{19}$$

where: - $Nu_{al}$ – the Nusselt criteria for the working agent;
- $\lambda_{al}$=0,0317 W/mK – the thermal conductivity of the working agent;
- $d_i$=0,006 m – the inner diameter of the pipe;

The Nusselt criteria for the working agent is calculated with the relation [4]:

$$Nu_{mai} = 0{,}021 \cdot \mathrm{Re}_f^{0,8} \cdot \Pr_f^{1/3} \cdot \left( \frac{\Pr_f}{\Pr_p} \right)^{0,25} \tag{20}$$

where: - $\mathrm{Re}_f$ – the Reynolds number for the working agent at the mean temperature in the heater ($t_{alm}$=95 $^0$C);
- $\Pr_f$ – the Prandtl criteria for the working agent at the mean pressure in the heater ($t_{alm}$=95 $^0$C);
- $\Pr_p$ – the Prandtl criteria for the working agent at the mean temperature in the pipe's wall ($t_{alm}$=130 $^0$C);

$$\mathrm{Re}_f = \frac{W_{al} \cdot d_i}{\nu_{alf}} \tag{21}$$

$$\Pr_f = \frac{\eta_{alf} \cdot c_{palf}}{\lambda_{alf}} \tag{22}$$

$$\Pr_p = \frac{\eta_{alp} \cdot c_{palp}}{\lambda_{alp}} \tag{23}$$

The mean global heat exchange coefficient of the heater:

$$k_s = \frac{1}{\dfrac{d_e}{d_i \cdot \alpha_i} + \dfrac{d_e}{2 \cdot \lambda_{Cu}} \ln\left(\dfrac{d_e}{d_1}\right) + \dfrac{1}{\alpha_e}} \tag{24}$$

The total pipes surface necessary to realize the imposed heat flux:

$$S_{it} = \frac{\dot{Q}_{inc}}{k_s \cdot \Delta T_m} \quad [\mathrm{m^2}] \tag{25}$$

The total perimeter of the heater's pipes:

$$P_{ti} = \pi \cdot d_e \cdot n_t \quad [\mathrm{m}] \tag{26}$$

The length of a pipe:

$$L = S_{it} / P_{ti} \ [\text{m}] \tag{27}$$

The adopted value is: L=580 mm;

The variation of the working agent's temperature along the heater is:

$$t_2 = t_2^{'} + \left(t_2^{''} - t_2^{'}\right) \cdot \frac{1 - \exp\left[-\frac{k_s \cdot S_x}{W_1} \cdot \left(1 - \frac{W_1}{W_2}\right)\right]}{1 - \exp\left[-\frac{k_s \cdot S}{W_1} \cdot \left(1 - \frac{W_1}{W_2}\right)\right]} \ [^0\text{C}] \tag{28}$$

where: - $t'_2$=70 $^0$C is the working agent's temperature at the entrance to the heater;

- $t''_2$=150 $^0$C is the working agent's temperature at the exit from the heater;
- k=183.297 is the global heat transfer coefficient of the heater;
- $W_1$ is the equivalent in water of the heating agent;

$W_1 = Q_{m1} \cdot C_{p1}$ [W/(kg·$^0$C)];

$Q_{m1}$=0.154167 kg/s is the mass flow of the heating agent;

$C_{p1}$=2300 K/(kg·$^0$C) is the specific heat at constant pressure of the heating agent;

$W_1$=354.5841 W/(kg·$^0$C);

- $W_2$ is the equivalent in water of the working agent;

$W_2 = Q_{m2} \cdot C_{p2}$ [W/(kg·$^0$C)];

$Q_{m2}$=0.1365 kg/s is the mass flow of the working agent;

$C_{p2}$=1000 K/(kg·$^0$C) is the specific heat at constant pressure of the working agent;

$W_1$=136.5 W/(kg·$^0$C);

- $S_x$ and S [m$^2$] are the heat exchange surface along the heater and the total heat exchange surface, respectively;

The variation of the working agent's temperature is presented in fig. 9.

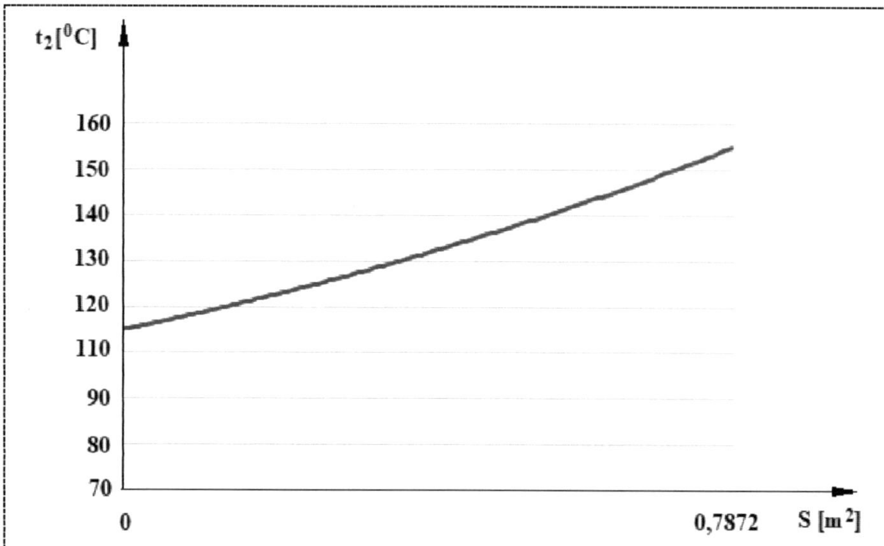


Fig. 9 The working agent's temperature variation along the heater

## 3. The simulation of the heat transfer process for the isochoric heating phase

For the simulation of the Stirling engine working agent's heating process the COMSOL Multiphysics software is used.

This is a finite elements analysis software, to solve and to simulate the electric, mechanic, fluid, acoustic and chemical phenomena. It uses partial differential equations coupled for different phenomena to be analyzed and offers an integrated development environment for applications in the most fields. COMSOL Multiphysics permits the simultaneous simulation of different aspects of a system or of some physical systems and of the interactions between these, the analysis can be done both static and dynamic.

To analyze the heater two coupled modules were used, one of turbulent flow, and the second of thermal transfer between the fluids.

To consider the conduction heat transfer, a thin layer defined by material wideness and material was introduced. The wideness and the material are the ones provided in the heater's design, the walls wideness is of 1 mm and its material is copper.

The 3D model of the heater was realized in the first phase It includes two domains: the one of the heating agent and the one of the working agent. The heater's 3D model is presented in fig. 10.

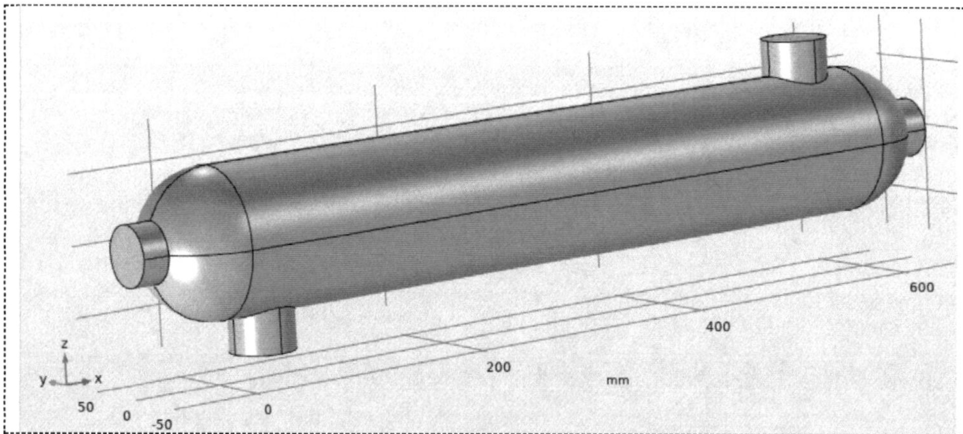


Fig. 10 The 3D model of the heater

To reduce the number of the finite elements of the model for the reduction of the running time, the software permits the use of a half 3D model sectioned with the plan of symmetry (ZX).

In fig. 11 is presented the sectioned heater's plan. The working agent of the Stirling agent is air, and the heating agent is Mono Ethyl Glycol. This agent was chosen because the liquid phase is stable in the interval of the temperatures and pressures in which must work.

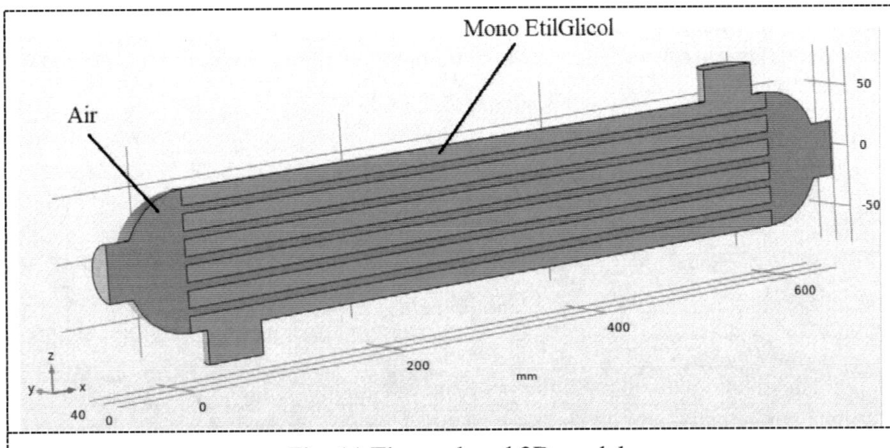


Fig. 11 The analysed 3D model

The entrance and the exit of the air from the heater has the section corresponding to the 54 pipes with the inner diameter $d_i$=6 mm, to have minimum pressure losses.

$$D_a = d_i \cdot \sqrt{54} \quad [\mathrm{m}] \tag{29}$$

Also, for the heating agent circuit the entrance and the exit diameter will be equal to the flow section inside the heater.

$$D_{MEG} = \sqrt{D_{mti}^2 - D_a^2} \quad [\mathrm{m}] \tag{30}$$

The heater's analysis is done for the case in which the flow of the two fluids is in counter current (fig. 12).

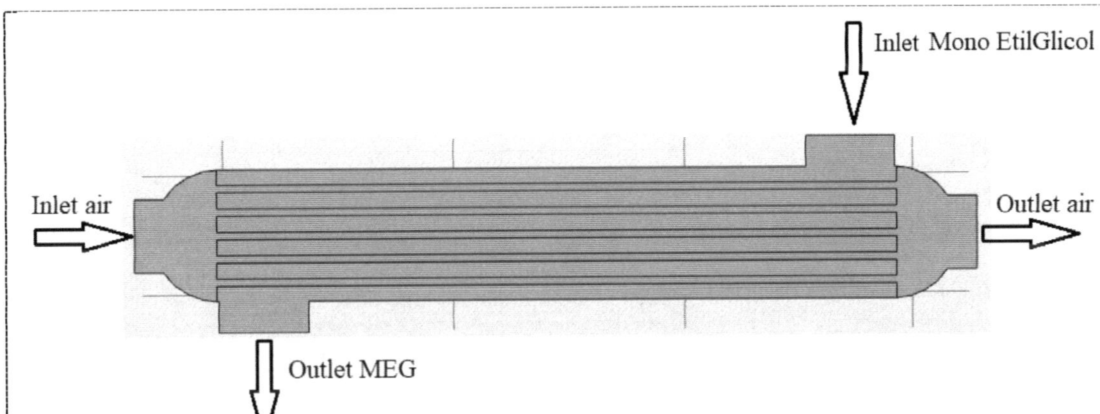


Fig. 12 The counter current flow of the fluids

The boundary conditions to simulate the heating process of the working agent are:
For the turbulent flow phenomenon:
- the inlet air speed=4 m/s;
- the inlet heating agent speed=0.1 m/s;
For the case of the heat transfer phenomenon:
- the inlet temperature of air=70 $^0$C;
- the inlet temperature of the heating agent=180 $^0$C;
For these data, the temperature variation of the working agent is presented in fig. 13. Thermal colors are used, and the temperature interval influences the maximum warm color and the maximum cold temperature, respectively.

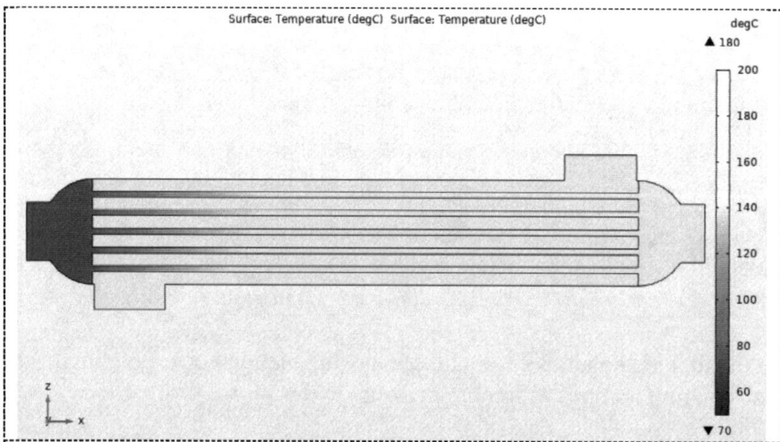


Fig. 13 The working agent's temperature distribution in the heater

The variation of the working agent's temperature along the heater obtained through simulation is presented in fig. 14.

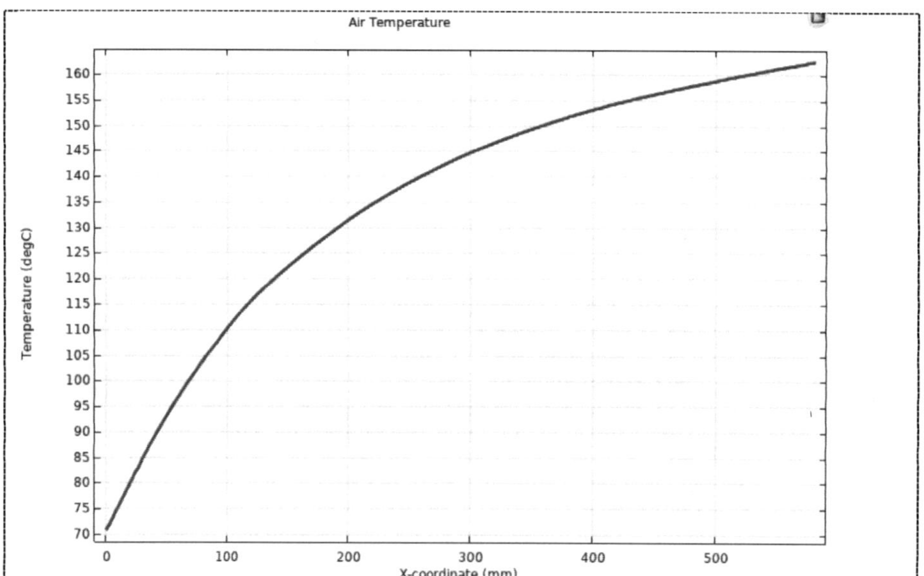


Fig. 14 The working agent's temperature along the heater

## 4. Conclusions

A comparison between the calculated variation of the working agent's temperature along the heater and the one obtained by simulation is presented in fig. 15.

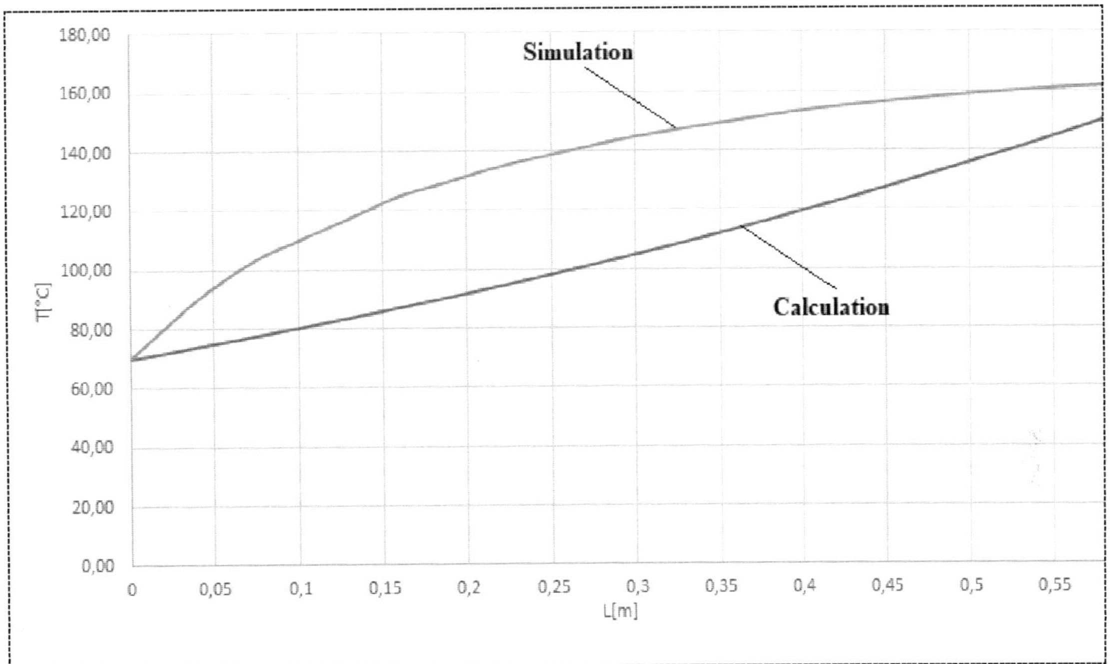


Fig. 15 Comparison between the calculation and the simulation

It can be observed that the values obtained by simulation for the variation of the temperature along the heater are up to 40% greater than the ones calculated. Also, the shapes of the curves are different.

The analytic calculation of the heat transfer is a very delicate problem. The formulas used in this case are empirical, or semiempirical, and were determined for specific cases and can produce errors up to 50%. But, in this case, more important is the temperature of the working fluid at the exit from the heater. The difference between the calculated temperature of the working agent at the exit from the heater and the one obtained by simulation is about 10%. That's quite reasonable in the case of the heat transfer. The variation of the temperature has no importance in the functioning of the Stirling engine.

The thermal efficiency of the reversible Stirling engine cycle is:

$$\eta = 1 - \frac{T_1}{T_3} \tag{31}$$

where: - $T_1$ [$^0$C] is the temperature of the cold source;

- $T_3$ [$^0$C] is the temperature of the warm source;

In this case, the temperature of the warm source is equal to the temperature of the working agent at the exit from the heater. If $T_1$=ct, from the (rel. 31) it results that the thermal efficiency of the Stirling engine is greater as this temperature rises. This temperature depends on the temperature of the heating agent at the entrance in the heater. The temperature of the heating agent depends on the intensity of the solar radiations. This means that the functioning conditions depend on the solar radiation.

The analytical calculus of the heater was based on a temperature of the heating agent at the entrance in the heater $t'_1$=180 $^0$C. Depending on the solar radiation, this temperature can rise to 200 – 220 $^0$C.

In this case, the Stirling engine is a part of a thermal hydraulic generator that was designed to store the mechanical work produced from the solar energy into hydrostatic energy inside a hydro-pneumatic piston accumulator [5]. It was designed to be used to replace the batteries as a storage device for the energy. Solar panels are used mainly to produce hot water. This solution intends to use much more of the disposable solar energy. The energy stored in the hydraulic accumulator that is produced in daytime can be used to produce electric current that can refill the batteries of an electric car during the night.

As mentioned before, solar panels can produce electric energy by using the Stirling cycle [2]. One of the particularities of the thermal-hydraulic generator is that the pistons have only an alternative rectilinear movement. This means that it is possible to produce electric current through a linear electric generator.

There was research to use linear electric generators for the propulsion of electric cars [6]. This possibility can be used for the propulsion of some small cars, for 1-2 passengers, such as electric cars used by elder persons to visit the center of the touristic cities.

The possibilities to use the Stirling cycle to produce energy from sustainable source are numerous. So, the proposed algorithm can be used for the dimensioning of such systems that can use sustainable energy sources for the propulsion of the road vehicles. After the dimensioning of the heat exchangers of a Stirling engine, specialized programs can be used to simulate its functioning.

## 5. References


[1] Homutescu C A, Savitescu Gh, Jugureanu E, Homutescu V M 2003, Intoducere in masini Stirling (Introduction in Stirling machines), CERMI Publishing, Iasi

[2] Artin Der Minassians 2007, Stirling Engines for Low-Temperature Solar-Thermal-Electric Power Generation, Technical Report No. UCB/EECS-2007-172, University of California Berkeley

[3] precer.com/Technology.html

[4] Stefănescu D., Leca A., at al 1983, Transfer de căldură și de masă. Teorie și aplicații, (Heat and mass transfer. Theory and applications) Editura Didactică și Pedagogică (Didactic and Pedagogical Plubishing), Bucharest

[5] Mitran T. Chioreanu N. Abaitancei H and Rus A. 2016 A Thermal-hydraulic system for the Conversion and the Sorage of Energy *Proceedings of the Annual Session of Scientific Papers „I.M.T. Oradea* Vol. XV p 67-70

[6] Toyota R&D developing free-piston engine linear generator; envisioning multi-FPEG units for electric drives unites 2014 *Green Car Congress*

IOP Conf. Series: Materials Science and Engineering **1311** (2024) 012004 doi:10.1088/1757-899X/1311/1/012004

# Conversion of a small size passenger car to hydrogen fueling: simulation of full load performance

**A Irimescu[1,*], B M Vaglieco[1], S S Merola[1], V Zollo[2] and R De Marinis[2]**

[1] Science and Technology Institute for Sustainable Energy and Mobility STEMS – CNR , Via G. Marconi 4, 80125 Napoli, Italy

[2] Demax SRL, Strada Statale 7 Appia km 251, 82014 Ceppaloni, Benevento, Italia

[*] adrian.irimescu@stems.cnr.it

**Abstract**. Hydrogen offers a valid choice when considering zero emissions transport. This alternative fuel is well suited for spark ignition (SI) engines and it can be implemented with relatively minor changes in terms of added components. Initial evaluations of feasibility with respect to available space for the fuel tank revealed that it is possible to convert a small size passenger car to hydrogen fueling and maintain a range comparable to the fully electric alternative. Peak power assessment showed that additional boosting was required compared to gasoline operation, close to the limit of compressor surge. As a result, the present study extended the engine speed range so as to evaluate full load performance levels that can be obtained when using hydrogen. 0D/1D simulation was applied for simulating power unit output as well as related control parameters. A dedicated laminar flame speed sub-model was implemented for including fuel chemistry effects. Predicted combustion phasing showed the required changes in ignition timing when switching fuels and revealed that the engine could be operated close to stoichiometry when using hydrogen. A reduction of torque was calculated at low engine speed, even if peak power ratings were practically the same as with gasoline fueling.

## 1. Introduction

Hydrogen fueled spark ignition (SI) engines fit well in a zero-emissions scenario and can provide a viable choice for short-term implementation [1]. Compared to other technologies such as fuel cells they are cost effectiveness thus further enhancing potential impact [2]. An intermediate step that involves hydrogen addition to hydrocarbon based fuels is also completely feasible, bringing immediate benefits in terms of efficiency and combustion stability [3].

Reduced volumetric energy density is one of main aspects to consider when using $H_2$ as fuel [4]; it further emphasizes the need to maximize efficiency so as to ensure increased vehicle range. Lower volumetric efficiency is also paramount, as this brings a drop in specific power if considering straight-forward implementation in SI units that were designed for gasoline [5]. High laminar flame speed is another interesting property of $H_2$ that ensures lean operation with good stability [6] and thus increases control margins. Switching between stoichiometric fueling at high load and lean at part load can be a valid strategy that ensures reduced emissions by using the well proven three way catalytic converter technology [7].

Within this context, the present work investigated specific requirements for converting a small size passenger car to hydrogen fueling. After an initial evaluation of achievable range and peak power, which had a positive outcome, a more detailed analysis was employed for studying the entire engine speed range at full load. This was achieved through 0D/1D modeling that highlighted the main issues to be resolved and the way for defining custom control strategies for $H_2$ fueling.

## 2. Simulation model

Previous evaluations revealed that the overall range that could be achieved with a 30 liter $H_2$ tank at 700 bar was around 100 km [8]. This figure is comparable to the 125-135 km rating specified by the manufacturer for the electric version of the vehicle [9]. Therefore, the overall goal of obtaining a product that can used in zero-emissions scenarios can be obtained even with a circular economy approach that further provides added value to society.

Table 1 shows the main engine specifications. The model used in the initial phase was modified for including a pressure ratio limit o as to avoid compressor stall. Fig. 1 shows an overview of the model. As an overall concept, the entire propulsion system is divided into components and each component may have multiple elements (e.g. the intake manifold is modeled as a series of T elements connected in series). Components have specific sub-models that can simulate different phenomena that range from simple filling-emptying to fuel evaporation and combustion.

Table 1. Engine specifications.

| | |
|---|---|
| Displacement | 599 $cm^3$ |
| Number of cylinders | 3 |
| Rated power | 40 kW @ 5250 rpm |
| Rated torque | 80 Nm @ 2000-4400 rpm |
| Bore x Stroke | 63.5 mm x 63.0 mm |
| Connecting rod length | 114 mm |
| Compression ratio | 9.5:1 |
| Number of Valves | 2 per cylinder |
| Intake valves opening/closure | 363/164 deg bTDC |
| Exhaust valves opening/closure | 157/349 deg a/bTDC |
| Fuel system | port fuel injection at 3.5 bar for gasoline and 5 bar for hydrogen |
| Ignition | inductive discharge, 2 spark plugs per cylinder |

Turbine and compressor maps were reconstructed by the software using the design point that features 2.3 pressure ratio at 230000 rpm and 7 lbs/min mass averaged flow [10]. The predictive EngCylCombSITurb [11] combustion model was employed for including fuel chemistry effects. A dedicated laminar flame speed correlation valid for engine conditions [12] was implemented for hydrogen, following the procedure described in [13]. Unfortunately, there was no in-cylinder pressure data available for validation. Nonetheless, the boost level was predicted to be at 1.5 bar for obtaining 40 kW at 5250 rpm with gasoline; it matches perfectly the OEM provided data and can be seen as a good starting point for drawing conclusions with respect to engine operation.

Engine speed was swept from 1000 to 5500 rpm at full load, i.e. by setting the target power at the level provided by OEM data. Target relative air-fuel ratio ($AFR_{rel}$) was set at 0.9 for gasoline and at stoichiometry for $H_2$. Spark timing was swept in 5 deg steps from -30 to 30 deg aTDC for each rpm value. The knock limited spark advance (KLSA) or the maximum MBT was chosen for every condition. Compared to the previous version of the model [8], an additional limit was imposed for the proportional integral derivative (PID) controller that actuates the wastegate. This was implemented so as to avoid compressor surge throughout the simulations.

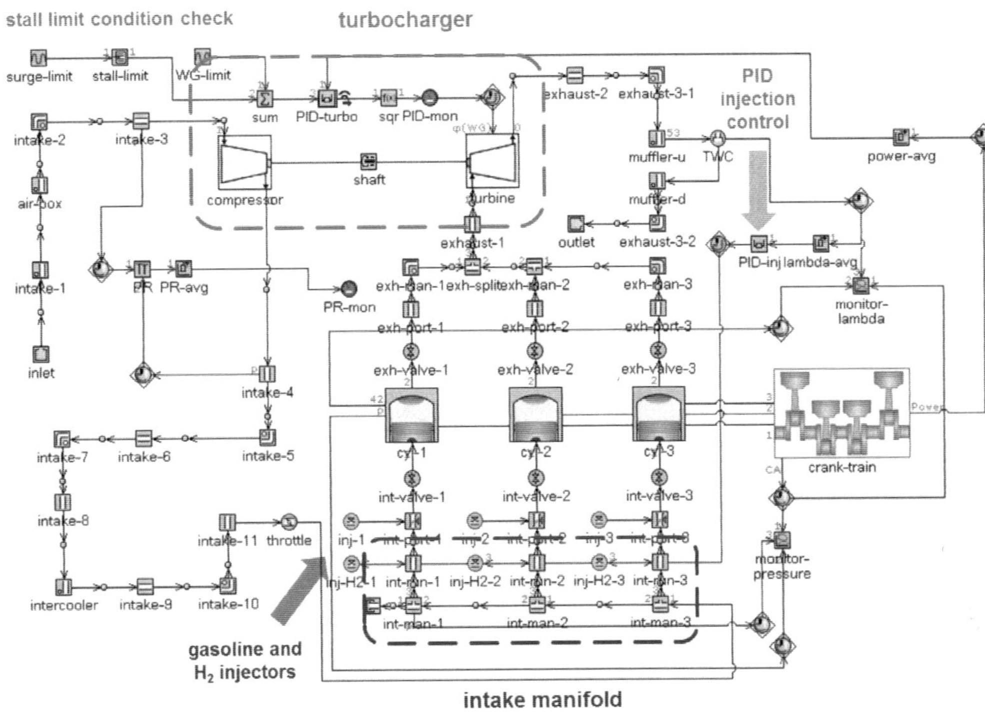


Fig. 1. Overview of the simulation model with several components highlighted.

## 3. Results and discussion

Peak power was already verified as completely compatible with hydrogen fueling [8]. Boost pressure needed to be increased by about 30%, from 0.5 $bar_g$ to 1.0 $bar_g$ when switching from gasoline to $H_2$. The model also predicted that gasoline was knock limited while hydrogen was not.

Fig. 2 shows the brake power curve when using the two fuel types. The model predicted that the reference power level can be achieved without any issues when using gasoline between 2500 and 5500 rpm. A slight difference was calculated for 1500 and 2000 rpm; achievable power was predicted to be around 8% lower at 1500 rpm and almost 5% less at 2000 rpm. This was due to the aforementioned surge limit imposed for the wastegate PID controller. One possible reason is that the compressor map was extrapolated by the simulation software based only on the design point. This could directly influence the surge line and thus result in lower predicted full load power. Another point that should be considered is that the engine was knock limited in these two points (the only condition that the KLSA setting was not reached was at 5500 rpm); for example, at 1500 rpm spark timing would need to be retarded to 10 deg aTDC so as to avoid knocking. More to the point, with this "safe" setting, only 9.25 kW (compared to the 10.10 kW reference) can be obtained. With the ignition point set at 5 deg bTDC, 10.01 kW power was predicted, but with over 20% of the charge still left unburned when knocking would occur. The first explanation seems to be a good starting point, but a combination of the two factors is most likely. Nonetheless, these observations will be recalled when analyzing the results obtained with $H_2$ fueling.

Gaseous fuel simulations suggest that knock limit would not be a problem. Only the 3500 and 4000 rpm conditions reached the KLSA setting. Whether knocking would be a problem at lower rpm, in the case that the required boost pressure would be available, is unclear. Nonetheless, induction time integral values were predicted to be around 40-80% between 1000 and 3000 rpm, thus suggesting sufficient knock mitigation margins.

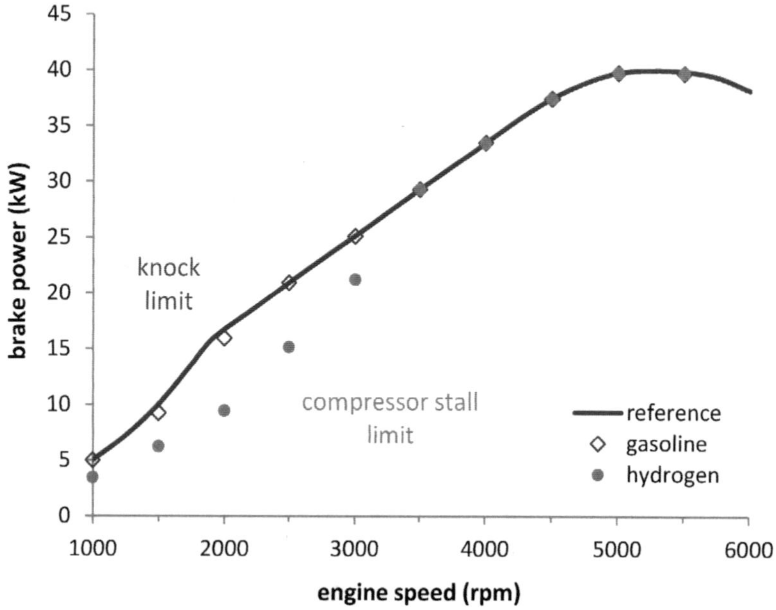


Fig. 2. Full load brake power simulated for gasoline (diamond symbols) and hydrogen (round symbols); the reference curve is also shown with a line.

Full load reference power levels were predicted to be reachable with $H_2$ only over 3500 rpm. Below this engine speed, as much as 40% loss in power was calculated, with 1500 rpm being the most critical condition. Overall a 30% drop in engine output was predicted from 1000 to 2500 rpm, due to compressor surge limitations. At 3000 rpm, the loss was only 15%. Again, the results obtained at 1500 and 2000 rpm with gasoline need to be kept in mind when considering these conclusions. Nonetheless, they highlight compressor surge as a major issue that needs to be considered when using $H_2$ as fuel.

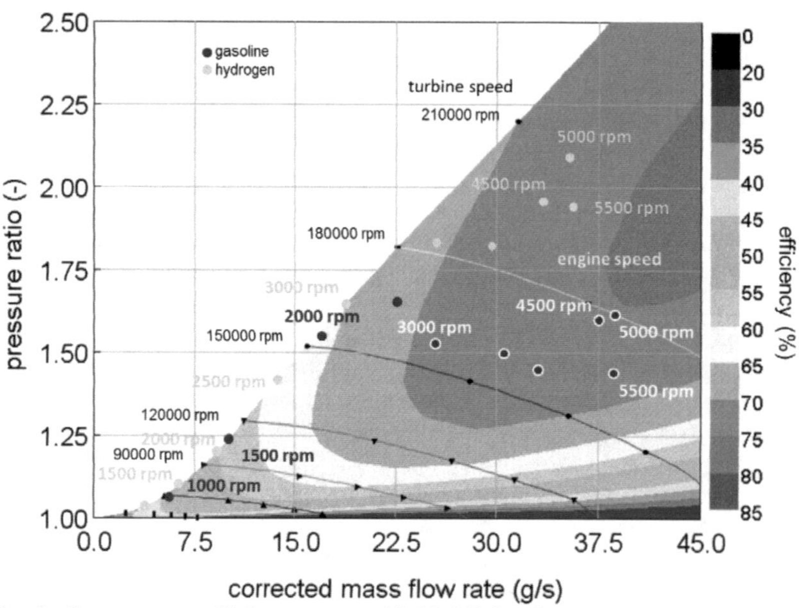


Fig. 3. Compressor efficiency map with highlighted operating points in the 1000-5500 rpm engine speed range at full load.

Fig. 3 gives a more detailed picture on the operating conditions that were simulated during full load operation. As expected, at low rpm the compressor was operated at the surge line for both fuel types. Above 2000 rpm gasoline fueling required a boost pressure around 0.5 $bar_g$, close to the OEM data. $H_2$ instead required higher pressure ratio (PR), close to 2; evidently, below 3000 rpm this target boost could not be reached. The hysteresis around the 1.5 and 2.0 PR values (far from the surge line) is due to the relatively coarse step used for sweeping ignition timing. A more interesting observation is that the surge limit is reached for a much more range of engine speed values, up to 3000 rpm when using $H_2$.

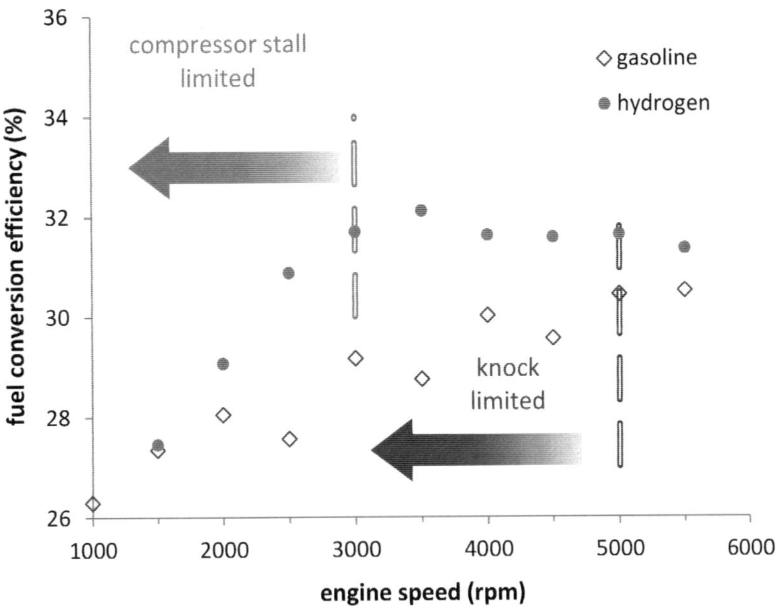


Fig. 4. Full load conversion efficiency for the two fuel types; compressor stall limited range and knock limit are also illustrated.

Gasoline featured around 3-10$\%_{rel}$ lower conversion efficiency compared to the gaseous fuel (Fig. 4). One of the main reasons is that the target air-fuel ratio was 0.9 for the liquid fuel and 1.0 for $H_2$ thus ensuring improved combustion efficiency for the latter. The improvement was most evident around 3500 rpm and was mostly due to the fact that gasoline operation at the KLSA setting featured a much more prominent effect. Fig. 5 shows the knock index (KI, a parameter that combines the unburned charge fraction at knock occurrence, cylinder volume, unburned gas temperature and equivalence ratio into one overall parameter) for cylinder 1 when using the two fuel types at 3500 rpm. Even if in both cases the KLSA setting was found to be between 0 and 5 deg bTDC, $H_2$ was closer to the point of MBT (at 0 deg bTDC spark timing); instead, much more consistent improvement could have been obtained with gasoline if ignition could have been advanced.

One reason for concern is that the KI seems to evolve in a much steeper manner for the gaseous fuel, suggesting that ignition may need to be set at a "safer" condition and thus not being able to exploit the full potential for improving efficiency. Cylinder-to-cylinder variation is another issue; e.g. at 0 deg bTDC spark timing cylinder 3 featured a KI around 20, while that for cylinder 1 was over 46. This is due to differences in volumetric efficiency noted in a previous study [8]. The fact that such large variations can be recorded when using $H_2$ means that knock mitigation through spark advance control needs to be more carefully considered.

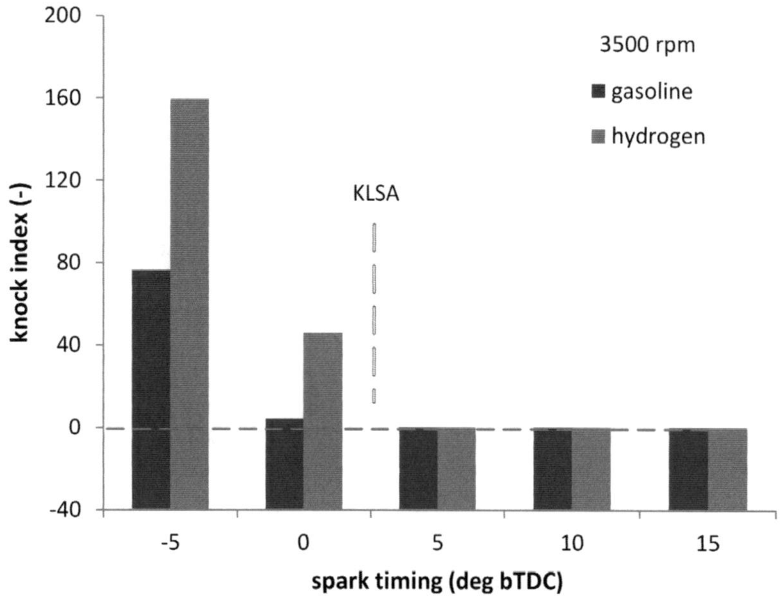


Fig. 5. Knock index for cylinder 1 at 3500 rpm with gasoline and hydrogen fueling.

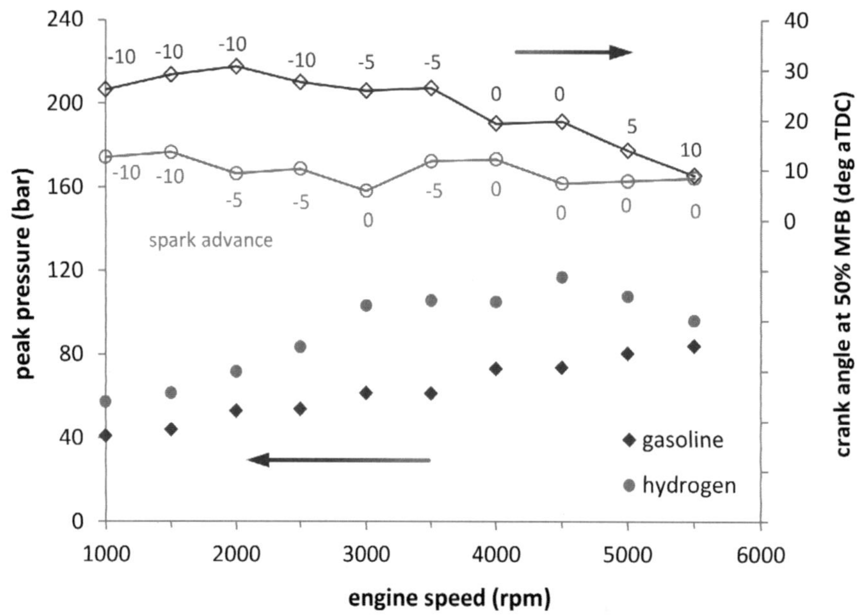


Fig. 6. Peak pressure and location of 50% MFB points for cylinder 1 at full load when using gasoline (diamond symbols) and hydrogen (round symbols); spark timing is also marked for each point close to the CA50 lines.

Peak pressure is another essential parameter, especially at full load. For $H_2$, this parameter was consistently higher compared to gasoline (Fig. 6) due to its higher laminar flame speed resulting in "faster" combustion but also because boost pressure needed to be augmented. This could represent another drawback when considering vehicle conversion to hydrogen fueling, given that changes to the piston-rod-crankshaft assembly require extensive modifications for enduring higher peak pressure.

Rapid combustion development determined CA50 (crank angle location of 50% mass fraction burned) values quite close to 10 deg aTDC that further emphasizes the conclusion that hydrogen was not knock limited. In fact, gasoline featured much more retarded combustion and only for the 5500 rpm condition reached the same CA50 as $H_2$. One interesting observation is that the spark timing was predicted to be within a relatively narrow range (please note that spark advance values listed in Fig. 6 are expressed as deg bTDC). This at least seems to suggest that the settings for gasoline could be used as a good starting point when remapping the ECU for $H_2$ use.

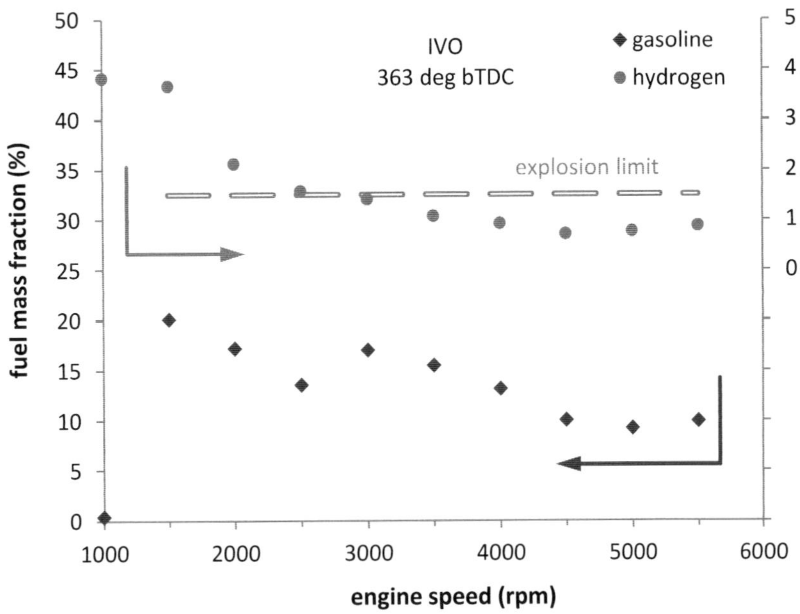


Fig. 7. Fuel mass fraction in the intake port of cylinder 1 with the start of injection fixed ad 270 deg aTDC.

Another aspect that needs to be considered is that hydrogen tends to feature backfire phenomena. Previous identified a start of injection setting of 270 deg aTDC as the best option for mitigating backfire [8]. This strategy was used for the present simulations as well.

Fig. 7 shows the concentration of fuel in the intake port of cylinder 1, at intake valve opening (IVO). It should be noted that the lower explosion limit is shown only for hydrogen. This is due to the fact that it features much lower ignition energy compared to gasoline. As evidenced in Fig. 7, the fixed timing of 270 deg aTDC ensures $H_2$ concentrations below the limit only over 3500 rpm. The SOI setting was optimized for the peak power point, i.e. at 5250 rpm. Using a SOI of 360 deg bTDC at 1000 rpm ensured the minimum concentration of fuel in the intake port at IVO and also the highest efficiency value. This further emphasizes the need for defining in-depth control strategies tailored for the use of hydrogen in SI engines.

## 4. Conclusions

An initial evaluation of peak power and overall range for a hydrogen fueled small size passenger car was extended to the entire rpm sweep at full load. One of the main conclusions was that compared to gasoline knock limitations are less likely. On the other hand, compressor surge limitations prevent the $H_2$ engine from reaching the same power level as with gasoline below 3500 rpm. A drop of 30-40% in power was predicted by the 0D/1D simulations at low engine speed. The numerical results obtained with gasoline at 1500 and 2000 rpm suggest that the actual drop in full load power could be less critical. Improving the compressor map used in the simulations was identified as an important step for ensuring more accurate results.

A relative improvement of up to 10% in fuel conversion efficiency was predicted for the gaseous fuel, mostly due to operation close to the point of optimized ignition settings and the possibility of using stoichiometric air-fuel ratio rather than rich fueling.

More sensitive knock variation and cylinder-to-cylinder differences were identified as possible issues that may hinder obtaining the full efficiency improvement potential of hydrogen. Higher peak pressure compared to gasoline is another factor that needs to be considered. This was due to increased boosting required for hydrogen, as well as rapid combustion development.

Backfire mitigation through optimized SOI settings was also found to be possible throughout the entire rpm sweep. Compared to high engine speed for which injecting $H_2$ during the exhaust was the best choice, at low rpm fuel delivery needs to be performed during intake.

All these aspects emphasize the complexity of converting passenger cars to hydrogen fueling and highlight the potential of obtaining efficient and close-to-zero emissions. Dedicated control strategies and careful consideration of multiple requirements are paramount for achieving the two main goals.

**References**

[1] S. Verhelst, T. Wallner. Hydrogen-fueled internal combustion engines. Progress Energ. Combust. Sci. 35(6):490-527, 2009, doi:10.1016/j.pecs.2009.08.001.

[2] Hydrogen Council, Hydrogen Insights 2021: A perspective on hydrogen investment, market development and cost competitiveness, February 2021.

[3] S. E. Hosseini, B. Butler. An overview of development and challenges in hydrogen powered vehicles. International Journal of Green Energy 17(1):13-37, 2020, doi: 10.1080/15435075.2019.1685999.

[4] Z.A. Stępień. Comprehensive Overview of Hydrogen-Fueled Internal Combustion Engines: Achievements and Future Challenges. Energies 14:6504, 2021, doi: 10.3390/en14206504.

[5] A. Irimescu. Study of Volumetric Efficiency for Spark Ignition Engines Using Alternative Fuels. Analele Univ. Eftimie Murgu Resita Fasc. Ing. 17(2):149-154, 2010.

[6] G.M. Kosmadakis, D.C., Rakopoulos, C.D., Rakopoulos. Assessing the cyclic-variability of spark-ignition engine running on methane-hydrogen blends with high hydrogen contents of up to 50%. Int. J. Hydrogen Energ. 46(34):17955-17968, 2021, doi: 10.1016/j.ijhydene.2021.02.158.

[7] T. Wallner, H. Lohse-Busch, S. Gurski, M. Duoba, W. Thiel, D. Martin, T. Korn. Fuel economy and emissions evaluation of BMW Hydrogen 7 Mono-Fuel demonstration vehicles. Int. J. Hydrogen Energ. 33(24):7607-7618, 2008, doi: 10.1016/j.ijhydene.2008.08.067.

[8] A. Irimescu, B.M. Vaglieco, S.S. Merola, V. Zollo, R. De Marinis. Conversion of a small size passenger car to hydrogen fueling: focus on rated power and injection phasing effects. SAE Technical Paper 2022-24-0031, 2022.

[9] *** www.smart.mercedes-benz.com, Smart EQ ForTwo.

[10] N. Cuturi, E. Sciubba. Design of a Tandem Compressor for the Electrically-Driven Turbocharger of a Hybrid City Car. Energies 14:2890, 2021, doi: 10.3390/en14102890.

[11] *** Gamma Technologies, GT-SUITE Flow Theory Manual Version 7.4, 2013.

[12] U. Gerke, K. Steurs, P. Rebecchi, K. Boulouchos. Derivation of Burning Velocities of Premixed Hydrogen/Air Flames at Engine-Relevant Conditions Using a Single-Cylinder Compression Machine with Optical Access. Int. J. Hydrogen Energ. 35 (2010): 2566-2577, doi:10.1016/j.ijhydene.2009.12.064.

[13] A. Irimescu, S. Di Iorio, P. Sementa. Laminar Flame Speed Based Optimization of Efficiency and Emissions for Methane-Hydrogen Fueled SI Micro-Generators. SAE Technical Paper 2021-24-0047, 2021, doi: 10.4271/2021-24-0047.

# The effects of water injection inside an internal combustion engine regarding $NO_x$ emissions

**A T Borborean[1*], V Stoica[1] and D Llelea[1]**

[1] Politehnica University of Timisoara, Mechanical Engineering Faculty, Mihai Viteazu Blv, 1, Timisoara, Romania

*Email: andrei.borborean@upt.ro

**Abstract.** In this paper we present the influence of water injection on NOx emissions and the performance of a gasoline engine subjected under light to medium loads. The test bench used for measurements has a 4-cylinder, 1.6-liter engine and operates with a wasted spark configuration. Water is injected upstream of the throttle valve through a single injector mounted in the intake runner. The injection timing and amount of water injected inside the engine have been controlled by NI CompactRIO hardware device by the aid of an in-house algorithm implemented in LabView. Water is injected twice at each rotation of the crankshaft in order to have water in all cylinders. The injection of water is done at a fix angle, one time at 0°CA and the other at 360°CA, only the amount is varied from 10% to 45% of the total fuel fraction injected. The results showed that at an amount of water of about 25% of the total fraction of fuel injected, it produced a considerable decrease in NOx emissions, while keeping engine performance unchanged.

## 1. Introduction

New legislation imposing new and new standards to reduce pollution from internal combustion engines, make manufacturers and researchers find new ways to reduce pollution. These pollution reduction methods can be both before the formation of emissions or after treatment, the so-called pre- and post-combustion emission reduction systems. For the moment, the most effective have proven to be the post-treatment ones.

However, in the last 10-15 years, a trend that was thought to be dead has been studied, namely the injection of water inside the cylinder of an internal combustion engine, which appeared during the Second World War in order to reduce the effect of detonation on engines. [1]

Having a high latent heat of vaporization, water has the substantial effect of reducing the temperature of the fresh charge. Furthermore, the water particles that do not vaporize inside the intake manifold reach the inside of the cylinder where they continue their vaporization process, reducing the overall temperature inside the combustion chamber. Managing in this way to reduce $NO_x$ emissions and reduce the knock limit in supercharged engines. [2]

This paper aims to present a way of implementing a water injection system inside a cylinder of an internal combustion engine but also to present a series of measurements on $NO_x$ emissions.

The first part will discuss the process of water injection and evaporation, followed by a study of the mechanism that occurs during the compression-combustion process when there are water particles in the cylinder in order to provide a better understanding on the unfolding phenomena.

Finally, a series of results on the effects produced by water injection on $NO_x$ emissions are presented, as well as a series of problems and challenges that need to be studied further.

## 2. Experimental test bench

As previously presented, the test bench is based on a spark ignition internal combustion engine with port fuel injection, having 4 valves per cylinder, manufactured by Daewoo. The engine has a displacement of 1598 [cc], a piston diameter and stroke of 79.0 x 81.5 [mm] at a compression ratio of 9.5 [-]. [3]

As can be seen in figure 1, the engine is connected to a hydrodynamic brake to be able to obtain various loads and operating regimes. In addition to these, the engine has numerous temperature and pressure sensors mounted on the intake and exhaust manifolds, to be able to monitor all the state parameters of the air-fuel-exhaust mixture. Fuel and water consumption is carefully measured using the gravimetric method. The air flow into the engine is measured in two ways, one using a double diaphragm and one using an electronic air flow meter.

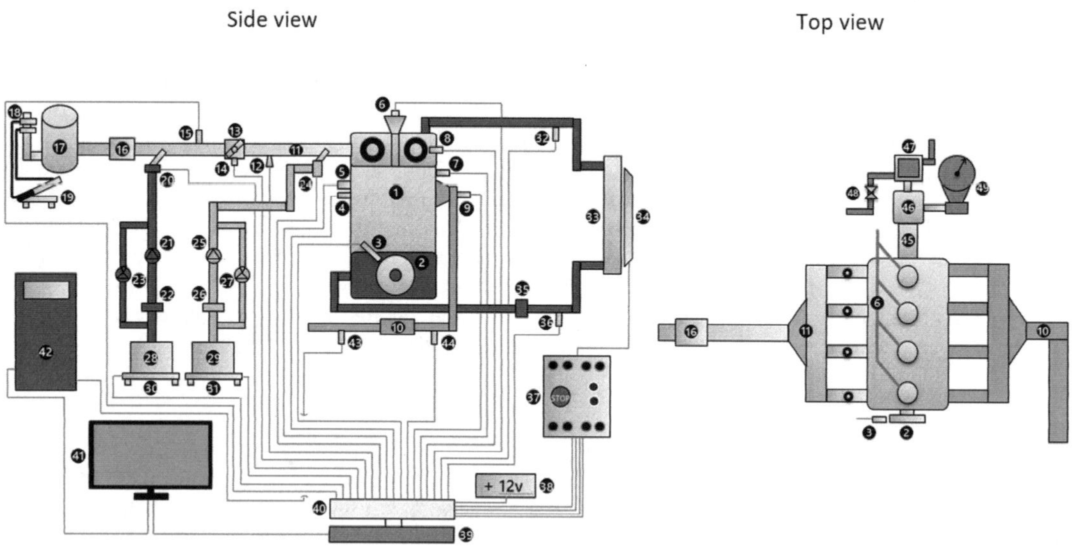


**Figure 1.** Schematic representation of the test bench

IOP Conf. Series: Materials Science and Engineering **1311** (2024) 012005 doi:10.1088/1757-899X/1311/1/012005

| No. | Description |
|---|---|
| 1 | Internal combustion engine |
| 2 | Cranckshaft pulley |
| 3 | Cranckshaft position sensor |
| 4 | Oil pressure sensor |
| 5 | Knock sensor |
| 6 | Ignition coils |
| 7 | Water temperature sensor |
| 8 | Camshaft position sensor |
| 9 | Exhaust temperature sensor |
| 10 | Exhaust manifold |
| 11 | Intake manifold |
| 12 | Intake manifold air pressure sensor |
| 13 | Throttle body |
| 14 | Throttle position sensor |
| 15 | Intake air temperature sensor |
| 16 | Mass air flow sensor |
| 17 | Settle tank |
| 18 | Double diaphragm |
| 19 | Pressure gauge |
| 20 | Water injector |
| 21 | Water pressure pump |
| 22 | Water filter |
| 23 | Water pressure regulator |
| 24 | Fuel injector |
| 25 | Fuel pressure pump |
| 26 | Fuel filter |
| 27 | Fuel pressure regulator |
| 28 | Water tank |
| 29 | Fuel tank |
| 30 | Water weighing scale |
| 31 | Fuel weighing scale |
| 32 | Coolant temperature sensor 1 |
| 33 | Heat exchanger |
| 34 | Cooling fan |
| 35 | Coolant flow meter |
| 36 | Coolant temperature sensor 2 |
| 37 | Command and control panel |
| 38 | 12V power supply |
| 39 | NI CompactRIO |
| 40 | Electronic signal conversion module |
| 41 | PC |
| 42 | CAPELEC 3201 |
| 43 | Exhaust gas analyzer probe |
| 44 | O2 Lambda sensor |
| 45 | Mechanical transmission |
| 46 | Hydrodynamic brake |
| 47 | Hydrodynamic brake cooling system |
| 48 | Adjusting valve of the hydrodynamic brake cooling system |
| 49 | Scale |

The engine is kept at a stable thermal regime using a larger heat exchanger than the original one fitted to the car dedicated to the engine.

All engine operating parameters are controlled via a real-time control system called Compact-RIO from National Instruments. In figure 2 we can see the graphical command-control interface and parameter recording implemented in the LabVIEW software program of the same company.

In order to study the effects of water injection on the emissions and performance of a spark ignition internal combustion engine, it was necessary to find a location for the water injector(s).

For this, three possible locations were studied:

- a single water injector mounted in the intake gallery upstream/downstream of the throttle body;
- multipoint injection, in the port valve;
- direct injection. [4]

The first option was chosen, being the simplest to implement and with the lowest costs. Although it is a weaker variant in terms of the uniformity of water particles in all cylinders, it still has a number of advantages that make it clearly superior to the other two variants. [4]

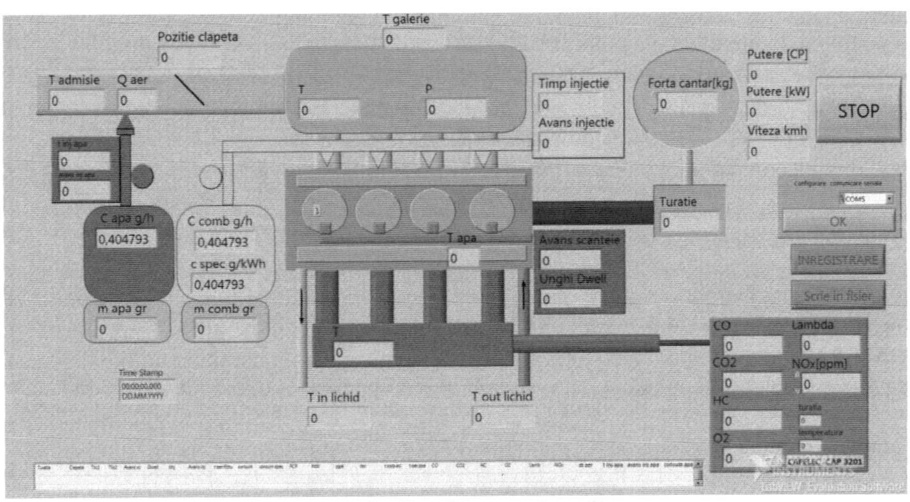

**Figure 2.** Graphic user interface

Next, for the thorough study of the effects of water injection on combustion and implicitly on pollutant emissions, it was necessary to implement an installation to be able to record and investigate the combustion process. Nowadays this is done by measuring and recording the variation of the pressure in the cylinder in relation to the angle of rotation of the crankshaft, the so-called p-α diagram. For this, the measuring installation shown in figure 3 was used.

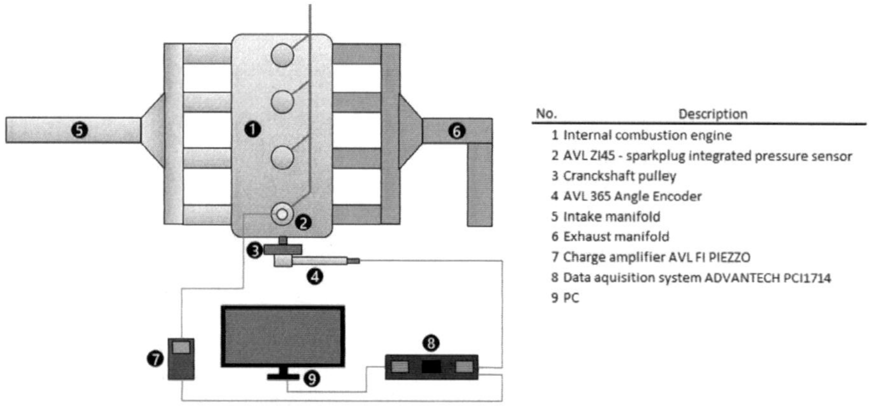


**Figure 3.** Schematic representation of the p-α diagram installation

The installation is based on two important sensors and a data acquisition system, namely:
- a sensor for recording the pressure in the cylinder, this sensor is a piezo type, embedded in a spark plug;
- a sensor for recording the angular position of the crankshaft;
- data acquisition system Advantech PCI1714 board.

The graphic interface for viewing and recording the diagram was made using the National Instruments LabVIEW software. For the analysis of the exhaust gases, a dedicated gas analyzer for spark ignition internal combustion engines, namely CAP 3201 of the Capelec company, was used.

## 3. Experimental procedure

On our stand, as can be seen in figure 1, water is injected with the aid of the water injector (20), inside the intake manifold (11), upstream of the throttle valve (13).

The installation that provides the water injection is a simple one, being composed of a water tank (28), a filter (22), a pump (21) that provides the pressure required for the injection and a pressure regulator (23), which maintains a constant pressure in the installation. During the experiments, the pressure in the water injection system was kept constant at 4 bar.

Before taking the measurements, the engine was brought to a stable thermal regime, namely the oil temperature of at least 80°C and the coolant temperature at least 75°C. All measurements were taken while maintaining a constant engine speed at around 2400rpm and at a throttle position of 25%.

The experimental measurements were divided into nine sets of measurements, each set performed three times. The first set of measurements was taken to obtain baseline measurements, i.e. the engine operating within normal parameters without injecting water into the cylinder. It was made only to be able to draw the indicated pressure diagram, shown under normal operating conditions. In addition, it was necessary to record an engine cycle without combustion. To achieve this, the spark was turned off for a fraction of the time, enough to be able to record a cycle without combustion.

The rest of the measurements were those to which the water injection is added. In order to be able to make a connection between the amount of water injected and that of fuel, the water injection was quantified as a mass percentage of the total mass of fuel injected. For a better understanding, you can follow the equation below.

$$\dot{m}_{inj_{water}} = \frac{\dot{m}_{inj_{fuel}} * k}{100}; \qquad k \in [5,10,15,20,25,30,35,40] \tag{1}$$

## 4. Experimental results

As stated above, we will begin by discussing the process of water vaporization. The graph below shows the temperature and pressure variations along the intake path. The intake air pressure is measured by the aid of the air pressure sensor (12). The temperature variation measured at the following points: inside the intake manifold before the water injector (16), at the port valve gate (15) and the temperature of the entire body of the intake manifold (50). Based on the graph shown in figure 4, we can observe that the air temperature inside the intake manifold measured before the water injector remains constant, and the temperature measured in the valve gate decreases as more water is injected. At the same time we can see that the temperature of the body of the intake manifold decreases as more water is injected.

Following the temperature variation, we can say that the water begins its vaporization process right from the moment of injection inside the intake manifold.

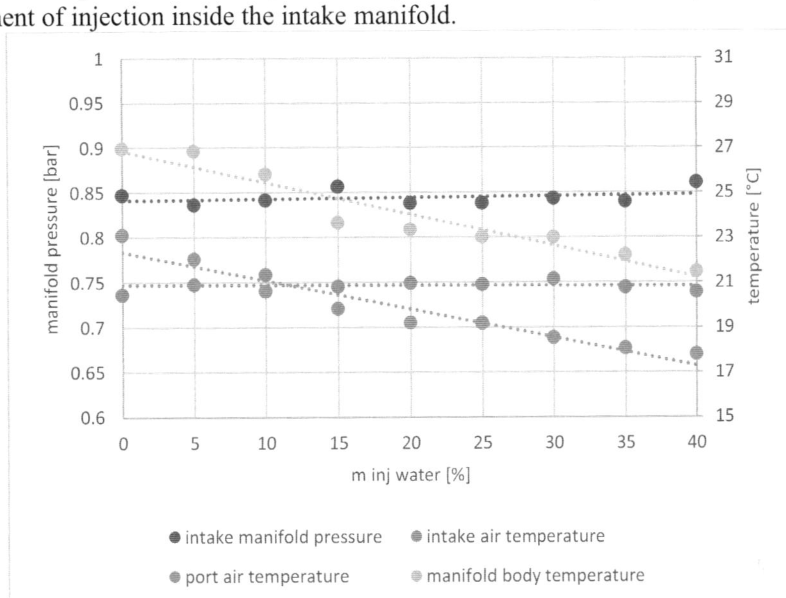


**Figure 4.** Variation of the air intake temperature and pressure

Continuing the study of the process of vaporization of water, we will shift our attention to the burning process inside the cylinder. Therefore, we will analyze the indicated pressure diagram obtained from the experimental measurements. We will start with the one in which we have normal combustion, without water injection and the one in which we have no combustion. In one of the ways to detect the moment of the start of combustion, the two diagrams, the one with and without combustion in the cylinder, will be superimposed [5]. Analyzing in more detail, we can say that the start of combustion takes place around 346 [°CA], which is also confirmed by a spark advance of 14 [°CA]. And for the normal combustion process we get the maximum pressure is recorded at 372 [°RAC] with a pressure of 49.5 [bar].

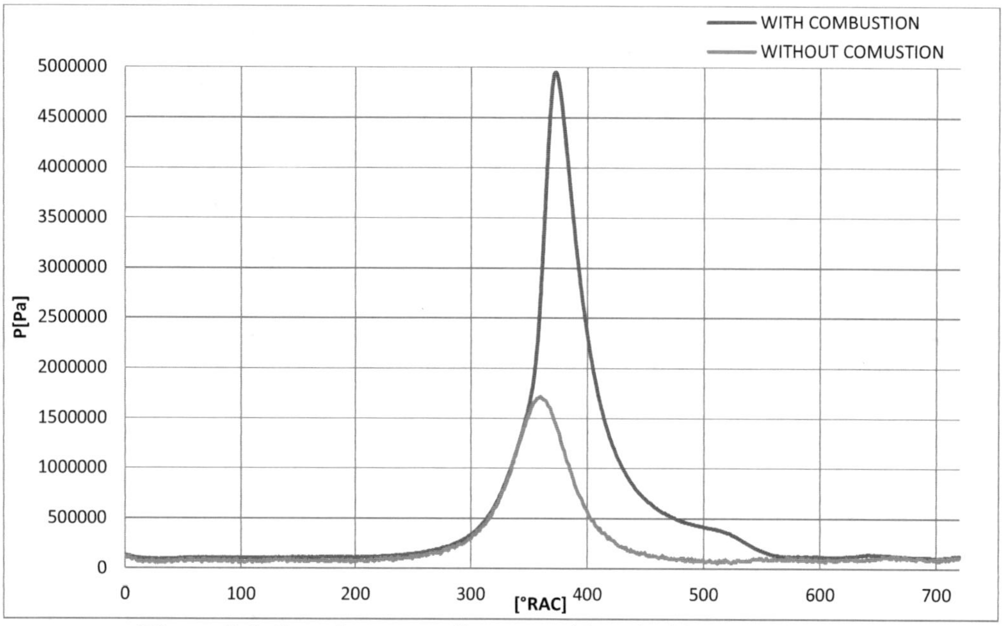


**Figure 5.** Indicated pressure diagrams with and without combustion

In the graph shown in figure 6, we will overlay all the recorded indicated pressure diagrams. At a first glance we can see that as we inject more water, the maximum pressure decreases by approximately 4.1 [bar] and moves towards the exhaust by 6 [°RAC]. This confirms a slower combustion with a lower temperature, which benefits the reduction of NOx formation, but slightly reduces the overall performance of the engine. This can be seen also in the recording of the measured brake power shown in figure 7.

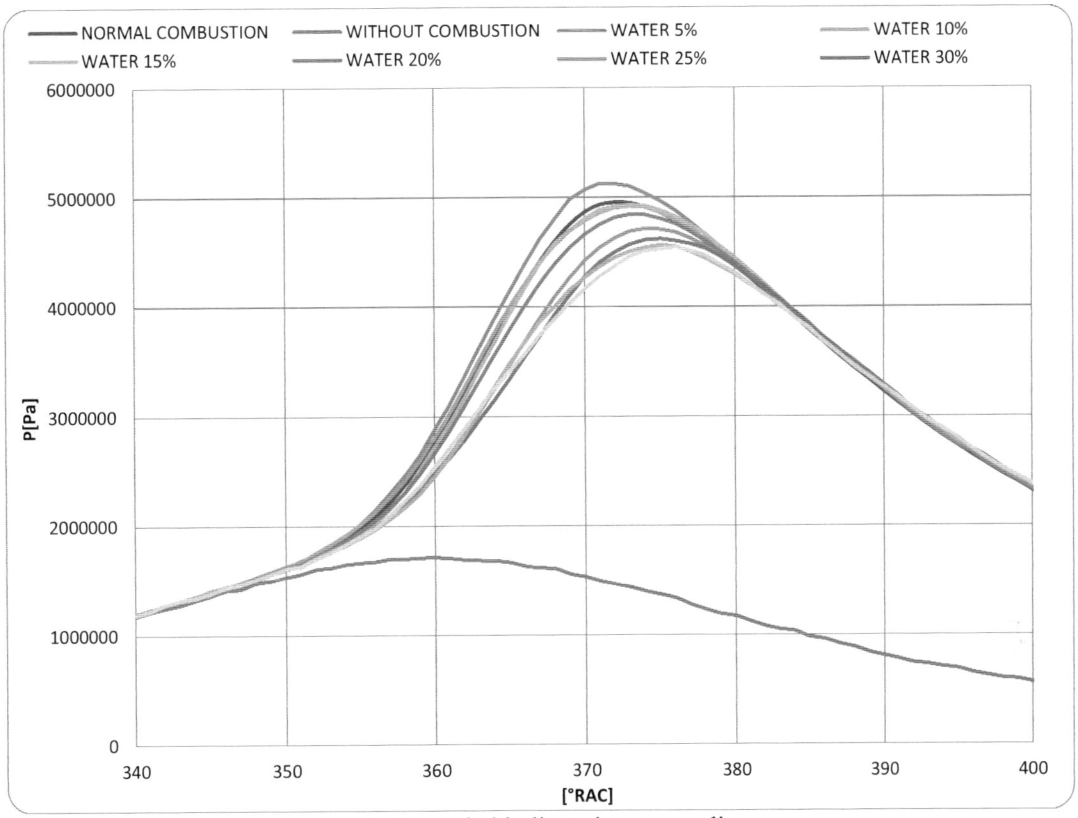


**Figure 6.** Recorded indicated pressure diagrams

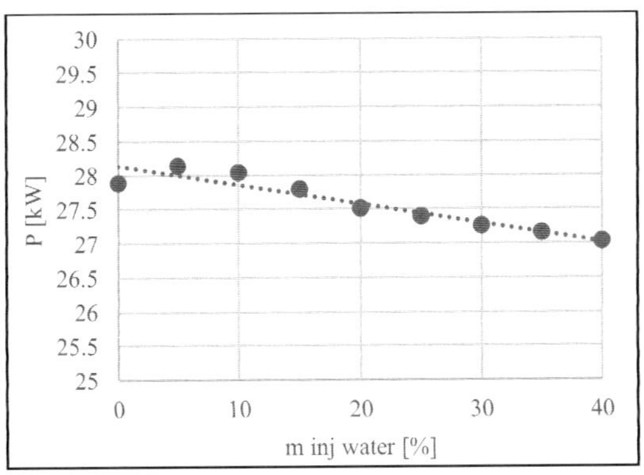


**Figure 7.** Measured brake power

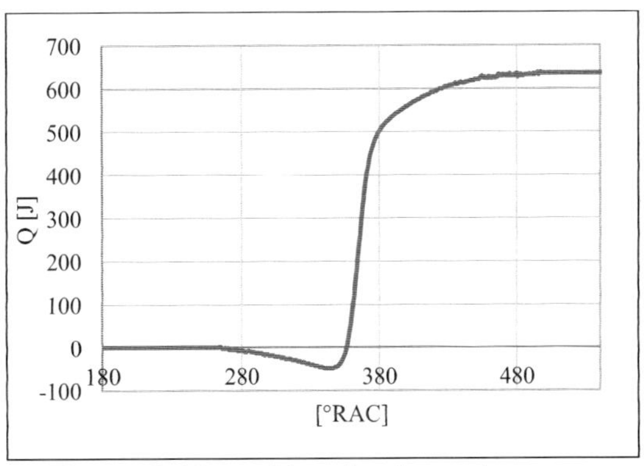


**Figure 8.** Normal burning heat release rate

In the above figure we can see the heat release law. This law was drawn without separating the heat released by combustion from the heat exchange at the walls. This can be seen at the beginning of the graph, where the graph descends into the negative zone, basically the working gas at that point is exchanging heat with the cylinder walls. All the heat release laws have been drawn based on the equation below, and they can be seen in the figure 9.

$$dQ = \frac{Vdp + kpdV}{(k-1)} \tag{2}$$

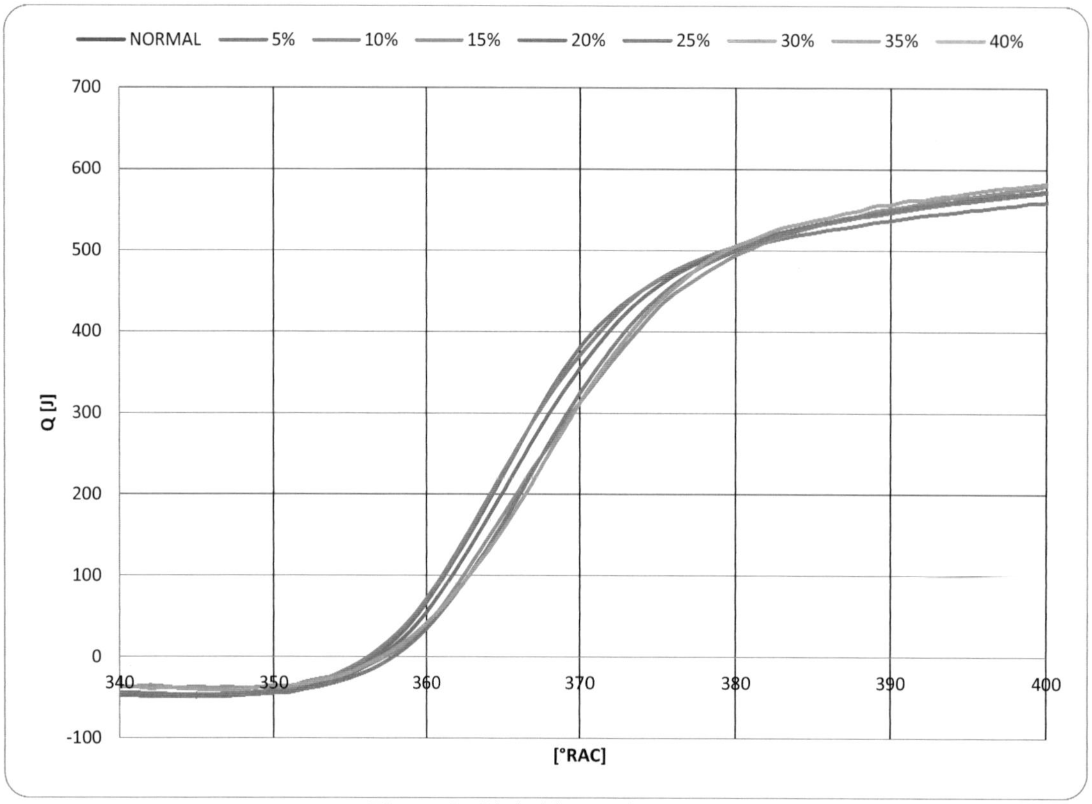


**Figure 9.** Global heat release rates

Next, we will analyze the variation of the resulting emissions from the measurements made with water injection. To begin with, we can observe that the hydrocarbons have a decreasing trend, which indicates a more complete combustion, this is also confirmed by the decreasing trend of recorded $O_2$ emissions, but also by the increasing trend of $CO_2$.

Finally, we cannot ignore the fact that emissions of nitrogen oxides, NOx, are drastically reduced as more water is introduced into the cylinder. At about 40% injected water mass relative to the total fuel mass, we get a $NO_x$ reduction of about 38%.

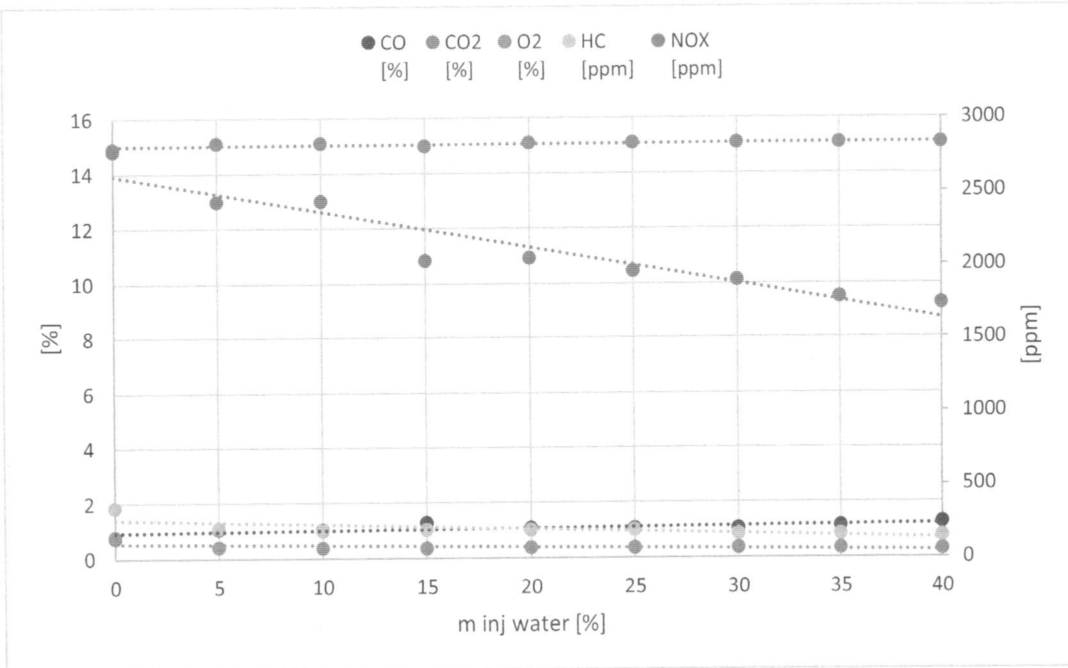


**Figure 10.** Recorded emissions

## 5. Conclusions

Based on the following measurements we can conclude that we can obtain a considerable reduction in $NO_x$ formation during water injection, with some insignificant power loss. But that loss of power can be compensated by adjusting the ignition timing beforehand to compensate for the shifting of the peak pressure, shown on the indicated pressure diagram towards the exhaust.

More detailed studies are necessary to find out the mechanism of the thermodynamic processes inside the cylinder, but also to find a recipe for some parameters, such as the spark advance and the amount of water injected. A wide series of experiments must be carried out to be able to obtain an exact relationship between the amount of water injected, the moment of injection but also the change in advance of the ignition timing. All this to achieve high efficiency and low $NO_x$ emissions.

Therefore, I consider this topic to be of interest for future studies in order to improve the performance of spark ignition internal combustion engines.

## References

[1] M. Rowe and G. Ladd, "Water injection for aircraft engines," *SAE Technical Paper,* no. 460192, 1946.

[2] A. Rohit, S. Satpathy, J. Choi, J. Hoard, G. Surnilla and M. Hakeem, "Literature survey of water injection benefits on boosted spark ignition engines," *SAE Technical Paper,* 2011.

[3] A. T. Borborean, V. Stoica and D. Lelea, "Test bench for studying the effects of water injection inside an internal combustion engine," in *International Conference on Applied Sciences - Materials Science and Engineering*, Baja Luka, 2018.

[4] I. Hermann, C. Glahn, M. Kluin, M. Paroll and W. Gumprich, "Water injection for gasoline engines-Quo Vadis?," *International conference on Knocking in Gasoline Engines,* vol. Springer, pp. 229-321, 2017.

[5] B. Grunwald, Teoria calculul și construcția motoarelor pentru autovehicule rutiere, București, 1980.

# CFD simulation of the interaction between water droplets on the GDL of a low temperature PEM FC in conditions of air cross-flow

**C Antetomaso[1], S S Merola[1*], A Irimescu[1], B M Vaglieco[1]**
**S Di Micco[2], E Jannelli[2], G Scarpati[2], E Simeoni[2]**
[1] CNR-STEMS, Naples, Italy
[2] University of "Parthenope", Naples, Italy

*Corresponding author: simonasilvia.merola@stems.cnr.it

**Abstract**. Droplet dynamics in low temperature proton exchange membrane fuel cells (LT-PEMFCs) are significantly important with respect to water management and removal. Three main phenomena are linked to water generation/transport in a LT-PEMFC: H2O production on the cathode side, electro-osmotic drag (anode to cathode) and back diffusion (cathode to anode). The excess/lack of water, depending on membrane humidity, strongly affects LT-PEMFC performance, involving its instability. In this work CFD simulation of droplets was performed when they were deposited in tandem on the gas diffusion layer (GDL) and subjected to 15 m/s parallel air flow. The volume of fluid (VOF) method was adopted and droplets with diameters that range from 0.3 mm to 0.8 mm were considered. Simulations demonstrate that deformations, oscillations and sliding of droplets were affected by their relative size and position. Results can contribute to define predictive models of water droplet dynamics on the GDL surface and their detachment towards the FC gas channel.

## 1. Introduction

The working principle of a PEMFC is quite simple: the user provides $H_2$ and $O_2$ respectively at the anodic and cathodic ends of the cell. Reactants will find their way through the porous Gas Diffusion Layer (GDL) and reach the Catalyst Layer (CL). The catalyst (usually Pt) will promote the redox reaction and the forming stream of H+ will pass through the membrane, while electrons will be forced in an external circuit. Problems with this device start to arise as we consider one of the by-products of the reactions: water. Water management in LT PEMFCs is studied in detail due to the challenges it features. Three main phenomena are linked to water generation/transport in a LT-PEMFC: H2O production on the cathode side, electro-osmotic drag (anode to cathode) and back diffusion (cathode to anode). The excess/lack of water, depending on the membrane humidity, strongly effects the LT-PEMFC performance, involving its instability.

In the last years experimental and numerical researchers focused their efforts towards a deep comprehension of what behaviour the water had under FC working conditions. One of the most diffused modelling techniques to resolve a multiphase flow is the Volume Of Fluid (VOF) [1]. Chen et al. [2] parameterized the working conditions of the cell (air velocity, water formation, wall wettability) and visualized critical points of the Gas Channel (GC). Their findings indicated that the wettability

angle and airflow intensity influenced flow patterns and critical droplet dimension, respectively. Polverino et al. [3] proposed and experimentally validated a simplified model to predict droplet oscillation and detachment from a GDL. Their results showed that droplet dimension and airflow had the biggest impact on the behaviour of the droplet. Other studies focused on the reciprocal influence of water droplet on GDL. The higher the load, the higher the water formation, the higher the probability that two (or more) water droplets formed on the GDL at comparable distance. Tsuji et al. [4] visualized the flow around two spheres; the obstacles were positioned at different reciprocal distance and the effect on vortexes formation and propagation was observed. Successively, Tsuji et al. [5] performed a 3D CFD simulation of the phenomenon. The results clearly related the vortex morphology with droplet distance and disposition (longitudinal or transverse with respect to the flow). Kotsev [6] deepened the study increasing the number of droplets and adding a triangular configuration. Razzaghi et al. [7] experimentally reproduced the detachment of two droplets in longitudinal and transverse configuration, confirming that the critical detachment velocity depended on the distance and disposition.
The aim of this paper was to investigate the relative influence of water droplets on a GDL surface when they were disposed longitudinally (in-tandem) with respect to the air-flow. Numerical simulation based on Volume Of Fluid, VOF, method was applied to study the dynamic of droplet pairs with diameters ranged from 0.3 mm to 0.8 mm. The droplets were invested by a 15m/s flow parallel to the line connecting the droplet centres. Numerical results were validated with experimental data obtained through optical visualization. The proposed approach represents a contribution to the definition of a predictive model of droplet dynamics in fuel cell components.

## 2. Experimental setup

Water droplets were lowered onto a GDL surface of a Celtec®-P membrane electrode assembly by using a NIPRO device equipped with stainless steel needle; the internal diameter of the cannula (450 μm ) represented the size limit of the deposited droplets. The GDL texture was analysed in previous investigations [8] that permitted to define the cross-section to use for CFD simulations. In absence of air-flow, the deposited droplets were visualized by using a 1920x1080 microscope CCD camera with 8-LED ring crown illuminator. The geometric characteristics of droplets were estimated by applying a customized image processing procedure. The optical set-up allowed the estimation of length with a high spatial resolution (5.8 μm/pixel). The contact angles resulted comparable (±2 deg) and matched those calculated from the Wenzel's relation [9][10]. Thus, the averaged value (123 deg) was used as boundary condition for the wettability parameter in the CFD simulations. Figure 1 shows the image of two droplets placed in-tandem with diameters of 600μm and 800μm, respectively.

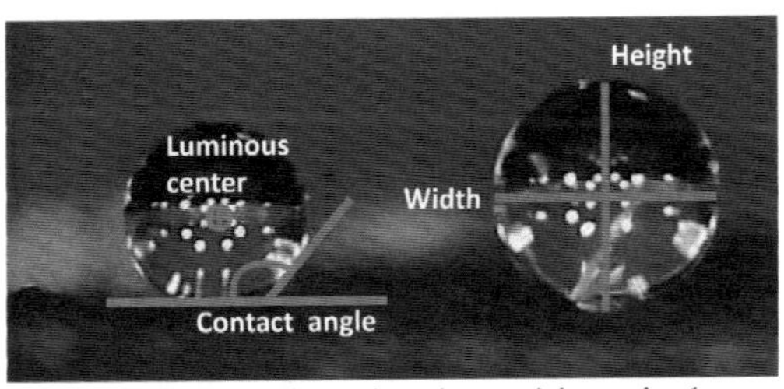

**Figure 1**. In-tandem droplets with typical morphologic parameters.

For the experimental investigations under dynamic conditions, a hot wire anemometer was used to measure with an accuracy of 0.1 m/s the velocity of the air flow produced by a fan that was placed at 50 mm from the upstream droplet. All investigations were performed at controlled temperature (25°C) and pressure (1 atm). Taking into consideration the setup characteristics, the evaporation effects were negligible. High-speed visualizations with 1000 fps were carried up by using a 512x512 CMOS camera (Optronis CR500x2). Previous studies on single droplets [8] demonstrated that the droplet started to oscillate even at flow velocities much smaller than the critical velocity that corresponded to the droplet detachment. Moreover, at fixed air velocity, the oscillation amplitude increased with the droplet size until the point that detachment was induced, while the frequency increased with the inverse of the volume up to a diameter comparable with the micro-groove structure of the GDL. In this work, to better investigate the effect of air-flow on the droplets, image processing developed by using

NI Vision 2020 was applied to identify the instantaneous position of the luminous center (LC) and estimate the associated shape factor as the ratio between the width and height of the bounding rectangle. Moreover, the LC oscillation frequencies in the direction of the flow were analysed by applying an FFT LabView 2020 routine.

## 3. Method

### *3.1. Governing equations*

The VOF method is used in this work. The primary phase is air, the secondary phase is water. A numerical value between 0 and 1 defines the quantity of water in each cell. In general, for n phases:

- $\alpha_q = 0$, the cell is empty (of the qth pahse);
- $\alpha_q = 1$, the cell is full (of the qth phase);
- $0 < \alpha_q < 1$, the cell contains the interface between phases.

The primary phase is evaluated with the following formula:

$$\sum_{q=1}^{n} \alpha_q = 1 \quad (1)$$

Representative properties of cells are calculated as a volume weighted mean. For a two phases flow:

$$\rho = (1-\alpha)\cdot\rho_1 + \alpha\cdot\rho_2 \quad (2)$$

$$\mu = (1-\alpha)\cdot\mu_1 + \alpha\cdot\mu_2 \quad (3)$$

where $\rho$ is the density and $\mu$ the viscosity of the mixture for the cell.

Formula (4) express mass conservation.

$$\frac{\partial\rho}{\partial t} + \nabla(\rho\cdot\overline{v}) = 0 \quad (4)$$

In VOF a single momentum equation is resolved for each cell in the domain.

$$\frac{\partial}{\partial t}(\rho\cdot\overline{v}) + \nabla(\rho\cdot\overline{v}\cdot\overline{v}) = -\nabla p + \nabla\left[\mu\left(\nabla\overline{v} + \nabla\overline{v}^T\right)\right] + \rho\cdot\overline{g} + \overline{F} \quad (5)$$

In this study we consider only two phases, so the Continuum Surface Force (CSF) model equation can be written as:

$$\overline{F} = \sigma\cdot\frac{\rho\cdot k\cdot\nabla\alpha}{\frac{1}{2}\cdot(\rho_1+\rho_2)} \quad (6)$$

where $\sigma$ is the surface tension coefficient, k is the surface curvature of the interface defined in terms of the divergence of the unit normal $\overline{n}$ as:

$$k = \nabla\overline{n} \quad (7)$$

with:

$$\overline{n} = \frac{\nabla\alpha_1}{|\nabla\alpha_1|} \quad (8)$$

The surface normal vector at the wall is given by:

$$\overline{n} = \overline{n}_w\cdot\cos\theta_w + \overline{t}_w\cdot\sin\theta_w \quad (9)$$

where $n_w$ and $t_w$ are the normal and tangential versor to the wall respectively and $\theta_w$ is the static contact angle, defined by the type of GDL.

### *3.2. Mesh, initial and boundary condition, simulation setup*

The computation domain is reported in Figure 2 together with the geometric details of the GDL cross-section obtained in previous investigations [8]. The general cell dimension was set at 0.05 mm, but this limit was lowered to 0.0125 mm on the right bottom half since it was the area in which we wanted to observe droplet oscillation and vortexes formation and development. The first droplet was initially deposited on the GDL at x = 15 mm upstream with respect to the air-flow. In order to understand the effect of the relative distance between the particles, indicated as length $L_i$, the deposition of the second droplet was performed at 1, 2 and 3 valleys of the GDL profile. Details on the boundary conditions and simulation setup are listed in Table 1 and Table 2, respectively. Following the approach adopted for the analysis of the sequences of images recorded on the experimental setup, the 2D CFD results were treated by applying a NI Vision and LabView procedure in order to evaluate the oscillation frequency and droplet shape deformation.

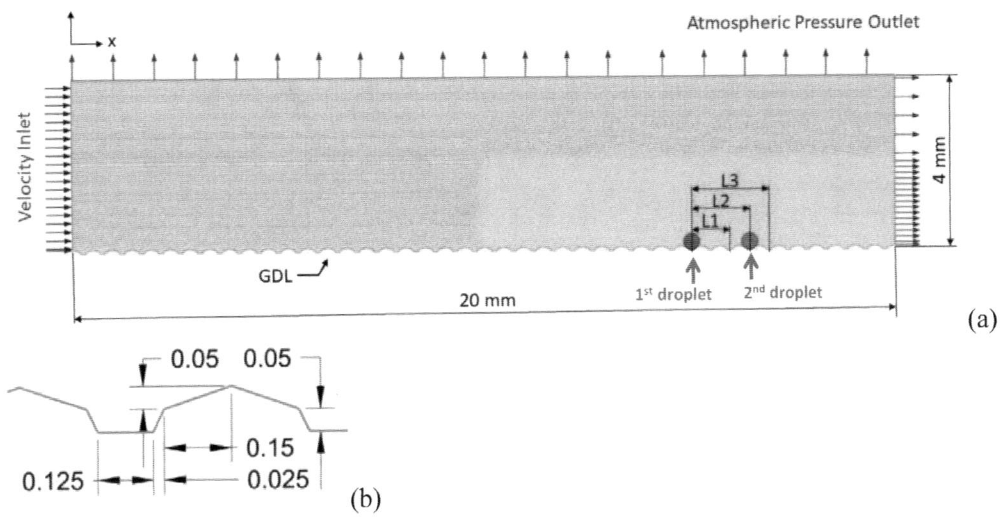


**Figure 2.** (a) Computational domain and (b) details of the GDL cross-section.

| | |
|---|---|
| Inlet Velocity [m/s] | 15 |
| Outlet Pressure [barg] | 0 |
| Droplet Diameter D [μm] | 300-800 |
| Droplet Distance L [mm] | $L_1$=0.96, $L_2$=1.4, $L_3$=1.9 |

**Table 1.** Boundary and initial condition

| | |
|---|---|
| Solver Type | Pressure-Based |
| Pressure-Velocity coupling | PISO |
| Spatial Discretization | |
| Pressure | PRESTO! |
| Momentum | Second Order Upwind |
| Volume Fraction | Geo-Reconstruct |
| Simulation duration [ms] | 100 |
| Simulation step | Adaptive Multiphase-Specific |

**Table 2.** Simulation setup

## 4. Results

Numerical results were aimed at the analysis of the horizontal oscillation mode. Several cases were under scrutiny, in which droplets with different diameters were deposited at the same distance (L1 = 0.96 mm) from a first droplet with fixed diameter of 300 μm. As shown in Fig. 3a, numerical simulations demonstrated that the presence of a first droplet affected the dynamic in the air-flow direction of the second one by increasing the oscillation frequency of its luminous center with respect to the single droplet case. The increase resulted more relevant for small particles and practically negligible when the diameter of the second droplet was double with respect to the first one. In terms of droplet shape, as it is shown in Fig. 3b, an increase in the distortion of the second droplet in the flow direction was observed (i.e. higher shape factors mean more extensive horizontal elongation). The distortion, that was more significant for larger droplets, corresponded to an increase in the amplitude of oscillations along the longitude. It should be noted that, in case of droplets larger than 800 μm, the larger elongation determined effects of coalescence with consequent reduction of the critical size for the detachment at fixed air velocity [3]. The CFD results were in very good agreement with experimental data measured in case of single droplets with average diameters of 600 μm and 800 μm, respectively.

The increase in the longitudinal elongation was explained by the analysis of the flow field. Fig. 4 shows the streamlines obtained at 10 ms from the cases of single and in-tandem droplets with 300 μm and 600 μm diameter. For the single droplet cases (Fig. 4a-4c), it is evident that the air flow generated a vortex behind the droplet. Because of the viscous force between the droplet and bottom wall, the velocity in the top part of the droplet was higher than that in the bottom part, leading to the stretching of the droplet along the flow direction. In case of droplet pairs (Fig. 4b-4d) at distance larger than sum of their diameters, vortices were formed upstream and downstream of the second droplet. For small droplet pairs, the size of both vortices was comparable with the first droplet diameter; this increased the drag forces on the second droplet sides inducing an increase in the oscillation frequency with no significant effect on the oscillation amplitude. At increasing diameter of the second droplet, the size of the downstream vortex increased. The unbalanced drag forces on the 2nd droplet sides reduced the effect on the oscillation frequency but increased the longitudinal distortion and thus the detachment potential.

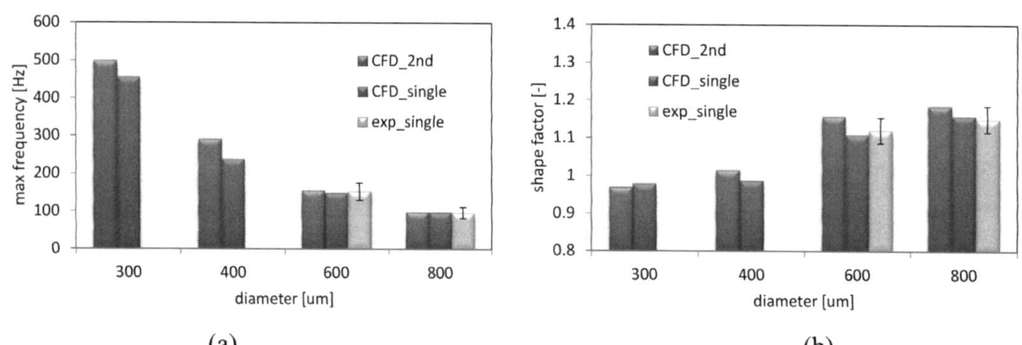


(a) (b)

**Figure 3**. Comparison between (a) highest intensity oscillation frequency and (b) shape factor for the second droplet of an in-tandem pair with respect to a single droplet as a function of the diameter. CFD results are validated with experimental data for 600μm and 800μm single droplets.

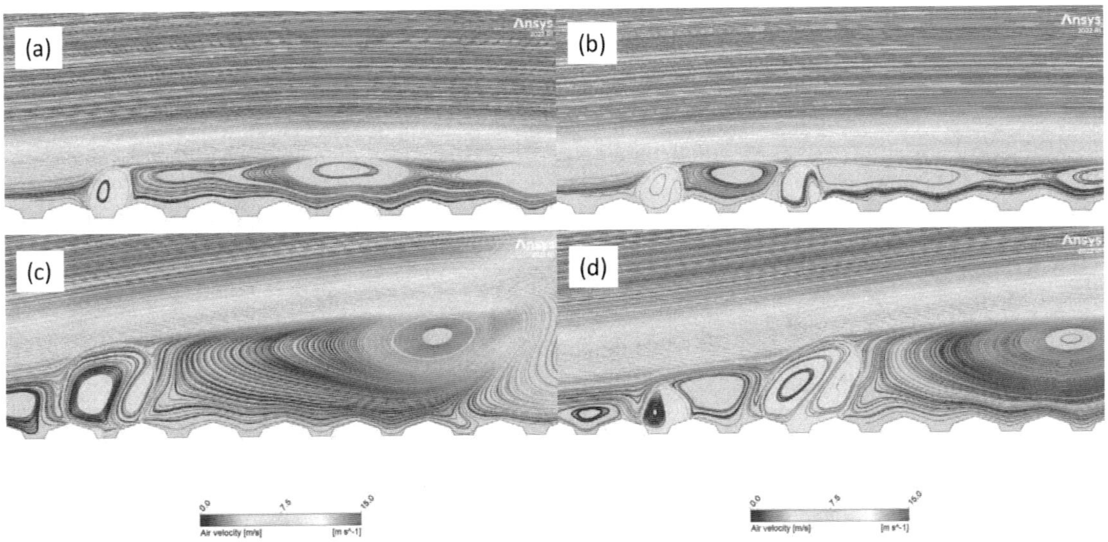


**Figure 4.** Streamlines obtained from CFD simulations at 10 ms from droplet deposition for the cases of (a) 300 μm single droplet, (b) 300μm - 300 μm droplets at distance $L_1$, (c) 600 μm single droplet and (d) 300μm - 600 μm droplets at distance $L_1$.

To better understand the effect of the longitudinal air-flow on the droplet pair, CFD results were compared with respect to the parameter S defined as the ratio of the distance L between the centres and the average diameters of the droplets:

$$S = \frac{2L}{(D_1 + D_2)}$$

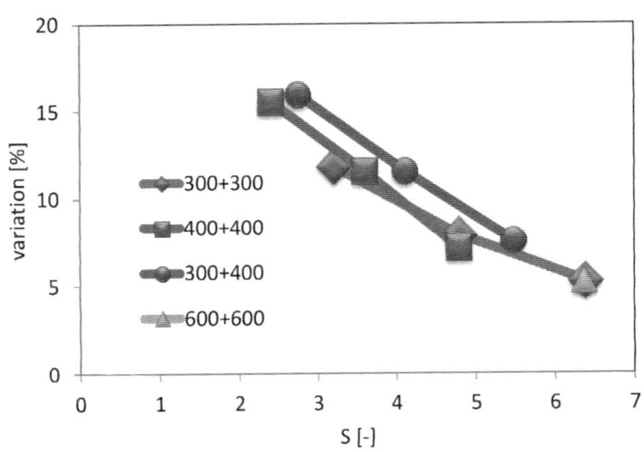


**Figure 5.** Percentage variation of the oscillation frequency relative to the second droplet with respect to the single droplet as a function of parameter S.

Figure 5 reports the percentage variation of the oscillation frequency relative to the second droplet with respect to the single droplet as function of the S parameter. It should be noted that for pairs with the same diameters in 600-800 μm range, no evaluation was statistically suitable for S values below 6 because the increase in the oscillation frequency induced detachment even if this was not observed in case of single droplets during 100 ms of simulation [8]. It should be noted that if the first droplet was larger than 600 μm, S=3 represented the minimal distance to avoid effects of coalescence and successive detachment. Anyway, despite of the size of the merged droplet no breaking up was observed. Nevertheless, for droplets smaller than 600 μm the increment in frequency decreased linearly with comparable slopes. It is interesting to note that by defining the non-dimensional parameter S, the correlations collapse on a seemingly linear trend that be directly used to predict oscillation variations with minimum computational effort. This of course can be used for defining optimized water management control strategies.

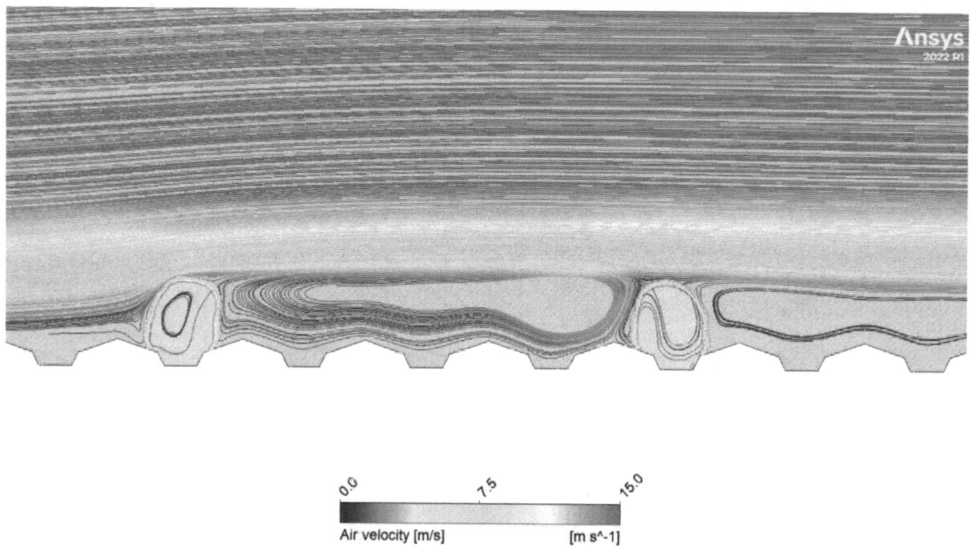


**Figure 6.** Streamlines obtained from CFD simulation at 10 ms from droplet deposition for the case of 300 μm - 300 μm droplets at distance $L_3$.

The influence of the first droplet continued at increasing S values until reaching S≅7 when the upstream side of the second droplet faced a strongly reduced drag compared to the downstream side of the first droplet, as shown in Fig. 6. For S≅10 the vortex between the droplets disappeared and open streamlines occurred. This also fits well with empirical observations of the effects of turbulence in geometries such as tubes.

## 5. Conclusions

CFD simulations of droplet interactions in PEMFC channels showed that there are specific differences with respect to the diameter of each body of liquid and their relative distance. For the cases of single droplets, predicted oscillation frequencies were found to be completely in line with the measurements performed on the experimental setup. The CFD simulations could be extended to a wide range of droplet size as well as distance-diameter combinations. As an overall conclusion, the oscillation frequency tended to increase as the distance between the droplets was shorter. The main reason was identified to be correlated to the presence of vortices that interact with the droplet pair. Coalescence and detachment effects were also found to be directly influenced by the combinations of diameters and distance between the droplets.

By defining a non-dimensional parameter that combines the distance between droplets and the sum of their diameters, it was found that the correlations with frequency variations collapsed roughly on the same linear trend.

The method and results can be used for defining correlations that define droplet interactions and can be applied in simulations that are less computationally intensive. In turn, these correlations can be directly used for setting up water management control strategies.

## References

[1] Ferreira, R. B., Falcão, D. S., Oliveira, V. B., & Pinto, A. M. F. R. (2015). Numerical simulations of two-phase flow in proton exchange membrane fuel cells using the volume of fluid method – A review. Journal of Power Sources, 277, 329-342.

[2] Chen, R., Qin, Y., Ma, S., & Du, Q. (2020). Numerical simulation of liquid water emerging and transport in the flow channel of PEMFC using the volume of fluid method. International Journal of Hydrogen Energy, 45(54), 29861-29873.

[3] Polverino, P., Esposito, A., & Pianese, C. (2013). Experimental validation of a lumped model of single droplet deformation, oscillation and detachment on the GDL surface of a PEM fuel cell. International journal of hydrogen energy, 38(21), 8934-8953.

[4] Tsuji, Y., Morikawa, Y., & Terashima, K. (1982). Fluid-dynamic interaction between two spheres. International Journal of Multiphase Flow, 8(1), 71-82.

[5] Tsuji, T., Narutomi, R., Yokomine, T., Ebara, S., & Shimizu, A. (2003). Unsteady three-dimensional simulation of interactions between flow and two particles. International journal of multiphase flow, 29(9), 1431-1450.

[6] Kotsev, T. (2018). Viscous flow around spherical particles in different arrangements. In MATEC Web of Conferences Vol. 145, p. 03008. EDP Sciences.

[7] Razzaghi, A., & Amirfazli, A. (2019). Shedding of a pair of sessile droplets. International Journal of Multiphase Flow, 110, 59-68.

[8] C. Antetomaso, A. Irimescu, S. S. Merola, B. M. Vaglieco, S. Di Micco, E. Jannelli, G. Scarpati, E. Simeoni (2022) CFD simulation of water droplet adhesion on the GDL of a low temperature PEM FC in air cross-flow conditions, ATI Conference, ID 88.

[9] Good, R. J., & Koo, M. N. (1979). The effect of drop size on contact angle. Journal of Colloid and Interface Science, 71(2), 283-292.

[10] Drelich, J., Miller, J. D., & Hupka, J. (1993). The effect of drop size on contact angle over a wide range of drop volumes. Journal of colloid and interface science, 155(2), 379-385.

## Acronyms

| | |
|---|---|
| CL | Catalyst Layer |
| CSF | Continuum Surface Force |
| FC | Fuel cell |
| FFT | Fast Fourier Transformation |
| GC | Gas Channel |
| GDL | Gas Diffusion Layer |
| LC | Luminous center |
| LT | Low Temperature |
| PEM | Proton Exchange Membrane |
| VOF | Volume Of Fluid |

IOP Conf. Series: Materials Science and Engineering **1311** (2024) 012007 doi:10.1088/1757-899X/1311/1/012007

# Simulation of the Parameters of a Lithium-ion Battery that Equips an Electric Vehicle

**A A Ancuța*, Ș Voloacă, G Frățilă, G Danciu**

[1] Department of Automotive Engineering, Faculty of Transport, University „Politehnica" of Bucharest, Romania

*Email: alex.ancuta96@yahoo.com

**Abstract**. The evolution in the field of electric vehicles is strongly influenced by the performances of the electricity storage sources. Thus the paper proposes to present some theoretical aspects related to the operation of lithium-ion batteries and factors that may influence their state of charge and of course the vehicle autonomy, as well as a simulation model to determine the variation of the main parameters of the battery when it is connected to a velocity variation profile according to WLTC and UDDS cycles. For this we want to show the influence of the discharge current on the battery over time. So, to realize the study the velocity variation over time for the two mentioned cycles will be converted into the current variation and the battery will be connected to this signal.

**Keywords:** Li-Ion Battery, Electric Vehicle, State of Charge.

## 1. Introduction

Currently, the electric vehicle is more and more publicized in terms of worldwide spread and this is also demonstrated by the higher number of sales of this type of vehicle among the population. It is well known, however, that one of the biggest problems actually preventing this new trend from capturing the market entirely is the vehicle's power source. The main source of electricity currently used by vehicles equipped with an electric propulsion system are batteries using lithium technology; they come with a series of disadvantages, at least for the time being, and we mention some of them: the reduced energy storage capacity on board an appearance vehicle that also generates a still reduced autonomy among electric vehicles; the large mass of the battery in relation to the stored energy; the high sensitivity of lithium batteries (here comes the need to use a permanent battery control and monitoring device called Battery Management System - BMS, a complex device from a constructive and functional point of view); the time required to recharge the batteries is still an issue and not least the recharging infrastructure. However, lithium batteries are the most used solution at the moment and also a starting point for research and development of new solutions to solve the previously listed problems.

In this paper, we aim to mainly present the SOC behaviour of the lithium battery in relation to two of the most important test cycles used worldwide, namely: WLTC (Worldwide Harmonised Light Vehicles Test Cycle) and UDDS (Urban Dynamometer Driving Schedule). The paper continues the studies on the behaviour of electric vehicle storage sources in relation to different influencing factors, in the present case the velocity of vehicle in relation to a test cycle and the discharge current profile obtained by converting the velocity profile.

## 2. Some theoretical aspects and data used for realize the simulation model

First of all, in order to carry out this study, it was necessary to choose an electric vehicle whose battery will be used in the proposed simulation model. In this sense, we have chosen the electric vehicle VOLKSWAGEN e-GOLF, which offers an electric vehicle intended mainly for driving in the urban environment, being a vehicle of small dimensions, with a hatchback type body that offers 5 seats. It is a model that falls into the middle area in terms of purchase price, so it can be available to several social classes ready to make the transition to an electric vehicle. After tens of millions of vehicles equipped with an internal combustion engine, the GOLF model continues its tradition under another variant in line with the trends in the automotive area.

The chosen model has a weight of 1510 kg and is equipped with a synchronous motor with permanent magnets that develops a power of 85 kW and a torque of 270 Nm. It uses a lithium-ion battery consisting of 27 modules, which presents the following characteristics, which were used in the realization of the model for simulation: battery mass 310 kg; 264 cells with a nominal voltage of 3.7 V; nominal battery voltage 323 V and nominal energy 21.2 kWh; battery capacity of 75 Ah; temperature range between -25°C and +55°C, [1].

As we specified at the beginning, the main parameter analysed in this paper is %SOC and the influence of velocity and discharge current on it. For this I used the velocity variation over time for the WLTC and UDDS cycles, which I first plotted in Excel, then imported into Simulink; also in Excel I converted the two velocity diagrams into current diagrams to be used later as a consumer for the lithium-ion battery that equips the e-Golf vehicle.

### *2.1. WLTC profile*

Using the data and tables from [2] and taking into account the fact that the chosen e-Golf vehicle reaches a top velocity of 150 km/h and has a specific power of approximately 56 W/kg, the velocity profile corresponding to the class was drawn for the simulations of 3b vehicles according to the WLTC classification.

We remind you that for this class of vehicles the cycle lasts 1800 seconds, during which the vehicle travels 23266 m (~23.3 km), and the maximum velocity is 131.3 km/h. The Excel program was used to draw the profile, where the data was entered according to the analysed class and the following result was obtained:

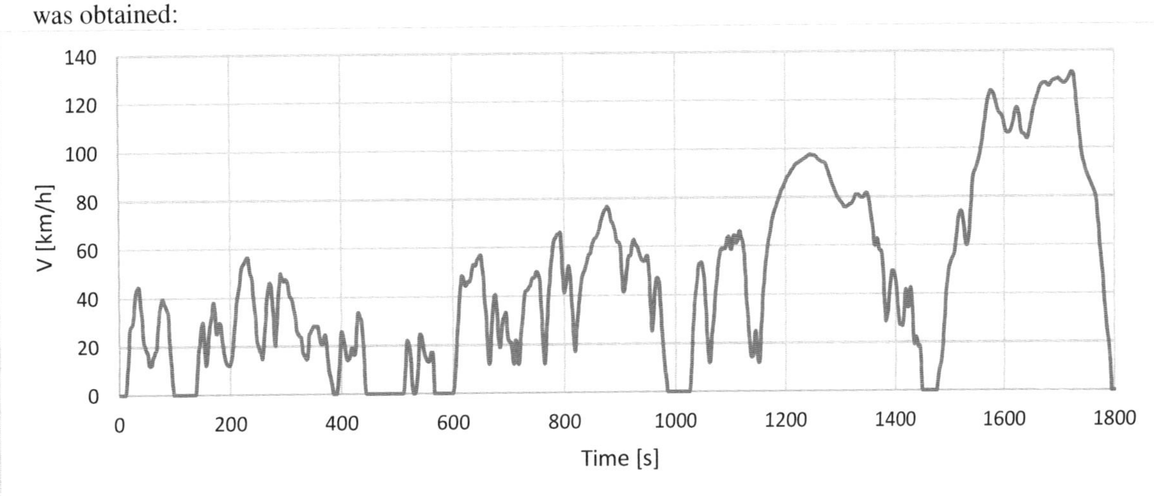


**Figure 1.** WLTC velocity profile for vehicle class 3b.

Later, in order to carry out the simulation and obtain the targeted results, this profile shown in figure 1 was converted into the electric discharge current profile of the e-Golf vehicle battery and we obtained the following diagram:

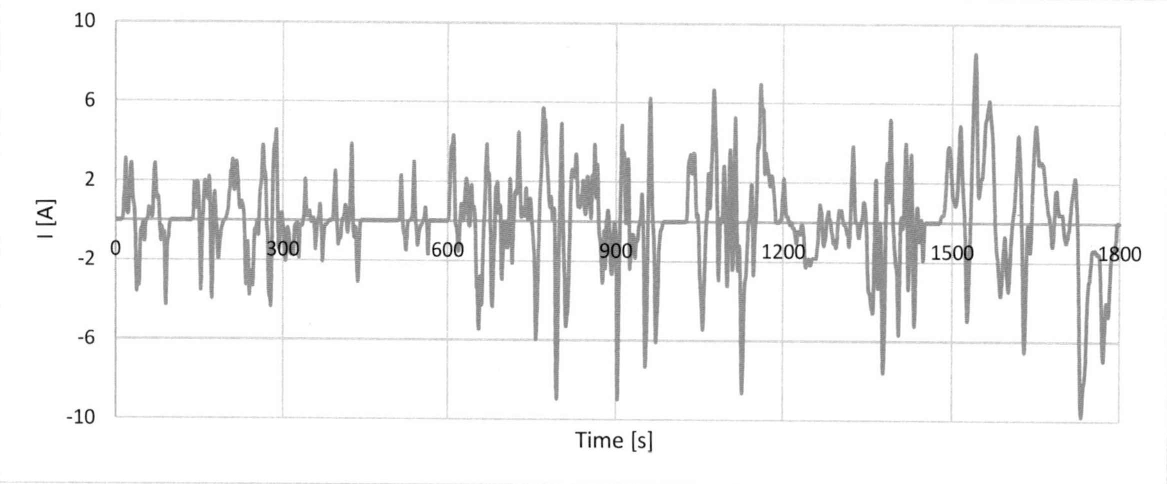


**Figure 2.** WLTC current profile used for simulation model.

### *2.2. UDDS profile*

For the UDDS cycle according to the information in [3], the data was retrieved and the velocity diagram was drawn and subsequently the profile of the electric current used as a consumer for the analysed lithium-ion battery.

Thus the following diagram was obtained:

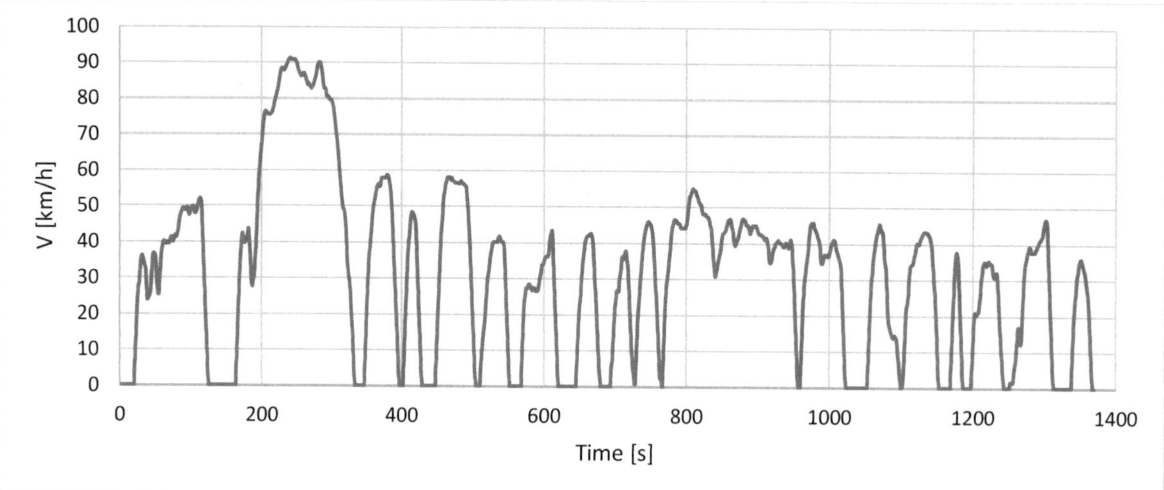


**Figure 3.** UDDS velocity profile.

According to the diagram from figure 3, we mention some characteristics of the UDDS cycle used in modeling and simulation: the travel time of this cycle is 1369 seconds, and the vehicle travels a distance of 16500 m (16.5 km) reaching a top velocity of 96.4 km/h, [4], about 35 km/h less than the previous cycle shown, WLTC.

The discharge current profile can be seen in figure 4:

IOP Conf. Series: Materials Science and Engineering **1311** (2024) 012007 doi:10.1088/1757-899X/1311/1/012007

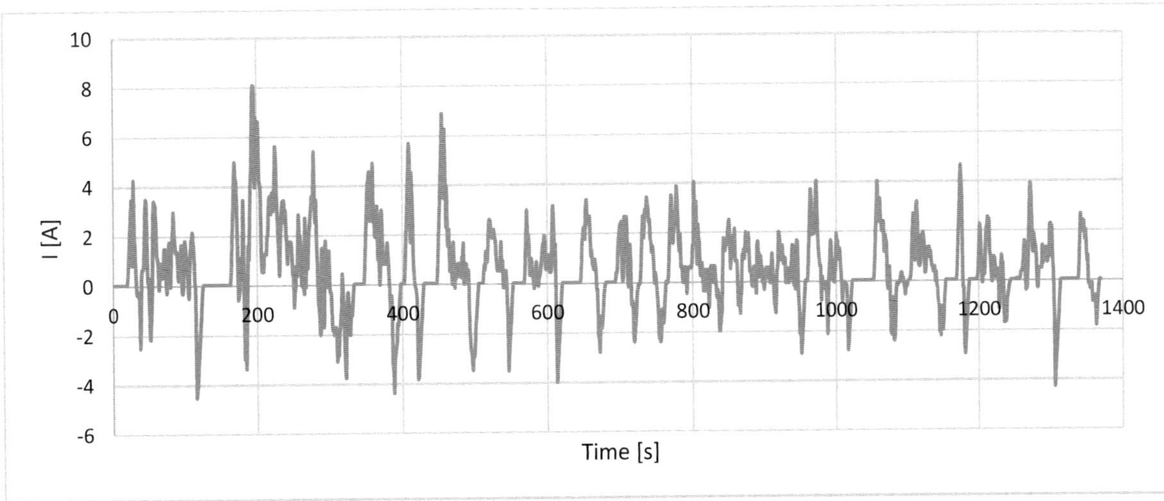


**Figure 4.** UDDS current profile used for simulation model, [5].

All these data are used in the next section to finalize the simulation model and obtain the targeted results.

## 3. Making the model used for simulation

To create the model used for the simulation, I used the SimScape library from Simulink where I took a battery model that I parameterized with the technical data of the e-Golf vehicle and obtained through the battery control block the theoretical discharge characteristic in relation with time, as we can see in the figure below:

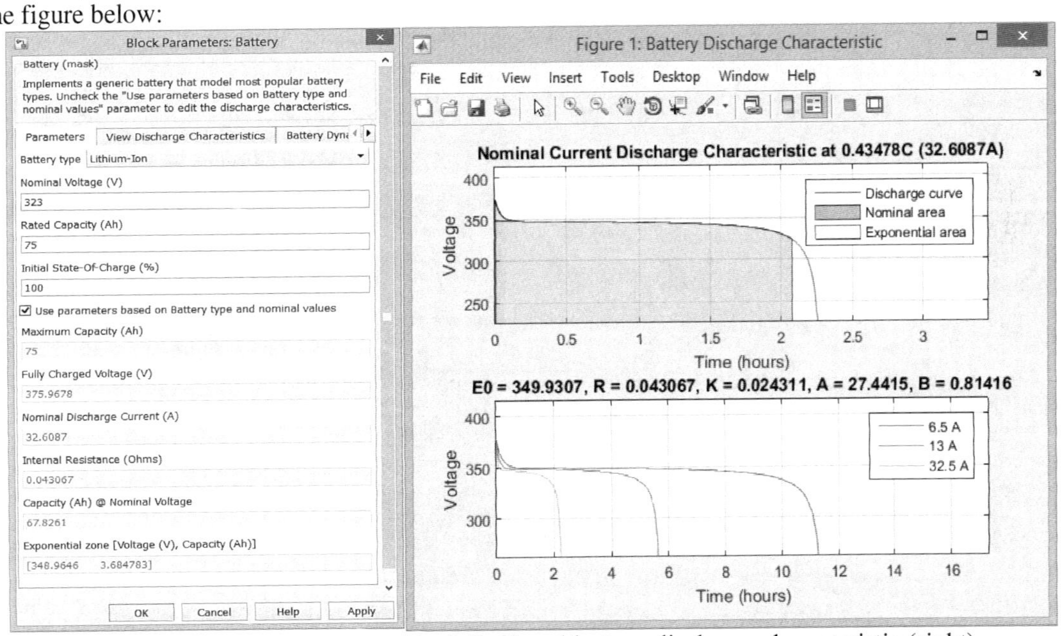


**Figure 5.** Battery parameters block (left) and battery discharge characteristic (right).

To import into Simulink the diagrams obtained in Excel, a block called "SignalBuilder" was used, and to use the signal emitted by it as one produced by a load connected to the battery, a "Controlled Current Source" block was used, which made the interconnection between signal source and battery terminals. In addition, in terms of simulation settings, we used the "ode45 (Dormand-Prince)" solver, and the simulation time was set based on the total time of each individual cycle, i.e. 0 - 1800 s for WLTC and 0 - 1369 s for UDDS and also used the "powergui" block with the continuous type variable. Thus, we obtained the following simulation model, also based on the explanations in [4], [5]:

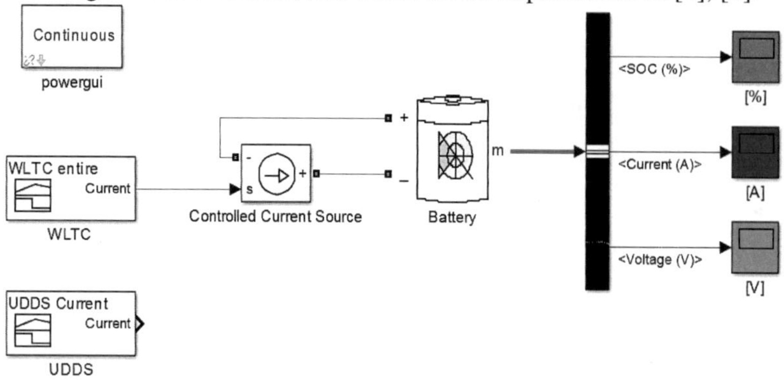


**Figure 6.** Simulation model for Lithium-Ion battery.

As for the signals used for simulation imported through the "SignalBuilder" block, they can be seen in the figures below:

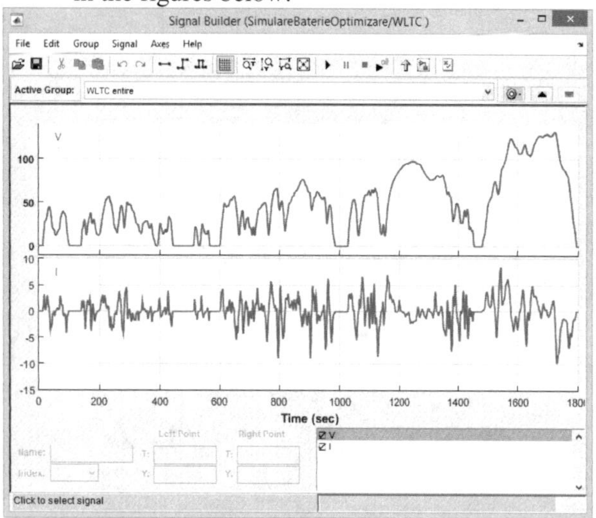


**Figure 7.** Signals from „WLTC SignalBuilder" block.

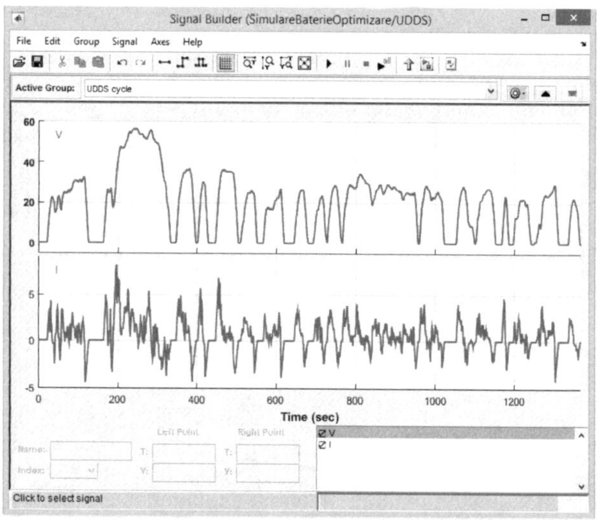


**Figure 8.** Signals from „UDDS SignalBuilder" block.

Thus, all the data necessary for the simulation were prepared for the two situations addressed. In the last part of the paper, the obtained results will be presented.

## 4. Results and conclusions

### *4.1. Simulation results for WLTC*

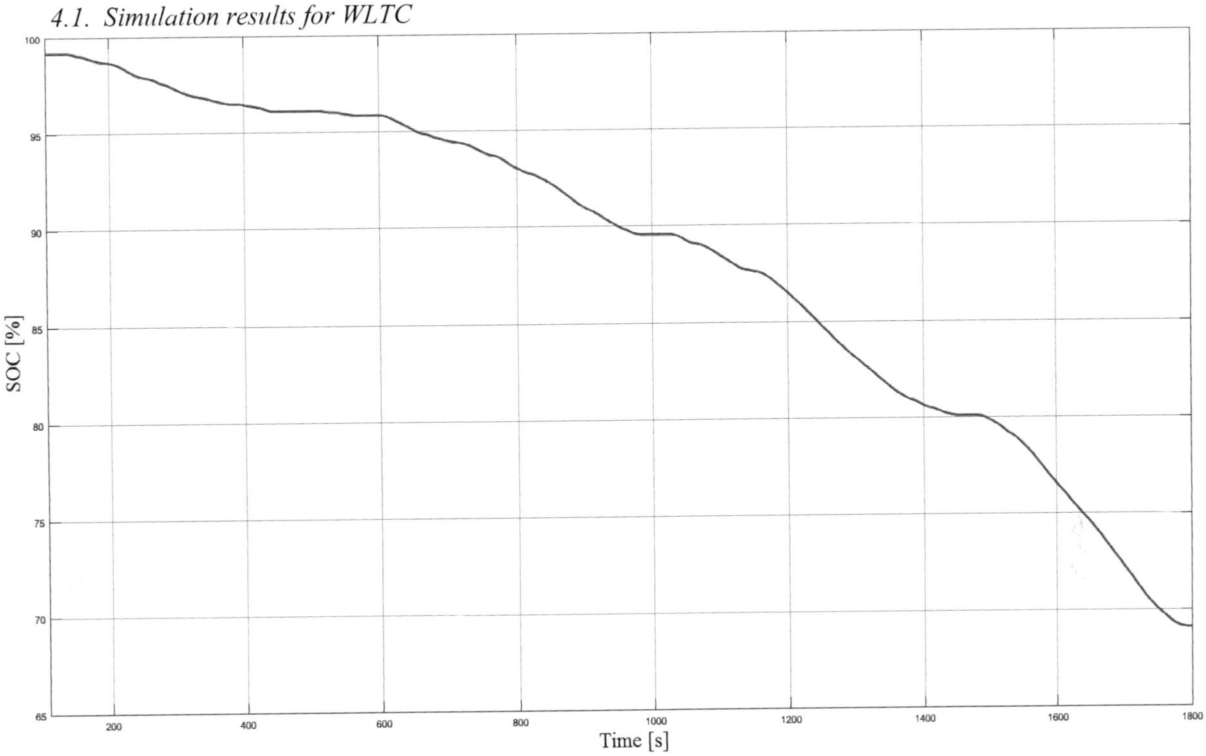


**Figure 9.** Battery SOC using WLTC profile.

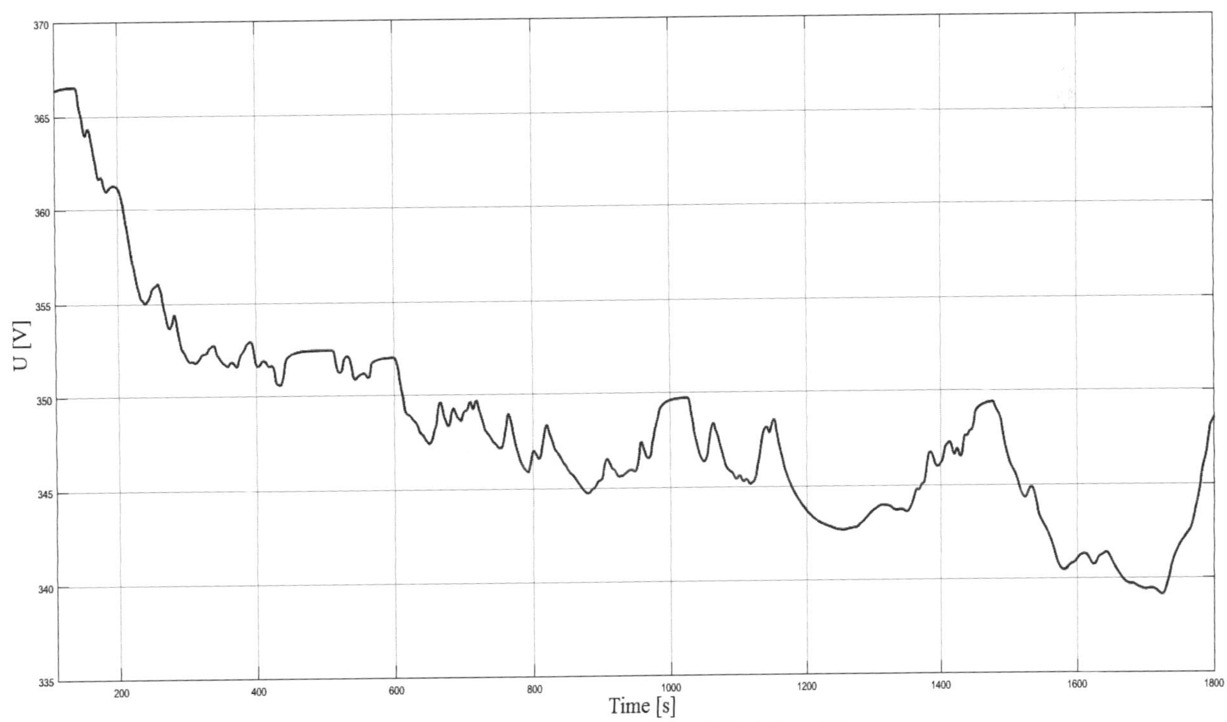


**Figure 10.** Battery voltage using WLTC profile.

IOP Conf. Series: Materials Science and Engineering **1311** (2024) 012007 doi:10.1088/1757-899X/1311/1/012007

### *4.2. Simulation results for UDDS*

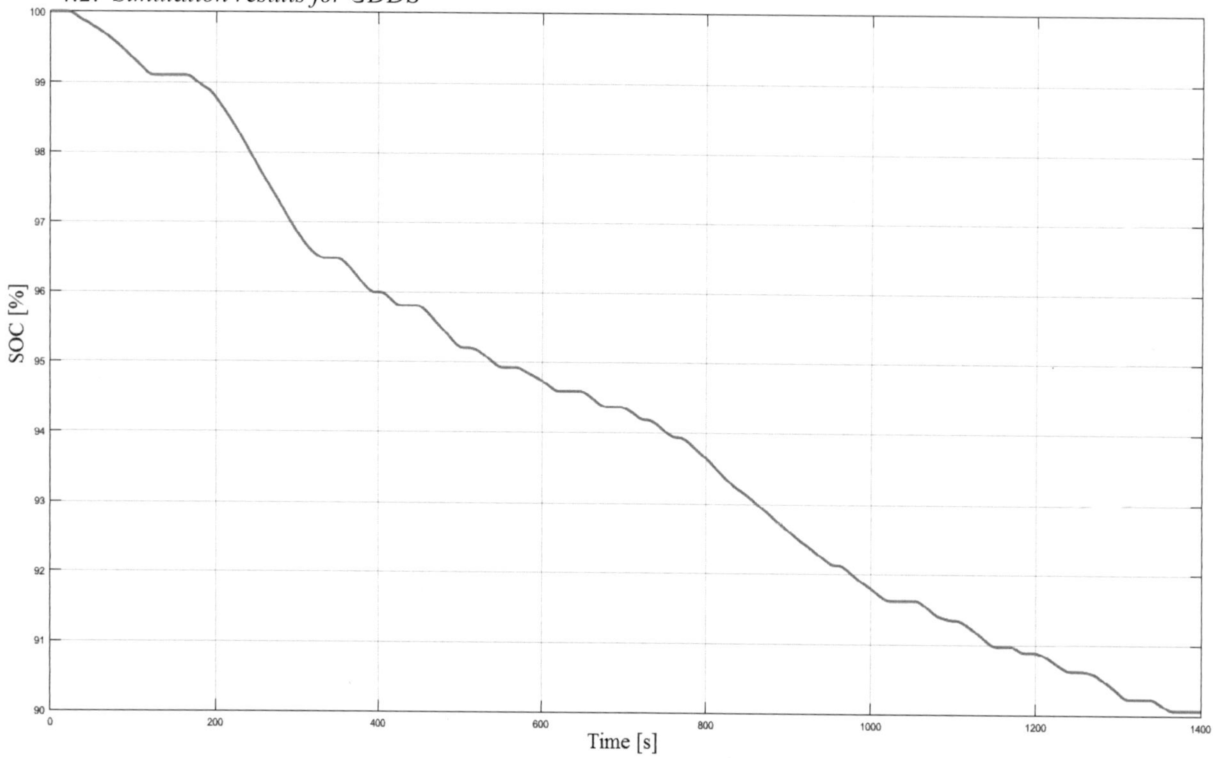


**Figure 11.** Battery SOC using UDDS profile.

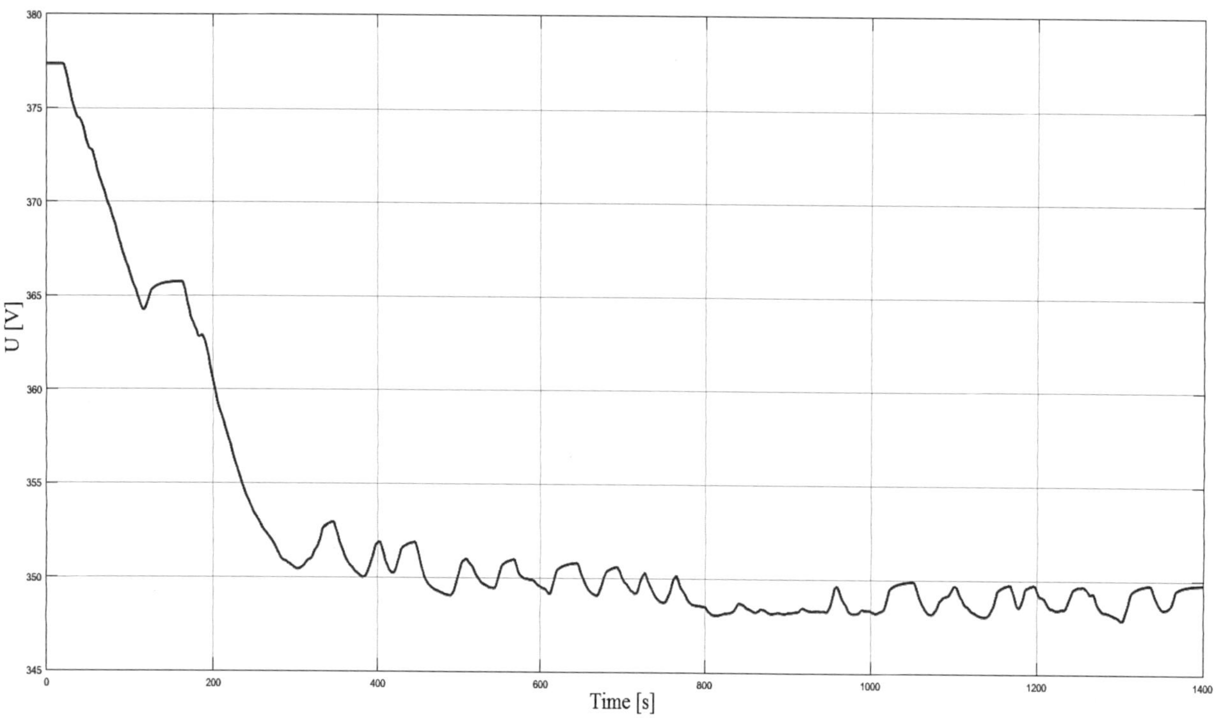


**Figure 12.** Battery voltage using UDDS profile.

### *4.3. Conclusions and comments*

- The SOC variation over time for the simulation using the WLTC cycle (figure 9) shows us that at the start of the simulation, at t=0 s, the SOC value=99.5%, an aspect that can be influenced by the current value at that time, which increases almost instantly once the vehicle starts running according to the velocity diagram; in other words, when the vehicle starts, the SOC gets a slight decrease;
- During the simulation, for 1800 s, the SOC reaches the value of 69.0752%, so it suffers a decrease of approximately 30.5%, with the observation that in the areas where the vehicle decelerates on the velocity diagram, the SOC stabilizes and remains constant until further acceleration; therefore, driving the vehicle in such a regime, when the battery is required by a variable discharge current, directly proportional to the driving velocity, will generate a complete discharge of the battery (SOC = 20...30%) in approx. 3600 s (one hour);
- The battery voltage when running on the WLTC cycle also suffers an obvious decrease during the 1800 s, but it must be specified that when the vehicle brakes, (see the areas where the speed drops suddenly or even reaches 0 value - figure 1), the voltage suffers a slight increase, due to regenerative braking, which manages to support the battery throughout the simulation;
- From second 1724, where the maximum velocity of the WLTC cycle is reached, i.e. 131.3 km/h, and where the voltage value is the lowest during the simulation, i.e. 339.1 V, there is a sudden increase in the voltage to around 350 V, which is due to the high deceleration of the vehicle at the end of the test cycle; in total, during the simulation, the voltage suffers a drop of 30 V;
- As for the SOC when running on the UDDS cycle, its decrease is much faster compared to the WLTC cycle, but the amount by which the SOC decreases during the 1369 s is also lower, about 10%, which is also due to the velocity running time, which is lower than the WLTC cycle; something that is transmitted to the discharge current applied to the battery, at the beginning of the run it is higher, later reaching lower values;
- The battery voltage on the UDDS cycle also has an accelerated decrease of about 25 V, in the time interval 200…350 s where the velocity also increases very quickly, after which the voltage tends to stabilize with small variations in the area of 340…350 V, where the positive effect of frequent decelerations when running according to the UDDS cycle is also observed, which produces an increase in voltage;

Thus, we can observe the influence of running on an imposed cycle on the battery and the differences that appear in terms of the influence on the main parameters of the battery, differences generated by the different characteristics and requirements of the two cycles used. In the following works, these cycles will be used for several types of electric vehicles and the forward resistances will also be taken into account, as well as other constructive parameters so that the electricity consumption can be determined in different situations and of course taking into account and environmental factors such as ambient temperatures.

**Acknowledgments**
I would like to thank first of all to my PhD supervisor Mr. Professor Gheorghe Frățilă for the constant support and continuous coordination they give me and of course to my Professor Grigore Danciu for all the support provided in conducting this research, for the advice and ideas he offered me all the time.

**References**

[1] SSP 530 Document – VOLKSWAGEN e-Golf catalog Service

[2] ECE/TRANS/WP.29/2020/78 (as amended by ECE/TRANS/WP.29/2020/93) – UN Regulation No. 154, pages 155 - 205

[3] P Chatterjee, J Singh, Y A R Avadh, S Kanchan – *Electric Vehicle Modeling in MATLAB and Simulink with SoC&SoE Estimation of a Lithium-ion Battery* – IOP Conf. Series: Materials Science and Engineering 1116 (2021) 012103, FSAET 2020

[4] D. Makarchuk, J. Kreicbergs, A. Grīslis, M. Gailis – *Analysis of Energies and Speed Profiles of Driving Cycles for Fuel Consumption Measurements* – Published 2015, ID:56063185

[5] https://skill-lync.com/student-projects/week-5-battery-characteristics-using-drive-cycle-21, last access 8/12/2022

IOP Conf. Series: Materials Science and Engineering **1311** (2024) 012008 doi:10.1088/1757-899X/1311/1/012008

# Research concerning air quality in the vicinity of a public transport stop

**D M Mustață[1], R Ionel[1], I Ionel[1]**

[1]Politehnica University of Timișoara, Piața Victoriei 2, Timișoara 300006, Romania

dan.mustata@student.upt.ro, ioana.ionel@upt.ro

**Abstract**. The research focuses on particulate matter and the results of their concentration in open air due to several impact factors like traffic intensity and weather conditions. The result of the 11 days episode measurement is provided in graphics. The study is part of a complex research involving the development of an integrated system for analysing air quality in the vicinity of public transport stops. Also, the number of passengers exposed to levels of poor air quality which are waiting in these stops will be recorded. The authors analysed a road section where future integration and applicability of such a system is presumed to become functional, recording the impact on both passengers and environment.

## 1. Introduction

Air quality is an important issue for all governments in the world, focusing, with great attention, on systemic methods to reduce air pollution levels, especially in urban agglomeration areas. Pollution and poor air quality in urban areas are responsible for a number of associated diseases, especially cardiovascular and respiratory. According to a report by the World Health Organization (WHO, 2014), premature death from heart attacks and severe heart diseases is associated 80% with the level of air pollution [1].

The high levels of air pollution and the effects on people's health began to be observed already at the beginning of the twentieth century, due to some instances of severe pollution on the environment, among which there can be mentioned the Meuse Valley (Belgium, 1930), London (Great Britain, 1952), and Donora (USA, 1948). In these situations, negative effects on human health have been observed with cases ranging from headaches, vomiting, and even deaths [2].

Among the main pollutants generated by the combustion process of internal combustion engines are suspended particles of different sizes (PM1.0, PM2.5, PM10) causing bronchitis, asthma, and inflammation of the respiratory tract [3], carbon oxides (CO, CO2) that cause cardiovascular diseases by reducing the ability of the blood to transmit oxygen [4], and nitrogen oxides (NO, NO2) that also cause respiratory diseases, especially asthma [5].

In general, governments and state institutions use conventional air quality measuring equipment, arranged in different positions throughout the studied area of an urban/peri-urban agglomeration. They are usually bulky, require large amounts of energy to operate, and are immovable. Recent advances in the field of air quality measurement have also brought new technologies through which measuring equipment becomes also portable, facilitating their moving, respectively their mobile use [6].

The introduction of an air quality monitoring system in public transport stations represents a punctual analysis, thus attempting a direct analysis on the recording of high pollution levels, which have a direct impact on a constant group of exposed passengers waiting in these public transport stops.

The state institutions ensure air quality monitoring at the entire level of urban areas with large agglomerations of people, but they do not specifically target situations of this nature. Passengers are exposed daily, waiting for public transport vehicles, to large amounts of pollution generated by the traffic present in the immediate vicinity of public transport stops. Obviously, passengers of personal vehicles do not feel the effects of pollution with the same intensity due to technological advances in the automotive industry. The presence of filters reduces the penetration of suspended particles of different sizes (PM2.5, PM10) into the passenger compartment of vehicles [7], but unfortunately passengers waiting in stations are constantly bombarded by vehicles that pass in front of them. In general, the elderly and children are the most exposed, as they represent the vast majority of passengers who choose to use public transport in their daily commutes.

Air quality measurement equipment will be used to measure both suspended particles and polluting gases resulting especially from the combustion process of motor vehicle thermal engines. For the analysis of the number of passengers which are exposed, advanced people counting systems will be used, arranged on a certain area, through video recording systems [8].

The main focus of the analysis presented in this article is to determine a variation of the concentration of the particles knowing that they have a high impact on human health. The implementation of the future measurement system will monitor continuously the air quality and the number of passengers waiting in public transport stops. This study represents a more detailed analysis of a previously performed study on the same road section. The episode duration for recording PM values was short compared to the present research covering more than 20 days which is detailed in the current article [9].

## 2. Methodology

Particulate matter represents the main focus. The measurement of PM2.5 and PM10 was performed next to a public transport stop on a frequently congested road section, located in the peri urban area of Timisoara in the town of Dumbravita, Timis County, Romania. The road is particularly congested due to several aspects, including a highly commuted area, the town is on the outskirts of Timisoara, one of the biggest cities in Romania, it is the main connection to the highway A1, and it is located in the vicinity of production facilities and logistics hubs. All these conditions present a high number of vehicles that are transiting this main road section from the city. Due to the small surface area of Dumbravita and being a residential area, the majority of public transport stops are placed next to this main road which has only one lane of traffic per travel direction. There are a total of 11 stops that border the main road section. The measurements where accomplished next to one of these stops, located approximately in the middle of the length of the road section at the geographical coordinates 45°47’54” N, 21°14’39” E.

The measurement method used is via gravimetric impactors, which use a glass fibre filter to capture the different sized particulate matter by sucking air at a specified air flow rate. The determination of the concentration for PMs was done by weighting the clean vs charged filters. An impactor measured PM2.5 concentrations and the second measured PM10 concentrations. Both impactors are calibrated, metrologically certified, and accredited by RENAR (Accreditation Association of Romania).

The measurement episode is from 15$^{th}$ of June to 25$^{th}$ of June 2022, representing 11 days of continuous measurement for PM2.5 and PM10. The number of vehicles passing in front of this public transport stop was measured with a recording device placed perpendicular to the axis of the main road so that it only recorded the vehicle type and amount. Personal identifiers were not recorded and cannot be extracted from the data, due to the position of the recording device. The number of vehicles was determined by measuring at every 15 min interval a 1 min interval counting for 24h.

## 3. Results and analysis

The 11-day continuous measurement determined the highest concentration at 36.84 μg/ Nm$^3$ for PM2.5, and 51.87 μg/ Nm$^3$ for PM10 on 19$^{th}$ of June 2022.

All measured values per day (24h measurement episode) are represented in Figure 1.

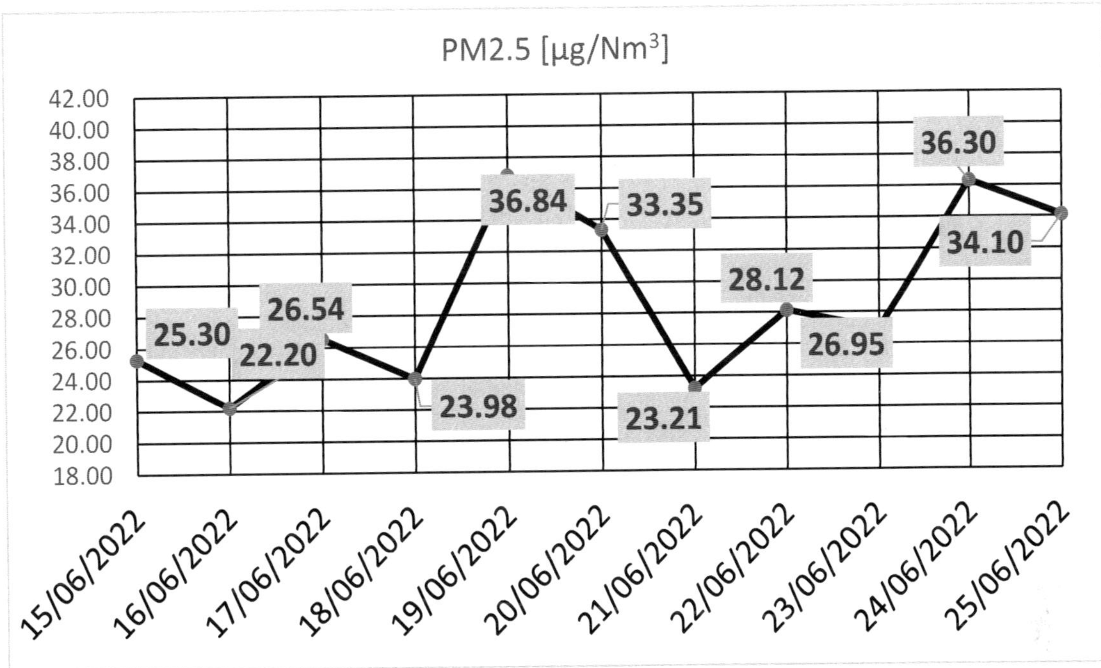


**Figure 1.** Concentration of PM2.5 recorded values for 11 days episode.

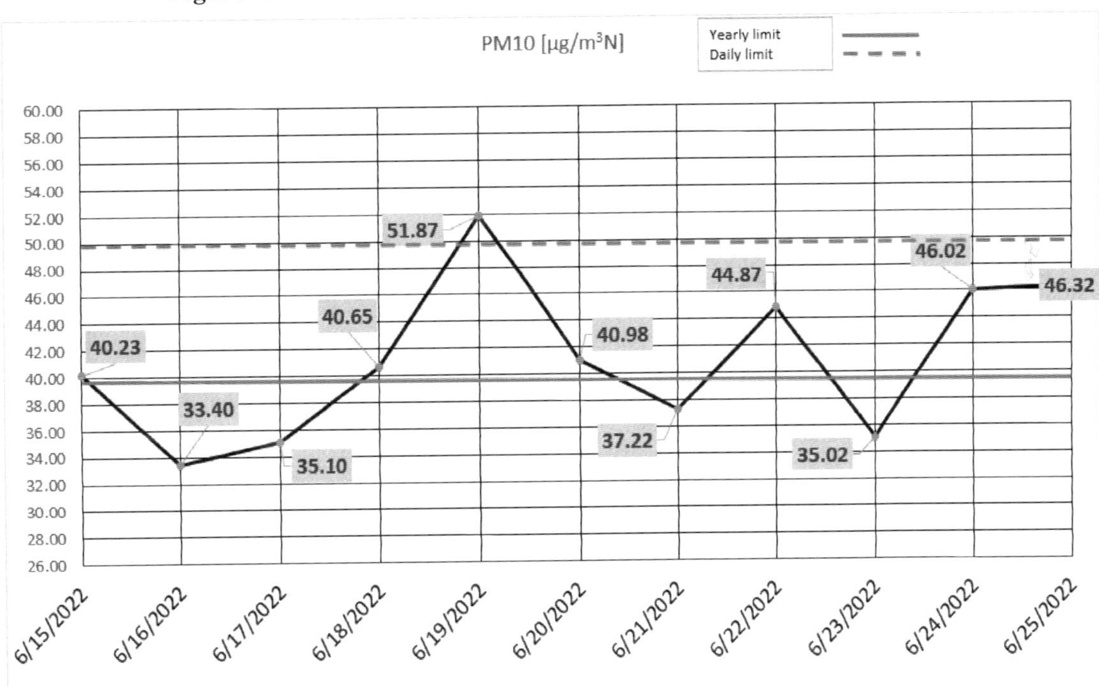


**Figure 2.** Concentration of PM10 recorded values for 11 days episode.

According to Romanian law, the accepted yearly limit for PM10 is 40 µg/ $Nm^3$, and additionally the daily limit is 50 µg/ $Nm^3$ and is not allowed to be exceeded for more than 35 days in a year. For PM2.5 the Romanian law is not providing limits [10].

Analysing both graphs, it can be recognized that the values are high above the accepted yearly limit and already exceeded for 1 day out of the 35-day maximum allowed daily limit of 50 µg/ $Nm^3$.

Multiple causes can be attributed to this peak and for the trends that are present in this 11-day episode.

One of this is due to the size of the traffic, where in Figure 3 it can be recognized that the trend of the number of vehicles is directly proportional to the values recorded for PM2.5 and PM10.

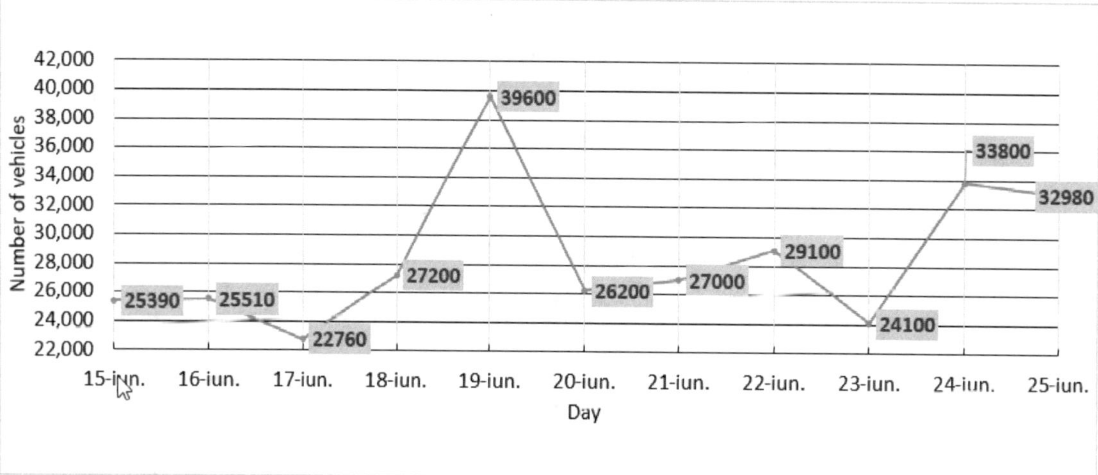


**Figure 3.** Traffic intensity for the 11 days episode.

Of course, more vehicles do not necessarily mean that the PM values are increasing as well; the meteorological conditions have to be considered also.

In the case of temperature measurement, in Figure 4 maximum and average recoded temperature are shown and it can be seen that the trend is again similar to that of the PM spread.

During heat waves and dry weather conditions, particulate pollution increases. The soil becomes very dry, which also results in an increase of dust particles getting raised into the atmosphere [11].

In addition to the heat waves that lead to an increase of the PM values, the wind conditions have also been ideal to allow this increase and high value.

Analysing the day of $19^{th}$ of June, where the highest values were recorded in both PM2.5 and PM10, the wind speeds were low, recorded at an average of 8 km/h and a wind direction from N-E [12]. The measurement device was placed perpendicular to the road section having the nozzles that sucked air toward the N-E direction, meaning that the wind was blowing more predominantly throughout the entire day towards the measurement device.

Related to air pressure, in which air quality is better in case of lower pressure conditions, because air being able to disperse easier with winds, it can be recognized that the air pressure measured during this 11-day episode was inversely-proportional to the PM values, the number of vehicles, and temperature trend. This normally should have led to a drop in PM values even if the number of vehicles was higher, but due to wind speeds, the air could not quickly disperse, being more stagnant. Wind speeds during the entire 11-day episode were recorded as an average value between 1 and 20 km/h [12].

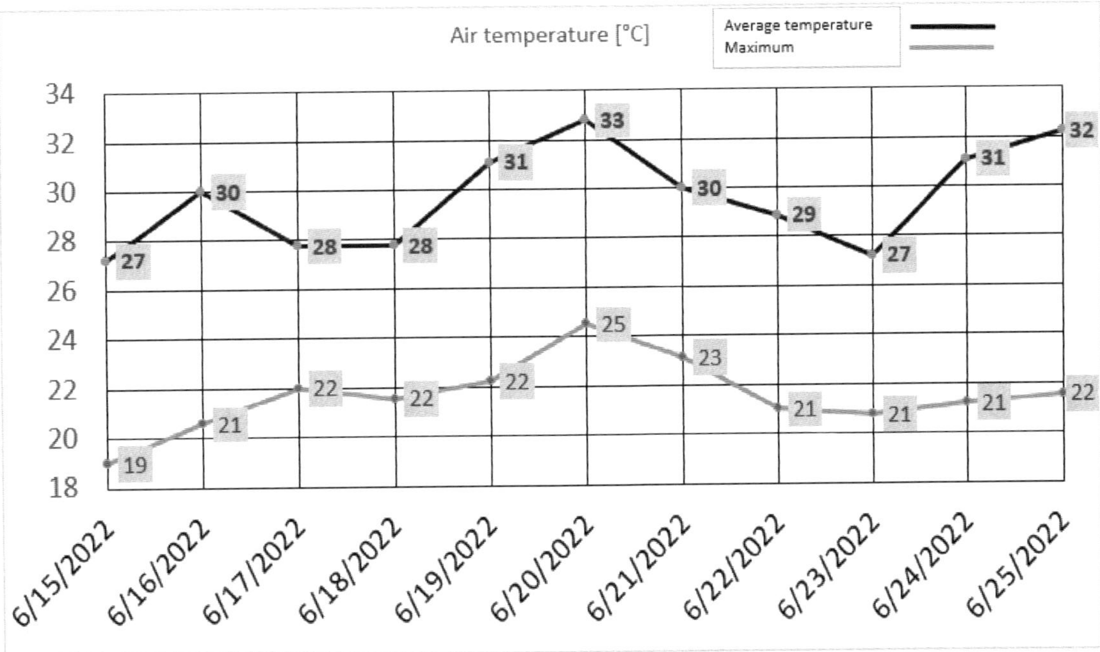


**Figure 4.** Air temperature for the 11 days episode.

## 4. Conclusions

The research presented in this article was to analyse a congested road section by continuous monitoring of particles PM2.5 and PM10 concentration. The concentrations are close to the yearly average limit of 40 µg/ Nm$^3$ and for 6 days, this has already exceeded, coming close to the threshold of the 50 µg/ Nm$^3$ of 35 days per year limit. According to the measurements and results, the road section is highly congested by the traffic. The meteorological data were also recorded. Both traffic intensity and weather conditions are well correlated to the variations of the particle concentration. The research supports a future development of a measurement system that will raise awareness and protect the passengers waiting in the public transport stops.

**Acknowledgements** The first author expresses his gratitude to the teaching staff of the Politehnica University of Timisoara, the LACIEDIN (www.mediu.ro) laboratory staff, and to the co-authors, especially to Prof.dr.ing.habil Ioana Ionel, the coordinating professor of his PhD study program.

## References

[1] Guerreiro C, Gonzalez Ortiz A and de Leeuw F 2017 Air quality in Europe — 2017 report European Environmental Agency 11-12 doi:10.2800/358908 https://www.researchgate.net/publication/323005604_Air_Quality_in_Europe_-_2017_report accessed on 29th Aug. 2022

[2] Metcalfe S, Derwent D 2005 Atmospheric Pollution and Environmental Change pp 84-85

[3] Zaheer J, Jeon J, Lee S, Kim J S 2018 *Effect of Particulate Matter on Human Health, Prevention, and Imaging Using PET or SPECT* Progress in Medical Physics p 81-91 DOI: https://doi.org/10.14316/pmp.2018.29.3.81

[4] https://www.epa.gov/co-pollution/basic-information-about-carbon-monoxide-co-outdoor-air-pollution#Effects accessed on 29th Aug. 2022

[5] https://www.epa.gov/no2-pollution/basic-information-about-no2#Effects, accessed on 29th Aug. 2022

[6] Agrawaal H, Jones C, Thompson J E 2020 *Personal Exposure Estimates via Portable and*

*Wireless Sensing and Reporting of Particulate Pollution* Int. J. Environ. Res. Public Health EISSN 1660-4601 **https://doi.org/10.3390/ijerph17030843**

[7] Yu N, Shu S, Lin Y, She J, Ip H S S, Qiu X, Zhu Y 2017 *High efficiency cabin air filter in vehicles reduces drivers' roadway particulate matter exposures and associated lipid peroxidation* PLOS ONE https://doi.org/10.1371/journal.pone.0188498

[8] Khan S H, Yousaf M H, Murtaza F, Velastin S 2020 *Passenger detection and counting for public transport system* NED University Journal of Research https://doi.org/10.35453/NEDJR-ASCN-2019-0016

[9] Mustață D M, Popa R M, Ionel I, Balogh R M, Streinu D R 2022 *Air quality analysis on a frequently congested road section* IOP Conference Series: Materials Science and Engineering https://doi.org/10.1088/1757-899X/1220/1/012058

[10] *Romanian Law no. 104 from 15 June 2011 in relation to air quality (original in Romanian: LEGE nr. 104 din 15 iunie 2011 privind calitatea aerului înconjurător)* Parliament of Romania http://legislatie.just.ro/Public/DetaliiDocument/129642, accessed on 29th Aug. 2021

[11] https://scied.ucar.edu/learning-zone/air-quality/how-weather-affects-air-quality , accessed on 29th Aug. 2022

[12] https://my.meteoblue.com/visimage/history_simple_hd?look=KILOMETER_PER_HOUR%2CCELSIUS%2CMILLIMETER&apikey=5838a18e295d&temperature=C&windspeed=kmh&precipitationamount=mm&winddirection=3char&city=Dumbr%C4%83vi%C8%9Ba&iso2=ro&lat=45.7969&lon=21.2425&asl=94&tz=Europe%2FBucharest&fcstlength=31&year=2022&month=7&lang=ro&params=20220731&ts=1662028614&sig=97886ec900617f3e3d5ad9100974c7cd , accessed on 29th Aug. 2022

IOP Conf. Series: Materials Science and Engineering **1311** (2024) 012009 doi:10.1088/1757-899X/1311/1/012009

# Experimental for depicting the influence upon gaseous emissions of an Otto engine by using hydrogen as fuel addition

**C M Suciu[1], S V Igreț[2] and I Ionel[1*]**

[1]Politehnica University Timisoara, Faculty of Mechanical Engineering, Piața Victoriei 2, Timișoara 300006, Timisoara, Romania
[2]Aurel Vlaicu University, Faculty of Engineering, Calea Aurel Vlaicu 43, 310141, Arad, Romania

*Email: ioana.ionel@upt.ro

**Abstract**. Due to the higher pollution restrictions implemented globally upon the internal combustion engines, for new but also older vehicles, suitable solutions must be found, for solving major, attested incriminations concerning pollution. The present research focuses on the possibility of the reduction of the emission concentration values emitted by a classic Otto spark ignition engine (Otto ICE), by adding hydrogen to the normal used fuel (gasoline). The hydrogen is generated via a water electrolytic conversion technology, applied, in symbiosis with the internal combustion engine and it is known as hydrogen air (HHA). The hydrogen quantity, as resulted from the HHA generating process and fuelled into the engine, represents 10% (by volume) of the total volume of the intake capacity. Being directly correlated to the stoichiometric ratio used to create the fuel mixture, the intake of hydrogen is finally controlled; therefore, the total calorific power of the new fuel mixtures is increasing by approximate 30%. All the measurements have been achieved in a lab using certified and approved gas analyzers and measurement instruments. The novelty of the research consists of *(i)* presenting results of measurements with the gas analyzer, one at an engine speed of 800 RPM and the second at 2000 RPM using only gasoline as fuel (normal or standard fuel); *(ii)concluding of* comparative measurements at the same engine speeds, but with addition of hydrogen in the fuel mixture. Since carbon dioxide makes up the vast majority of greenhouse emissions, the results indicate also a more reduced $CO_2$ concentration in the exhaust. One presents that the electrolytic convertor is capable of producing 1800 litters of HHO gas from 1 litre of water. By upgrading the oxyhydrogen system, one succeeded to produce more HHO gas and replacing the amount of conventional fuel. As conclusions one demonstrated that lower emission values can be obtained with the addition of hydrogen air (HHA).

## 1. Scope of the research

Over the last decade, automobile constructors are confronting more severe pollution restrictions. The global direction is going towards electrification. But vehicles, equipped with internal combustion engines (ICE), are still being produced, sold and used on public roads. The batteries' low range characteristics for the electric vehicles and the high price on vehicles equipped with electric motors are two major reasons presently that an ICE vehicle still can represent a first choice for the majority of the population, when it comes to buy a new vehicle. The production of batteries implies an extremely polluting process. Energy consumption necessary for producing electric powered vehicles is mainly

IOP Conf. Series: Materials Science and Engineering **1311** (2024) 012009 doi:10.1088/1757-899X/1311/1/012009

based on fossil fuels and carbon is emitted [1], [2], [3]. Worth to mention are the high prices of the batteries, their large volume and also the difficulty in reusing the waste batteries.

The possibility of emission reduction is investigated by the present automobile engine manufacturer industry. A turn towards greenhouse effects technologies vehicle manufacturers starts considering utilizing unconventional synthetic fuels [4], [5] as potential solution.

Hydrogen is to be considered an alternative fuel source. Hydrogen has a superior calorific power to all conventional fuels and costs are lower for obtaining [6], [7], [8].

## 2. Measurements and results

An ICE vehicle, equipped with a gasoline engine on which it was mounted an electrolytic converter, was used for the measurements.

Table 1 presents the engine specifications.

Table 1. Engine specifications.

| Characteristic | Value |
|---|---|
| Number of cylinders | 4 |
| Cylinder's layout | Straight/ inline |
| Cylinder bore [mm] | 79 |
| Cylinder stroke [mm] | 81.5 |
| Compression ratio | 10.5 |
| Number of valves | 16 |
| Fuel type | Gasoline |
| Engine's position | Front, Transversal |
| Engine's displacement [$cm^3$] | 1,598 |
| Intake | Aspiration |
| Max power output [kW] | 77 @ 6,200 RPM |
| Max torque output [N·m] | 150 @ 3,900 RPM |
| Transmission layout | Front wheel drive |
| Gearbox | Manual |
| Number of gears | 5 |
| Fuel tank volume [l] | 58 |
| Emission class | Euro 3 |

Capelec 3200 gas analyser, presented in figure 1, had been used for the measuring of the emission values. There were two measurements, one at engine's idle speed and the other at an engine speed of 2000 rotations per minute (RPM).

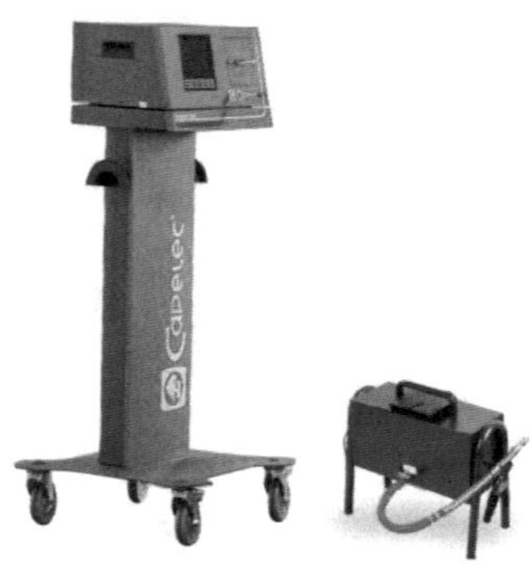


**Figure 1.** Capelec 3200 stand-alone opacimeter

In order to produce gas from water, using the electrical power supply from the vehicle, water is dissociated into hydrogen and oxygen. The gas mixture created is also called Oxy-hydrogen gas (HHO gas) [9], [10].

According to [11], in the future years' hydrogen is to be expected to become a valuable fuel, contributing also to the quality of the atmospheric air. Authors' [11] arguments consist in the fact that hydrogen is the most widespread has on Earth. It can be obtained via both renewable and nonrenewable by various processes. Furthermore, the results can be promising solution for the future retrofit of the ICE trains, by using only HHO gas as fuel.

Figure 2 represents the electrolytic converter system. This consists of a generator, a pulse modulator, a water tank and a filter. The HHO generator is characterized by positive, negative and neutral plates. Continuous voltage is applied to the electrode. At the anode (the positive charged plate) water molecules loose two electrons, forming an oxygen molecule and four hydrogen ions. The oxygen obtained is vented into the atmosphere. The hydrogen ions obtained pass through the neutral plates being attracted by the cathode (the negative charged plate). This converts the four ions into two hydrogen molecules. The water tank contains water and potassium hydroxide in proportion of 5-15%. The mixture is further used as an electrolyte in the generator. The water tank is connected to the generator via a hose. The pulse modulator using voltage between 12-14 Volts controls the amount of HHO to be produced. Maximum quantity of HHO that can be produces from 1 liter of water is of 1800 liters after the system is powered on and starts producing HHO. Before passing to the vehicle's intake system, the HHO passes through the bubble filter with the purpose of stopping the small gas particle and vapours. The HHO gas reaches temperature up to 50 degrees Celsius and thus vapour may occur. The resulted vapour quantity is reduced within the bubble filter. The HHO gas that passes the bubble filter enters the intake system at a pressure equal to the one of the atmosphere' due to the chosen engine being naturally aspirated. The electrical current required for the whole system is between 2 and 20 Amperes. It is able to produce between 2 and 4 litters of HHO gas per minute.

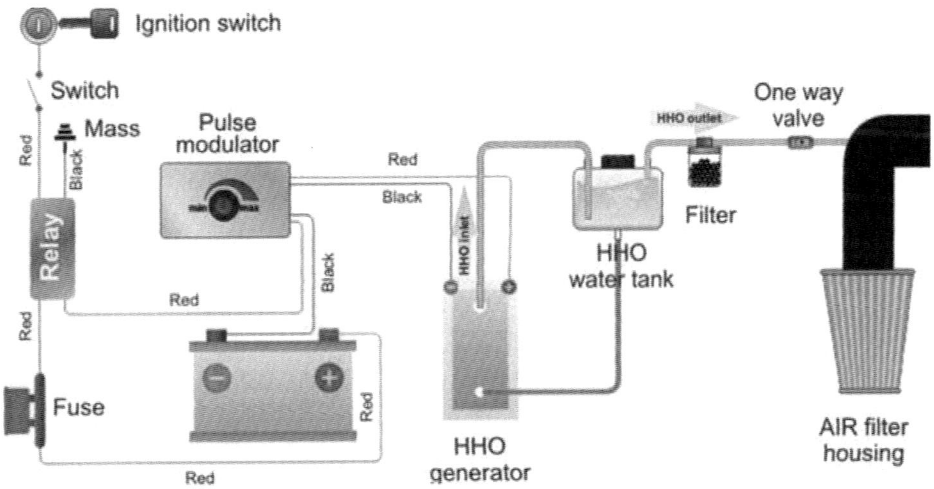


Figure 2. Installation scheme of the HHO system as used in the research, according https://images.app.goo.gl/2QdheyWmAwZeJXpC8.

The heating value of the hydrogen and thus its energy density is higher than that of the gasoline by about 3 times. The resulted hydrogen from the process represents 10% of the total volume of intake's volume per stroke.

Measure values have been obtained using Capelec 3200 gas analyzer. Table 2 represents measured values by using only gasoline as fuel. For table 3 the engine had been exploited in the exact same environmental conditions. Values measures represent the emissions after using gasoline fuel with addition of HHO gas. Both tables represent measurements at two engine speeds of an Otto engine, one at 800 rotations per minute (RPM) and one at 2,000 RPM.

Table 2. Measurement results using only gasoline as fuel.

| No. | Engine speed [RPM] | CO[%vol] | $CO_2$ [%vol] | HC [ppm] | $O_2$ [%vol] | Temperature [°C] | Air excess ratio |
|---|---|---|---|---|---|---|---|
| 1 | 800 | 0.72 | 10.90 | 113 | 2.86 | 92 | 0.996 |
| 2 | 2,000 | 0.52 | 11.50 | 142 | 2.56 | 92 | 1.030 |

Table 3. Measurement results using gasoline and HHO as fuel

| No. | Engine speed [RPM] | CO [% vol] | $CO_2$ [%vol] | HC [ppm] | $O_2$ [%vol] | Temperature [°C] | Air excess ratio |
|---|---|---|---|---|---|---|---|
| 1 | 800 | 0.22 | 4.10 | 60 | 13.2 | 92 | 1.816 |
| 2 | 2,000 | 0.11 | 5.60 | 80 | 12.6 | 92 | 1.612 |

Charted representations of the tables are presented in figures 3 and 4. Figure 3 overlays results from the measurements made at an engine speed of 800 RPM while figure 4's measurements had been made at an engine speed of 2,000 RPM.

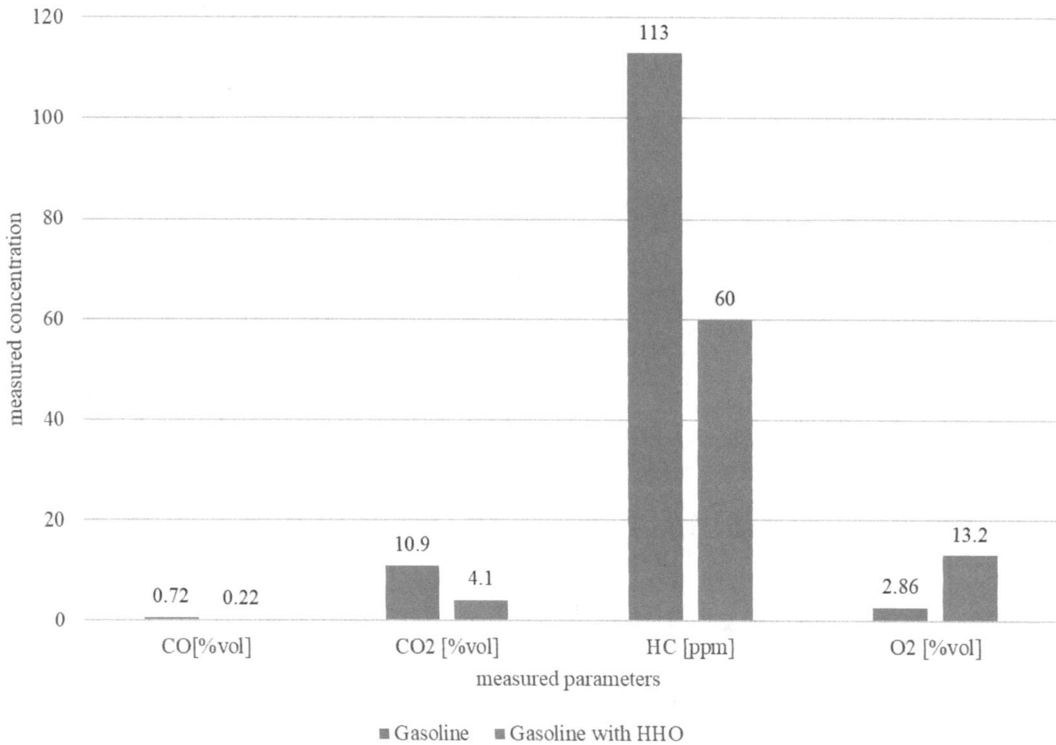


Figure 3. Comparative concentration measurement results overview at an engine speed of 800 RPM

At the idle engine speed of 800 RPM comparing the two measurements, with and without the addition of oxyhydrogen gas, results conclude that when using HHO together with gasoline, the monoxide carbon (CO) volume is lower by 106.3% compared to only using gasoline as fuel, carbon dioxide's ($CO_2$) value is lower by 90.6% compared to using only gasoline as fuel, hydrocarbon's (HC) particles are 61.27% lower compared to using only gasoline as fuel, more oxygen is produces, by 128.7% comparted to when only gasoline is used as fuel, this can also be noticed by the higher values of the oxygen sensor's measurements

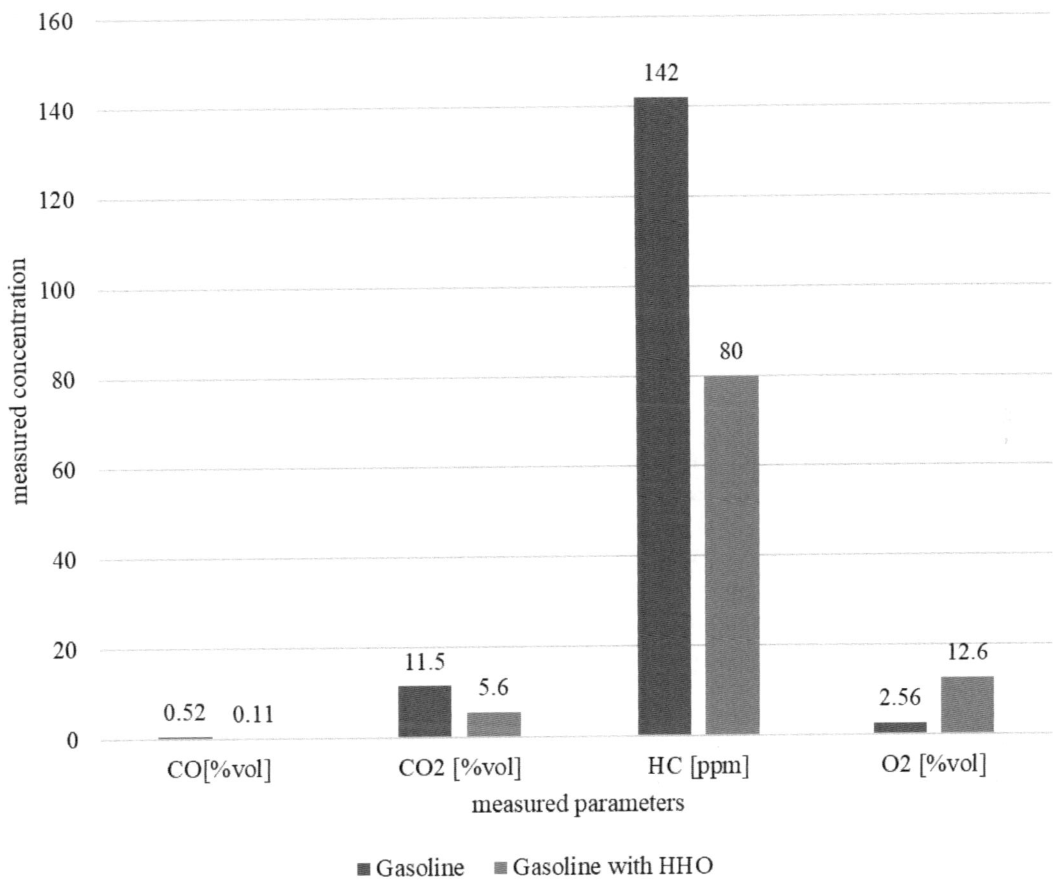


Figure 4. Comparative concentration measurement results overview at an engine speed of 2,000 RPM

The advantages of HHO are also noticeable when engine's speed is increased from 800 to 2,000 RPM. At 2000 RPM, CO and $CO_2$ values are lower by 69% compared to using only gasoline as fuel. Analysing the concentrations for HC particles are 55.8% lower compared to using only gasoline as fuel, more oxygen is produces, by 132.4% comparted to when only gasoline is used as fuel, this can also be noticed by the higher values of the oxygen sensor's measurements. For both measurements the oxygen sensor read higher values with the addition of HHO. The higher oxygen values are due to water being formed by two molecules of hydrogen and one of oxygen. Replacing 10% of the intake air which contains around 80% nitrogen with HHO increases combustion efficiency thus lowering fuel consumption and increasing engine power. The addition of hydrogen is also responsible for lowering the amount of resulted CO and $CO_2$, which are the specific pollutants when using hydrocarbon-based fuels for powering ICE. The solution presented is a good alternative for electrically powered units in the automotive industry, knowing that Li-ion batteries manufacturing process is pollutant and harmful for the environment.

## 3. Conclusion

The paper presents the benefits of adding oxyhydrogen to an internal combustion engine. One proved that the proposed system is responsible of the reduction of engine's emissions level. Initially, meaning before retrofitting, the vehicle was categorized as a Euro 3 engine, after the proposed retrofit it passes the Euro 6 pollution class-imposed limits.

Since hydrogen is quite easy to be obtained and available, the presented solution can be a retrofit solution for all ICE. The presented electrolytic converter can produce up to 1800 litters of HHO gas from 1 liter of water. An upgrade to the system can lead to a higher gas production. The results conclude around 3 time more effective power from the engine, cheaper fuel and really low emissions.

**Acknowledgement**

The research is part of the PhD program of the first author at the Politehnica University of Timisoara. The authors acknowledge for the substantial support. Also, the Romanian Automotive Register from Arad, where all experiments occurred.

## 4. References

[1] H S Das, M M Rahman, S Li, C W Tan 2020 Electric vehicles standards charging infrastructure and impact on grid integration: A technological review. *Renewable and Sustainable Energy Reviews* **120(C),**

[2] W W Jacob, J J Michalek, C Samaras 2021 *Environmental Science & Technology* 55,

[3] A Mayer, H Burtscher, S Loretzand, M Kasper, J Czerwinski 2018 *High air pollution in vehicle cabins due to traffic nanoparticle emission exposure and a solution for in-use vehicles* **421(3),**

[4] R Peters, M Decker, L. Eggemann et al. 2020 Thermodynamic and ecological preselection of synthetic fuel intermediates from biogas at farm sites. *Energy, Sustainability and Society* **10(4),**

[5] J Szczygieł, M Kułażyński 2020 Thermodynamic limitations of synthetic fuel production using carbon dioxide: A cleaner methanol-to-gasoline process. *Journal of Cleaner Production* **276.**

[6] Y Haseli 2018 Maximum conversion efficiency of hydrogen fuel cells. *International Journal of Hydrogen Energy* 43(18) 9015-9021,

[7] S Jungwoo, H Won-Sik, C Hyundoi 2019 Can hydrogen fuel vehicles be a sustainable alternative on vehicle market?: Comparison of electric and hydrogen fuel cell vehicles. *Technological Forecasting and Social Change* **143** 239-248,

[8] C Acar, I Dincer 2020 The potential role of hydrogen as a sustainable transportation fuel to combat global warming. *International Journal of Hydrogen Energy* **45(5)** 3396-3406,

[9] S Balaji, I Saleel 2018 Production and use of HHO gas in IC engines. *International Journal of Hydrogen Energy* **43(14)** 7140-7154,

[10] S Balaji, T Venugopal 2020 Analysis of onsite HHO gas generation system. *International Journal of Hydrogen Energy* **45(28)** 14218-14231,

[11] C M Kalamaras, A M Efstathiou 2013 Hydrogen Production Technologies. Current State and Future Developments. *Conference paper, special issue Power Options for the Eastern Mediterranean Region* 1-9

IOP Conf. Series: Materials Science and Engineering **1311** (2024) 012010 doi:10.1088/1757-899X/1311/1/012010

# The emission of off-road vehicles and their reduction options

**N Bíró [1*], P Kiss [2] and D Szőllősi [3]**

[1] IBIDEN Hungary Kft., Technical Center, Test Engineer

[2] Hungarian University of Agriculture and Life Sciences, Institute of Technology, Department of Vehicle Technology, professor

[3] IBIDEN Hungary Kft., Technical Center, Senior Engineer

*Email: Biro.Norbert.5@phd.uni-mate.hu

**Abstract.** The reduction of harmful emissions is a popular topic in both scientific and public circles. Contrary to popular belief, emissions from transport are only a small part of global emissions. As shown in the figure, the average of the EU 27 countries attributes only 15% of emissions from transport. From common perspective, vehicles with internal combustion, including diesel engines, will soon be replaced by alternative drivetrains. However at least at the commercial and off-road transportation, the diesel technology it is not so easy, and also the environmental reasons are questionable. Although the local emissions of electric vehicles are zero, their global emissions are, according to some research, more significant than their counterparts with internal combustion engines. The total electricity produced in the world in 2016 was about 25 PWh, two-thirds of which came from fossil energy carriers, which seriously damages the ecosystem, so it cannot be labelled as green energy.

## 1. Introduction

Despite its many advantages, there are several downsides of the diesel engines. Like most fossil fuels, diesel and hydrogen are essential components. In an ideal thermodynamic process, only carbon dioxide ($CO_2$) and water vapor ($H_2O$) would be produced in the diesel engine. [13]Factors interfering with ideal conditions: inadequate air-fuel mixing ratio, particle content of intake air, change in combustion temperature, inaccuracy of ignition timing, turbulence in the combustion chamber, non-ideal combustion chamber shape (note: ideal for spherical shape, with centric combustion). Due to these factors, significantly more harmful substances are formed. [1]

In order to reduce the amount of harmful particles in the atmosphere, the European Union began to continuously reduce the amount of harmful emissions to the environment (harmful ambient air load) with standards developed for diesel commercial vehicles in 1990. The most significant of these are nitrogen oxides ($NO_x$), carbon monoxide (CO), hydrocarbons (HC), and particulate matter (PM), as well as sulfur dioxide ($SO_2$). (Figure 1) [3]

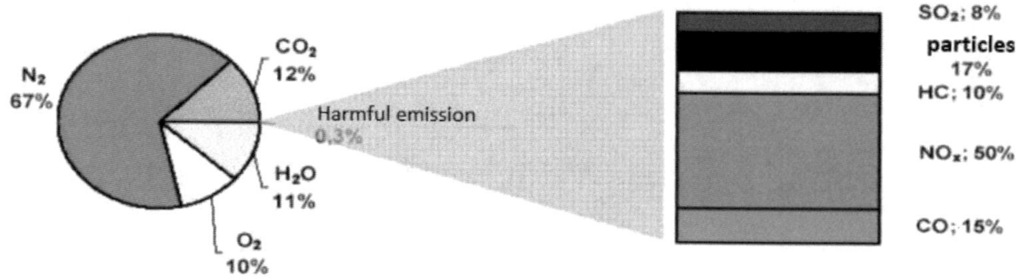


**Figure 1.** Average exhaust gas composition of diesel engines. [3]

Normally, the nitrogen in the air does not react with the oxygen during combustion, however, as a result of the inhomogeneous combustion characteristic of diesel engines, from a temperature of 1600 °C, the nitrogen already reacts with the oxygen and produces $NO_x$ emissions.

The two main components of $NO_x$ emissions are NO and $NO_2$. Significantly more toxic NO is slowly converted to $NO_2$ under normal atmospheric conditions. Road traffic accounts for 40-70 % of anthropogenic $NO_x$ emissions, of which about 85 % is related to diesel vehicles. [4] Nitrogen oxides also promote the formation of acid rain and smog.

CO is the result of imperfect combustion and is minimal in diesel engines running on high excess air. A colorless and odorless gas, the blood binds to the hemoglobin responsible for transporting oxygen orders of magnitude more than oxygen. It causes concentration problems, slowed reflexes and confusion, and eventually death.

HC emissions can be derived from unburned fuel. It is formed in the engine at lower temperature parts of the combustion chamber due to improperly high temperatures. It contributes to the increase in ground-level ozone, which can cause long-term mental problems in children and several other illness. [5]

As a result of combustion, the particulate emissions (PM) of diesel engines are also significant. These compacted particles (agglomerates) can be formed by the following materials:

• carbon particle,
• dust particles from inadequate air filtration,
• partially burned fuel,
• partially burned lubricating oil,
• ash from fuel and lubricating oil,
• sulphates,
• water molecule,
• metal particles [6].

Due to the many components, the particles are very complex. The materials listed above usually compress into a central carbon core to form particles typically 15-40 nm in size, approximately 90 % of which are smaller than 1 μm. [7] To express the number of particles: a diesel engine emits ~1014 particles, which is 3 orders of magnitude higher than the particulate emissions of a gasoline engine. Emissions that are harmful to human health can cause lung and respiratory problems and, according to the current k of science, can cause a major contribution to the development of cancer. [8]
In response to health concerns, more and more stringent emission values have been introduced for new diesel engines worldwide. Current emissions are forcing everyone in the world to continually improve and system-wise, where we need to use and combine advanced engine management in parallel with a variety of sophisticated exhaust after-treatment technologies.

There are several well-functioning exhaust gas treatment technologies for diesel engines today.
The most significant PM reduction technologies:
diesel oxidation catalyst (DOC),

diesel particulate filter (DPF),
closed crankcase ventilation (CCV).

Oxidation catalysts, which were also responsible for the oxidation of hydrocarbons (HC) and CO in early petrol engines, continue to play an important role in diesel engines. Although originally developed to reduce gaseous emissions, they have also been shown to be effective in reducing particulate emissions, which can be as high as 20-50 % (total particle weight). [8]

Diesel particle filter is also a proven technology. They have a honeycomb structure, with the cell always closed on the opposite side. The filtration thus takes place on the pores of the cell walls. (Figure 2)

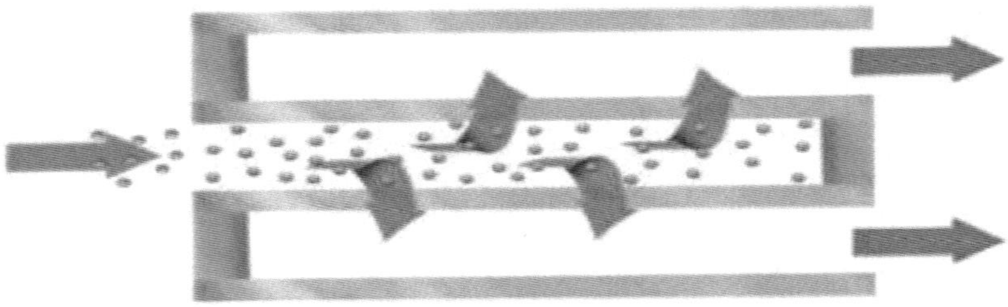

**Figure 2.** Exhaust gas flow at cell wall pores and accumulation of PM particles. [2]

The two most common raw materials for the particulate filter ceramic body are Cordierite ($2MgO - 2Al_2O_3 - 5SiO_2$) and silicon carbide (SiC). Cordierite filters are significantly cheaper due to their manufacturing technology, with very good particle filtration properties. This is also true for SiC filters, but this technology is an investment that is offset by its ability to withstand significantly more robust, extreme engine conditions. Today's modern particulate filters, whether Cordierite or SiC, can filter out 99.9% of engine particulate emissions. [1]

## 2. Equipments and methods

The test is performed on a dynamometer using a test engine using the following equipments:
the exhaust gas treatment system to be tested with the associated diesel engine,
gas concentration measuring device (AVL AMAi60),
DPF filtration efficiency measuring machines (AVL Particle Counter Plus and AVL Smart Sampler).

The engine brake pad is capable of real vehicle simulation, the gas, brake, clutch and torque converter of an imaginary vehicle are controlled by the simulation software. The engine used for the test was a 4-cylinder 2000 cm3 EURO5 compliant diesel engine braked by a 220 kW electric brake generator (vehicle acceleration, constant speed travel) or driven (vehicle engine start, vehicle engine brake operation). (Figure 3)

**Figure 3.** Test cell with measuring equipment. [source: Ibiden Hungary]

As emission regulations have emerged, there has been a need for equipment that can be used to measure emissions from vehicle engines in a repeatable and comparable way around the world.
In our test room there is an AMAi60 multi-channel continuous emission measuring device from AVL List GmbH. (Figure 4) This equipment uses 2 independent sampling channels to simultaneously

IOP Conf. Series: Materials Science and Engineering **1311** (2024) 012010 doi:10.1088/1757-899X/1311/1/012010

measure emissions at 2 different points in the exhaust system (before and after the catalyst). Each channel measures, displays and records the amount of 7 different gas components. The measured gas components are THC (total hydrocarbon), $CH_4$ (methane), $NO_x$ (nitrogen oxides), NO (nitrogen monoxide), $CO_2$ (carbon dioxide), CO (carbon monoxide), $O_2$ (oxygen).

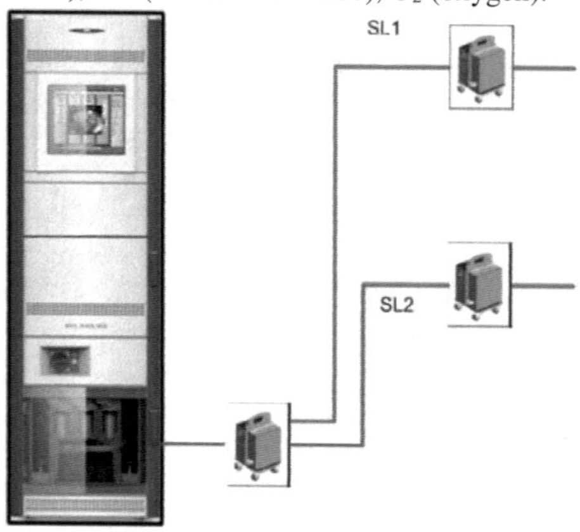


**Figure 4.** AVL AMAi60 exhaust gas analyzer. [9]

DPF filtration efficiency can be measured and calculated with two devices. One of our devices counts the particle number (PN) emitted by the vehicle. This device is a Particle Counter Plus from AVL List GmbH. Principle of operation: Exhaust gas enters (sample inlet, red marked duct, Figure 5) the primary dilution unit (chopper diluter, PND1) where it is diluted with HEPA filtered compressed air (blue marked duct, Figure 5). It then passes through a heated tube (channel marked in green, Figure 5) to the evaporator (ET), where the diluted gas is heated to such an extent that all the volatiles evaporate and only the solid particles remain in the diluted sample gas. A second diluent (PND2) is then used to further dilute the sample gas and enter the condensation particle counter (PNC). In PNC, butanol condenses on particles to make the particles larger diameter so that the optics of the equipment can visually detect them. The enlarged particles are then counted based on the scattered light pulses as the particles pass through the laser beam. This makes it possible to determine the number of particles per unit volume (#/cm3).

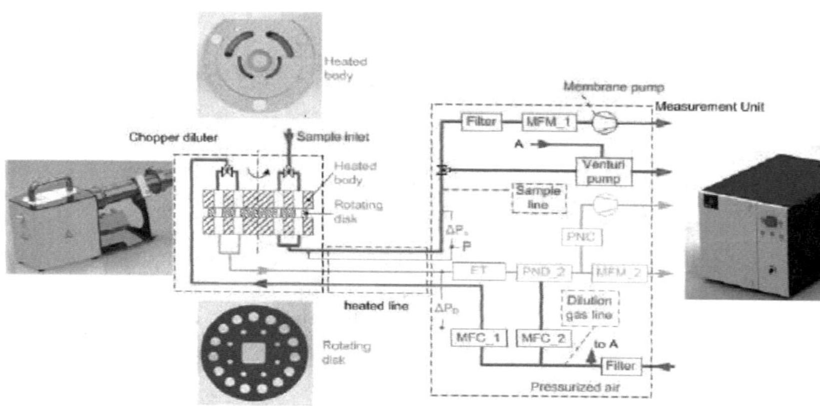


**Figure 5.** Construction of AVL Particle Counter Plus particle counter. [10]

Another device that measures DPF filtration efficiency uses a gravimetric method, but the Smart Sampler from AVL List GmbH is used to collect particles (diesel soot). With the exception of condensed water, all particles fall into the "PM" category, which is maintained at controlled

IOP Conf. Series: Materials Science and Engineering **1311** (2024) 012010 doi:10.1088/1757-899X/1311/1/012010

temperatures of 51.7 °C or less. This device collects these particles on a disc-shaped (47 mm outside diameter) filter paper from the diluted exhaust gas during a vehicle test cycle, the mass of which is measured after conditioning. Because the amount of soot collected is very small, each filter.

**Figure 6.** Filter paper with particles. [source: Ibiden Hungary]

The only officially approved measurement method for diesel vehicles is the gravimetric method, where the weight of the diluted PM is measured. The sampling system must operate at a constant dilution ratio (Rratio = 300) during the transient cycle, so the amount of gas passing through the filter paper (Gtot) must be precisely controlled. In addition, the amount of dilution air (Gdil) must be added accurately and quickly for constant dilution. The sample gas mass flow rate (Gp) is calculated from the difference between the mass flow rate on the filter paper and the dilution air mass flow rate. [11]

$$Gp = Gtot - Gdil \tag{1}$$

The exhaust gas mass flow rate (Gexh) of the engine is calculated from the mass of air intake by the engine per unit time (Gair) and the mass of fuel consumed per unit time (Gfuel) by the test bench computer and forward it to the SPC. (Figure 7) [11]

$$Gexh = Gair - Gfuel \tag{2}$$

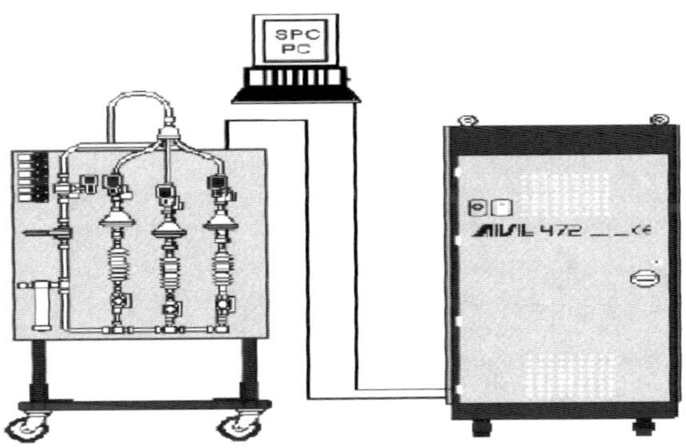


**Figure 7.** Structure of the AVL Smart Sampler measuring system. [11]

The SPC PC calculates the instantaneous dilution air mass flow rate (Gdil) from the exhaust mass flow rate (Gexh) from the dilution value (Rratio = 300).

Test cycles:

Particle collection in the DPF filter (preconditioning): EUDC (Extra Urban Driving Cycle),

Filter efficiency for measurement: NEDC (New European Driving Cycle),

## 3. Measurement

DPF regeneration with vehicle engine, post-injection (Post Injection).

Particulate collection: During the 400 second EUDC test cycle, the vehicle travels 6954.9 meters at a maximum speed of 120 km/h. Wait is 39 seconds at idle before and after the start of the cycle for the vehicle engine to warm up or cool down. (Figure 8)

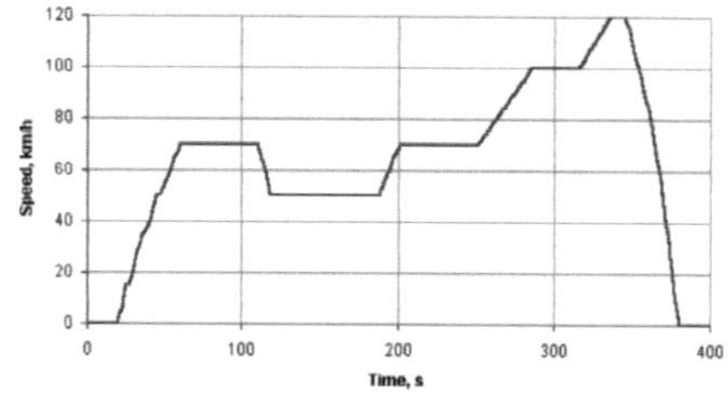


**Figure 8.** EUDC test cycle flow chart. [12]

Filtering efficiency measurement: The NEDC cycle is a complex cycle consisting of 4 ECE15 cycles and 1 EUDC cycle. The total driving time is 1180 seconds while the vehicle engine is idling for 267 seconds. (Figure 9)

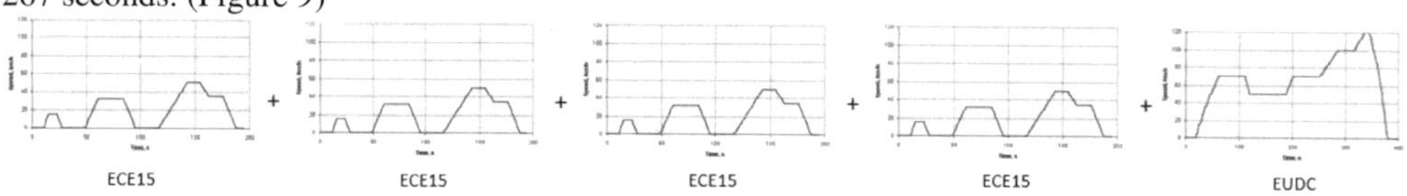


**Figure 9.** Flow diagram of a NEDC complex test cycle. [12]

Regeneration of the DPF filter: During test periods, the DPF filter must be regenerated (burned out) so that the accumulated diesel particles (diesel soot) burn without residue, or the filter returns to its original state. Therefore, the DPF filter is heated to 650 °C for 30 minutes to allow combustion to take place. This condition is achieved by modifying the engine's fuel injection process. Post injection is used. (Figure 10) Near the end of the stroke (while still burning), fuel is injected into the cylinders before the exhaust stroke, which escapes through the exhaust system into the diesel oxidation catalyst (DOC) and oxidises there (enough), thereby raising the exhaust gas temperature DPF before filter. This temperature can be adjusted by the amount of fuel injected in the central engine control (ECU) electronics.

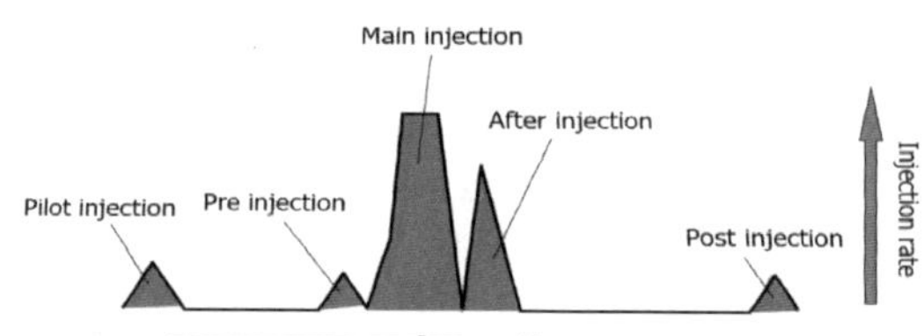


**Figure 10.** Fuel injection timing diagram. [13]

The Diesel engine uses more air during operation than is needed for perfect combustion (excess air condition), the excess gets out into the exhaust system, and this oxygen helps the DOC to operate, or the DPF to regenerate.

Procedure for measuring the change in DPF filtration efficiency:

During continuous measurement series, it is found that the filtration efficiency of DPF reaches the maximum efficiency (99.9 %) after 10-20 hours of operation. This phenomenon is demonstrated in a series of tests in this publication. A series of tests were performed that measure the initial DPF filtration efficiency in the first two steps and then the in-process efficiencies of the increasing particle. (Figure 11)

The amount of particles increases in proportion to the operating hours of the DPF. Thus, it shows the change in filtration efficiency depending on the operating time of the DPF.

| Step | Process |
|---|---|
| 1 | NEDC cycle without soot accumulation |
| 2 | Active Regeneration |
| 3 | NEDC cycle without soot accumulation |
| 4 | Active Regeneration |
| 5 | NEDC cycle with 1 times soot accumulation |
| 6 | Active Regeneration |
| 7 | NEDC cycle with 2 times soot accumulation |
| 8 | Active Regeneration |
| 9 | NEDC cycle with 3 times soot accumulation |

**Figure 11:** Test series to verify the change in DPF filtration efficiency. [source: Ibiden Hungary]

Before step 0, the initial weight of the DPF filter is measured and used as the initial reference weight. After each test period (test step), the weight of the DPF filter is measured and the difference in the initial weight shows the actual weight of the particles remaining in the filter.

## 4. Results

During the evaluation of the series of measurements, it became apparent how the initial filtration efficiency of the DPF filter (55%) increases as the operating time progresses. It can be seen that with increasing particle load, the number of particles passing through the DPF filter decreases strongly, so the filtration efficiency increases. (figure 12)

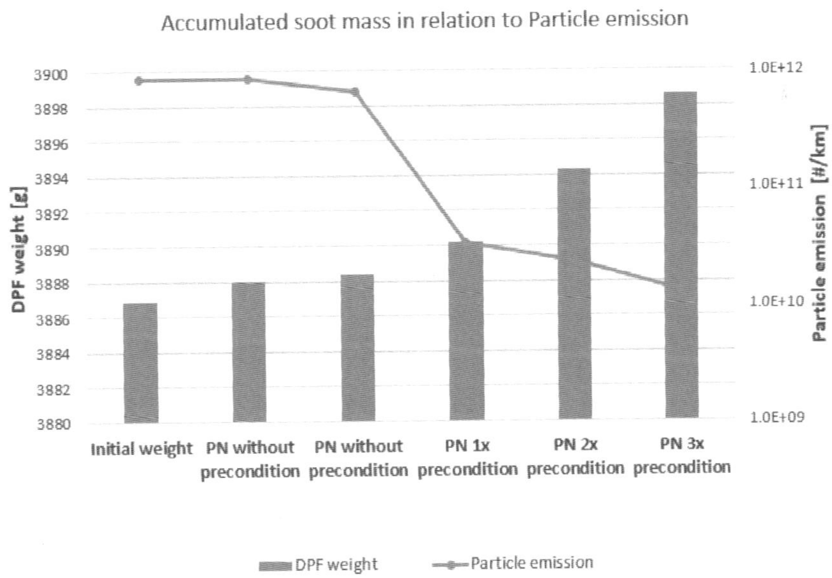


**Figure 12.** Accumulated soot and particle number.

Filtration efficiency increases from initial 55 % to final 99.9 % (over 16 hours of operation) as DPF uptime progresses. (Figure 13)

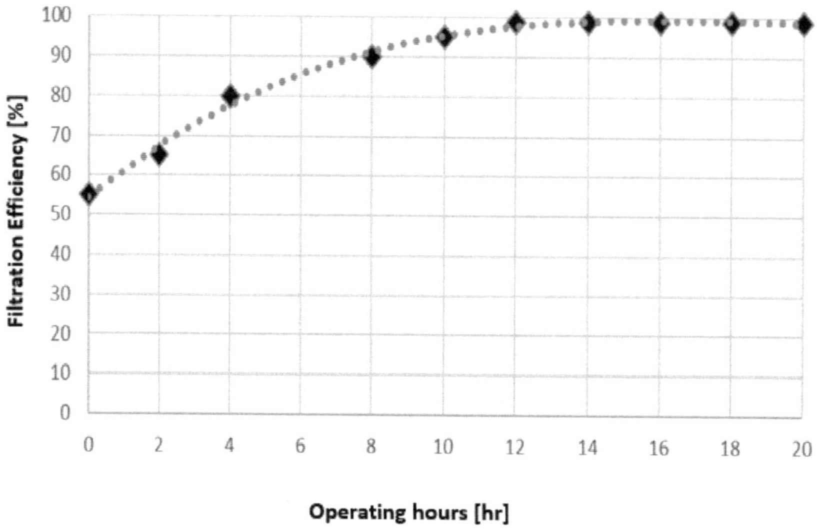


**Figure 13.** Change of filtration efficiency over operating hours.

If we further examine the result of Figure 12, we can see that the initial mass of DPF could not be recovered in any way after the tests. That is, there is always a particle (soot) left inside the filter. These soot particles adhered to the DPF filter particles (SiC), thereby reducing flow gaps on the inner surface of the filter. Therefore, the filtration efficiency increased with increasing operating time. (Figure 14)

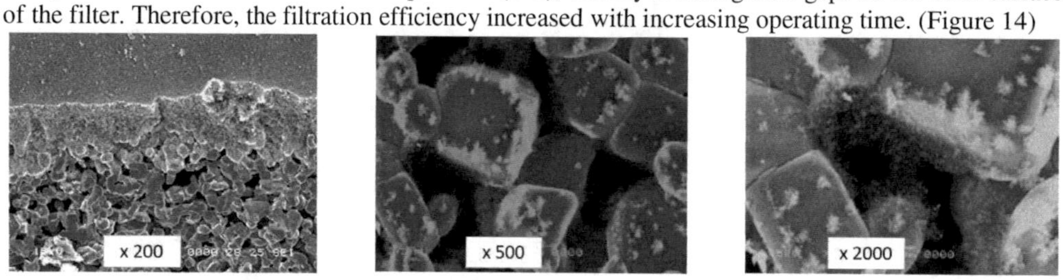


**Figure 14.** Electron micrograph of the cell wall with the deposited carbon black layer. [source: Ibiden Hungary]

## 5. Summary

A layer of soot adhering to DPF particles (white layer at the edges of the particles on the electron microscope image, Fig. 14) results in improved long-term filtration efficiency, thereby also helping diesel vehicles to operate more environmentally friendly.

Conclusions/Summary:

The test has shown that the DPF filter unit's initial filtration efficiency (55 %) after 16 hours of operation close to 100 % and maintains it for the rest of its life span.

During unlocking the background of this phenomenon, we have discovered that a thin layer of soot adheres to the particles of the filter material (SiC) and narrows the passages and gaps between the particles of the filter substrate, thus increasing the filtration efficiency.

This creates a residual, useful soot layer that cannot be removed by regeneration, so this filtration efficiency can be considered final.

In this experiments with DPF, we have observed a phenomenon which shows the initial filtration efficiency of DPF is lower than the filtration efficiency after a few hours of operation.

## References

[1] Szőllősi D, Bíró N, Kiss P: *A dízel részecskeszűrő (DPF) koromszűrési hatékonyságának megállapítása.* Mezőgazdasági Technika 2020. szeptember: LXI. Évfolyam 2-5. (2020).

[2] Biró N: *Egyedi Adblue befecskendező rendszer tervezése és tesztelése*. Diplomadolgozat, Szent István Egyetem (2020) pp. 10-15.
[3] Szabó J Z: *Gépjárművek üzemanyag ellátó berendezései*. Óbudai Egyetem, Mechatronikai és Autótechnikai Intézet, (2013).
[4] Lee T, Park J, Kwon S, Lee J, Kim J: *Variability in operationbased NOx emission factors with different test routes, and its effects on the real-driving emissions of light diesel vehicles*. Sci Total Environ 461–462:377–385, (2013).
[5] Demers D, Walters G: *Guide to exhaust emission control options*. BAeSAME, Bristol (1999).
[6] Agarwal AK *Biofuels (alcohols and biodiesel) applications as fuels for internal combustion engines*. Prog Energy Combust Sci 33:233–271 (2007).
[7] Burtscher H: *Physical characterization of particulate emissions from diesel engines: a review*. Aerosol Sci 36:896–932 (2005).
[8] Manufacturers of Emission Controls Association. (MECA) - *Emission Control Technologies for Diesel-Powered Vehicles*. (2007).
[9] *AMAi60, User's guide. AT2520E Rev. 2.9 AVL iGEM 1.6 AMA*, (12/2016).
[10] *APCplus - AVL Particle Counter™ Product Guide. AT2858E, Rev. 14* , (12/2017).
[11] *SPC472 SMART SAMPLER INTEGRATION, User's Guide. AT3198E, Rev. 11*, (3/2010).
[12] Dieselnet.com, Standards, Cycles. *ECE 15 + EUDC / NEDC*.
[13] Csankó Cs, Szőllősi D, Bíró N, Kiss P: *Dízelmotor NOx kipufogógáz-emissziójának mérése*. Mezőgazdasági Technika 2021. szeptember: LXII. Évfolyam 2-5. (2021).

# Experimental concerning the influence upon gaseous emissions by using HHO addition to an LPG powered Otto engine

**Cosmin Constantin Suciu[1], Sorin Vlad Igreț[2] and Ioana Ionel[1]**

[1]Politehnica University Timisoara, Faculty of Mechanical Engineering, Piața Victoriei 2, Timişoara 300006, Timisoara, Romania
[2]Aurel Vlaicu University, Faculty of Engineering, Calea Aurel Vlaicu 43, 310141, Arad, Romania

ioana.ionel@upt.ro

**Abstract**. Because of the increased pollution regulations imposed on internal combustion engines across the world as well to preserve the air quality, effective solutions for new and older cars must be found in reference to their emission levels. This paper focuses on the impact of adding oxyhydrogen (HHO) fueled into an Otto engine working with liquefied petroleum gas (LPG), in reference to the emission concentration values. HHO is produced using a water electrolytic conversion technique that works in tandem with the Otto engine. The electrolytic convertor can generate 1,800 litters of HHO gas from 1 liter of water. Due to this addition, the total calorific power of the fuel mixture is growing by around 30 % in comparison to only LPG fueled situation, considered as basic one. All measurements were conducted in a laboratory with accredited and approved gas analyzers and measuring devices. The study presents comparative results of the basic measurements, at an engine speed of 800 rotations per minute (RPM) and respectively at 2,000 RPM, when using an LPG as fuel, respectively a mixture LPG and HHO. As conclusion, one proves that the addition of HHO can generate as effect the decreased of particular emission values.

## 1. Scope of the research

Internal combustion engines (new or older) must produce low pollutant emission values, in respect to the euro class where they fit. Most manufacturers are aiming towards electric or hybrid vehicles. Due to battery's charging time, availability and range of electric cars, internal combustion engine (ICE) vehicles are still being produced and sold, and they represent still the first choice for majority of users. High energy consumption and the pollutant manufacturing process of batteries used in electric vehicles are still mainly based on burning fossil fuels, therefore carbon is invested in a large amount during the production process [1], [2], [3].

High prices, large volumes and little options on reusing the waste batteries motivate automobile manufacturers to investigate further on the emission reduction. Solutions like using unconventional synthetic fuels represent a turn towards greenhouse effect technologies [4], [5]. Older cars must also be checked during their utilisation, mainly by regular checks in authorised units, but retrofitting them with modern solution might also be an option, mainly for price reducing of the utilisation but also for diminishing the emission levels.

Lower costs for the fuel supply and superior calorific power of the LPG and HHO mixture turn this solution to a feasible one. Costs of obtaining HHO are lower compared to the technologies for producing conventional fuels [6], [7], [8]. Multiple studies on HHO addition into spark ignited engines (Otto ICE) using gasoline or LPG as main source of fuel [9], [10], [11], [12] demonstrated an increase in the engine's output power, therefore a better engine performance together with a decreased fuel consumption are resulting. Hydrogen's thermal efficiency through combustion is three times higher than petrol's, thus higher peak pressure therefore a higher engine efficiency [12] are expected. Engine's thermal efficiency increases even at extreme engine operating as increased compression ratio to 300:1 [12]. Moreover, all studies have showed that HHO addition improves HC and CO emissions but increased $NO_x$ concentrations, due to the higher combustion chamber temperatures. One notes that this $NO_x$ levels could be reduced using for example an exhaust gas recirculation technology [9], [10], [11], [12]. Another aspect, worthy of mentioning, is the air excess ratio under lean-burn conditions achieved, which maintain a better combustion environment due to the addition of HHO [9]. Furthermore, the results could be an intermediate solution towards fuel cells for future ICE retrofits using only hydrogen as fuel.

## 2. Measurements and results

For the experiment, one used a spark-ignition engine (SIE) fuelled with LPG retrofitted with an electrolytic converter. Engine specifications are presented in Table 1.

Table 1. Engine specifications.

| Characteristic | Value |
|---|---|
| Number of cylinders | 4 |
| Cylinder's layout | Straight/ inline |
| Cylinder bore [mm] | 79 |
| Cylinder stroke [mm] | 81.5 |
| Compression ratio | 10.5 |
| Number of valves | 16 |
| Fuel type | LPG |
| Engine's position | Front, Transversal |
| Engine's displacement [$cm^3$] | 1,598 |
| Intake | Aspiration |
| Max power output [kW] | 77 @ 6,200 RPM |
| Max torque output [N·m] | 150 @ 3,900 RPM |
| Transmission layout | Front wheel drive |
| Gearbox | Manual |
| Number of gears | 5 |
| LPG tank volume [l] | 30 |
| Emission class | Euro 3 |

LPG is known to be one of the most potent fuels for application in ICE driven vehicles, meeting all the requirements for the SIE's exhaust emission levels. Among the many advantages, one mentions its non-toxicity, non-corrosiveness and an octane number higher than that of gasoline, being up to 112 octane number [13]. Higher octane values result in a possibility to operate at higher compression ratio, higher thermal efficiency and no knocking events. The lower heating value (LHV) per unit mass is slightly higher for LPG compared to that of gasoline, as it is 50 MJ/kg for LPG, whereas for gasoline it is 44 MJ/kg [13]. The volume energy density is lower for LPG by 16.2 % than that of gasoline (LPG value is 28 MJ/l and for gasoline is 33.4 MJ/l) [14]. Since LPG contains flammable mixture of hydrocarbon gasses, more precise propane ($C_3H_8$) and butane ($C_4H_{10}$) in order to obtain an improved mixture formation, LPG is distributed in a gaseous state, therefore achieving better idling and acceleration ability of the engine. Another benefit worth to be mentioned is referring to the fact that the engines' cylinder wall, piston rings and spark plugs, in case of being fuelled with LPG mixture, are free of carbon deposits. Due to LPG's lower volumetric energy density than that of gasoline and because it is delivered in a gaseous state, engines fuelled have lower volumetric efficiency and engine performance, as the fuel consumption increases by 15-20 %, in comparison to the case when the engine is gasoline fuelled [14], [15], [16].

Emissions values have been determined using Capelec 3200 gas analyzer presented in figure 1. This stand-alone multigaz unit measures the emissions of all gasoline, LPG and compressed natural gas (CNG) vehicles. With a response time of under 5 seconds, it measures Hydrocarbons (HC), Oxygen ($O_2$), Carbon monoxide (CO) and Carbon Dioxide ($CO_2$). Multiple measurements took place, two at 800 RPM meaning the engine's idle speed and two at 2,000 RPM, under no engine load. The first set of measurements used (LPG as main (single, basic) fuel and the second set of measurements used LPG and HHO as fuel mixture. For the exhaust flue gas measurements, the gas analyzer's probe has been inserted into the vehicle's exhaust.

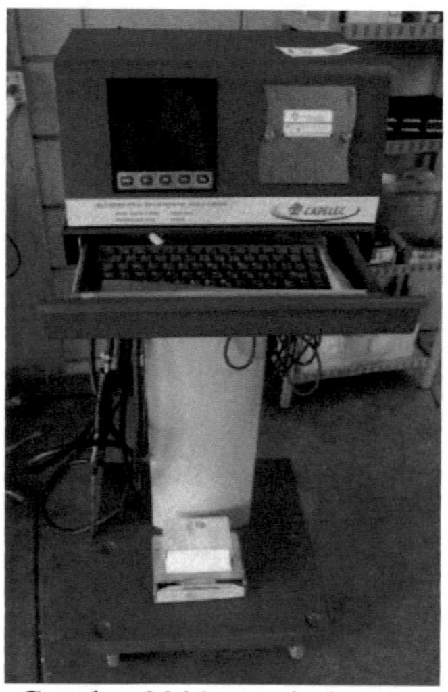

Figure 1. Capelec 3200 stand-alone opacimeter.

The HHO gas mixture is produced from water. Dissociation of water into hydrogen and oxygen is made using the electrical power supply of the vehicle on which the system is installed. [17], [18].

Quality of atmospheric air is very important and a goal for manufacturers for the future years is to improve it. Hydrogen is obtainable by renewable and non-renewable processes [18]. Hydrogen can be used in both spark ignition and compression ignition engines, having a self-ignition temperature of around 585 degrees Celsius [19] therefore needing external source to initiate the combustion process. Table 2 presents hydrogen's properties in comparison with other fuels.

Table 2. Comparison of hydrogen with other fuels [16].

| Fuel | Methane | Propane | Octane | Methanol | Hydrogen | Gasoline | Diesel |
|---|---|---|---|---|---|---|---|
| Lower heating value [MJ/kg] | 50 | 45.6 | 47.9 | 18 | 119.9 | 44.5 | 42.5 |
| Higher heating value [MJ/kg] | 55.5 | 50.3 | 15.1 | 22.7 | 149 | 47.3 | 44.8 |
| Air/Fuel Ratio | 17.2 | 15.6 | 0.31 | 6.5 | 34.3 | 14.6 | 14.5 |
| Combustible Range [%] | 44696 | 2.1-9.5 | 0.95-6 | 6.7-36 | 27485 | 1.3-7.1 | 0.6-5.5 |
| Flame Temp[°C] | 1914 | 1925 | 1980 | 1870 | 2207 | 2307 | 2327 |
| Min. Ignition Energy [MJ/kg] | 0.3 | 0.3 | 0.26 | 0.14 | 0.017 | 0.29 | - |
| Auto Ignition Temp. [°C] | 540-630 | 450 | 415 | 460 | 585 | 260-460 | 180-320 |

IOP Conf. Series: Materials Science and Engineering **1311** (2024) 012011 doi:10.1088/1757-899X/1311/1/012011

According to table 2 for burning 1 kg of hydrogen, 34,3 kgs of air are needed in comparison to gasoline, where the air/fuel ratio is 14.6:1 [19], [20]. One critical factor to be taken in consideration when adding hydrogen to the mixture is now created mixture's air/fuel ratio thus backfire may occur. Direct injection is one solution for a better control to avoid backfiring [21]. Since hydrogen's form of substance is gaseous, it displaces more of the combustion chamber than a liquid fuel for the same mass thus less volume inside the combustion chamber for air.

Hydrogen, which offers low carbon flexibility and storage necessity, can play a significant role in enabling complementary routes to deeper decarbonization alongside electricity. As a result of their concern for the benefits to the environment, several European countries are attempting to define green hydrogen standards [22].

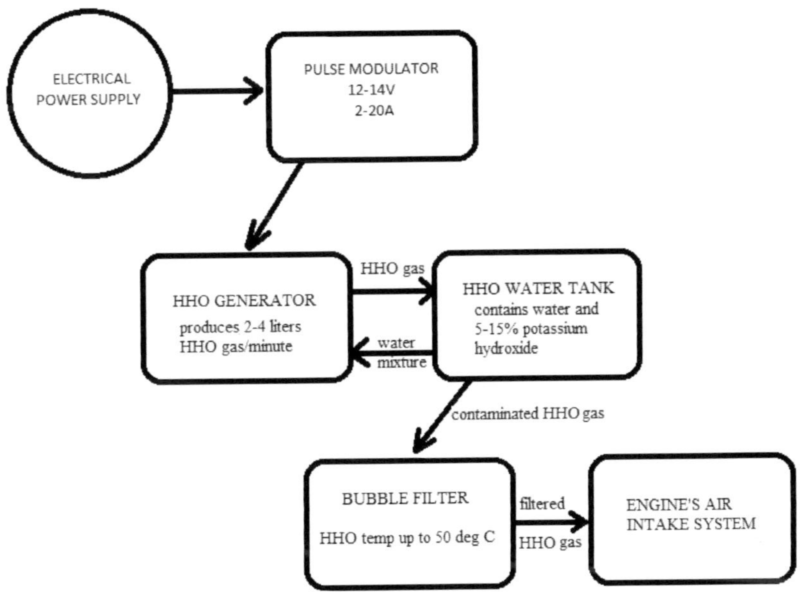


Figure 2. Installation scheme of the HHO system as used in the research, according. https://i.pinimg.com/736x/ec/0d/8a/ec0d8a247ba9bec9337ef109d7404b6e.jpg

Figure 2 represents the electrolytic converter system. Starting with the pulse modulator, in order to function as intended the electrical power supplies it with an electrical tension between 12 and 14 Volts. Depending on the amount of HHO gas needed to be produced, the pulse modulator controls the HHO generator with an electric current between 2 and 20 Amperes. Inside the HHO generator (figure 3) are multiple plates with positive, negative and neutral charges. While continuous electrical tension passes through the electrodes, the anode is where water molecules loses two electrons. The split of a water molecule creates two hydrogen molecules and one oxygen molecule thus resulting HHO gas. The generator can produce up to 1800 liters of HHO from 1 liter of water at a speed of 2 to 4 l/min.

IOP Conf. Series: Materials Science and Engineering **1311** (2024) 012011 doi:10.1088/1757-899X/1311/1/012011

Figure 3. HHO generator mounted on the vehicle used for measurements.

The HHO generator and the HHO water tank are linked together (figure 4), exchanging fluids and gasses via hoses.

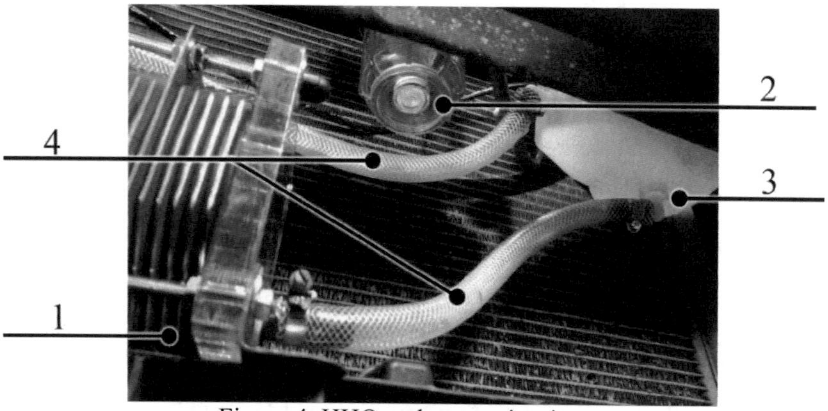


Figure 4. HHO and water circuit.
(1) HHO generator (2) bubble filter (3) HHO water tank (4) Hoses.

In the first phase the HHO gas created in the generator, receiving fluid from the HHO water tank. The HHO gas is passing through the HHO water tank (figure 5). At this moment it contains small gas particles and vapors. The gas reaches temperatures up to 50°C before entering the bubble filter. The role of the bubble filter (figure 5) is to reduce the concentration of gas particles, vapor and temperature before the HHO gas enters the engine's air intake system (figure 6).

Figure 5. (1) Bubble filter (2) HHO water tank.

The resulted HHO represents 1% of the total intake's volume/stroke and in the case of the selected engine is 39.95 $cm^3$. Since at engine's idle speed the cylinder is filled with fuel in a proportion of around 30% of the cylinder's total volume and the engine does 3523.17 cycles each second from which 14 parts are air and one part is fuel (due to stoichiometric ratio of 14:1), 1% of 14 is HHO. The percentage of produced HHO is then calculated to 32 $cm^3/s$ or 2 l/min.

Figure 6. HHO being added into the intake system.
(1) engine's intake system (2) Hose from bubble filter.

Table 3 represents the obtained measured values via Capelec 3200 from the vehicle using LPG as fuel. Data in table 4 were obtained in the same manner as for table 3 but HHO gas had been added to the LPG. Disseminated results are presented in figures 7 and 8. In both tables, two measurements are presented, one at the engine's idle speed of 800 RPM and another at an engine speed of 2,000 RPM.

Table 3. Measurement results using LPG as fuel.

| Engine speed [RPM] | 800 | 2,000 |
|---|---|---|
| CO [%vol] | 0.1 | 0.07 |
| $CO_2$ [%vol] | 3.8 | 4.3 |
| HC [ppm] | 47 | 56 |
| Coolant temperature [°C] | 92 | 92 |

Table 4. Measurement results using LPG and HHO as fuel.

| Engine speed [RPM] | 800 | 2,000 |
|---|---|---|
| CO [%vol] | 0.03 | 0.01 |
| $CO_2$ [%vol] | 0.8 | 0.6 |
| HC [ppm] | 0.3 | 0.38 |
| Coolant temperature [°C] | 92 | 92 |

By comparing the basic measurement results from Table 3 graphically represented in Figure 7 and Figure 8, obtained at an engine speed of 800 RPM with HHO addition, concentrations values changed by: i) CO is lower by 70 % ii) $CO_2$ reduced by 99,4 % iii) HC particle concentration diminished by

99.36 %. The impact of HHO on emission concentrations is significant, especially on HC, bringing its concentration to a value very close to 0. This determines an improved engine efficiency and a lower fuel consumption and thus the possibility to face a cleaner air in rush hour traffic jams scenarios.

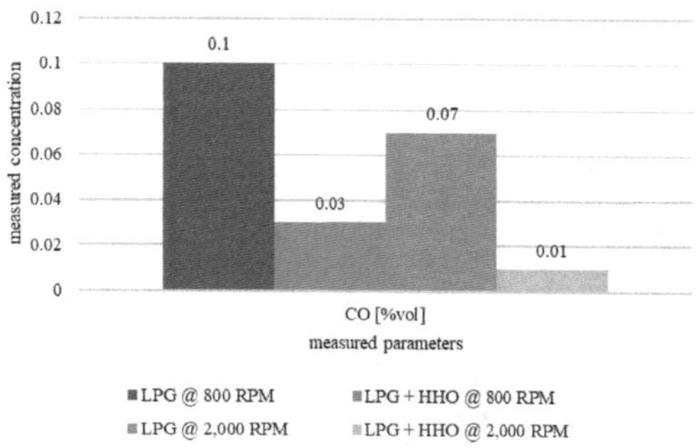


Figure 7. Measurement results for CO.

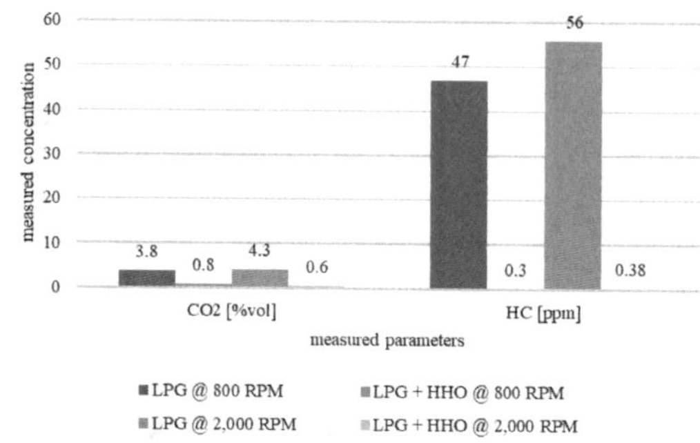


Figure 8. Measurement results for $CO_2$ and HC.

Other noticeable advantages of HHO addition are at the engine speed of 2,000 RPM, according to the values from Table 4 represented as in both graphs in figures 7 and 8, compared to standard situation $CO_2$ concentration are lower by 86 %. HC particles concentration dropped by 99,3%. A similar pattern regarding the influence of HHO between the two tables can be noticed. Knowing that manufacturing process of Li-ion batteries is pollutant and harmful for the environment, the solution presented is a good alternative for electrically powered units in the automotive industry.

## 3. Conclusion

The paper presents the advantages of adding HHO to an LPG powered ICE. One proved that the engine's emission levels are impacted by the proposed system. A Euro 3 initially categorized engine was retrofitted with the presented solution and finally achieved the Euro 6 pollution class-imposed limits. Since hydrogen could become highly available, the presented solution can be a feasible intermediate solution as retrofit option for all ICE. The presented electrolytic converter can produce from 1 liter of water up to 1,800 liters of HHO gas. The results conclude a more effective engine, cheaper fuel and very low emissions.

## 4. References

[1] H S Das, M M Rahman, S Li, C W Tan 2020 Electric vehicles standards charging infrastructure and impact on grid integration: A technological review. *Renewable and Sustainable Energy Reviews* **120(C)**,

[2] W W Jacob, J J Michalek, C Samaras 2021 *Environmental Science & Technology* 55,

[3] A Mayer, H Burtscher, S Loretzand, M Kasper, J Czerwinski 2018 *High air pollution in vehicle cabins due to traffic nanoparticle emission exposure and a solution for in-use vehicles* **421(3)**,

[4] R Peters, M Decker, L. Eggemann et al. 2020 Thermodynamic and ecological preselection of synthetic fuel intermediates from biogas at farm sites. *Energy, Sustainability and Society* **10(4)**,

[5] J Szczygieł, M Kułażyński 2020 Thermodynamic limitations of synthetic fuel production using carbon dioxide: A cleaner methanol-to-gasoline process. *Journal of Cleaner Production* **276.**

[6] Y Haseli 2018 Maximum conversion efficiency of hydrogen fuel cells. *International Journal of Hydrogen Energy* **43(18)** 9015-9021,

[7] S Jungwoo, H Won-Sik, C Hyundoi 2019 Can hydrogen fuel vehicles be a sustainable alternative on vehicle market?: Comparison of electric and hydrogen fuel cell vehicles. *Technological Forecasting and Social Change* **143** 239-248,

[8] C Acar, I Dincer 2020 The potential role of hydrogen as a sustainable transportation fuel to

IOP Conf. Series: Materials Science and Engineering **1311** (2024) 012011 doi:10.1088/1757-899X/1311/1/012011

combat global warming. *International Journal of Hydrogen Energy* **45(5)** 3396-3406,

[9] Z Zhe, H Yan, Y Xiumin, L Ming, Z Zhihong, G Zezhou, W Tianqi, L Decheng 2022 Comparative study on combustion and emission of SI engine blended with HHO by different injection modes of gasoline. *International Journal of Hydrogen Energy* 0360-3199,

[10] V Balu 2020 HHO gas kit as a petrol additive in a 4 stroke IC engine. *IJARET Z. Zhe, H. Yan, Y. Xiumin, L. Ming, Z. Zhihong, G. Zezhou, W. Tianqi, L. Decheng 2022 Comparative study on combustion and emission of SI engine blended with HHO by different injection modes of gasoline, International Journal of Hydrogen Energy, 0360-3199* **11(11)** 2090-2095,

[11] S Kharde, P Khodade, R Nagare, G Nagare, H Rajule 2019 Review on efficiency increasing HHO kit for IC engine. *International journal of research in engineering, science and management* **2(2)**,

[12] R Franzi, S Macchello, L Roche, N Galante, N A Galante 2021 Determinación de la performance y emisiones en motores de combustión interna al suministrar hidrógeno bajo el modo del gas HHO (de Brown) en el proceso de combustión. *FRBA – Revista Proyecciones* **19(1)**,

[13] K N Duc, V N Duy 2016 Study on performance enhancement and emission reduction of used fuel-injected motorcycles using bi-fuel gasoline-LPG. *Energy for Sustainable Development* **43** 60-67,

[14] D K Nguyen, D V Nguyen, L Hoang-Dinh, V T Nguyen, T Le-Anh 2019 Performance and emission characteristics of a port fuel injected, spark ignition engine fueled by compressed natural gas. *Sustainable Energy Technol Assess* **31** 383-389,

[15] T Suthisripok, N Phusakol, N Sawetkittirut 2017 Bi-fuel system - Gasoline/LPG in A used 4-stroke motorcycle - fuel injection type. *IOP Conf Ser: Mater Sci Eng* **241** 012021,

[16] V N Duy, K N Duc, N C Van 2021 Real-time driving cycle measurements of fuel consumption and pollutant emissions of a bi-fuel LPG-gasoline motorcycle. *Energy Conversion and Management: X* **12(100135)** 2590-1745,

[17] S Balaji, I Saleel 2018 Production and use of HHO gas in IC engines. *International Journal of Hydrogen Energy* **43(14)** 7140-7154,

[18] S Balaji, T Venugopal 2020 Analysis of onsite HHO gas generation system. *International Journal of Hydrogen Energy* **45(28)** 14218-14231,

[19] C M Kalamaras, A M Efstathiou 2013 Hydrogen Production Technologies. Current State and Future Developments. *Conference paper, special issue Power Options for the Eastern Mediterranean Region* 1-9,

[20] V Kumar, D Gupta, N Kumar 2015 Hydrogen use in internal combustion engine: A review. *International Journal of Advanced Culture Technology* **3(2)** 87-99,

[21] A Onorati, R payri, B M Vagelieco et al. 2022 The role of hydrogen for future internal combustion engines. *International J of Engine Research* **23(4)** 529–540,

[22] I Staffell, D Scamman, A V Abad, P Balcombe, P E Dodds, P Ekins, N Shahd, K R Warda 2019 The role of hydrogen and fuel cells in the global energy system. *Energy Environ. Sci.* **12** 463-491

**Acknowledgement**
The research is part of the PhD program of the first author at the Politehnica University of Timisoara.

# Study of energy consumption improvement for electric vehicles in platoon driving

**C Constantin[1], C Renţea[2*], M Băţăuş[2] and G Frăţilă[2]**
[1]Faculty of Transports, University Politehnica of Bucharest, 313 Spl. Independentei, Bucharest, Romania
[2]Aurel Vlaicu University, Faculty of Engineering, Calea Aurel Vlaicu 43, 310141, Arad, Romania

*Email: rentea.cristian@yahoo.com

**Abstract.** It is known that platoon driving has a large impact on fuel economy for conventional vehicles and using current available technologies, could be implemented especially on highway driving. Since the transition to electric vehicles is in progress, it is important to evaluate its applicability to this type of vehicles which have different characteristics than conventional ones equipped with an internal combustion engine. This paper aims to investigate the energy consumption improvement for a platoon of up to 7 electric passenger vehicles using a validated detailed model created in LMS Imagine.Lab AMESim, an industrial modelling and simulation environment. The energy consumption of individual vehicles and also the platoon's will be calculated for several stabilized speeds and also on a standardized high-speed cycle (HWFET). The analysis will show how the special characteristics of an electric vehicle (lower drag coefficient, increased mass, regenerative braking, etc.) have an influence on the vehicle's energy consumption when driving in a platoon. This will allow us to also evaluate the platoon driving efficacity in increasing the vehicle's driving range.

**Keywords:** Platoon driving, platooning, energy consumption, electric vehicles

## 1. Introduction

Nowadays every car manufacturer has at least one electric vehicle on the market and plans to expand in this direction. With the European Union ban on new combustion engines vehicles sales from 2035 this market is on the rise like never before.

Electric vehicles are for the moment and for the most part the upper region of the price point and customers expect the newest technologies and the best driving safety and comfort at this level. Most of them come with semi-autonomous driving and efforts are being made to introduce fully autonomous vehicles on the market.

Using these technologies platoon driving can be implemented as a mean to increase fuel efficiency. In the case of electric vehicles, the power consumption is the highest at highway driving where the vehicles drives at constant high speed so the benefits of regenerative braking are almost cancelled. Since platooning is the most efficient in these exactly conditions it could be used as a mean to reduce power consumption and complement the lack of regenerative braking.

With the improvements in the advanced driver-assistance systems (ADAS) and vehicle-to-vehicle (V2V) communication the leading vehicle intentions can be know by the following cars so they can make predictive actions to smooth the traveling speed by avoiding unnecessary braking and to maintain a short distance between vehicles to decrease the air drag coefficient.

The influence of platoon driving on the drag coefficient was studied in the past with experiments on real vehicles [1], on models in wind test tunnels [2], [3], but also by the mean of computational fluid dynamics (CFD) simulations [4], [5]. The potential reduction in $CO_2$ emissions was studied in [6] on passenger cars in different configurations from 1 and up to 7 vehicles driving in a platoon.

IOP Conf. Series: Materials Science and Engineering **1311** (2024) 012012 doi:10.1088/1757-899X/1311/1/012012

This paper investigates the energy consumption reduction that can be achieved thru platoon driving at stabilized high speed and also on the Highway Fuel Economy Cycle (HWFET) for a passenger electric vehicle and to see how the vehicle range can be improved using this method. The gains will also be measured in comparison to an uncoordinated convoy.

Following the market trend, the study is done for a SUV class electric vehicle by dynamic modelling and simulation using a model developed using LMS Imagine.Lab AMESim, an industrial modelling and simulation environment.

For the purpose of this paper the vehicle specifications were chosen to resemble the average for the electric SUV market with a specific power of 81W/kg. The characteristics of the Nissan Ariya were selected: vehicle mass: 2200kg, aerodynamic drag coefficient (Cd): 0.297, frontal area: 2.6m$^2$ and usable battery capacity: 87kWh.

## 2. Model development

For the simulation of the electric vehicle, the model from Figure 1 was developed using LMS Imagine.Lab AMESim.

The model is based on the submodels from the IFP Drive library: the mission profile and ambient data (DRVMP2A), the driver (DRVDRVA00A), the 1D vehicle (DRVVEH4A), the control unit (DRVCUE0A), the electric motor (DRVELMT0A), the power electronic converter (DRVPET0B) and the battery (DRVBAT04). From the Mechanical library, a rotary load submodel (RL01), a reducer submodel (RN011) and a clutch submodel (FR2R000) are used. The clutch submodel is introduced only as a causality converter and the parameters are chosen to not influence the energy consumption results. Transducers for motor rotary velocity, vehicle linear velocity and displacement are also employed.

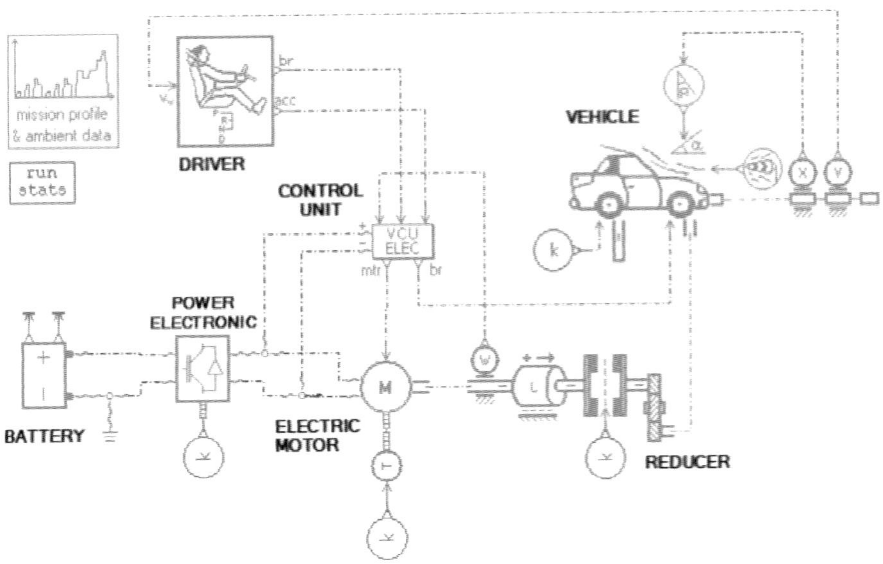


**Figure 1.** Model design of an electric vehicle for the energy consumption study.

The energy consumption was determined for the WLTC cycle, and the vehicle range was calculated and compared with the range declared by the manufacturer. Using the 87kWh battery capacity the range of 448km was obtained which is 10% lower than the official range of 499km declared by the manufacturer for the ARIYA E-4ORCE 87 KWH version on the WLTC cycle.

The vehicle velocity in a convoy of vehicles can be determined using a leader-follower model. There are numerous theoretical models of the driver based on different methods: differential equations, conditional constructs, algorithmic, fuzzy logic, neuronal networks [1]. Figure 2 shows the diagram of the system driver-vehicle for the front vehicle tracking.

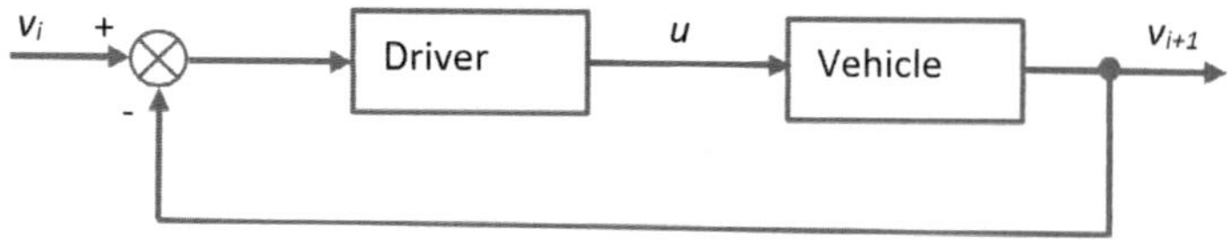


**Figure 2.** Diagram of the system driver-vehicle for the front vehicle tracking.

Starting from the vehicle equation of rectilinear motion the following equation can be deduced [7].

$$\dot{v_{i+1}} = k \cdot [v_i(t-T) - v_{i+1}(t-T)] \tag{1}$$

where: $v_i$ is the velocity of the vehicle in front; $v_{i+1}$ is the velocity of the vehicle in rear; $t$ is the current time, $T$ is the driver reaction time and $k$ is a constant.

The equation (1) can be integrated with the Signal&Control Library from AMESim (Figure 3), and, by successive integrations, the vehicle velocity profiles of a non-platoon vehicle convoy are computed starting from the first profile.

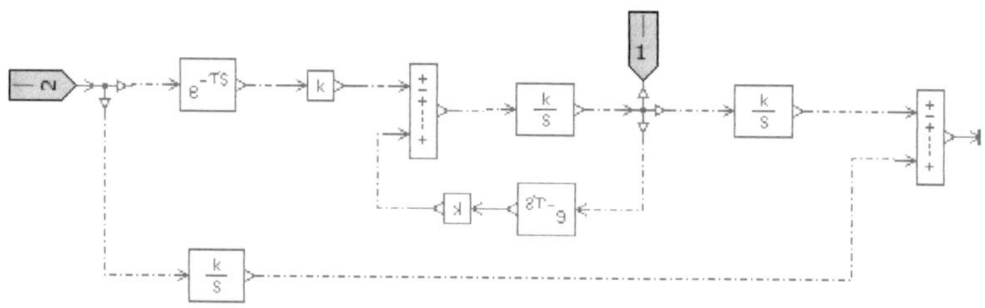

**Figure 3.** Linear model for leader-follower for a single vehicle.

The evolution of the vehicle's velocities during a high-speed perturbation for a non-platoon vehicle convoy of 6 vehicles is shown in Figure 4.

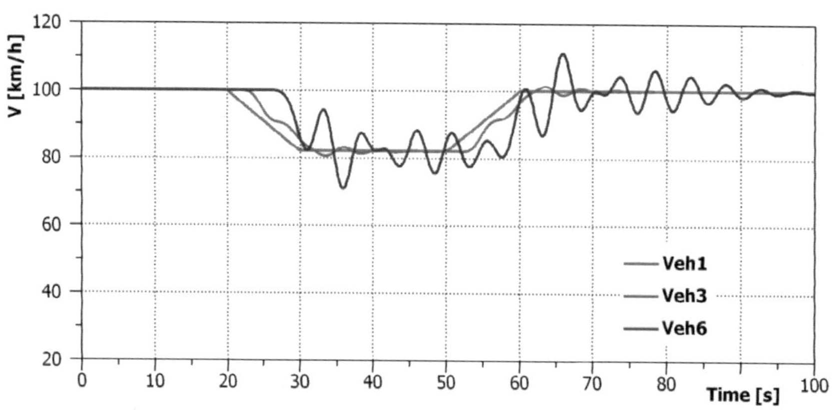


**Figure 4.** The velocity profiles of vehicles 1, 3 and 6 from the uncoordinated convoy.

## 3. Results

The simulations were done for a single electric vehicle and for short spacing (1-meter inter-vehicle distance) platoons consisting of 2 and up to 7 identical vehicles. The variation of the aerodynamic drag (Cd) is taken from [4] and applied to our vehicle.

Three constant vehicle velocities were considered for the simulations: 70 km/h, 100 km/h and 130 km/h. The results are shown in Table 1 and Figure 5.

**Table 1.** Energy consumption for stabilized speed.

| No. of vehicles | Average energy consumption [Wh/km] | | | Improvement over the isolated vehicle [%] | | | |
|---|---|---|---|---|---|---|---|
| | 70km/h | 100km/h | 130km/h | 70km/h | 100km/h | 130km/h | $C_d$ |
| **1** | 163.3 | 216.9 | 314.5 | 0 | 0 | 0 | 0 |
| **2** | 158.5 | 207 | 297.8 | 2.9 | 4.5 | 5.2 | 8.8 |
| **3** | 151.9 | 193.7 | 275.4 | 6.9 | 10.6 | 12.4 | 20.6 |
| **4** | 152.5 | 194.9 | 277.4 | 6.5 | 10.1 | 11.7 | 19.6 |
| **5** | 152.1 | 194.1 | 276 | 6.8 | 10.4 | 12.2 | 20.2 |
| **7** | 150 | 189.7 | 268.5 | 8.1 | 12.5 | 14.6 | 24.2 |

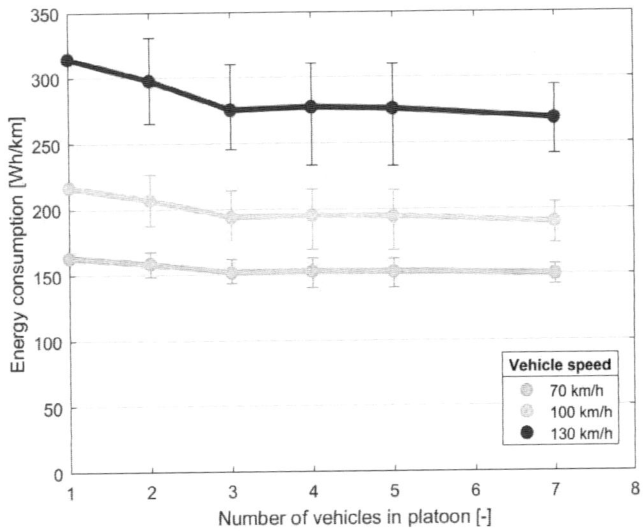


**Figure 5.** Energy consumption for platoon driving at high stabilized speed.

From the Table 1 data, we can observe that we cannot simply convert the drag coefficient gain to energy consumption gains as the drag coefficient's influence varies with vehicle speed. As expected, the biggest gains in energy consumption are at higher vehicle speed and can reach 14.6% for a 7-vehicle platoon over an isolated vehicle.

There is also worth mentioning that the vehicle position in the platoon has a large influence on energy consumption. The biggest gains are obtained for the first vehicle at 26% gains over the isolated vehicle while the last vehicle can only obtain 6.5%. This combined with the fact that some vehicle design features may work better for isolated vehicles than for a platoon as stated in [3], makes vehicle positioning in the platoon harder since we must increase the efficiency of the platoon while benefiting as many traffic participants as possible. The position in the platoon could be dynamically chosen by considering the vehicle's platoon driving history or the disadvantages of a position with a lower benefit could be mitigated by taxes benefits.

Using the leader-follower model on a standardized high-speed cycle (HWFET) a non-platoon vehicle convoy of up to 7 vehicles was simulated. The HWFET cycle is a driving schedule, developed by the U.S. EPA (Environmental Protection Agency) for the determination of fuel economy of light duty vehicles and has a mean speed of 78.2 km/h with a vehicle speed fluctuation of 19.2%.

In Figure 6 we can see the variation of the energy consumption for a convoy of 7 uncoordinated vehicles and the influence of the regenerative braking in the HWFET cycle.

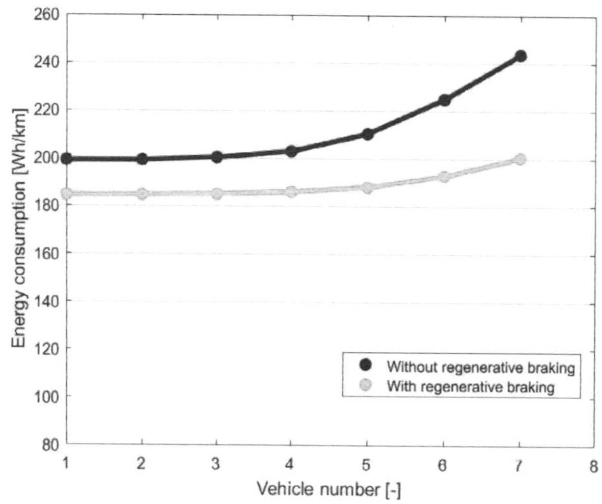


**Figure 6.** Energy consumption for uncoordinated high traffic density driving.

In Table 2 the energy consumption for HWFET cycle using regenerative braking for every participant to a 7-vehicle platoon and similar uncoordinated convoy is shown. The data is used to compute the improvements over the isolated vehicle and the uncoordinated convoy. Since as seen in Figure 6, the uncoordinated convoy has a negative influence on the energy consumption, we obtained the highest gains of up to 11.8% when comparing the platoon to other driving methods.

**Table 2.** Energy consumption for HWFET cycle with regenerative braking.

| No. of vehicles | Average energy consumption [Wh/km] | | Improvement [%] over | |
|---|---|---|---|---|
| | Platoon | Uncoordinated convoy | Isolated vehicle | Uncoordinated convoy |
| 1 | 184.6 | 184.6 | 0 | 0 |
| 2 | 178 | 184.7 | 3.6 | 3.6 |
| 3 | 169.2 | 184.8 | 8.4 | 8.4 |
| 4 | 170 | 185.1 | 7.9 | 8.2 |
| 5 | 169.4 | 185.6 | 8.2 | 8.7 |
| 7 | 166.5 | 188.8 | 9.8 | 11.8 |

Similar in Table 3, we can see the average energy consumption when not using the regenerative braking. As expected, the improvement is higher over the uncoordinated convoy.

**Table 3.** Energy consumption for HWFET cycle without regenerative braking.

| No. of vehicles | Average energy consumption [Wh/km] | | Improvement [%] over | |
|---|---|---|---|---|
| | Platoon | Uncoordinated convoy | Isolated vehicle | Uncoordinated convoy |
| 1 | 199.1 | 199.2 | 0 | 0 |
| 2 | 192.8 | 199.3 | 3.1 | 3.2 |
| 3 | 184.3 | 199.6 | 7.4 | 7.7 |

| | | | | |
|---|---|---|---|---|
| 4 | 185.1 | 200.5 | 7 | 7.7 |
| 5 | 184.6 | 202.4 | 7.3 | 8.8 |
| 7 | 181.8 | 211.5 | 8.7 | 14 |

In Figure 7 we can observe, the improvement in energy consumption is higher in platoon driving when benefiting from regenerative braking, since less kinetic energy is lost due to the aerodynamic drag which allows the generator to recuperate more energy.

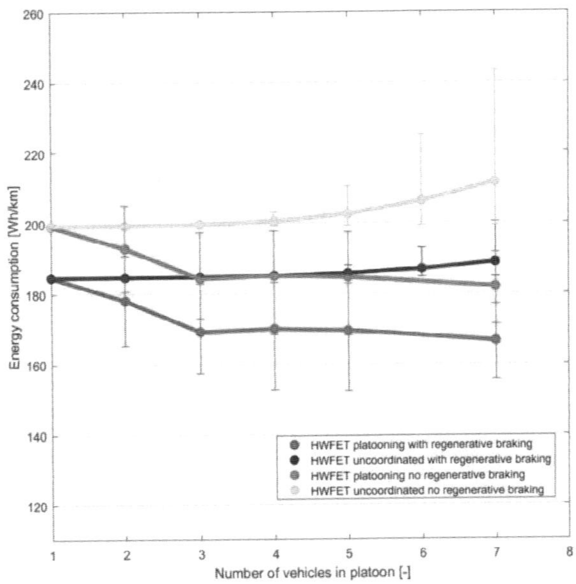


**Figure 7.** Comparison of energy consumption for HWFET cycle.

Platoon driving benefits over energy consumption are considerable but drivers need to have more tangible results, so we translated the gains into range improvements.

In Table 4 we can observe a 10.3% improvement over an isolated vehicle for the WLTC cycle and 17.6% improvement for stabilised highway driving, which increases the range by almost 48km and breaks the 300km highway driving barrier.

**Table 4.** Platoon driving influence on range.

| | Range [km] (with regenerative braking) | | | | | |
|---|---|---|---|---|---|---|
| **No. of vehicles** | **1** | **2** | **3** | **4** | **5** | **7** |
| **WLTC** | 448 | 466 | 488 | 486 | 487 | 494 |
| **130km/h** | 277 | 296 | 319 | 317 | 318 | 325 |

The gains in range are shown in Figure 8, where we can observe that the most efficient platoon is the one made from 3 vehicles.

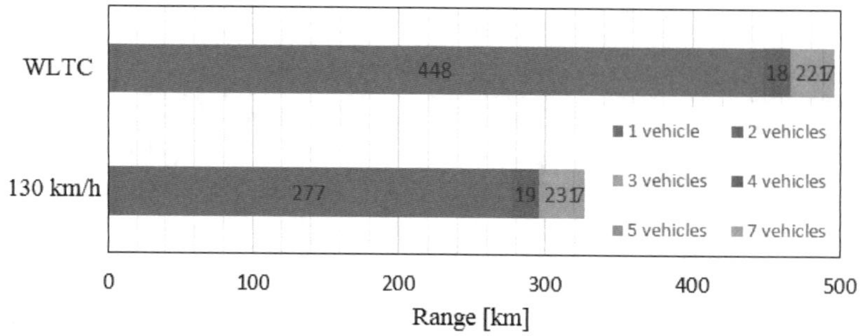


**Figure 8.** Range gains during platoon driving.

## 4. Conclusions

It was shown that in order to evaluate the benefits of platoon driving we must look beyond the mean drag coefficient and also evaluate the impact of the platoon speed and number of involved vehicles.

For the SUV electric vehicle that was studied, for stabilised speeds, the improvements range from 2.9% for 2 vehicles at 70km/h and reach 14.6% for 7 vehicles at 130km/h. These gains are lower than the ones obtained for a conventional powertrain that starts from 3.5% for 2 vehicles at 70km/h and reach 18.7% for 7 vehicles at 130km/h as shown in [6].

For a standardised high-speed cycle (HWFET) using regenerative braking the improvements start from 3.6% for a 2-vehicle platoon and go up to 9.8% for a 7-vehicle platoon.

These gains are comparable with improvements obtained using other technologies. As stated in [8], by using a two gears transmission instead of a single gear one reductions in energy consumption of 14.5% for the WLTC cycle and of 8.1% for the HWFET can be obtained. Also, by using a two motors architecture the energy consumption can be further improved by 4.1% for the WLTC cycle or 7.9% for the HWFET cycle as stated in [9].

Most of the improvements are obtained when using 3 vehicles platoon so this is the most efficient setup using the least number of vehicles. Further small improvements can still be obtained but it remains to be seen if short platoons which also have to coordinate between them are a better solution than a longer platoon.

Platoon driving offer real benefits where electrical vehicles have their weak spot, driving at high stabilised speed and offers tangible results like increased range of up to 17.6% at 130km/h. Also, by decreasing the drag coefficient it provides better regenerative braking conditions.

## References

[1] Mark Michaellan, Fred Browand 1999 *Quantifying Platoon Fuel Savings*, 1999 Field Experiments, University of Southern California,

[2] M. Zabat, N. Stabile, S. Frascaroli, F. Browand 1995 *The Aerodynamic Performance of Platoons*, Final Report, California PATH Research Report UCB-ITS-PRR-95-35,

[3] M. Zabat, N. Stabile, S. Frascaroli, F. Browand 1995 *The Aerodynamic Performance of Platoons*, Final Report, California PATH Research Report UCB-ITS-PRR-95-35,

[4] Renata Gnatowska, Marcin Sosnowski 2017 *The influence of distance between vehicles in platoon on aerodynamic parameters*, The European Physical Journal Conferences, Renata Gnatowska,

[5] Marcin Sosnowski 2017 *The influence of distance between vehicles in platoon on aerodynamic parameters*, The European Physical Journal Conferences,

[6] C. Constantin, C. Rențea, M. Bățăuș, G. Frățilă 2022 *Study of $CO_2$ emissions in platoon driving*, International Conference of Mechanical Engineering,

[7] Stoicescu, A.P. 2007 *Proiectarea performanţelor de tracţiune şi de consum ale autovehiculelor*, Editura Tehnică, Bucureşti,

[8] C. Renţea, M. Oprean, M. Băţăuş, G. Frăţilă 2019 *The influence of multi-speed transmissions on electric vehicles energy consumption*, IOP Conference Series: Materials Science and Engineering Volume 564,

[9] C. Renţea, M. Băţăuş, S. Maican, M. Oprean, G. Frăţilă 2020 *Comparative study of different electric vehicle configurations in terms of energy consumption*, IOP Conference Series: Materials Science and Engineering Volume 1037

IOP Conf. Series: Materials Science and Engineering **1311** (2024) 012013 doi:10.1088/1757-899X/1311/1/012013

# Study on the efficiency of Battery Electric Vehicle

**O Chiver[1],N Burnete[2], I R Şugar[3*] and L Neamt[1]**
[1] Technical University of Cluj-Napoca, Electrical, Electronic and Computer Engineering Department, Str. Dr. Victor Babeş Nr. 62A, 430083BAIA-MARE, Romania
[2]Technical University of Cluj-Napoca, Automotive Engineering and Transports Department, B-dul Muncii Nr. 103-105, 400641, CLUJ-NAPOCA, Romania
[3] Technical University of Cluj-Napoca, Engineering and Technological Management Department, Str. Dr. Victor Babeş Nr. 62A, 430083BAIA-MARE, Romania

*Email: radusugar@yahoo.com

**Abstract.**The paper deals with aspects that influence the efficiency of electric vehicles and, respectively, their efficiency compared to hybrid or ICE vehicles.Different driving cycles are considered, especially referring to driving on the highway, or urban driving with repeated starts and stops, with average speed relatively low, and the acceleration not too high.Besides these, a comparison will be made between the situations in which the BEV has one speed or two speeds, since the weight of the vehicle is greater in the latter case.Then a hybrid and a conventional model will be tested for the same driving cycles and the same vehicle body.

## 1. Introduction

The problems related to the pollution produced by internal combustion engine (ICE) vehicles are well known, the harmful effects being especially clear in large urban agglomerations, where, because of them, the air becomes unbreathable. Although researchers are continuously looking for solutions to improve ICEVs, [1], [2], [3], [4], [5], [6], things seem to have taken a major turn once United Nations Climate Change Conference, Glasgow 2021, where the COP26 declaration has been adopted. According to this, automotive manufacturers, will work towards reaching 100% zero emission new car sales in leading markets by 2035 or earlier. Of course, not all manufacturers or all governments have signed this declaration, but it can still be considered as future directives to reduce the impact of human activity on the environment. Therefore, the possible solutions to be considered at this time exclude even plug-in hybrid vehicles (PHEV).Zero emission vehicles include only battery electric vehicles (BEV) and fuel cell (or hydrogen) vehicles. Basically, the differences consist in the way of obtaining/storing electricity that powers the motor. In all cases, the goal is to replace ICE with electric motors to propel the vehicle. There are some significant advantages that the latter confer concerning the former, such: much higher yield, usually over 90%, but reaching a maximum of 96-98%; lack of polluting emissions during travel; the possibility of recovering braking or coasting energy; the simplicity of stopping/restarting while the vehicle is not moving. Overall, depending on how the electricity is generated (needed to power the vehicle), it is possible to reach a situation where its operation has zero emissions (if the electricity is generated from renewable sources).The biggest disadvantage, which imposes some restraint on the purchase of BEVs, is the limited autonomy [7], while ICEVs have large autonomy as well as a very short time needed to fill the tank. Also, the use in cold weather, when it is necessary to heat the vehicle, still reduces the discrepancies between ICEVs and BEVs. However, it seems that the return of electric vehicles instead of ICEVs will be faster than expected. Hybrid vehicles arose with the idea of getting ICEVs' and BEVs' advantages. There are many types of hybrid vehicles from the least electrified ones (only with start/stop function and braking energy recovery) to those Pug-in hybrid (PHEV) that use ICE only when the battery energy is below a certain level, thus increasing the autonomy of the vehicle.

The types of used motor for electric vehicles depends on the required power, the type of vehicle (2, 3, or 4-wheels), and technology, many papers refer this [9], [10].The classic DC motors are rarely used now due to expensive maintenance. They can still be found in some countries like India, especially in the case of vehicles for urban public transport.

Brushless DC motors (BLDCM) have the perfect characteristics for use in electric traction, just like the classic DC. Unlike the classic ones, which have the big disadvantage of expensive maintenance due to the brushes and commutator, BLDCMs are electronically commutated and are maintenance-free. They have higher efficiency, up to 96% and they are suitable for high power density. The main drawback is the high cost because of permanent magnets, but also, they have ripple torque and noise. They are found in two forms, with the outer or inner stator. Also called Hub motors because the wheel directly connection to the rotor, the two-wheeler manufacturers prefer the BLDC motors with the outer rotors. This type also does not require a special space for mounting the motor. There is a restriction on the motor dimensions which limits the power output in the out-runner configuration. Motors with inner rotors require mounting space and a power transmission system to the wheels, unlike the first ones. They are used especially in three-wheeled vehicles.

Permanent magnets synchronous motors (PMSM) are like BLDC, they have the right characteristics for traction, high specific power, and very high efficiency, up to 98%.Unlike BLDC, they do not present ripple torque and noise, but they are more expensive. Most of the manufacturers (BMW, Chevrolet, Ford, Nissan, Toyota, etc.) use PMSM motors for their electric vehicles and there are also available for higher power ratings which allow them using for heavy vehicles as well.

The PMSRM (permanent magnets synchronous reluctance motor, called also IPMM – interior permanent magnets motor) along with the PMSM are among the best motors on the market. The PMs (permanent magnets) make the motor behave well at low speeds, with high torque, allowing good starting and ramp running. Furthermore, PMs offer advantages in the generator regime of the electric machine. At high speed, when the negative effect of the induced back EMF of the PMs occurs, through the control mode of the supplying voltage, the PMs magnetic field become negligible and the motor works as a synchronous reluctance motor (variable reluctance). At high speeds SRM has good performances, so the PMSRM properly controlled, gives good characteristics both at low and high speeds (it is used in Tesla 3).

Finally, IM (induction motor) it is still one of the widely used motors for electric vehiclesand both two-wheeler and four-wheeler manufacturers use it(Tesla, Tata, Mahindra).Among its main advantages are the low price, robustness, and competitive efficiency. The main disadvantage is the somewhat complicated control, the higher specific weight, but also the lower efficiency compared to motors with PMs.

## 2. Objectives

The main goal of the paper is analyzing the overall efficiency of BEV in different situations, but also comparatively with hybrid and ICE vehicles. The considered situations refer to different driving cycles (representative for urban and highway respectively); 1-speed and 2-speed gearbox for BEV; IM and PMSM. Because BEVs usually have only a 1-speed gearbox, the authors highlight the way a 5-speed gearbox is changed into a 2-speed gearbox. After that, the authors will analyze the efficiency of the main components of every type of vehicle for considered driving cycles.

## 3. Methodology

### *3.1. Vehicle model*

The total resistance force, taking into consideration only longitudinal forces, is the sum of the aerodynamic drag – $F_{aero}$, the rolling resistance -$F_{rr}$, and the grade, $F_{grade}$. The tractive force - $F_t$, imposes the acceleration and the speed. The mechanical model of the vehicle is [8],

$$m\frac{dv}{dt} = F_t - F_{aero} - F_{rr} - F_{grade} \tag{1}$$

The resistance forces are expressed as,

$$F_{aero} = \frac{1}{2}C_d\rho A v^2, F_{rr} = C_{rr}mgcos\theta, F_{grade} = mgsin\theta \tag{2}$$

We noted: air density, $\rho$; frontal vehicle area, $A$; vehicle velocity, $v$; drag coefficient, $C_d$; resistance coefficient, $C_{rr}$; vehicle mass, $m$; gravity acceleration, $g$; road grade, $\theta$.
The mechanical model is valid for all vehicle types, what differs is the way the tractive force is produced. The overall vehicle mass results as sum of the main components (e.g., for BEV: vehicle body, motor, energy storage system, transmission including gearbox and cargo).

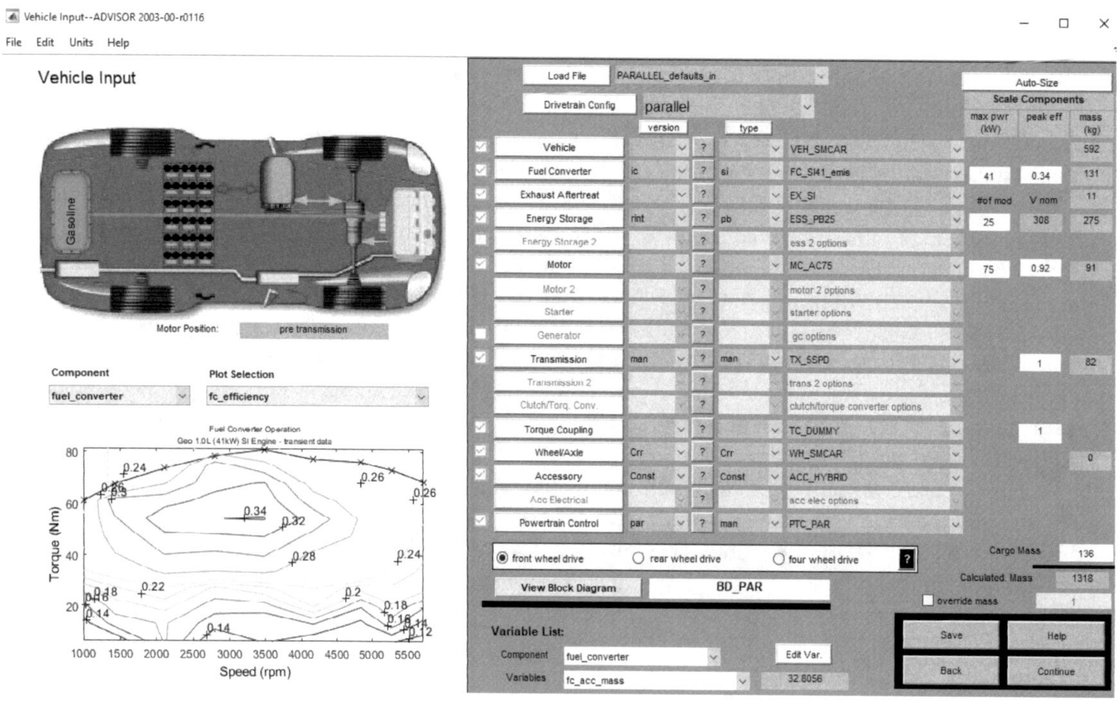


**Figure 1.** ADVISOR main preset interface

*3.2. Models' parameters*

IM1 and IM2 models have been configured as BEV, AC75 motor was set. This is an induction motor with 75 kW continuous power, 320V, the maximum torque is 200 Nm, over-torque factor is 1.8, maximum allowed current, 480A, minimum allowed voltage, 120V, maximum speed is 10000 RPM, and the mass is 91 kg. Fig. 2a shows the motor characteristics and the operating points when HWFET driving cycle has been imposed for IM1.While the IM1 has a single gear ratio, the IM2 has a gearbox with two gear ratios. For IM1 the self-determined gear ratio is about 6.7. For IM2, after many simulations the two gear ratios of 13.45 and 3.77 have been chosen from the existing ones in a 5-speed gearbox. Also, the gear change was set at a vehicle speed of 60 km/h. The same speed has been considered for the PMSM2 model. The vehicle mass is 1176 for IM2 and 1144 for IM1.

For PMSM models (also BEVs), a 58-kW continuous power motor has been set. Other parameters: maximum torque around 400 Nm; over torque factor, 1; maximum allowed current, 480A; minimum allowed voltage, 120V; maximum speed, 4000 RPM, and the mass, 70 kg. Fig. 2b shows the motor characteristic and operating points for PM1 running in HWFET cycle. The self-determined gear ratio for PM1 is about 2.7 and for PM2 the set two gear ratios are 3.77 and 1.67.Overall vehicle mass is 1155 kg for PM2 and 1123 kg for PM1.
The used hybrid model (Hy) has been the preset one. It has a parallel configuration (fig. 1), the gasoline engine, 1 liter, 41 kW and characteristics as in fig. 1. Beside this, the model has an induction motor (AC75) which is the same as for IM1 and IM2. In fact, the entire electrical system is the same. The 5-speed gearbox has the gear ratios: 13.45, 7.57, 5.01, 3.77 and2.84. The overall vehicle mass is 1318 kg.
The conventional model (ICE) has the same engine and gearbox as the hybrid model. The total mass is 952 kg.
All models have the same vehicle body, 592 kg, cargo mass, 136 kg, and a constant electric load for accessories, of 700 W.

### *3.3. Vehicle configuration*
The authors carried out the study using the ADVISOR (ADvanced VehIcle SimulatOR) software, created by the U.S. Department of Energy's National Renewable Energy Laboratory's (NREL) Center for Transportation Technologies and Systems. This software is free and runs in Matlab/Simulink environment. In ADVISOR the dynamic of the vehicle is approximated as a series of discrete steps, during each of these, the components are assumed to be in a steady state, so the transient quantities (e.g., current, voltage, motor speed) are not taken into consideration. This allows the use of efficiency or power loss tables, which are generated by testing a powertrain component at a given torque and speed (or current and voltage), to find the required power of the various components at each moment. The ADVISOR authors made the necessary components for vehicle configuration, based on the Simulink library components. Many needed parameters are found by tests, or they have the values according to the component manufacturer's specifications and industry partners' data.
After starting ADVISOR, the main interface from fig.1 appears. The first two dialog boxes refer to the file that holds the configuration data of a specific model (generic or of a specific manufacturer) - Load File box and to the type of drivetrain (electric, hybrid, ICE, etc.) - Drivetrain Config box. The two are in correspondence, so if EV is set in Drivetrain Config, in Load File we can have, e.g., EV defaults_in (generic electric vehicle) or gm_ev1_in (which is electric vehicle from General Motors). Of course, even after the user has chosen a particular preset type, one can make reconfigurations by accessing the active boxes and selecting other structure models (among those that can be set in the chosen configuration) according to his own desire. Since the user has access to the program codes of the existing components/structures in ADVISOR, they can make (if has the necessary programming knowledge) changes to the control logic or certain parameters. In this way, the existing and modified component will behave as a new one, according to the user's wishes.
In this paper, for study the BEV, the authors made the following changes: a 5-speed gearbox was defined as a 2-speed one, selecting only two gear ratios, and changing the mass accordingly; the authors also altered the BEV control logic (PTC_EV file)to be able to work with a 2-speed gearbox, imposing the change of gears depending on the vehicle's speed. Finally, four models of BEV will be studied, as follows: the first being with an induction motor, with a single transmission ratio (self-determined according to the maximum speed of the motor, so that the vehicle can reach a maximum speed of approximately 145 km/h, considering a slip of 10%.) - called IM1; the second is the same vehicle with the IM but with a gearbox with two transmission ratios - IM2; the other two types resulted from replacing the IM with a PMSM - PM1 and PM2.Besides the four types of electric vehicles, a hybrid model - Hy, and a conventional model - ICE, were also configured to make some comparisons.

All the calculations that ADVISOR performs are based on the speed-time dependence (driving cycle) imposed on the vehicle, this being the main input characteristic. The driving cycles imposed on the models were HWFET (specific to highway driving, with data: time - 765s, maximum speed – 96.4 km/h, average speed – 77.58 km/h, maximum acceleration – 1.43 m/s$^2$, maximum deceleration – 1.48 m/s$^2$, average acceleration – 0.19 m/s$^2$, distance – 16.51 km, idle time – 6s, 1 stop) and FTP (for urban driving, having data: time - 2477s, maximum speed – 91.25 km/h, average speed – 25.82 km/h, maximum acceleration – 1.48 m/s$^2$, maximum deceleration – 1.48 m/s$^2$, average acceleration – 0.51 m/s$^2$, distance – 17.77 km, idle time – 360 s, 22 stops).In addition to imposing a specific driving cycle, one can set a constant road grade (zero has been chosen) or can assess a grade ability, and acceleration tests.

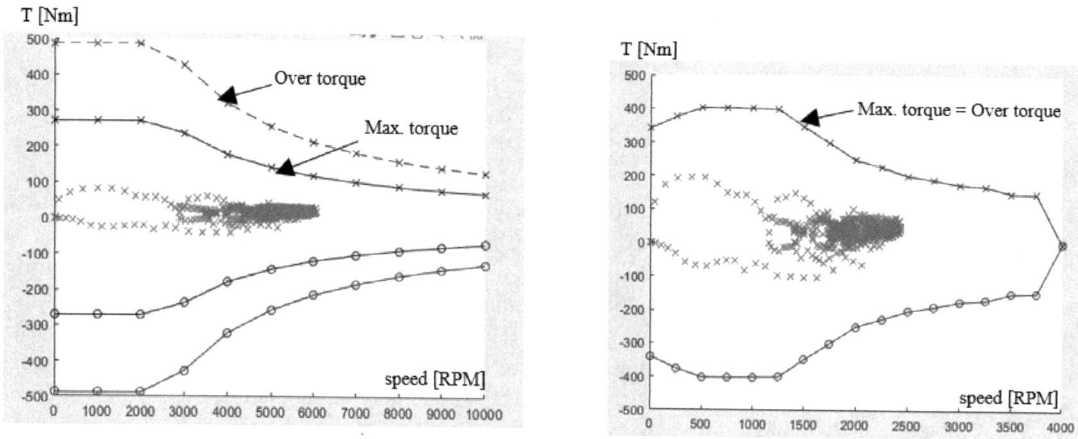


a) IM1 b) PM1

**Figure 2** Motor characteristic and operating points in HWFET cycle

### *3.4. Results*

We considered the overall efficiency the most important parameter of the efficiency of all vehicle models because it is calculated in terms of the consumed energy and necessary energy. First, we will compare the efficiency of all models for each driving cycle, fig. 3 and 4 show these results.

IOP Conf. Series: Materials Science and Engineering **1311** (2024) 012013 doi:10.1088/1757-899X/1311/1/012013

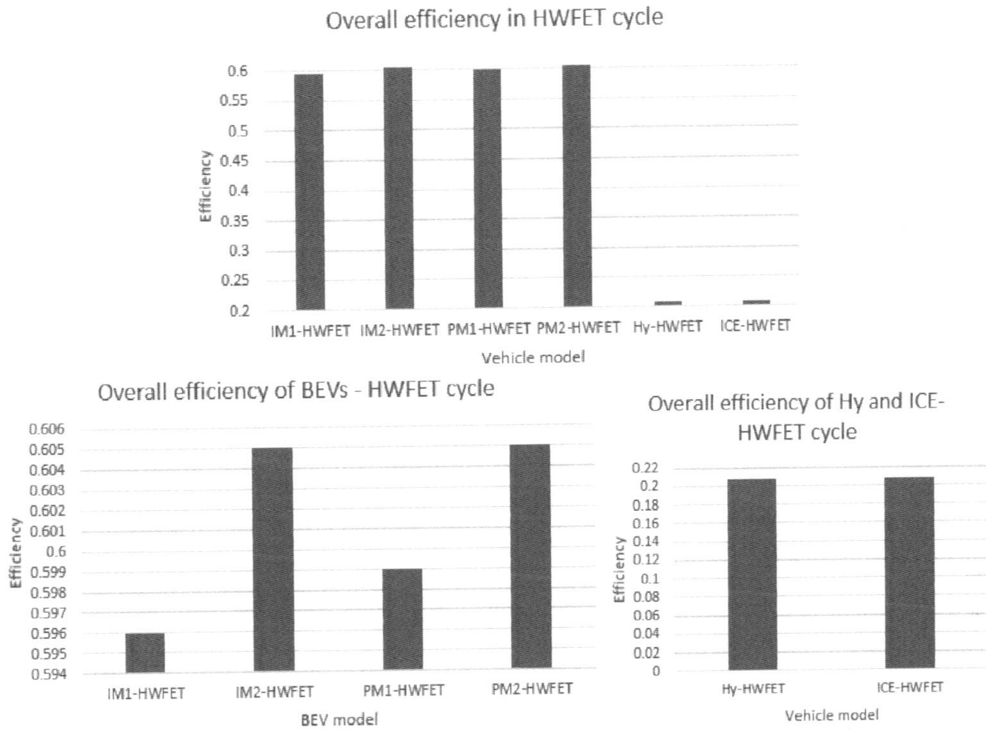


**Figure 3** Overall efficiency of the models for HWFET driving cycle

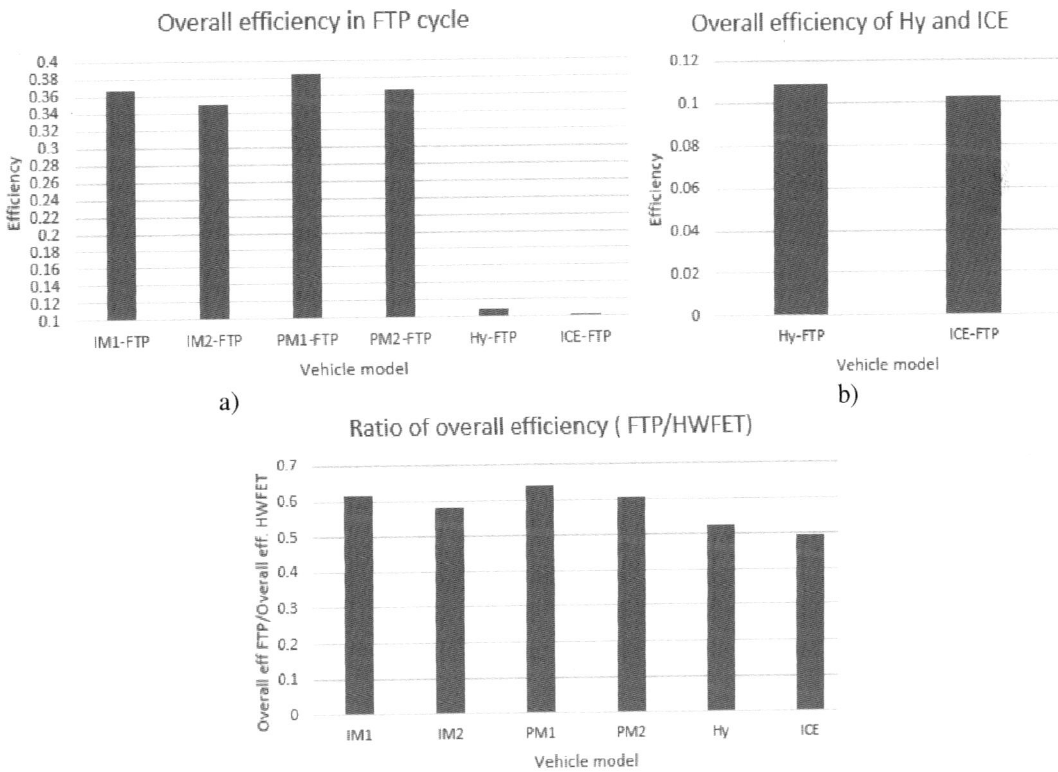


**Figure 4** Overall efficiency of the models for FTP driving cycle (a, b) and FTP/HWFET efficiency

Looking at figure 3, it can be noted that in case of driving on the highway (HWFET), the difference in efficiency between BEVs and the other types of vehicles is major. When BEVs have only one gear ratio, the vehicle with PMSM is the best. The maximum efficiency is for BEVs (both IM and PMSM) with 2-speed gearbox. Using this gearbox, the efficiency of IM vehicle increases more than efficiency of PMSM vehicle. This can be explained by the fact that, percentage wise, the weight of the vehicle with IM increases less (2.79%) than the weight of the vehicle with PMSM (2.85%), when 2-speed gearbox is used. There is no difference in efficiency between the hybrid (Hy) and the conventional (ICE) vehicle when running on the highway. The advantages offered by regenerative braking are annihilated by the extra weight of the electrical system, in the motoring regime.
When the vehicles running in urban cycle (FTP) the efficiency is lower for all models. In reported values, ICE and Hybrid efficiency decreases more than BEVs efficiency, when running on urban cycle, fig. 4c.Another important thing to highlight is the fact that in the case of the FTP cycle, BEVs with a single gear ratio are more efficient than those with a 2-speed gearbox. This is because the cycle speed is low, and the second gear is used very little, but the vehicle mass is higher.
Regarding the efficiency of components, we will refer only to the motor-controller and engine. The results are show in fig. 5.
It is noted that in the highway driving cycle, in the motor regime, IM2 (0.861) has the best yield followed very closely by PM2 (0.86). In the case of driving in the urban cycle (FTP), the best yield is at PM1 and PM2 (0.806 and 0.805) followed closely by IM1 (0.795).Although we are talking about the same motor-controller as in the IM models, in the case of the hybrid model the efficiency is much lower.

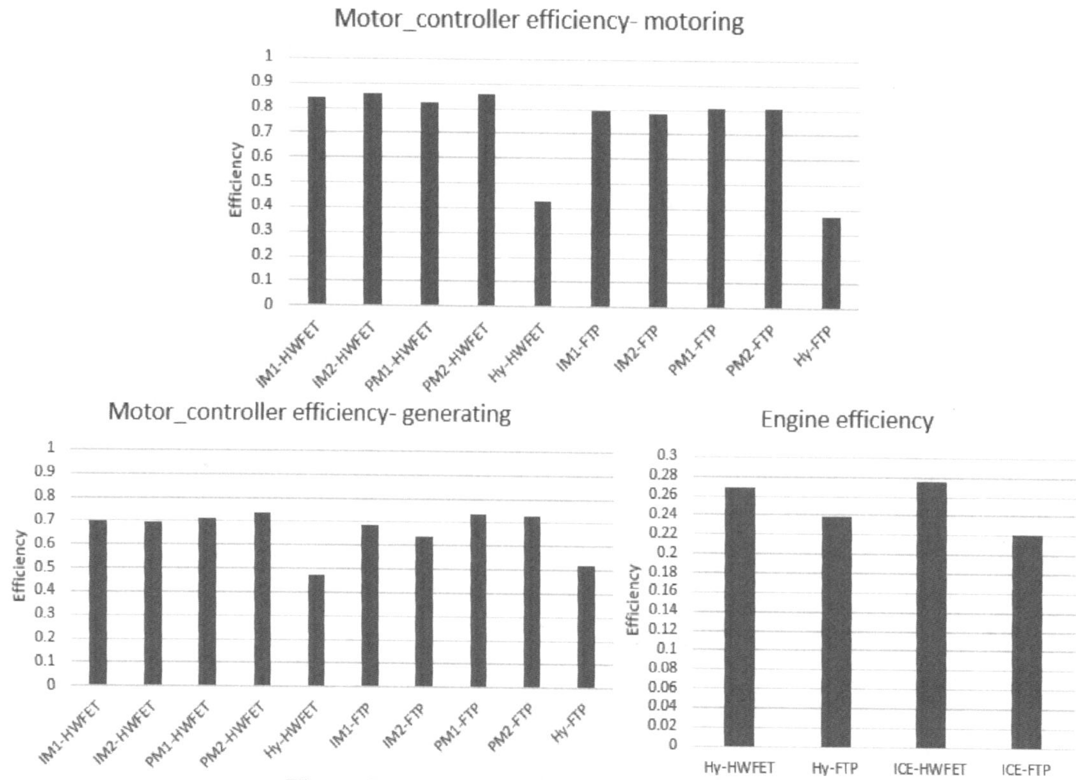

**Figure 5** Motor-controller and engine efficiency

For the generator regime, of course, the PM models have the highest efficiency. In the generator regime, the PM models have the highest efficiency, and the hybrid model has the lowest efficiency. We will highlight that for this regime, the hybrid model is more efficient when running in the urban

cycle than when running on the highway. Regarding engine efficiency, the ICE model has the best efficiency (0.276) when running on the highway but also the minimum (0.22) when running on the urban cycle.
Some data about models' performances are shown in table 1.

**Table 1**. Models' performances

| **Model**<br>**Parameter** | **IM1** | **IM2** | **PM1** | **PM2** | **Hy** | **ICE** |
|---|---|---|---|---|---|---|
| Time (0-100 km/h) [s] | 11.3 | 11.8 | 12.5 | 12.4 | 12.8 | 18.4 |
| Time (65-100 km/h) [s] | 6.4 | 6.5 | 6.1 | 6.5 | 7.2 | 9.7 |
| Time (0-137km/h) [s] | 25.6 | 25.7 | 24.2 | 25.5 | 25.6 | 41.8 |
| Max. acc. [m/s$^2$] | 5 | 5 | 3.1 | 4.2 | 5 | 2.8 |
| Max. speed[m/s] | 157.6 | 163.8 | 150.7 | 171.6 | 199.5 | 155.8 |
| Max. grade at 88.5 km/h [%] | 16.5 | 16 | 15.6 | 16.1 | 14.6 | 9.7 |

Analysing the data in the table, the following can be found: IM1 has the shortest time to reach 100km/h; from 65 to 100 km/h, and also from 0 to 137 km/h, PM1 need the shortest time; Hy model reach the highest maximum speed, and between BEVs, the PM2; the maximum grade at 88.5 km/h can be maintain by IM1, but PM2 and IM1 also have good values; finally, all BEV models have closely performances.

## 4. Conclusions

- When the vehicles run on the highway, the total efficiency of BEVs is around 0.6, while the hybrid and conventional models have an efficiency of around 0.2. Also in this cycle, a two-speed gearbox offers BEVs a higher efficiency than a single-speed gearbox;
- In the FTP cycle, overall efficiency is around 0.35 for BEVs and about 0.1 for Hy and ICE models. A two-speed gearbox has no advantage for BEVs in this cycle;
- The hybrid model is more efficient than the classic one only in the FTP cycle (urban);
- In the HWFET cycle, there is no difference between PM2 and IM2 models regarding efficiency, while in the FTP cycle, PM1 is the best followed by IM1;
- IM1 has the shortest time to reach 100km/h; from 65 to 100 km/h, and from 0 to 137 km/h, PM1 need the shortest time; Hy model reach the highest maximum speed, and between BEVs, the PM2; the maximum grade at 88.5 km/h can be maintain by IM1, but PM2 and IM1 also have good values; usually, the PM machine offers the best efficiency in all regimes (motor and generator).

## References

[1] Aleonte M, Cosgărea, R, Jelenschi L, and Cofaru C 2011 *Technical solutions for improving the efficiency of a two stroke SI engine* (Bulletin of the Transilvania University of Braşov, Series I: Engineering Sciences • Vol. 4 (53) No. 2) p. 1,
[2] Aliasand A E and Josh F T 2020 *Selection of Motor for an Electric Vehicle: A Review, Materials Today* (Proceedings, Volume 24, Part 3, ISSN 2214-7853) p. 1804,
[3] Burnete N V, Burnete N, Jurchis B and Iclodean C 2017.*Influences of diesel pilot injection on ethanol autoignition - a numerical analysis*,(IOP Conf. Ser.: Mater. Sci. Eng. 252,012066),
[4] Ehsani M. Singh K V Bansal H O and Mehrjardi R T 2021 *State of the Art and Trends in Electric and Hybrid Electric Vehicles*, (Proceedings of the IEEE, vol. 109 no. 6,) p. 967,
[5] Jafari, M Gauchia A. Zhang K and Gauchia L 2015 *Simulation and Analysis of the Effect of*

*Real-World Driving Styles* (EV Battery Performance and Aging, in IEEE Transactions on Transportation Electrification, vol. 1, no. 4) p. 391,
[6] Gutarevych Y Mateichyk V Matijošius J Rimkus A GritsukI Syrota,O and Shuba Y 2020 *Improving Fuel Economy of Spark Ignition Engines Applying the Combined Method of Power Regulation*(Energies 13, no. 5) p. 1076,
[7] Jurchiș B M Burnete N Burnete N V and Iclodean C D 2018 *Particulate matter emission characteristics for a compression ignition engine fueled with a blend of biodiesel and diesel* (IOP Conf. Ser.: Mater. Sci. Eng. 444, 072012),
[8] Martins J BritoF P Costa,T and Costa J 2018 *Thequest for improving efficiency in internal combustion engines* (Conference: 3ª Conferência Internacionalem Produção do Conhecimento Científicoem Timor-LesteAt: Dili, Timor-Leste),
[9] Nagasawa T Okura Y Yamada R Sato S Kosaka H Yokomori T and Lida N 2021 *Thermal efficiency improvement of super-lean burn spark ignition engine by stratified water insulation on piston top surface* (International Journal of Engine Research, Volume: 22 issue: 5) p. 1421,
[10] Simpson T Bousfield G Wohleb A and Depcik C 2022 *Electric Vehicle Simulations Based on Kansas-Centric Conditions* (World Electric Vehicle Journal 13, no. 8) p. 132.

IOP Conf. Series: Materials Science and Engineering **1311** (2024) 012014 doi:10.1088/1757-899X/1311/1/012014

# INFLUENCE OF DUAL FUELING ON THE PERFORMANCE OF A COMPRESSION IGNITION ENGINE

**S Izvorean[1*], V Stoica[1], D Lelea[1]**

[1]Politehnica University Timisoara, Faculty of Mechanical Engineering, 1 Mihai Viteazu Blvd., Timisoara, Romania

*Email: stefan.izvorean@student.upt.ro

**ABSTRACT** The diesel engine although offering a high efficiency and low fuel consumption is slowly being marginalized regarding thermal engines due to high pollutant like NOx and smoke particles. Although there are alternative propulsion systems, they are not proper for industrial use. For this reason, we need to find a solution to improve this engine and to obtain lower emissions and a higher efficiency.
In this paper we propose a solution that can improve the efficiency and reduce the pollutants produced by diesel engine, using a secondary fuel. The idea is to inject the secondary fuel in the intake manifold of the engine, creating a mix between the air and the fuel. This mixture will fill the cylinders and a pilot injection will be used to ignite the content of the cylinder and to provide the additional fuel needed to obtain the targeted performance, while maintaining the emissions at a minimum.
In this scope an experimental test bench with a diesel engine, equipped with a secondary injection system was created. The secondary fuel is injected, in the intake manifold, separately than the main diesel fuel, via a single point injection system. The entire system was created in the laboratory and is driven with a in house software solution that allows us to control every aspect of the engine such as timing, injection duration, fuel pressure and so on. As a result of the experiment, was observed that the engine performance is not diminished. The NOx emission presents a decrease in the same operating conditions. The smoke opacity is reduced significantly and the HC emissions present a slight increase for the same conditions

## 1. Introduction

When thinking about dual fueling a compression ignition engine, multiple variants in which a secondary fuel can be introduced in the combustion chamber comes to mind, also multiple types of secondary fuel such as gases or liquids. If the secondary fuel is liquid, it can be introduced as a mix with the main fuel as shown by Shaoji Chen et al. [1]. A mathematical simulation algorithm was created, and tested on a real engine, in order to determine the effects of methanol substitution of diesel by mixing the two fuels., The results show a reduction of smoke particles with the increase of methanol content, the CO increases with load but decreases with methanol substitution. The engine efficiency is improved with the presence of methanol in the fuel content, the NOx increases in the case of this simulation. In a similar fashion Xinling Li et al [2] performed tests on a 4-cylinder engine with a series of mixes between different types of alcohols and diesel. The results show that HRR is increasing, the ignition delay is increasing, and the burn duration is decreasing. It is observed that the smoke particles are decreasing but the CO and HC are increasing.

An alternative to this method is to introduce the secondary fuel in the intake manifold. Here different types of fuels can be used, like gas or alcohol. Mirosław Karczewski et al. [3] uses a test bench with a truck engine, equipped with a secondary fuel system in the intake manifold. The CNG was used as secondary fuel. The measurements were concentrated on the on-road performance of the truck. It was observed a slight decrease in performance of the engine but also a reduction of CO2. The NOx presented an increase and so did the smoke particle. The authors consider that the mix of CNG and air from the cylinder was homogeneous and the injected fuel did not have enough air to properly burn. Another perspective about gas as a secondary fuel is offered by Swarup Kumar Nayak et all [4]. A Kirloskar

TAF1 engine was modified to work in a dual fuel configuration with diesel and biogas, in addition with a system that allows the increase of injection timing. The result are excellent and show an improvement in all key criteria, the performance of the engine is increased, the CO, HC, NOx and smoke particle are reduced when compared with running only on diesel.

One more way to introduce a secondary fuel in the combustion chamber is to inject a liquid fuel in the intake manifold of the engine, this is also the way used for the measurement in this paper. The effects of this method for dual fueling are present also by V.D Caudhari et al [5]. The presented results are for a single cylinder engine that is running dual fueled, by diesel in the cylinder and methanol in the intake. Compared with standard single fuel, an increase in engine performance is observed alongside with a reduction in some conditions of HC and smoke particles. The CO and CO2 values are high compared to single fuel. Good result are shown also by the Zhanming Chen at all [6]. A 4-cylinder engine was equipped to run dual fuel. The secondary fuel used was methanol, ethanol and n-butanol. The result show that the burn process is shorter and leads to a higher peak in cylinder pressure. The presence of secondary fuel increases the engine efficiency, the measurement with ethanol shows a 0.9% increase. The smoke particle are reduced in dual fuel mode and the NOx is reduce with methanol use, increased with n-butanol and for ethanol are not conclusive. Also, a series of mathematical simulation model for the dual fuel are created. Zengna Weic et al. [7] created a model and validated with a real engine. The results after simulation with different substitution of diesel with ethanol show that the in cylinder pressure is higher but the rate with which the pressure rises for higher substitution is dangerous and is consider that the engine runs with knock. The emissions show a reduction of smoke and CO but an increase of HC and NOx. Zhiyong Li et al [8] created another mathematical model, and validated by measurements. The simulation aims to compare the port injection of methanol with direct injection of methanol. The results show that the direct injection offers a better response for high substitution ratios and better burn characteristic for low loads, but the system proposed need an injector with 2 sets of offset holes that can inject 2 separate fuels. The port injection is easier to implement and still offers a reduction is particulate matter, but the NOx and CO are increased in this simulation.

Changing the timing of diesel injection in the case of alcohol port injection can improve the burn process. Yanqing Cui et al. [9] uses an optical engine to observe the flame formed inside a combustion chamber of a dual fuel engine, fueled with methanol in the intake manifold and n-heptane or n-dodecane in the cylinder. It is observed that the lover reactivity fuel (n-heptane) that is injected directly does not manage to ignite the charge in the cylinder, but a positive trend is observed in the case of injection timing advance. When a higher reactivity fuel (n-dodecane) is introduced in the cylinder alongside the increased injection timing multiple flame kernels are formed and the burn process takes place.

## 2. Experimental test bench Introduction

### *2.1 Generalities*

The experimental test bench is composed out of 3 main elements, the thermal engine, the engine hydraulic brake, and the electronic control unit with the software solution.

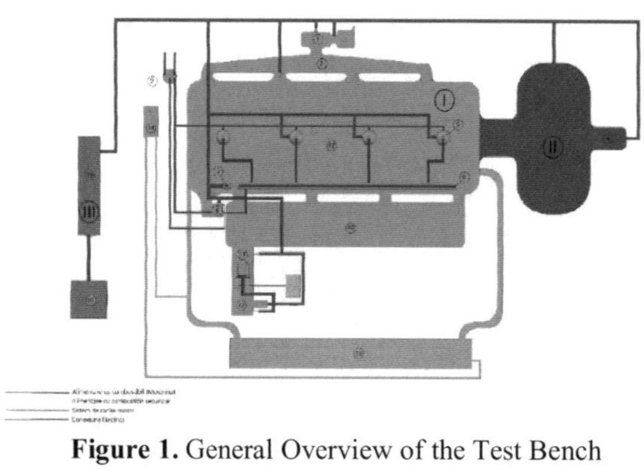

**Figure 1.** General Overview of the Test Bench

IOP Conf. Series: Materials Science and Engineering **1311** (2024) 012014 doi:10.1088/1757-899X/1311/1/012014

In the figure 1 the experimental test bench is depicted. The main components are as follow: 1-Turbocharger; 2-Exhaust Manifold; 3-Whater Brake for Engine Load; 4-Adjustabla Valve for Engine Load; 5-Diesel Injectors; 6-Common Rail; 7-Common Rail Pressure Sensor; 8-High Pressure Diesel Pump; 9-Diesel Filter; 10-Intake Manifold; 11-Secondary Fuel Injectors; 12-intake Adaptor; 13-Collant Heat Exchanger; 14-Expansion Tank; 15-Signal Conditioning Box; 16-Computer for engine control

### *2.2 Thermal Engine*

The engine used for the measurement is a turbocharged compression ignition Ford engine. It has a displacement of 1998 cm3, the bore and stroke are equal and measure 86 mm. The compression ratio of the engine is 18,2:1. The main injection system, is a Common-Rail system, able to reach 1600 Bar of pressure. For the secondary fuel, an injection system that uses 2 gasoline fuel injectors was created. They inject the secondary fuel into the intake manifold, where it is mixed with the intake air creating a mixture.

The fueling of the engine is realized from external tanks, mounted on electronic scales that report the fuel consumption, to the data centralization program, every minute.

The engine cooling is realized with water-to-water heat exchangers, the coolant temperature is kept around 102 °C and is controlled by the engine thermostat.

**Figure 2.** Test Bench with Diesel Engine

### *2.3 Engine Dynamometer*

In order to test different load scenarios, the engine is connected to a water engine dynamometer., The dynamometer is produced by the Junkers company and it is an impact screw style dynamometer. This type of dynamometer uses bars inside of the housing. These bars are dragged through the water inside the brake, creating a torque between the rotor connected to the motor and the stator connected to the load cell used to report the engine torque.

### *2.4 Electronical Control*

In order to control the engine without being limited by an off the shelf control solution, an entire hardware and software solutions, from sensors readout to engine position determination and output control was created. For the injectors a peak-hold control circuit was created, the circuit is based on a LM1949 integrated circuit from Texas Instruments, the designed circuit for main injectors allows for a 26 Ampere peak current and a 10 Ampere hold current, this allows the injector to open in 400 μs. For the secondary fuel injector due to the lower pressure a much smaller amount of current is necessary, so a 3-ampere peak and 1 ampere hold current is used, the open time for this injector is 600 μs.

The common rail pressure is control with a metering valve in the high-pressure pump. This valve is controlled with a 1KHz PWM signal, the duty cycle can be control manually or with a PID regulator implemented in the software.

The sensor reading, engine position calculation and output commands are realized by a compact-RIO real time control system from National Instrument, the software is realized in LabView, a programing tool also from National Instruments.

The engine position is determined from 2 sensors, one sensor mounted on the flywheel, this provides a signal pulse every 6 °CRK, the angle in between teeth is estimated based on the crankshaft radial speed. In order to differentiate between the 2 engine rotations, a secondary sensor is mounted on the camshaft, this sensor gives a signal when the piston from the cylinder 1 is on the compression stroke. For the injection advance, injection time, rail pressure (in case of PID use) numeric controls are implemented in the software, so the user can set them to the desired value.

Because of the dedicated software the sensors used on the test bench needed to be calibrated. Therefore, the characteristic of each sensor was determined.

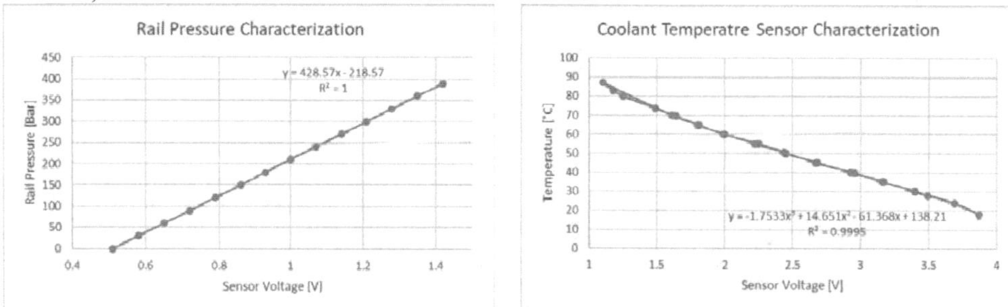


Figure 3. Sensor Calibration Curves

## 3. Experimental results

### *3.1 Diesel Only Injection Optimization*

As the experimental test bench was created with a dedicated software solution some preliminary measurement and characterization were necessary. In order to understand the optimal injection pressure and optimal injection timing for the engine, but also possibly some limitation of the test bench, a characterization at a set engine state was necessary. The parameters set were: engine speed of 1800 rpm ±1% and an output power of 17.75 kW±2% intake air pressure and temperature are at ambient conditions. These parameters represent approximately the necessary output for the vehicle from which the engine is obtained to drive with 100 Km/h. A measurement matrix was created to obtain the full picture. The measurements started form an injection pressure as low as 400 BAR and reached 1100 BAR. The intention was to go even higher but because of the small size of the fuel tank, the temperature of diesel fuel was high even at 1100 BAR. In order to protect the engine and especially the high pressure components the decision to limit the pressure at 1100 BAR was taken. The second parameters that was altered during the testing was the injection timing, at each rail pressure used the injection timing was varied in order to find the optimal value. The optimal value was considered to be where the specific fuel consumption was at the minimum. During the measurement as the parameters were changed, the injection time was reduced or increased in order to maintain the engine speed as close as possible to the desired RPM. The engine dynamometer was not tempered with so the load was constant at a set engine speed.

The test matrix and the results obtained are presented in the table 1 and figure 4:

**Table 1.** Test Matrix for Engine Performance on Diesel Only

| | Rail Pressure [BAR] | | | | | | |
|---|---|---|---|---|---|---|---|
| | 400 | 500 | 600 | 700 | 800 | 900 | 1100 |
| Injection Timing [° BTDC] | - | - | - | - | - | -5 | -5 |
| | - | 0 | 0 | 0 | 0 | 0 | 0 |
| | - | 2.5 | 2.5 | 2.5 | 2.5 | - | - |
| | - | 5 | 5 | 5 | 5 | 5 | 5 |
| | - | 7.5 | 7.5 | 7.5 | 7.5 | - | - |
| | - | 10 | 10 | 10 | 10 | 10 | 10 |
| | - | 15 | 15 | 15 | - | 15 | 15 |
| | - | 20 | 20 | 20 | - | 20 | 20 |
| | 25 | - | 25 | - | - | - | - |

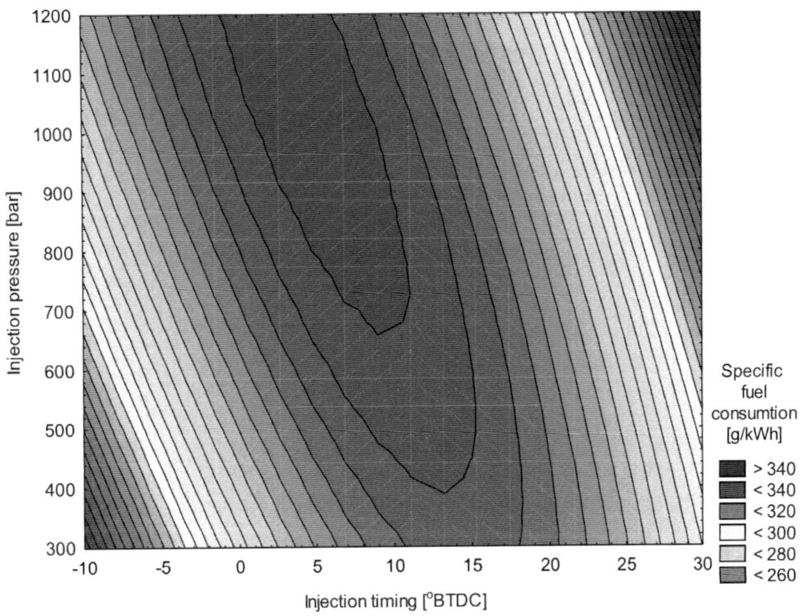


**Figure 4.** Optimal injection Timing for engine speed 1805±32 rpm and load 17.75±0.66 kW

The values obtained show that the optimal injection pressure is above 700 BAR and the optimal timing varies with pressure. The test with alcohol were started at the pressure of 900 BAR as the specific consumption is not different than the consumption at a higher pressure and the lower pressure is easier to manage.

### *3.2 Alcohol Performance Measurements*

The alcohol used is Technical Alcohol which is a homogeneous mix between Metilic alcohol, Ethylic alcohol and Isopropyl alcohol. The dose of alcohol was increased until an increase in the voltage read of the knock sensor with which the engine is fitted. The literature states that in the case of alcohol injected in the intake, there is a maximum amount that can be used for safe operation. After that point the pressure rise rate is to big and the engine runs in conditions similar to knock. A maximum of 36% from the total energy introduced was substituted with alcohol. The percent of alcohol variated slightly, as in order to find the optimal timing with alcohol, the diesel dose was reduced or increased, in order to maintain the engine speed as in the measurement for diesel optimal settings. The average value for alcohol substitution is 34%.

The measurement with alcohol were performed at a single injection pressure of diesel, set to 900 BAR and alcohol pressure was set to 4 BAR. The results for optimal timing while using alcohol in the intake are presented in figure 5.

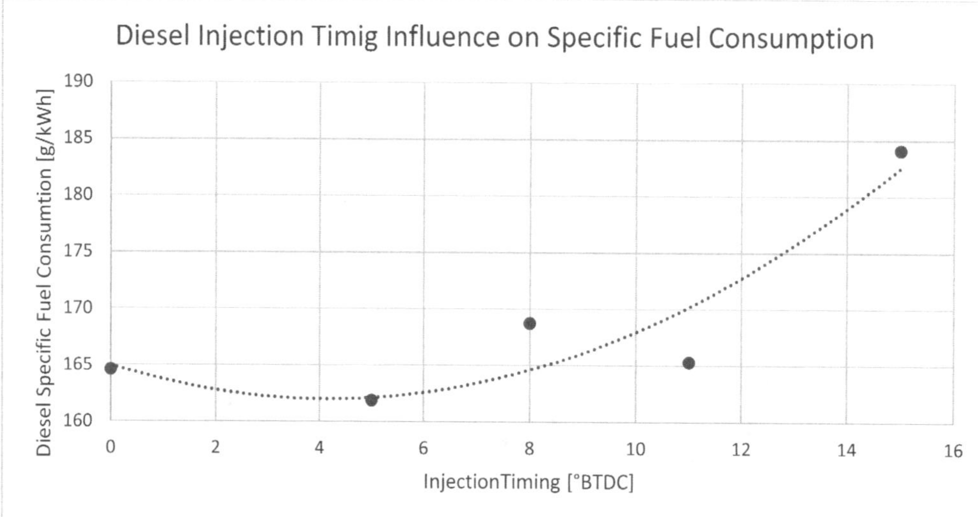


**Figure 5**. Optimal Injection Timing With Alcohol Injection for engine speed 1804±36 rpm, diesel fuel pressure 900±14.44 Bar and load 17.31±0.54 kW and alcohol hourly consumption 3133±376 g/h

*3.3 Alcohol Injection Influence on the Emissions*

The emissions of the engine were monitored with a Capelec CAP 3201., The CO, CO2, HC, O2, Lambda, NOx and smoke opacity were measured. The base values for diesel only are presented in the table 2.

**Table 2.** Diesel Only Emissions

| Diesel Only Emissions | |
|---|---|
| CO [% vol.] | 0.01 |
| CO2 [% vol.] | 9.2 |
| HC [ppm vol.] | 94 |
| 02 [% vol.] | 8.4 |
| Lambda [-] | 1.63 |
| NOx [ppm vol.] | 733 |
| K [1/m] | 0.13 |
| Diesel Injection Timing [°BTDC] | 11 |

The main difference with alcohol use is presented in the figure 6 as a function of diesel injection timing for engine speed 1804±36 rpm, load 17.31±0.54 kW and alcohol hourly consumption 3133±376 g/h

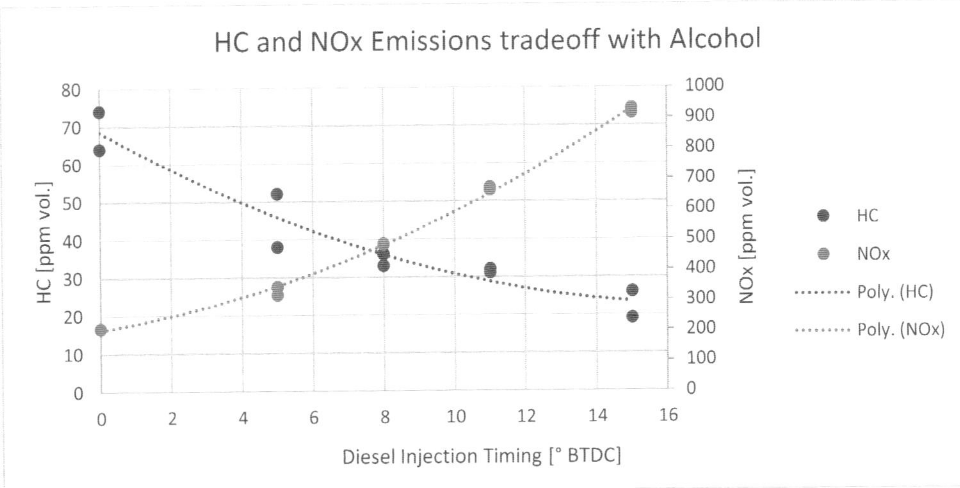


**Figure 6.** HC and NOx Emissions tradeoff with Alcohol Injection

## 4. Conclusions

The result obtained show us that about 1/3 of the energy introduced in the engine can be substituted with alcohol injected in the intake manifold, without any performance loss and with an improvement on the emission of the engine. From the information presented we can conclude that:

- The smoke particles are reduced to ¼ compared to the diesel only operation
- The NOx is reduced for the same operation conditions and can be reduce even more by optimizing the injection
- The HC is slightly increasing with alcohol presence and can be reduce more by altering the injection, but in order to obtain a low amount of HC the NOx value increases
- The Carbon Monoxide CO shows a ten time increase with alcohol presence but the value is still low at 0.1 % vol.

## References

[1] S. Chen, J. Tian, J. Li, W. Li and Z. Zhang, "Investigation of the Performance and Emission Characteristics of a Diesel Engine with Different Diesel–Methanol Dual-Fuel Ratios," *Processes,* vol. 9, no. 1944, 2021.

[2] X. Li, C. Guan, K. Yang, C. S. Cheung and Z. Huang, "Impact of lower and higher alcohol additions to diesel on the combustion and emissions of a direct-injection diesel engine," *Environmental Science and Pollution Research,* 2019.

[3] M. Karczewski and M. Wieczorek, "Assessment of the Impact of Applying a Non-Factory Dual-Fuel (Diesel/Natural Gas) Installation on the Traction Properties and Emissions of Selected Exhaust Components of a Road Semi-Trailer Truck Unit," *Energies,* 2021.

[4] S. K. Nayak, A. T. Hoang, S. Nizetic, X. P. Nguyen and T. H. Le, "Effects of advanced injection timing and inducted gaseous fuel on performance, combustion and emission characteristics of a diesel engine operated in dual-fuel mode," *Fuel,* vol. 310, 2022.

[5] V. Chaudhari and D. Deshmukh, "Fuel flexibility study of various fuels with charge dilution and high compression ratio for medium-load operating RCCI engine," *Fuel,* vol. 310, 2022. A reference.

[6] Z. Chen, J. He, H. Chen, L. Geng and P. Zhang, "Comparative study on the combustion and emissions of dual-fuel common rail engines fueled with diesel/methanol, diesel/ethanol, and diesel/ n-butanol," *Fuel,* vol. 304, 2021.

[7] Z. Weic, Z. Yuxin, X. Qi, L. Yichi and X. Yuanli, "A simulation of ethanol substitution rate and EGR effect on combustion and emissions from a high-loaded diesel/ethanol dual-fuel engine," *Fuel,* vol. 310, 2022.

[8] Z. Li, Y. Wang, Z. Yin, Z. Gao, Y. Wang and X. Zhen, "To achieve high methanol substitution ratio and clean combustion on a diesel/methanol dual fuel engine: A comparison of diesel

methanol compound combustion (DMCC) and direct dual fuel stratification (DDFS) strategies," *Fuel,* vol. 304, 2021.

[9] Y. Cui, H. Liu, M. Wen, L. Feng, C. Wang, Z. Ming, Z. Zhang, Z. Zheng, H. Zhao, X. Wang, L. Liu and M. Yao, "Optical diagnostics and chemical kinetic analysis on the dual-fuel combustion of methanol and high reactivity fuels," *Fuel,* vol. 312, 2022.

IOP Conf. Series: Materials Science and Engineering **1311** (2024) 012015 doi:10.1088/1757-899X/1311/1/012015

# The study on the engine pressure variation of utilizing n-butanol-gasoline blends at fueling spark ignition engines

**Cr Sandu[1], C Pană[1], N Negurescu[1], Al Cernat[1], Cr Nuțu[2] and R Georgescu[1]**

[1]Thermotechnics, Engines, Thermical and Frigorific Equipment Department, University Politehnica of Bucharest, Bucharest, Romania
[2]Automotive Engineering Department, University Politehnica of Bucharest, Bucharest, Romania

Corresponding author: sandu_cristian2007@yahoo.com

**Abstract.** Alternative fuel solutions are highly regarded as potential solutions to more emissions regulation as well as more efficient energetic consumption requirements. The more efficient combustion engines become, the lesser resources they require to propel vehicles thus resulting in lower carbon footprint and lower pollutants such as CO, HC and NOx. Alcohols are promising alternative fuels that can be used to fuel vehicles, on their own or in blends with conventional fuels such as gasoline and diesel. A lot of studies have been done on alcohols such as ethanol, butanol and methanol. Each of the mentioned solutions have their advantages and disadvantages. Butanol looks a promising solution due to its high miscibility properties making it possible to blend in higher percentages with other fuels such as gasoline for example. Compared to ethanol, butanol is also less corrosive, making it ideal for older vehicles as well without any modifications to the fuel systems. Many studies have shown that the high oxygen content of butanol can help improve the combustion process in spark ignition engines, may improve thermal efficiency resulting in lower emissions and better overall engine stability. An improved engine stability can be measured by a lower variability between engine cycles during engine functioning. The objective of this paper is to determine the impact of fueling a spark ignition engine with a blend of 15% vol. n-butanol – 85% vol. gasoline on pressure variability and combustion. No engine adjustments/optimizations are done at fueling with the blended fuel vs pure gasoline. The baseline was set at an engine speed of 2500 min-1 and load of 55% while fueling with pure gasoline. After setting the baseline, the same measurements are done at fueling with the blended fuel.

## 1. Introduction

The target for most industries and companies today is the reduction of their carbon footprint, meaning the reduction of pollutants, greenhouse gas emissions and more efficient use of energy. The automotive sector has been focusing on searching for more green and regenerable solutions to power vehicles. One possible solution to overcome the challenges mentioned above is the use of alcohols, some of which are already in use today (ethanol for example). A study conducted in 2020 by Cristian Sandu et al. [1] has shown improvements in combustion stability and variability by blending 10% vol. butanol with gasoline at fueling a conventional spark ignition engine. A 13% reduction in variation was observed for the maximum pressure. Regarding the production of corn ethanol, a study concluded that it failed to reach its greenhouse emissions target by 24%. The carbon footprint of ethanol production is no better than that of gasoline production [2]. A contradictory study was published in 2020 concluding that corn ethanol production produces less emissions compared to gasoline by around 39 to 43% [3]. Alcohols have shown an improvement of engine efficiency, lower fuel consumption and emissions, several studies have concluded [4][5][6]. A study by Wu. M et al [7] indicates that the use of butanol in the transportation sector may reduce conventional fuel usage between 39 and 56 percent and greenhouse gases by 48 percent. Another added benefit of butanol is a lower corrosion level and may aid component manufacturers by reducing the need of redesigning components for vehicles [8]. A study by Zswaja and Naber [9] concluded that the use of butanol may result in lower coefficient of variation and improved engine stability. Alasfour [10] blended 30% butanol with gasoline and observed a 7% decrease of power while fueling with the mixture. Another study found a small difference in emissions between gasoline and 10% butanol-gasoline mixture [11]. Alasfour [12] noticed a 10% increase in $NO_x$ emissions, by increasing the intake air temperature, while fueling with a 30% blend of butanol and gasoline. Engine miss-fire and knock were also observed. Higher $NO_x$ emissions can also be observed by increasing the spark timing [13]. Dernotte J. et al. [14] observed a reduction in variability for the indicated mean effective pressure by adding butanol to gasoline in small percentages. A variation higher than 10% for the mean indicated effective pressure may impact the drivability characteristics of the vehicle according to Heywood JB [15].

The objective of this paper is to study the impact of fueling a turbocharged 1.5L spark ignition engine (A15MF) with a blend of 15% vol. n-butanol and gasoline on the coefficients of variation of the maximum pressure, mean indicated effective pressure and mass fraction burned. The engine functions at a load of 55% and speed of 2500 $min^{-1}$ at different air-fuel ratios. Measurements are captured across 250 consecutive engine cycles.

## 2. Experimental engine setup

The experimental setup is presented in figure 1.

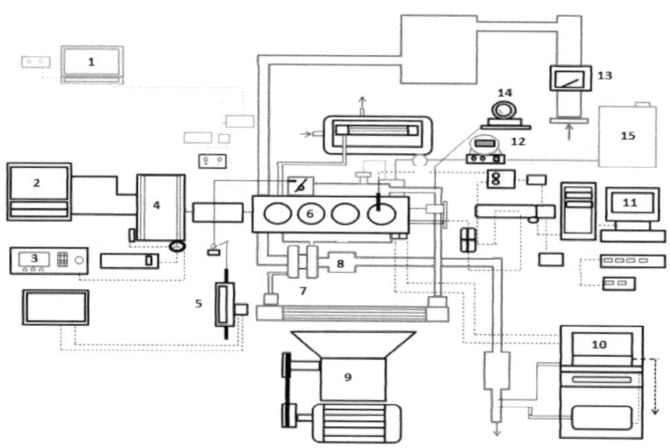


1 – ECU, 2 – dyno cooling, 3 – dyno control, 4 – Eddy dyno, 5 – throttle actuator, 6 – SI engine, 7 – turbocharger, 8 – catalytic converter, 9 – radiator cooling system, 10 – gas analyzer, 11 – PC, 12 – fuel mass flow transducer, 13 – air flow transducer, 14 – boost pressure transducer, 15 – fuel reservoir

**Figure 1.** Test bench schema [1]

## 3. Results

*3.1 Rich mixture λ = 0,89*

Below in figure 2, the maximum pressure dispersion is presented. The results of pure gasoline are presented with blue dots (G) and with the blended fuel with red (GB15). The coefficient of variability is calculated with the following formulas [16]:

$$COV_p = \frac{\sqrt{\frac{1}{N_c - 1} \cdot \sum_{k=1}^{N_c} (p_k - p_m)^2}}{p_{Max_m}} \quad (1)$$

$$p_m = \frac{\sum_{k=1}^{N} p_k}{N_c} \tag{2}$$

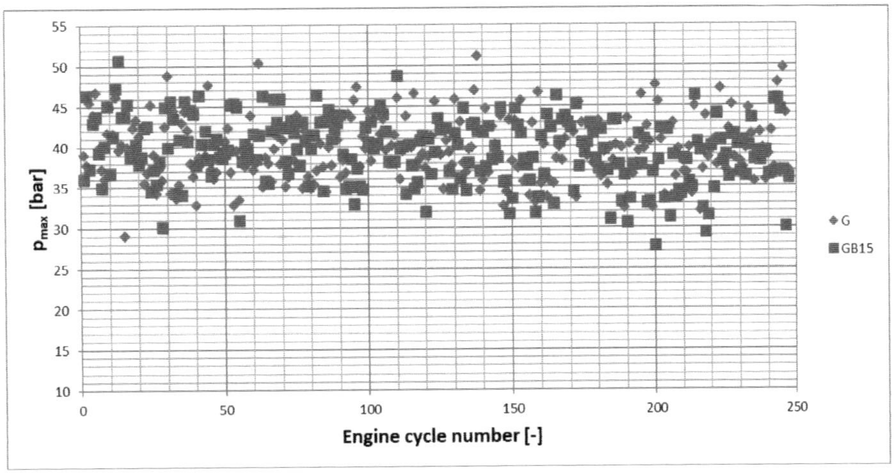


**Figure 2.** Max pressure dispersion at rich mixture

The addition of 15% vol. n-butanol to gasoline shows a slight degradation in maximum pressure variation. The $(COV)_{Pmax}$ for neat gasoline is 9,59% while de $(COV)_{Pmax}$ for the blended fuel is 10,23% (figure 3). The degradation in variation for maximum pressure is almost 6,5% while fueling with the blended mixture. This may be explained by the fact that the engine uses the same adjustments for gasoline. The average maximum pressure for gasoline is 39,75 bar VS 39,47 bar for gasoline-butanol, so no noticeable difference. $X_c$ is the coefficient of substitution of gasoline for butanol, and in this case it's 15%.

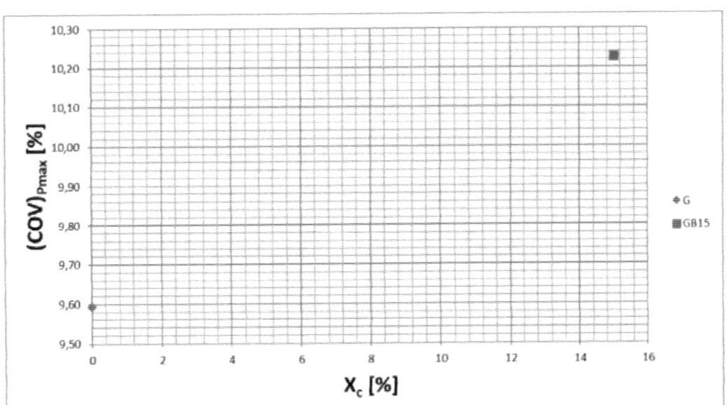


**Figure 3.** Max pressure coefficient of variation at rich mixtures

The indicated mean effective pressure (IMEP) is presented in figure 4. In both cases, the percentage of dispersion is low, with an advantage for gasoline. The $(COV)_{IMEP}$ for gasoline is 2,76% VS 3,19% for the blended fuel. An increase of almost 14,5% variation is observed for the blended fuel vs pure gasoline. The $(COV)_{IMEP}$ values are presented in figure 5. The mean values for IMEP are the same in both cases (19,45 bar for gasoline VS 19,43 bar for 15% vol. butanol-gasoline), indicating no loss of power at fueling with the blended fuel. The same is also true for maximum pressure. A better representation of pressure diagrams variation across 250 consecutive engine cycles is presented in figure 6.

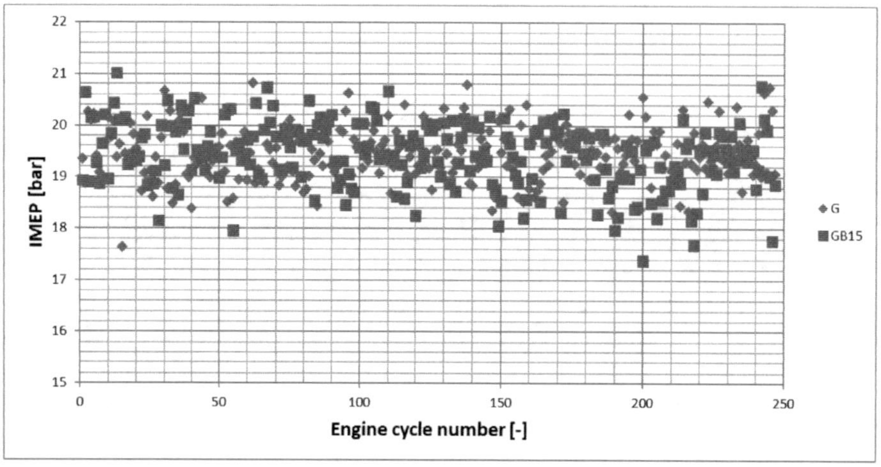


**Figure 4**. Indicated mean effective pressure dispersion at rich mixtures

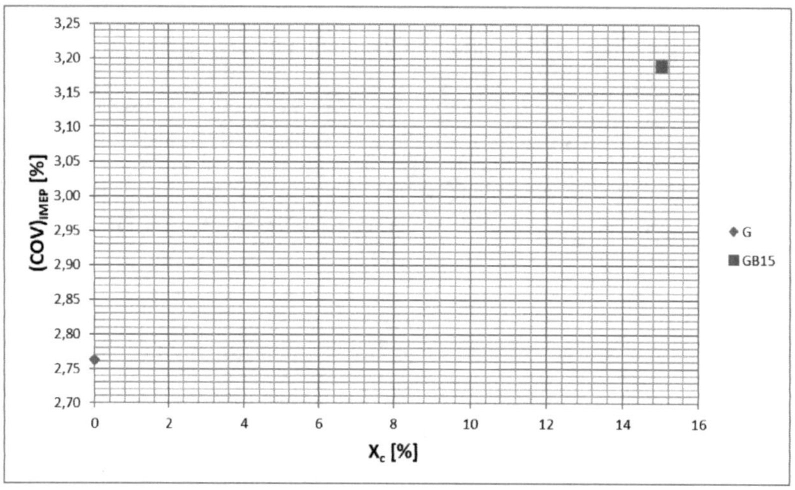


**Figure 5.** Indicated mean effective pressure coefficient of variation at rich mixtures

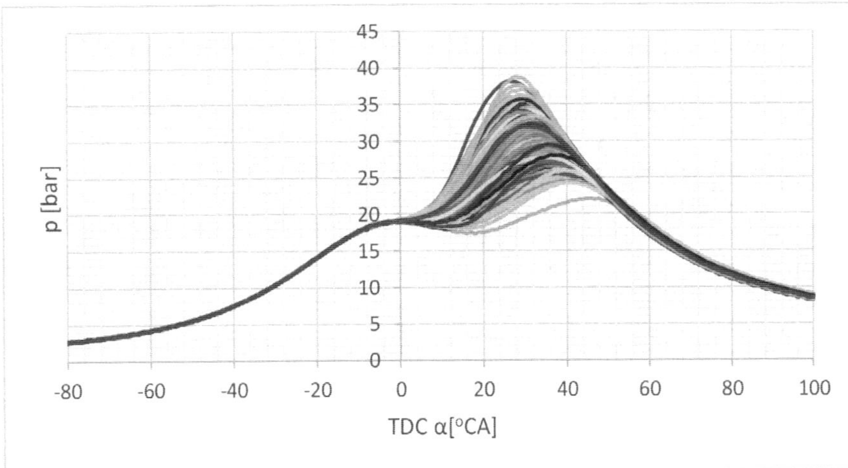


a.

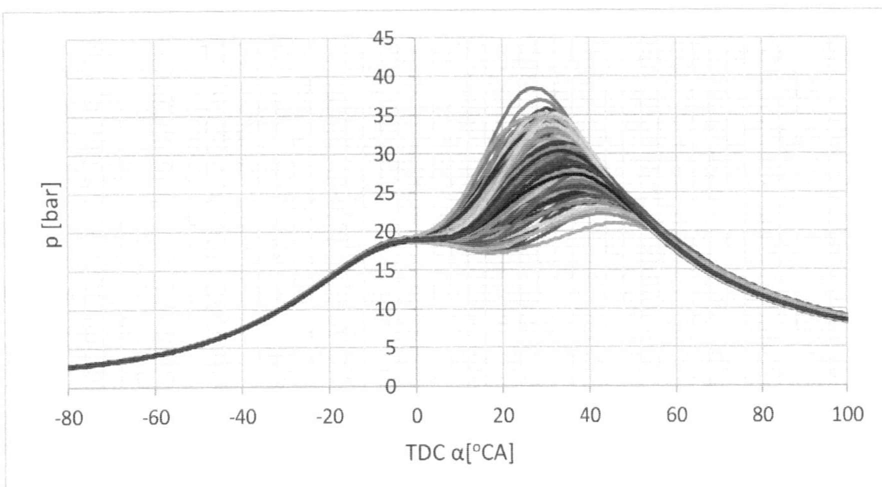


b.

**Figure 6.** Pressure diagrams across 250 engine cycles at rich mixtures for a) gasoline and b) blended fuel

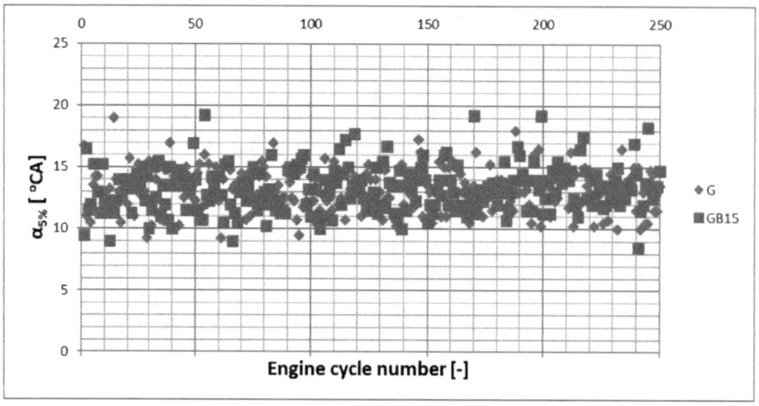


**Figure 7.** $\alpha_{5\%}$ at rich mixtures

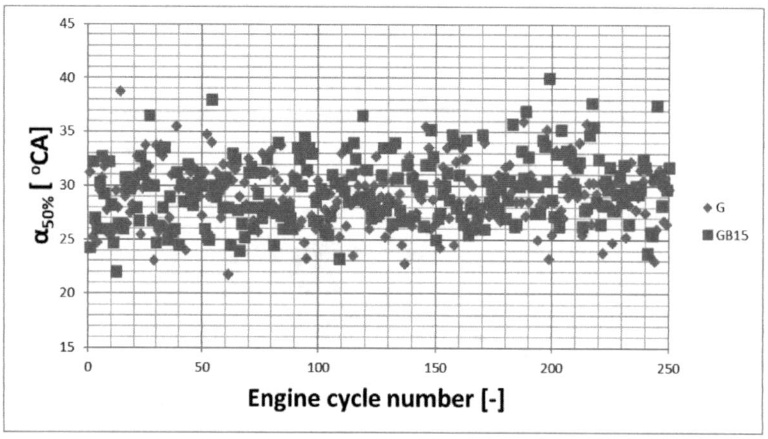


**Figure 8.** $\alpha_{50\%}$ at rich mixtures

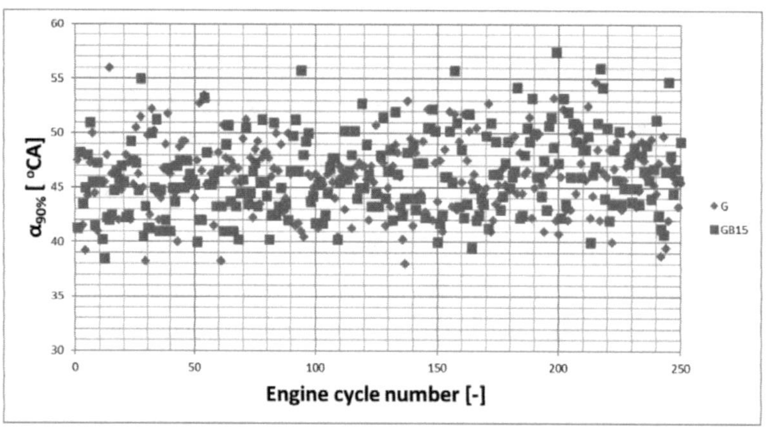


**Figure 9.** $\alpha_{90\%}$ at rich mixtures

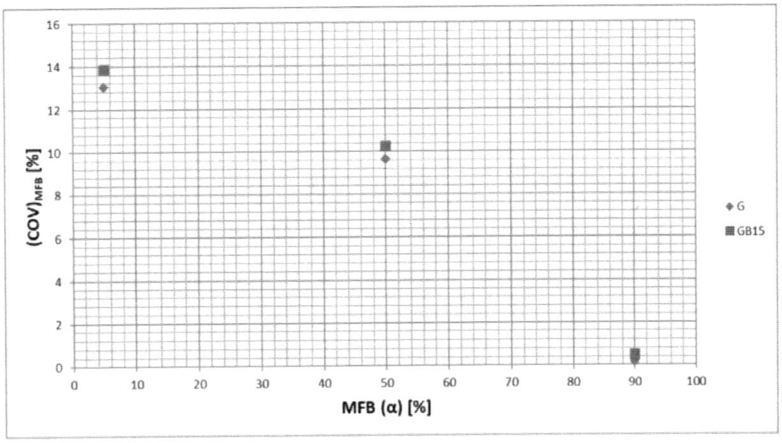


**Figure 10.** Coefficient of variation at different MFB at rich mixtures

The angles of the mass fraction burned for 5%, 50% and 90% heat released are presented in figures 7 to 9. For 5% heat released the coefficient of variation $(COV)_{\alpha 5\%}$ is 13,05% for pure gasoline and 13,87% for the blended fuel. An 6% increase in variation is observed at fueling with the blended mix at 5% heat released during combustion. The average angle achieved for gasoline was 13,188 °CA VS 13,262 °CA for the blended fuel (0,56% increase).

The average angles at which 50% heat is released are 29,21 °CA for gasoline and 29,54 °CA for the blended fuel. The $(COV)_{\alpha 50\%}$ is 9,31% for gasoline and 10,24% for the blended fuel, a 6% degradation.

The average angles for 90% heat released is similar in both cases, 46 °CA. Again, there is an 8,5% advantage for gasoline, with a $(COV)_{\alpha 90\%}$ of 0,45% vs 0,49% for the blended fuel. A comparison of variation coefficients is presented in figure 10. The additional oxygen content doesn't improve the engine stability at rich mixtures; however, no significant degradation is observed either. Regarding maximum pressures and mean indicated pressures, the blending of n-butanol presents no noticeable difference compared to gasoline, although a slight decrease was expected due to butanol's lower calorific value.

### *3.2 Lean mixture λ = 1,28*

The maximum pressure dispersion at lean mixtures is presented in figure 11 where a more unstable engine combustion is observed when compared to rich mixtures. The variation coefficient $(COV)_{Pmax}$ is higher in both cases for lean mixtures, with 15,34% for pure gasoline and 16,02 for the blended fuel. Gasoline is 4,33% more stable compare to the n-butanol-gasoline blend. Also, the average values for maximum pressure are 6% higher for gasoline (34,92 bar) when compared to the blended fuel (32,82 bar). This could be explained by the lower calorific values for n-butanol and by using the same engine adjustments (optimized for gasoline). The coefficient of variation values for the maximum pressure are presented in figure 12.

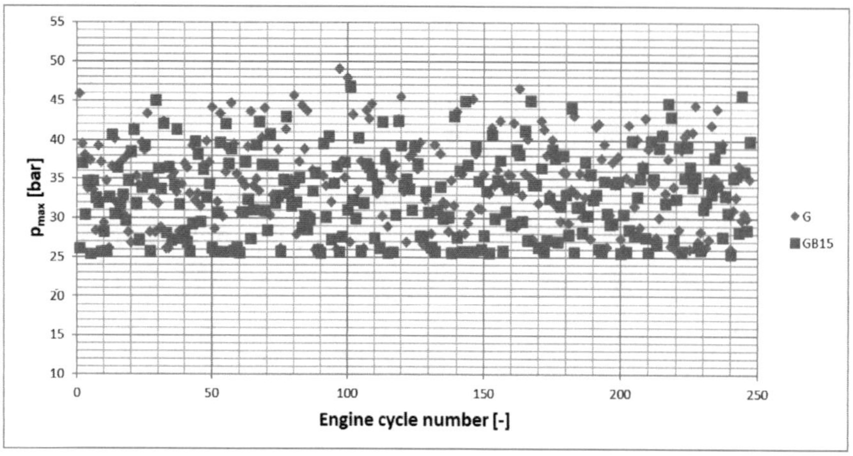


**Figure 11.** Max pressure dispersion at lean mixtures

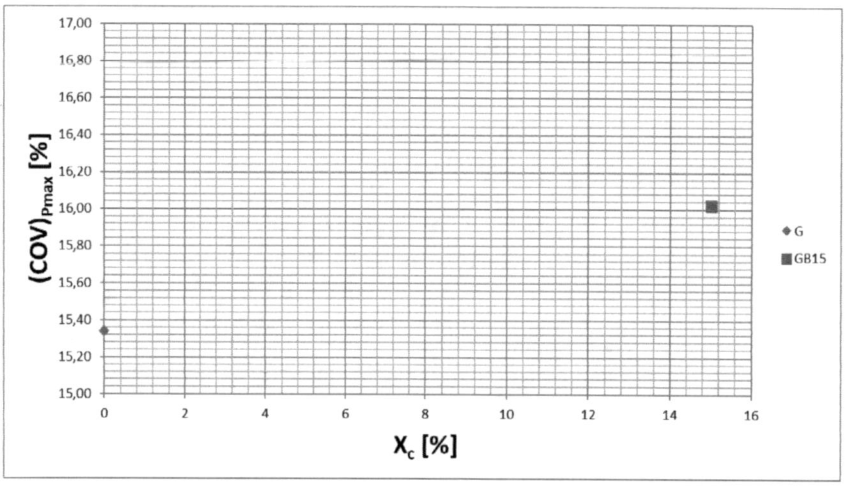


**Figure 12.** Max pressure coefficient of variation at lean mixtures

The mean indicated effective pressure values at lean mixtures are presented in figure 13. The dispersion is slightly higher at lean mixtures when compared to rich mixtures. There is no improvement in variation at fueling with the blended fuel when compared to gasoline. The $(COV)_{IMEP}$ for gasoline is 4,78% and 5,29% for n-butanol-gasoline blend, a 10% increase in variation. These values can be seen in figure 14. Regarding absolute median values, there is a drop of 3% for the blended fuel, from 18,76 bar to 18,2 bar.

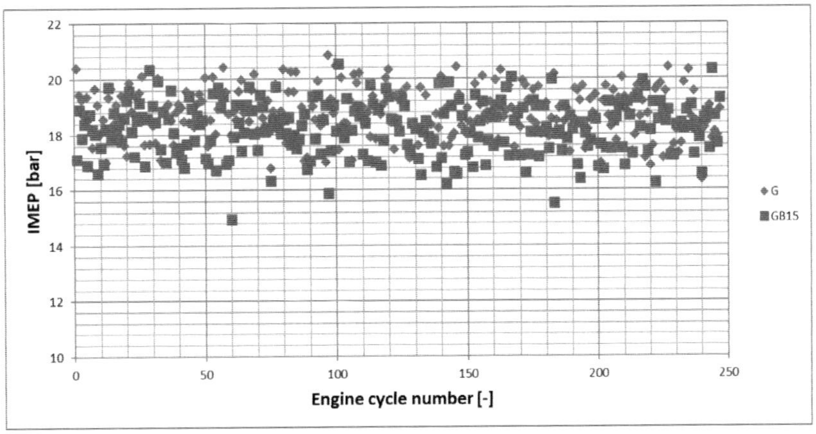


**Figure 13.** Indicated mean effective pressure dispersion at lean mixtures

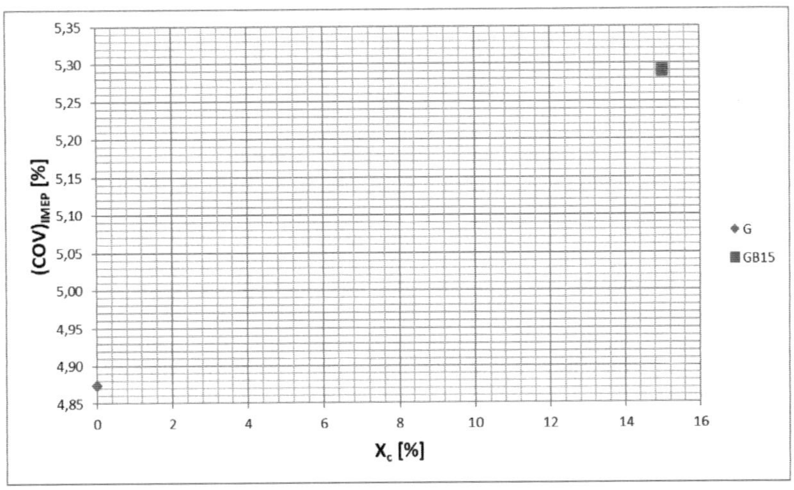


**Figure 14.** Indicated mean effective pressure coefficient of variation at lean mixtures

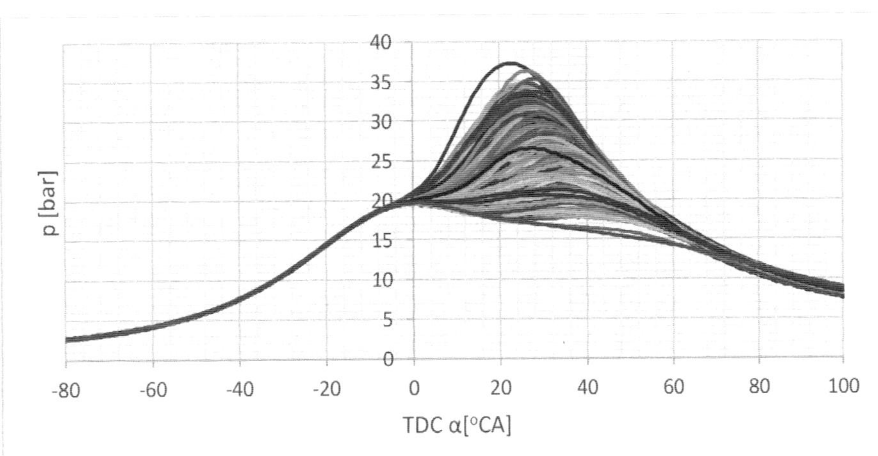


a.

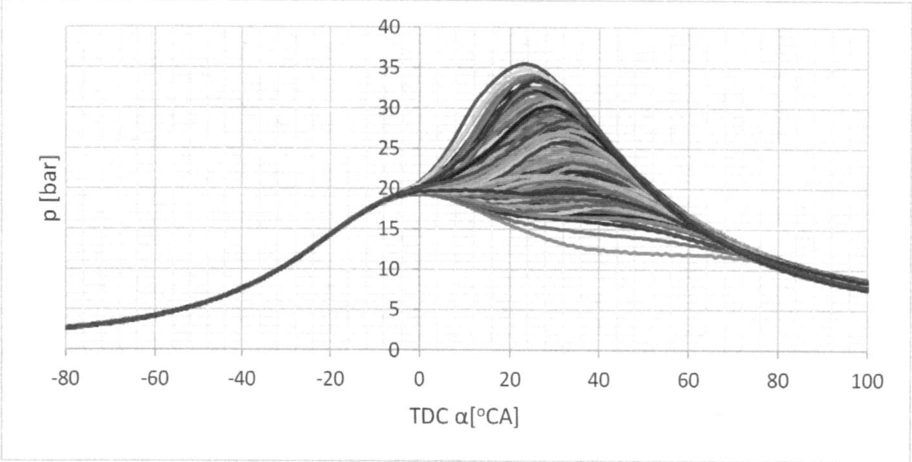


b.

**Figure 15.** Pressure diagrams across 250 engine cycles at lean mixtures for a) gasoline and b) blended fuel

The indicated diagrams for both fuels are presented in figures 14-15. As a comparison between the two diagrams, it is observed that there is lower dispersion in values at fueling with the blended mix indicating a small improvement in combustion but also lower maximum pressure values. The combustion of the blended mix moved further away in to the expansion. The same engine tunning/adjustments were used in both cases and should be taken into consideration.

The mass fraction burned for 5%, 50% and 90% at lean mixtures are presented in figures 16 to 18. The average angle at which 5% heat is released is 12,18 °CA for gasoline and 12,85 °CA for the blended mix. The variation coefficient for gasoline is 25,23% VS 25,89% for the blended mix. Gasoline has a 2,58% advantage in this case.

Regarding 50% heat released (figure 17), $(COV)_{\alpha 50\%}$ is 16,66% for pure gasoline and 17,83% for the blended fuel (6,8% difference). For 90% heat released (figure 18) the average angle is around 54 °CA in both cases with $(COV)_{\alpha 90\%}$ = 0,77% for gasoline, and 0,81% for n-butanol-gasoline blend. The overall representation of COV values for different heat released percentages is presented in figure 19.

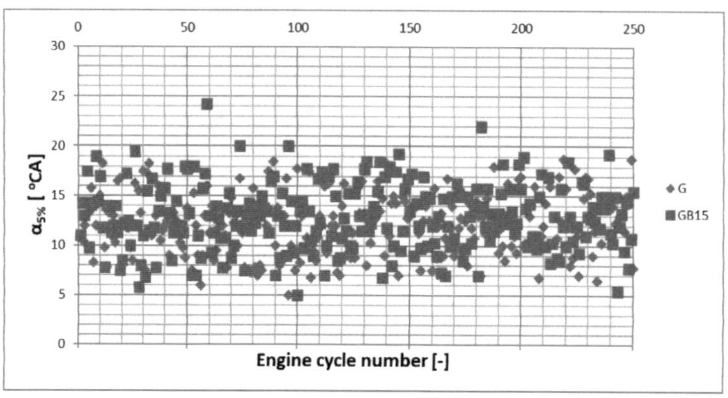


**Figure 16.** $\alpha_{5\%}$ at lean mixtures

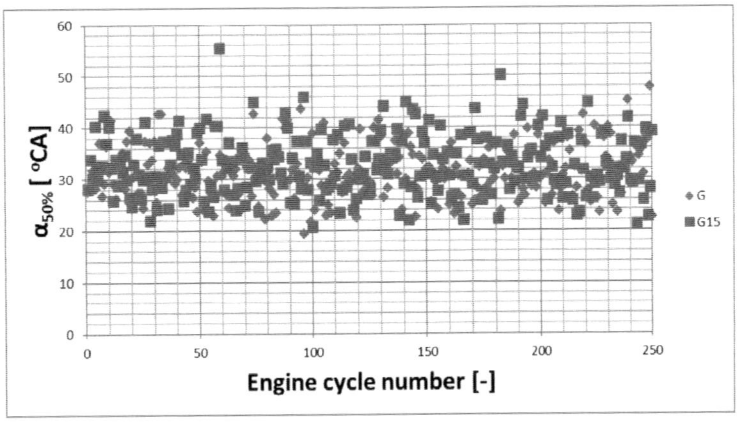


**Figure 17.** $\alpha_{50\%}$ at lean mixtures

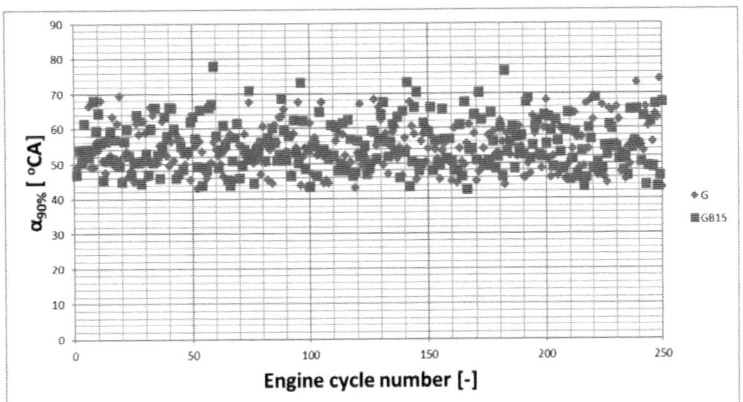


**Figure 18.** $\alpha_{90\%}$ at lean mixtures

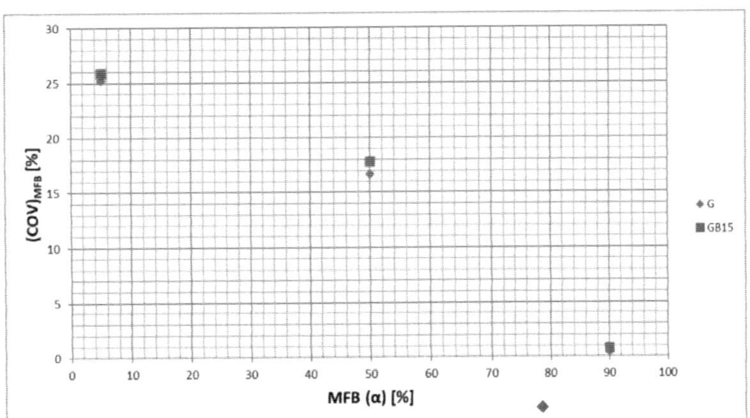


**Figure 19.** Coefficient of variation at different MFB at lean mixtures

**Table 1.** Results

| Parameter | | $(COV)_G$ [%] | $(COV)_{GB15}$ [%] | Diff. VS gasoline [%] |
|---|---|---|---|---|
| $P_{max}$ | λ=0,89 | 9,59 | 10,23 | +6,45 |
| | λ=1,28 | 15,34 | 16,02 | +4,33 |
| IMEP | λ=0,89 | 2,76 | 3,19 | +14,45 |
| | λ=1,28 | 4,87 | 5,29 | +8,26 |
| $MFB\alpha_{5\%}$ | λ=0,89 | 13,05 | 13,87 | +6,1 |
| | λ=1,28 | 25,23 | 25,89 | +2,58 |
| $MFB\alpha_{50\%}$ | λ=0,89 | 9,63 | 10,24 | +6,13 |
| | λ=1,28 | 16,66 | 17,83 | +6,78 |
| $MFB\alpha_{90\%}$ | λ=0,89 | 0,45 | 0,49 | +8,51 |
| | λ=1,28 | 0,77 | 0,81 | +5,06 |

## 4. Conclusions

At the investigated engine conditions, the addition of 15% -vol n-butanol to gasoline shows no improvement of the combustion process nor is the degradation too bad considering the gasoline engine adjustments were used. All results are presented in table 1. Overall, considering all parameter variations, the average degradation at fueling with n-butanol-gasoline blend is around 6,7%. One possible reason was already mentioned (engine adjustments) and another possible reason could be the lower calorific heat of butanol compared to gasoline. Looking at the pressure diagrams, the combustion isn't moved into the expansion at fueling with the blended fuel even though it burns slightly slower than pure gasoline.

Regarding engine pressures, at rich mixtures the blended mix shown no decrease in maximum pressure nor indicate mean effective pressure. The variation coefficients do increase in the case of the blended mix by 6,45% for max pressure and 14,45% for IMEP. At lean mixtures, there is a decrease of 6% maximum pressure and a 3% drop in IMEP for the blended fuel when compared to gasoline. Higher engine instability is noticed for both fuels at lean mixtures, with higher COV values. The highest degradation is noticed for IMEP with an 8,26% increase at the use of butanol-gasoline.

Regarding mass fraction burned, at all investigated heat release percentages, the average crank angle is slightly higher at the use of the blended fuel. At both rich and lean mixtures, the coefficients of variation are lower for gasoline, ranging from 2,58% to 8,51% better performance.

All things considered, with some engine optimization, butanol can be considered a good fuel alternative for spark ignition engines as there is little impact on engine pressures but with slightly higher variation. Adapting spark ignition engine to use a blend of butanol and gasoline require no modifications to the current fuel and ignition systems design as it's less corrosive than ethanol.

## References

[1] Cr. Sandu, C. Pana, N. Negurescu, Al. Cernat, D. Fuiorescu, R Georgescu, C. Nutu 2020 The study of cyclic variability at a n-butanol spark ignition engine fuelled, IOP Conference Series: Materials Science and Engineering 997 012132.

[2] Tyler L., Nathan P.H. 2022 Environmental outcomes of the US Renewable Fuel Standard Proceedings of the National Academy of Sciences 119(9): e2101084119.

[3] Lewandrowski J, Rosenfeld J, Pape D, Hendrickson T, Kirsten J & Moffroid K 2020 The greenhouse gas benefits of corn ethanol – assessing recent evidence Biofuels 11 pp 361-375.

[4] Gautam M., Martin II D.W. (2000) Combustion characteristics of higher-alcohol/gasoline blends, Proceedings of the IMechE Part A: J. Power and Energy 214, 5, 497-511.Yoon J, Ryu J and

Lim K B 2005 A novel reconfigurable ankle rehabilitation robot for various exercises Proceedings of the 2005 IEEE International Conference on Robotics and Automation ICRA 2005 pp 2290-2295

[5] Gautam M., Martin II D.W., Carder D. (2000) Emission characteristics of higher-alcohol/gasoline blends, Proceedings of the IMechE Part A: J. Power and Energy 214, 2, 165-182.

[6] Yacoub Y., Bata R., Gautam M. (1998) The performance and emission characteristics of C1-C5 alcohol-gasoline blends with matched oxygen content in a single-cylinder spark-ignition engine, Proceedings of the IMechE Part A: J. Power and Energy 212, 5, 363-379.

[7] Wu M, Wang M, Liu J, Huo H. Assessment of potential life-cycle energy and greenhouse gas emission effects 293 from using corn-based butanol as a transportation fuel. Biotechnol Progr 2008;24:1204-1214.

[8] E. Galloni, G. Fontana, S. Staccone, F. Scala, 2016 Performance analyses of a spark-ignition engine firing with gasoline-butanol blends at partial load operation Elsevier Energy Conversion and Management 110 pp 319-326.Doğan O. The influence of n-butanol/diesel fuel blends utilization on a small diesel engine performance and emissions. Fuel 2011;90:2467–2472.

[9] Szwaja S, Naber JD. 2010 Combustion of n-butanol in a spark-ignition IC engine Fuel 89 pp 73–82Liu H, Bi X, Huo M, Lee CF, Yao M. Soot Emissions of Various Oxygenated Biofuels in Conventional Diesel Combustion and Low-Temperature Combustion Conditions. Energ Fuel 2012;26:1900-1911.

[10] Alasfour FN. 1997 Butanol – a single-cylinder engine study: availability analysis Appl Therm Eng 17 pp 37–49Yuqiang L, Lei M, Karthik N, Timoty H L, Yilu L, Shengming L, Chia-Fon L, Combustion, performance and emissions characteristics of a spark-ignition engine fueled with isopropanol-n-butanol-ethanol and gasoline blends. Elsevier 2016; 184: 864-872.

[11] Wallner T, Miers SA, McConnell S. 2009 A comparison of ethanol and butanol as oxygenates using a direct-injection, spark-ignition engine J Eng Gas Turbines Power 032802:9He B-Q, Yuan J, Liu M-B, Zhao H. Combustion and emission characteristics of a n-butanol HCCI engine. Fuel 2014;115:758–64.

[12] Alasfour F.N. (1998) NOx emission from a spark-ignition engine using 30% iso-butanol – gasoline blend: Part 1: Preheating inlet air, Appl. Therm. Eng. 18, 5, 245-256.d

[13] Alasfour F.N. 1998 NOx emission from a spark-ignition engine using 30% iso-butanol – gasoline blend Part 2: Ignition timing, Appl. Therm. Eng. 18 pp 609-618.

[14] Dernotte J, Mounaim-Rousselle C, Halter F, Seers P. 2010 Evaluation of butanol-gasoline blends in a port fuel-injection, spark-ignition engine Oil Gas Sci Technol 65 pp 45–51.

[15] Heywood JB (1988) Internal combustion engine fundamentals. McGraw-Hill series in mechanical engineering. New York: McGraw-Hill.

[16] N. Negurescu, C. Pană, M.G. Popa, *Spark ignition engines. Processes, Matrix* ROM, Bucharest 2013

**Acknowledgments**

The authors address special thanks to AVL GmbH Graz Austria for providing the necessary equipment's. This work was supported by a grant of the Ministry of Research, Innovation and Digitization, CNCS - UEFISCDI, project number PCE 5/2022, PN-III-P4-PCE-2021-0777, within PNCDI III.

IOP Conf. Series: Materials Science and Engineering **1311** (2024) 012016 doi:10.1088/1757-899X/1311/1/012016

# Comparative dynamic analysis for a ventilated brake disk

**I Vieru[1], A A Boroiu[1], V Nicolae[1], D G Marinescu[1]**

[1] University of Piteşti, Târgul din Vale 1, 110040 Piteşti, Romania

E-mail: viorel.nicolae@upit.ro

**Abstract**. The paper presents a comprehensive dynamic analysis of the ventilated brake disc that fitted on the braking system of a car. There were taken into account several variants of material for the brake disc, determining the own modes of vibrations and also there was performed a frequency response calculation that allows us to estimate the dynamic behavior of the constructive variants, with direct influences on the noises and vibration (NVH).

The analysis of the frequency response consists in the validation of the structures in relation to a vibration level in order not to ensure that the frequency response is not too high during a forced stress. Thus was calculated the inertia that characterizes the dynamic rigidity of the analyzed models at certain points, according to the specified directions.

The finite element method was used as the calculation method, and CATIA V5 was used as the application, both for modeling and analysis. The CATIA application is an easy environment and already familiar to designers, so it can be used successfully in making preliminary calculations on estimating dynamic behavior in some models used in new projects.

## 1. Introduction

Traffic noise is one of the most widespread pollutants, in congested urban areas being considered by the WHO (World Health Organization) as having a major negative impact on health [1]. Recently, in the field of automobiles the noise level has been continuously reduced, an important element in this approach being the study of the noises produced by the components of the braking system and in particular by the brake disc.

Countless studies have been carried out in order to reduce the noise produced at the level of the brake disc, in an attempt to understand, predict and prevent the occurrence of this phenomenon.

In the beginning, the experimental methods represented a more credible way of approaching the study of these complex phenomena [2].

When it comes to designing new products or improving existing ones, the finite element method is by far the most widely used tool for numerical analysis, having the main advantage that it can simulate any changes to disc brake components much faster and easier than the experimental methods. In addition, the experimental methods are expensive due to the hardware costs and long lead times for design iterations [3].

Of course, there are remarkable developments in both directions, both in terms of experimental determinations and related to the realization of more and more complex calculation models, which together allow car manufacturers to manufacture models that have a noise level more diminished.

## 2. Theoretical considerations

The calculation of the frequency response allows the estimation of the dynamic behavior of the structures subjected to a forced excitation. Inertness characterizes the dynamic stiffness of a system at certain points, being defined by the acceleration induced to a point by a unitary excitation at this point. Inertness is also an important parameter as it allows the evaluation of the stiffness induced in the structure.

Knowing it is therefore important at the level of excitation points or of the fixing points of a functioning structure, where the main vibratory transfer takes place.

The inertness of a point is the ratio of the local acceleration to the excitation force ($\frac{\gamma_i}{F_i}$) for all directions (i ).

The inertness at this point is:

$$\frac{\gamma_i}{F_i} = \frac{\omega^2 .x}{k.x} = \frac{\omega^2}{k} \tag{1}$$

For practical reasons, the inertness graph is represented in decibels and is expressed with the relation:

$$R\gamma(dB) = 20.\log\left(\frac{\gamma}{F}\right) \tag{2}$$

where $\gamma$ represents the acceleration in [m.s-2] and F the excitation force in [N].

As a rule, to interpret the calculated (or measured) inertness values, it is compared with the equivalent stiffness curve (ISO-stiffness). This curve corresponds to an average level calculated for a structure, which has a variable stiffness depending on the frequency.

This curve is drawn with the following formula [8]:

$$R_k(dB) = 20.\log\left(\frac{\omega^2}{K}\right) \tag{3}$$

where:

Rk is the stiffness expressed in dB
$\omega = 2\pi\nu$ is the pulsation expressed in [Hz]
K represents the stiffness expressed in [N/m].\

## 3. The calculation model

The finite element method was used as the calculation method, and CATIA V5 was used as the application, both for modeling and analysis. Figure 1 shows the 3D model of the calculated brake disc.

a) The geometrical model. b) The discretized model.

**Figure 1** The brake disc model.

In a first phase, a dynamic modal analysis was carried out using the FEM module, considering three types of materials whose properties are presented in Table 1.

**Table 1.** Material properties used in the analysis

| Material / Properties | Cast iron GL09 variant 1 | Cast iron GL09 variant 2 | Cast iron GL11 |
|---|---|---|---|
| Young module [MPa] | 121000 | 121000 | 101000 |
| Material density, ρ [ Kg/m3] | 7200 | 7300 | 7100 |
| Poisson coefficient | 0,291 | 0,291 | 0,291 |

As it is known, the results of the modal analysis, realized by using the numerical analysis, are influenced by the shape of the part analyzed and by the material properties: the density of the material, the value of the poisson coefficient and especially the value of the Young's modulus [4], [5], [6],[7].

Next, the calculation of inertness was performed for cast iron GL09 (variant 2) and cast iron GL11.

The working mode in CATIA V5, for the calculation of inertia, includes the following stages [9]:

a. A frequency analysis of the structure is carried out;

b. It is inserted a static loading case, with the same constraints as in point a, applying a distributed force equal to 1 N at the point of interest and in the direction in which the analysis is to be carried out (Figure 2);

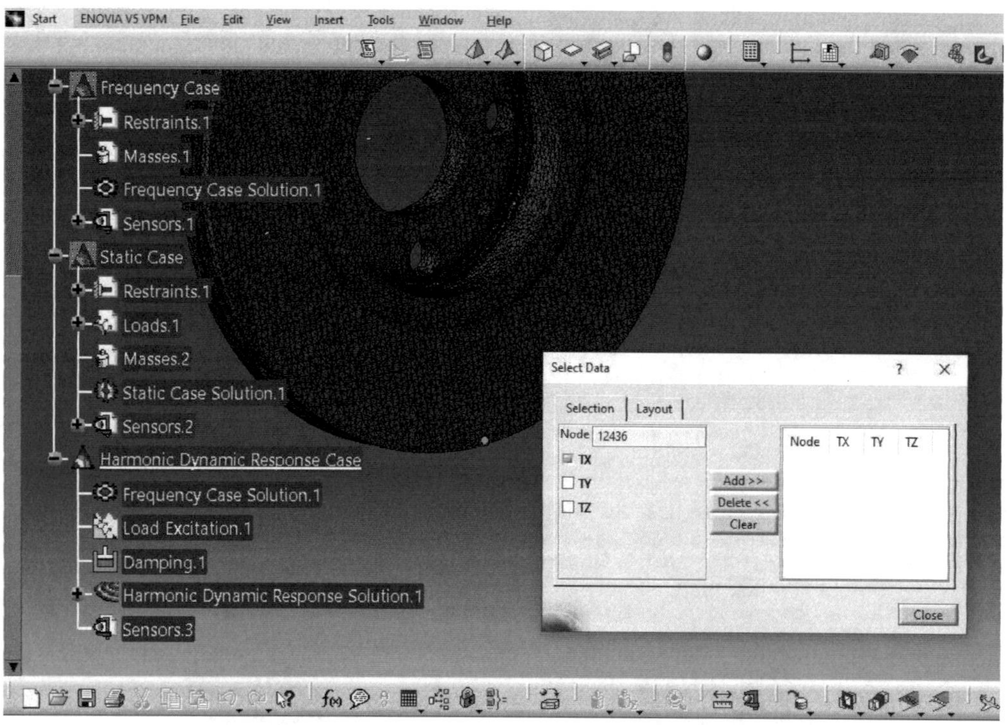


**Figure 2.** CATIA interface - the unit force application point.

c. Definition of the harmonics by indicating their parameters. The parameters are the frequency (Frequency Case Solution), the load (Load Excitation), the damping (Damping) and the harmonic excitation for the considered frequency range (Harmonic Dynamic Response Case).

IOP Conf. Series: Materials Science and Engineering **1311** (2024) 012016 doi:10.1088/1757-899X/1311/1/012016

Thus, each of the unitary forces defined in the "Static Load Case" will generate, in one direction of interest, a harmonic acoustic response.

## 4. The research results

The results obtained using FEM analysis for the first 10 vibration modes are presented in Table 2 and represent a first image of the dynamic behavior of the brake disc in the case of its realization from the materials with the properties shown in Table 1. It is obvious that the results obtained have relatively close values for the three analyzed cases.

**Table 2.** Frequencies of eigenmodes for the ventilated disk made of cast iron GL09 with different densities and cast iron GL11

| No. | Ventilated disk made of cast iron GL09 E=121000[MPa] ρ= 7200 [ Kg/m$^3$] | Ventilated disk made of cast iron GL09 E=121000[MPa] ρ= 7300 [ Kg/m$^3$] | Ventilated disk made of cast iron GL11 E=101000[MPa] ρ= 7100 [ Kg/m$^3$] |
|---|---|---|---|
| | **Frequency [Hz]** | | |
| 1 | 1304.8 | 1295.8 | 1314.0 |
| 2 | 1313.5 | 1304.5 | 1322.8 |
| 3 | 2148.9 | 2134.1 | 2164.0 |
| 4 | 2815.6 | 2796.3 | 2835.4 |
| 5 | 2820.3 | 2800.9 | 2840.0 |
| 6 | 2861.8 | 2842.1 | 2881.8 |
| 7 | 2862.6 | 2842.9 | 2882.7 |
| 8 | 3252.7 | 3230.4 | 3275.6 |
| 9 | 3312.6 | 3289.8 | 3335.8 |
| 10 | 4686.6 | 4654.4 | 4719.5 |

Next, the results obtained in the case of the inertness calculation for the brake disc made of cast iron GL09 (variant 2) and cast iron GL11 are presented.

Figures 3 and 4 show the acceleration amplitudes as a function of frequency for the brake disc made of the mentioned materials. The force was applied at the point indicated in Figure 2 and the direction of force application was considered perpendicular to the friction surface of the disc (parallel to its axis of rotation).

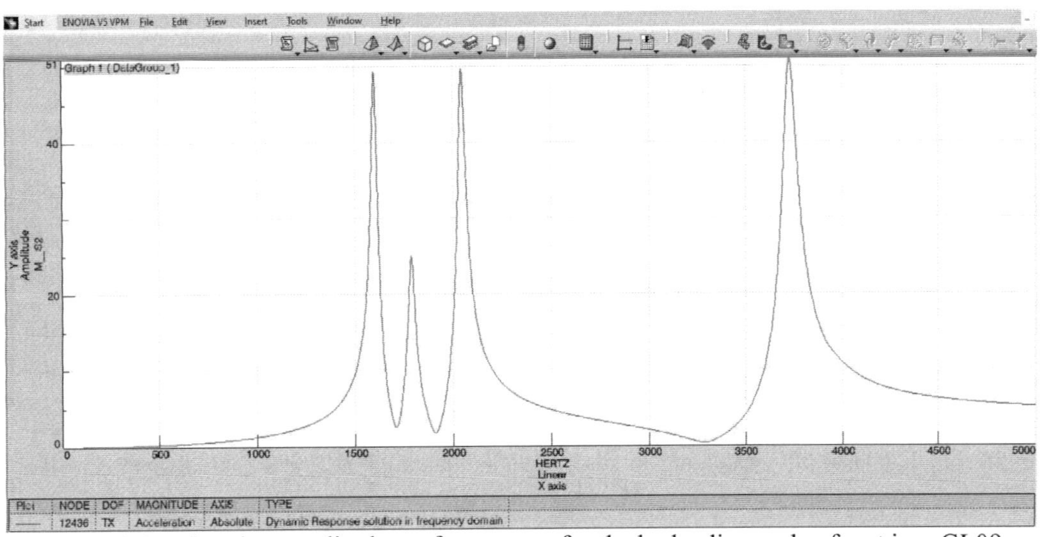

**Figure 3** Acceleration amplitude vs. frequency - for the brake disc made of cast iron GL09.

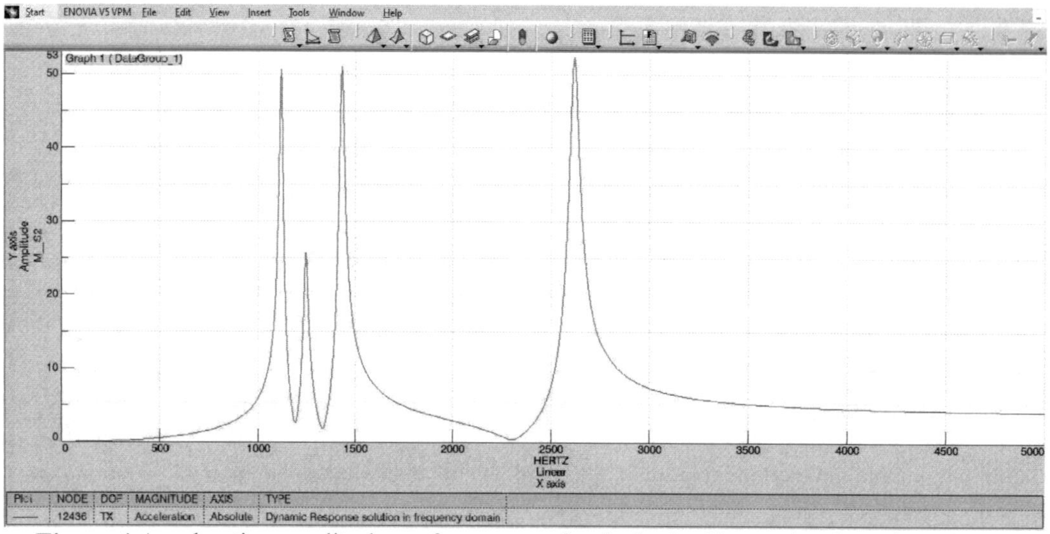


**Figure 4** Acceleration amplitude vs. frequency - for the brake disc made of cast iron GL11.

The CATIA application can perform several ways of representing these values, but for a comparative analysis the obtained results were imported into a Microsoft Excel file, where the two graphs represented in decibels were superimposed as well as the equivalent stiffness graph that was calculated with the relation (3). The resulting graph is presented in Figure 5.

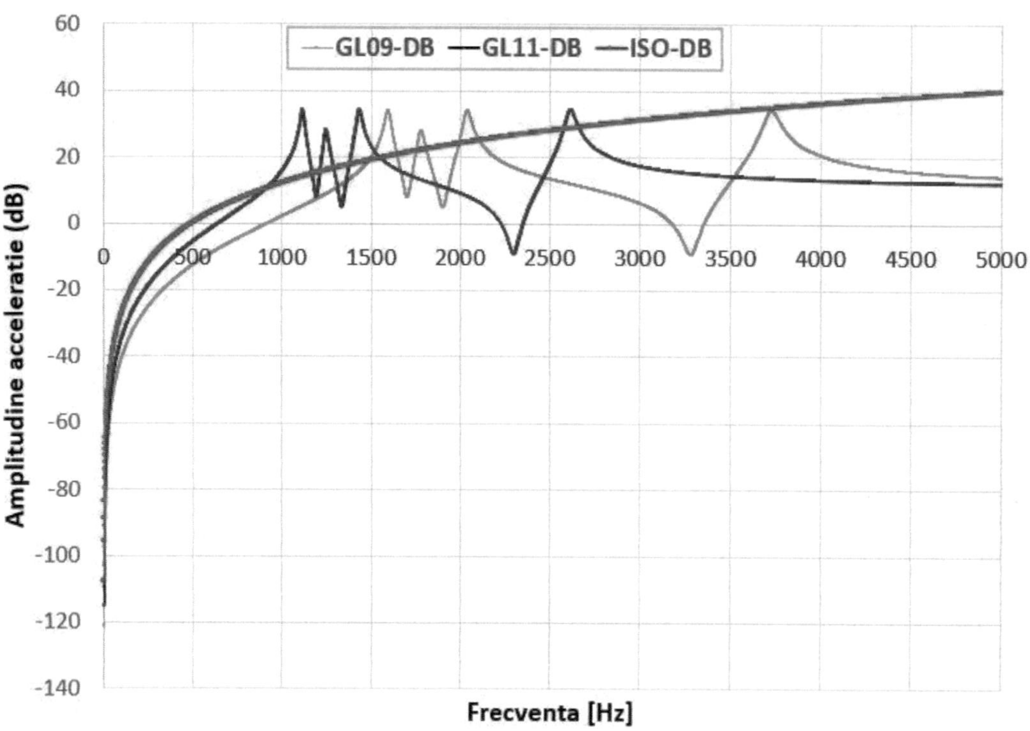


**Figure 5.** The amplitudes of accelerations (in dB) for GL09, GL11 and the equivalent stiffness (ISO) vs. frequency.

The ideal situation is when the acceleration graph does not exceed the equivalent stiffness curve. Thus, it is found that between 1000 and 1500 Hz the behavior of the brake disc made of GL11 is completely deficient, the acceleration graph far exceeding the equivalent stiffness curve. Between the frequencies of 1500 and 2500 Hz, a better dynamic behavior is observed for the brake disc made of GL11 cast iron. In order to modify the dynamic behavior, the designer has two possibilities: to change the geometric shape of the analyzed part or to use other materials with modified properties.

## 5. Conclusions

The analysis of the frequency response therefore consists in validating the structure in relation to a vibration level in order to ensure that the frequency response is not too high for a forced request. The CATIA application represents an easy and already familiar environment for designers, so it can be used successfully in making preliminary calculations regarding the estimation of dynamic behavior for newly designed parts. Thus, any identified deficiencies can be corrected from the beginning of the design, mainly through changes made to the constructive form of the parts.

The methodology developed in this work can be successfully used in the comparative analysis between two or more new projects, thus establishing which solution is better in terms of dynamic behavior, with direct implications on the production of noises and vibrations.

## References

[1] European Environment Agency, Noise in Europe 2014, Luxembourg: Publications Office of the European Union, 2014, ISBN 978-92-9213-505-8.

[2] Nishiwaki, M.R., Review of study on brake squeal. JSAE Review, 1990, Vol. 11, No. 4, 48¬54.

[3] Ouyang, H., Nack, W. V., Yuan, Y., & Chen, F, Numerical analysis of automotive disc brake squeal: a review. International Journal Vehicle Noise and Vibrations, 2005, Vol. 1, Nos. 3-4, 207-230.

[4] Yang, S., & Gibson, R. F, Brake vibration and noise: review, comments and proposals. International Journal of Materials and Product Technology, 1997, Vol. 12, Nos.4-6, 496-513.

[5] Shodh Ganga, A reservoir of Indian Theses; Available from: http://shodhganga.inflibnet.ac.in/bitstream/10603/11551/8/08_chapter%203.pdf.

[6] Abu-Bakar, A.R., Disc brake squeal - A prediction Methodology, AV Akademikerverlag, ISBN 978-3-639-45271-6, Stuttgart, Germany, 2012.

[7] Stojanovic, N, et. All, Experimental and numerical modal analysis of brake squeal noise, MVM 2018, Vol. 44, No. 4, DOI: 10.24874/mvm.2018.44.04.05.

[8] ***, *ANAMODE version 4.9, Manuel theorique by RENAULT*, Systemes informatiques Vehicule & Process Numeriques.

[9] ***, CATIA V5 Release 20, User's Documentation, Dassault Systemes,1999-2009.

IOP Conf. Series: Materials Science and Engineering **1311** (2024) 012017 doi:10.1088/1757-899X/1311/1/012017

# The influence of warning lights on the truck cabins on aerodynamic performance

**E M Vieru[1], M D Iozsa[1], L Ilea[1] and G Frățilă[1]**

[1]Automotive Engineering Dept., Faculty of Transports, POLITEHNICA University of Bucharest, ROMANIA

vieru_mihai_e@yahoo.com, daniel_iozsa@yahoo.com, ghe_fratila@yahoo.com

**Abstract.** In recent development of goods transportation by road, there has been a growing interest in truck aerodynamics as a consequence of increased desire to reduce fuel consumption and emissions. Commercial vehicles have undergone continuous modifications in terms of body design (cabin) and trailers or semi-trailers. There have been and still are areas that disrupt air flow and can be optimized for an improved aerodynamic performance. As part of this areas, the paper will study the influence of beacon lights and light ramps positioned on cabin roof for European type trucks (tractor units). These warning equipment's are imposed by transportation laws, reason for which eliminating them for achieving described purpose is not an option. In this context, the reduction of aerodynamic coefficient (Cd) can be pursued by analysing and optimizing the dimensions, shape and position for warning lights, in order to reduce the magnitude level of swirling effect. Different configurations of these elements will be tested and their impact in terms of air flow on the truck body will be analysed, with focus on roof area. The study will be done by simulating the interaction between airflow and truck in a virtual aerodynamic wind tunnel. The results will be analysed with the help of parameters like velocities, streamlines and pressures from a theoretical and experimental point of view to find the most viable option for each of the presented elements. The CFD (computational fluid dynamics) numerical simulations for chosen tests will be performed at a speed of 110 km/h considering that it is an often-encountered speed for trucks and also that aerodynamic phenomenology can be BETTER observed. By analysing the results obtained from the performed CFD simulations, the best positions and shapes for the beacon and ramp lights are desired to be identified. The paper conclusions will contain best configurations of warning lights and emissions improvement.

## 1. Introduction

Trucks are directly associated with the transport of goods, which is what they were initially designed for [1]. Whether we are talking about simple trucks with chassis, trailer, semi-trailers, flatbed, covered or others, dimensions of these vehicles in terms of length, height and width are usually high. This is imposed by the fact that a larger volume is required for the transport of goods. As a consequence, from an aerodynamic point of view these vehicles have the disadvantage of generating a fairly high forward resistance.

Over time, various areas were subjected to optimisation so to reduce the aerodynamic coefficient, and with success, an increased performance was obtained [2]. However, the potential for optimization remains increased in this domain, that is why the research continues to find the best configurations regarding the body of trucks. There are still areas that impact the aerodynamic performance in a negative way, and this study will focus on the area of the roof of the tractor unit, and highlight the influence of the light ramps and the beacon lights. Given that there are different regulations imposed by different states and countries, by law, certain elements such as beacon lights are mandatory, but they can be

IOP Conf. Series: Materials Science and Engineering **1311** (2024) 012017 doi:10.1088/1757-899X/1311/1/012017

optimized to obtain the best performance in terms of air flow [3]. However, there are also elements that are applied to the truck body to improve visual quality of the truck and also to increase driving efficiency. For example, the reflectors applied on the roof of the cabin truck generate more visibility during the night, thus increased driving safety. Most of the time, these elements are applied by owners without talking into account the constraints in the domain, such as the degradation of aerodynamic performance.

The images in Figure 1, 2 and 3 highlight, as an example, some of these configurations both on private trucks and trucks that belongs to various state institutions.

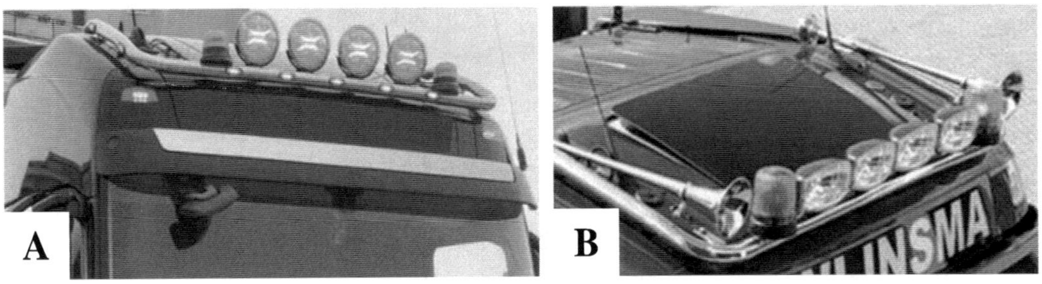


**Figure 1.** Models of light ramps on private trucks (A – round shape, B – rectangular shape). [4]

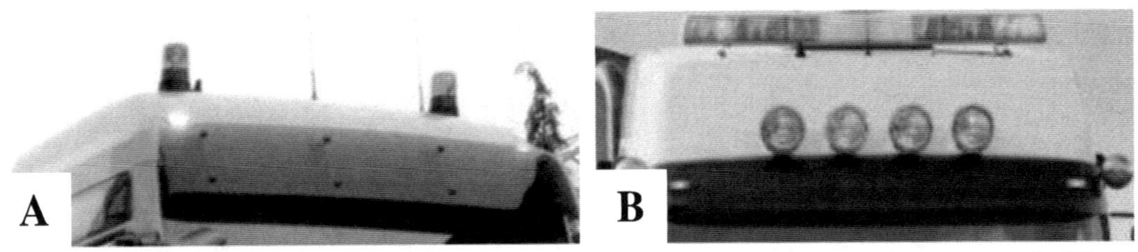


**Figure 2.** Models of beacon lights on private trucks (A – sides beacons, B – beacon ramp). [5], [6]

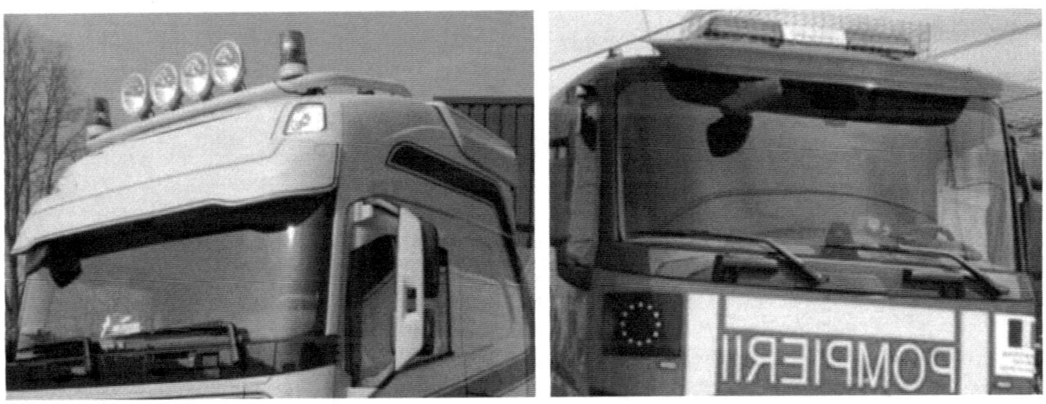

**Figure 3.** Models of warning lights on institution trucks. [4], [7]

As can be seen in the images above, there is a diversity in terms of warning lights, from the point of view of geometry but also of positioning.

For this study, an analysis will be made on several geometric models tested in the virtual environment and by observing the air flow, they impact will be highlighted and different improvements will be proposed.

## 2. Geometric model

In order to test the configurations to observe and analyse the different air flows and to be able to express an opinion related to the impact on the aerodynamic performance, 11 cases were tested for the roof elements. It was chosen to use a tractor unit without elements as the starting point, as indicated in Figure 4, so that the impact can be expressed much more easily by comparison. The tested accessories are described in figures 5, 6, 7, 8 and 9.

**Figure 4.** Configuration 1.

Figure 4 shows the first configuration of the tractor unit. In this case, as previously mentioned, there are no elements on the roof.

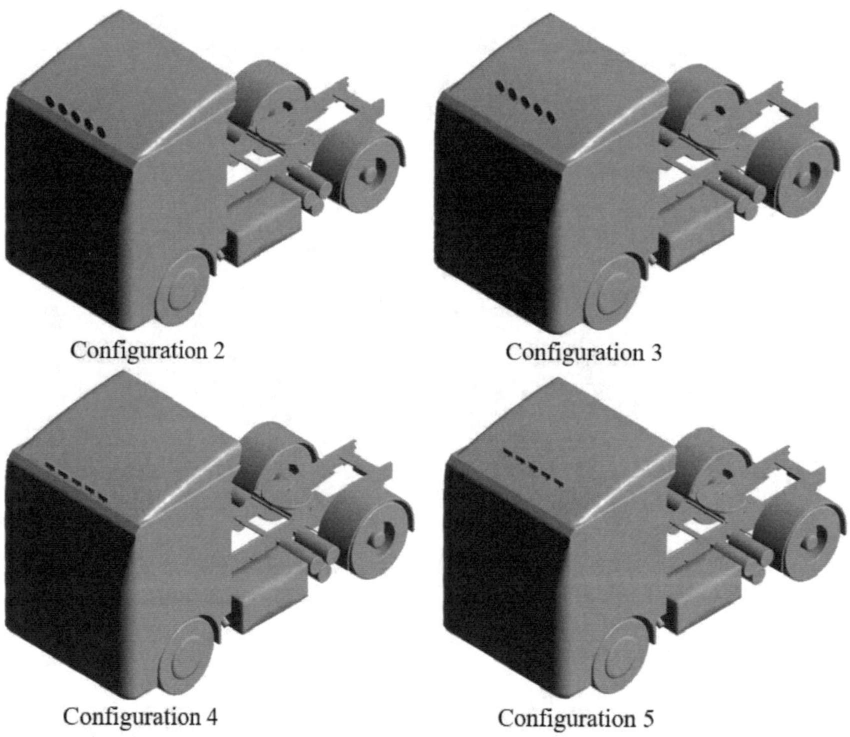


**Figure 5.** Configurations 2, 3, 4, 5.

Figure 5 shows the 4 configurations for the ramp light. Two different models were chosen for the rail ramp lights: round and rectangular. In addition, the impact was checked when the rail is positioned in -X compared to its initial, usual, position (in the front of the cabin). Next, are the details for the 4 configurations from Figure 5:

- Configuration 2: Ramp lights with round type elements positioned in the front of the cabin.
- Configuration 3: Ramp lights with round type elements positioned with 470 mm in -X from Configuration 1.
- Configuration 4: Ramp lights with rectangular type elements positioned in the front of the cabin.
- Configuration 5: Ramp lights with rectangular type elements positioned with 470 mm in -X from Configuration 3.

**Figure 6.** Configuration 6.

In order to be able to see the aerodynamic impact on a specific case, Figure 6 shows configuration when a ramp light with round type elements is positioned in front of the cabin and with beaconlights on sides.

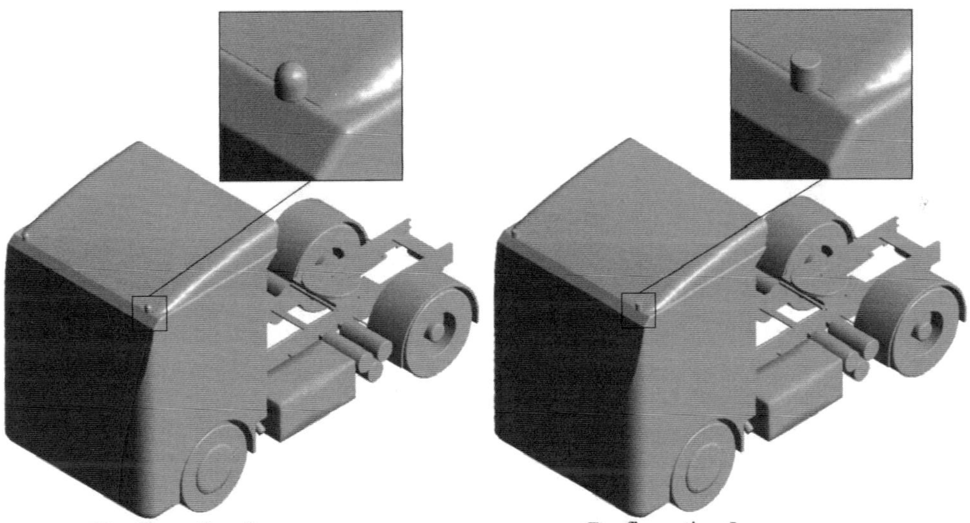


**Figure 7.** Configurations 7 and 8.

Taking into account the diversity of traffic lights on the market, two types of beacons were tested: cylindrical and rounded, in the upper part.

**Figure 8.** Configuration 9

According to Figure 3, there are many ramps type beacons traffic lights on the market and this led to the need to test this equipment. In figure 8 it is presented Configuration 9 when beacon ramp it is positioned in the front of the cabin.

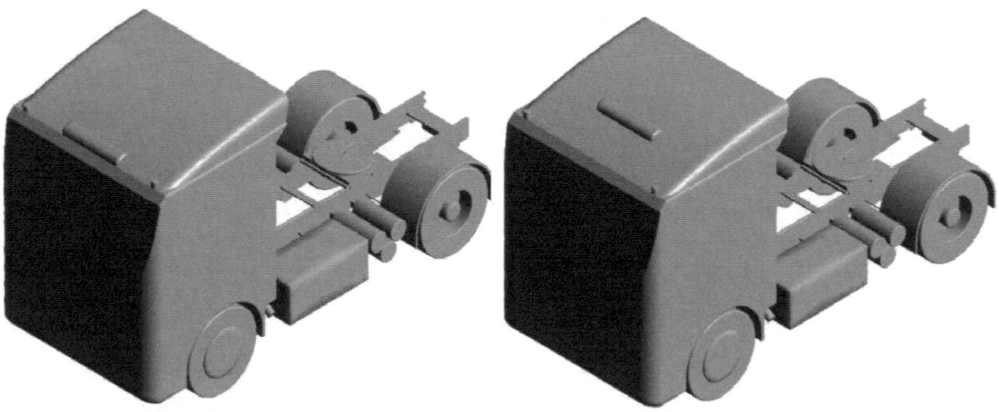

Configuration 10 Configuration 11

**Figure 9.** Configurations 10 and 11

The impact of the beacon ramp lights was verified and together with the cylindrical side beacons, these designs are shown in Figure 9. In addition, the beacon ramp was positioned both in the front part of the cabin (Configuration 10) and also translated by 470 mm in -X (Configuration 11).

## 3. Simulation

The CFD simulations will be done in the virtual environment and only the tractor unit will be used without the application of a semi-trailer as a result of the fact that the area of interest is on the cabin roof. In addition, being an aerodynamic study and because the geometric model is 1:1 scale, a large wind tunnel was simulated to avoid errors during processing as much as possible. The image below shows the sketch of the virtual wind tunnel and the positioning of the model in relation to it.

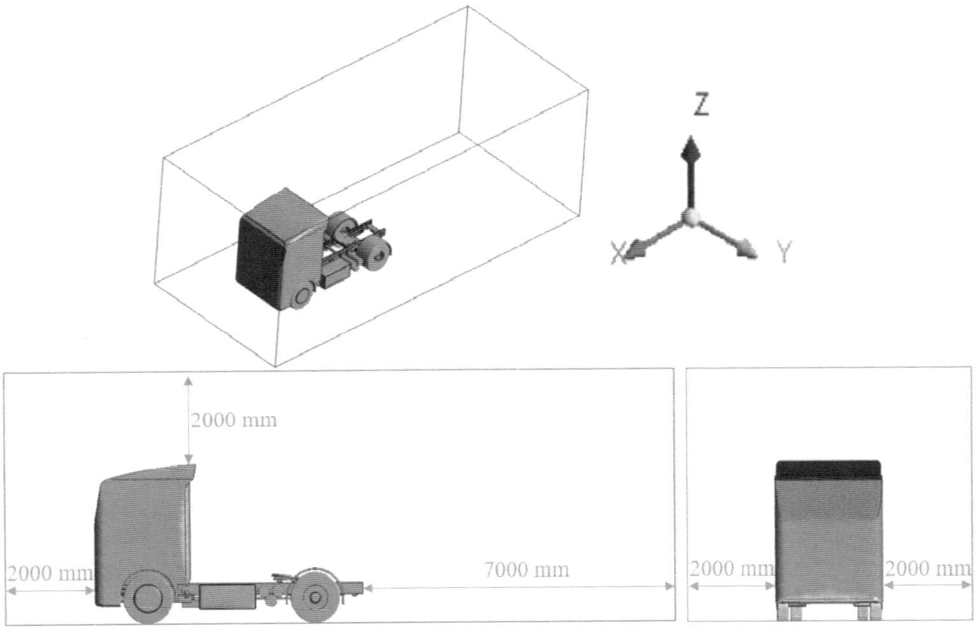


**Figure 10.** Virtual wind tunnel.

*3.1. Discretization domain*

Considering that the air flow will be analysed for elements of, relatively, small dimensions of the beacons in relation to those of the truck, it is necessary to observe the current lines very well. Therefore, a finer discretization of the models will help to generate these results as accurately as possible [8], [9].

Figure 11 shows, as an example, the meshing of the model without roof elements and for which the finest discretization method that the calculation program could provide was used. For this model the used elements were Tetrahedra and has 1942478 nodes and 10660528 elements.

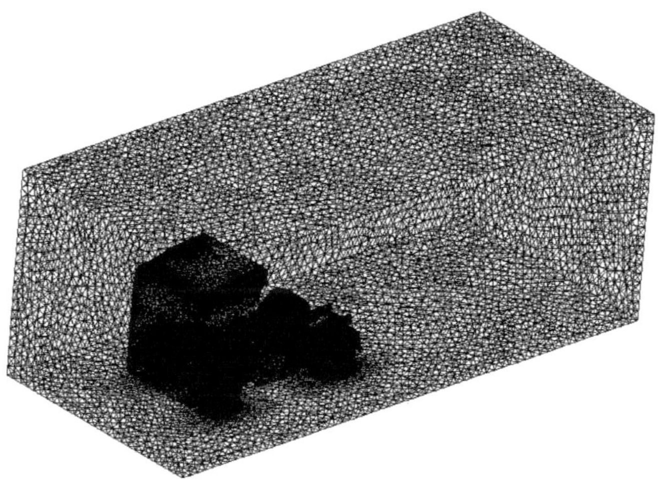

**Figure 11.** Discretization of the geometric model and wind tunnel.

*3.2. Set-up*

In order to prepare the launch of simulations in the virtual environment, it was chosen that the walls to have condition of free slip so, at the interference between the moving fluid and the stationary wall, the normal component of the fluid velocity field it is equal to zero [10]. The air speed at which the geometric models were tested is 110 km/h, a value at which the air flow can be observed quite well and precisely. The pressure of 1 atm was chosen as the exit parameter for the tunnel.

Based on these preparations, the obtained results and their analysis will be presented below.

## 4. Results

For the interpretation of the results, it was chosen to use streamlines, velocity vectors, and pressures. Also, for a good visibility of the analysed parameters such as streamlines and velocity vectors, it was chosen to exemplify them on a ZX plane at Y0 coordinate where in almost all cases there are elements that disturb the air flow.

So, as shown in Figure 12, the air flow is represented in the form of velocity vectors and streamlines. Based on the vector representation, the direction of the air flow after its interaction with a component can be easily observed. In this first simulation, the base case is presented in which the tractor unit has no elements on the roof. So, it can be seen the fluid flow of air on upper part of the cabin. This quasi-laminar flow can be considered ideal given the fact that the air does not meet any disturbing element in its path on the roof. Therefore, for the following cases, this configuration can be used as a basic model for comparison.

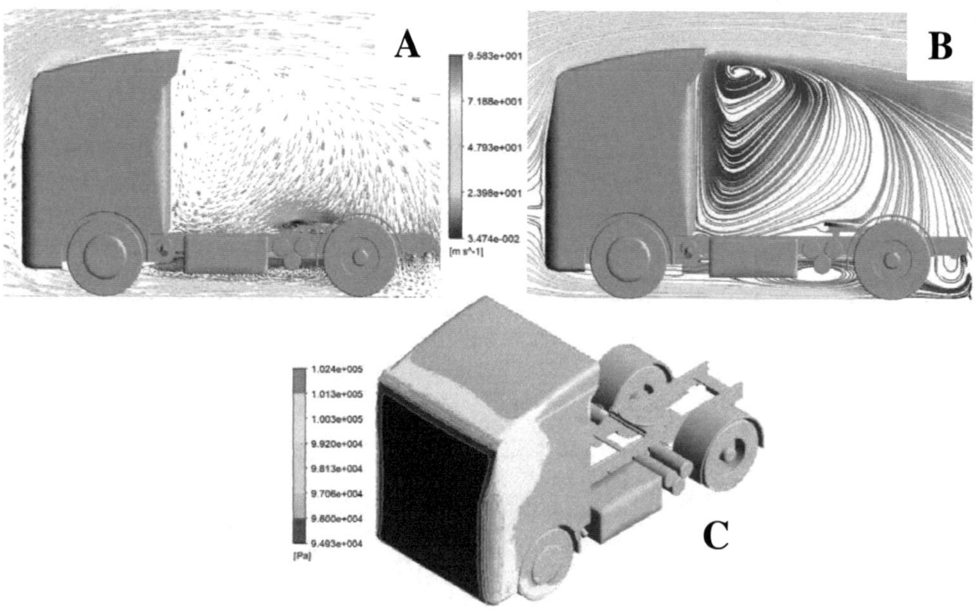


**Figure 12**. Velocity vectors (A), streamlines (B) and pressure (C) for Configuration 1.

Figure 13, shows the air flow for configurations 2 and 3. The air flow coming from the front of the truck comes into contact with the ramp of lights, which degrades the normal air flow and generates vortices. They contribute to the global degradation of the aerodynamic performance by increasing the resistance to the forward movement acting as a brake. The difference between the two cases is given by the fact that the light ramp positioned in '– X' with 470 mm generates smaller turbulences, so we can conclude that it is better placed on the cabin. In the case of Configuration 2, an area with large turbulences can be observed behind the light reflectors that generate recirculation, so it can be considered air with quasi-null velocity. In this situation, the air flow must bypass this air gap, which generates an unbalanced and turbulent flow.

IOP Conf. Series: Materials Science and Engineering **1311** (2024) 012017 doi:10.1088/1757-899X/1311/1/012017

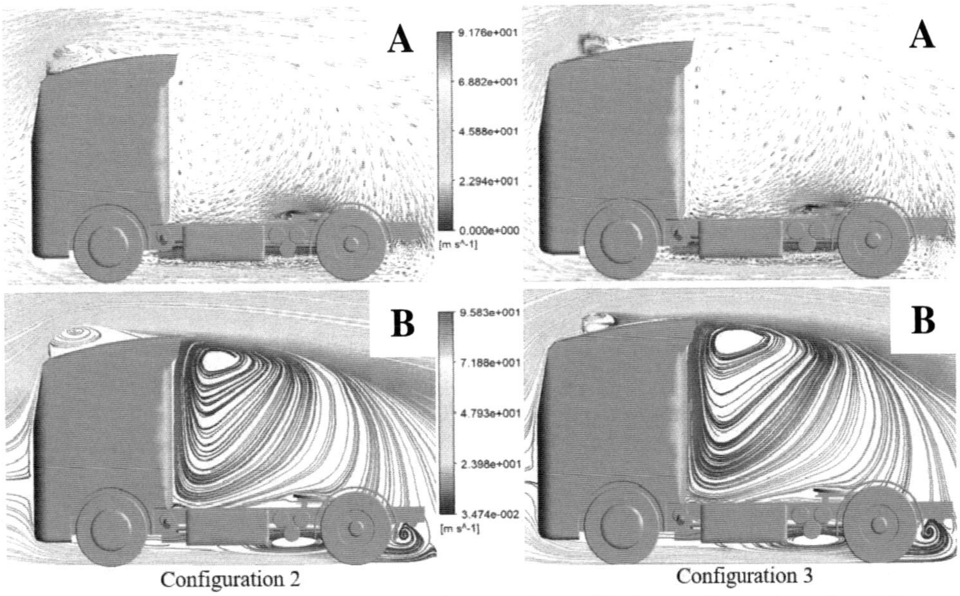


**Figure 13.** Velocity vectors (A) and streamlines (B) for configurations 2 and 3

For configurations 4 and 5, as shown in Figure 14, the representation of the air flow is also done in the form of vectors and streamlines. And in this case, as in the previous comparison, the pack of lights positioned towards the middle of the cabin has a better flow with smaller recirculation. At the same time, taking into account the geometric shape of the reflectors of Configuration 5 it has, according to the air flow, the best aerodynamic performance compared to configurations 2, 3 and 4. Simulation number 4 keeps the same behavioural pattern as configuration 2 in terms of air flow.

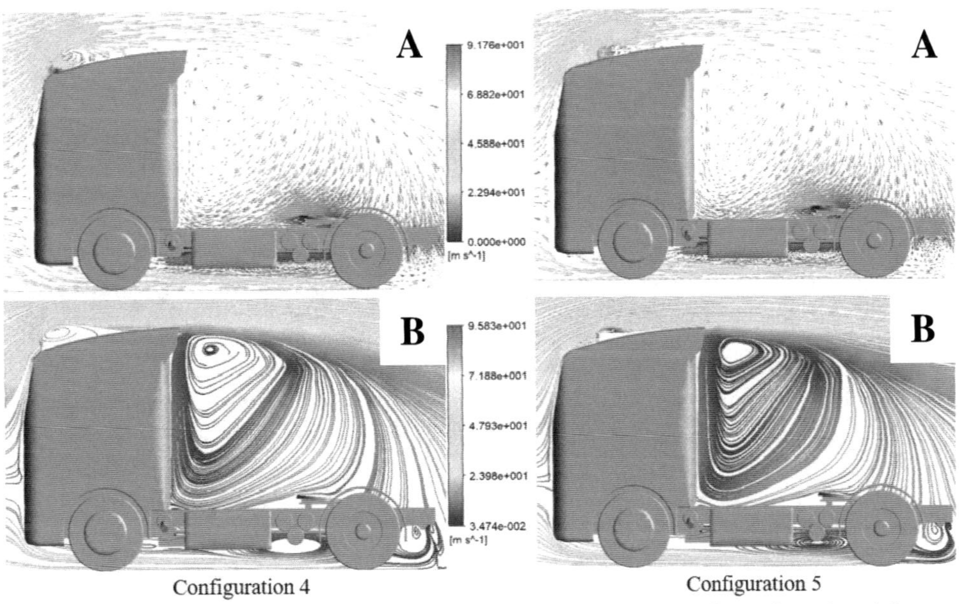


**Figure 14**. Velocity vectors (A) and streamlines (B) for configurations 4 and 5

Configuration 6, presented in Figure 15, adds for the frontally positioned light ramp, a set of warning beacon lights both on the left and on the right side. From the point of view of the air flow at Y0 section, it can be said and observed that there are no changes compared to Configuration 2.

IOP Conf. Series: Materials Science and Engineering **1311** (2024) 012017 doi:10.1088/1757-899X/1311/1/012017

However, for the streamlines in the ZX plane in the middle of the beacon light, it can be observed that, although the air has a strong first contact with its cylindrical surface, the streamlines reattached very quickly to the roof of the cabin. This will generate kind of a flatter profile on the sides, then in the case without beacon traffic lights. Therefore, it can be approximated that a slightly improved flow is presented.

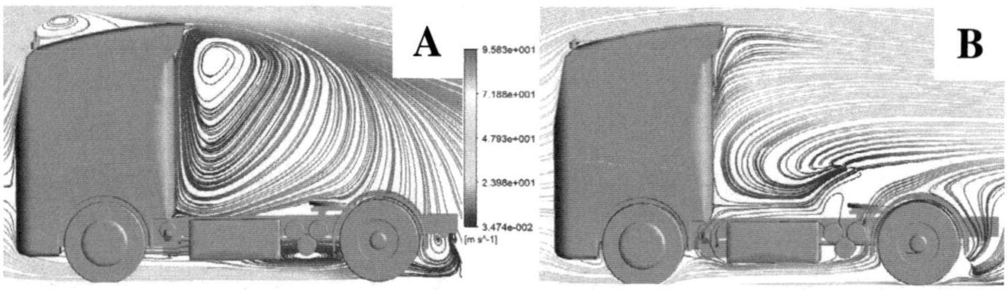


Configuration 6

**Figure 15.** Streamlines for Configuration 6. (A - Section Y0, B - Section through beacon light)

The comparison between configurations 7 and 8 in Figure 16, has the main purpose of presenting the difference between the beacon traffic lights with two different geometric shapes. It is observed that the behaviour of the current lines is the same: the interaction of the air with the beacon light surface, the apparition of local vortices, the detachment from the surface of the beacon and the rapid reattachment on the roof. In the case of Configuration 8, when the element has a cylindrical shape in the upper part, a more aggressive detachment can be distinguished. However, the upper base of the cylinder generates the approximately flat profile that was mentioned previously. So, it can be considered that, although the spherical beacon light has a better behaviour from the overall air flow point of view, in the present case the cylindrical beacon generates the best aerodynamic performance.

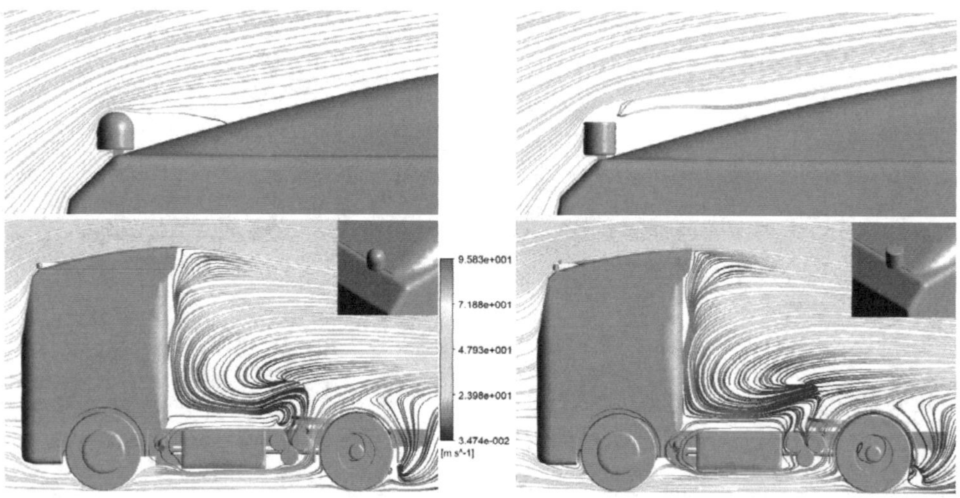


Configuration 7 Configuration 8

**Figure 16.** Streamlines for Configuration 7 and Configuration 8.

The results for Configuration 9 are presented in Figure 17 and describes the impact for the beacon light ramp. In both images, with the velocity vectors and in the one with streamlines, the relatively fluid flow on this element can be observed. Given the nature of the ramp's geometry, the well-defined profile in the Y0 section can be observed, which helps to reattach the air on the cabin roof and generate relatively small turbulence.

IOP Conf. Series: Materials Science and Engineering **1311** (2024) 012017 doi:10.1088/1757-899X/1311/1/012017

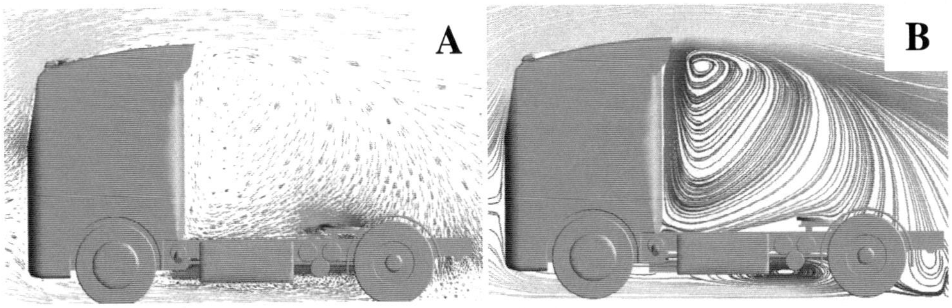


Configuration 9

**Figure 17.** Velocity vectors (A) and streamlines (B) for Configuration 9.

For configuration 10 and 11, the simulations described in Figure 18, the influence of the traffic beacon light ramp and the side beacons is presented. From the point of view of the air flow at Y0, the ramp, by nature of the geometry, generates a good flow that can be observed in the streamlines of Configuration 10. The reattachment of the air on the roof is done quickly and it can be seen that compared to configuration 11 there are fewer areas with zero velocity. At the same time, the behaviour of the side beacons is similar to one from the previous configurations (6 and 8).

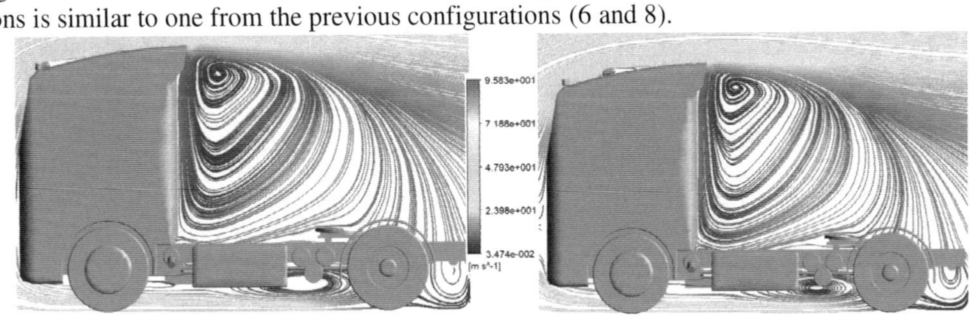


Configuration 10 Configuration 11

**Figure 18.** Streamlines for Configuration 10 and Configuration 11.

To have a numerical basis from point of view of the results obtained, refer to the Table 1 which shows the forces on X axis for each configuration. Comparing these force values, it can be seen that the minimum is obtained by Configuration 8 and the maximum is obtained by Configuration 2. According to the air flow of Configuration 1, this should have had the best performance. However, for the current configuration of the truck's geometry, it can be seen that the cylindrical side lights bring an improvement in aerodynamic performance.

**Table 1.** Values of force on X axis

| | F [N] |
|---|---|
| Configuration 1 | 4563 |
| Configuration 2 | 4880 |
| Configuration 3 | 4735 |
| Configuration 4 | 4766 |
| Configuration 5 | 4702 |
| Configuration 6 | 4584 |
| Configuration 7 | 4311 |
| Configuration 8 | 4274 |
| Configuration 9 | 4296 |

| | |
|---|---|
| Configuration 10 | 4316 |
| Configuration 11 | 4325 |

For a better overall visibility of the results, all the forces values from Table 1 are arranged in the graph presented in Figure 19. Except for Configuration 1, it can be seen that configurations 2, 3, 4, and 5 have the highest force values. It should be noted that for the previously mentioned configurations, only the reflectors were placed on the roof. Therefore, based on the numerical values and the interpretation of the air flows, it can be said that the light ramps degrade the aerodynamic performance of the trucks. Consequently, their integration in the roof deflector can be considered as a viable option. The configurations from 7 to 11 are represented by the models with beacon lights, which by the nature of the construction allow a better flow of air therefore a more aerodynamic performance than in the previous cases.

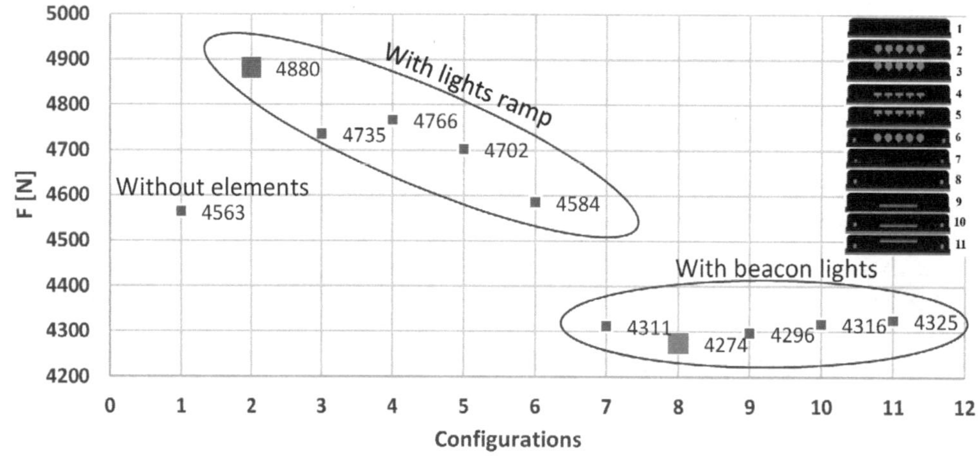


**Figure 19.** The values of the forces on the X axis depending on the nature of configurations.

Consulting the arrangement of the values from the graph above, it can be seen that Configuration 2 (light ramp, round type elements) has the highest force value on the X axis, a value of 4880 N. This describes the fact that this model has degraded aerodynamic performance. In addition, Configuration 3, when the ramp is positioned towards the middle of the cabin, has a better performance, the numerical difference between 2 and 3, in terms of force on X axis is 145 N. This shows that the positioning of the lights rail must be correlated with the geometry of the roof deflector of the cabin, which together, impacts the global air flow. For the ramp of lights with rectangular elements (Configuration 4 and 5), also in this case its positioning towards the middle of the cabin brings a greater gain compared to the one positioned at the front. Among these 4 configurations (2, 3, 4 and 5), the best force value is 4702 N and is representative for the case where the light rail has a rectangular profile and is positioned towards the middle of the cabin.

From the point of view of the beacon lights, for configurations 7, 8, 9, 10 and 11, it can be observed that the lowest and best value of force on X axis is 4274 N and is represented by Configuration 8. For this configuration only the side beacons of cylindrical type are presented. And in these cases, the necessity of corelating beacon lights with the roof deflector is shown. Between configurations 7, 9, 10 and 11 there are not very big differences, but a notable thing is the impact of the beacon rail between Configuration 10 (frontal positioned) and 11 (positioned to the middle of the cabin). When the ramp is translated on -X axis by 470 mm, together with the side beacons, degrades the aerodynamic performance, this by comparing it with the case in which it is located frontally with side beacons.

As a global view, it can be observed that the highest value of the force on the X axis is 4880 N for Configuration 2 (light ramp with round elements) and the lowest value is 4274 N for Configuration 8 (lateral beacons of cylindrical type), the difference between them being of 606 N.

Although Configuration 1 should represent the ideal case of air flow but, it is between the maximum and minimum value of the tested models, but it can be deduced that for the tractor unit modelled in this study the ramp of lights degrades the aerodynamic performance, and from the numerically point of view, it can be said that it is done with 317 N. On the other hand, the side beacons bring a numerical gain of 289 N for the modelled tractor unit (compared to Configuration 1).

In addition, for a clearer image of the values from an aerodynamic point of view, it was chosen to express the aerodynamic coefficient on the X axis ($C_x$), and these values can be found in Table 2. Equation (1) was used to calculate these values [10].

$$C_x = \frac{F_x}{\frac{1}{2}\rho v^2 A}\ [-] \tag{1}$$

For the $F_x$ values, the forces expressed in Newton from Table 1 were used, the air density, ρ, was considered 1.225 kg/m$^3$, the air speed has a value of 30.555 m/s and the frontal area of the tractor unit is 7.94 m$^2$ (in all cases because the elements do not exceed the edges of the truck) [10]. Therefore, adding these parameters in the formula above, the values from the following table are obtained.

**Table 2.** Values of $C_x$

| | $C_x$ [-] |
|---|---|
| Configuration 1 | 1.004 |
| Configuration 2 | 1.074 |
| Configuration 3 | 1.042 |
| Configuration 4 | 1.049 |
| Configuration 5 | 1.035 |
| Configuration 6 | 1.009 |
| Configuration 7 | 0.949 |
| Configuration 8 | 0.941 |
| Configuration 9 | 0.946 |
| Configuration 10 | 0.950 |
| Configuration 11 | 0.952 |

As shown in Figure 19, the best configuration is number 8, which, as it is presented in Table 2, has the lowest value of Cx, so the minimum rolling resistance. Next, and based on these values, it is concluded that the auxiliary elements degrade or can even improve aerodynamic performance

## 5. Conclusion

The purpose of this study is to present the influence of different auxiliary elements added to the roof of the truck cabin in terms of the impact on aerodynamic performance. As noted, certain components added to the cabin increase the resistance to forward movement, but there is also the possibility that some of them improve the air flow. For the simplified geometric model used in the cases of this study, it was observed that the addition of some extra elements brings a gain in global aerodynamic performance. This gain represents a lower forward resistance, therefore a lower fuel consumption so, a reduction in $CO_2$ emissions. This describes the fact that a better corelation of the truck body with the previously added auxiliary equipment is necessary right from the design point but also their highlighting through their integration into the bodywork.

## References

[1] Erik Gregersen, The complete history of wheeled transportation: From cars and trucks to buses and bikes, Britannica Educational Publishing, New York, 2012
[2] Adithya Hariram, Thorsten Koch, Björn Mårdberg , Jan Kyncl, A study in options to improve aerodynamic profile of heavy-duty vehicles in Europe, 2019
[3] Addendum 64: Regulation No.65, Uniform previsions concerning the approval of special warning lamps for power-driven vehicles and their trailers, United Nations, 7 December 2011
[4] https://www.dauto.ro/bara-proiectoare-volvo-fh4-euro-6-globetrotter-xl-3.html?
[5] https://www.paul-trans.ro/img/56921agabaritic.jpg
[6] https://www.obiectivdesuceava.ro/local/la-noapte-si-maine-trafic-ingreunat-din-cauza-a-doua-transporturi-agabaritice-care-vor-strabate-judetul-suceava/
[7] https://opiniabuzau.ro/wp-content/uploads/2022/05/incendiu-780x470.jpeg
[8] Liu, Y., & Glass, G. (2013). Effects of Mesh Density on Finite Element Analysis. SAE Technical Paper Series. doi:10.4271/2013-01-1375 [9] Liu, Y., & Glass, G. (2013). Effects of Mesh Density on Finite Element Analysis. SAE Technical Paper Series. doi:10.4271/2013-01-1375
[9] Suresh, K., & Regalla, S. P. (2014). Effect of Mesh Parameters in Finite Element Simulation of Single Point Incremental Sheet Forming Process. Procedia Materials Science, doi: 10.1016/j.mspro.2014.07.048
[10] Mihail Daniel Iozsa, Caroserii pentru autovehicule rutiere, Editura Politehnica Press, Bucuresti, 2016

# Optimization of the handling behaviour of an electric car by adjusting the kinematic gains of suspension

**U. Deryal, H. A. Dikmen, B. Uzuner**

[1] Chassis Systems and Analysis of Department, CDMTech Engineering Solutions

ugur.deryal@cdmtech.com.tr

**Abstract**. In this paper, within the scope of development of an electric car on a vehicle platform that is borned as internal combustion engine is presented. The vehicle dynamics behaviour changes due to the change of weight distribution and centre of gravity. Replacing the internal combustion engine with an electric motor, plus the addition of the battery moves the center of gravity towards the rear of the vehicle. Due to the increase in the rear axle load, the tendency of the vehicle to oversteer increases. To overcome this phenomenon, the rear suspension kinematic gains (camber, toe change due to vertical movement of the wheel) are adjusted thanks to the multi-link suspension. In the fist part of the study the platform is measured and Admas Car model is correlated depends on measured handling metrics. Subsequently, centre of gravity changes are applied to the verified Adams Car model. Also main platform handling metrics is defined as design targets and rear axle kinematic gains are defined as design parameters. A design of experiments (DOE)-based optimization study was performed via Adams®/Insight commercial software to achieve required handling metrics which satisfy design targets. Kinematic gains of the suspension system is analyzed via Adams car and sensitivity deepens on harpoints is determined. Design space for hardpoints is determined depends on mechanical design. The lowest and highest points are determined on the x, y and z axis and experiment list are designed according to DOE. Maneuvers that are for measuring the handling metrics are performed one after the other by changing the parameters according to the boundary conditions comes from DOE. A regression mathematical model is obtained and optimum parameters are calculated for the handling behaviour optimisation. Optimum parameters are applied on the Adams Car model again and handling manoeuvres are repeated. Results show that, although weight distribution changes, handling behaviours remain the same as based platform.

## 1. Introduction

Independent suspensions have several advantages for optimizing the handling behaviour of vehicles. In recent years most of the vehicle manufacturers uses independent front suspension and semi-independent twist beam axles for the rear. But, as the vehicle segment rises, manufacturers prefer to use independent suspension for rear axle too. Multi-link suspensions are the commonly used rear independent suspension for higher segments. Thanks to multi-connection geometry, kinematic and compliance parameters are able to be adjusted.

Common type of two different rear suspension types is shown in the Figure 1 is the general type of twist beam axles and Figure 2 is general type of multi-link suspensions. As seen in the Figure 2, multi-link suspension has several connection points that can be adjusted independently.

**Figure 1.** Twist Beam Axle [3]

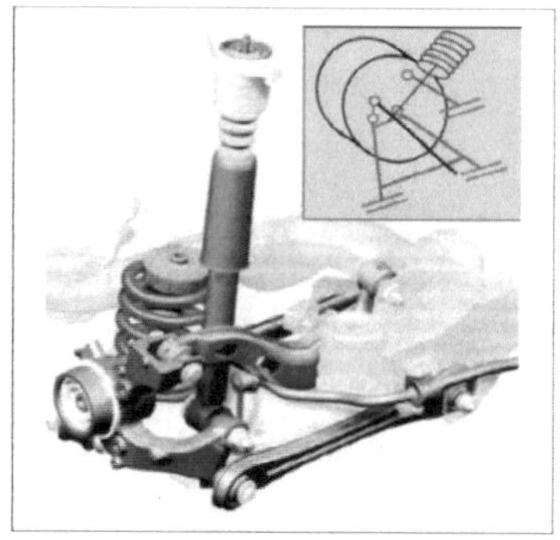

**Figure 2.** Multi-link suspension [2]

Another use case of the recent years automotive engineering is the electrification. While some of manufacturers designs innate electric vehicles, some of the changes the propulsion system internal combustion engine-based powertrain to electric drive units. If a vehicle is born electric, the entire design is carried out according to the electric driveline and it brings most of advantages for adjusting handling behaviour. But if propulsion unit changes to the electric driveline, centre of gravity moves to the back of car and handling behaviour is changed. Main problem, when the end user buys the electric of the same brand vehicle, they expect the same behaviour. So, making changes to the main running gear is a critical decision.

**Figure 3.** A electrical borned vehicle Tesla [6]

In this study, as defined in the abstract, electrical drive unit is integrated to a vehicle that has an internal combustion engine-based powertrain. Because of high voltage battery package and electrical drive unit, centre of gravity is shifted towards to the rear axle as seen in Figure 4.

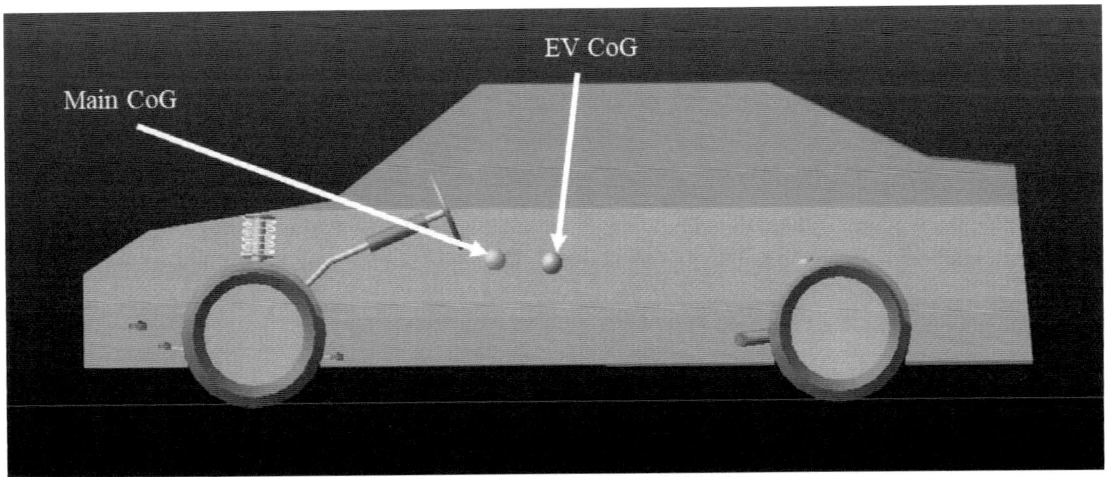


**Figure 4.** CoG Representation of Vehicle

Because of shifting the centre of gravity towards to the rear axle, rear tires' slip angle increases and vehicles behaves more oversteer. To achieve that, an optimization study is occurred on the rear multi-link suspension.

## 2. Theory Background

One of the most important metrics of the handling dynamics is understeer gradient. Understeer gradient is dependent to the steer angle on the front wheels. To understand understeer gradient, firstly steady-state cornering equation should be defined as seen in (1) [1].

$$\delta = 57.3\frac{L}{R} + \left(\frac{W_f}{C_{\alpha f}} - \frac{W_r}{C_{\alpha r}}\right)\frac{V^2}{gR} \quad (1)$$

where:

| | |
|---|---|
| $\delta$ | = Steer angle at the front wheels |
| L | = Wheelbase |
| R | = Radius of turn |
| V | = Forward speed |
| g | = Gravitational acceleration constant |
| $W_f$ | = Load on the front axle |
| $W_r$ | = Load on the rear axle |
| $C_{\alpha f}$ | = Cornering stiffness of the front tires |
| $C_{\alpha r}$ | = Cornering stiffness of the rear tires |

As seen in (1), steer angles on the front tire are effected by the lateral acceleration, axle loads and tire cornering stiffness. The (1) can be written in a shorthand form as follows [1]

$$\delta = 57.3\frac{L}{R} + Ka_y \quad (2)$$

where:

| | |
|---|---|
| K | = Understeer gradient (deg/g) |
| $a_y$ | = Lateral acceleration (g) |

Taking the K alone;

$$K = \left(\frac{W_f}{C_{\alpha f}} - \frac{W_r}{C_{\alpha r}}\right) \tag{3}$$

If it's clearly seen that in (3) understeer gradient will decrease by the mass transferred to the rear axle if the cornering stiffness remains the same. When the understeer gradient decreases, vehicle behaves more oversteer. But increasing the camber gain of the rear suspension, makes cornering stiffness increased and eliminates the mass effect on K.

## 3. Modelling and Model Verification

To investigate vehicle handling behaviour, a Multibody Dynamics model is created via Adams Car. (Figure 5)

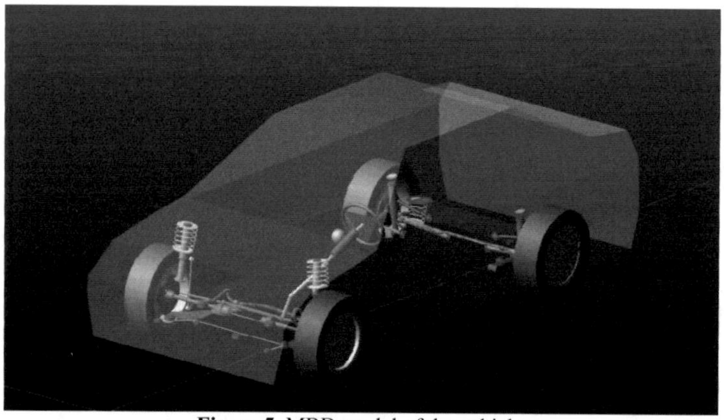

**Figure 5.** MBD model of the vehicle

And, original vehicle equipped with internal combustion engine-based powertrain is measured to get handling metrics.

**Figure 6.** Vehicle measurement

Most of handling metrics are correlated with the Adams Car model, as roll gradient, understeer gradient, yaw gain and roll velocity on slalom. The steering wheel angle result can be seen in Figure 7 (Numerical results cannot be seen because of confidentiality)

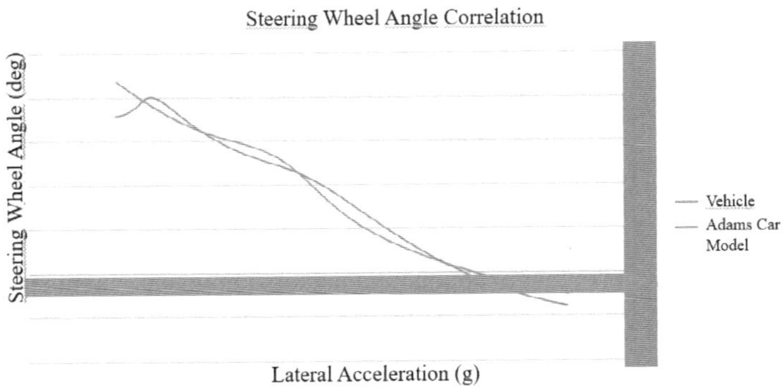


**Figure 7.** Steering Wheel Angle Correlation

Correlation results show the Adams Car model has enough accuracy to integrate centre of gravity changes and optimize handling behaviour.

## 4. Simulation and Optimization

After the correlation of the model, effect of centre of gravity changing is integrated. While making the simulations, steering wheel angle against the lateral acceleration and yaw rate against lateral acceleration is investigated. Because, the last user feels the steering wheel angle and yaw response of the vehicle mostly.

### *4.1. Steady-State Cornering*

To investigate differences, firstly a ramp steer simulation is occurred. Steering wheel angle input increases ramply 6 deg/s and vehicle's lateral acceleration increases linearly. After the simulation steering wheel angles of the cars are shown in graphics against lateral acceleration.

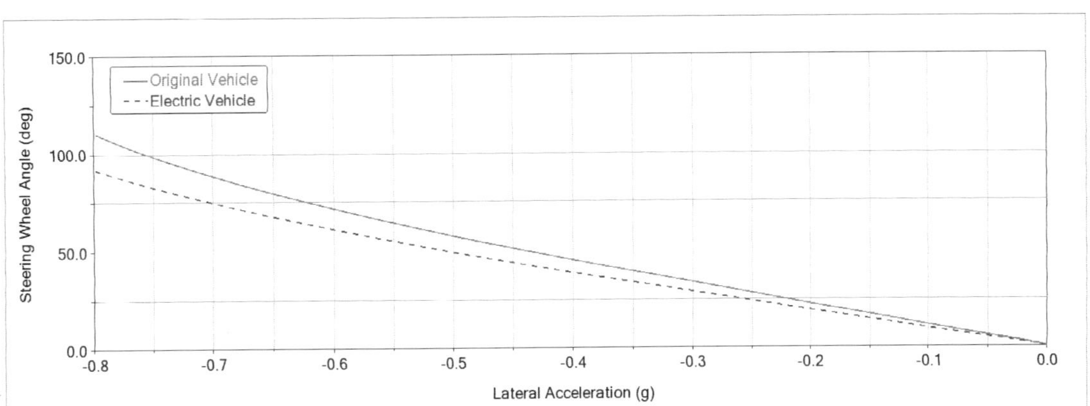


**Figure 8.** Ramp Steer Output

IOP Conf. Series: Materials Science and Engineering **1311** (2024) 012018 doi:10.1088/1757-899X/1311/1/012018

As seen in Figure 8, electric vehicle model needs less steering wheel angle for the same lateral acceleration. If we look at from the driver side, driver should turn steering wheel less, and it means more oversteer behaviour.

When we look at the yaw rate vs lateral acceleration in Figure 9, it's clearly seen electric vehicle gets more yaw rate.

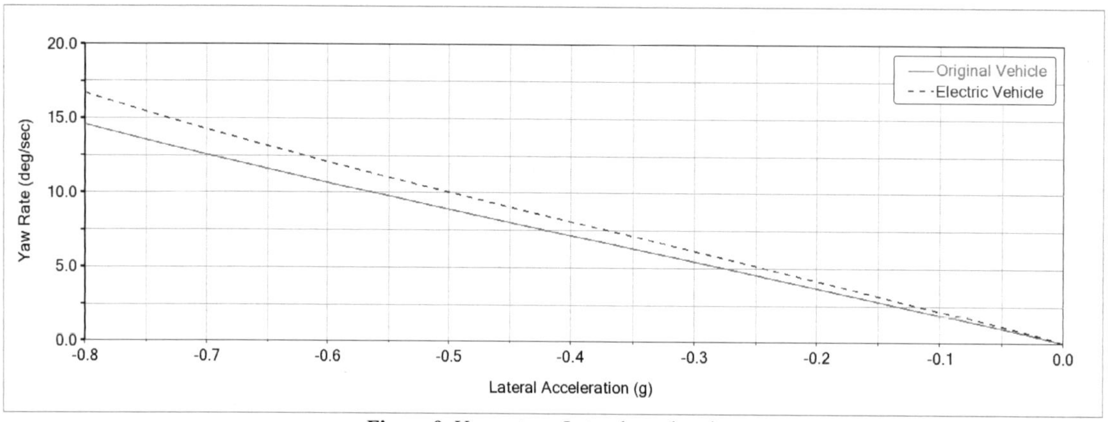


**Figure 9.** Yaw rate vs Lateral acceleration

### 4.2. *Transient Cornering*

Electric vehicle behaviour is clearly seen in steady-state manoeuvres. Also, a step steer analysis is done to see transient behaviour (Figure 10).

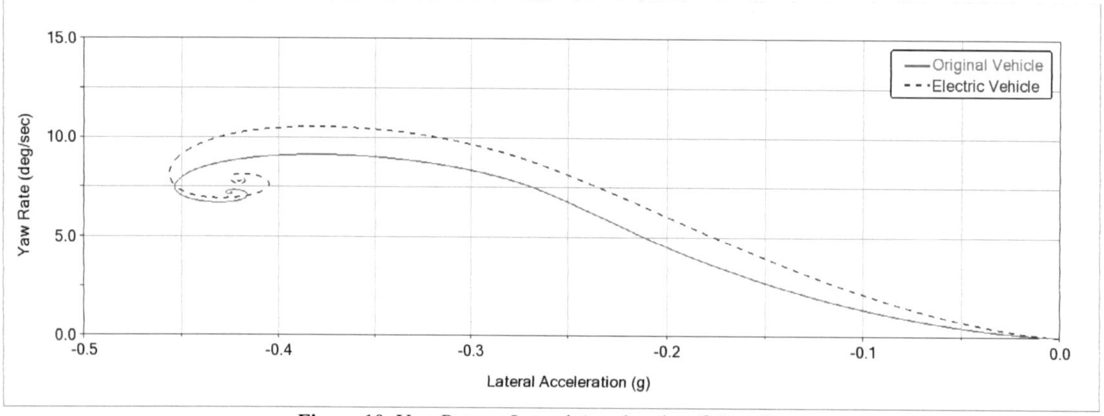


**Figure 10.** Yaw Rate vs Lateral Acceleration @Step Steer

### 4.3. *Rear Suspension Geometry*

Cornering simulations are done and seen that electric vehicle has more oversteer behaviour. To achieve that rear suspension geometry is analysed and points that can be changed determined. (Figure 11)

IOP Conf. Series: Materials Science and Engineering **1311** (2024) 012018 doi:10.1088/1757-899X/1311/1/012018

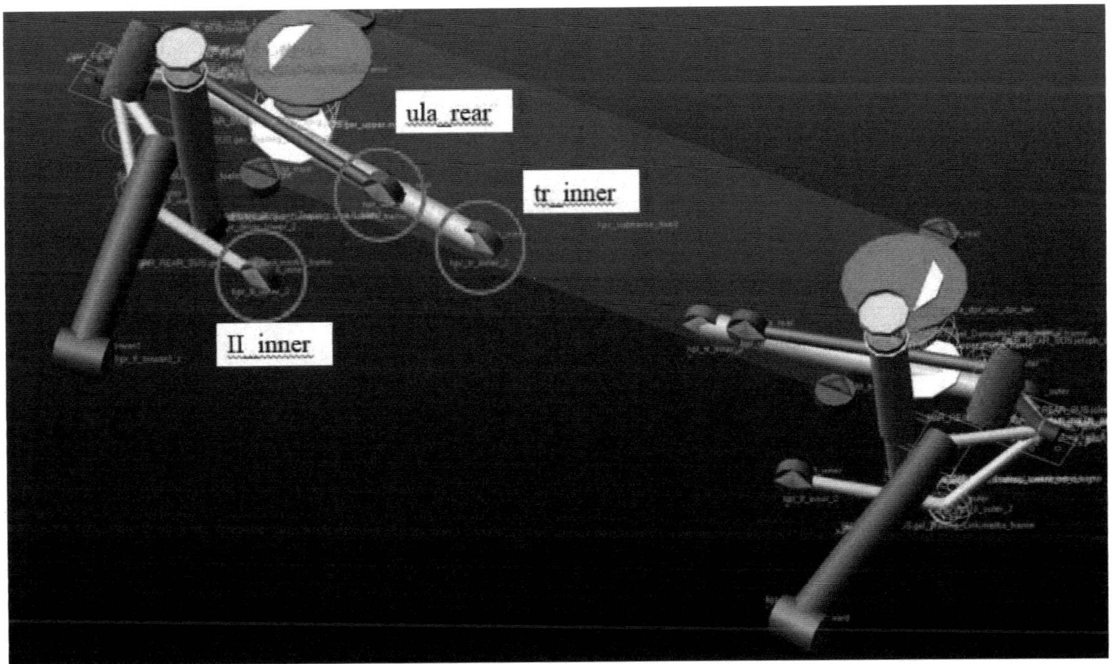


**Figure 11.** Rear Suspension Geometry

### 4.4. *Optimization*

As seen in the Figure 10, three of hardpoints can be changed during the handling optimization. It means, three hardpoints in three dimensions, 9 parameters. Increasing the number of parameters makes the problem always complicated. To achieve that complication, at firstly a design of experiment (DOE) study is done and Pareto diagram is investigated. Most effective parameters are selected due to Pareto and DOE is studied again.

Pareto result of the first DOE study is seen on Figure 12.

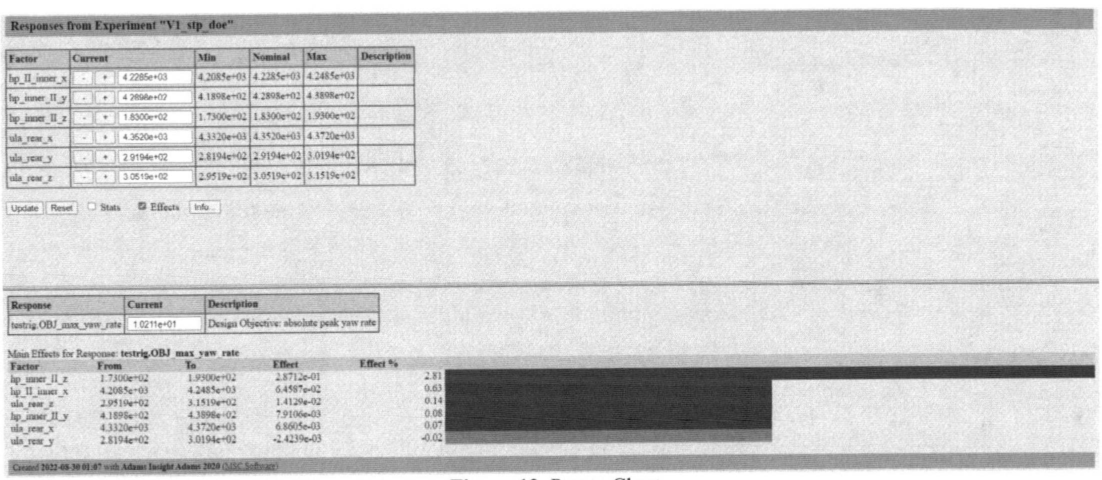

**Responses from Experiment "V1_stp_doe"**

| Factor | Current | Min | Nominal | Max | Description |
|---|---|---|---|---|---|
| hp_II_inner_x | 4.2285e+03 | 4.2085e+03 | 4.2285e+03 | 4.2485e+03 | |
| hp_inner_II_y | 4.2898e+02 | 4.1898e+02 | 4.2898e+02 | 4.3898e+02 | |
| hp_inner_II_z | 1.8300e+02 | 1.7300e+02 | 1.8300e+02 | 1.9300e+02 | |
| ula_rear_x | 4.3520e+03 | 4.3320e+03 | 4.3520e+03 | 4.3720e+03 | |
| ula_rear_y | 2.9194e+02 | 2.8194e+02 | 2.9194e+02 | 3.0194e+02 | |
| ula_rear_z | 3.0519e+02 | 2.9519e+02 | 3.0519e+02 | 3.1519e+02 | |

Update Reset Stats Effects Info...

| Response | Current | Description |
|---|---|---|
| testrig.OBJ_max_yaw_rate | 1.0211e+01 | Design Objective: absolute peak yaw rate |

Main Effects for Response: **testrig.OBJ_max_yaw_rate**

| Factor | From | To | Effect | Effect % |
|---|---|---|---|---|
| hp_inner_II_z | 1.7300e+02 | 1.9300e+02 | 2.8712e-01 | 2.81 |
| hp_II_inner_x | 4.2085e+03 | 4.2485e+03 | 6.4587e-02 | 0.63 |
| ula_rear_z | 2.9519e+02 | 3.1519e+02 | 1.4129e-02 | 0.14 |
| hp_inner_II_y | 4.1898e+02 | 4.3898e+02 | 7.9106e-03 | 0.08 |
| ula_rear_x | 4.3320e+03 | 4.3720e+03 | 6.8605e-03 | 0.07 |
| ula_rear_y | 2.8194e+02 | 3.0194e+02 | -2.4239e-03 | -0.02 |

Created 2022-08-30 01:07 with Adams Insight Adams 2020 (MSC Software)

**Figure 12.** Pareto Chart

As seen in Figure 11, II inner hardpoint has the most effect. So main study is done with II inner hardpoint in x and z direction.

To optimization maximum yaw rate value on the step steer simulation is used as objective.

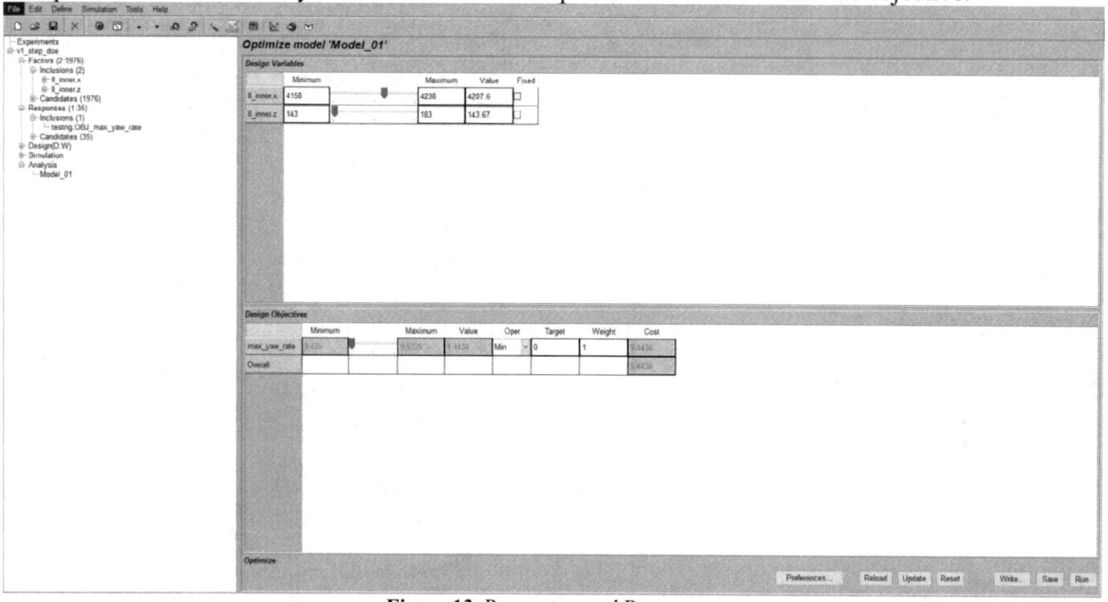


**Figure 13.** Parameters and Responses

The target is about 9,5 deg/s maximum, when we run the optimization solver we can see that by changing the hardpoint it is achievable.

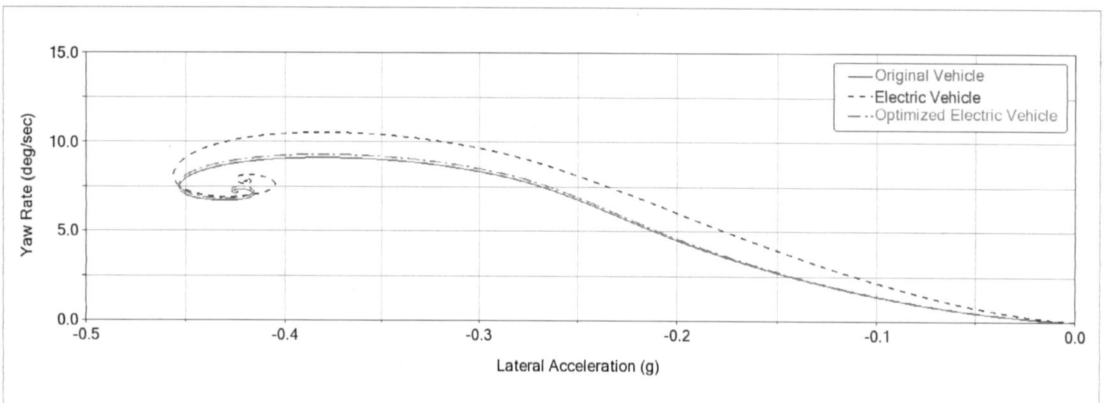


**Figure 14.**

## 5. Conclusion

Most of the time, planar dynamics is rarely enough for vehicle dynamics calculations but thanks to multibody dynamics always there are tuning opportunities. As seen in the study as a phenomenon is achieved by optimization approach on multibody dynamics models.

It's clearly seen that, by the changing kinematics of the suspension vehicle behaviour can be tuned. In this study camber gain of the suspension is increased and rear tires carry more lateral loads. It makes the rear tire slip angle smaller and being more oversteer is prevented.

**References**

[1] Gillespie, T.D., Fundamentals of vehicle dynamics / T.D. Gillespie. p. cm. Includes bibliographical references and index. ISBN 1-56091-199-9 1. Motor vehicles--Dynamics. I. Title. TL243.G548 1992 629.2--dc20ERsoy
[2] Heißing, B., Ersoy, M., Chassis Handbook
[3] Reimpell, J., Stoll, H., Betzler, J.W., The Automotive Chassis: Engineering Principles
[4] Adams Car User Manual
[5] Adams Insight User Manual
[6] www.tesla.com

IOP Conf. Series: Materials Science and Engineering **1311** (2024) 012019 doi:10.1088/1757-899X/1311/1/012019

# Comparative study on rolling resistance force in tire-ice interaction with a variation of tread rubber compound

**M N Shenvi* and C Sandu**
TMVS Laboratory, Virginia Tech, Blacksburg, VA 24060
* Author to whom any correspondence should be addressed

mshenvi@vt.edu*, csandu@vt.edu

**Abstract**. The complex phenomenon involved in the tire-ice interaction can be better understood by studying the tire and ice properties. A significant force at the contact patch is the rolling resistance force. To study the effect of tire characteristics on rolling resistance forces, data collected from a free-rolling tire can be used as there is no tractive or braking torque applied to the wheel and thus the data representing the longitudinal forces is only related to resistive forces viz. friction and rolling resistance force. Although a lot of research and literature has been developed for tire-ice interaction over the years, there is a significant lack of work in the subsection of free-rolling tires specifically to study the effect of ice parameters on rolling resistance forces. To fulfill the aim of this study, several sets of experiments for 16 winter tires of 8 different rubber compounds in free-rolling conditions were conducted. The experimental data was compared with predictions of 4 well-known formulations from the literature. Based on the findings and the relative error between the predicted and measured values, it was proposed that the high error percentage could be due to the variation in the viscoelastic properties of the tread rubber. The longitudinal force generated during those tests would thus be a combination of frictional forces due to adhesion between the tread rubber and its viscoelastic nature in addition to the contribution from the remaining sections of the tire. The data collected from the experiments and results from implementing the existing models to obtain frictional force will further help in analyzing the relation between the rolling resistance force with variation in normal load, inflation pressure, and change in camber angle once a generic formulation considering the rubber compound effects is incorporated. Another significant finding of this work is that formulations in literature used for finding the rolling resistance coefficient of radial passenger car tires were found to highly underestimate the coefficient when compared to the equivalent coefficient of friction found from experimental results. The future scope may include the development of a model which will incorporate the viscoelastic properties, especially the loss moduli and loss tangent to evaluate the coefficient of friction and rolling resistance force in the free-rolling condition of a winter tire on ice as the rolling resistance losses are found to be highly correlated to these two properties.

## 1. Introduction

The interaction between the tire and the terrain is a complex phenomenon that affects the overall performance of the vehicle from a safety as well as economical perspective. The complexity is further escalated when the terrain is (non-deformable) ice, due to the lower friction available at the ice surface and the possibility of localized melting of ice leading to water formation. The interaction between the tire and ice with respect to tractive and braking performance has been studied extensively in recent years [1], [2], [3], [4], [5], [6], [7]. The free-rolling performance of the tire especially on ice has not received much attention.

The free-rolling motion of a tire is defined as the condition in which no external torque (either driving or braking) is applied to the wheel [8]. Thus, the tire is expected to face resistive forces in the longitudinal direction. This phenomenon commonly occurs in all vehicles as the energy spent in the compression of the rubber entering the contact patch cannot be fully recovered and also causes stress distortion in the contact patch area causing the resultant normal force to be ahead of the center of the contact patch [9]. The rolling resistance force is directly proportional to the normal load. The most common method used for the evaluation of rolling resistance is on a drum-testing machine according to SAE J1269 [10] at high speeds. Generally, the longitudinal forces acting on the tire consist of air drag, friction force, and rolling resistance. The Terramechanics Rig is operated at lower speeds and air drag could be neglected. This work presents a hypothesis that due to the high pressures at the contact patch and the adhesion between the tire and the ice when the tread enters the contact patch, the resistive force experienced by the tire could be separated into two components viz. the rolling resistance force and another force of friction/adhesion at the contact patch, as shown in equation (1).

$$F_x = \mu_{adh} * F_z + \mu_{rr}\, F_z \qquad (1)$$

One of the major reasons the rolling resistance force is so important is that it starts acting as soon as the tire begins to rotate. Another reason, which may be a concern during operation on ice, is that it also contributes to the temperature rise of the tire which in turn would affect the water film formation on ice. This can affect the properties of the tire material for all ambient conditions. In general, the factors that affect the rolling resistance of the tire [11] are the temperature of the tire, the inflation pressure, the normal load, the velocity of the tire, the material and design of the tire, as well as the slip ratio. From a tire manufacturer's perspective, the most important factors are the dynamic viscoelasticity properties of various regions of the tire. In an earlier study conducted at the lab [12], efforts were made to benchmark three commonly used methods to evaluate the effective rolling radius of the tire. This work also attempts to calculate the rolling resistance coefficient using empirical methods from the literature and evaluate the relative accuracy to lay the foundation for future research on the effect that the variation in tread rubber can have on the rolling resistance force on the ice.

Section 2 of this work summarizes the relevant literature followed by a description of the testing program in Section 3. Section 4 discusses the significant findings of the testing program and relevant conclusions and scope for future work are presented in Section 5.

## 2. Methodology

Rolling resistance in tires is highly undesirable and its reduction is considered to be one of the goals in solving the 'magic triangle' of tires. Though the exact contribution of the individual sections of the tire to the rolling resistance losses is disputed in different studies [13] it could be estimated to be as high as around 80% of the total losses purely due to the tread region losses. Experiments were conducted in [14] with a vehicle in traction mode on snow and on ice to analyze the tire-ice interaction with a concentration on friction coefficients. They found the dependence of the friction coefficient on the temperature (T) and the aggregate material (A) as in equation (2) for a car tire on ice. However, the applicability of this equation to the current testing regime would not be accurate as it does not consider any operational parameters or properties of the tire and the ice.

$$\mu = 0.11 - 0.0052\,.\,T + 0.0002\,.\,A \qquad (2)$$

Another method was implemented [15] for real-time estimation of the tire friction coefficient for electric vehicles. The model in that work was a modification of Burckhardt's model [16] and is depicted in the equation (3). In equation (3), '$s$' is slip, and '$A_i$' are regression coefficients while '$C_i$'

are constants. A drawback of this model is that for the condition of free rolling of a tire on ice, it cannot be applied, as the friction estimated is zero.

$$\mu(s) = C_1 - A_1 e^{-23.99s} - A_2 e^{-33.82s} - A_3 e^{-94.13s} - A_4 e^{-306.39s} - C_3 \tag{3}$$

A concept of 'Mean Equivalent Rolling Force' was devised [10] to compare the rolling resistance of tires. The test practice they used was SAE J2452. The rolling resistance model developed from test data is shown in equation (4). The drawback of utilizing this in our current regime was the variation in the normal load and inflation pressure required to provide a better correlation model and the difficulty in performing high-speed testing on ice.

$$RR = P^{\alpha} Z^{\beta} (a + b.V + c.V^2) \tag{4}$$

where RR is the rolling resistance force, P is the inflation pressure, Z is the normal load, V is the longitudinal velocity, and (α,β) and (a,b,c) are coefficients to be found out using regression.

In the work on vehicle dynamics [9] several formulations for evaluation of the rolling resistance coefficient ($\mu_r$) are presented as shown in equations (5-7). Equations (5) and (6) evaluate the rolling resistance coefficient based on velocity ($v_x$) while equation (7) evaluates the rolling resistance coefficient based on tire type (radial/bias-ply), normal load ($F_z$), inflation pressure (p), and velocity. Of these, equations (6-7) were chosen to be utilized to find the rolling resistance coefficient for further analysis. The values for the coefficients were used from the book [9] authored by Jazar Reza.

$$\mu_r = \mu_0 + \mu_1 v_x^2 \tag{5}$$

$$\mu_r = C_0 + C_1 v_x + C_2 v_x^4 \tag{6}$$

$$\mu_r = \frac{K}{1000}\left(5.1 + \left(\frac{5.5 \text{ x } 10^5 + 90F_z}{p}\right) + \left(\frac{1100 + 0.0388F_z}{p} v_x^2\right)\right) \tag{7}$$

As the speed of the rig is low, the rolling resistance coefficient can also be found [11] using equations (8-9), however equation (9) is based on empirical data tested on a concrete surface, and hence equation (8) was chosen as a possible candidate for analysis, instead of equation (9) which considers more factors.

$$C_{rr} = 0.01\left(1 + \frac{V}{100}\right) \tag{8}$$

$$C_{rr} = f_0 + 3.24 * f_s \left(\frac{V}{100}\right)^{2.5} \tag{9}$$

where V is the speed in mph, $f_0$ and $f_s$ are factors dependent on inflation pressure.

Another possibility of evaluating the rolling resistance coefficient on ice was through the use of the ERDC VTI model [17] which was used to validate a winter driving simulator that predicts rolling resistance on hard surfaces according to equation (10), where inflation pressure is in PSI.

$$C_{rr} = 0.007 + \frac{0.0939}{p_{inf}} \tag{10}$$

## 3. Testing Program

The testing program was designed to focus on the effect of the tread rubber compound as a part of a larger study [18], [19]. The detailed design of experiments for the experimental testing program is published in [18]. For this work, the points of interest are the testing of sixteen tires of identical design and construction with the variation being in the composition of tread rubber compound (2 tires per rubber compound). The applied normal load was 4000 N and the inflation pressure was 28 psi. The camber and toe angles were kept at 0° while the ice was maintained at a temperature of -1°C. Further, the tires were cooled to around -27°C before every test to account for the mounting and calibration procedure so that once the tests began the tire would be cooler or at least match the temperature of the ice.

This was done in the Terramechanics rig [20] at the TMVS lab. The possible variation in the operational parameters includes changes in the normal load, inflation pressure, slip ratio, camber and toe angles, as well as compaction in case of deformable terrain testing. The rig consists of two motors viz. the carriage motor which provides a linear speed of 0.06 m/s to the carriage, and the wheel motor which can be used to apply rotational velocity to simulate a specific slip ratio. The results of the current work disengaged the clutch connection between the wheel and the wheel motor in order to avoid any external torque application. The rig also consists of an active normal load controller system [21] which is useful to keep the normal load in a small range of the target normal load [12].

The rig is instrumented with the Kistler P650 hub, which measures different types of data, but most importantly the forces and moments in three mutually perpendicular directions. The sampling frequency of the hub is 160 Hz requiring filtration of noise and data averaging before looking at the results in the post-processing stage. The longitudinal force measured by the Kistler sensor represents the drawbar pull if the calibration is performed according to proprietary standards. In a previous study at the lab [22] the Kistler sensor calibration was performed when the tire was free rolling on ice causing a considerable reduction of the resistive forces and providing insights only into the tractive forces. Though this method could be useful for having the tractive forces under a microscope, it would be difficult to analyze the free-rolling behavior. In addition, the tire in real-life conditions faces a combination of tractive and resistive forces i.e., the drawbar pull, and hence the proprietary method of calibration of the Kistler sensor was used in this work.

### *3.1. Experimental Testing Setup*

The creation of the ice-bed for testing was done in several steps

1. Placing the C-channels inverted on the side brackets of the upper level of the chamber.
2. Applying two layers of plastic tarp and fixing them to the sides of the chamber while ensuring no obstruction to the movement of the carriage.
3. Place a layer of insulation foam to fit the inner dimensions of the chamber on top of the tarp. Further, the foam sheets are fixed with respect to each other, and the inner walls of the chamber are followed by another layering of plastic tarp.
4. Unrolling the tubular heat exchanger on top and make connections with the chiller. The open sections of the hose pipes are covered with Rockwool insulation to prevent heat loss. The chemical used for heat exchange is a 50% ethylene glycol solution. The chiller is then set to a certain target temperature and has a feedback mechanism via a thermocouple that needs to be placed on top of the ice surface to control the operation of the chiller.
5. The system goes through a warm-up stage of a day and then tap water is sprayed at a rate of approximately 2 mm every 2 hours for about 4-5 days in order to achieve an ice bed thickness of about 3 inches. This rate was found optimal for the compressive strength of ice during prior investigations [23], [24] at the lab instead of creating a water pool and allowing it to freeze gradually.
6. To have a uniform surface roughness of ice, which is known to impact friction [25], any ice flakes or previous run tire traces on the ice surface were cleared using the tools (metal scraper and rubber squeegee) available in the lab.

The tires went through a preparation stage. In the first stage, the tires had their vent spews removed and were worn out by mounting them on a vehicle and running them for about 100 km and intermediate tire rotations for the break-in that ensures uniform wearing of tires. In the second stage, K-type thermocouples were chosen and fixed in the grooves of the tires to measure the rise in temperature that could occur in the contact patch. The advantages and disadvantages of this have been documented [26].

### *3.2. Rubber compound properties*

Although the exact constitution of the rubber compounds was unknown, some mechanical and thermal properties were known as below. The values of the properties have been published in other works by the authors and found in the literature [18], [19], [26].

1. Rubber compound density
2. Thermal conductivity of the rubber
3. Rubber's specific heat capacity
4. Stiffness-related properties: The viscoelastic properties of rubber are a prime contributor to the rolling resistance effects in a tire. The viscoelastic properties like loss modulus, storage modulus, dynamic modulus, and the tangent of the phase difference at various temperatures were available for many of the rubber compounds.

## 4. Results & Discussion

The equivalent coefficient of friction (resistive forces normalized with respect to the normal load acting on the tire) [12] for tires of all the rubber compounds was in the vicinity of each other [26]. For this work, the analytical evaluation of the rolling resistance coefficient was done using 4 methods viz. Empirical equations from Jazar's work considering normal load and inflation pressure (equation (7)), empirical equation from Jazar's work considering variation in longitudinal speed (equation (6)), the method for rolling resistance on hard undeformable surfaces by ERDC (equation (10)), and the empirical equation from Gillespie's work (equation (8)). The choice of these 4 methods was based on the availability of operational parameters as well as use in literature. The goal was to analyze the effectiveness of these equations in predicting the rolling resistance coefficient (RRC) for the experimental program of testing winter tires on ice. The predicted values found using these methods are:

1. Jazar's equation (7): 0.0079
2. Jazar's equation (6): 0.0099
3. ERDC VTI code equation (10): 0.0104
4. Gillespie's equation (8): 0.01

Upon calculation of the relative error in the prediction of the rolling resistance coefficient using these equations with respect to the experimental findings, the results could be summarized as shown in Table 1.

**Table 1:** Relative Error in the estimation of RRC with respect to experimental results

| Rubber Compound | Equivalent coefficient of friction [26] | Percentage Error with respect to equation. (7) | Percentage Error with respect to equation. (6) | Percentage Error with respect to equation. (10) | Percentage Error with respect to equation. (8) |
|---|---|---|---|---|---|
| **A** | 0.0338 | 76.60 | 70.68 | 69.20 | 70.39 |
| **B** | 0.0345 | 77.11 | 71.32 | 69.87 | 71.03 |
| **C** | 0.035 | 77.39 | 71.66 | 70.23 | 71.38 |
| **D** | 0.0332 | 76.12 | 70.08 | 68.57 | 69.77 |
| **E** | 0.031 | 74.32 | 67.82 | 66.19 | 67.49 |
| **F** | 0.0257 | 69.25 | 61.47 | 59.52 | 61.08 |

| | | | | | |
|---|---|---|---|---|---|
| **G** | 0.0347 | 77.24 | 71.48 | 70.04 | 71.19 |
| **H** | 0.0319 | 75.16 | 68.87 | 67.29 | 68.55 |

Based on the results from the experiments, it was found that tires of rubber compound F had the least value of resistive forces experienced, while rubber compound C had the highest amount of resistive forces per unit of normal load. However, on evaluating the percentage errors and plotting as a histogram using the 4 methods, it was found that from a prediction perspective, rubber compound F had the least amount of prediction error in comparison to the other compounds as shown in Figure 1. For the equivalent coefficient of friction, an order of resistive forces per unit of normal load will be

**C>G>B>A>D>H>E>F**

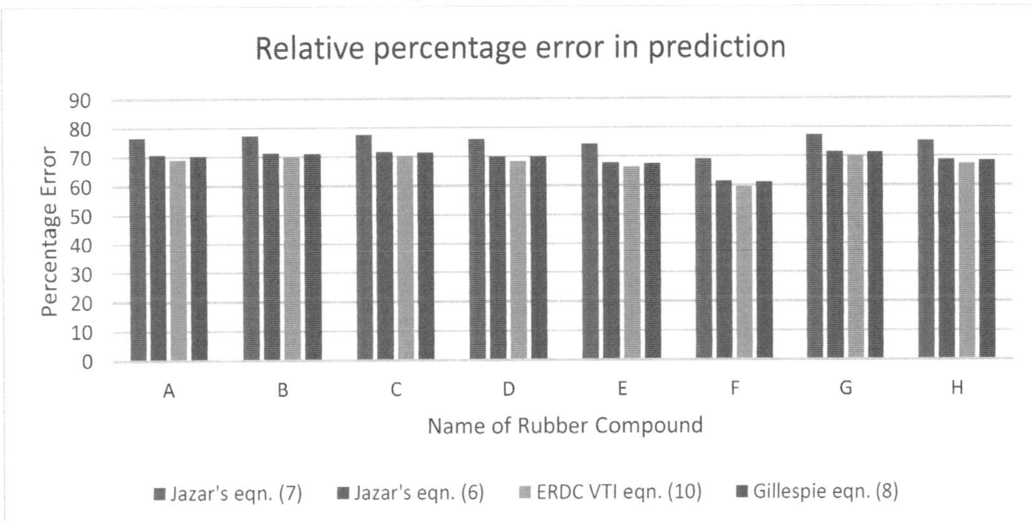


**Figure 1**: Relative percentage error in prediction of rolling resistance coefficient

Another aspect worth considering here is that except for the 1st method, all the remaining methods tend to have a nearly similar prediction error percentage. Further, another method based on equation (5) was considered for this study but given the low speed of testing that was used on the rig and the low value of $\mu_1$, it was found to always produce a constant value equal to the rolling resistance coefficient if the longitudinal velocity was zero i.e., the equation does not seem to hold at the low-velocity values that are being tested on the rig.

Based on these findings, it is hypothesized that although existing analytical or empirical relations would indeed be subject to some inherent assumptions, the amount of percentage error is too high to be quantified purely based on assumptions of a specific model or test setup. Although this difference in prediction could be due to the specific tread design of winter tires, it will still not completely explain the difference in the measured equivalent coefficient of friction. Thus, the mechanical properties of the tread rubber and its adhesive behavioral differences in interaction with ice would be a major factor involved in the equivalent coefficient of friction experienced by the tires in the experimental tests. In the case of rubber, the loss of energy between the compression and release cycles is generally considered to be a function of normal load and thus it is proposed that the current formulations need terms that could account for the viscoelastic properties of rubber. One such equation that is proposed is equation (1), which would separate the contributions of the tread rubber and the other layers of the tire to the rolling resistance.

## 5. Conclusions and Future Work

This work attempted to evaluate the relative accuracy of the existing formulations that are used for predicting the rolling resistance coefficient in the case of radial passenger car tires with respect to experimental results of free-rolling winter tires. Further, it was found that 4 such well-known formulations have a huge difference when compared to the current testing regime. Thereby, it was proposed that these differences would be primarily due to two reasons. Firstly, the inherent assumptions in the case of the four formulations, and secondly the viscoelastic property differences in the tread. As the exact composition of the tread rubber was not known, it would be difficult to quantify the contributions of the adhesive behavior between the tread rubber and the ice.

From a future work perspective, one step to quantifying the differences would be analyzing the composition of the tread rubber. For example, it was found in studies [27], [28], [29] that using the filler material silica and improving its dispersion in the matrix reduces rolling resistance. Another possibility would be regression analysis of the current data by using the viscoelastic properties of tread rubber as variables to quantify the individual contributions as it was found that the loss tangent and loss modulus have a relation with the rolling resistance [13], however, evaluating that relation for this work was not possible as all values of loss tangent and loss modulus were not known.

**References**

[1] Gießler M, Gauterin F, Wiese K, Wies B. Influence of friction heat on tire traction on ice and snow. Tire Sci Technol. 2010;38(1):4–23.
[2] Sokolovskij E. Automobile braking and traction characteristics on the different road surfaces. Transport. 2007;22(4):275–8.
[3] Zhang Y, Gao J, Li Q. Experimental study on friction coefficients between tire tread rubber and ice. AIP Adv. 2018;8(7).
[4] Hyun Chul Jung, Woo Cheol Park KMJ. Finite Element Analysis of Tire Traction Using a Rubber-Ice Friction Model. Open Journal of Applied Sciences. 2018;8:495–505.
[5] Jimenez E, Sandu C. Towards a real-time pneumatic tire performance prediction using an advanced tire-ice interface model. J Terramech. 2018;81:43–56.
[6] He R, Jimenez E, Savitski D, Sandu C, Ivanov V. Investigating the Parameterization of Dugoff Tire Model Using Experimental Tire-Ice Data. SAE International Journal of Passenger Cars - Mechanical Systems. 2017;10(1):83–92.
[7] Waluś KJ, Olszewski Z. Analysis of tire-road contact under winter conditions. Proceedings of the World Congress on Engineering 2011, WCE 2011. 2011;3:2381–4.
[8] He R, Sandu C, Mousavi H, Shenvi MN, Braun K, Kruger R, et al. Updated standards of the international society for terrain-vehicle systems. J Terramech. 2020;91:185–231.
[9] Jazar RN. Vehicle Dynamics: Theory and Application. Springer. Springer US; 2017.
[10] Grover PS, Bordelon SH. New parameters for comparing tire rolling resistance. SAE Technical Papers. 1999;108(May):1527–34.
[11] Gillespie Thomas D. Fundamentals of Vehicle Dynamics. Warrendale, PA: SAE International; 1992.
[12] Mousavi H, Shenvi, M.N., Sandu, C. Experimental study for free rolling of tire on ice. In: ASME 2019 International Design Engineering Technical Conferences and Computers and Information in Engineering Conference IDETC/CIE 2019. 2019.
[13] Lou AY. Relationship of Tire Rolling Resistance to the Viscoelastic Properties of the Tread Rubber. Tire Sci Technol [Internet]. 1978;6(3):176–88. Available from: www.archive.org
[14] Navin F, Macnabb M, Nicolletti C. Vehicle Traction Experiments on Snow and Ice. SAE Technical Paper Series. 1996;1(412).
[15] Lin C, Wang G, Cao WK, Zhou FJ. Real-time estimation of road friction coefficient for the electric vehicle. Proceedings of the 2012 3rd World Congress on Software Engineering, WCSE 2012. 2012;172–5.
[16] Burckhardt M. Fahrwerktechnik: Radschlupf-Regelsysteme. Wurzburg: Vogel-Verlag. Vogel; 1993.
[17] Parker MW, Shoop SA, Coutermarsh BA, Wesson KD, Stanley JM. Verification and validation of a winter driving simulator. J Terramech. 2009 Aug;46(4):127–39.

[18] Shenvi MN, Verulkar A, Sandu C, Mousavi H. Tread rubber compound effect in winter tires: An experimental study. J Terramech [Internet]. 2022;99:57–73. Available from: https://doi.org/10.1016/j.jterra.2021.11.004

[19] Shenvi MN, Mousavi H, Sandu C. Tread rubber compound effect in winter tires: Benchmarking ATIIM 2.0 with classical models. J Terramech. 2022 Jun 1;101:43–58.

[20] Sandu C, Taylor B, Biggans J, Ahmadian M. Building Infrastructure for Indoor Terramechanics Studies: The Development of a Terramechanics Rig at Virginia Tech. In: Proceedings of 16th ISTVS international conference, Turin, Italy. Turin, Italy; 2008.

[21] Naranjo SD. Experimental investigation of the tractive performance of an instrumented off road tire on a soft soil terrain [Internet]. Virginia Polytechnic Institute and State University; 2013. Available from: http://hdl.handle.net/10919/51165.

[22] Mousavi H, Sandu C. Experimental study of the tread rubber compound effects on Tire Performance on Ice. SAE Int J Commer Veh. 2020;13(2):99–101.

[23] Bhoopalam AK, Sandu C, Taheri S. Experimental investigation of pneumatic tire performance on ice: Part 1 - Indoor study. J Terramech [Internet]. 2015;60:43–54. Available from: http://dx.doi.org/10.1016/j.jterra.2015.02.006.

[24] Jimenez E, Sandu C. Experimental Investigation of the Tractive Performance of Pneumatic Tires on Ice. Tire Sci Technol. 2020;48(1):22–45.

[25] Lahayne O, Pichler B, Reihsner R, Eberhardsteiner J, Suh J, Kim D, et al. Rubber Friction on Ice: Experiments and Modeling. Tribol Lett. 2016;62(2):1–19.

[26] Shenvi MN. Comparative Study of the Effect of Tread Rubber Compound on Tire Performance on Ice [Masters]. Virginia Polytechnic Institute and State University; 2020.

[27] Hiramatsu K, Kamamizu K, Komai M, Okazaki T, Takino H. Rubber and collagen-fiber blends for studless winter tire applications. Rubber World. 1991;204(2):38–43, 45.

[28] Okel TA, Rueby JA. Silica morphology and functionality: Addressing winter tire performance. Rubber World. 2016;253(5):21–33.

[29] Weng P, Tang Z, Guo B. Solving “magic triangle” of tread rubber composites with phosphonium-modified petroleum resin. Polymer (Guildf) [Internet]. 2020;190(January):122244. Available from: https://doi.org/10.1016/j.polymer.2020.122244

**Acknowledgments**

This study has been partially supported by Sumitomo Rubber Industries Ltd. and Terramechanics, Multibody, and Vehicle Systems Laboratory at Virginia Tech. The authors would like to thank the guidance and support provided by Dr. Toshio Tada.

# Multibody Dynamics and Terramechanics-based modeling and simulation of Quarter-Car with Suspension

**E Chaturvedi[1*], J Mukherjee[2], and C Sandu[3]**

[1,2] Graduate student, TMVS Laboratory, Virginia Tech, Blacksburg, 24060, USA
[3]Robert E. Hord Jr. Professor, Department of Mechanical Engineering, Director of TMVS Lab, Virginia Tech, Blacksburg, 24060, USA
*Author to whom any correspondence should be addressed
ekanshchat96@vt.edu[1*], jmukherjee@vt.edu[2], csandu@vt.edu[3]

**Abstract**. While tires only compress when acted upon by a load in on-road scenarios, in off-road conditions, on the other hand, the tires not only undergo compression but also sinkage into the ground, in a manner that depends on the terrain. The tire-ground contact, in turn, affects the forces acting on the suspension system. Hence, it is necessary to study the vertical dynamic response of a system in off-road conditions by accounting for the terramechanics phenomenon (tire-terrain interface). The forces acting at the tire and terrain interface in off-road conditions have been studied in terramechanics separately; however, their integration with multibody systems is relatively less explored. In this paper, a multibody dynamic-based modeling and simulation of a quarter car with a linear spring-damper suspension system for off-road vehicles is presented. The generalized external forces are obtained using the Bekker-Wong based soil pressure-sinkage approach for two different kind of multibody systems interacting with sandy loam terrain. This model is integrated with the Tangent Space Ordinary Differential Equations of the quarter car and suspension system and solved using numerical integration techniques. Thus, by accounting for the terramechanics in the multibody system, the technique provides a more realistic response of the system in off-road scenarios.

## 1. Introduction

Many research studies have studied the dynamic response of suspensions using quarter car modeling. This is fundamental to developing control algorithms as well as designing suspension systems for vehicles. Rahnejat [1] outlined a method to model a two-dimensional McPhearson strut suspension using multibody systems approach. Hegazi et. al [2] further developed multibody models for full vehicle handling. Inman [3] developed a base excitation model for simulating effects of external excitations on a quarter-car model. Goncalves [4], in an experimental study, tested five semi-active control policies on a two-degree-of-freedom (DOF) quarter-car system using magneto-rheological fluid damping. Levesley et. al. [5] further used the experimental road inputs for designing control strategies for a semi active suspension using multibody dynamics. Moreover, Sandu et. al. [6] simulated a quarter car model through index-1 differential algebraic equations (DAEs) and demonstrated the effectiveness of this approach by comparing the results with a physical test on the rig.

However, a recurrent observation with the existing literature is that these exclude the effects of terrain parameters especially for deformable terrain, which govern tire-ground interaction and hence the dynamic response of the suspension itself. The existing literature on terramechanics models emphasizes well on the nonlinear nature of the wheel-soil interface. Taheri et. al. [7] reviewed major empirical, semi empirical and finite element based terramechanics models in a comprehensive review paper. Ersal et. al. [8] developed a reduced-order nonlinear terramechanics model as a surrogate of the

IOP Conf. Series: Materials Science and Engineering **1311** (2024) 012020 doi:10.1088/1757-899X/1311/1/012020

Soil Contact Model (SCM) through extending a state-of-the-art Bekker model [9] to account for additional dynamic effects. Mocera et. al. [10] developed a semi-empirical model and simulated the interaction of grousers on deformable terrain.

There is substantial literature available for the design of suspension systems for on-road vehicles as well as for the field of terramechanics. It is necessary to bridge the gap between terramechanics and multibody system formulation of vehicles in order to better model off-road vehicle systems. Previous studies conducted in the TMVS lab [16], [17], [18] showed that it is important to assess the performance of off-road vehicles in their operational environment. This becomes especially important when it comes to off-road vehicle suspensions where terrain profiles and terrain type impact the drivability of the vehicle. Therefore, the presented approach attempts to link the multibody dynamics modeling approach with Bekker-Wong pressure-sinkage model [9], [11] and Janosi-Hanamoto shear stress-shear displacement model [12], which are well-established terramechanics models.

## 2. Methodology

### *2.1. Multibody formulation using Tangent Space ODE formulation*

The multibody dynamic systems considered in this study are made of rigid links constrained by ideal joints. Such constrained dynamic systems are usually formulated using Index 3, Index 2, or Index 1 DAEs. However, these DAE formulations consider only one of the position, velocity, or acceleration constraints, respectively [13]. Further, Haug [14], [15] proved that the tangent space ODE formulation captures the highly nonlinear nature of constrained multibody dynamic systems and exhibits better error control with all the dynamic parameters after integration. This is because it includes constraints on position, velocity, as well as acceleration simultaneously. Hence, the presented work was conducted using Haug's tangent space ODE formulation. In this section, the general formulation of a multibody system using tangent space ODE is provided. The formulation is then applied to a case study in section 3.

Let $\mathbb{I}$ be an inertial frame of reference centered at $O$ with axes $X, Y, Z \in \mathbb{R}^3$, represented as $I \triangleq \{O, X, Y, Z\}$ and a body fixed frame $\mathbb{B} \triangleq \{O', x, y, z\}$ centered at O' with axes X',Y',Z' $\in \mathbb{R}^3$. The generalized coordinates are given by $\boldsymbol{q} = [\boldsymbol{r}\, \boldsymbol{p}]^T$ where $\boldsymbol{r} = [\boldsymbol{x}\, \boldsymbol{y}\, \boldsymbol{z}]^T$. The Euler parameters, $\boldsymbol{p} = [\boldsymbol{e}_0\, \boldsymbol{e}_1\, \boldsymbol{e}_2\, \boldsymbol{e}_3]^T$ and the constraint equations are given by $\boldsymbol{\Phi}(\boldsymbol{q}, t) = 0$. For a constrained multibody system, the general equation of motion in the DAE form is given by,

$$\boldsymbol{M}(\boldsymbol{q})\ddot{\boldsymbol{q}} - \boldsymbol{S}(\boldsymbol{q}, \dot{\boldsymbol{q}}) + \boldsymbol{\Phi}_q^T(\boldsymbol{q}, t)\boldsymbol{\lambda} - \boldsymbol{Q}^A(\boldsymbol{q}, \dot{\boldsymbol{q}}, t) = 0 \tag{1}$$

The equivalent tangent space ODE formulation derived in [13,14] is given by,

$$\boldsymbol{D}^T\boldsymbol{M}\boldsymbol{D}\ddot{\boldsymbol{v}} = \boldsymbol{D}^T\left(\boldsymbol{M}\boldsymbol{U}\boldsymbol{B}\boldsymbol{\gamma} + \boldsymbol{S} + \boldsymbol{Q}^A\right) \tag{2}$$

where,

$$\gamma = -\left(\boldsymbol{\Phi}_q(\boldsymbol{q})\dot{\boldsymbol{q}}\right)_q\dot{\boldsymbol{q}} - \boldsymbol{\Phi}_{tt} \tag{3}$$

$$\boldsymbol{U} = \boldsymbol{\Phi}(\boldsymbol{q}_0)_q^T \tag{4}$$

$$\boldsymbol{B} = \boldsymbol{\Phi}(\boldsymbol{q})_q\boldsymbol{U}^{-1} \tag{5}$$

$$\boldsymbol{D} = \boldsymbol{I} - \boldsymbol{U}\boldsymbol{B}\boldsymbol{\Phi}(\boldsymbol{q})_q\boldsymbol{V} \tag{6}$$

The term $\boldsymbol{M}(\boldsymbol{q})$ is the combined mass-inertia matrix, $\boldsymbol{V}$ is the null space of the constraint Jacobian $\boldsymbol{\Phi}_q(\boldsymbol{q})$ and $\boldsymbol{q}_0$ is the orientation of the system at a given time step where reparameterization happens. The terms $\boldsymbol{U}$ and $\boldsymbol{V}$ together form the tangent space planes. For details on the tangent space ODE formulation, the reader is referred to [13], [14].

### *2.2. Terramechanics model*

In deformable terrain, the forces acting between the tire and the surface depend on several parameters, such as soil moisture, soil cohesiveness, tire inflation pressure, tire normal load, etc. The state-of-the-art terramechanics models which capture these aspects of tire terrain interaction could be classified as analytical, empirical, and semi-empirical. The normal force acting on the tire due to its interaction on the deformable terrain depends on both the normal stress and the shear stress. Based on the superposition principle, the stresses are evaluated by integrating the stresses in an infinitesimal element over the contact patch. The variable of integration is the angle subtended at the wheel center and the integration limits are the entry and exit angles. In this study, the Bekker-Wong (7-9) equations provide the relationship to obtain normal stress. The Janosi-Hanamoto equations (10-12) provide the relationship to obtain the shear stress, which is also a function of the normal stress. The following equations summarize how the stresses are obtained in order to calculate the vertical force acting on the tire $F_z$:

$$z = h(\theta) \tag{7}$$

$$h(\theta) = \begin{cases} r(cos\theta - cos\theta_f, \theta_m \le \theta \le \theta_f \\ r(cos\theta_e - cos\theta_f, \theta_r \le \theta \le \theta_m \end{cases} \tag{8}$$

$$\sigma(\theta) = \left(\frac{k_c}{b} + k_\phi\right) h(0)^n \tag{9}$$

$$\tau(\theta) = \left(c + \sigma(\theta)tan\phi_f\right)\left(1 - e^{\frac{-j(\theta)}{K}}\right) \tag{10}$$

$$j(\theta) = r\left(\theta_f - \theta - (1 - i)\left(sin\theta_f - sin\theta\right)\right) \tag{11}$$

$$i = \frac{\omega r - V_x}{\omega r} \tag{12}$$

$$F_z = rb \int_{\theta_r}^{\theta_f} (\tau(\theta)sin + \sigma(\theta)cos\theta)d\theta \tag{13}$$

In the above equations, the terms, $z$ represents the sinkage, $c$ is the soil cohesion, $b$ is the tire width in the contact area, $k_c$ *and* $k_\phi$ are Bekker-Wong parameters, $\phi_f$ is the internal friction angel, $\theta$ is the angular displacement from the point of entry at the tire-terrain interface, $\sigma(\theta)$ is the normal stress, $\tau(\theta)$ is the shear stress, $j(\theta)$ is the shear displacement, $i$ is the slip ratio, $\omega_r$ is the wheel spin velocity, $V_x$ is the linear velocity of the wheel and $r$ is the undeflected geometric radius of the tire. For the case study in section 3, the terrain was assumed to be sandy loam. The following parameters for sandy loam were used in the terramechanics model.

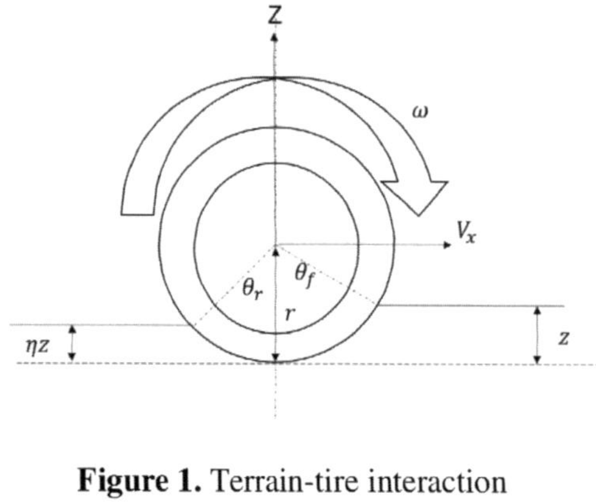


**Figure 1.** Terrain-tire interaction

**Table 1**: Terrain parameters

| **Parameter** | **Value** |
|---|---|
| $c$ | $1700\ Pa$ |
| $k_c$ | $5300$ |
| $k_\phi$ | $1515000$ |
| $\phi_f$ | $29^o$ |
| $\omega$ | $3\ rad\ s^{-1}$ |
| $V_x$ | $0.35\ ms^{-1}$ |
| $b$ | $0.075\ m$ |
| $r$ | $0.25\ m$ |

## 3. Case study

Two case studies were considered: a quarter-car model with linear suspension and a quarter-car model with McPherson strut suspension, which is a nonlinear model.

### *3.1. Linear Suspension Model:*

The linear suspension model comprises a sprung mass $M$, unsprung mass $m$ connected by a damper and spring with damping constant $C$ and stiffness $K$ respectively, as shown in Figure 2. Since in suspension test rigs the input profile resembles the road profile, an actuator plate is modeled which also mimics a deformable terrain. The movement of the actuator plate (virtual deformable terrain) excites the system assembly. The system, when formulated using a multibody approach, consists of three joints. The sprung mass and unsprung mass are connected by a translation joint. The sprung mass and ground are connected by a translational joint. The actuator plate, of negligible mass, is connected to the ground through a translational joint. The system can be formulated using the tangent space ODE formulation for reasons specified in section 2. The number of generalized coordinates, $\boldsymbol{q} \in \mathbb{R}^3$, where $\boldsymbol{q} = [\,x\ y\ \theta]^T$. Since there are three translational joints, the constaint manifold, $\boldsymbol{\Phi}(\boldsymbol{q}) \in \mathbb{R}^6$. Hence, the number of degrees of freedom of the system is 2. For further details into modeling the constraint manifold, readers are directed to [13], [14].

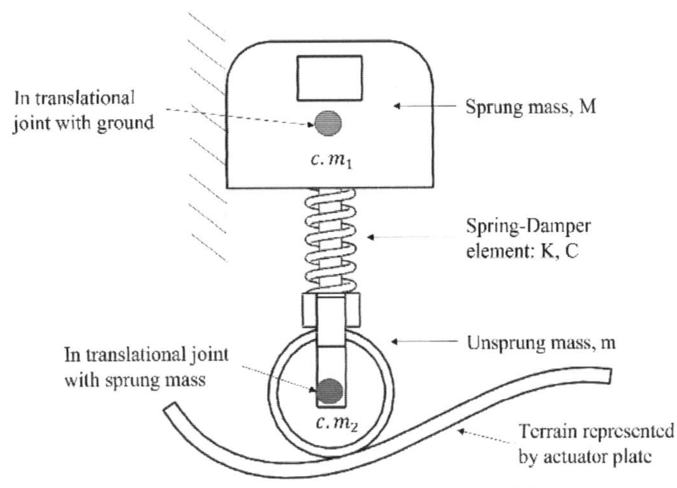


**Figure 2.** Linear suspension model

**Table 2**: Linear model study parameters:

| **Parameter** | ***Value*** |
|---|---|
| $M$ | $200\ kg$ |
| $m$ | $71\ kg$ |
| $K$ | $27600\ Nm^{-1}$ |
| $C$ | $1000\ Nsm^{-1}$ |
| mass of actuator plate | $0.001\ kg$ |

### *3.2. Nonlinear Suspension Model: McPherson Strut Suspension*

The nonlinear suspension model comprises a sprung mass $M$, unsprung mass $m$ connected by a McPherson strut suspension and spring with damping constant $C$ and stiffness $K$ respectively as shown in the Figure 3. The nonlinear suspension assembly is modeled as five bodies (including the actuator plate) connected by six constraints. Of the six constraints, there are three translational joints and three revolute joints. The actuator plate is connected to the ground through a translational joint.

The sprung mass is connected to the ground through a translational joint. Body 2 (cylinder) is connected to the unsprung mass through a translational joint. The Body 2 is connected to the sprung mass through a revolute joint. Body 4 is connected to the sprung and unsprung mass through revolute joints. The system can be formulated using Tangent Space ODE formulation for reasons specified in section 2. The number of generalized coordinates, $\boldsymbol{q} \in \mathbb{R}^3$, where $\boldsymbol{q} = [\,x\ \ y\ \ \theta]^T$. Since there are three translational joints and three revolute joints, the constaint manifold, $\boldsymbol{\Phi}(\boldsymbol{q}) \in \mathbb{R}^{12}$. Hence, the number of degrees of freedom of the system is two.

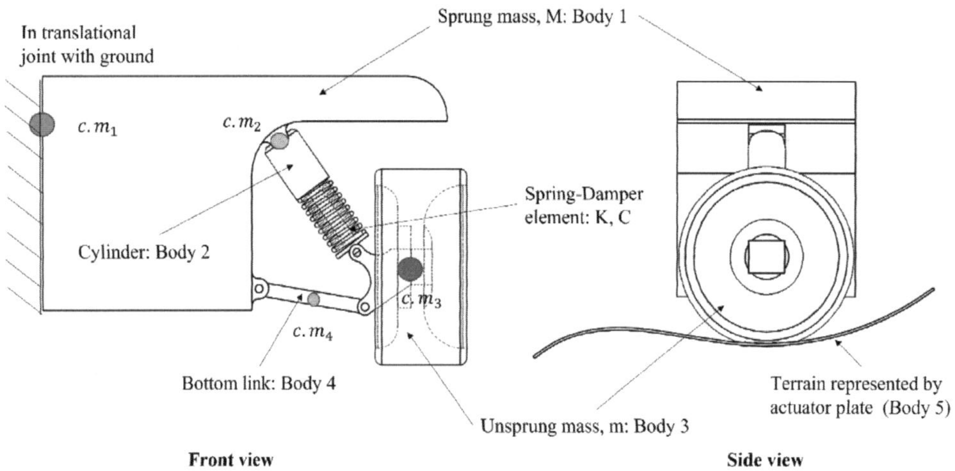


**Figure 3.** Nonlinear model: McPherson strut suspension.

**Table 3**: Nonlinear model study parameters

| **Parameter** | **Value** |
|---|---|
| $M$ | 200 Kg |
| $m$ | 20 Kg |
| $K$ | 37600 N/m |

| | |
|---|---|
| $C$ | 1000Ns/m |
| mass of actuator plate | 0.001 kg |
| mass of body 2 | 3 Kg |
| mass of body 4 | 5 Kg |

### *3.3. Link between multibody formulation and terramechanics model*

The multibody system formulated using the tangent space ODE is linked with the Bekker-Wong pressure-sinkage terramechanics model during the formulation of the vector containing external forces acting on the system.

The following algorithm explains this link between two models:

1. At time $t_n$, the sinkage $z$ is evaluated using the difference in the coordinates of actuator plate and unsprung mass.
2. The sinkage $z$ is used to evaluate the compaction force using equations (7-13).
3. The force value obtained in step 2, is then entered into $\boldsymbol{Q}^A$ in equation (2).
4. The acceleration of generalized coordinates is obtained from equation (2).
5. The equation (2) is then integrated and the generalized coordinates and velocities are evaluated for time $t_{n+1}$.

## 4. Results and discussion

The multibody formulation in combination with the terramechanics force models were employed to a linear suspension model. The position, velocity, and acceleration plots are shown in the figure along with the soil contact force plot. The system was solved as an initial value problem, solely on its inherent dynamics excited by the terrain profile.

### *4.1 Linear suspension model*

As observed in Figure 4, the system response converges to a steady state, after a brief initial transient phase. This can be seen from the relative positions of the terrain, sprung mass and unsprung mass in the position-time plot. The multibody equation was numerically solved using implicit SDIRK 54 integrator. It was ensured that there was no violation of the constraint manifold. The norm of the constraint manifold was found to be $||\boldsymbol{\Phi}(\boldsymbol{q})|| = 3.6 \times 10^{-10}$ which is well below the set threshold of $10^{-6}$.

For the simulation study, the suspension assembly was dropped on the sandy loam terrain and then made to follow a terrain profile. Initially, the wheel sinks into the sandy loam, however, after a finite amount of time, the energy in the suspension dissipates which results in the steady-state behavior. Sandy loam is a mixture of sand and clay, hence it is expected for a compaction behavior when the tire traverses through such a terrain, that is, there should be a finite amount of sinkage during traversal. The difference between the terrain profile and terrain contact point gives the sinkage. It can be observed that the terramechanics forces in the vertical direction is directly proportional to the sinkage raised to the exponent $n$ as predicted by the Bekker's pressure-sinkage relationship. However, the compaction is minimum after 2.5s which can be inferred from the sinkage. The sprung mass displacement also is shown to converge to a steady state behavior, the harmonic nature arises due to the terrain profile input. These results demonstrate the applicability of terramechanics based force models in a tangent space ODE formulation of multibody systems.

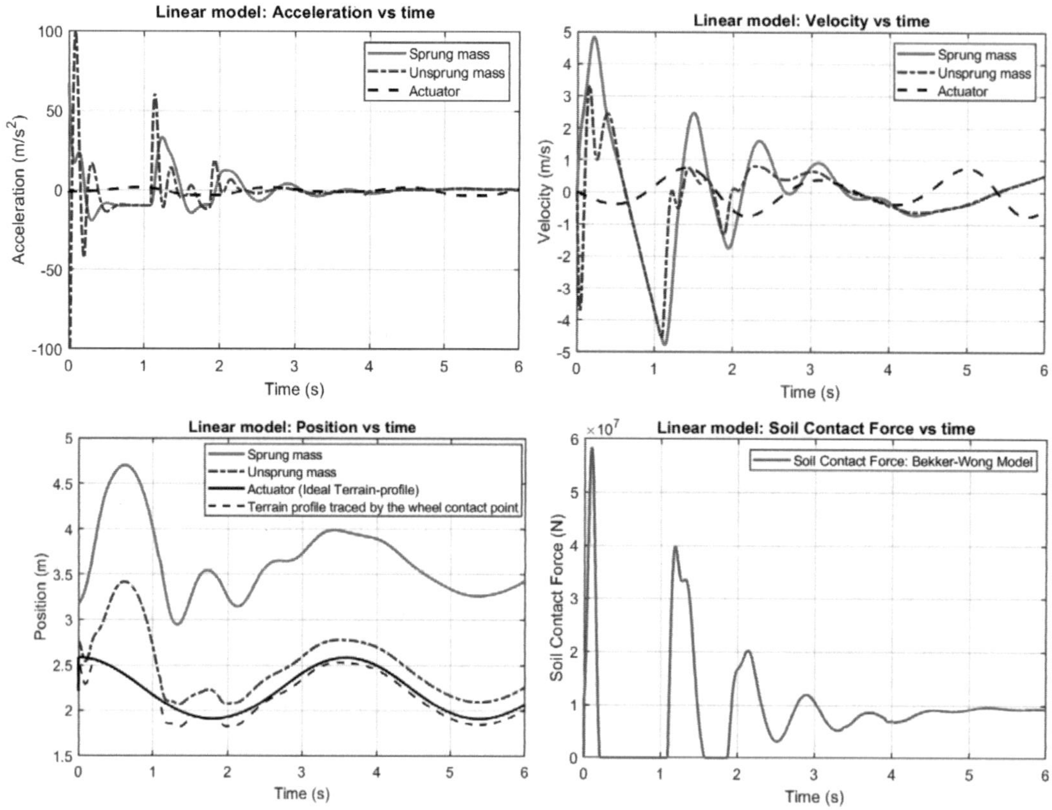


**Figure 4.** Linear suspension model results

*4.2 Nonlinear Model: McPherson Strut Suspension*

As in the linear suspension model, the simulation was carried out for a nonlinear model for McPherson strut suspension. Unlike the linear model, the nonlinear suspension system accounts for the different joints between the links, which implies that several of the constraint forces have been accounted for in the system. The system was simulated as an initial response system as in the previous case and the position, velocity and acceleration plots are shown in Figure 5. It can be observed in this case as well, that after a brief initial transient phase, the system converges to a steady state behavior. The tangent space ODE was integrated using an implicit numerical integration scheme, SDIRK 54. The simulation was run for 6 seconds and there were no singularities observed during the simulation. It was ensured that there was no violation of the constraint manifold. The norm of the constraint manifold was found to be $||\boldsymbol{\Phi}(\boldsymbol{q})|| = \mathbf{5.8 \times 10^{-8}}$ which is well below the set threshold of $\mathbf{10^{-6}}$.

For the simulation of the nonlinear suspension, the system assembly was dropped on the sandy loam terrain (for the dynamic settling, in order to obtain the initial configuration parameters), and then made to follow a terrain profile. It can be seen from the position plot that the wheel sinks into the sandy loam initially, however, after a finite amount of time, the energy in the suspension dissipates which results in the steady state behavior. In the case of the nonlinear model, the sinkage is larger than that of the linear model. It can again, be observed that the terramechanics forces in the vertical direction is directly proportional to the sinkage raised to the exponent $n$ as predicted by the Bekker's pressure-sinkage relationship. The sprung mass displacement also is shown to converge to a steady state behavior, the harmonic nature arises due to the terrain profile input. These results demonstrate the applicability of terramechanics based force models in a tangent space ODE formulation of multibody systems.

IOP Conf. Series: Materials Science and Engineering **1311** (2024) 012020 doi:10.1088/1757-899X/1311/1/012020

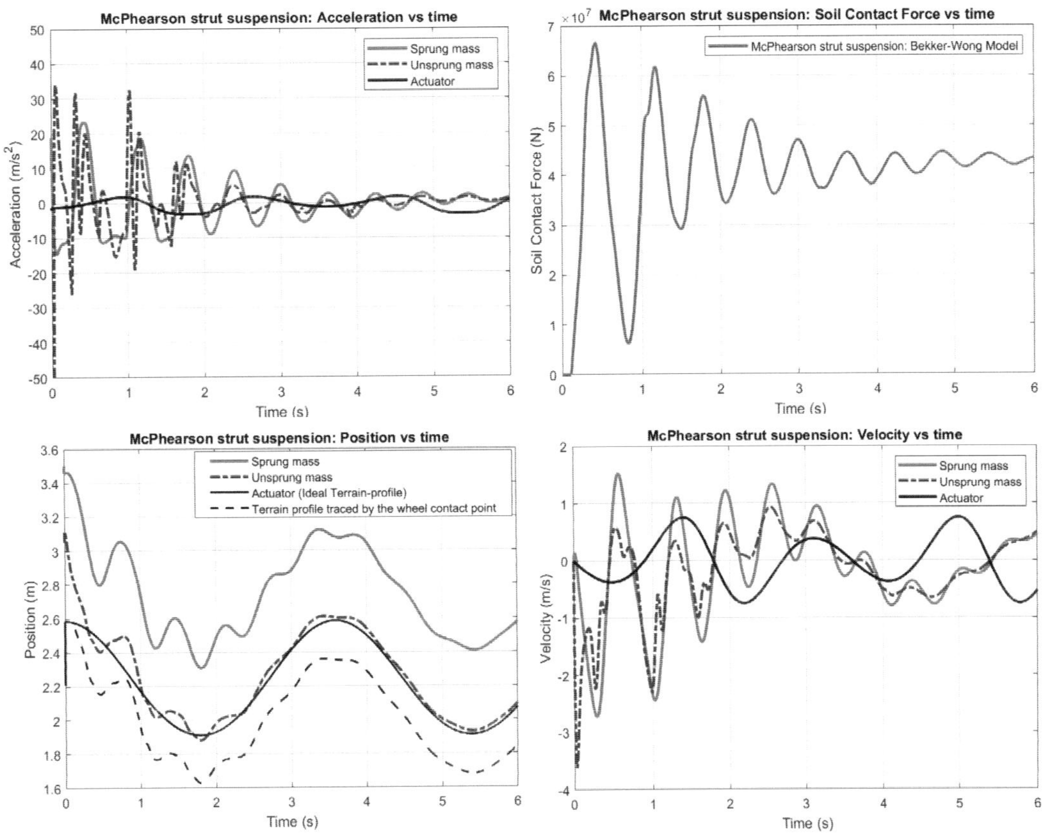


**Figure 5.** McPherson suspension model results

## 5. Conclusions

Two different types of suspension systems were modeled to study vertical dynamics of a quarter-car vehicle model on deformable terrain. Analysis of such systems require modeling not only the suspension system but also the forces acting on the tire due to its interaction with deformable terrain. To this end, a linear and a nonlinear suspension system with sprung and unsprung masses were modeled. The tangent space ODE multibody dynamics formulation was employed to model the suspension system. The forces acting at the tire-terrain interface were modeled using Bekker-Wong and Janosi-Hanamoto terramechanics forces. The combined formulation using multibody dynamics approach and terramechanics were simulated for a sandy loam terrain. The position, velocity, acceleration and soil contact force plots were obtained. This study demonstrated the applicability of terramechanics based force models in multibody systems. Such a modeling framework could be leveraged to understand design parameters in off-road vehicle suspension systems. Future work includes validating the simulation results on a suspension rig with deformable terrain. Further, the terramechanics models can be used with neural network surrogates that can be directly incorporated in the multibody dynamic system formulation.

## References

[1] Rahnejat H 1998 *Multi-Body Dynamics Vehicles, Machines, and Mechanisms* (Professional Engineering Publishing Limited, London).
[2] Hegazy S, Rahnejat H and Hussain K 2000 Multi-Body Dynamics in Full-Vehicle Handling Analysis under Transient Manoeuvre *Veh. Sys. Dyn* **34** 1-24.

[3] Inman D 2000 *Engineering Vibration 2nd ed.* (Prentice Hall, Upper Saddle River, NJ).
[4] Goncalves F D 2001 *Dynamic analysis of semi-active control techniques for vehicle applications, in Mechanical Engineering* (Virginia Polytechnic Institute and State University, Blacksburg).
[5] Levesley M, Ramli R, Stembridge N and Crolla D 2007 Multi-body co-simulation of semi-active suspension systems. *Proc. Inst. Mech. Eng.: JMBD* **221** 115 – 99.
[6] Sandu C, Andersen E R, and Southward S 2011 Multibody dynamics modeling and system identification of a quarter-car test rig with McPherson strut suspension, *Veh. Sys. Dyn* **49** 153-179.
[7] Taheri Sh, Sandu C, Taheri S, Pinto E, Gorsich D 2015 A technical survey on Terramechanics models for tire–terrain interaction used in modeling and simulation of wheeled vehicles, *J Terramech* **57** 1-22.
[8] Dallas J, Jain K, Dong Z, Sapronov L, Cole M P, Jayakumar P and Ersal T 2020 Online terrain estimation for autonomous vehicles on deformable terrains, *J Terramech* **91** 11-22.
[9] Bekker M G 1962 *Theory of Land Locomotion* (The University of Michigan Press, Ann Arbor, MI).
[10] Mocera F, Somà A and Nicolini A 2020 Grousers Effect in Tracked Vehicle Multibody Dynamics with Deformable Terrain Contact Model *Appl. Sci.* **10** 6581, 2020.
[11] Wong J Y and Reece A 1967 Prediction of rigid wheel performance based on the analysis of soil-wheel stresses part I. Performance of driven rigid wheels. *J. Terrramech.* **4** 81–98, 1967.
[12] Janosi Z, Hanamoto B and Ferraris G 1961 The analytical determination of drawbar pull as a function of slip for tracked vehicles in deformable soils. In: *International Conference of the Mechanics of Soil-Vehicle Systems* (Torino, Italy).
[13] Haug E J 2022 *Computer-Aided Kinematics and Dynamics of Mechanical Systems*, Vol. II, Modern Methods, 3rd ed.
[14] Haug E J 2016 An Ordinary Differential Equation Formulation for Multibody Dynamics: Holonomic Constraints *ASME. J. Comput. Inf. Sci. Eng* **16** (2).
[15] He R, Sandu C and Osorio J E 2019 Systematic Tests for Study of Tire Tractive Performance on Soft Soil: Part I – Experimental Data Collection *J. Terramech.* **85** 59-76 .
[16] He R, Sandu C and Osorio J E 2019 Systematic Tests for Study of Tire Traction on Soft Soil: Part II — Parameterization of Terramechanics Models and Tire Models *J. Terramech.* **85**. 77-88.
[17] Li L and Sandu C 2007 On the Impact of Cargo Weight, Vehicle Parameters, and Terrain Characteristics on the Prediction of Traction for off-Road Vehicles *J. Terramech.* **44** 221-238.

**Acknowledgements**
This research did not receive any specific grant from funding agencies in the public, commercial, or not-for-profit sectors. The research has been supported by the Terramechanics, Multibody, and Vehicle Systems (TMVS) Lab at Virginia Tech.

IOP Conf. Series: Materials Science and Engineering **1311** (2024) 012021 doi:10.1088/1757-899X/1311/1/012021

# Behaviour of wet clay due to off-road traffic loading

**V S Swamy[1], R Pandit[2], A Yerro[2], C Sandu[1], B Castellanos[2], D Rizzo[3], and K Sebeck[3]**

[1] Terramechanics, Multibody, and Vehicle Systems (TMVS) Laboratory, Department of Mechanical Engineering, Virginia Tech, Blacksburg, VA, United States
[2] Department of Civil and Environmental Engineering, Virginia Tech, Blacksburg, VA, United States
[3] U.S. Army CCDC Ground Vehicle Systems Center, 6501 E. 11 Mile Road, Warren, MI 48397

varshasswamy@vt.edu

**Abstract**. The presence of water in the soil can greatly affect the trafficability of off-roading vehicles, making it difficult to operate. As the water content increases, it can eventually lead to sinkage and immobility. Despite the relevance of this topic in many applications, such as military, agricultural, and earthmoving vehicles, the studies available on tire and moist soil interaction are limited. The numerical modeling of such a system is very challenging because it requires accounting for hydro-mechanical effects, large deformations, non-linear behavior, and complex interactions between the tire and the soil. This work contributes to the understanding of the behavior of moist clay under the influence of traffic loading. In particular, the preliminary results of numeric simulations to study the interaction between moist clay and a passing tire using physics-based techniques are presented. A great emphasis is given to the development of an accurate soil model that can capture the fundamental characteristics of moist clays in saturated conditions.

## 1. Introduction

Over the years, tire-soil interaction studies have been essential to determine the trafficability of different vehicles in different off-road conditions. The modeling of tire-soil interaction is complex, with many aspects to be considered. The responses of the soil and the tire-soil interface are highly dependent on the load transferred from the tires, the inflation pressure of the tire, the tire characteristics (e.g., geometry, construction, material), and the different factors directly related to the soil characteristics, including moisture content, soil type, and density, among others [1]. Numerical techniques have been used to model this complex system, including the tire, the soil, and their interaction. The Finite Element Method (FEM) has been employed to simulate and validate the soil-tire interaction. Along with studies concentrating on modeling and validating the tire model, studies on the soil response have also become relevant because accurately capturing the terrain's performance is essential to predict off-road trafficability. The behavior of dry soil has been modeled in the literature using simple models like Von-Mises [2], Mohr-Coulomb [3], Drucker-Prager [4], [5], and advanced models like the Drucker – Prager cap models [6], [7], [8]. Based on the compiled literature, it was noticed that there had been more studies on coarse-grained soils (sands and gravels) than fine-grained soils (clays and silts). Plastic fine-grained soils are highly non-linear, and their compression and shearing behavior is dependent on the loading

DISTRIBUTION A. Approved for public release, distribution unlimited. OPSEC #6910

IOP Conf. Series: Materials Science and Engineering **1311** (2024) 012021 doi:10.1088/1757-899X/1311/1/012021

history of the soil. For this reason, the characterization of such behavior usually requires the use of advanced elastoplastic constitutive laws.

The study of saturated soil behavior is also critical for the prediction of off-road trafficability since it decreases with increasing water content. The introduction of pore water in the system further complicates the numerical models, and, in the literature, there is not much information on this subject. The existing models typically consider simple constitutive laws that are calibrated under very specific soil conditions and cannot be easily generalized. The use of well-established advanced soil models, commonly used in geotechnical engineering (e.g., the Cam Clay model), could potentially improve the understanding of the physics behind the effects of pore water pressure in the interaction of a tire with saturated soils.

## 2. Fundamental soil behavior

The behavior of fine-grained soils is not the same as of coarse-grained materials because of factors such as the difference in particle size, the electrochemical interactions between soil particles, and the low hydraulic conductivity of fine-grained soils. The addition of water further complicates modeling these materials. Key factors controlling the behavior of fine-grained soils are (a) the stresses carried by the soil grains (i.e., effective stresses), (b) the drainage conditions (i.e., drained vs. undrained conditions), and (c) the stress history (i.e., normally consolidated vs. overconsolidated). These concepts are explained herein.

The behavior of soils depends on the stresses carried by the solid particles through the contacts, so-called "effective stresses." In dry conditions, effective stresses are equal to the total stresses applied to the soil resulting from self-weight and external loading. However, when the soil is saturated (space between particles is fully filled with water), the effective stresses are not equivalent to the total stresses because the water in the porous can exert pressure on the soil particles reducing the contact stress. In this case, effective stresses equal the difference between the total stress and the pore water pressure. Upon loading, especially in fine-grained soils, as the soil skeleton is compressible, but water is not, the load is first taken by the water in the porous. As time passes, depending on the permeability of the soil, water migrates from the porous, reducing the pore pressure until the soil particles take all the applied load. This process is known as consolidation. The effective stress concept is the most important concept of soil mechanics, and it was proposed by Karl Terzaghi in 1936 [9].

From the definition of effective stress, it is easy to perceive the importance of predicting the evolution of pore water pressure to understand the behavior of soils under different loading conditions. For saturated soils, a loading process changes (increases or decreases) the water pressure in the pores. As a result, the stress carried by the solid particles (i.e., effective stress) varies. Among other things, the pore water pressure developed resulting from external loading (excess pore water pressure) depends on the hydraulic conductivity (or permeability) of the soil and the rate of loading. Very permeable soils dissipate excess pore pressure much faster than low permeable soils because the liquid can move faster through pores. In this framework, two hydro-mechanical conditions can be defined: drained and undrained. The drained condition corresponds to the scenario when no excess pore pressure is induced by the loading and only hydrostatic pore pressure is acting on the soil (e.g., when the soil is loaded very slowly compared to its ability to drain) or when enough time has passed for the excess pore pressure caused by the applied loads has dissipated. Contrarily, an undrained condition is considered when the loading induces excess pore pressure (e.g., when soil is loaded rapidly compared to its ability to drain). Overall, the drainage and loading conditions control the effective stress evolution and, ultimately, the soil strength.

The behavior of fine-grained soils is also highly dependent on the stress history. In particular, fine-grained soils can be normally consolidated (NC) or overconsolidated (OC). If the present effective overburden pressure to which the soil is being subjected is the maximum it has seen (preconsolidation pressure), the soil is deemed NC. On the contrary, if the present effective overburden pressure is less than the preconsolidation pressure (i.e., the soil has been through an unloading process), the soil is OC. It should be considered that a loading process can change an OC soil into NC if the stress level reaches

IOP Conf. Series: Materials Science and Engineering **1311** (2024) 012021 doi:10.1088/1757-899X/1311/1/012021

and exceeds the preconsolidation pressure. At the same mean effective stress, the shear strength for OC soils is generally higher than that of NC soils. These differences can be explained by the process of consolidation and its consequences (e.g., reduction in void ratio). The consolidation process happens in saturated soils and consists of the reduction in soil volume due to a reduction in the volume of water in the porous (note that the consolidation process differs from compaction, in which the reduction in volume is caused by a reduction of pores due to air migration). The migration of pore water results in volumetric deformation of the soil (settlement), particles rearrange, and the material densifies. Therefore, the same soil subjected to different stress histories will have different particle arrangements (i.e., packing), which will result in different behaviors for the same loading conditions.

A well-accepted framework in soil mechanics for describing the above soil responses by bringing together the consolidation, volumetric strains, stress history, shear stress, and pore water pressure is the critical state soil mechanics [10]. The non-linear stress evolutions before failure can be predicted. A unique failure state called the critical state is expected for the soil to attain at higher deformations.

## 3. Behavior of clay

Different laboratory tests like the incremental stress consolidation and controlled rate of strain consolidation tests are used to characterize the compressibility behavior of soils, while others like the triaxial, direct shear, direct simple shear, and ring shear are used to characterize the strength and some, are also used to characterize the stress-strain behavior of soils. Strength tests can be performed in different ways, including consolidated drained (CD) and consolidated undrained (CU). The first descriptor refers to whether the specimen is consolidated or not before shearing, while the second one refers to the conditions during shear. In a drained test, the specimen is allowed to change in volume change during shear, and the excess pore water pressure is maintained close to zero by applying the loads at a very slow rate. Drained tests are used to obtain effective stress shear strength parameters. In undrained tests, the load is applied more rapidly, the volume change is not permitted during the shearing stage, drainage is not allowed, and pore water pressures are generated. If pore water pressures are measured, then effective stress shear strength parameters can be obtained.

Figure 1 shows an idealized behavior of NC and OC fine-grained soils in a CD triaxial loading according to the critical state framework. Both specimens are assumed to be consolidated to the same mean stress (Figure 1a). OC and NC soils will tend to have similar strength at larger deformations (Figures 1a and 1b). The main difference is that OC soil tends to present a more pronounced strain-softening behavior with a higher peak at smaller strains (Figures 1a and 1b). NC soil is typically more ductile. In terms of volume change, NC soil compresses (contractive in nature) during shearing, while OC tends to expand at larger strains (dilative in nature). (Figure 1c). It is important to highlight that the behavior described using CD testing is representative of saturated fine-grained soils in drained conditions (i.e., under long-term conditions, very slow loading rates) and is also valid for dry conditions.

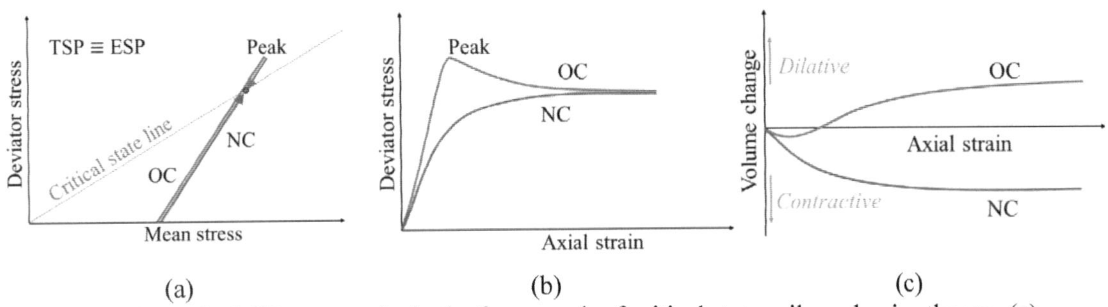


**Figure 1.** Theoretical CD test results in the framework of critical state soil mechanics theory. (a) Total and effective stress paths (TSP and ESP). (b) Deviatoric stress vs. strain curve. (c) Volume change vs. strain.

DISTRIBUTION A. Approved for public release, distribution unlimited. OPSEC #6910

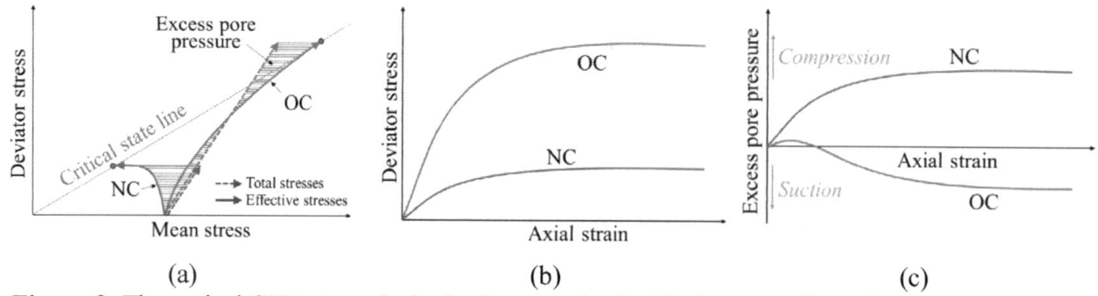


**Figure 2.** Theoretical CU test results in the framework of critical states soil mechanics theory. (a) Total and effective stress paths (TSP and ESP). (b) Deviatoric stress vs. strain curve. (c) Excess pore pressure vs. strain.

Figure 2 schematically represents the results from a CU triaxial loading performed on saturated NC and OC fine-grained soils. The results from CU tests are especially relevant to determining the behavior of saturated fine-grained soils in undrained conditions (or rapid loading). In this case, although the same total loading conditions are applied in both specimens (Figure 2a), the overall behavior of NC and OC soils is very different. The evolution of effective and total stresses does not converge because of the accumulation of excess pore pressure (indicated with a shaded zone in Figure 2a) as the loading increases. The evolution of the pore pressure generated during shear in both clays is also presented in Figure 2c. It is well known that NC soils develop excess pore pressures in compression (resulting from their tendency to contract). OC soils tend to develop a small positive excess pore water pressure in compression followed by suction (negative excess pore water pressures), resulting from their tendency to dilate. As a result, for NC soil, the mean effective stress tends to decrease, while for OC soil, it tends to increase. The direct implication of this behavior is that the strength of NC soils, when loaded in undrained conditions, is smaller than the one from OC soils and smaller than the strength for the same material subjected to drained conditions.

In addition to this fundamental behavior, other conditions influence the undrained behavior of fine-grained soils; for example, they depend on the strain rate (i.e., viscous effects) [11].

## 4. Clay behavior when a tire passes

The behavior of clay subjected to a tire traversing above it is complex. As the wheel passes, two types of loads are acting on it; the weight typically applied vertically on the ground surface (increase in confining mean stress in the soil mass) and the shear loads due to the tire rotation. This shear force exerted on the soil gives the tire traction to move forward. In dry conditions, the soil behavior is equivalent to drained conditions. Total and effective stresses are the same (i.e., soil particles carry all the load). The load increase tends to trigger particle rearrangement, and the void ratio decreases as the soil compacts (air does not offer any resistance and escapes from the soil voids).

In saturated conditions, because of the low permeability of fine-grained soils and the speed of loading, the pass of one tire can be considered an undrained problem, and excess pore pressures are expected to develop. During rapid loading, the shear stresses induced on the soil (weight of the vehicle and the traction transmitted through the tire) are taken by the solid skeleton, while the total mean stress is carried by the pore water trapped in the voids. In the hypothetical case of perfectly elastic clay, the mean effective stresses would not change. After the wheel passes, the total stress and pore water pressure would both return to the initial value as before the tire passed. However, the reality is much more complex; clays are history-dependent materials (e.g., different responses for NC or OC soils) that accumulate plastic deformation, and the development of excess pore pressure is controlled by both mean and deviatoric components of the loading (i.e., shear stresses can also induce excess pore pressure). Considering the fundamentals of soil behavior in undrained conditions (Figure 2), one would expect that a tire passing on NC fine-grained soil would trigger a sudden decrease of the effective stresses in the soil and, in turn, a reduction of the shear strength would immediately difficult its trafficability. On the

IOP Conf. Series: Materials Science and Engineering **1311** (2024) 012021 doi:10.1088/1757-899X/1311/1/012021

other hand, a tire passing on OC fine-grained soil might experience an increase in effective stress, the strength would be enhanced, and the trafficability would improve.

If short-term multi-pass of vehicles is expected, the pore water pressure can accumulate and, in NC fine-grained soil, it can highly worsen the trafficability in the area. At the same time for long-term conditions, resulting from the consolidation process, excess pore pressure can dissipate (at a slow rate), and the soil condition in NC soils will generally improve. On the contrary, in OC soils, the trafficability can improve during short-term conditions and worsen for long-term conditions with the dissipation of suction. Herein, we focus on the study of fine-grained soils due to a single-pass

## 5. Modeling in LS-DYNA

Fine-grained soils are stress-history-dependent materials, and using cap models, such as the Cam-Clay model [12], is one way to incorporate such non-linearities required to simulate the stress-strain behavior of dry and wet, NC and OC fine-grained soils in a unified effective-stress framework. The model selected in this work is the two-invariant Drucker-Prager cap model available on LS-DYNA R12.0 [13]. It is a volumetric hardening cap model. It consists of one failure envelope and an associated elliptical yield surface (cap) that limits the elastic region (Figure 3). The intersection between the cap with the horizontal axis represents the preconsolidation pressure. Therefore, the stress state of NC soils before failure is expected to fall on the cap surface, while OC soils are expected to be in the elastic region. Once the cap is activated, the model hardens, and volumetric and plastic strains are developed. Soil failure is reached when the effective stress path (ESP) reaches the failure envelope. This model also has the capability to incorporate strain-rate effects, but in this work, this feature has not been considered.

The input parameters for the fine-grained soil model tested herein were obtained from a clayey soil previously parameterized by NASA [14]. Other inputs were selected based on values used in the Cam-Clay model. All the input parameters have been provided in Table 1. It is important to highlight that the Drucker-Prager cap model is an effective-stress based model, and the same set of effective strength parameters can be used to model both dry and saturated soil behavior.

In order to investigate the capabilities of the model to reproduce the basic soil behavior discussed in Section 3, two 3D FEM models were implemented in LS-DYNA to reproduce a CD and CU triaxial test. Consolidated triaxial testing consists of two phases, the hydrostatic loading phase, where the confinement pressure is applied on the boundary, and the shear phase, where the top nodes are displaced down with a constant velocity for strain-based tests. The CD test does not account for excess pore pressure to reproduce the drain condition in the laboratory. In the CU test, the analysis is drained in the first phase with permeable boundaries (to allow excess pore pressure to dissipate) and undrained in the second phase with impermeable boundary conditions. The responses for both NC and OC soils for each type of test were simulated. The OC model was implemented by hydrostatically loading the soil to a higher pressure followed by hydrostatic unloading before the deviatoric phase.

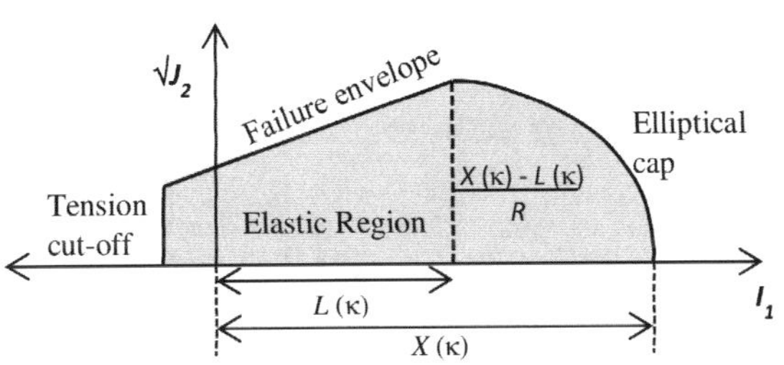


Where
$I_1$ is the first invariant of the stress tensor
$J_2$ is the second invariant of the deviator tensor

**Figure 3.** Two invariant Drucker-Prager cap yield surfaces, modified from [13].

DISTRIBUTION A. Approved for public release, distribution unlimited. OPSEC #6910

**Table 1.** Input parameters of the model LS-DYNA [13].

| Parameter | RO | Bulk | G | Alpha | Theta | R | D | W | TOFF |
|---|---|---|---|---|---|---|---|---|---|
| **Value** | 1453 kg/m$^3$ | 23.166 MPa | 17.23 MPa | 0.0606 | 0.285 | 8 | 0.08 | 0.5 | 13 kPa |

After evaluating the capabilities of the Drucker-Prager cap model, two more FEM models were created in LS-DYNA to investigate the influence of water on tire-soil interaction. The two 3D models are half model consisting of a tire rotating on a horizontal bed of fine-grained soil (1m wide, 8m long, and 2 m deep). The first model is in dry conditions and the second one is in saturated undrained conditions. These models require modeling the soil, the tire, and the contact. In both cases, the material is NC fine-grained soil, and the same parameters used for the triaxial tests are used. The tire is represented by a simple ring with a rigid rim with the dimensions of a P225/55R16 tire. An elastic material modeled with a Young's Modulus of 4.4 MPa and a Poisson's ratio of 0.49 is employed to model the tire. A vertical load of 1000 N is applied to the tire and is rotated kinematically. The soil-tire contact is modeled using surface-to-surface contact from the penalty-based methods readily available in LS-DYNA. The friction coefficient at the contact, obtained from the literature, is 0.68 peak friction and 0.65 sliding friction for the dry case and 0.55 peak friction and 0.4 sliding friction for the saturated case [15]. Non-reflecting infinite boundary conditions are imposed to mitigate spurious reflections along the edges. The model calculation consists of three phases: (1) stress initialization of the soil, (2) tire placement, and (3) rotation phase. In the dry model, pore pressures are zero during the whole process. In the saturated model, drained conditions are considered for the first phase of the model, which means that excess pore pressure dissipates and hydrostatic pore pressure distribution is maintained. The placement and rotation phases are conducted in undrained conditions. In all cases, the constitutive model is evaluated in terms of effective stresses.

## 6. Results

### *6.1. Consolidated drained (CD) triaxial test*

The CD test results (shear phase) are provided in Figure 4. As expected, in Figure 4a, the I1-√J2 plot is the same for both NC and OC clay. This is because both systems are subjected to the same Total Stress Path (TSP, controlled by the loading conditions) and TSP and effective stress path (ESP) are equivalent, given that excess pore pressure is zero. However, in the axial strain vs. √J2 plot (Figure 4b), the NC clay experiences the highest mean effective stress it has ever experienced and rapidly accumulates plastic deformation from the start of the shear phase. In the case of the OC clay, the loading would be elastic initially till the stress state reaches the failure, represented by the initial straight line in the red plot in Figure 4b. After that, it starts yielding, and the deformation now has both elastic and plastic parts. This can be seen as the second part of the red curve in Figure 4b, where the slope has reduced.

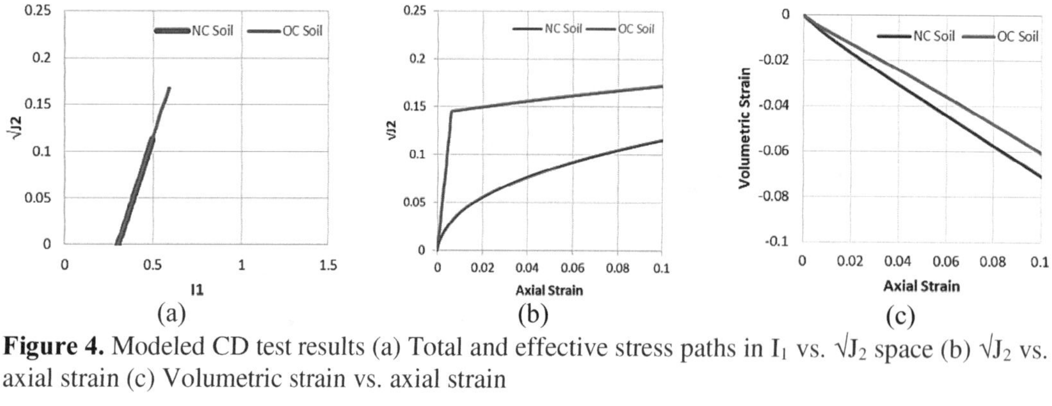


**Figure 4.** Modeled CD test results (a) Total and effective stress paths in $I_1$ vs. $\sqrt{J_2}$ space (b) $\sqrt{J_2}$ vs. axial strain (c) Volumetric strain vs. axial strain

DISTRIBUTION A. Approved for public release, distribution unlimited. OPSEC #6910

### *6.2. Consolidated undrained (CU) triaxial test*

In the CU test, the ESP of both soils is no longer the same (Figure 5a). For the NC soil, the ESP tends to the left. On the other hand, for OC soil, the stress path is overall much more vertical (first, it goes to the right, and when it reaches the cap, it tends to the left until it reaches the failure envelope) reaching the failure envelope at much larger confinement. The differences between the ESP are controlled by the pore water pressure evolution, presented in Figure 5c. These differences are also observed in Figure 5b where the strength mobilized by the NC soil is much lower than the one mobilized by the OC clay.

Comparing Figure 5 with Figure 2, it can be concluded that the behavior of NC clay is generally well captured. Some features of OC clay are also qualitatively captured (e.g., increased strength, reduced excess pore pressure); however, as seen in the CD case, the model cannot predict dilation behavior, and suction is not developed at higher strains. In practice, this might imply that the model might underestimate the trafficability of OC clay.

### *6.3. Tire on dry and saturated soil*

Loading the tire on the NC soil causes a settlement of 26.9 mm in the dry case and 68.8 mm in the saturated case (Figures 6a and 6b). Figures 6c and 6d show the shear stress profile for both cases. It can be seen that the shear stress is positive in the front and negative in the back of the tire, thus agreeing with work by other researchers [16].

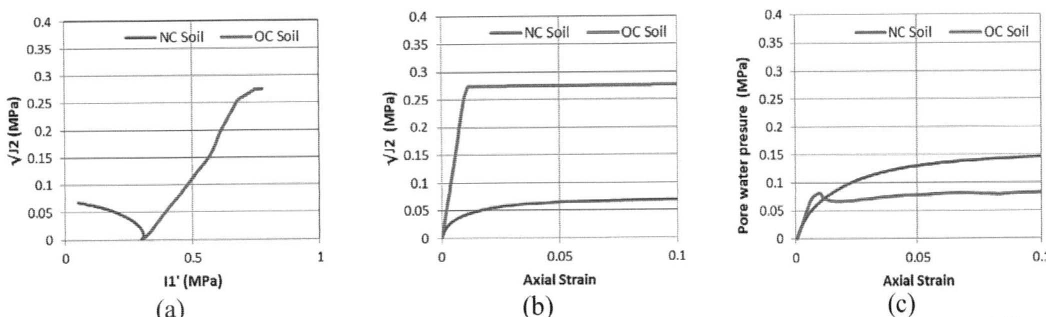


**Figure 5.** Modeled CU test results (a) Total and effective stress paths in $I_1$ vs. $\sqrt{J_2}$ space (b) $\sqrt{J_2}$ vs. axial strain (c) Volumetric strain vs. axial strain

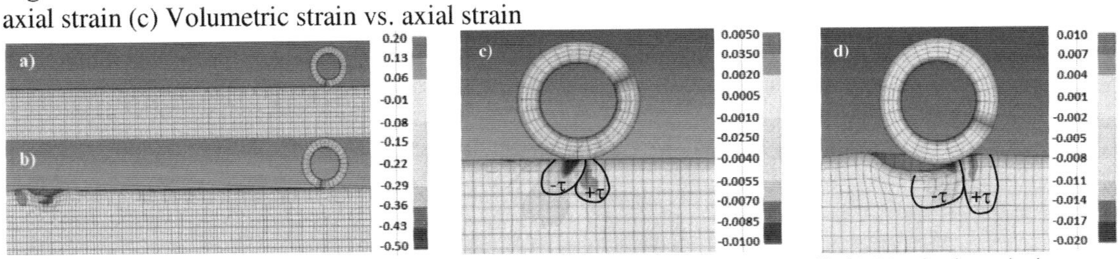


**Figure 6.** Results from the tire-soil simulation: (a) Vertical strain in dry soil (b) Vertical strain in saturated soil (c) Shear stress in dry soil (d) Shear stress in saturated soil. Units in MPa.

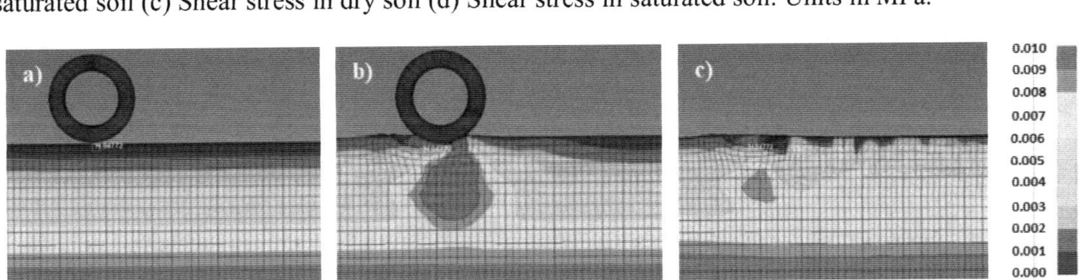


**Figure 7.** Pore water pressure distribution at different times: (a) just before the rolling starts, (b) during the rolling, and (c) after the wheel passes. Units in MPa.

DISTRIBUTION A. Approved for public release, distribution unlimited. OPSEC #6910

For the saturated case, the pore water pressure distribution at different instants is shown in Figure 7. When gravity is loaded, the pore pressures increase with depth, with no excess pore water pressure resulting from the drained analysis. On loading the tire, we see a sudden increase in the pore water pressure just below the tire. After passing, we see that the pore water distribution is not the same as before the tire was placed; this is because it is not allowed to dissipate. This final condition would represent the initial pore water pressure distribution for a second tire passing shortly after the first one, which will encounter a more challenging condition.

## 7. Conclusion

This work brings to light the importance of understanding the pore water pressure evolution when modeling saturated soils and plastic fine-grained soils like clays in particular. From this preliminary analysis, it is concluded that, overall, this soil model can capture the most relevant characteristics of NC in saturated conditions better than OC soils. The tire passing over dry and saturated NC soil is modeled. Shear stress contours are consistent with results from the literature. It is observed that the tire sinkage is higher in the saturated condition, and excess pore water pressure is accumulated as a result of the passing. These changes in the pore water distribution are important to note, especially if multi-passes are expected.

## References

[1] Hallonborg U. Super ellipse as tyre-ground contact area. J Terramechanics. 1996;33(3):125–32.
[2] Hambleton JP, Drescher A. Development of Improved Test Rolling Methods for Roadway Embankment Construction. Minnesota Department of Transportation. MN/RC 2008-08. 2008.
[3] Saliba JE. Elastic-viscoplastic finite-element program for modeling tire/soil interaction. J Aircr. 1990;27(4):350–7.
[4] Farhadia P, Golmohammadia A, Malvajerdib AS, Shahgholi G. Finite element modeling of the interaction of a treaded tire with clay-loam soil. Comput Electron Agric. 2019;162:793–806.
[5] Cueto OG, Coronel CEI, Bravo EL, Morfa CAR, Suarez MH. Modelling in FEM the soil pressures distribution caused by a tyre on a Rhodic Ferralsol soil. J Terramechanics. 2016;63:61–7.
[6] Shoop SA. Finite Element Modeling of Tire – Terrain Interaction. 2001.
[7] Xia K, Yang Y. Three-dimensional finite element modeling of tire/ground interaction. Int J Numer Anal Methods Geomech [Internet]. 2012;36:498–516.
[8] Li H, Schindler C. Investigation of tire-soil interaction with analytical and finite element method. Mech Based Des Struct Mach. 2013;41:293–315.
[9] Terzaghi K. A fundamental fallacy in earth pressure computations. Harvard University. 1936.
[10] Wood DM. Soil behaviour and critical state soil mechanics. Cambridge university press; 1990.
[11] Karlsson R. On cohesive soils and their flow properties. Sweedish Geotech Inst Rep No 5 Stock. 1963;
[12] Roscoe KH, Burland JB. On the generalized stress-strain behavior of “wet” clay. J Heyman FA Leckie (eds), Eng Plast (Cambridge Cambridge Univ Press. 1968;535–609.
[13] Livermore Software Technology (LST). LS-DYNA Keyword user ’s manual volume II - Material Models.
[14] Thomas MA, Chitty DE, Gildea ML, T’Kindt CM. Constitutive soil properties for Cuddeback Lake, California and Carson Sink, Nevada. 2008.
[15] Jin H, Zhou M. On the road friction recognition based on the driving wheels deceleration. IEEE Transp Electrif Conf Expo, ITEC Asia-Pacific 2014 - Conf Proc. 2014;
[16] Wong JY. Theory of Ground Vehicles. John Wiley & Sons; 2001.

## Acknowledgements

The authors wish to acknowledge the technical and financial support of the Automotive Research Center (ARC) in accordance with Cooperative Agreement W56HZV-19-2-0001 U.S. Army DEVCOM Ground Vehicle Systems Center (GVSC) Warren, MI.

IOP Conf. Series: Materials Science and Engineering **1311** (2024) 012022 doi:10.1088/1757-899X/1311/1/012022

# Conceptual design of an electric vehicle chassis using topology optimization method

**I L Scurtu, D Moldovanu**

Department of Automotive Engineering and Transports, Technical University of Cluj-Napoca, Romania

dan.moldovanu@auto.utcluj.ro

**Abstract**. Nowadays, the chassis represents the most important structural part of the vehicle, that has a great influence on the passenger's safety. Lately, topology optimization represents a vital part of the automotive design process. The chassis design can improve the stability and rigidity of the vehicle for accurate handling. In this paper is presented the conceptual design of an EV (electric vehicle) chassis generated by using the topology optimization method. The introduction of the paper presents the latest method used in automotive optimization and the theoretical aspects. In the second part, the basic chassis structure is modelled using CAD software. The objective of this optimization study is to decrease the mass of the chassis to a minimum, without affecting the mechanical properties. Two optimization cases are studied using the steel and aluminium allow material at different boundary conditions. At the final of this research, all results are compared, and the conclusions are presented.

## 1. Introduction

Nowaday, the vehicles design process combine multi-disciplinary requirements, such as acoustics, dynamic stability or crashworthiness. The computer integration in the design process using dedicated software, completed by the cooperation beetwen of different experts lead to find the optimal design solutions [1].

The study presented by Matsimbi begins with a pertinent review of a structural topology optimization formulations used to provide an optimal automotive body structure. One of the first optimized chassis structure, it stands out after the 1930s. By substituting the separate chassis frame structure into a fully unitary design, Shallow (1939) obtain 16% weight saving and increased by approximately 50% of torsional stiffness, compared to the previous chassis structure. It is worth to mention that the author did not use a specific optimization method, based on the overall experience, he demonstrates that a can be obtained a lighter chassis structure by changing the connectivity points of the individual elements [6]. By combining the steel and aluminium materials, Fenyes (1981) found that is possible to develop a lighter beam model of a chassis structure. To generate the optimal design of the chassis was used on optimization program to find the best materials of each part. As a drawback of a method can be mentioned that the optimization technique can be used on the known geometry structure, this being in accordance with the structural sizing optimization [6].

The topology optimization method was applied to improve the structural design of the race car chassis, designed for Formula Varsity Malaysia competition. The main goal is to design the race care structure by reducing the weight, without compromising the driver ergonomics and chassis strength. The monocoque body was designed by using an Asian-size mannequin, following the competition requirements. Initially the chassis was made from steel plate which is converted to a tubular space frame. The constrain of the optimization process is to find the optimal shape by using maximum 30% of monocoque body design space volume. The proposed body frame is analysed for torsional, vertical, and lateral loading, and results was compared to the similar structure of the 2015 electric vehicle model. It is found that the mass of the new optimized model is about 11% lower than of the previous competition model [2].

The topology optimization and additive manufacturing are studied to increase the stiffness of the frontal vehicles structure in the crash events. The studied model includes the main component of the vehicles: the engine, transmission, cooling pack, fuel tank, passengers compartment volume, etc. By using the Optistruct CAE software, the frontal crash loadings are applied, resulting the optimal distribution of the material in the design volume [3].

Another optimization study applied on electric body frame have the goal to reduction the weight by determining the optimal material distribution in design space. The optimal shape is generated applying four conditions of statical analysis: horizontal bending, emergency braking, ultimate torsion, and emergency turn. Appling the topology optimization algorithm in Optistruct software for a basic chassis model reduce the initial weight with approximately 18.96% for improved design [4].

## 2. Optimization process

Nowadays, the optimization process plays an important role in the products design, becoming an indispensable method to generate the manufacturing products. Structural design is based on a basic concept model of a structure. The main structural optimization methods used are topological optimization, shape optimization and topographic optimization.

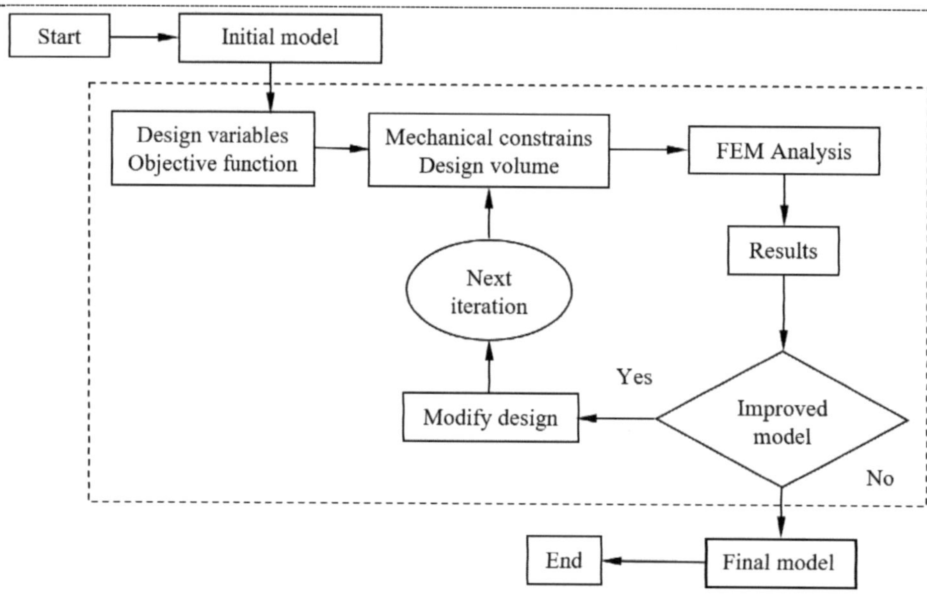


**Figure 1.** Flowchart of the optimization process [8]

The topological optimization process involves the determination of the optimal material distribution in a design space defined by the user, specification of the boundary conditions and assigned material proprieties.

IOP Conf. Series: Materials Science and Engineering **1311** (2024) 012022 doi:10.1088/1757-899X/1311/1/012022

Another formulation of the topological optimization process is defined as the minimization of the structural compliance with the condition of volume reduction. This optimization method is applied at the initial stage of the product design, and it is a particular case of the shape optimization.

Figure 1 shows the general flowchart that describe the necessary steps used to provide a topological optimization of a mechanical structure using dedicated software [8]. The start point of the optimization process is given by the import of the geometrical model into optimization software. In the second stage, the design variables and the objective function are defined, which can minimize the mechanical structure compliance, maximizing the rigidity at a minimum material usage. The definition of the mechanical constrains, and the load case creation are applied in third stage of the optimization process.

In the fourth stage of the process, the basic model is analysed by the dedicated solver of the iterative process. The obtained results are analysed, if the new geometry of the models fulfil the required conditions, then the model is the optimized model. If the model does not meet the required conditions, then the structural model begin an iterative calculation process, until the final optimized structure is generated.

## 3. Topological optimization methodology

### *3.1. CAD chassis design*

The proposed chassis frame model for topology optimization process is modeled in SolidWorks modelling environment [5]. It consists of the base part of the chassis on which the components will be assembled.

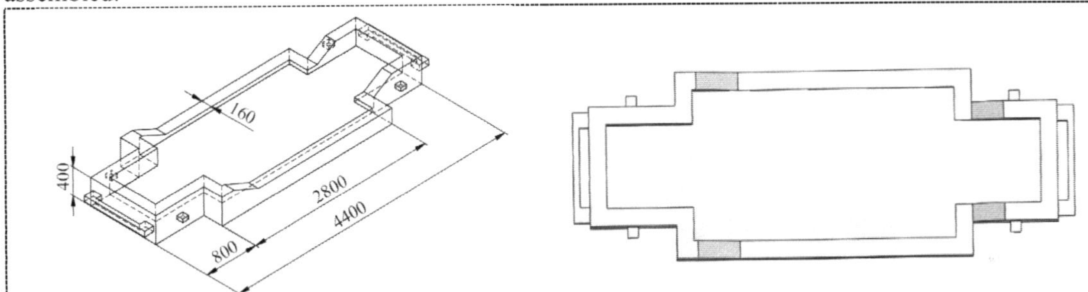


a) Overall dimension of the base model b) Top view of the base model

**Figure 2.** Base model of the chassis frame

Using the modelling techniques, the shape and the overall dimensions of the proposed chassis frame are presented in figure 2.

### *3.2. Chassis discrete model*

To perform the topological optimization, the model volume is divided into two parts: non-design volume and design volume.

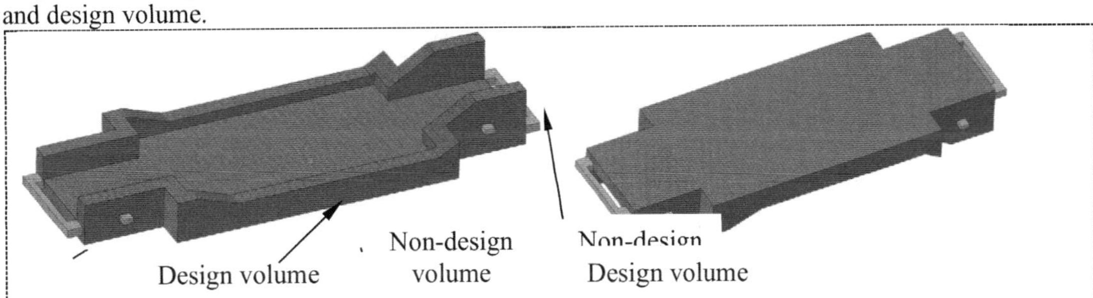


a) Overall dimension of the base model b) Top view of the base model

**Figure 3.** Discrete model of the chassis frame

IOP Conf. Series: Materials Science and Engineering **1311** (2024) 012022 doi:10.1088/1757-899X/1311/1/012022

The non-design volume is defined in area connection zone where the defined geometry is still unchanged, in this case the non-design volume is defined in suspension area connection and on the frontal and rear side of the frame, as can be viewed in figure 3. The geometrical model is meshed into 116950 tetrahedral elements, with a 40 mm target length.

### *3.3. Material model*

In this study, the base model is topological optimized in two configurations: first configuration in which the material used was steel, and second configuration made of aluminum alloy. For both materials are assigned MAT 1 material card available in Optistruct interface [7]. The mechanical properties of the assigned materials are presented in table 1.

**Table 1.** Mechanical properties of the materials.

| Mechanical properties | Steel | Aluminum |
|---|---|---|
| Initial density | 7.85e-6 Kg/mm$^3$ | 2.7e-6 Kg/mm$^3$ |
| Poisson ratio | 0.3 | 0.33 |
| Young modulus | 210 GPa | 70 GPa |

### *3.4. Load case creation*

In this section are presented the load case created and applied on the frame structure. To generate the optimized shape of the frame, the chosen forces are applied linear static. The loads are defined as a linear combination of load sets, as can be seen in figure 4.

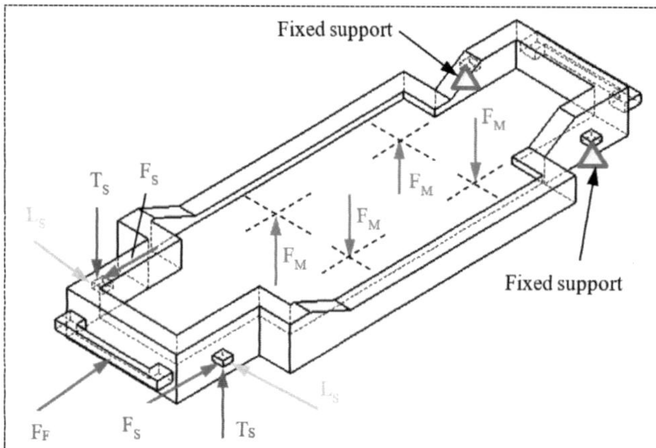


| |
|---|
| $F_F$ – Frontal load = 40000N |
| $F_S$ – Frontal suspension load = 2x2000N |
| $L_S$ - Side suspension load = 2x2000N |
| $T_S$ - Torsional suspension load = 2x2000 N |
| $F_M$ – Auxiliary loading = 4x2000 N |

**Figure 4.** Load case creation

The frontal load ($F_F$) is placed in front of the chassis on the non-design component and show the crash load on the mechanical structure. The frontal suspension loads ($F_S$) are pairs placed at the region of suspension connection. The lateral stiffness loadings ($L_S$) are pairs placed on the left and the right side of the frame. For torsional stiffness two vertical force ($T_S$) are acting in opposite direction, placed on rigid component (non-design) which correspond to the suspension connections.

For the first steel frame optimization the imposed geometrical parameters are in the range of 30 and 90 mm, and the material distribution in the design volume is relative to three symmetrical planes. The chosen parameters for the second aluminum frame optimization are in the range of 20 and 60 mm, and the design is generated depending on two symmetrical planes.

IOP Conf. Series: Materials Science and Engineering **1311** (2024) 012022 doi:10.1088/1757-899X/1311/1/012022

## 4. Results overview

The chassis modelling and the optimization process calculation were conducted on a graphic workstation, with an Intel Xeon CPU, 2.5 GHz, 32 GB RAM memory and Nvidia Quadro K620 graphical card.

### *4.1. Topology optimization of the steel frame*

Because a large design space was used, the several optimization design solutions are obtained for different material amount, as can be seen in figure 5.

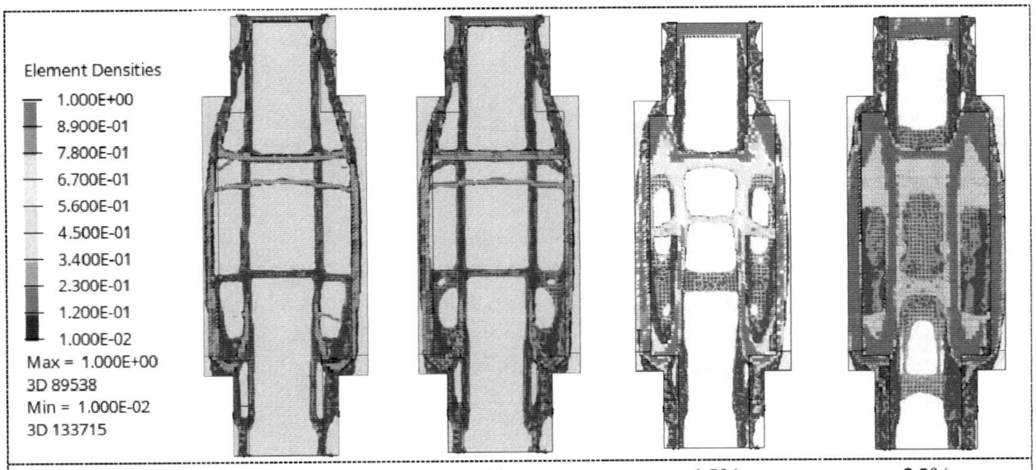


| 5% | 8% | 15% | 25% |
|---|---|---|---|

**Figure 5.** Topology results for different material amount used - steel frame

Element density distribution of the material in design volume is determined by the parameters range imposed by the user.

### *4.2. Topology optimization of the aluminium frame*

The resulted geometry shapes in the range of 5% and 25% are presented in figure 6, were can be observed the element density distribution in design volume.

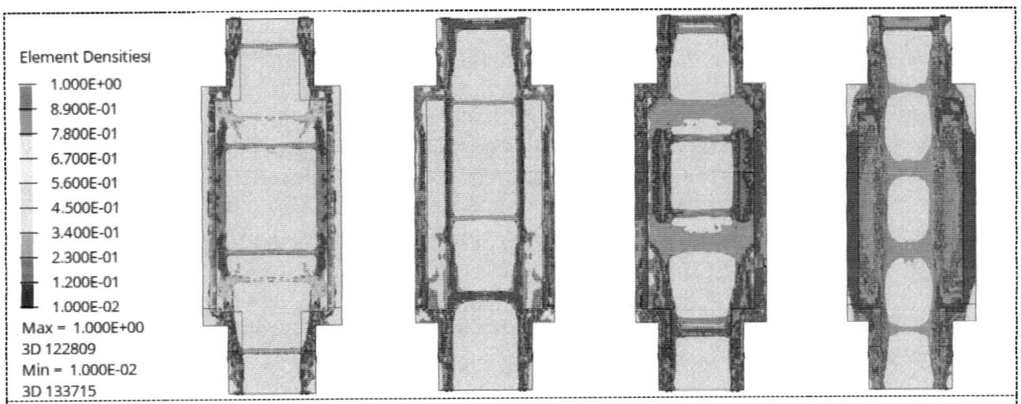


| 5% | 8% | 15% | 25% |
|---|---|---|---|

**Figure 6.** Topology results – aluminum

For a better view of an optimized geometry at 5% distribution material, in figure 7 is presented the axonometric view of the resulted shape for both optimization cases.

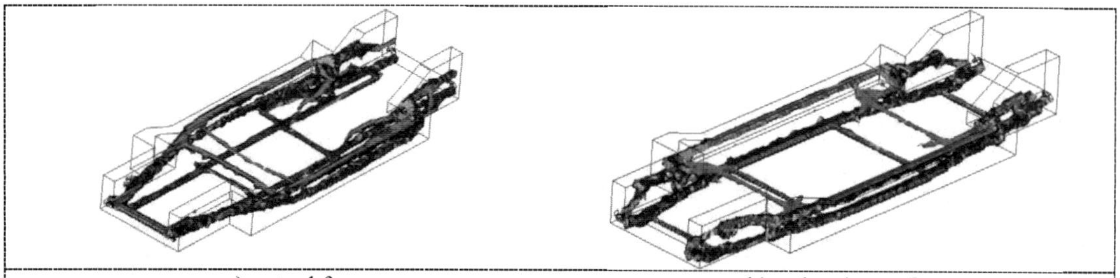

a) - steel frame b)- aluminum frame

**Figure 7.** Topology results 5% material distribution

## 5. Conclusions

This works represents the starting point for the determination of an optimal EV frame geometry. Two topological optimization studies, applied on the base frame chassis geometry, are presented. The alternative options to generate lightweight mechanical structure are given by the utilisation of different materials (steel and aluminium). The material distributions in the design space generate shapes, that can be adapted and designed using the weldments profiles. However, for the final frame design, it is important to provide a higher factor of safety which involves increasing the mechanical performance of the frame in a situation of the crash impact. Unfortunately, by applying the topology optimization method is impossible to determinate the wall thickness. Using solid beams lead to an increasing acquisition cost.

## References

[1] Aulig, N., Nutwell, E., Menzel, S., & Detwiler, D. (2018). *Preference-based topology optimization for vehicle concept design with concurrent static and crash load cases.* Structural and Multidisciplinary Optimization, 57(1), 251-266, DOI 10.1007/s00158-017-1751-z

[2] bin Ab Razak, M. S., bin Hasim, M. H., & bin Ngatiman, N. A. (2017). *Design of electric vehicle racing car chassis using topology optimization method.* In MATEC Web of Conferences (Vol. 97, p. 01117). EDP Sciences.

[3] Chuang, C. H., Chen, S., Yang, R. J., & Vogiatzis, P. (2018). *Topology optimization with additive manufacturing consideration for vehicle load path development.* International Journal for Numerical Methods in Engineering, 113(8), 1434-1445.

[4] Ge, D., Zhu, L., & Xuan, D. (2017). Topology Optimization in Electric Car Body Frame Based on Optistruct. In MATEC Web of Conferences (Vol. 100, p. 01016). EDP Sciences, DOI: 10.1051/ matecconf/20710001016

[5] Kiraly,A., 2002 Grafica Inginereasca (Engineering Graphics), Editura U.T.PRESS Publishing, ISBN 973-8335-35-3, Cluj-Napoca.

[6] Matsimbi, M., Nziu, P.K., Masu, L.M., & Maringa, M. (2021). *Topology Optimization of Automotive Body Structures: A review*,* International Journal of Engineering Research and Technology. ISSN 0974-3154, Volume 13, Number 12 (2020), pp. 4282-4296

[7] Optistruct – User manual, Trial Version.

[8] Scurtu, L. I., Bodea, S. M., & Jurco, A. N. (2016). Design optimization method used in mechanical engineering. Journal of Industrial Design and Engineering Graphics, 11(2), 13-1

IOP Conf. Series: Materials Science and Engineering **1311** (2024) 012023 doi:10.1088/1757-899X/1311/1/012023

# Aerodynamic evaluation of a commercial bus structure

**M C Balcau**

Department of Automotive Engineering and Transports, Technical University of Cluj-Napoca, Romania

monica.balcau@auto.utcluj.ro

**Abstract**. This article focuses on the aerodynamics evaluation of a commercial intercity bus used on the highway. The research begins with a brief introduction to the state-of-the-art vehicle aerodynamic assessment using the iterative CFD (computational fluid dynamics) method. The second part presented the method to generate the basic CAD model of a commercial bus using the blueprint images. The boundary conditions and the discretization of computational volume are presented in the third part of this document. The aerodynamic evaluation is performed at a velocity of 33.34 m/s, highlighting the velocity distribution around the bus structure and the air pressure on the bus and road surface. For high accuracy of the simulation results, a movable road surface and a rotation zone is introduced around the wheels, according to the simulation case. After analysing the results of the original model is proposed a geometrical improvement by adding the air deflectors to the bus model. The improved bus model is simulated under the same boundary conditions as the base model. The last part of the document presents the comparison study and its conclusions.

## 1. Introduction

Due to the computer science development, engineers have developed numerical applications that facilitate the aerodynamic evaluation using CFD simulation software. In the last time, CFD has been a frequently applied method to evaluate the vehicles aerodynamics. Using the CFD simulation tool the aerodynamic performance of vehicles can be predicted before of the vehicle manufacturing process, reducing the production costs of the final product. In addition to generate a new aerodynamic model, applying the CFD method, the older vehicle structure can be improved by adding the air deflectors on their structure to reduce the relative pressure on the vehicle surface and obtain a lower drag coefficient.

The start point of this study is given by the analysis of the latest studies in which CFD aerodynamics software and structural device was used to improve the aerodynamic performance [1], [4]. In the last time many CAD development companies have integrated solvers for FEA and CFD computing which makes easy the all design process. A useful CFD software tool used by many users is FlowWorks solver integrated in SolidWorks design environments [11]. Scurtu and Gheres perform an CFD evaluation of two vehicles in platoon establishing the optimal distance between two vehicles that travel on the road. To obtain an accurate result in the CFD simulation was added the moving road surface and the rotational effects on the wheel housing. When the road surface is added in simulation appear the “ground effect”, decreasing the lift coefficient. Another CFD study was performed by Bodea et all applied on the trailer truck structure in two configurations, the base model, and improved model. The improved model has mounted the air defectors on cabin the roof side, in the space between the cabin and trailer and in the rear side of the trailer. The results show an improvement of the drag and list coefficients.

## 2. Work description

This study is focused to improve the aerodynamic performance of an intercity bus used on highway. The study was validated by the CFD simulation performed using dedicated software at a speed of 33.34 m/s.

### *2.1. Bus model*

The bus geometry was designed by using the SolidWorks software through blueprints pictures of a Mercedes Travego bus model, as can be seen in figure 2, [10]. The sketches of the body bus are positioned on the projection planes in the working area of the modelling software [3], [11].

The overall dimensions of the CAD model designed in this study are in concordance with the real bus model, see figure 1.

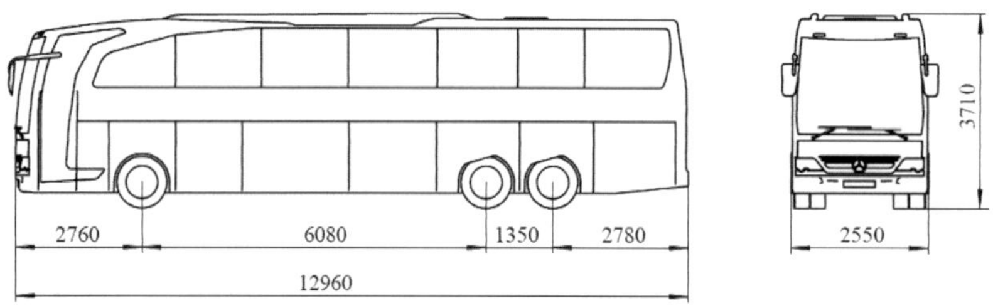


**Figure 1.** The overall dimensions of the body bus

The wheels housings, rims, windshield wipers, mirrors and other details related were modeled in minutiae, as can be seen in figure 3. The front side of the bus is designed with the closed radiator grill so that there is no airflow in the engine compartment. The external details of the bus body were reproduced as best as possible to obtain accurate results. The cross-sectional area of the bus body value used for determining the drag coefficient is 8.63 m$^2$ and it is measured in CAD software environment. Area of the bus body, used to determine the lift coefficient is 32.437 m$^2$.

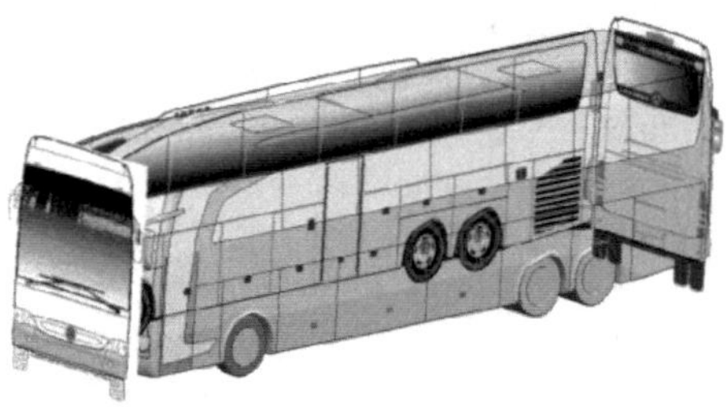

**Figure 2.** Modelling stage of the bus

**Figure 3.** Final CAD model of the original bus

### *2.2. Aerodynamic forces*

Several studied on this topic have been studied to establish the aerodynamics force that acts on the vehicle body [8], [9]. The aerodynamic forces determined in this study are drag force and lift force. The drag coefficient and the lift coefficient are also calculated. It is known that the distribution of aerodynamic force on the bus body is as follows: 40-45 % on the external surface, 25-30 % on wheels housing and 20-25 % from the underbody of the body and engine area [7]. The traction force is given by de relation:

$$F_T = F_D + F_R + m \cdot \frac{dV}{dt} + m \cdot g \cdot \sin\alpha \tag{1}$$

where:
$F_D$ – drag force, N
$F_R$ – friction force at rolling, N
m – mass of the vehicle, kg
α – angle of road inclination, °
Expression 2 presents the equation of the rolling resistance force.

$$F_R = f_R \cdot G \tag{2}$$

where:
$f_R$ – friction coefficient
G – gravitational force (G=m·g), N

*2.2.1. Drag force.* The main force applied on the bus body while driving is the drag force, which is given by the property of the fluid to resist the vehicle's forward motion. Conventionally in automotive aerodynamics, the reference axis on which aerodynamic force acts is the X-axis.

The magnitude of this force is influenced by the vehicle velocity, air density and the design shape of the bus. The relationship used for the aerodynamic force is expressed by relation 3:

$$F_D = \frac{1}{2} \cdot \rho \cdot C_D \cdot A \cdot V^2 \tag{3}$$

where:
$C_D$ – drag coefficient
A – cross-sectional area of the bus body, $m^2$
V – bus velocity, m/s
ρ – air density, $kg/m^3$

The aerodynamics performance of a vehicle is associated with the drag coefficient. The value of the drag coefficient expresses the degree of intensity between the vehicles body and surrounding fluid. The expression of the drag coefficient deduced from the drag force relationship have the form:

$$C_D = \frac{2 \cdot F_D}{\rho \cdot A \cdot V^2} \tag{4}$$

*2.2.2. Lift force.* During the road forwarding the bus structure is influenced by the lift force which can be defined as the sum of all forces generated around the vehicle body that are projected on a perpendicular plane to the air flow direction during the travel [2]. The expression of the lift force is presented in relation 7.

$$F_L = \frac{1}{2} \cdot \rho \cdot C_L \cdot A_L \cdot V^2 \tag{5}$$

where:
$C_L$ – lift coefficient
$A_L$ – area of the bus body on the ground, $m^2$

Thus, lift coefficient is given by the relation:

$$C_L = \frac{2 \cdot F_L}{\rho \cdot A_L \cdot V^2} \tag{6}$$

The lift force results from the interaction between a fluid and a solid body, in this case between bus and airflow that get from under the bus structure. The lift force may have a negative value, in this case being oriented toward the road. In case of the high magnitude load an additional pressure is exerted on the vehicle tires, leading to premature tire wear and an increase in fuel consumption. In the opposite case, if this force is too high positive, the risk of losing control of the vehicle is high [2].

*2.2.3. Side force.* When the bus is travelling, there is a lateral force that can be defined as the sum of all the forces generated around the bus body that are projected on a plane of the air flow direction. The lateral force is given by the relation:

$$F_S = \frac{1}{2} \cdot \rho \cdot C_S \cdot A_S \cdot V^2 \quad (7)$$

where:
$C_S$ – side coefficient
$A_S$ – longitudinal section area of the bus body, $m^2$

The relation of the side coefficient is:

$$C_S = \frac{2 \cdot F_S}{\rho \cdot A_S \cdot V^2} \quad (8)$$

*2.3. Mathematical approach*

The mathematical approach uses the Reynolds Averaged Navier Stokes *(RANS)* equations, a model used in an engineering application that use the fluid flow simulation [6]. Equations can be written in Cartesian tensor form, as follows:

$$\frac{\partial U_i}{\partial x_i} = 0 \quad (9)$$

where:
U – velocity tensor
$x_i$ – vector position
$\overline{u'_j u'_i}$ – the average of fluctuating velocity
P – the mean static pressure
μ – the molecular viscosity

The strain rate expression is given by relation:

$$\rho = \frac{\partial}{\partial x_j}\left(U_j U_i + \overline{u'_j u'_i}\right) = -\frac{\partial P}{\partial x_i} + \frac{\partial}{\partial x_j}\left(2\mu S_{ij}\right) \quad (10)$$

The standard turbulence model k-ε is applied in this study and is given in the following equation:

$$S_{ij} = \frac{1}{2}\left(\frac{\partial u_i}{\partial x_j} + \frac{\partial u_j}{\partial x_i}\right) \quad (11)$$

The standard k-ε turbulence model is applied in this study and in the following equation:

$$\rho U_j \frac{\partial k}{\partial x_j} = \tau_{ij} \frac{\partial U_i}{\partial x_i} - \rho\varepsilon + \frac{\partial}{\partial x_j}\left[\left(\mu + \frac{\mu_T}{\sigma_k}\right)\frac{\partial \varepsilon}{\partial x_j}\right] \quad (12)$$

where:
$\tau_{ij}$ – the Reynolds stresses

$\mu_T$ – turbulent viscosity
$\sigma_k$ – closure coefficient
k – turbulent kinetic energy
ε – dissipation rate of turbulent energy

In the next relation is given the dissipation rate of turbulence kinetic energy equation:

$$\rho U_j \frac{\partial \varepsilon}{\partial x_j} = C_{\varepsilon 1} \frac{\varepsilon}{k} \tau_{ij} \frac{\partial U_i}{\partial x_j} - C_{\varepsilon 1} \rho + \frac{\varepsilon^2}{k} + \frac{\partial}{\partial x_j}\left[\left(\mu + \frac{\mu_T}{\sigma_\varepsilon}\right)\frac{\partial \varepsilon}{\partial x_j}\right] \quad (13)$$

The turbulent viscosity expression is given by the relation:

$$\mu_T = \rho C_\mu \frac{k^2}{\varepsilon} \quad (14)$$

### *2.4. Reynolds number*

The Reynolds number is a dimensionless number used to characterize the flow regime. The value of this number determines the relationship between the vacuum forces and the inertial forces of a moving fluid. Inertial forces are determined by Newton's second law and are responsible for the maximum acceleration of the fluid and viscous forces [7]. The expression that determines the value of this number is presented in relation 15:

$$R_e = \frac{\rho \cdot V \cdot L}{\mu} \quad (15)$$

where: $V$ – represents air velocity, (m/s), $L$ – represents length of the vehicle, (m), $\rho$ – represents air density, (kg/m$^3$), $\mu$ – represents dynamic air viscosity, (kg/m·s).

The value of the air density is ρ = 1,225 kg/m$^3$, corresponding to a temperature of 15°C, the length of the bus is l = 12,960 m and the dynamic air viscosity μ = 1,802 · 10$^5$ kg/m·s. In this study the airflow regime is established considering the airflow over a plane surface [5]. In the case of external flows when the value of Reynolds number is less than 5x10$^5$ it is a laminar flow, where viscous forces are dominant and is characterized by smooth, constant fluid motion. If Reynolds number is in range of 5·10$^5$ and 10$^7$ the flow regime is a transient one. When its value exceeds 10$^7$ the flow regime will be turbulent, which tend to produce flow instabilities, chaotic eddies, and vortices. In this case the calculated Reynolds number is 2.973 ·10$^7$, that include the airflow in a turbulent flow regime.

## 3. CFD simulation

### *3.1. Boundary conditions*

In figure 4 are presented the overall dimensions of a given computational domain in which the bus model is introduced. The bus is placed on the road surface, to which has been applied a translation movement corresponding to the inlet air velocity. The bus model is placed at six meters from the air inlet which allows the analysis and visualization of the pressure effect that occur on the road surface when the vehicle is on the road. The base model computational domain is divided into 161716 cells arranged in layers with higher density in the near area of the vehicle model to a better view of the airflow around the vehicle. The 22355 of these cells are in contact with the vehicle and road surface.
The computational domain of the improved bus model is divided into 138051 fluid cells, of which 19281 cells are in contact with the bus, and road surface. The CFD simulation was performed for two case studies.

First simulation, when is established the aerodynamic evaluation of the original model of the bus and second simulation, where the aerodynamic evaluation is applied on the geometry of the improved bus that has mounted the air deflectors to improve the aerodynamics capabilities.

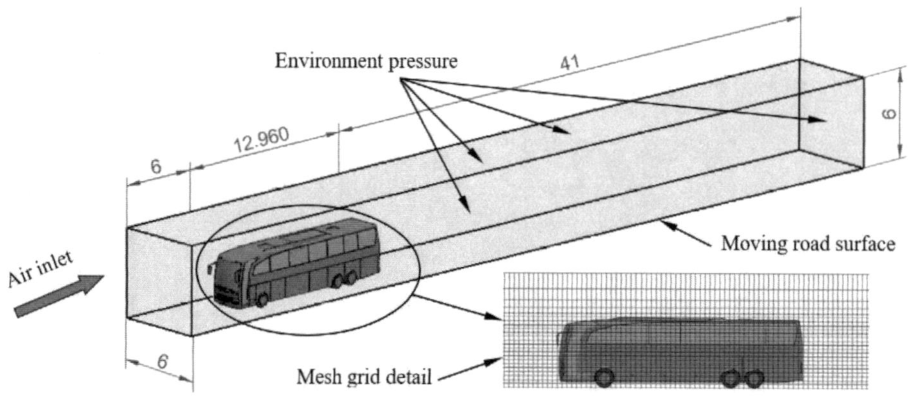


**Figure 4.** Boundary condition in computational volume

*3.2. Results – original model*

The air pressure distribution on the original bus model is presented in figure 5. The high-pressure value is applied in the front side of the bus, on the windshield and on the mirror support.

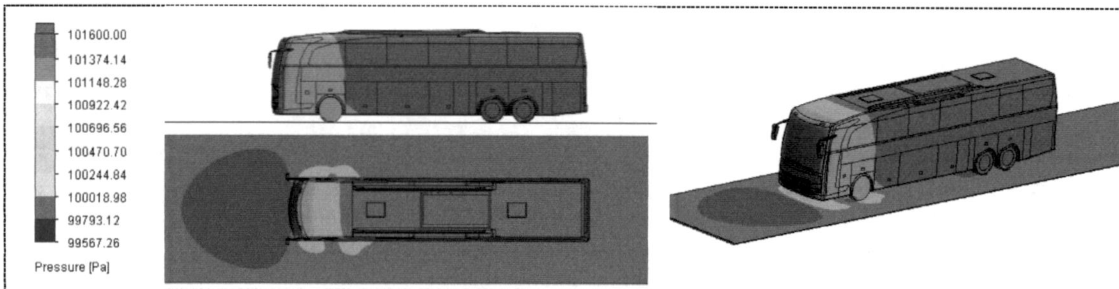


**Figure 5.** Air pressure distribution on the original bus model and road surface at 33.34 m/s velocity

In figure 6 is presented the air flow velocity distribution on the original bus model at velocity of 33.34 m/s. The low velocity fields are situated in the rear side of the bus, where a turbulent airflow was created. Also, on the rear side of the mirror and the wheels housing the airflow velocity is very low.

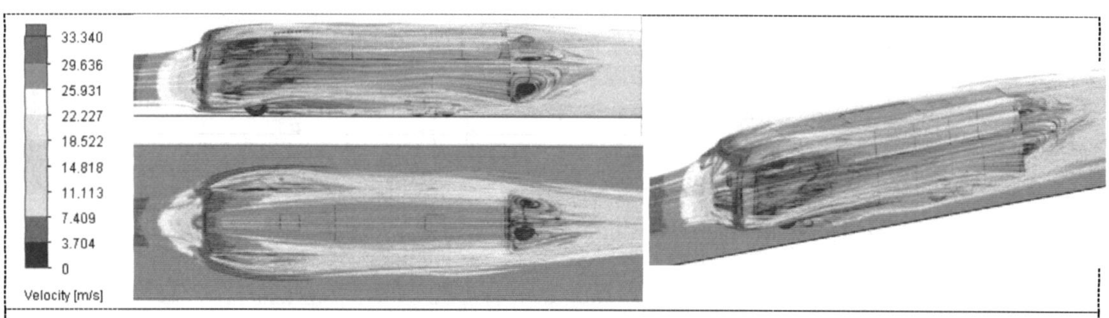


**Figure 6.** Air flow velocity distribution on the original bus model and road surface at 33.34 m/s velocity

*3.3. Results - improved model*

The improved bus model has added two air deflectors mounted on the bus roof and in the rear side of the bus. The roof air deflector covers the existent gaps from de original bus and makes the smoother airflow due to the curved form. The heigh of the bus remain unchanged. As an inconvenient, the rear side deflector increases the original length of the bus with 0.86 m. The tapered shape of this air deflector directs the airflow decreasing the low velocity field generated at the original bus model. This deflector can be designed to work automatically when the bus exceeds a chosen speed. Or it can be designed as an extra space for luggage. In the figure 7 is represented the pressure distribution on the improved bus and road surface in range of 99567 to 101600 Pa. The high-pressure field is situated on the front side of the bus as in the original model, but in the front side of the roof the pressure is lower comparing to the original model.

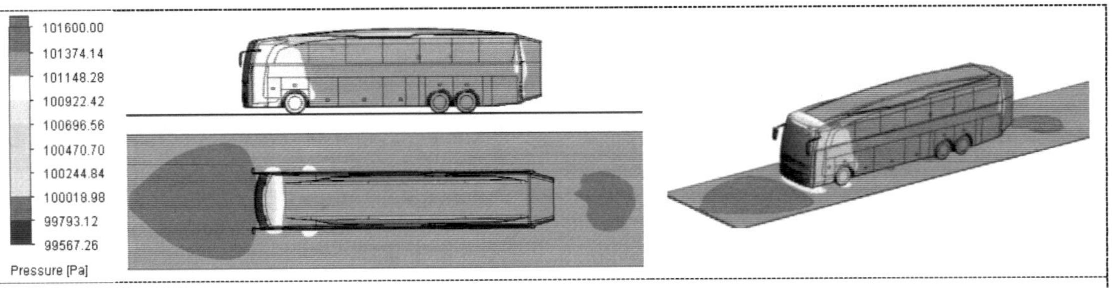


**Figure 7.** Air pressure distribution on the improved bus model and road surface at 33.34 m/s velocity

The airflow velocity distribution around the improved model at 33.34 m/s velocity is presented in figure 8. The low air velocity field from the rear side of the bus is more lower than the original version. The air deflector mounted on the rear side of the bus has a signifiant impact to the aerodymic performance of the bus. The air deflector mounted on the roof makes the airflow smoother.

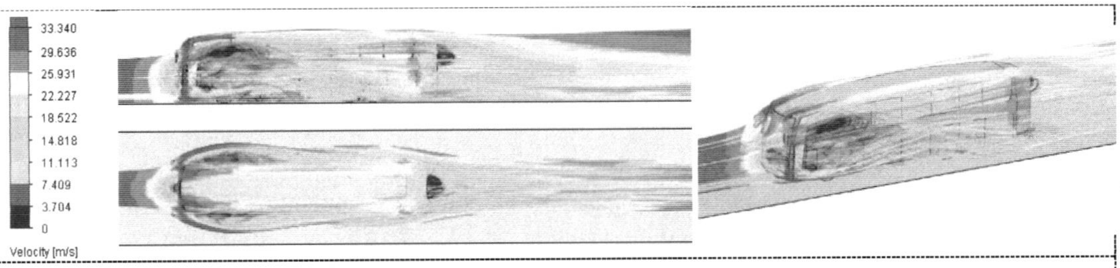


**Figure 8.** Air flow velocity distribution on the improved bus model and road surface at 33.34 m/s velocity

The numerical results of both simulations are presented comparatively in table 1. Analysing the obtained values of the drag coefficient it is established that their values is lower because the undercarriage of the bus is represented by a smoother surface, while the underbody of the real vehicle is a surface with many bumps.

**Table 1.** Numerical results at 33.34 m/s velocity

| | Base model | *Improved model* | *Percentage difference* |
|---|---|---|---|
| **Drag coef.** | 0.45 | 0.380 | 15.5 |
| **Drag Force [N]** | 2614 | 2195.39 | 16.01 |
| **Lift coef.** | 0.045 | 0.049 | 8.89 |
| **Lift force [N]** | 978.29 | 1058.01 | 8.15 |

The lower values of the lift coefficient are influenced by the road surface in the study. It can be observed that the percentage difference between these models is considerable.

IOP Conf. Series: Materials Science and Engineering **1311** (2024) 012023 doi:10.1088/1757-899X/1311/1/012023

## 4. Conclusions

This paper presents an aerodynamic evaluation of the commercial model of an inter-city bus based on CFD simulation. An important aspect of this evaluation was performed by adding the movable "ground effect" represented as a road surface and the rotational field around the wheels housing. This study was realised to design the air deflectors that can be mounted on the older buses structure to improve the aerodynamics performance. The error that can occur between the CFD simulation and the reality can be caused by the reason that the bus is in motion and the air is fixed, while in the CFD simulation the air is in motion and the bus is fixed. Also, another error can be the meteorologic condition, these phenomena are not applied exactly in the CFD simulation.

## References

[1] Balcau, M. 2021 *Aerodynamic study of a car towing a motorcycle trailer* Journal of Automotive Engineering, ISSN 2457 – 5275, Vol. 27-1, pp 31-38.

[2] Bodea S.M., Prodan C.V., Scurtu I.L. 2019 *The Aerodynamic Study of a Body Truck* Proceedings of the 4th International Congress of Automotive and Transport Engineering (AMMA 2018). AMMA2018 2018. Proceedings in Automotive Engineering. Springer, Cham.

[3] Kiraly,A., 2002 *Grafica Inginereasca*, Publisher U.T.PRESS, ISBN 973-8335-35-3, Cluj-Napoca.

[4] Jurco A.N. 2021 *Study of the influence of roof luggage box on a vehicle aerodynamics* Automotive engineering, ISSN 1842-4074 vol 58 pp 15-18

[5] Scurtu I.L. 2021 Aerodynamic performance evaluation for a vehicle structure equipped with a bicycle rack Automotive engineering ISSN 1842-4074 vol 58 pp 10-14

[6] Scurtu, I. L., Gheres, M. I. 2021 *Numerical evaluation of vehicles aerodynamics in platoon using CFD simulation* IOP Conference Series: Materials Science and Engineering (Vol. 1220, No. 1, p. 012024). IOP Publishing.

[7] Sebben, S., Walker, T., Landström, C. 2014 *Fundamentals, Basic Principles in Road Vehicle Aerodynamics and Design* Encyclopedia of Automotive Engineering, 1-16.

[8] Todoruţ, A., Cordoş, N. 2018 *Evaluation of the vehicle sideslip angle according to different road conditions* International Congress of Automotive and Transport engineering (pp. 814-819). Springer, Cham.Engineering, Vol. 58(3), pg. 349-360, Ed. U.T.PRESS, ISSN 1221-5872, 2015

[9] Todoruţ, A., Cordoş, N., Burdea, M.D., Bălcău, M., 2015 *The evaluation of normal load redistribution on the static axles and on the wheels, when the vehicle is in motion* Cluj-Napoca, Acta Technica Napocensis, Series: Applied Mathematics, Mechanics, and Engineering, Vol. 58(3), pg. 349-360, Ed. U.T.PRESS, ISSN 1221-5872,

[10] Site: *https://alldrawings.ru/en/pictures/item/mercedes-travego-bus-2007-drawings-dimensions-pictures-of-the-car-1*, 2022

[11] Solidworks 2021 - *Software help*

# Implementing Model Predictive Control (MPC) in Steer-by-Wire Systems for Future Automated Vehicles

**V Changoski**[1,], **A Vasileva**[1], **K Jakimovska**[1] **and D Danev**[1]

[1] Ss. Cyril and Methodius, University in Skopje, Faculty of Mechanical Engineering-Skopje, Institute of Engineering Design, Mechanization and Motor Vehicles, North Macedonia

vasko.changoski@mf.edu.mk

**Abstract.** The need for future green sustainable transportation represents a motivation for the researchers in order to create a better society. In this paper a proposed Steer-by-Wire system for future automated vehicles is analysed by implementing model predictive control (MPC) controller that would optimise the desired behaviour of the vehicle. By using MATLAB/Simulink, a bicycle vehicle model is simulated in standardised test manoeuvres according to standard ISO:7401. These results are used as referent data for the desired trajectory of the vehicle and as input data for the MPC controller. The idea of using these results for the controller is generated by the desire of creating a shared information intelligent systems. Shared information and control in the future intelligent systems would be crucial for creating the desired sustainable transport. By traveling, the vehicle would collect data of the desired measured variables and those data would be send to every other vehicle. Therefore, the next vehicle would possess the previous data and would try to improve the vehicle response and safety. Every optimised trajectory would be shared with the rest of the vehicles and this cycle would result in further optimization and in creation of a safer, more comfortable and more sustainable transport. The optimisation is done by the MPC controller which is responsible for defining the steering wheel angle and therefore creating the Steer-by-Wire system that is entirely defined by the MPC controller without driver commands. By comparing the trajectory of the automated vehicle and the desired trajectory gained from the previous simulations, the desired steering wheel angle is defined. This results in creation of a Steer-by-Wire system that improves the vehicle dynamics, safety and dynamic response of the vehicle. The mathematical modelling, results of the simulations and advantages of using the MPC controller for the Steer-by-Wire system in automated vehicle are presented.

**Keywords:** automated vehicles, steer-by-wire system, model predictive control (MPC), vehicle bicycle model, sustainable transport

IOP Conf. Series: Materials Science and Engineering **1311** (2024) 012024 doi:10.1088/1757-899X/1311/1/012024

## 1. Introduction

The need for research is supported by a review of previous surveys in the field of green sustainable transportation, and the need for environmentally friendly, safe, and high performance vehicles. First, in reference [1] a 'automation' is defined as "the use of electric or mechanical devices to replace human labour" (Oxford English Dictionary (OED)). Furthermore, in reference [2] is presented that steering systems transferring the drivers' steering commands to tires have experienced several evolution levels and have a trend from a pure mechanical toward a purely electrical system. The mechanical parts of a vehicle are being exchanged by electrical sensors and actuators with the aim of improved performance [3]. As is shown in reference [3] a Steer-by-Wire system (SBW) purpose to remove the physical connection between the steering wheel and the wheels of a car by using electrically controlled motors to swap the direction of the wheels and to give feedback to the driver. Reference [4] presents the control strategy for wheel synchronization and the variable steering ratio. The research paper [5] proposes a novel adaptive hierarchical control approach for Steer-by-Wire (SbW) vehicles to improve handling stability. The reference [6] presents an approach to vehicle stabilization that leverages: the steering torque and reliably estimates vehicle sideslip and the tire-road friction coefficient the estimates, to define state boundaries that exclude unstable vehicle dynamics and utilizes a model predictive envelope controller to bound the vehicle motion within this stable region on the state space. In a research paper [7] Model Predictive Control (MPC) is presented as a set of advanced control methods, which explicitly use a model to predict the future behaviour of the system.

Authors in [8] develop MPC controller for trajectory following autonomous vehicles and also propose trajectory optimisation. In [9] an analysis is made on the impact of trajectory optimization for autonomous vehicles when they are on their limits of endangering the vehicle stability. Also in [10] by using MPC an algorithm is proposed for controlling and optimizing the lateral vehicle trajectory. The paper [11] presents an MPC that determines which combination of trucks, trains, and ships to use for transporting the containers and what routes empty and full trucks should use as one integrated problem. The results indicate an improved vehicle utilization. Additionally, the integrated problem approach allows for more realistic constraints and costs.

In [12], [13], [14] the vehicle dynamics is analysed and control methods are proposed for its improvement, while [15], [16] the MPC concept is analyzed and in [17] efficient MPC method is developed for urban traffic network.

The usage of conventional vehicles and steer-by-wire vehicles would overlap at certain period in the future, together with the automated vehicles. In order to utilize the advantages of the autonomous vehicles and increase their reliability, data sharing connection between every vehicle should be implemented. The idea of the research is to use conventional or steer-by-wire vehicle that would be controlled by a driver as a reference model and after successful maneuver, the data would be send to the automated vehicles controlled by MPC controller in attempt to complete the maneuver successfully and improve the vehicle performances. Previously gained information would be send to the next vehicles and optimization of trajectory would be mace in order to achieve even further improvement in the vehicle performances and stability.

The aim of this research paper is to analyze the advantages of implementation of the MPC in the Steer-by-Wire systems in 2 wheel steering (2WS) and 4 wheel steering (4WS) systems, in order to improve the vehicle stability and performances which will lead to safer future road transportation.

## 2. Analytical modelling of the vehicles and MPC controllers

### *2.1. Vehicle bicycle model*

In this paper, a 2DOF bicycle model is used (Figure 1). This model would be used as a reference model where it is assumed that the information from the reference model is obtained from a vehicle commanded by a driver and as a steer-by-wire model where the command for steering wheel angle is given by the MPC controller.

The model is defined using the equations (1) and (2) where its lateral velocity ($V_y$) and yaw rate ($\omega_z$), while the longitudinal velocity ($V_x$) is considered to be constant.

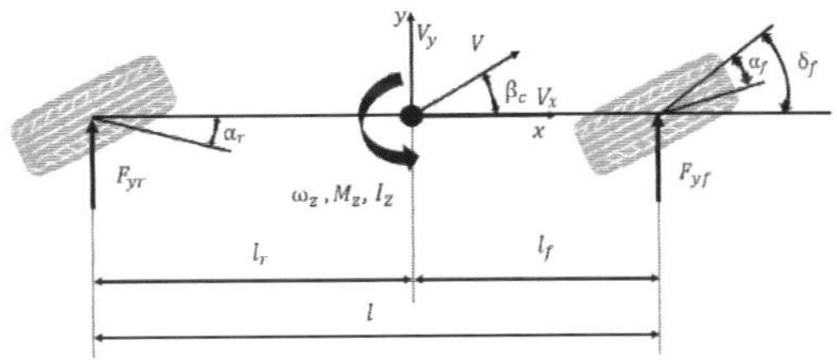

**Figure 1.** Vehicle bicycle model

$$\dot{V}_y = \left(\frac{-K_{\alpha f}-K_{\alpha r}}{mV_x}\right)V_y + \left(\frac{K_{\alpha r}l_r-K_{\alpha f}l_f}{mV_x} - V_x\right)\omega_z + \frac{K_{\alpha f}}{m}\delta_f \quad (1)$$

$$\dot{\omega}_z = \left(\frac{K_{\alpha r}l_r-K_{\alpha f}l_f}{I_zV_x}\right)V_y + \left(\frac{-K_{\alpha f}{l_f}^2-K_{\alpha r}{l_r}^2}{I_zV_x}\right)\omega_z + \frac{K_{\alpha f}l_f}{I_z}\delta_f \quad (2)$$

Parameters of the vehicle are presented in Table 1 which, are defined as future automated passenger vehicle form D - segment.

**Table 1.** Bicycle vehicle model - parameters

| | |
|---|---|
| Mass ($m$): | 1700 kg |
| Wheelbase ($l$): | 2875 mm |
| Distance from centre of mass to front axle($l_f$): | 1380 mm |
| Inertia radius ($i$): | 1250 mm |
| Front wheel cornering stiffness($K_{\alpha f}$): | 56000 N/rad |
| Rear wheel cornering stiffness($K_{\alpha r}$): | 56000 N/rad |

The idea and goal of implementing the steer-by-wire system in the future automated vehicles would result in increasing the free space of the vehicles for implementing other systems or maybe shortening the vehicle length. In this study, an approximation is made. In other words, the wheelbase of the vehicle would remain the same, and reconfiguring the vehicle system's position would allow for the implementation of a steer-by-wire system for the rear-wheel steering system. Hence, the maximum turning angles would be the similar or same as the front wheels steering angles.

Therefore, to analyze the previous approximation and to further improve the vehicle dynamics, stability and safety, a 4-wheel steering (4WS) linear bicycle model is used as a test vehicle (Figure 2). This model is also a 2DOF model, defined using equations (3) and (4).

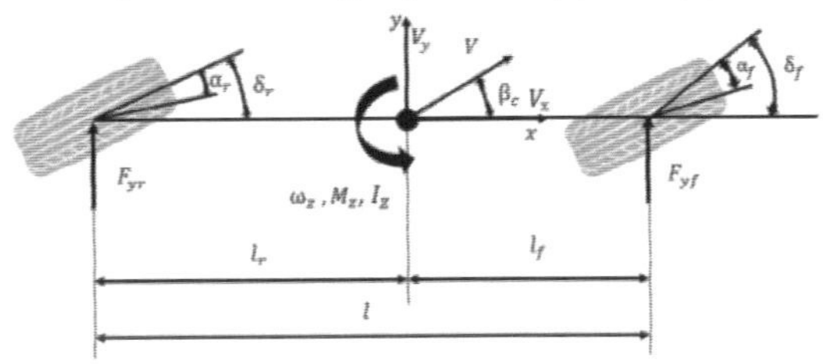

**Figure 2.** 4WS bicycle model

$$\dot{V}_y = \left(\frac{-K_{\alpha f}-K_{\alpha r}}{mV_x}\right)V_y + \left(\frac{K_{\alpha r}l_r-K_{\alpha f}l_f}{mV_x} - V_x\right)\omega_z + \frac{K_{\alpha f}}{m}\delta_f + \frac{K_{\alpha r}}{m}\delta_r \quad (3)$$

$$\dot{\omega}_z = \left(\frac{K_{\alpha r}l_r-K_{\alpha f}l_f}{I_zV_x}\right)V_y + \left(\frac{-K_{\alpha f}{l_f}^2-K_{\alpha r}{l_r}^2}{I_zV_x}\right)\omega_z + \frac{K_{\alpha f}l_f}{I_z}\delta_f - \frac{K_{\alpha r}l_r}{I_z}\delta_r \quad (4)$$

IOP Conf. Series: Materials Science and Engineering **1311** (2024) 012024 doi:10.1088/1757-899X/1311/1/012024

*2.2. MPC controllers*

In this paper as a control method, a MPC controller was used (Figure 3). The MPC controller is responsible for commanding the steering wheels angles of the vehicle models, without drivers command. The command of the controller is obtained from controller base on the difference (error) between the values of the side-slip-angle ($\beta_c$) and lateral trajectory displacement ($y$) of the automated vehicle and the reference vehicle. These parameters were chosen in order to track the trajectory of the reference vehicle, and to control the stability of the vehicle and consciousness of the passengers about the direction of travel of the vehicle by analyzing the value of the side-slip angle.

Based on the previous statement the cost function (J) of the MPC controller is presented in equation (5).

$$J = \sum_{i}^{N}[w_{ei}(X_d - X)^2 + w_{ui}U^2] \tag{5}$$

Where the $X_d$ represents a matrix of the parameters of the reference model, $X$ matrix of the parameters of the automated vehicle, $U$ represents the control parameter of the MPC controller, while $w_{ei}$ and $w_{ui}$ the associated weight coefficients. The values of the coefficients and the shape of the matrixes differ depending on the automated vehicle. In table 2 the weight parameters are defined.

**Table 2.** Cost function – weight coefficients

| | |
|---|---|
| 2WS – Side-slip angle ref. value ($w_{e11}$): | 1 |
| 2WS – Lateral displacement ref. value ($w_{e21}$): | 1 |
| 4WS – Side-slip angle ref. value ($w_{e11}$): | 1 |
| 4WS – Lateral displacement ref. value ($w_{e21}$): | 0.7 |
| 2WS– optimised – Side-slip angle ref. value ($w_{e11}$): | 0.5 |
| 2WS – optimised - Lateral displacement ref. value ($w_{e21}$): | 1 |
| 4WS – optimised – Side-slip angle ref. value ($w_{e11}$): | 1 |
| 4WS – optimised – Lateral displacement ref. value ($w_{e21}$): | 0.1 |

The parameters were chosen in order to emphasize the importance of the certain reference values. For example, it can be concluded that the primary objective of the 4WS is to lower the side-slip angle (where the desired reference value is 0), while due to the limits of the only front wheel steering of the 2WS, the primary objective is set to following the reference trajectory. The weight coefficients for the inputs are equal with value of 1.

The control algorithm of the models and the general idea of the research is presented in Figure 3.

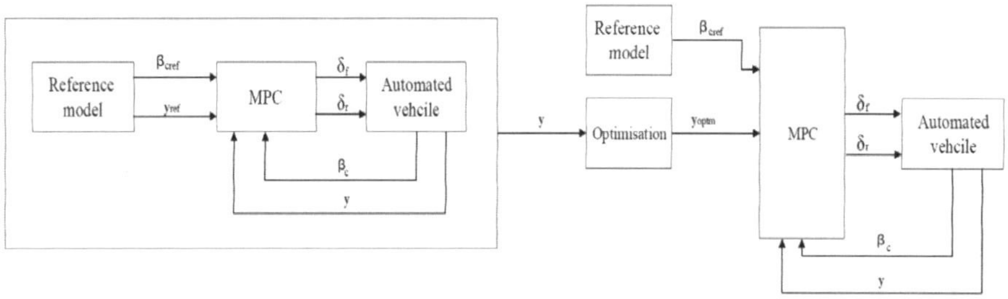


**Figure 3.** MPC controller

IOP Conf. Series: Materials Science and Engineering **1311** (2024) 012024 doi:10.1088/1757-899X/1311/1/012024

## 3. Simulation results

In this section the results of the conducted simulation are presented. A driving scenario is defined where a 2WS vehicle, commanded by driver (Passive vehicle) is performing the single lane change manoeuvre according to the standard ISO 7401 which defines the lateral transit test methods for road vehicles. The idea of the driving scenario is that the driver detects an obstacle on the road and successfully avoids it. This information is transferred to thc 2WS and 4WS steer-by-wire vehicles (2WS – MPC and 4WS - MPC) where it is used as reference data for the MPC controllers. Afterwards, when the alternative vehicles reach the same obstacle they are supposed to complete the manoeuvre successfully.

The passive vehicle is travelling with constant velocity of 80 km/h, while the command for the steering wheel angle is defined as sine function with frequency of 0.5 Hz. The value of the steering wheel angle is defined in order to achieve the necessary lateral acceleration of $4\ m/s^2$ in the first amplitude, according to the standard (Figure 4).

Due to the geometry of the 2WS – MPC vehicle and its parameters, this vehicle would try to follow the parameters of the passive vehicle, while because of the rear wheel steering system of the 4WS – MPC, larger improvements are possible to be made. In this vehicle, reference data for the side-slip angle is defined as a constant value of 0 which forces the controller to achieve the desired goal, and with that accomplished, the vehicle stability and passenger's consciousness of the traveling direction would be improved.

In Figure 5, the steering wheels angles of the vehicles and the commands from the MPC controllers can be observed. The most significant difference is the steering wheels angles of the 4WS – MPC vehicle where, the rear wheels achieve steering angle almost as the front wheels steering angle of the passive and the 2WS – MPC vehicles.

While the 2WS – MPC vehicle is trying to follow the trajectory and by that the performances of the passive vehicle, the biggest improvements in the implementation of the 4WS MPC system can be observed in Figures 6, 7 and 8. The 4WS – MPC possesses smaller values of the yaw rate and the goal of achieving a side slip angle of value 0 is achieved, thus improving the vehicle stability of the vehicle.

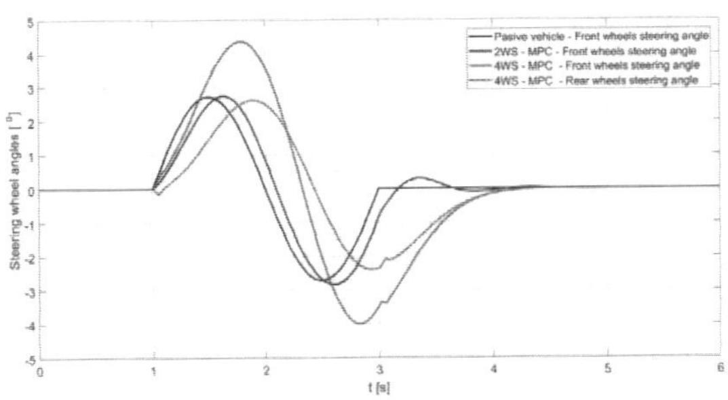


**Figure 4.** Steering wheel angles

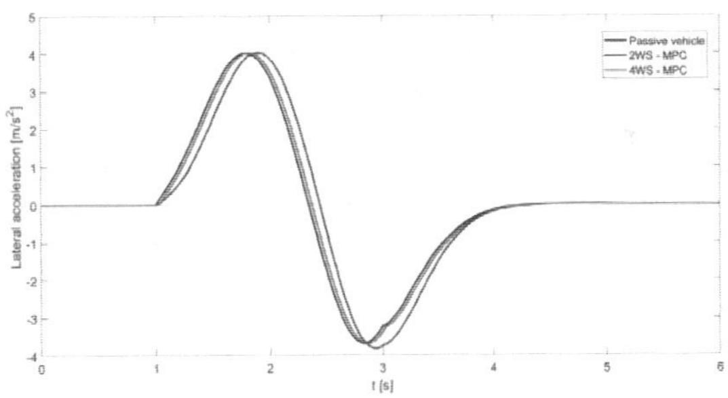


**Figure 5.** Lateral acceleration

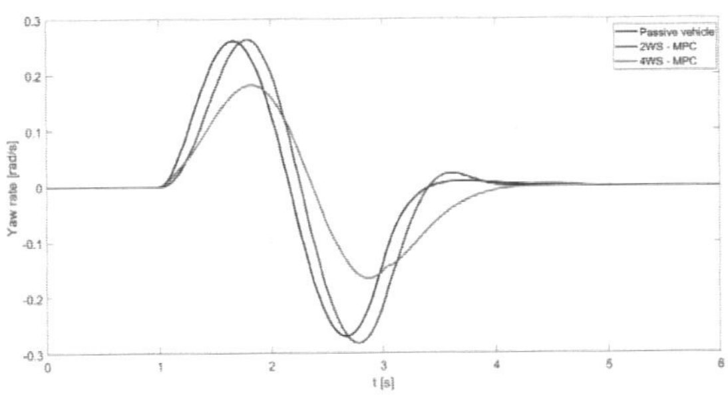


**Figure 6.** Yaw rate

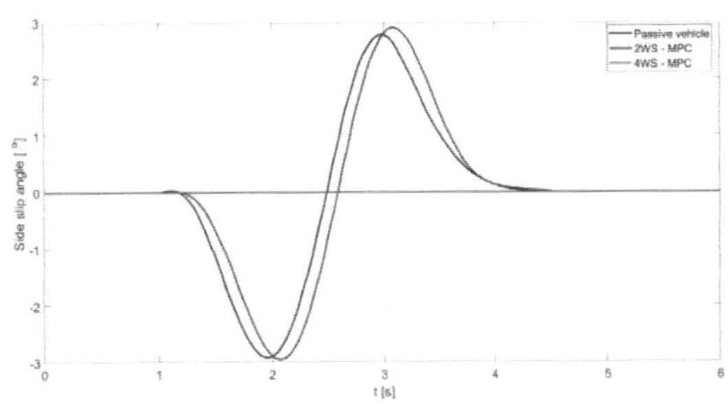


**Figure 7.** Side slip angle

IOP Conf. Series: Materials Science and Engineering **1311** (2024) 012024 doi:10.1088/1757-899X/1311/1/012024

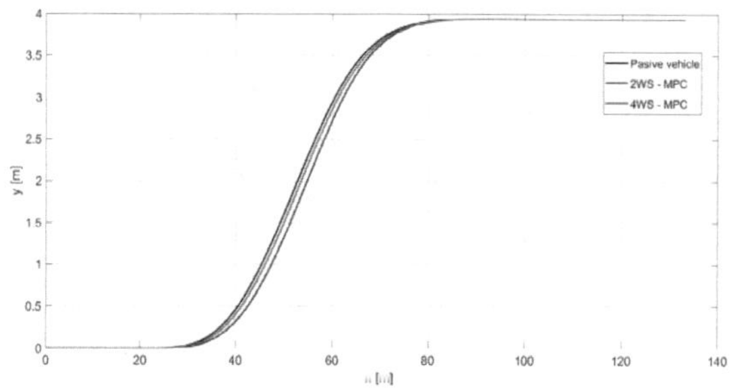


**Figure 8.** Trajectory

## 4. Optimisation of the trajectory of the vehicles

After it was confirmed that the reference trajectory satisfies the vehicle performances of both 2WS and 4WS vehicles, and if the information about the already mentioned obstacle is transferred from the passive vehicle in time, an optimization of the trajectory is suggested in order to improve the vehicle performances. Given the previous defined trajectory from the passive vehicle, the idea is to start the manoeuvre sooner and to follow the smoother trajectory. Instead of traveling in a straight line at the beginning of the manoeuvre, the proposed optimised trajectory will force the vehicles to turn in opposite direction at the beginning of the manoeuvre and to turn in direction of avoiding the obstacle (Figure 9).

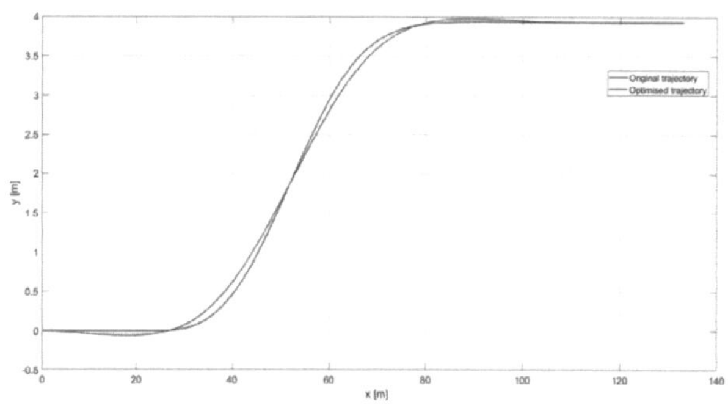


**Figure 9.** Comparison between the original and optimised trajectory

Using the reference optimised trajectory as the reference trajectory, this information is used as in input in the MPC controllers for the next vehicles. Comparisons between the 2WS – MPC and 4WS – MPC vehicles and the optimised ones are presented in the next analysis. In Figure 10, the steering wheels angle that represents the output from the MPC controllers is presented. From Figure 11 it can be observed that the optimised vehicles start the manouver sooner, and possess smaller amplitudes of the vehicle parameters. By analyzing Figure 12 for the yaw rate and the side slip angle it can be concluded that both of the optimised vehicles possess smaller values for the yaw rate, and side-slip angle (Figure 13), where for the 4WS the side slip angle possesses value close to 0. This concludes that the vehicle stability and performance are further improved over the previous vehicles. In Figure 14 it is shown that the optimised vehicles follow the optimised trajectory.

IOP Conf. Series: Materials Science and Engineering **1311** (2024) 012024 doi:10.1088/1757-899X/1311/1/012024

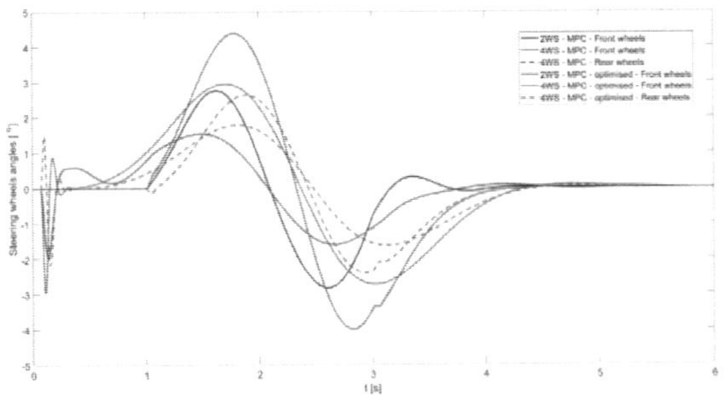


**Figure 10.** Steering wheel angles

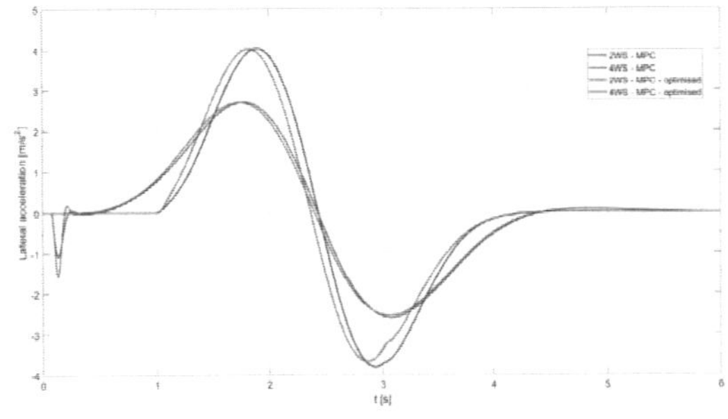


**Figure 11.** Lateral acceleration

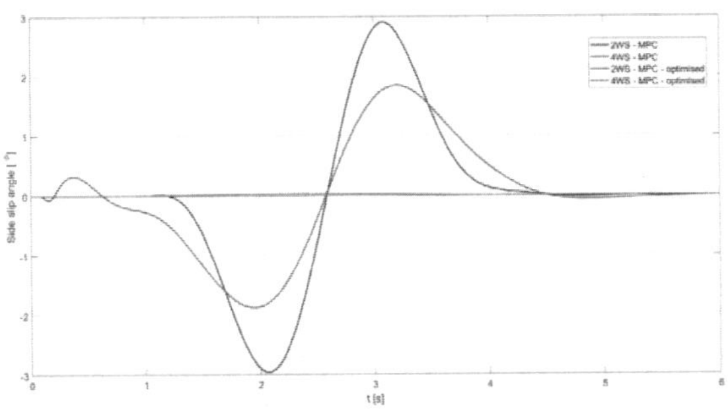


**Figure 12.** Yaw rate

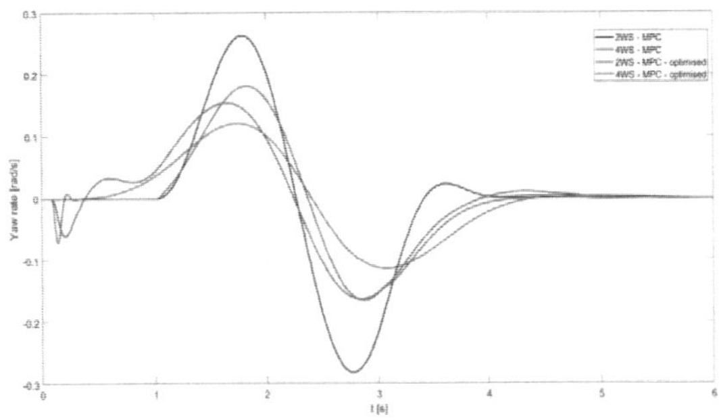


**Figure 13.** Side slip angle

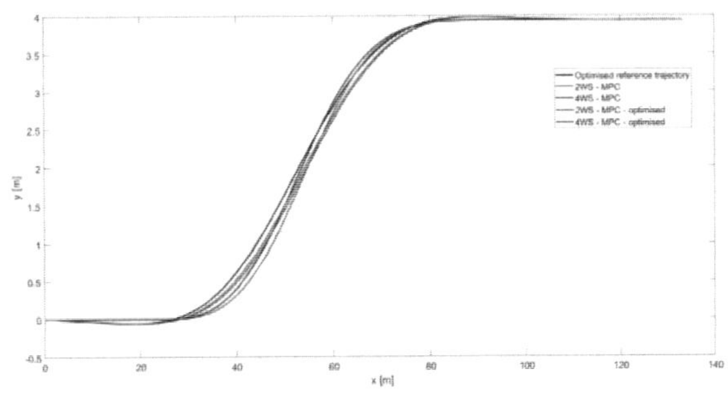


**Figure 14.** Trajectory

## 5. Conclusion

By conducting the simulations the MPC controller had shown that its application as a steer-by-wire system can bring certain improvements for future automated vehicles. These improvements are especially prominent for the 4WS vehicles. The MPC controllers allow better vehicle stability, improved consciousness for the passengers while traveling and improved vehicle safety and road transportation.

The importance of the reference trajectory also showed that an optimized trajectory will further improve the previously mentioned performances and characteristics of the vehicles.

The complexity of the considered issues indicates the requirement for further investigation and deepening of the study. Additionally, a more complex reference model and testing vehicles could be created where the number of input reference parameters and controlled variables of the MPC controller would be increased. Beside the steering wheels angles, those parameters would include the control of longitudinal speed, braking torque, adaptive suspension etc.

IOP Conf. Series: Materials Science and Engineering **1311** (2024) 012024 doi:10.1088/1757-899X/1311/1/012024

**References**

[1] Shladover, S. E. (2018). Connected and automated vehicle systems: Introduction and overview. Journal of Intelligent Transportation Systems, 22(3), 190-200.

[2] Mortazavizadeh, S. A., Ghaderi, A., Ebrahimi, M., & Hajian, M. (2020). Recent developments in the vehicle steer-by-wire system. IEEE Transactions on Transportation Electrification, 6(3), 1226-1235.

[3] Kader, A. (2006). Steer-by-wire control system. Swarthmore College Department of Engineering.

[4] Fahami, S. M. H., Zamzuri, H., Mazlan, S. A., & Zakaria, M. A. (2012, June). Modeling and simulation of vehicle steer by wire system. In 2012 IEEE Symposium on Humanities, Science and Engineering Research (pp. 765-770). IEEE.

[5] Yih, P., & Gerdes, J. C. (2005). Modification of vehicle handling characteristics via steer-by-wire. IEEE transactions on control systems technology, 13(6), 965-976.

[6] Beal, C. E., & Gerdes, J. C. (2012). Model predictive control for vehicle stabilization at the limits of handling. IEEE Transactions on Control Systems Technology, 21(4), 1258-1269.

[7] Oldewurtel, F., Parisio, A., Jones, C. N., Gyalistras, D., Gwerder, M., Stauch, V., ... & Morari, M. (2012). Use of model predictive control and weather forecasts for energy efficient building climate control. Energy and Buildings, 45, 15-27.

[8] Tosolin, G., Cartró, J., & Sharma, V. (2020). Development of model predictive motion planning and control for autonomous vehicles. In 10th International Munich Chassis Symposium 2019 (pp. 323-340). Springer Vieweg, Wiesbaden.

[9] Subosits, J. K., & Gerdes, J. C. (2021). Impacts of model fidelity on trajectory optimization for autonomous vehicles in extreme maneuvers. IEEE Transactions on Intelligent Vehicles, 6(3), 546-558.

[10] Gutjahr, B., Gröll, L., & Werling, M. (2016). Lateral vehicle trajectory optimization using constrained linear time-varying MPC. IEEE Transactions on Intelligent Transportation Systems, 18(6), 1586-1595.

[11] Larsen, R. B., Atasoy, B., & Negenborn, R. R. (2021). Model predictive control for simultaneous planning of container and vehicle routes. European Journal of Control, 57, 273-283.

[12] Kong, J., Pfeiffer, M., Schildbach, G., & Borrelli, F. (2015, June). Kinematic and dynamic vehicle models for autonomous driving control design. In *2015 IEEE intelligent vehicles symposium (IV) (pp. 1094-1099). IEEE.*

[13] Guo, H., Cao, D., Chen, H., Lv, C., Wang, H., & Yang, S. (2018). Vehicle dynamic state estimation: State of the art schemes and perspectives. IEEE/CAA Journal of Automatica Sinica, 5(2), 418-431.

[14] Aripin, M. K., Md Sam, Y., Danapalasingam, K. A., Peng, K., Hamzah, N., & Ismail, M. F. (2014). A review of active yaw control system for vehicle handling and stability enhancement. International journal of vehicular technology, 2014

[15] Rossiter, J. A. (2017). Model-based predictive control: a practical approach. CRC press.

[16] Kocijan, J., Murray-Smith, R., Rasmussen, C. E., & Girard, A. (2004, June). Gaussian process model based predictive control. In Proceedings of the 2004 American control conference (Vol. 3, pp. 2214-2219). IEEE.

[17] Lin, S., De Schutter, B., Xi, Y., & Hellendoorn, H. (2012). Efficient network-wide model-based predictive control for urban traffic networks. Transportation Research Part C: Emerging Technologies, 24, 122-140.

IOP Conf. Series: Materials Science and Engineering **1311** (2024) 012025 doi:10.1088/1757-899X/1311/1/012025

# The electric transmission stress behaviour investigation while being used for a middle-class road vehicle

**F Visan, V Croitorescu**

University POLITEHNICA of Bucharest, Faculty of Transport, SRCTCSAT, 313 Splaiul Independentei st., $6^{th}$ sector, Bucharest

valerian.croitorescu@upb.ro

**Abstract.** The paper presents the design for an electric transmission used for a middle-class road vehicle. The transmission gears are designed, calculated and virtually developed. A middle-class road vehicle is chosen for virtual implementation. The geometrical models for the gears are designed and developed. The transmission integration into the chosen vehicles is defined. The stress behaviour for the transmission is investigated following different torque performances demands. Different materials for the gears are considered.

## 1. Introduction

The road vehicles state of the art engages all the manufacturers to develop and to manufacture more and more hybrid and electric propulsion vehicles. Every car manufacturer directs a large part of its design activities towards innovative solutions of propulsion systems achieve the imposed requirements and regulations regarding pollutant emissions. Different regulations are applied for each continent regarding pollutant emissions coming from road vehicles. Both, the outlet & consumer markets are being directed toward solutions to reach the consumers demands. Therefore, most of the road vehicles development activities are aiming to reach dedicated development approaches for the innovative powertrains. One step forward consists of several conventional powertrain updates, when the engine is using different innovative systems (stratified fuel mixtures, extended combustion chamber, stratified fuel injection etc.) and well-known technical solutions (downsizing, cylinder deactivation etc.) to achieve lower fuel consumption and harmful emissions. Another step forward consists of converting a conventional powertrain into a hybrid and even into an electric powertrain. These are the most met practices.
Nowadays, the road vehicles manufacturers offer also fuel cell powertrains beside the already optimised hybrid and electric powertrains that already equip many of their models. The power transmission from the power source to the wheels represents one of the main goals for all the road vehicles manufacturers. In addition to these goals, the infotainment and the entertainment systems desire to come close in the optimised functionalities, which have an important role to propulsion.
The viable solution represented by the hybrid vehicles is competing with the electric vehicles for a better energy and environment management to reduce the greenhouse gasses. However, the key to low-carbon mobility offers real environmental benefits when time and space resources are well managed, but also limited autonomy, which have proven to be the main challenges.
During recent years, the transport sector has done its part to reduce carbon dioxide emissions. But the growth of two-thirds of fossil fuels consumption comes from the transport industries, which is extremely unfavourable for the sustainable development of human society.

Following the Paris Climate Agreement, the electric vehicles represent a very important step for the automotive industry. In 2021, more than 1 in 5 cars registered in the EU was electrically chargeable. This positive trend can only be sustained if governments start making investments in charging infrastructure and put in place meaningful and sustainable incentives. [1]
In the second quarter of 2022, sales of battery electric vehicles continued to expand in the EU, accounting for 9.9% of total passenger car registrations. The European market reached a grow for the registrations of battery electric vehicles (BEV) in the EU of 11.1%, reaching 233413 cars sold, during the second quarter of 2022. Plug-in hybrid electric vehicles (PHEV) expanded their market share during the period, despite a double-digit drop in units sold (-12.5% to 206501 cars). Sales of hybrid electric vehicles (HEV) in the European Union slipped back by 2.2% during the second quarter of the year. However, due to the significant drop in sales of conventionally fuelled cars, HEVs expanded their overall market share (22.6%). [2]
China Association of Automobile Manufacturers statistics says that the production of New Energy Vehicles (NEVs = BEVs & PHEVs & FCV) increased, reaching almost 755000 units in September 2022 from 4717000 units during the first nine months of this year (Jan-Sept.2022) [3], while the sales also increased, reaching almost 708000 units in September 2022 from 4567000 units sold between January and September 2022 [4].
A total of 77687 plug-in vehicles (63243 BEVs and 14444 PHEVs) were sold during September 2022 in the United States, up 42.4% from the sales in September 2021. PEVs captured 6.98% of total LDV sales this month. Cumulatively, 643858 PHEVs and BEVs have been sold in 2022. [5]
Between 2021 and 2027, the size of the global electric vehicle market is expected to increase over four-fold to reach an estimated global market size of some 1.4 trillion U.S. dollars by 2027. This translates to a notable compound annual growth rate (CAGR) of more than 19.19 percent between 2022 and 2027. [6]. As shown above, the number of electric vehicles is increasing continuously, while the batteries are seen by many to be by far the most cost-effective form of energy storage. Frequent starts and stops, high acceleration/deceleration rate, high torque at low speed, high power density, large and highly efficient operating area, low vibration and noise, high reliability and a high price-performance ratio are few of the demands the consumers have for the automotive industry. Motors and electronic motor power controllers are the basic parts for converting electro-mechanical energy. Therefore, the requirements for electric powertrains mainly include the following: good torque control capability for vehicle dynamics performances, reliability and durability for safety and secure usage, higher lifetime of the vehicle (not including the battery), high efficiency and a high performance/cost ratio for energy savings and users' capital [11].
This paper aims to present the stress behaviour of an electric transmission used for a middle-class road vehicle, after being converted from a conventional powertrain to an electric powertrain. The middle-class road vehicle has been chosen after several investigations where the transmission should better integrate, considering from the beginning the limitations imposed. The functional and construction particularities were considered, like the body shape, low weight, proper exterior dimensions for easily usage in crowded cities. In addition, energy limitations were also considered, like the autonomy and the charging scenarios, together with the power provided by the batteries and the torque developed by the electric machine.

## 2. The functional architecture of the powertrain

The flexibility of choosing the vehicle functional architecture represents a high-quality demand for each road vehicle.
Four different solutions have been designed and investigated before and after starting the conversion: front driven axle (Figure 1), both driven axles using different transmission for each axle (Figure 2), rear driven axle with planetary gearboxes and electric machines for each wheel (Figure 3), all wheels driven by an independent electric machine and transmission near each wheel (Figure 4). The electric machine takes place of the engine, the batteries take place of the fuel reservoir. The main components presented for the following architectures are the batteries (B), the variator (VT), the electric machine (M), the transmission (R, R+D, TP+D).

All the four solutions have pros and cons in terms of dynamic performances and energy management, well justified by the transmission integration approaches with the electric machine and the wheels.

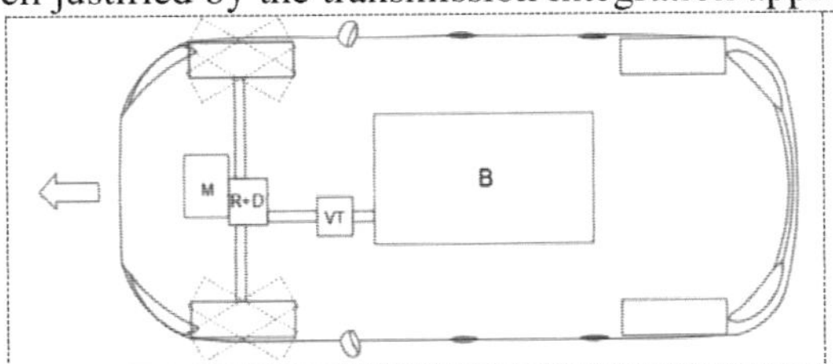


**Figure 1.** Front driven axle

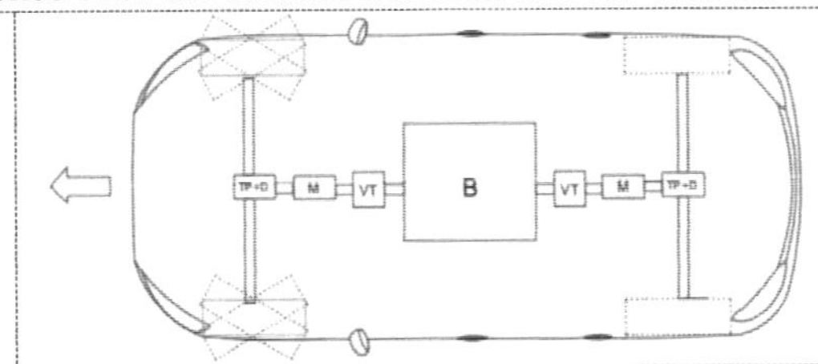


**Figure 2.** Both driven axles using different transmission for each axle

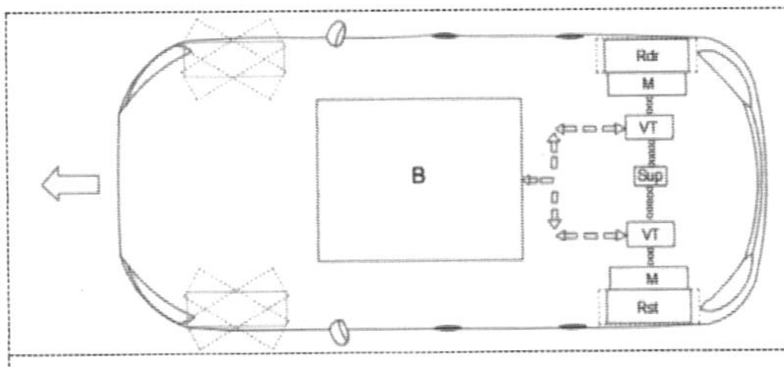


**Figure 3.** Rear driven axle with planetary gearboxes and electric machines for each wheel

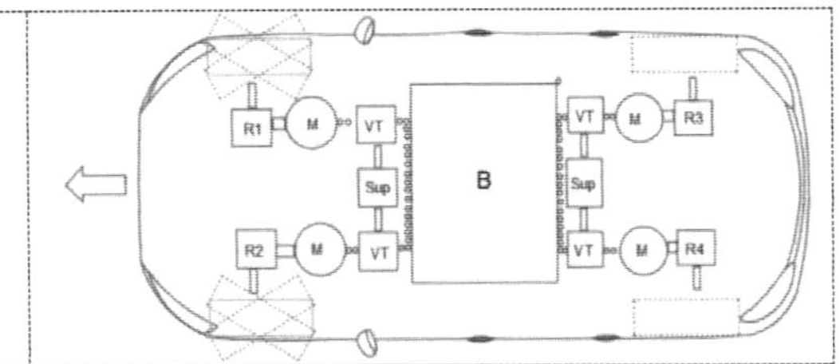


**Figure 4.** All wheels driven by an independent electric machine and transmission near each wheel

The implemented solution consists of the front driven axle, where the transmission is coupled directly to the electric machine. Considering the car autonomy, the costs, the maintenance, the on-road car behaviour, the stability and the ability while running on different roads, power losses etc. , three different solutions have been designed and investigated, during the implementation stages: 1- simple transmission that includes the differential in the same housing and has groove coupling (Figure 5), 2-the input transmission axle includes of the gearbox axle (Figure 6), 3-a planetary gearbox in the same housing with the differential (Figure 7).

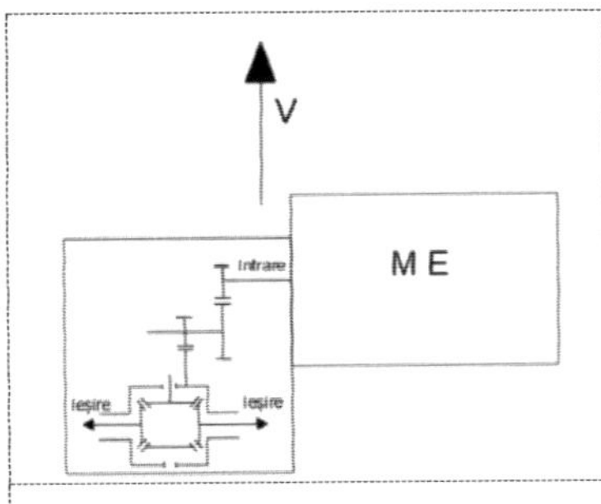


**Figure 5.** Simple transmission that includes the differential in the same housing

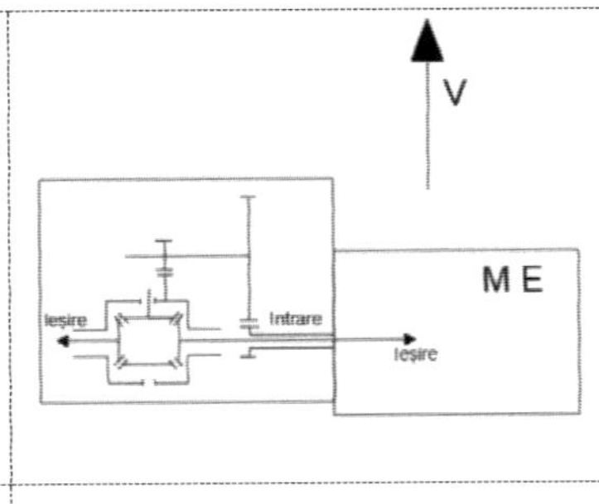


**Figure 6.** The input transmission axle includes of the gearbox axle

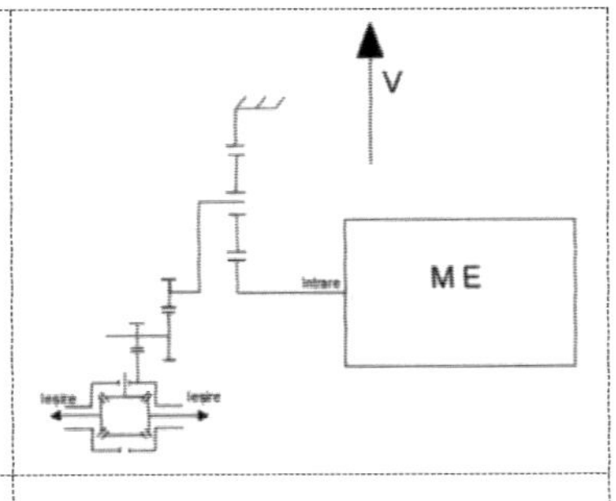


**Figure 7.** Planetary gearbox in the same housing with the differential

## 3. The specifications for the main powertrain components

Based on different selection criteria, that included both the vehicle class and the needed traction requirements, the electric machine and the inverter were chosen from a known manufacturer. Their technical specifications are presented in Table 1.

The battery package has an important role for designing the powertrain. The battery technology was chosen to be Li-Ion. Its starting state of charge was set to be 80%. The battery was calculated starting from the needed traction power at the defined vehicle speed of 60 km/h, to supply the vehicle auxiliar equipment (limited to 10% of the maximum power) and to reach the vehicle required autonomy between 300 to 350 km (estimated for different traffic conditions, different temperature values, different road conditions, auxiliar equipment operation etc.). The maximum voltage for which the electric machine and inverter can properly operate was also considered. The number of cells and their arrangement (series or/and parallel) were determined and set. The cell and the battery technical specifications are presented in Table 1.

The transmission was calculated considering the maximum vehicle speed to be reached, the tyres specifications, the rotary velocity of the wheels. It was chosen from different solutions already manufactured starting from the calculated transmission ratio that is 9.19. Its shape and external volume to be easily integrated with the electric machine for the chosen architecture were considered. The chosen transmission has a compact design and can be used on front or rear axle or as well an all-wheel drive. Its technical specifications are presented in Table 2.

**Table 1.** The technical specifications for the electric machine, inverter, and battery. [7][8][9]

| **Electric Machine** | | **Inverter** | | **Battery Cell** | | **Battery Package** | |
|---|---|---|---|---|---|---|---|
| Parameter | Value | Parameter | Value | Parameter | Value | Parameter | Value |
| Rated speed | 4,200 rpm | LV input voltage for operation (ISO16750) | 12 V | Nominal Capacity & Minimum Capacity | 5000 mAh (0.2C discharge) | Total number of cells | 4104 |
| Max. Torque (at max. Converter current) | 220 Nm | HV range – input voltage for operation with max. electricity | 250 – 450 V | Charging / Discharge Cut-off Voltage | 4.2 V / 2.6±0.05V | Series / parallel cells | 108 / 38 |
| Permanent moment | 130 Nm | Continuous current RMS (IACcont) | 350 A | Max. Charge / Discharge Current | 5000 mA / 10 A | Operating Temperature | Charge: 0 to 45° C |
| Continuous output | 70 kW | Max current RMS (IACmax) | 600 A | Standard Charge Current | 2500mA | | Discharge: -20 to 60°C |
| Max. power | 96 kW | Temperature 1* | -40 to +85 ° C | Standard / Rapid Charge time | 3 hours / 1.5 hours | | |
| Max. number of revolutions | 12,000 rpm | Temperature 2** | -20 to +60 ° C | Internal Resistance | the measured 30 mΩ | Energy | 65.7 kWh |
| Efficiency | 95% | Type. Efficiency | 97% | Nominal Voltage | 3.7 V | Capacity | 190.43 Ah |
| Weight | 51.5 kg | Weight | 15.3 kg | Weight | 94g | Weight | 385 kg |

* Temperature 1: Ambient temperature range during operation, **Temperature 2: Cooling water temperature range at the entrance

**Table 2.** Transmission technical specifications. [10]

| **Power** | **Drive torque** | **Output torque** | **Gear ratio** | **Weight** | **Dimensions** | **Number of revolutions** | **Additional functions** | **Motor type** |
|---|---|---|---|---|---|---|---|---|
| up to 150 kW | up to 270 Nm | up to 2,500 Nm | 1 : 9.59 | 51.5/23.8 kg (engine / gearbox) | 456 x 501 x 325 (L/W/H) mm | up to 13,000 rpm | Mechanical differential, parking lock | Hybrid Synchronous Motor (HSM) |

## 4. The transmission verification and static stress approach

The chosen transmission was mathematical verified and virtually developed, both functional and geometrical models being created.

The transmission axles are designed starting from the distance between them that must be as small as possible. The output torque is used to determine the distance. The gears teeth number are calculated depending on the gears type and their layout angles, together with some coefficients (specific tooth displacements) and other geometric characteristics (thickness, height, width of the gear, frontal roll angle, angle of the profile of the teeth in the frontal plane etc.).

The forces between the teeth for both gears were determined and are presented in table 3. Based on the determined values, the needed verification was achieved.

IOP Conf. Series: Materials Science and Engineering **1311** (2024) 012025 doi:10.1088/1757-899X/1311/1/012025

Both primary and secondary shafts geometrical characteristics were determined. The stress behaviour (bending, strength) for both shafts was investigated. The corresponding graphic diagram is presented below (Figure 8).

**Table 3.** The forces of the gear teeth

| Power | Tangential force | Radial Force | Axial force | Normal force |
|---|---|---|---|---|
| Main | 8.33 kN | 3.14 kN | 2.23 kN | 9.18 kN |
| Secondary | 19.61 kN | 7.38 kN | 7.13 kN | 22.13 kN |

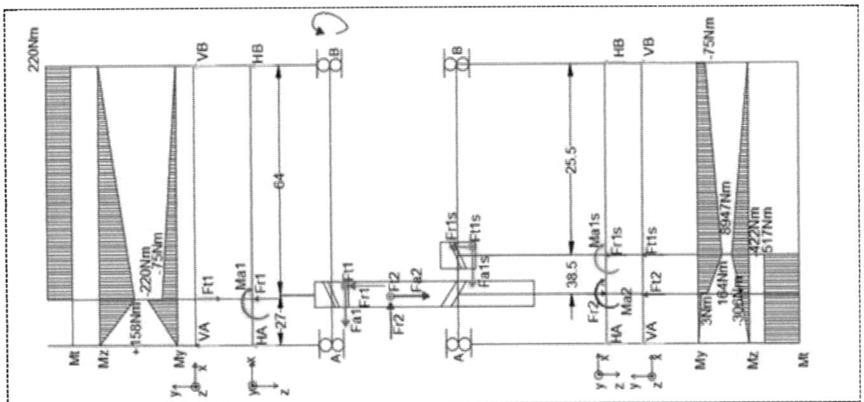

**Figure 8.** The stress behaviour diagram

In addition, the bearings from inside the gearbox were investigated. The sustainability and the loads of the bearings on the primary and secondary shafts was achieved.

The lubricant inside the gearbox was chosen basically limited by the gears kinematic and dynamic parameters, the gear types and the characteristics of the materials they are made off, including also the working environment and the design solution. An important parameter in choosing the type of lubricant is the peripheral speed of the gears.

The transmission case was also designed, where the following parameters were calculated and verified: wall thickness, body and cover flange thickness, thickness of the body and the cover ribs, all the screws used for assembly housing. The transmission case was also checked for overheating and fast cooling.

### 5. The transmission modelling approach

The transmission 3D model was developed using the Fusion 360 platform. Each 3D component with constant cross-section was made using Extrude sub-menu facility, especially for elements whose sketches were projected in the longitudinal plane, to establish the length of the element with the XZ median plane. The Extrude sub-menu facility was used throughout the modelling, as it is a basic element alongside the Hole and Loft functions. Figure 9.a. shows the primary shaft model using Extrude/Hole function.

The validation of the mathematical model was performed using the same modelling platform. It included the checking steps for both the pinion and the driven wheel. The obtained values were certified and prepared to be used for the geometrical modelling stages. Figure 9.b. shows the validation of the mathematical model by certifying the pinion and driven wheel parameters.

The components integration was made using the Joint function. All the components are integrated together, including the wheels, the shafts, the elastic rings and the bearings, with respect to the verified dimensions, as shown in figure 10.a.

The final assembly, without the case, is presented in figure 10.b., including all the major components used for the better evaluation of the stress behaviour.

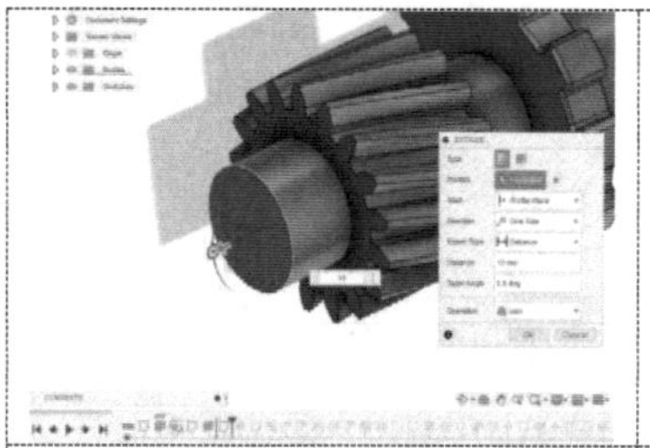

**Figure 9.a.** The 3D modelling for the primary shaft using Extrude/Hole function

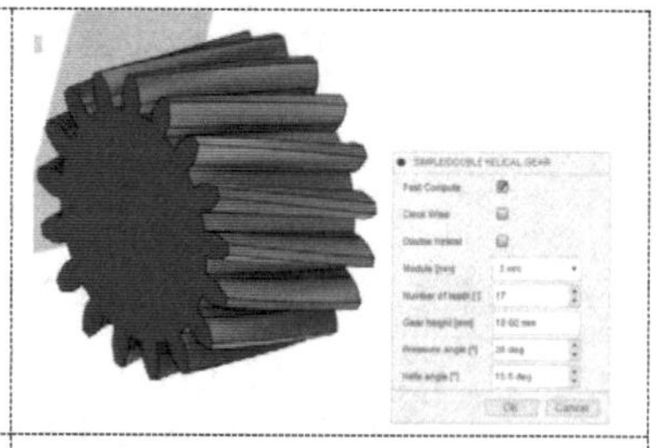

**Figure 9.b.** Validation of the mathematical model by checking the wheels parameters

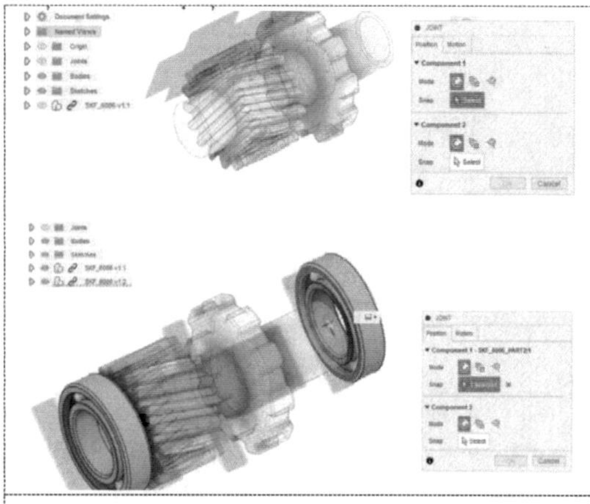

**Figure 10.a.** The integration of the components using Joint function

**Figure 10.b.** The transmission geometrical model (without the case)

## 6. The stress behaviour investigation

Transmissions of modern vehicles are very complex assemblies consisting of a considerable number of components made of different materials, by different technologies and assembled by several assembly methods and processes. For this reason, the structural analysis of a transmission is very difficult to achieve without using computational models.

Simplified models must consider that not all components are equally important in taking the stresses acting on the gears. That is why these components can be neglected in the calculations aimed at analysing the resistance of the gear under normal operating conditions.

The stress analysis was performed using the Ansys software (Figure 11.a.). Each component of the gearbox was simplified, to be easily integrate and analyse their geometrical shapes, assuring that this approach cannot bring possible changes of the wanted results. For example, the irregular shape of the case in cross-section will be replaced by a section of a simple geometric shape of a rectangle but the dimensional proportions will be preserved.

The contact areas are at the level of the gear teeth, both for the drive and the driven wheels. For continuing the analysis, the made of material for the shafts and the gears was chosen and their technical properties were introduced. The shafts and the gears are using different made of materials. For the shafts\` material, the yield stress is 835 MPa, while the breaking stress is 1800 MPa. For the gears\` material, the yield stress is 735 MPa, while the breaking stress is 930 MPa. The maximum allowable stress based on the yield stress was determined to be 500 MPa.

The assembly of the shafts and gears was discretized, 50278 elements and 90318 nodes were used. The bearings and the elastic rings were not considered, this approach being evaluated to generate minimal errors, if any. (Figure 12.a)

The motion constrains and the contact conditions were applied (Figure 12.b). The contact between the drive and driven wheels was simulated for different scenarios. For example, one scenario consists of the torque of the electric machine applied to the transmission primary shaft. The coupling between the electric machine shaft and the transmission primary shaft is made by using grooves. The two shafts cannot move in the normal direction of the applying force (in theory they no longer move on the grooves, remains joined, with no backlash). It allows to concentrate the input torque to the primary shaft and to apply the stress only the teeth. Another scenario consists of the primary shaft blocked (with no operation) and the stress for the gears` teeth are investigated (worst case).

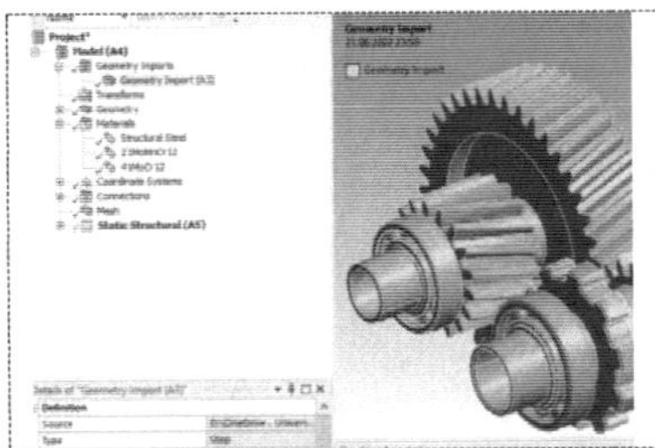

**Figure 11.a.** The model to be investigated in Ansys

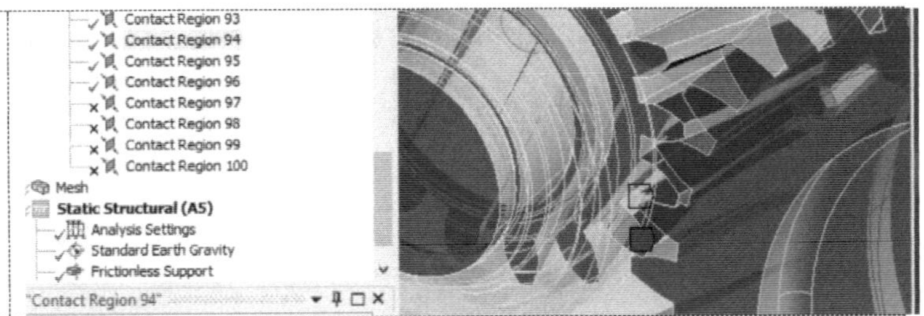


**Figure 11.b.** The coupling contact area of the gear teeth under investigation

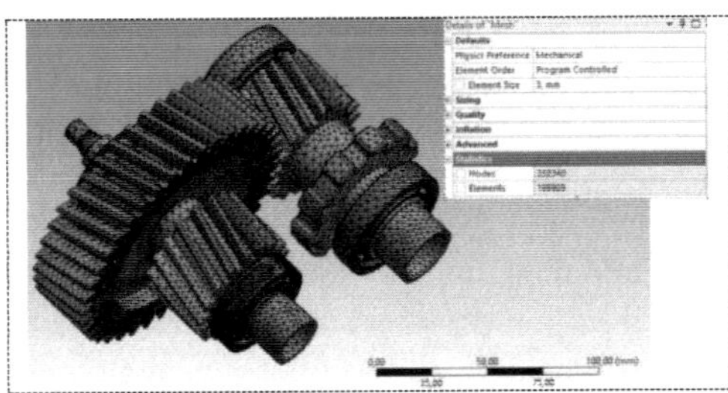

**Figure 12.a.** The finite element model discretization

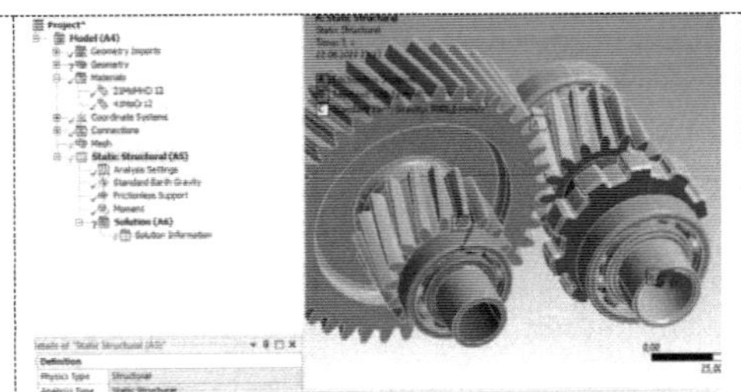

**Figure 12.b.** The constrains and the loads for the gears

Bending and shearing demands on the gears` teeth are continuously present when the motor torque is applied. The output torque from the electric motor is 220Nm. Friction is reduced in the central area of the shafts, because of the grooves coupling, allowing tangential direction displacement of the primary shaft. The deformation is maximum for the teeth head areas of the primary shaft (Figure 13.a, b), but also for the teeth of the parking mode wheel. The pitch diameter defines the teeth displacement because the meshing of the drive and drivel wheels is continuous.

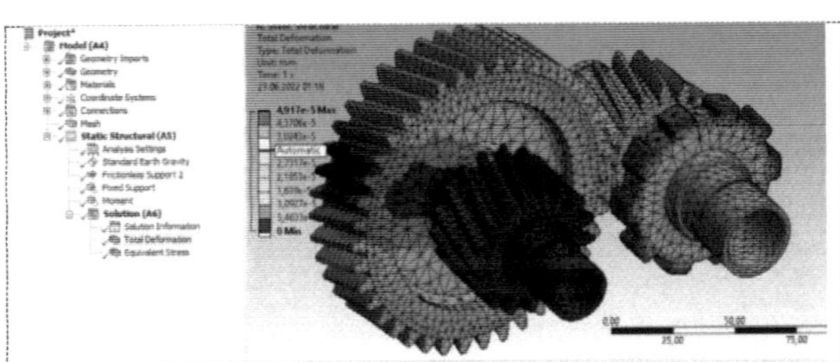

**Figure 13.a.** The deformation of the gears` teeth

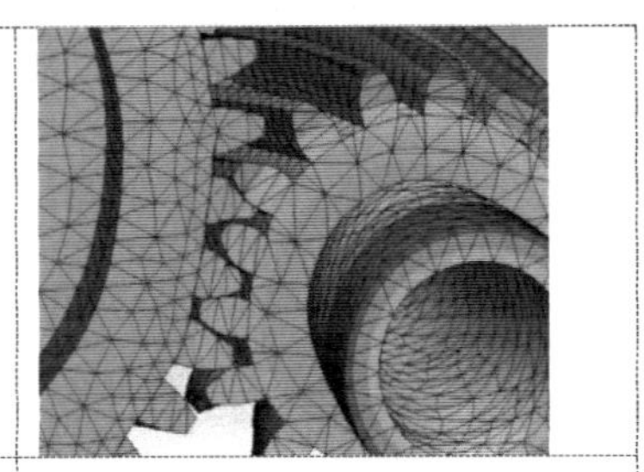

**Figure 13.b.** The detail of the deformation

The equivalent stress is met when the drive and the driven wheels are in contact, when not more than two or three teeth are simultaneously pressing on their homologous teeth. The maximum stress encountered in the drive wheel teeth was about to 150 MPa (Figure 14). The obtained value met the admissible value, being below the maximum admissible stress of 500 MPa. Therefore, another material can be chosen, maybe cheaper, but keeping the equivalent stress demands.

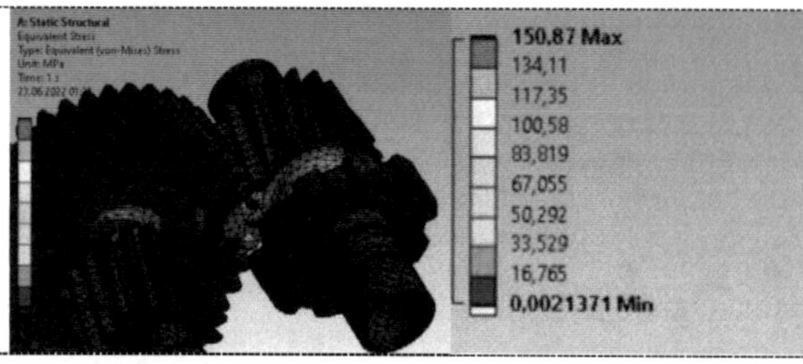


**Figure 14.** The maximum stress between the teeth in contact

## 7. Conclusions

The fast dynamics of road vehicles evolution bring together many different powertrain solutions, with high density on hybrid electric and electric propulsion systems. The current potential customers are pursuing the vehicle manufacturers with demands that are aiming fuel economy and dynamic performances. The authorities from most of the continents are continuously working on requirements and regulations to be applied for each vehicle model from each manufacturer. The solutions are quickly coming from all the vehicle manufacturers, to achieve both the customers needs and the authorities` directions. With this respect, the hybrid electric and electric vehicles are using innovative solutions for power and torque efficiency, including the transmissions design and optimization.

Within this paper a transmission for a middle-class electric vehicle was designed and investigated. Starting from the powertrain architecture, its each main component was chosen, the traction force and the maximum torque were calculated. The stress behaviour for the drive and driven wheels was investigated. Based on the obtained results, the initial chosen material can be changed with a cheaper one. In addition, the transmission design and its functional behaviour for being used for middle-class road vehicles were validated.

## References

[1] https://www.acea.auto/publication/2022-progress-report-making-the-transition-to-zero-emission-mobility/
[2] https://www.acea.auto/fuel-pc/fuel-types-of-new-cars-battery-electric-9-9-hybrid-22-6-and-petrol-38-5-market-share-in-q2-2022/
[3] http://en.caam.org.cn/Index/show/catid/59/id/1887.html
[4] http://en.caam.org.cn/Index/show/catid/60/id/1899.html
[5] https://www.anl.gov/esia/light-duty-electric-drive-vehicles-monthly-sales-updates#:~:text=A%20total%20of%2077%2C687%20plug,have%20been%20sold%20in%202022
[6] https://www.statista.com/statistics/271537/worldwide-revenue-from-electric-vehicles-since-2010/
[7] https://www.brusa.biz/portfolio/hsm1-6-17-12/?lang=en
[8] https://www.brusa.biz/portfolio/dmc544/?lang=en
[9] https://www.eleteks.com/sale-8325159-original-3-7v-panasonic-26650a-26650-5000mah-li-ion-rechargeable-battery-max-10a-discharge-battery-n.html
[10] https://www.brusa.biz/portfolio/dtso1-096/?lang=en
[11] Maria Hernandez, Maarten Messagie, Omar Hegazy, Luca Marengo, Oliver Winter, Joeri Van Mierlo, 2017, Environmental impact of traction electric motors for electric vehicles applications, *The International Journal of Life Cycle Assessment*, Vol.22, pag.54-65

IOP Conf. Series: Materials Science and Engineering **1311** (2024) 012026 doi:10.1088/1757-899X/1311/1/012026

# Research regarding the simulation of the autonomy for the electric vehicles in the case of the WLTC test cycle

**Corneliu Minzatu**[1], **Valentin Nişulescu**[1], **Gheorghe Bancă**[1], **Marius Toma**[1], **Cristian Renţea**[1]

[1]Politehnic University Bucharest, Romania.

corneliu_minzatu@yahoo.com

**Abstract**. In current mainstream research, energy consumption and autonomy are just a few of the the major objectives anticipated for electric road vehicles. The production of electric vehicles is continuously expanding, offering to the market a large variety of constructive variants in order to satisfy the requirements for their exploitation. Also, in the current energy context, it is expected from the electric car manufacturers, to deliver high quality models which can be characterized as the actual alternatives to actual existing solutions equipped with an internal combustion engine. Under these circumstances, the continuous research it is a high priority for everyone electric vehicle producers, where simulation on the computer is basically not absent as a go-between step prior to the validation of the final draft proposals. According to these assumptions, in the first part of the article you will find the AMESim simulation platform points, where it describes the model made with it and the simulated test cycle. In the second part of the article, as a consequence of the simulation of the experimental electric vehicle, are highlighted some aspects of the autonomy behaviour in the case of simulation on the World-wide harmonized Light duty Test Cycle (WLTC), for all four speed zones specific to it.

## 1. Introduction

The vehicles powered by an electric power supply are becoming a major interest in modern society, so the producers of the electric vehicles have suggestions the different design variants that can be the result meets the highest expectations [1]. In addition to these, another component that needs to be developed is the charging network for traction batteries, which unfortunately, both in Europe and worldwide, is unable to cope with the volume of vehicles produced. Thus the development of this type of vehicle, through the permanent evolution of their volume, is highly dependent on the level of implementation of the charging network [2].

Two performance criteria are perhaps among the most important expectations users have of electric cars: the distance traveled after a single charge of the traction battery and the electricity consumption. The range, represented by the total number of kilometers traveled after a single charge of the traction battery, varies from one manufacturer to another and is obviously lower or higher depending on the intended use of the vehicle [3]. It is influenced by certain technical constructional conditions of the traction battery, so it is closely dependent on the amount of energy stored in the traction battery and on the energy consumed by the comfort electrical system [5]. Also, a specific characteristic of the traction battery that influences the distance traveled after a single charge is its state of health, which should be maintained at a satisfactory level for long periods of time [4].

IOP Conf. Series: Materials Science and Engineering **1311** (2024) 012026 doi:10.1088/1757-899X/1311/1/012026

The simulation on the autonomy of the electric car submitted for research using the AMESim platform is based on the European legislation applicable to this type of vehicle by the authorities responsible for granting type-approval, namely Commission Regulation (EU) 2017/1151 of June 1, 2017 supplementing Regulation (EC) No 715/2007 of the European Parliament and of the Council on type-approval of motor vehicles [8]. The speed profile to be followed by the simulated electric vehicle is the one regulated by the WLTC (World-wide harmonized Light duty Test Test Cycle), shown in the following figure.

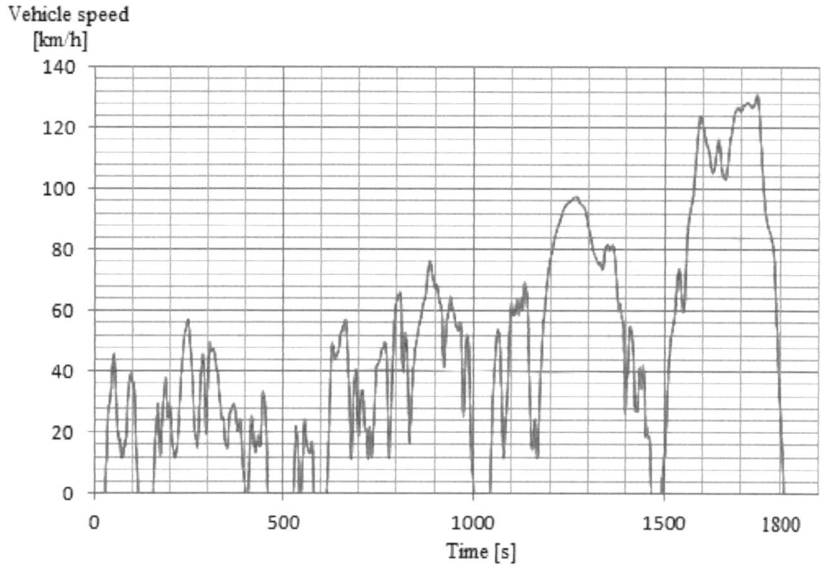


**Figure 1.** WLTC test cycle

The WLTC speed allure is composed of the four zones characterized by different dynamics and different allocated time, as follows: zone 1 represented by low speed and 600 seconds time, zone 2 represented by medium speed and 400 seconds time, zone 3 represented by high speed and 500 seconds time and zone 4 represented by very high speed and 300 seconds time. The cycle profile indicates that the vehicle covers a wide range of dynamic loads in order to faithfully reproduce the circumstances it may encounter under real driving conditions. The total cycle time is 1800 seconds, the distance traveled about 23 kilometers, the maximum speed 131 km/h, the average speed about 47 km/h and the stationary percentage 13%..

The AMESim platform, one of the most popular simulation and modeling platforms used by vehicle manufacturers prior to the deployment phase, is used for simulation. AMESim Platform incorporates a multitude of the simulation functionality utilized by the electric vehicle's producers and by different other industrial branches, like as the construction of machinery, ships, aviation, etc., being able to respond for a very widely diversified range of clients, regarding the conception of the virtual and calculational models furthermore, the efficiency is simulated through the assisted tests.

The AMESim platform can be used to simulate vehicle types, engines, mechanical and automatic gearboxes, axles, differentials, mechanical and electrical components of the listed systems, emission treatment equipment, injection equipment, cooling and lubrication equipment. Another special characteristic is the integration of specific test means and test types related to the type or system approval legislation, which really provides particular benefits at the time of research, the phase prior to the development of systems and components [7]. Simulation platforms therefore provide sufficient technical advantages to reduce production costs, making simulation an indispensable application in both research and manufacturing [6].

## 2. Experimental

For the realization of the modelling and the simulation was used the AMESim platform. This has been performed for a mid-class vehicle which is equipped with an electric engine, and which is equipped with a reducer. All the modeled and simulated vehicle data is identical to that of a real vehicle that will be tested at the roller stand to confirm the results obtained by simulation.

The created model has the objective of highlighting on the basis of mathematical algorithms the depletion mode regarding the state of charge of the traction battery as well as the electricity consumption using established relationships, subjecting the vehicle to the specific driving conditions of the WLTC cycle, in the present work being presented only results on the way of decreasing the state of charge of the battery.

The main technical characteristics of the simulated vehicle were:

- Electric Vehicle Mass: 1137 kg;
- Over inertia: 1253 kg;
- Front wheel - drive;
- Type of electric motor: synchronous with wound rotor ;
- Maximal Power: 67 kW;
- Maximal Torque: 210 Nm;
- Power of the traction battery: 23 kWh;
- Battery Traction Technology: Lithium Ion, 405V.

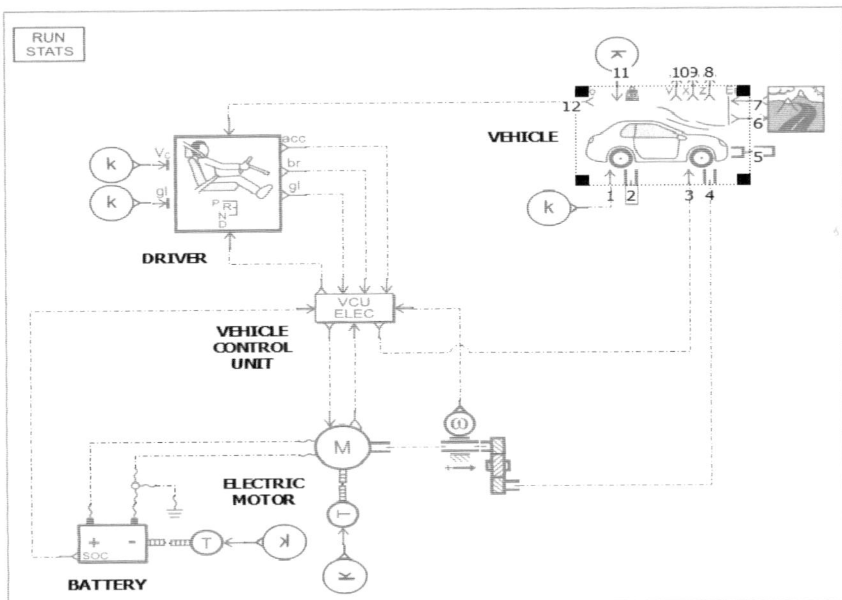


**Figure 2.** Representing of the AMESim model for an electric vehicle

Was used the predefined data in order to simulate the inertia and loading requirements of the simulated roll stand:

- The load and power supported by the roller stand at 82 km/h: 6.9 kW.
- Coefficients of rolling resistance: a ( N ) 6,9; b [ N / (km/h) 2 ] 0,0459.

The figure above shows the model of the electric vehicle for simulation purposes, the visually identifiable elements as follows:

- an element which simulates the car under simulation;
- an element named "driver" which simulates the behavior of the driver and provides control of the car when simulated on the WLTC cycle;
- an element which simulates the electric motor, which equips the modeled car;
- an element simulating the electric vehicle electronic control unit - VCU;
- an element simulating the gear-type transmission;
- an element simulating the traction battery.

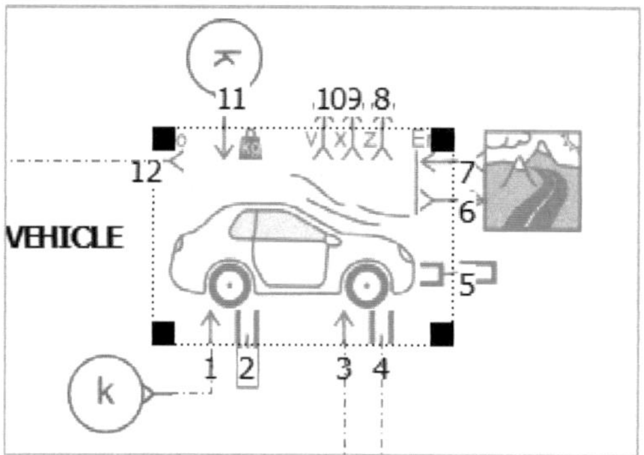


**Figure 3.** Electric vehicle component

The components listed above represent the main input parameters for the simulated model, but it also uses other subcomponents to obtain information about inertia, torque, engine speed, temperatures, which are the basis of the internal calculation algorithms [9]. In the figure above we have highlighted the main elements that make up the simulated car model, the purpose of the physical and mathematical input data displayed at its level being the following:

- input data 1 and 3: torque signals from the braking action of the front/rear axles are received via these input data.
- input data 2 and 4: through these input data the drive torque information is received and returns the angular speed information, specific to each axle.
- input data 5: this input data is strictly mechanical, through which the external forces that can be applied to the car are input, through which acceleration and speed information is returned.
- input data 6 and 7: through this input data, information on the wind speed is received, as well as information on the ramp and slope of the roadway.

Some of the computational relations that the model uses to achieve the simulated rolling are presented below:

- For wheel radius:

$$R_R = \frac{1}{2} \cdot D_r + 0.01 \cdot H \cdot l \qquad (1)$$

Where $D_r$ is the rim diameter of the wheel $[m]$; $l$ is the tire width $[m]$.
$H$ is the tire height [%];

- For the overall mass of the electric vehicle taking also into account the effect of inertia of the wheel in linear movement:

$$m_v = m + 4 \cdot \frac{J_r}{R_{r2}} \cdot l \tag{2}$$

Where $m$ is the overall mass of the electric vehicle $[kg]$; $l$ is the radius of the wheel$[m]$.
$J_r$ is Inertia of the wheel $[kgm^2]$;

- The running on the roller stand, without longitudinal slippage of the tire, the driving force is determined by the relation:

$$F_m = \frac{T_4 + T_2}{R_d} \tag{3}$$

Where $R_d$ is the dynamic wheel radius $[m]$; $T_4$ şi $T_2$ are the torque inputs in the ports 4 and 2, which are specific to the two front and rear axles $[Nm]$.

The resisting forces simulated from the electric vehicle model is determined using relationship as follow:

- The force created by by the ramp resistance $F_p$ $[N]$.

$$F_p = m \cdot g \cdot \sin(\arctan(0.01 \cdot \alpha)) \tag{4}$$

Where $m$ is the overall mass of the electric vehicle $[kg]$; $\alpha$ is the rolling road inclination [%]; $g$ is the gravitational acceleration 9,81 $[m/s^2]$

- The force that is triggered by the aerodynamic resistance $F_a$ $[N]$

$$F_a = \frac{1}{2} \cdot \rho_a \cdot C_x \cdot S \cdot (v + v_v)^2 \tag{5}$$

Where $\rho_a$ is the density of the air specified in the sub model $[kg/m^3]$; $S$ is the section of the vehicle $[m^2]$; $v_v$ is the speed of the wind $[m/s]$; $v$ is the speed of the vehicle $[m/s]$; $C_x$ is the aerodynamic coefficient [/];

- The rolling resistance force $F_{rul}[N]$

$$F_{rul} = m \cdot g \cdot (f + k \cdot v + vânt \cdot v^2) \tag{6}$$

Where $vânt, f$ şi $k$ are constant values.

## 3. Results and discussions

Based on the simulation, it was possible to plot the evolution of the state of charge of the traction battery for each specific speed zone simulated on the WLTC cycle. In this way, the simulated behavior of the state of charge of the traction battery was analyzed as a function of the load and the speed of the electric motor, evidently generated by the speed of displacement and the dynamics of the cycle. The simulation was run until the state of charge of the traction battery was fully exhausted. On the basis of this feedback, it had also been possible to identify the speed zones in which the electric car performs best economically, characterized by low electricity consumption.

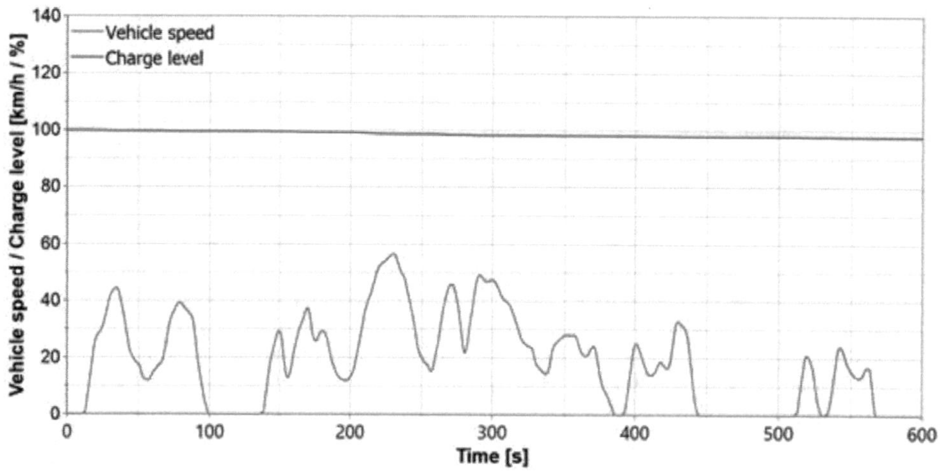


**Figure 4**. State of charge evolution of the traction battery in the low speed range

In the figure above, the shape of the speed profile realized by the simulated car in the low speed zone is illustrated, as well as the evolution of the state of charge in the traction battery. The simulated low speed driving zone is attributed to the urban exploitation of the vehicle characterized by intense traffic, several stops and starts, with a maximum speed of approximately 56.5 km/h and an average speed of 18.9 km/h.

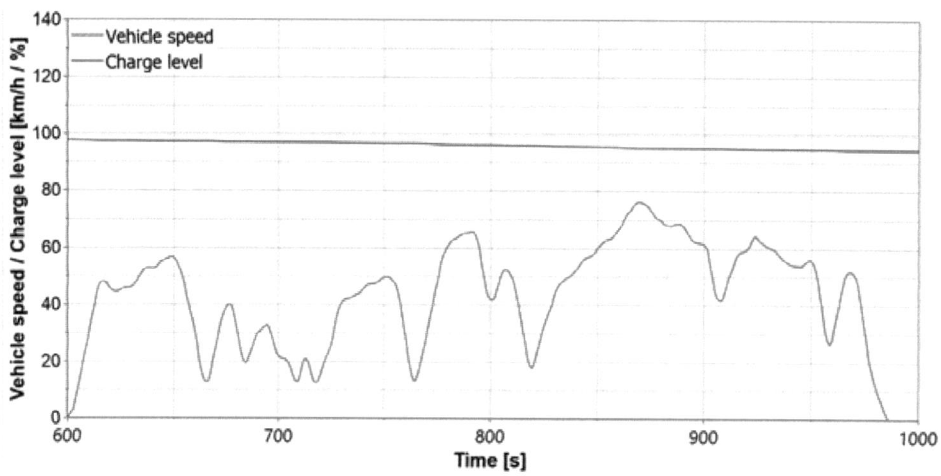


**Figure 5**. State of charge evolution of the traction battery in the medium speed range

Considering that the dynamic loads during the operation of the vehicle at low speed are not exceptionally high, the behavior of the traction battery in regarding to the maintaining of the state of charge is very good. An important positive contribution in this area can also be attributed to the fact that the energy consumption is low and the recuperative energy brakes are multiple. Also, through simulation, it has been evidenced that, in such situations, the state of the charge from the battery does not deplete quickly, having a smooth regressive allure. This trend is also maintained in the case of the medium speed zone, shown in Figure 5, which is characterized by a maximum speed of 76.6 km/h and an average speed of 39.5 km/h, which also disposes of a significant number of decelerations with a beneficial role of the energy recuperation.

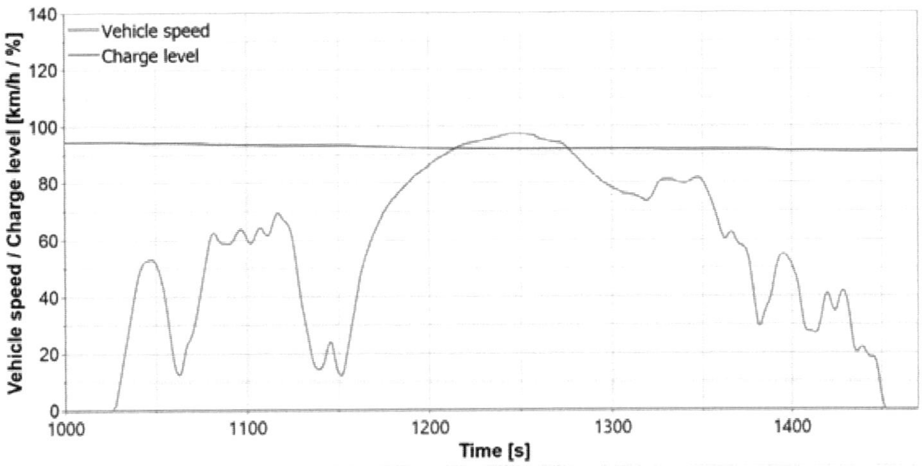


**Figure 6**. State of charge evolution of the traction battery in the high speed range

The high speed simulated driving zone, presented in the figure above, is characterized by a maximum speed of 97.5 km/h and by the average speed of 56.6 km/h. In such conditions, without a recuperative braking the strains on the traction battery are significant, the state of charge decreases rapidly, as a consequence the high-speed driving has a negative impact on the electric autonomy.

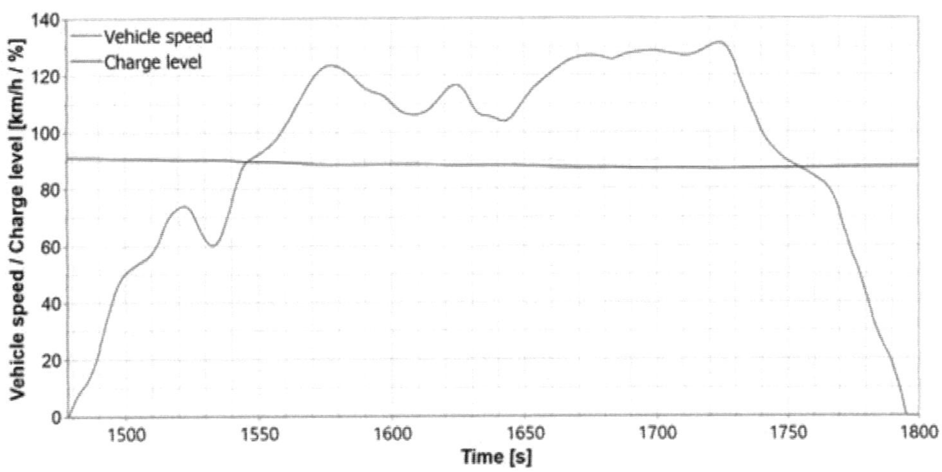


**Figure 7**. State of charge evolution of the traction battery in the verry high speed range

For the simulated very high speed driving described in the figure above, where the maximum speed is 131.3 km/h and the average speed is 92 km/h, the behavior of the battery state of charge decrease has an identical behavior as in the high speed range. As a result, the simulation showed that the state of charge of the traction battery decreases rapidly with increasing speed, such an operation of the cars having a major impact on the electric range.

## 4. Conclusions

Through simulation, it was revealed that in the event of the driving of the vehicles at the medium and the low speeds characterized by the low and by the moderate dynamic loads, the traction batteries maintain their state of charge, the effect on the electrical autonomy being reduced. In the event of simulated driving on the WLTC cycle in the low and the medium speed zone, we observe the usage profile of the vehicles in the urban mode, characterized by the multiple stops and starts and by the sufficient deceleration zones, these having a positive contribution regarding to the electric energy recuperation by the traction battery.

The driving simulations have shown that when the vehicles are operated at high and very high speeds, typically characterized by high and very high dynamic loads, the preservation of the state of charge of the traction battery is hardly achievable, resulting a very high effect on the electric autonomy of the traction battery. As for the high and very high speed aspect of the simulated WLTC cycle, it can be compared with extra-urban and highway driving. In these conditions, with virtually no regenerative braking, the energy recovered by the traction battery is non-existent, so the impact on electric range is very high.

The simulated evolution of the state of charge of the battery with effect on the electric range during the WLTC cycle indicates that vehicles with an electric power source have a very good utility in urban mode and a satisfactory utility in extra-urban or highway mode. Technological advances in the manufacture of traction batteries and their increased power output may allow vehicles to increase their battery range in extra-urban or highway mode.

**References**

[1] Buchmann, I.: Batteries in a Portable World, 2nd end. Cadex Electronics Inc. (2001). ISBN 978- 0968211823
[2] Paul, D., Stănescu, O., Electric, hybrid and fuel cell cars, Politehnica, Timişoara (2015). ISBN: 978-606-35-0011-4.
[3] Jeong, N.T., Yang, S.M.,: Urban driving cycle for performance evaluation of electric vehicles. Int. J. Automot. Technol 17(1), 145-151 (2016).
[4] Bamban, V., et al.: Analysis of range extender electric vehicle performance using vehicle simulator. Energy Procedia 68, 409-418 (2015).
[5] Shi, Q., et al.: The study of a new method of driving cycles construction. Procedia Eng. 16,79-87 (2011).
[6] Ababei, S., Culea, G. Electrical measurements and data acquisition, Tehnica-Info, (2003) ISBN 9975-63-214-9
[7] Barlow, T. J. et al.: A referance book of driving cycles for use in the measurement of road vehicle. Published Project Report PPR354, TRL Limited (2009).
[8] EUR-Lex Homepage, http://data.europa.eu/eli/reg/2017/1151/oj, last accessed 2022/08/20.
[9] Help AMESim Rev12.

IOP Conf. Series: Materials Science and Engineering **1311** (2024) 012027 doi:10.1088/1757-899X/1311/1/012027

# Virtual Development of a Challenger type, single-seater vehicle for the World Solar Challenge Competition

**B Roman, V Croitorescu**

University POLITEHNICA of Bucharest, Faculty of Transport, SRCTCSAT, 313 Splaiul Independentei st., 6th sector, Bucharest

valerian.croitorescu@upb.ro

**Abstract.** The paper presents the evaluation of the feasibility of a single-seat solar prototype vehicle for the World Solar Challenge competition, by estimating its autonomy. The research carried out contains several stages of study, including the development of mathematical models that are intended to be accurate enough to provide realistic predictions on the functional behaviour of the battery pack, so that it can be determined whether the designed vehicle will be able to complete the stages of the proposed competition. The paper also presents the CFD modeling approach, and the results obtained after simulating different body shapes advancements. In addition, the functional modeling approach is presented to evaluate the vehicle battery range, including the possibility to use different designs and types for the battery pack and to verify if the designed vehicle can achieve the race finish line.

## 1. Introduction

In the field of competitions with solar vehicles, two competitions stand out, namely Bridgestone World Solar Challenge – BWSC (in Australia) and Ilumen European Solar Challenge – IESC (in Belgium), their regulations imposing the constructive characteristics of the vehicle that is the subject of the paper. These are international competitions for solar vehicles, aimed at **participants from the academic environment**, supported by companies from various industries like: electronics, automotive, energy, IT etc. Their aim is to stimulate research in the field of solar vehicles, to develop the technologies used in their design and manufacturing, thereby providing new technical solutions for the eco-friendly of transport of the future.

The competition is divided into two (at IESC) or three classes (at BWSC): Challenger, Cruiser and Adventure (only BWSC). Challenger class vehicles are the aim of this paper. They are single seaters that want to achieve maximum performance.

## 2. The body shape design

Two body shapes are approached considering their standing out during the past competitions that took place. The vehicle behaviour is influenced by its functional and construction particularities, in addition to the dimensions, weight and energy performances. The large wheel track, the increased cornering stability, the large side-area section and the reduced wheelbase characterize the Catamaran vehicle type (Figure 1.a). In addition, the reduced longitudinal area section, the body shape, the position for centre of gravity, the reduced wheel-track and the reduced cornering stability characterize the Bullet vehicle type. (Figure 1.b) The air resistance for both vehicle types have a large proportion of the total drag. The aerodynamic design represents the most important and it is crucial to be studied and improved. One proposed approach consists of using the steering system for the rear axle, because the disruptive effect of crosswind and the generation of top drag force is reduced. This strategy has been approached by teams that have competed during previous years competitions using Catamaran vehicle type. Another proposed approached is met for the Bullet vehicle type, regarding the cornering stability, when the negative effects of the small track width are limited by actively tilting the vehicle body while cornering. The present study is approaching the BULLET vehicle type.

**Figure 1.a.** The CATAMARAN vehicle type [4]

**Figure 1.b.** The BULLET vehicle type [5]

*2.1. The applied regulations*

The BWSC regulation were considered for the next steps. The driver position inside the vehicle must achieve all the safety requirements, considering the driver seat geometry (Figure 2.a) and the roll-cage design (Figure 2.b). The heel must be below the hip level and the distance between the pilot's helmet and any neighbouring element must be at least 50 mm. [7] The visibility is checked by the pilot's ability to understand images of 75 mm height at four meters away, between 0.4 meters below his eye level, respectively 0.7 m above it, at a maximum angle of 100° to the direction of driving. The roll cage structure must protect the driver following different impact scenarios. For the frontal impact a five g load, opposing the direction of travel, applied to the front of the occupant roll-cage in an area less than 250 mm high and less than 600 mm wide is accepted. For the side impact a five g load into the side of the occupant roll-cage, applied adjacent to the driver's torso in an area less than 250 mm high and less than 600 mm wide is accepted. For the on top impact a load with components five g down, 1.5 g sideways and four g backwards, applied at each possible area of contact between the occupant roll-cage and the ground when the occupant roll-cage is upside down; the contact area for each test load must have a diameter less than 150 mm. [7]

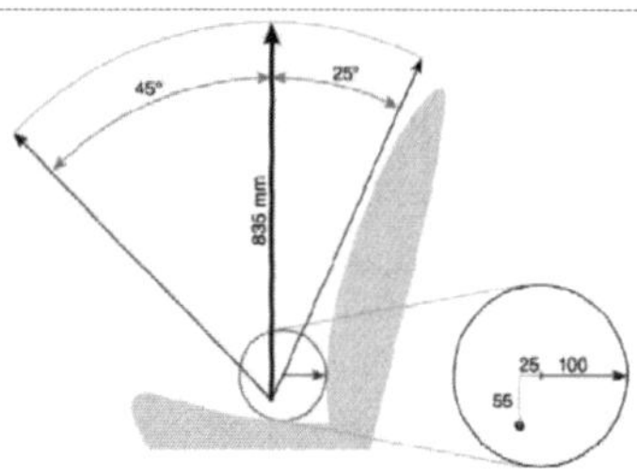


**Figure 2.a.** The driver seat geometry [7]

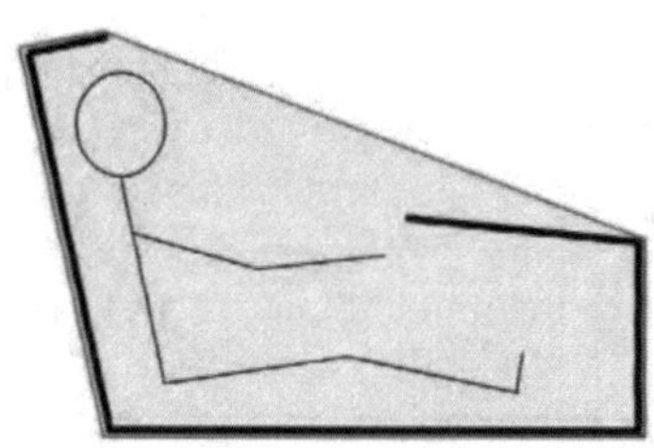

**Figure 2.b.** The roll-cage design [7]

*2.2. The vehicle body and the wheels*

The 3D model for the vehicle body shape was developed (Figure 3).

Since there is no statistical data for vehicles of this type, some components masses were determined by measuring the volume of 3D modelled parts. For the body and rims, the density of 1750 kg/m3, corresponding to carbon fibre was considered. In addition, for the removable roof, a density of 1200 kg/m3 was associated to the polycarbonate, while for the tubular protective frame and roll-cage, the titanium with a density of 4420 kg/m3 was chosen.

The battery and the solar panels maximum allowed masses were approximated. The photovoltaic panels were placed on the removable upper part of the body, the available surface being measured directly on the 3D model (4.35 m2). The wheels should be integrated under the body shape.

IOP Conf. Series: Materials Science and Engineering **1311** (2024) 012027 doi:10.1088/1757-899X/1311/1/012027

The chosen tires were Bridgestone Ecopia, for which the free diameter is 19.09 inches, equivalent to 484.89 mm, and the rolling resistance coefficient, measured for the speed of 100 km/h, corresponds to the following range: 0.0045-0.005. [8]

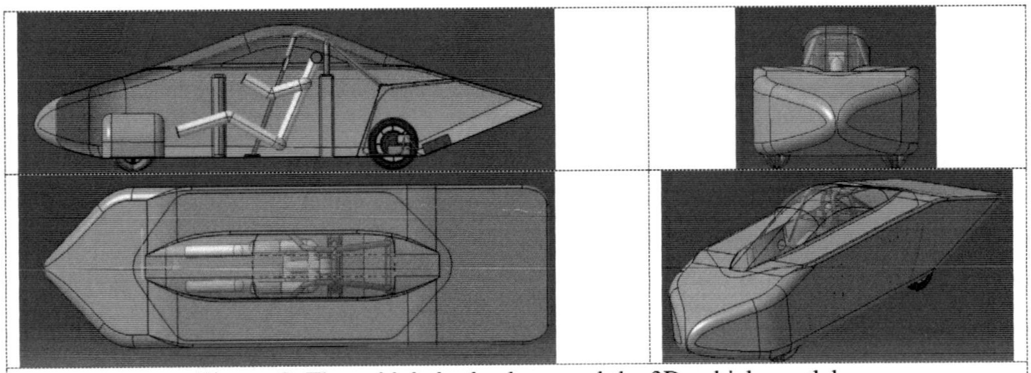

**Figure 3.** The vehicle body shape and the 3D vehicle model

### *2.3. The body shape aerodynamics*

The Computational Fluid Dynamics (CFD) is a branch of fluid mechanics that uses algorithms, numerical methods and computers to model and solve problems that arise during fluid flow. [14] Modeling is based on solving a set of differential conservation equations, supplemented by numerous additional equations, often semi-empirical models, to treat turbulence, pressure, cavitation, heat exchange, transport of chemical species or dispersed phases. [14]
A two-dimensional model (Figure 4) was developed for reducing the processing time of the simulation. The wheels were not introduced because areas of extremely low pressure were formed behind them, because of the existence of a small gap between the contact path and the road surface, that does not exist in real scenarios. The vehicle longitudinal section and the area of the front section were used as input parameters. Its meshing was carried out. The resolution decrease in areas away from the vehicle profile. The first rows in the vicinity of the contour are parallel to it.

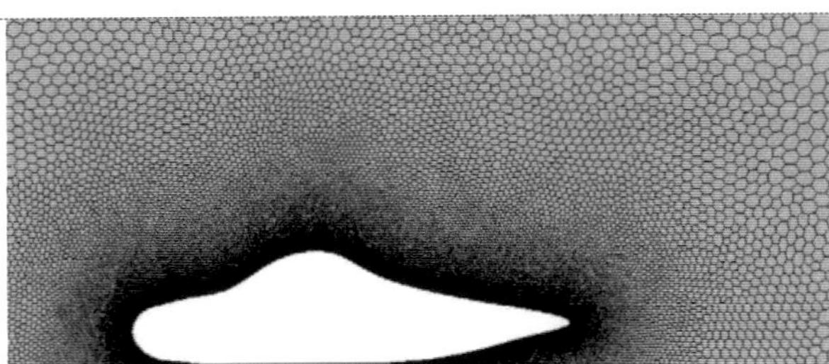

**Figure 4.a.** The 2D model meshing

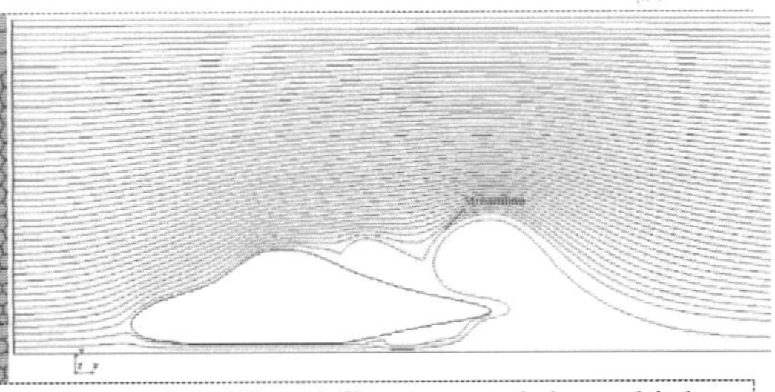

**Figure 4.b.** The airflow around the vehicle

The body shape was slightly improved by analyzing the vector visualization. Two problems were identified, for the roof area and for the front vehicle section. The frontal area was calculated, being 1.6 square meters. Simulations of 1000 iterations were performed, for the speeds of 6, 11, 17, 22, 28 and 33 m/s (Figure 5).
The vector representations of the airflow showed that the vortices were reduced in size and tend to follow the shape of the body shape more closely. For the speed values of 17 m/s and 33 m/s a large vortex behind the vehicle appeared because of the air flows below the vehicle that merged with each of the one above. During the simulation, its spin direction alternated, a phenomenon known as "vortex shedding" appeared around 20 m/s (Figure 4.b).
Because the vehicle is extremely light and its aerodynamic shape tends to generate high lift, these characteristics must be monitored to ensure operational safety. The lift and aerodynamic coefficients were calculated, being presented in figure 6.a&b together with the lifting generated force in figure 6.c.

IOP Conf. Series: Materials Science and Engineering **1311** (2024) 012027 doi:10.1088/1757-899X/1311/1/012027

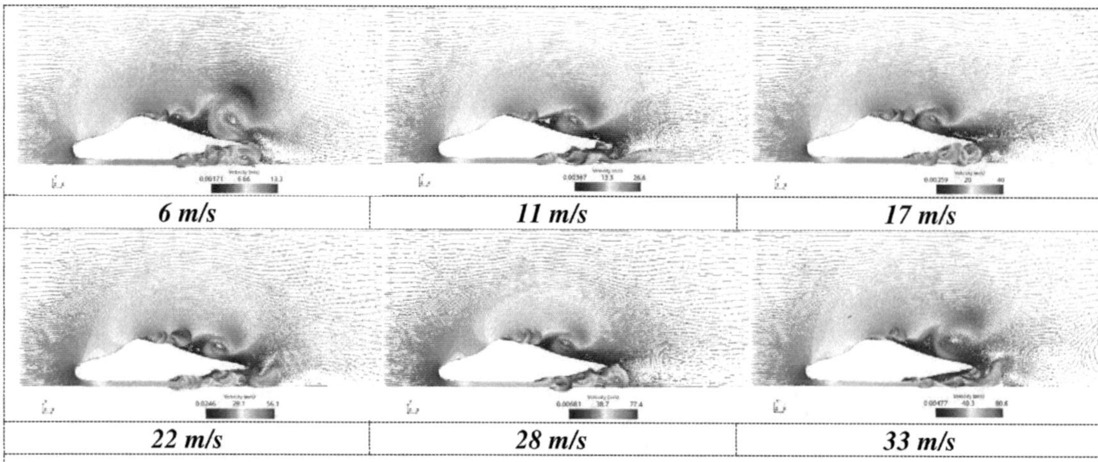


**Figure 5.** Vectorial visualization at different speed values: 6, 11, 17, 22, 28, 33 m/s

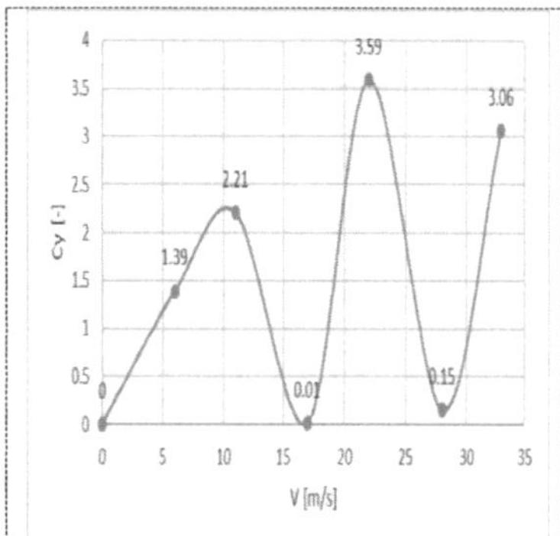


**Figure 6.a.** The lift coefficient

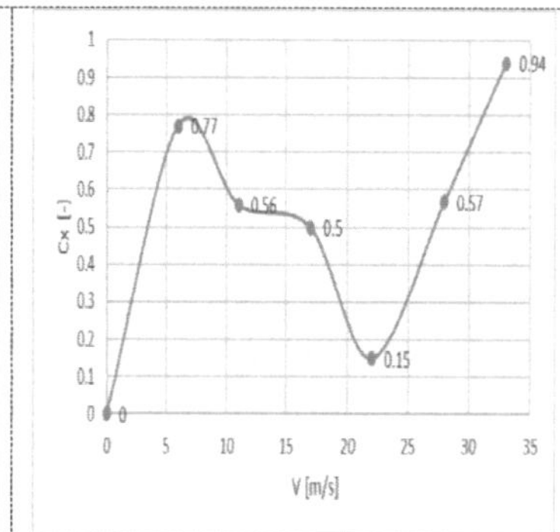


**Figure 6.b.** The aerodynamic drag coefficient

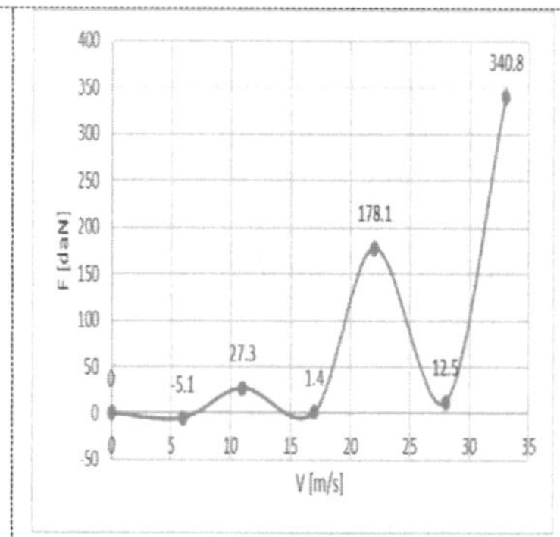


**Figure 6.c.** The lifting generated force

The lifting generated force reached very high values, exceeding the vehicle weight for the speed of 33 m/s. For the speed of 22 m/s, the lifting generated force reached around 178 daN. Given the fact that the air flow angle varies due to the bumps from the road, it is mandatory to monitor the vehicle behavior.
The simulated model is not exactly like the real one (the below presented assumptions were applied), being narrower for the top side. The aerodynamic drag coefficient varied drastically, that indicated the existence of some errors. Even if the drag coefficient unwanted variations appeared, its value for driving the vehicle with 22 m/s is promising, being around 0.15.

## 3. The powertrain design

The powertrain design includes the calculation for the rolling resistance, grade resistance and air resistance. They consist of the combination of forces that work against the forward motion of your vehicle. The weight of the vehicle, gravity and inertia, the amount of friction between the tires and the road surface, air drag and the vehicle speed (that varies from o to 30 m/s) were used for the calculation steps.
The air resistance has the largest influence, its hyperbolic trend being transferred also to the characteristic of total resistance (Figure 7.a). The power required for the propulsion system must be around 5 kW (Figure 7.b).

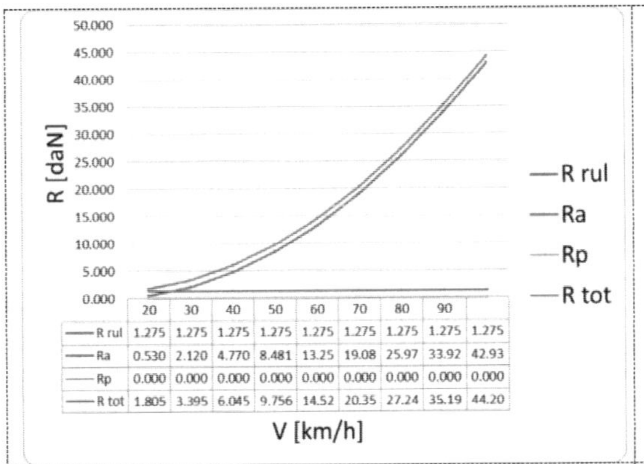


| | 20 | 30 | 40 | 50 | 60 | 70 | 80 | 90 | |
|---|---|---|---|---|---|---|---|---|---|
| R rul | 1.275 | 1.275 | 1.275 | 1.275 | 1.275 | 1.275 | 1.275 | 1.275 | 1.275 |
| Ra | 0.530 | 2.120 | 4.770 | 8.481 | 13.25 | 19.08 | 25.97 | 33.92 | 42.93 |
| Rp | 0.000 | 0.000 | 0.000 | 0.000 | 0.000 | 0.000 | 0.000 | 0.000 | 0.000 |
| R tot | 1.805 | 3.395 | 6.045 | 9.756 | 14.52 | 20.35 | 27.24 | 35.19 | 44.20 |

**Figure 7.a.** Resistances to forward the vehicle

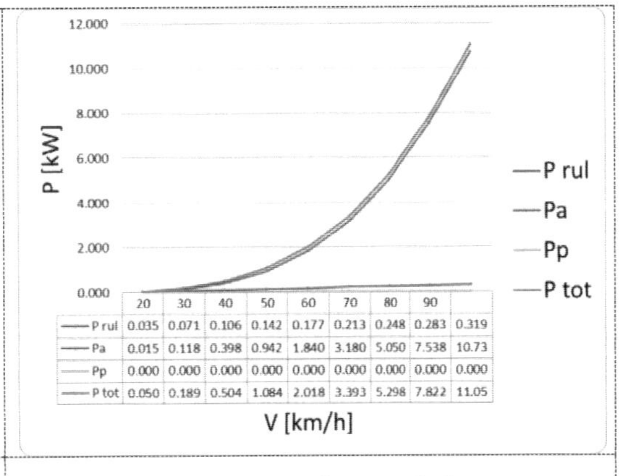


| | 20 | 30 | 40 | 50 | 60 | 70 | 80 | 90 | |
|---|---|---|---|---|---|---|---|---|---|
| P rul | 0.035 | 0.071 | 0.106 | 0.142 | 0.177 | 0.213 | 0.248 | 0.283 | 0.319 |
| Pa | 0.015 | 0.118 | 0.398 | 0.942 | 1.840 | 3.180 | 5.050 | 7.538 | 10.73 |
| Pp | 0.000 | 0.000 | 0.000 | 0.000 | 0.000 | 0.000 | 0.000 | 0.000 | 0.000 |
| P tot | 0.050 | 0.189 | 0.504 | 1.084 | 2.018 | 3.393 | 5.298 | 7.822 | 11.05 |

**Figure 7.b.** The power characteristics

### *3.1. The powertrain components*

The powertrain design consists of choosing its components carefully, to reach the needed efficiency for the race purpose. The chosen electric motor is a hub motor (BLDC), having compact shape, high efficiency (95%) and high specific power. Its controller is linked both to the electric motor and to the battery management system because the battery pack is connected to the solar panels. The battery package is limited by its weight and external volume. It must be integrated directly to the vehicle structure, but with an easy-to-extract solution in case of any possible accidents. The battery cells are using Li-Ion technology, having high energy density. The battery nominal voltage is 72 Volts, but with a minimum of 50 Volts for safely operation. The battery current will reach 200 Ampers. The maximum power provided is 14.4 kW. The minimum capacity of the battery package is 66 Ah measured for the 25 degrees Celsius. The solar panels technology is the crystalline silicone.

To reach the 22 m/s speed, the vehicle should develop at least 2 kW power, being optimal for the chosen competition. However, the maximum powertrain obtained power of 5 kW allows driving the vehicle with speeds around 30 m/s.

### *3.2. The powertrain functional model*

The powertrain functional model consists of A MATLAB Simulink model (Figure 8) that was created to provide the estimated range value. A battery of 400 cells, a solar panel with an area of 4 m2 and a power output of 160 W/m2 were considered.

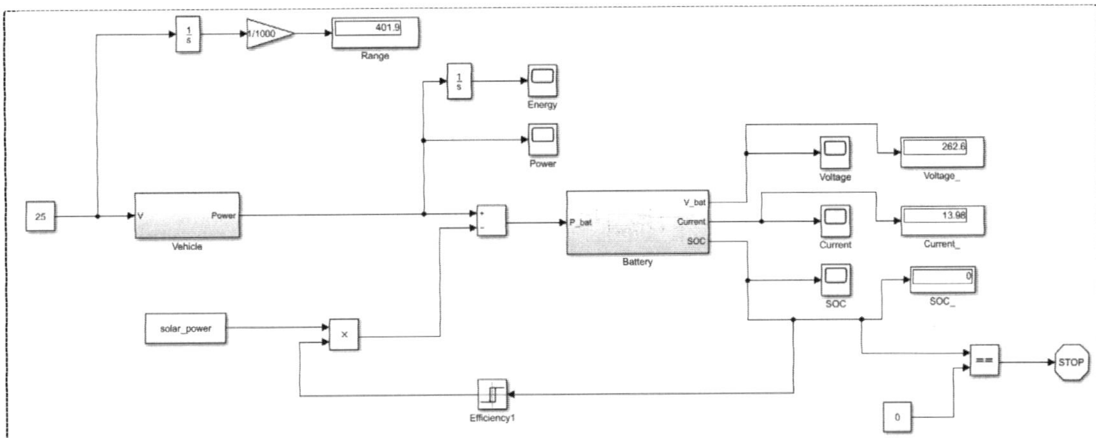


**Figure 8.** The functional powertrain model

IOP Conf. Series: Materials Science and Engineering **1311** (2024) 012027 doi:10.1088/1757-899X/1311/1/012027

After simulation, the resulted range was about 401.9 km, enough to cover a stage of the competition. Large variations for the obtained aerodynamic coefficients were also met. However, for the speed of 80 km/h, the obtained value is reasonably accurate. The simple fact that we can visualize the flow of air around the body allows us to identify problem areas and optimize its shape.

## 4. Conclusions

The World Solar Challenge competition reaches several teams for competing with their technical achievements in terms of developing newly and more efficient solution for the participating vehicles. The participants come from the academic environment, supported by companies from various industries like electronics, automotive, energy, IT etc. Their aim is to stimulate research in the field of solar vehicles, to develop the technologies used in their design and manufacturing, thereby providing new technical solutions for the eco-friendly of transport of the future.

Within this paper several research stages were presented, starting from the body shape design to powertrain design and their simulation while operating. A CFD modelling was performed, together with a functional model simulation. The CFD simulation highlights the complexity of fluid motion for this kind of vehicle, while the functional model simulation highlights the approach for determining the autonomy of the vehicle providing values within real limitations and the ability to be a quick tool for comparing later versions of the battery pack or solar panels implemented.

## References

[1] https://teamsolaris.com/wp-content/uploads/2020/03/iesc.png.webp
[2] https://lh3.googleusercontent.com/a2OGB8k6zxDulI_s3yDZHXfdgFQA wTYGFBebKL8riLCHIGqg372MuSfUFo_cf1heEfKYnMJWoy06MoGkQUK_bkw=s1500
[3] https://en.wikipedia.org/wiki/World_Solar_Challenge
[4] https://www.materialise.com/en/blog/materialise-helps-drive-agoria-solar-team-through-outback
[5] https://www.bridgestone.com/bwsc/stories/article/2019/10/28-1.html
[6] https://www.autoblog.com/2013/11/04/testing-toyotas-three-wheel-leaning-i-road-trike-in-japan/
[7] 2021 Bridgestone World Solar Challenge Regulations Release version 1.0 issued 19 June 2020
[8] http://images.wikia.com/solarcar/images/3/35/NGM_Bridgestone_Ecopia_Tire_Specs.pdf
[9] MITSUBA Special motor for solar car instruction manual, MITSUBA CO.,LTD, 2014
[10] https://www.dioda.ro/fotovoltaice/30960-panou-fotovoltaic-solar-monocristalin-flexibil-40w-18v-545-350mm.html?utm_source=2performant&utm_campaign=2Performant&utm_source=a7c80ffd2&utm_medium=CPS&gclid=CjwKCAjwsfuYBhAZEiwA5a6CDE_aDxDUt40wng0pj1ntNGroe30Qa-SLqgvpLVmX-roDvtTbMEZZsBoCv5IQAvD_BwE
[11] https://grabcad.com/library/mit-motorsports-2013-formula-sae-frame-1
[12] https://www.flickr.com/photos/cambridgeuniversity-engineering/18673519140
[13] https://www.utoronto.ca/news/u-t-s-blue-sky-solar-racing-team-unveils-its-newest-vehicle
[14] https://ro.wikipedia.org/wiki/Mecanica_fluidelor_numeric%C4%83
[15] https://www.bridgestoneamericas.com/en/press-release-details.en.2019.bridgestone-to-supply-solar-car-tires--using-fuel-efficient-tire
[16] https://www.ferrari.com/en-EN/competizioni-gt/articles/tech-insight-cfd-design-in-the-488-gte
[17] https://scientificgems.wordpress.com/2019/09/18/2019-world-solar-challenge-revised-teams-list/
[18] https://batterybro.com/products/panasonic-sanyo-ga-ncr18650ga-3450mah-10a#:~:text=%EF%BB%BFMax.,Minimum%20capacity%3A%203350%
[19] https://s13emagst.akamaized.net/products/14774/14773413/images/res_3d8657a453cd952791d1e3baab542f87.jpg

# Artificial Intelligence Supporting Early Automotive Engineering Processes

**A Kreis[1], M Hirz[2]**

[1]Graz University of Technology, Institute of Automotive Engineering, Research Group Virtual Product Development, Inffeldgasse 11, 8010 Graz, Austria
[2]Graz University of Technology, Institute of Automotive Engineering, Department Automotive Mechatronics, Inffeldgasse 11, 8010 Graz, Austria

email: alexander.kreis@tugraz.at

**Abstract**. Due to progressive increase of complexity, the automotive industry is subject to constantly changing trends varying from the introduction of greener products and components to the deployment of technological advances in development and engineering processes. In relation to both, sustainable automotive products as well as the deployment of technological advances, the integration of AI (Artificial Intelligence) approaches in combination with virtual products in automotive development and engineering processes is of great importance. In combination with knowledge-based CAx (Computer-Aided engineering), the integration of AI approaches delivers an enormous potential to enhance automotive development processes. In addition to process optimizations, the integration of various AI approaches considers sustainability (e.g., optimization of component geometries and materials, reduction of emissions over the entire life cycle, $CO_2$ reduction through improved development) and economical aspects (e.g., resources savings throughout the entire development process, time and cost savings through earlier error detection, avoidance of unnecessary process steps). The present approach deals with the integration of AI and knowledge-based engineering methods in the early phases of automotive development and engineering processes. Furthermore, the paper points to the time, cost and resources reduction potential, leading to earlier market entries and a greener industry. Finally, the paper demonstrates the integration of AI technologies into industrial development and engineering processes based on selected application scenarios.

## 1. Introduction

Because of rising costs, increase in competition, external influencing factors, as well as political regulations demanding more sustainable products, development and engineering processes have to be designed and optimized more efficiently. The increase of automation required to implement efficiency improvements affects both engineering service providers and production companies in a wide range of industries. In the presented paper, the focus is on automotive development and engineering processes in order to develop more enhanced and sustainable products by the integration of appropriate AI approaches. The demands on sustainable development and production methods have increased enormously in recent years to face the different environmental-related factors in automotive industry [1], [2]. In addition to the already known emission reductions (e.g., the limitation of the average fleet $CO_2$ emission to 95g/km in the European Union [3], [4]), the emission balance of products over the entire life cycle will be increasingly considered in the future. To improve this emissions balance, a wide range of actions must be applied, including but not limited to the further development and optimization of development processes. To achieve the goals of more efficient, optimized and sustainable development processes and automotive products, this paper provides an insight into the integration of appropriate AI methods into automotive engineering processes and the associated potentials for improving automotive products, both from an economical and sustainability perspective. Furthermore, this article briefly introduces both state-of-the-art processes in automotive development and production as well as AI applications used in the automotive industry.

## 2. State-of-the-Art

### *2.1. Automotive Engineering Processes*

The automotive industry is influenced by variate changing internally and externally shaped boundary conditions, which leads to the fact that automotive development and production processes must be adapted continuously. Especially in the field of development processes, the shift from 2D drawings created on drawing boards towards the use of 3D-CAD models in combination with optimization approaches (e.g., simultaneous engineering [5], [6] and frontloading [6]) in virtual product development has been a game changer during the past decades [7]. However, these approaches reach their limits in terms of optimization capabilities and resource saving potential today. For this reason, the introduction of AI-based technologies in automotive development processes is the next logical step to reach further optimizations.

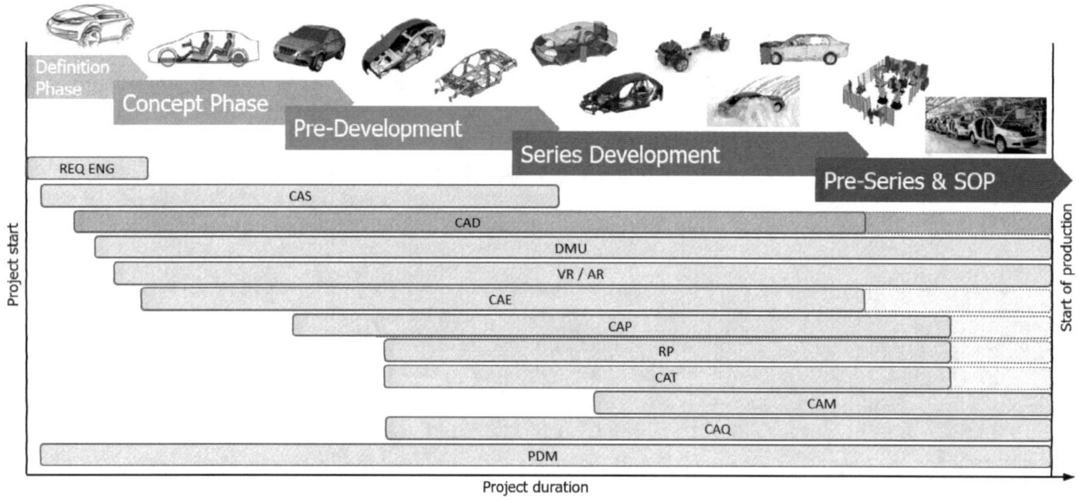


**Figure 1.** Individual phases of an automotive development process, exemplary focussing on body-in white (BIW), referring to [7], [8].

Figure 1 gives an overview of a comprehensive automotive development schedule and the different integrated individual phases. Regarding [7], [8] and [9], the development process starts with the phase “Definition” in which the first ideas and specifications are elaborated. The next phase in the development process is the phase “Concept”. In this phase the elaborated specifications are used to create the first styling and packaging drafts of the vehicle to be developed. Both phases, “Definition” and “Concept” are supported by different CAx methods, like CAS (Computer-Aided Styling), CAD (Computer-Aided Design), CAE (Computer-Aided Engineering), etc. The two initial phases are followed by the phases “Pre-Development” and “Series Development”. In these two phases – where the main development effort takes place – there is a closer cooperation of several CAx methods. Based on the developed specifications as well as styling and packaging drafts, 3D-CAD models are created. With the help of CAE and CAM methods, the 3D-CAD models of the automotive products are refined and optimized in several simulation loops. The last phase to round up figure 1 is the phase "Pre-Series & SOP", which initiates production processes with the Milestone "SOP" (start of production). Due to the fact, that in the early phases, both the phase "Definition" and the phase "Concept" include a large number of unknown internal and external factors with variables as well as parameters not yet clearly defined, there is an enormous potential to integrate AI approaches into the development process (c.f. figure 3). Besides the high potential for optimizations in the early phases, there is still the greatest potential for detecting errors and thus keeping the effort required for necessary changes to a minimum (c.f. figure 4).

### 2.2. *Artificial Intelligence Approaches Used in Automotive Industry*

Due to the ongoing technical progress in the automotive industry and beyond, there is already a variety of application areas that rely partially or entirely on AI applications. Especially in cases where large amounts of data have to be processed (e.g., in the field of autonomous driving), AI approaches deliver an advantage compared to conventional development methodologies. In addition to the use of AI methods for processing large amounts of data, AI is also used for optimization measures in product development. According to [10] and [11], research on AI methods is intended to accelerate time-consuming CAD-CAE simulation loops to evaluate them independently. Like for most AI-based methods, a database with existing simulation results is used for this purpose. Using ML (Machine Learning) algorithms, the AI is to perform complex evaluations with regard to future simulation results from existing digitally stored expert knowledge and to assess the results. This is expected to help engineers optimizing products more effectively, reducing and speeding up typical CAD-CAx process loops [10]. Consequently, this approach aims at improving the effectiveness of computer-aided layouts, however, individual components and modules of vehicles are considered for this purpose instead of holistic systems. Further examples are the AI applications from Amazon Web Services (AWS). As the world's most widely used cloud platform, AWS offers ML applications as well as some DL-based (Deep Learning) services and products. AI algorithms for image and video classification are used, for example, in Amazon Recognition, Amazon Prime Photos, and Amazon's Firefly services. These algorithms are capable of orienting themselves on a large data volume of images or videos and recognizing certain motifs from them (e.g., face recognition). Another example is the AI-based speech recognition of Amazon Alexa. Here, DL algorithms help to continuously evolve the application to correctly recognize even incomprehensible speech input (e.g., pronunciation errors, dialects). Probably the largest use case of AWS DL algorithms today, however, is in Amazon's online shopping system, which recommends the best possible products to individual customers [12], [13].

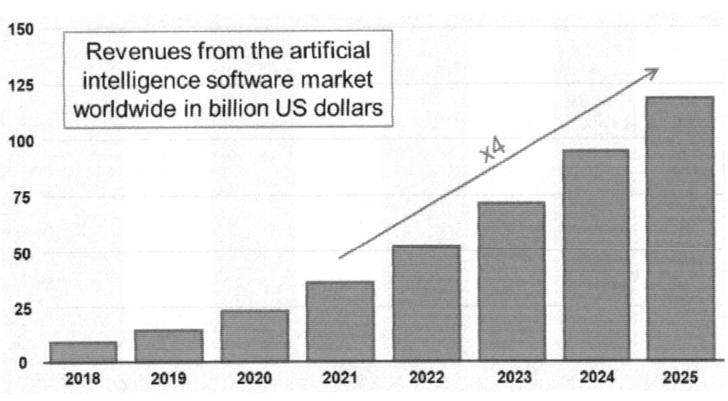


**Figure 2.** Revenue in billion U.S. dollars from the AI software market worldwide (2018-2025), referring to [14].

Another example, according to [15], is AI-based research efforts that will design the outer shape of future automotive bodies, incorporating big amounts of data. The focus here is clearly on styling development while technical tasks are not addressed. According to [16], AI-based methods support development tasks to automatically generate optimized workpieces for additive manufacturing. A variety of possible solutions for complex geometries are generated from a wide range of input data, including functional requirements, materials, manufacturing processes, and other constraints. These variants can be optimized according to different specifications (e.g., performance, weight).

These approaches pursue the support of the development of individual, often complex, components. Although this variety of different AI-based applications does not take into account the overall vehicle development, it clearly shows the potential to integrate AI methods into the development process of complex products such as those found in the automotive industry. Figure 2 shows the increase in revenue and thus the derived potential for the use of AI software systems in the upcoming years. With increasing trend, revenue for AI software applications will quadruple from 2021 to 2025 [14].

## 3. Potentials and Limits of Integrated AI Methods in Automotive Industry

The integration of AI methods into automotive development processes offers a high potential to optimize and accelerate existing process steps. This is of particular interest in phases where there is still a high degree of uncertainty about variables and parameters of the components as well as the end-products. Especially for automotive engineering providers and system suppliers, who have to deal with a number of different customers, there is an enormous potential. Figure 3 shows possible savings potentials by the integration of AI methods in the early phases of automotive product development. Preliminary studies carried out on industrial projects have shown that a resource and time saving potential in the range of 15-25% could be achieved. The early phases of the automotive development process are defined as phases "Definition" and "Concept". For consideration in figure 3, the phase "Definition" is further divided into the two sections "Product Vision" and "Feasibility".

Besides the high time and resource saving potentials in the early phases of automotive product development, there is also a not negligible potential in the subsequent phases "Pre-Development", "Series Development" and "Pre-Series & SOP". The potential of optimization measures in these phases is estimated to approximately 5%, which can be achieved for example by supporting CAD-CAx simulation loops or by supporting data exchange processes between several CAx environments (c.f. [17], [18]). Since the approach presented here is limited to automotive development processes, which ends with the phase "Pre-Series & SOP", other areas (e.g., "Production", "Logistics", "After-sales") are not specified.

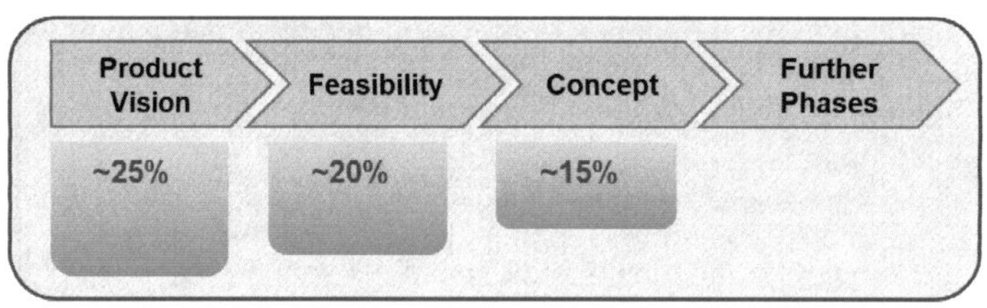


**Figure 3.** Expected optimization potential by use of AI applications in early product development phases.

The characteristics of possible savings and optimization potential that can be achieved with the help of the integration of AI methods is illustrated in Figure 4. Here, three curves are used to show the general impact of CAx and AI support over the project duration, contributing to earlier error detection. The dotted curve shows the option of product development without the use of CAx applications. It can be clearly seen that the general influence on the automotive development process is high at the start but decreases considerably with project progress. The dashed curve sketches the potential influences, which are given with the support of CAx systems. Here again, it can be clearly seen that these decreases slightly over the entire project duration. The further the project has progressed, the lower the possible influence on serious improvements. If compared the two curves – with AI support (continuous curve) and with CAx support – it can be clearly seen that the integration of AI methods already intervenes in the phase

"Research & Product Vision", i.e. even before product development is initiated. This already offers some possibilities in the early phases of development and before, thus showing the higher potential.

In addition to the introduced advantages and potentials of the use of AI methods in automotive development, there are also some limitations to be considered. First of all, it should be mentioned that AI methods are primarily able to support the development processes and no finished product can be achieved at the push of a button solution. This is also due to the fact that automotive development is very complex. In order to create a complete vehicle, several thousand parameters must be defined. However, at the start of development, only a few, typically less than hundred parameters are available. Here, the AI methods can intervene and support the development and guarantee a faster and enhanced product creation. For this, however, the boundary conditions must also be available and restricted, e.g., by implementation of large databases, so that sufficiently exact predictions can be made. This facilitates the work of the involved engineers and contributes to a faster and less error-prone development of the products.

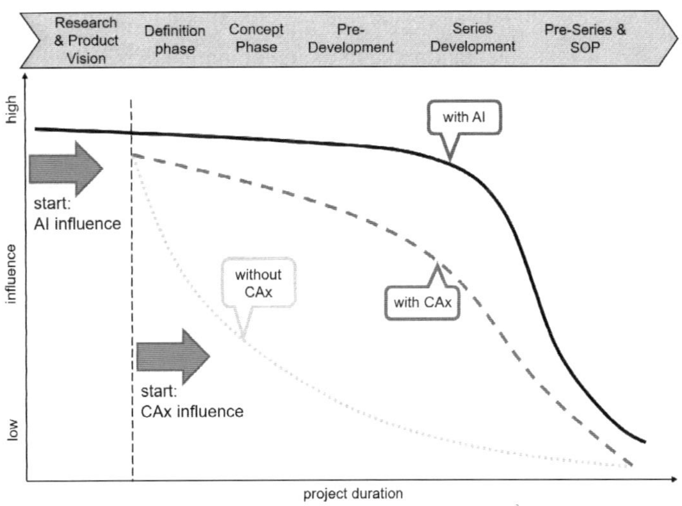


**Figure 4.** Influences of the use of CAx and AI systems over the course of product development processes, referring to [19].

The integration of AI methods can additionally make significant contributions to the development of sustainable products and systems. High levels of complexity and the associated challenges to ensure sufficient transparency in automotive development processes often lead to increased risks of insufficiently optimized products and workflows resulting in resource wastage. AI-supported optimizations of products and their development processes can achieve high levels of detail by analyzing and evaluating complex interdependencies, whereas the use of traditional development methods would come to their limits. This means that AI is able to generate and transparently present certain best-in-class solutions for specified optimization targets (e.g., utilization of energy efficient manufacturing processes, use of sustainable materials, man-hour saving potentials, etc.), which can be considered in subsequent working steps in order to enhance future automotive development processes [20], [21], [22], [23], [24].

## 4. Integration of an AI Development Assistant in Automotive Development Processes

As mentioned above, there is a high potential of AI methods to improve product development. Especially in the early phases of automotive product development processes, the use of AI approaches can be advantageous since the degrees of freedom for new products are numerous in these phases. In this context, figure 5 shows a systematic workflow of a comprehensive AI approach, the so-called AI development assistant (AIDA). The AIDA consists of different areas, which are processed partly simultaneously and partly in sequence. The basic process includes the three areas "Automated Data Acquisition and Preparation", "AI Training Database" and "Output of the AI Development Assistant". These three phases are run through in sequence, while input information is provided simultaneously in each area. The structural workflow of the entire AI methodology consists of two main paths, which can be linked together. One of these paths represents the process of parameter definition and evaluation up to the output of an x-dimensional parameter solution space with the optional inclusion of analyzed image data. The second path involves automated parameter derivation and generation of polygon models using AI-assisted object recognition from imagery originating from various sources. The definition of initial information and specifications for the program start includes initial parameters, boundary conditions, as well as the definition of development criteria (target parameters of optimizations) and the desired output information.

An AIDA is developed in such a way that certain development strategies are applied. Thus, the analysis of available as well as the definition of additionally required data is important. Furthermore, specific AI development guidelines regarding sustainability aspects are considered. After comprehensive definition of task and start parameters, an automated data acquisition for both paths is performed. In this process, the data (image material, parameter sets, etc.) are selected from different databases and prepared for the AI training session. The data sources used can include information from already completed product development projects (e.g., existing CAD models), benchmark data as well as information freely available in the internet. However, depending on the sources and data types used, preparation and evaluation might be necessary. For example, information taken freely available from the internet must be structured via a preparation process and provided with meta information before it can be used for subsequent AI-based processing.

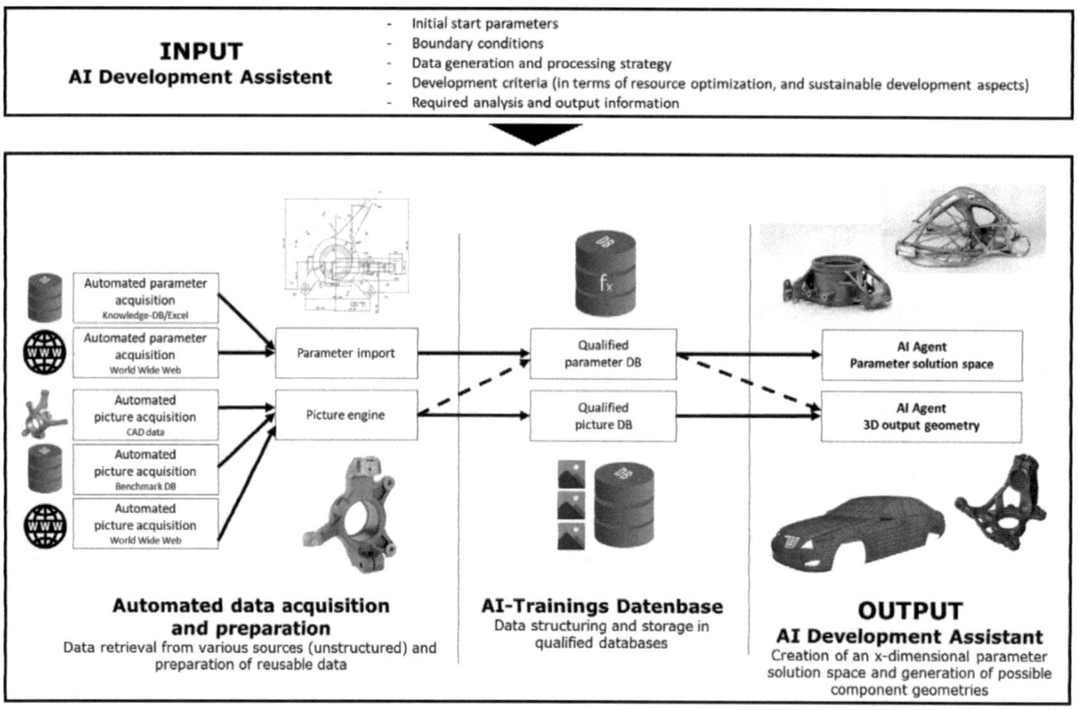


**Figure 5.** Systematic workflow of the introduced AI development assistant, including picture sources: [26], [27], [28], [29], [30], [31], [32], [33], [34], [35].

In case of the integration of images, the corresponding data are provided in the form of predefined ISO view types of vehicles, modules and components, for example. This image and parameter data, linked with meta information, are provided to the AI training algorithm as learning material. From this image analysis, system and component suggestions based on polygon modeling programs (e.g., Blender [25]) are automatically generated using a suitable AI algorithm. After evaluation of the output quality, the automation of data acquisition can be improved iteratively. The goal is to optimize the required amount and form of data input such that an efficient process can be presented to meet the specified quality criteria. In addition to the processing of image data and the output of polygon models, purely parameter-based data sets are also to be processed, e.g., in order to describe functional product properties. An exemplary application of the AIDA includes the conceptual layout of new vehicle architectures, and the definition of corresponding parameter sets. The multi-dimensional solution parameter space contains various aspects, such as legal framework conditions, ergonomic specifications, sustainability, customer specifications, technological-functional specifications, as well as aspects from production, service and maintenance. According to the specifications, the parameter space is filled with data required for the development of the solution and, with the help of AI, solution proposals are developed and evaluated with regard to their optimality, e.g., according to the Pareto principle. Depending on the task at hand, the data used are drawn from different sources. In addition to model and image-based data material, purely parameter-based databases and proprietary available information are also processed. This form of analysis is used, for example, to generate suggestions for further dependent, but not yet determined parameters, as well as, in addition, to determine the possible solution areas per parameter.

## 5. Conclusion

The introduction of an AI development assistant in combination with knowledge-based CAx method support in automotive development and engineering processes is of great importance for both, the optimization of automotive products and the acceleration of automotive processes. Besides, the introduced approach is able to consider economical and sustainable boundary conditions, so that more efficient and sustainable products can be achieved. The AI development assistant can also be applied to other phases of the automotive development and beyond. The application in the early phases, as shown in this publication, mainly refers to the high uncertainty of boundary conditions and specifications and thus offers the greatest potential for optimization measures. However, since the phases "Definition" and "Concept" are shorter in time than the subsequent development and product life-cycle phases, there is a high potential for improvements, too.

The use of AI approaches enables more efficient and resource-saving product development processes and thus brings significant advantages for automotive development. The paper presented has addressed the fact that there is enormous potential for the use of AI-based approaches, particularly in the early stages of product development. However, some challenges have to be considered – for example the large amount of sufficiently qualitatively processed data available, as well as the fact that AI approaches in product development should serve as a support and cannot create the finished product (no push of a button solution).

## References

[1] Cioca L I, Ivascu L, Turi A, Artene A and Găman G 2019 Sustainable Development Model for the Automotive Industry *Sustainability* 11 22 6447 https://doi.org/10.3390/su11226447
[2] Kreis A 2020 *Tailored Data Exchange Processes for Automotive Body Development* Ph.D. Thesis Graz University of Technology Austria
[3] Niikura A 2018 The impact of the application of aluminum *ATZ Extra Issue – Motor Fan Special Edition illustrated – Aluminum Technology* 4 pp 2-3
[4] Krollmann J and Wei B 2019 Lightweight design – Do we need lightweight engineering in battery electric vehicles? *Mobiles* pp 6-9
[5] Eigner M and Stelzer R 2009 *Product Lifecycle Management* (Berlin: Springer) ISBN 978-3-540-

68401-5
[6] Dankwort C W, Ovtcharova J and Weidlich R 2003 *A concept of engineering objects for collaborative virtual engineering: automotive case study* (Dresden: Fraunhofer, I.R.P. (ed.) Proceedings of the ProSTEP iViP Science Days) ISBN 978-3-816-76389-5
[7] Hirz M, Gfrerrer T, Lang J and Dietrich W 2013 *Integrated Computer-Aided Design in Automotive Development – Development Processes, Geometric Fundamentals, Methods of CAD, Knowledge Based Data Management* (Heidelberg, New York, Dordrecht, London: Springer) https://doi.org/10.1007/978-3-642-11940-8
[8] Kreis A, Hirz M and Rossbacher P 2021 CAD-Automation in Automotive Development – Potentials, Limits and Challenges *Computer-Aided Design and Applications 18* 4 pp 849-863 https://doi.org/10.14733/cadaps.2021.849-863
[9] Weber J 2009 *Automotive Development Processes – Processes for Successful Customer Oriented Vehicle Development* (Munich: Springer) https://doi.org/10.1007/978-3-642-01253-2
[10] Daimler *CAx goes AIAx-Künstliche Intelligenz lernt von Experten* https://media.daimler.com/marsMediaSite/de/instance/ko/CAx-goes-AIAx-Kuenstliche-Intelligenz-lernt-von-Experten.xhtml?oid=41324165 date of access: 19.07.2022
[11] Liebscher M SCALE GmbH – IT solutions for CAE *Big data project VAViD successfully completed* https://www.scale.eu/en/news/corporate-news/newsitems/big-data-project-vavid-successfully-completed?set_language=en date of access: 19.07.2022
[12] Amazon Web Services (AWS) *What is Artifical Intelligence?* https://aws.amazon.com/machine-learning/what-is-ai/?nc1=h_ls date of access: 19.07.2022
[13] Amazon Web Services (AWS) *Cloud Computing with AWS* https://aws.amazon.com/what-is-aws/?nc1=h_ls date of access: 19.07.2022
[14] Statista 2021 *Revenues from the artificial intelligence software market worldwide from 2018 to 2025, by region* https://www.statista.com/statistics/721747/worldwide-artificial-intelligence-market-by-region/ date of access: 19.07.2022
[15] Sreejith 2016 *Nissan's Idea: Let an artificial intelligence design our cars* https://www.drivespark.com/four-wheelers/2016/nissan-build-artificial-intelligence-017266.html date of access: 19.07.2022
[16] Technology and operations management 2018 *Project Dreamcatcher: Can Generative Design Accelerate Addititve Manufacturing?* https://digital.hbs.edu/platform-rctom/submission/project-dreamcatcher-can-generative-design-accelerate-additive-manufacturing/ date of access: 19.07.2022
[17] Kreis A and Hirz M 2020 Tailored Data Exchange Process to Support Production Engineering *IEEE Xplore* IEEE 7th International Conference on Industrial Engineering and Applications (ICIEA) Bangkok Thailand pp 979-983 https://doi.org/10.1109/ICIEA49774.2020.9101937
[18] Kreis A, Hirz M and Stadler S 2020 A Contribution to Optimized Data Exchange Supporting Automotive Bodywork Engineering *Computer-Aided Design and Applications 17* **1** pp 178-189 https://doi.org/10.14733/cadaps.2020.178-189
[19] Vajna S, Weber C, Zeman K, Hehenberger P, Gerhard D and Wartzack S 2017 *CAx für Ingeneure – Eine praxisbezogene Einführung* (Wiesbaden: Springer Vieweg) ISBN 978-3-662-54624-6 (eBook)
[20] Luckow A, Cook M, Ashcraft N, Weill E, Djerekarov E and Voster B 2017 Deep Learning in the Automotive Industry: Applications and Tools *IEEE Xplore* IEEE International Conference on Big Data pp 3759-3768 https://doi.org/10.1109/BigData.2016.7841045.
[21] Glenn G 2020 *Forbes – Environmental Sustainability and AI* https://www.forbes.com/sites/glenngow/2020/08/21/environmental-sustainability-and-ai/?sh=fcb5c7e7db3a date of access: 19.07.2022
[22] Publication office of the EU 2019 Directorate-General for Communications Networks, Content and Technology (European Commission): Ethics guidelines for trustworthy AI https://doi.org/10.2759/346720

[23] Van Wynsberghe A 2021 Sustainable AI: AI for sustainability and the sustainability of AI *AI and Ethics* **1** pp 213-218 https://doi.org/10.1007/s43681-021-00043-6.
[24] Seebo 2020 *Artificial Intelligence – The Driving Force of Industry 4.0* https://www.seebo.com/industrial-ai/ date of access: 19.07.2022
[25] Blender https://www.blender.org/ date of access: 19.07.2022
[26] Ebalstudio *Tutorials – Modeling Cars in Polygons* https://www.ebalstudios.com/blog/modeling-cars-polygons date of access: 19.07.2022
[27] Autodesk *Generative Design* https://www.autodesk.de/solutions/generative-design date of access: 19.07.2022
[28] PNG All http://www.pngall.com/database-png/download/2285 date of access: 19.07.2022
[29] PNG Wing https://www.pngwing.com/en/free-png-nhrpu date of access: 19.07.2022
[30] Kamble A V, Patil S S and Harugade M L 2016 Design and Development of Steering Knuckle with Spindle and its Optimization – A Review *International Journal of Engineering Research & Technology (IJERT) 3* **5** pp 43-45 https://doi.org/10.17577/ijertv5is030068
[31] Vijayarangan S, Rajamanickam N and Sivananth V 2013 Evaluation of metal matrix composite to replace spheroidal graphite iron for a critical component, steering knuckle *Materials & Design* **43** pp 532-541 https://doi.org/10.1016/j.matdes.2012.07.007
[32] Autodoc https://www.auto-doc.at/ridex/14351339?gshp=1&gclid=EAIaIQobChMIt7nM7tOW7wIVGZ3VCh2 mgATsEAQYAyABEgK9PvD_BwE date of access: 19.07.2022
[33] ProgrammerSought *How catia generative shape design combines two sides_defining structure with function | Dassault Systèmes CATIA function drives generative design...* https://www.programmersought.com/article/73507460901/ date of access: 19.07.2022
[34] Hexagon *MSC Apex Generative Design* https://www.mscsoftware.com/en-uk/product/msc-apex-generative-design date of access: 19.07.2022
[35] Winaero https://winaero.com/colorful-windows-10-icons-photos-app-again/ date of access: 19.07.2022

IOP Conf. Series: Materials Science and Engineering **1311** (2024) 012029 doi:10.1088/1757-899X/1311/1/012029

# Topology optimization with the redesigned suspension system of the internal combustion vehicle converted to an electrical vehicle

**H A Dikmen[1], U Deryal[1], G Yağlı[1] and Ş Günenç[1]**

[1] Chassis Systems and Analysis of Department, CDMTech Engineering Solutions

**Abstract**. In this paper, the topology optimization of lower wishbone for a passenger car which is the converted of an internal combustion engine vehicle to an electrical vehicle was presented. On the other hand, the need to study on a suspension system arose for electrical car that was redesigned. In study, topology optimization of the structural design of the lower wishbone to be used in the front suspension was carried out by Finite Element Analysis (FEA) Method. For this purpose, firstly, the suspension spring curves and damper curves that will improve the handling, ride and comfort behaviour of the vehicle were determined, suspension system was modelled with the Multi Body Dynamics (MBD) approach. Hardpoints is necessary for vehicle dynamics and that is also an input for FEA were obtained from new suspension system via MBD analysis. In the second step, The preliminary design of the lower wishbone was created, taking into account together with the positions of the wheel drive shaft, suspension spring and brake system structural elements according to hardpoints. After designing the structure suitable for the joint types, initial design was studied on the remaining region that is for weight reduction region except the joint connection surfaces. Then, topology optimization was applied to the preliminary design for the selected load types that is come from MBD analysis. Thus, the excess volume on the preliminary design was determined for each load type. In the last stage, the results of topology optimization analysis were evaluated to reach out target weight and the producible design of the wishbone was made. Finally, Finite Element Analysis were applied to the final design for validation purposes within topology optimization work was done with the forces coming from Multi Body Dynamics. After statically solved Finite Element Analysis, a modal neutral file(.mnf) was created via FEA solver and was integrated to Adams Car as a flexbody. Flexbody was attahced to the other parts which are rigid bodies like nuckle according to Multi Body Dynamics approach. Dynamics durability cases was carried out via Adams Car. Stress results was compared to FEA results and evaluated depends on yield stress. Study results shows that the new design is lighter and competitive design and also safe in dynamics analysis for redesigned passenger car.

## 1. Introduction

Driving comfort and handling capabilities generally affect the suspension system in automotive industry. There are two wishbones in suspension as upper and lower control arms. While driving, vehicle is exposed to shock loads come from unsmooth road conditions. In this case, the role of these wishbones is to absorb the shock loads. The upper and lower wishbone have different design and structures based on the model and purpose of using the vehicle. This study presents lower wishbone of a transformed passenger car.

Patil et al. They studied to the lower wishbone is the better shock absorber than the upper arm because of its position and load bearing capacities [1]. Viqaruddin et al. They presented topology optimization with material layout based on load and boundary conditions. The geometry is finalized by keeping the strength is constant, and reducing the weight by 30% [2]. Rahman et al. They used the I-beam cross section gives higher stiffness and a moment for the part to sustain the higher bending moment [3]. Gadade et al. They created a typical A-type suspension arm of same material of AISI 1040 as per the consideration in FEA. The load goes on increasing and at last they found the stress in material at maximum stressed area is of lower [4].

In this passenger car has two lower wishbone which is a kind of triangular component. One of these wishbones has bushings in two ends, pivot in one end. The preliminary geometry was designed taking into account together with the positions of the wheel drive shaft, suspension spring and brake system structural elements according to hardpoints. Topology optimization is applied to designed wishbone by (Finite Element Method) FEM with forces come from Multi Body Dynamics (MBD) analysis and same geometry analysed in ADAMS as a flex-body. Reducing the weight is the main for the purpose of without affecting the durability and safety.

## 2. Design

The wishbone is one of the important parts of the suspension system was modelled via the Multi Body Dynamics (MBD) approach. The suspension spring curves and damper curves that will improve the handling, ride and comfort behaviour of the vehicle were determined in Adams. Hardpoints is necessary for vehicle dynamics and that is also an input for FEA were obtained from new suspension system via MBD analysis. In the second step, the preliminary design of the lower wishbone was created, taking into account together with the positions of the wheel drive shaft, suspension spring and brake system structural elements according to hardpoints. After designing the structure suitable for the joint types, preliminary design was studied on the remaining region that is for weight reduction region except the joint connection surfaces. In this case, lower wishbone was cut into parts for avoid destroying the joint connection surfaces shown in Figure 1.

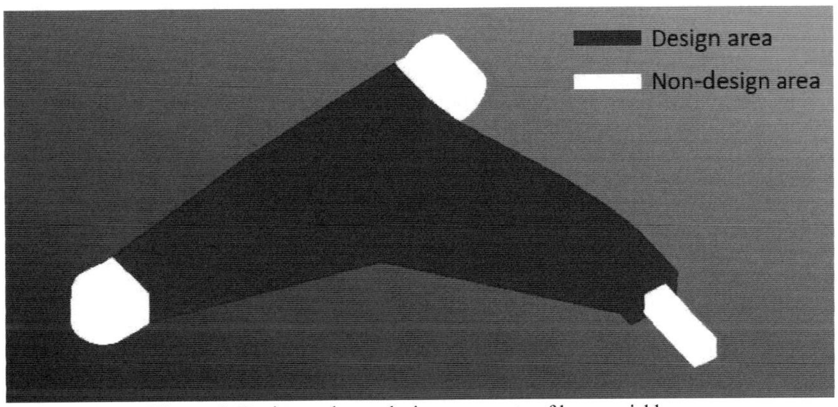


**Figure 1.** Design and non-design area parts of lower wishbone

## 3. Theory and Method

Topology optimization is kind of design optimization subject to load and boundary conditions. This study presents the reducing weight of the lower wishbone design by topology optimization to be used in the transformed vehicle. Simulation gives us a load distribution as stress spreads across the boundaries. In this stress path, if it is observed which elements are carrying the load lower than yield strength value and bigger load than the legible ones, this path gives us about new configurable design. Every single element has different behaviour about stress, strain, deformation but Elements are related with every single adjacent element. The main idea in base on this topology optimization which is geometrical optimization eliminate the elements which is not significantly carrying the stress in the model. All remaining elements will be enough for working this lower wishbone with loads come from dynamic analysis. In formulation, elastic energy should be minimized and following the eq. 1-4 respectively [5],

$$c = \frac{1}{2} U^T K U \quad (1)$$

$$KU = F \quad (2)$$

$$\rho_i \epsilon [0,1], \forall i \quad (3)$$

$$g = \sum_i \rho_i - V_0 \leq 0 \quad (4)$$

| | |
|---|---|
| c | = Elastic energy |
| K | = Matrix rigidity |
| U | = Displacement |
| ρ | = Density |
| V | = Volume |
| F | = Force |

## 4. Modelling

Model was imported in MSC Apex for geometrical clean up and meshing. All lower wishbone is meshed hexahedron elements as shown Figure 2 and AISI 1040 Steel which has high carbon rate is applied to geometry as a material. Design area is modelled as deformable part and the other three parts are rigid structure in white colour as shown Figure 1 because of the connection surface is not wanted to be eliminate.

There are three important load conditions which are come from MBD analysis while vehicle is driving on road as shown Table 1, these load cases applied to joints as forces and moments inputs

Wishbone is such an unconstrained structure part of vehicle so with inertia relief theory which is advanced option is simulated in this static analysis.

**Table 1.** Load Cases

| Analysis Cases |
|---|
| Braking |
| Lateral |
| Vertical Bump |

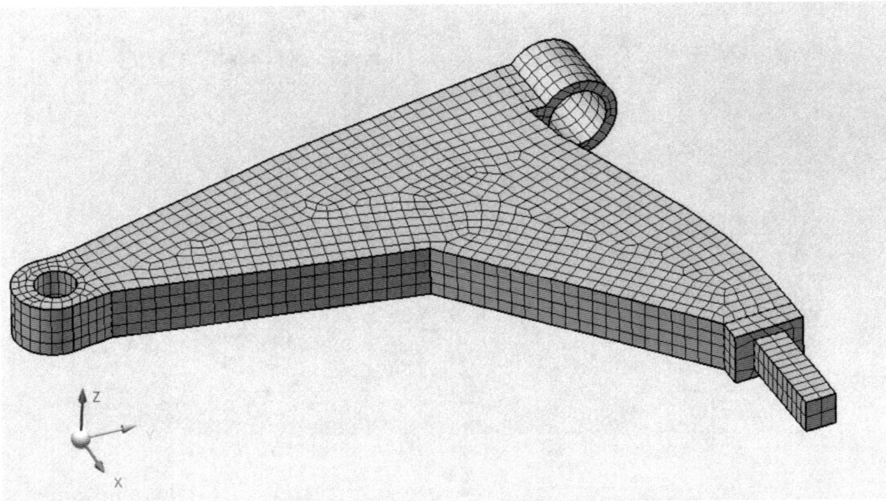


**Figure 2.** Mesh model in MSC Apex

**Table 2.** Mechanical properties of material

| Material Name | Elasticity (MPa) | Poisson Ratio | Density (kg/mm^3) | Yield Strength (MPa) |
|---|---|---|---|---|
| AISI 1040 | 210000 | 0.3 | 7.86e-06 | 361 |

## 5. Topology Optimization

Preliminary model is topology analysed in MSC Patran with three different load cases which are braking, cornering, vertical-bump after modelling MSC Apex. For purpose of weight reduction, Constraint target is %43 mass fraction in these analyses.

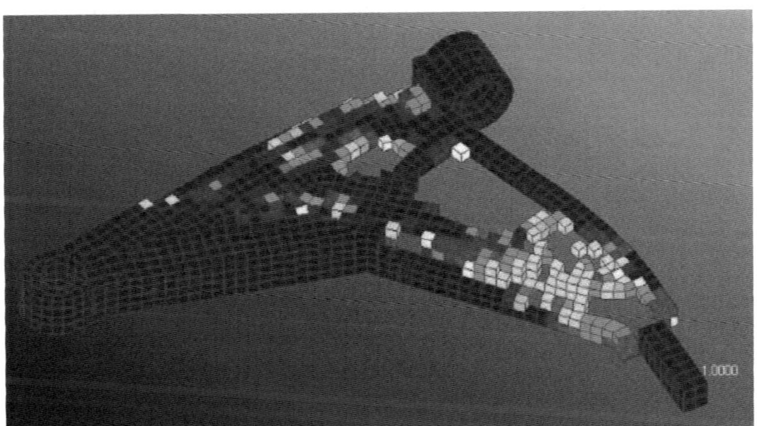


**Figure 3.** Geometry after topology optimization for braking load case

IOP Conf. Series: Materials Science and Engineering **1311** (2024) 012029 doi:10.1088/1757-899X/1311/1/012029

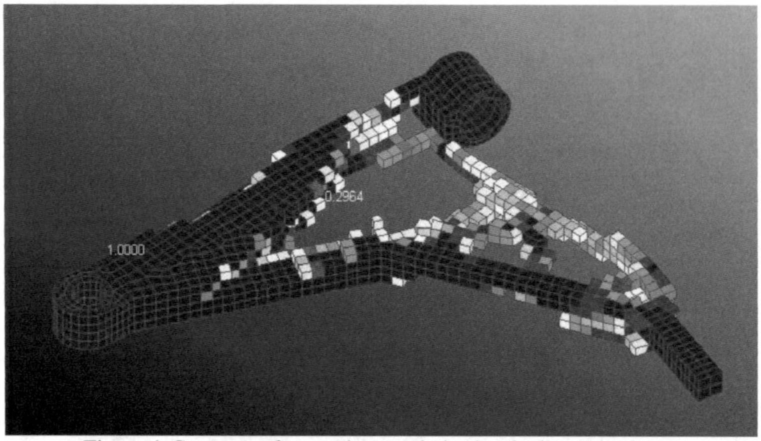

**Figure 4.** Geometry after topology optimization for lateral load case

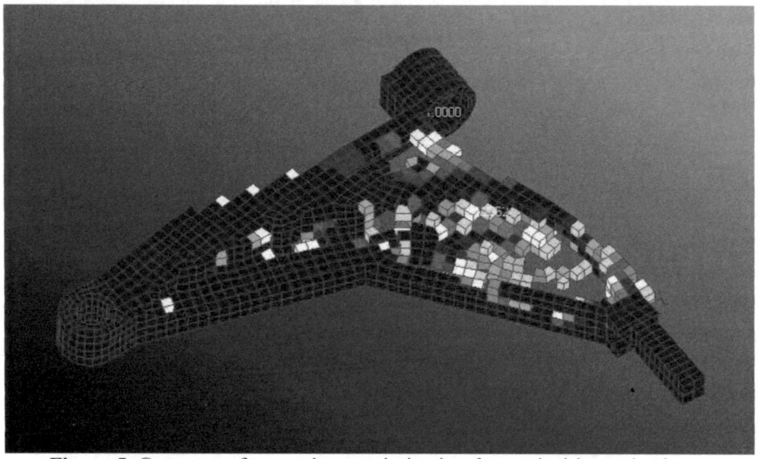
**Figure 5.** Geometry after topology optimization for vertical-bump load case

After topology optimization for each load case, these geometries are exported as a .stl file and combination of these three new geometries is transformed to a new wishbone geometry in Catia. To validate the new geometry, all load cases are applied to new design in MSC Apex again. On the other hand, the new geometry is exported as a flex-body to import Adams and to be analysed in dynamic conditions.

## 6. Results

In this study, model is created on CATIA V5 software. Modelling is done on MSC Apex, Patran was use for topology optimization and Nastran was used to solve this FEA problems as shown in Figure 6,7,8 and last part of these analysis Adams Car is used for flex-body analysis as shown Figure 9. The model is analysed with three important load cases under inertia relief theory.

IOP Conf. Series: Materials Science and Engineering **1311** (2024) 012029 doi:10.1088/1757-899X/1311/1/012029

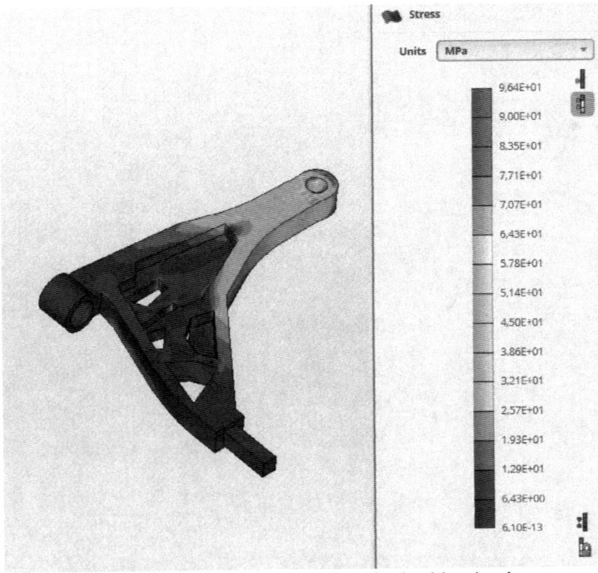


**Figure 6.** New design stress values for braking load case

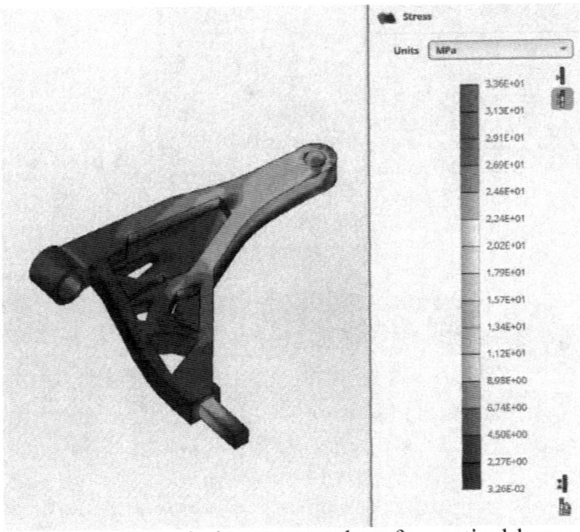


**Figure 7.** New design stress values for vertical-bump load case

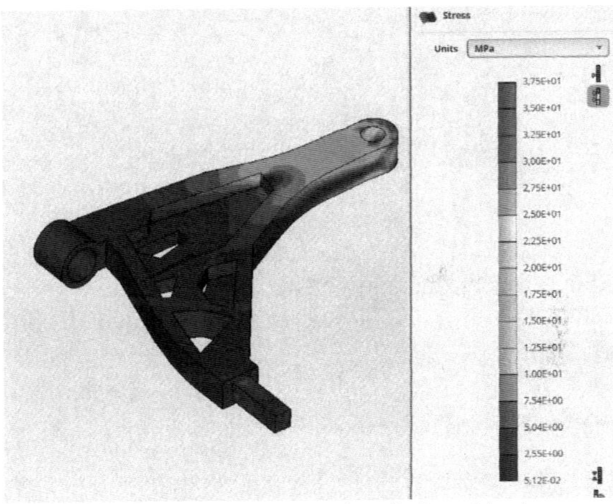


**Figure 8.** New design stress values for lateral load case

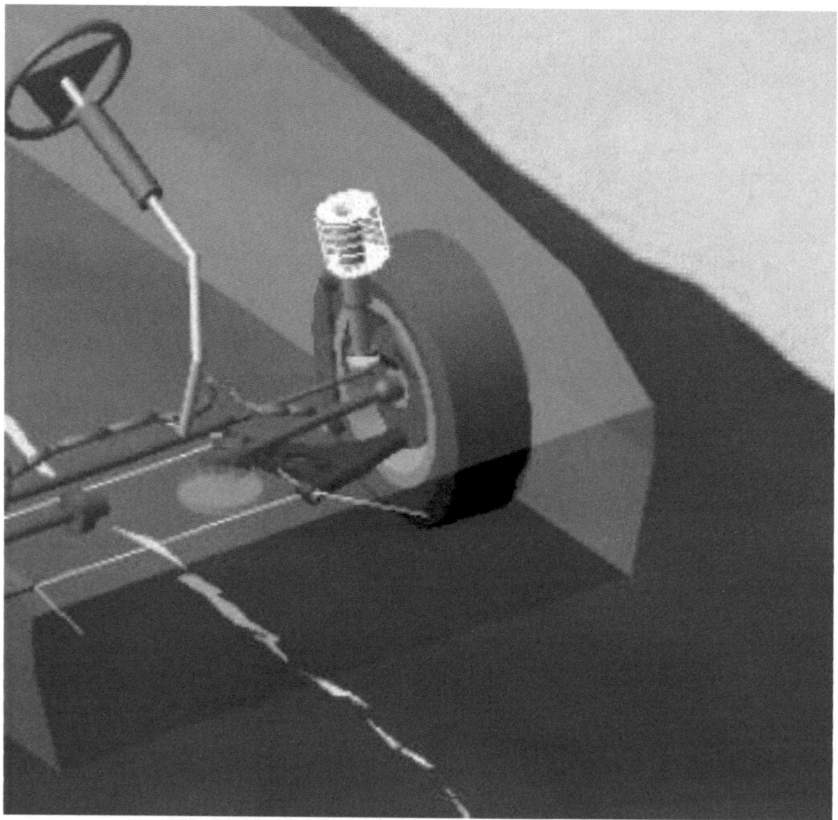

**Figure 9.** Flex-Body analysis under dynamic condition

## 7. Conclusion

In this project, a new lower wishbone design is presented to be used in the transformed passenger car. We reached out to target weight with this analysis procedure. Static analysis is compared with the flex-body analysis which is done in Adams Car for validation. This comparison gave us the accuracy of the model. In sense of durability, these analysis results show us the new model is in safety region according to chosen material for those load cases.

## References

[1] Patil S P, Birwatkar S R, Phadke P D, Pawar K R and Pawar A R 2021 *Journal of Analysis and Computation* vol 15 p 91-100
[2] Viqaruddin M, Reddy D 2017 *Materials Today:Proceedings* vol 4 p 9230-9236
[3] Rahman M H A, Salleh M S, Razak M S A, Kamal M R M, Marjom Z, Anuar L and Saad N A M 2018 *International Journal of Engineering & Technology* vol 7 p 71-75
[4] Gadade M B and Todkar R G 2015 *International Resarch Journal of Engineering and Technology* vol 02 p 759-766
[5] Bendsoe M P and Sigmund O 2003 *Topology Optimization Theory, Methods and Applications* p 4

IOP Conf. Series: Materials Science and Engineering **1311** (2024) 012030 doi:10.1088/1757-899X/1311/1/012030

# Experimental and numerical analyses on the modal behaviour of an aluminium alloy sheet

**A R Moisescu and G Anghelache**

University Politehnica of Bucharest, Splaiul Independentei 313, Bucharest, Romania

raluca.moisescu@upb.ro

**Abstract**. The study of modal behaviour for vehicle body panels represents a field of interest for automotive engineering, having a major influence on automotive comfort, with respect to vehicle noise, vibration and harshness, but also concerning the durability of vehicle body. The investigation of natural frequencies and vibration mode shapes is crucial in the design phase, because essential information can be obtained concerning the behaviour of the components in dynamic conditions. The current paper focuses on the investigation of vibration modes and frequency response function on an aluminium alloy sheet. The aim of the paper is to perform the study of modal behaviour both experimentally and theoretically, through measurements using the impact hammer method and respectively through finite element analysis. The CAD model of the aluminium alloy sheet has been developed in Catia V5 environment in view of elaborating the finite element model, which has been used for investigation of modal behaviour, taking into account the case of free vibrations. Through the modal finite element analysis performed in Ansys Workbench, the natural frequencies of the aluminium alloy sheet, as well as the corresponding mode shapes have been obtained. Experiments have been performed on an aluminium alloy sheet, mounted on an elastic support with low natural frequency. The impact hammer method has been used for obtaining the natural frequencies, through frequency analysis of accelerations measured with two piezoelectric accelerometers mounted on the aluminium alloy sheet. The frequency response function has been computed from the response acceleration measured on the aluminium alloy sheet and the impact force measured using the force transducer of the impact hammer. Virtual instruments have been developed in LabVIEW environment for data acquisition, time domain and frequency analysis, as well as for determining the frequency response function using the measured signals. The natural frequencies obtained from the finite element analyses have been compared to the results of measurements and very good similarities have been ascertained. The experimental investigation of modal behaviour contributed to the validation of the finite element model, but further research is required to investigate the possibilities of improving the modal behaviour of the aluminium alloy sheet.

## 1. Introduction

The study of modal behaviour for vehicle body panels represents a field of interest for automotive engineering, having a major influence on automotive comfort, with respect to vehicle noise, vibration and harshness, but also concerning the durability of vehicle body [1], [2], [3]. The vehicle dynamic performances and fuel economy improvement also imply mass reduction requirements that can be achieved using light alloy or composite materials [4], [5], [6], [7], [8], [9]. The investigation of natural frequencies and vibration mode shapes is crucial in the design phase [10], [11], [12], because essential information can be obtained concerning the behaviour of the components in dynamic conditions.

The current paper focuses on the investigation of vibration modes and frequency response function on an aluminium alloy sheet. In a mechanical system, a mode is a standing wave state of vibration, in which all the components of the system have sinusoidal movement at a fixed frequency corresponding to that mode [13]. For a real system, linear superposition of these vibration states can be performed, and therefore several modes can be achieved simultaneously when an impact is applied on the system to induce vibrations. The aim of the paper is to perform the study of modal behaviour both experimentally and theoretically, through measurements using the impact hammer method and respectively through finite element analysis.

## 2. Experimental investigation of the modal behaviour

The modal behaviour of an aluminium alloy sheet, mounted on an elastic support with low natural frequency, has been investigated experimentally through the impact hammer method. This method consists of measuring the vibrations of a mechanical system, such as the metal sheet, generated by applying an impact with a special hammer equipped with a force transducer [14]. The vibratory response of the metal sheet is measured simultaneously with the impact force, so that the frequency response function can be ascertained:

$$H_{ij}(\omega) = \frac{\bar{y}_j(\omega)}{\bar{x}_i(\omega)}$$

where $x_i(t) = \bar{x}_i(\omega) \cdot e^{i\omega t}$ is the impact signal, $y_j(t) = \bar{y}_j(\omega) \cdot e^{i\omega t}$ is the response signal, $\bar{x}_i(\omega)$ and $\bar{y}_j(\omega)$ are complex spectra of impact and response signals [15]. From the frequency response function of the experimental results, natural frequencies of the aluminium alloy sheet can be determined.

Measurements have been performed on an aluminium alloy sheet that has been suspended on elastic support to allow free vibrations, as shown in figure 1, and has been instrumented with two piezoelectric accelerometers, each having one measuring direction and a very small mass of 0.6 g. The measuring points have been chosen near an edge midpoint and a corner respectively, in view of detecting the peaks and nodal points and thus studying the mode shapes by comparing magnitudes and phases of vibrations detected by the transducers. The accelerometers have been mounted with the measuring direction normal to the metal sheet surface, using dedicated adhesive supplied by the manufacturer, as can be seen in figure 2. The impact force has been measured using a specialized impact hammer equipped with piezoelectric force transducer. The signals from the accelerometers and from the force transducer fitted on impact hammer have been processed using a 3-channel signal conditioner for piezoelectric transducers, shown in figure 3. The impact force has been applied at different locations on the aluminium alloy sheet, oriented on direction normal to the surface. In addition to the impact hammer, transducers and signal conditioning device, the measuring system also included a data acquisition card with multiple analog input channels and a computer.

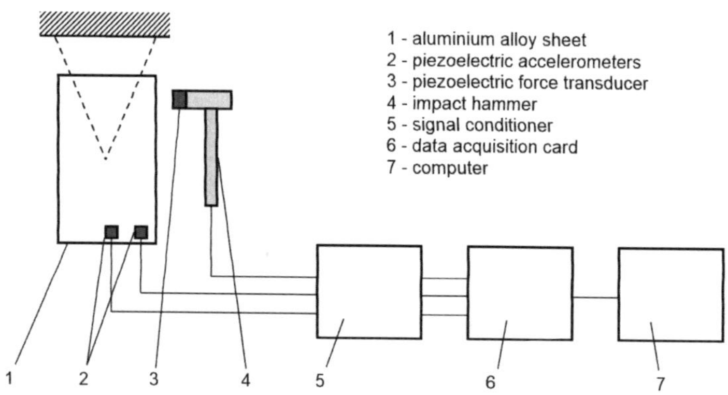


**Figure 1.** Experimental layout for modal behaviour investigation

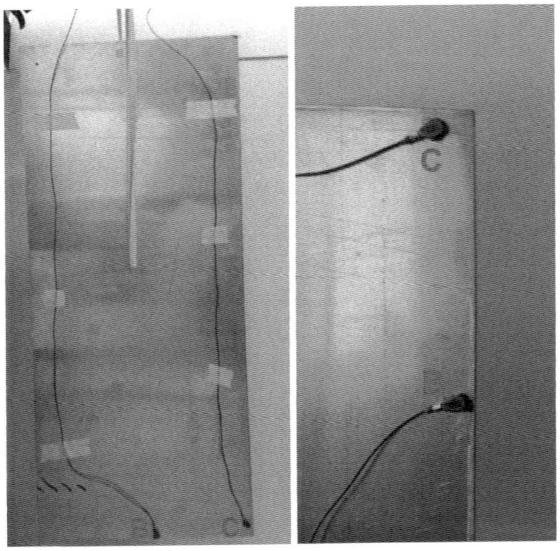


**Figure 2.** The aluminium alloy sheet instrumented with piezoelectric accelerometers

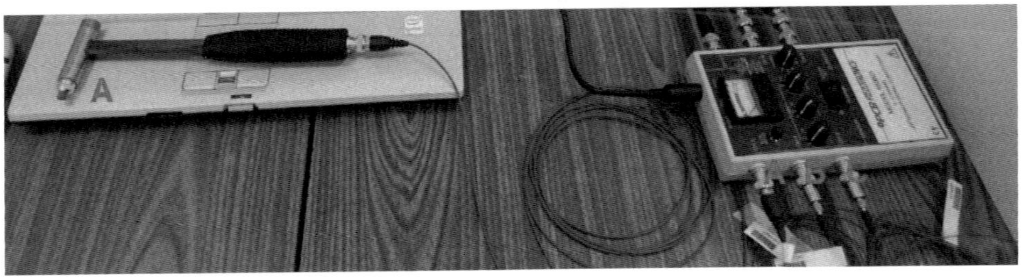


**Figure 3.** The impact hammer with piezoelectric force transducer and the signal conditioner

The applications for data acquisition and data processing have been developed in LabVIEW software [16]. The data acquisition application, shown in figure 4, uses three analog input channels of the data acquisition card to measure the impact force (channel A) and the accelerations of response vibrations (channels B and C), then converts the voltages to force and accelerations using the respective sensitivity constants of the transducers, and finally stores the results files on a computer. The data processing application, shown in figure 5, retrieves the stored data, performs the frequency analyses, computes the frequency response functions, and plots the results for the two measuring locations, so that the natural frequencies and mode shapes can be identified.

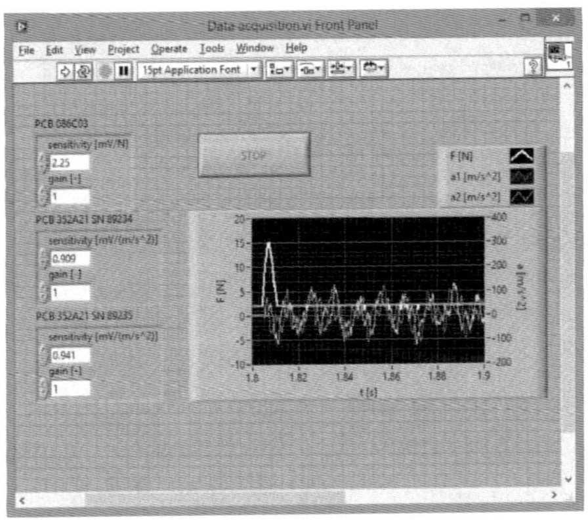

**Figure 4.** The application for data acquisition

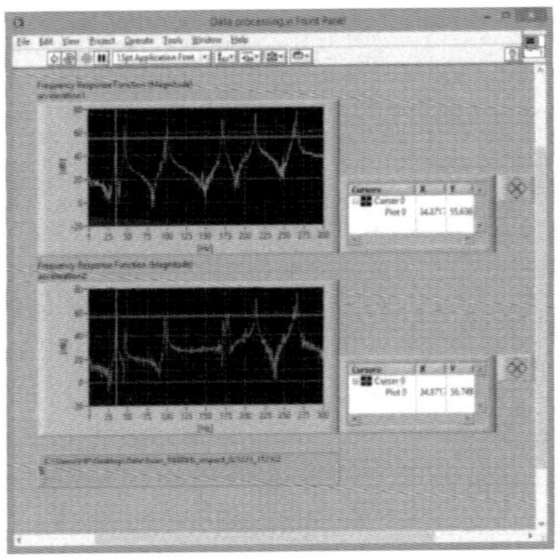

**Figure 5.** The application for data processing

An example of measured signals is presented in figure 6 and data processing through the frequency response functions is illustrated in figure 7.

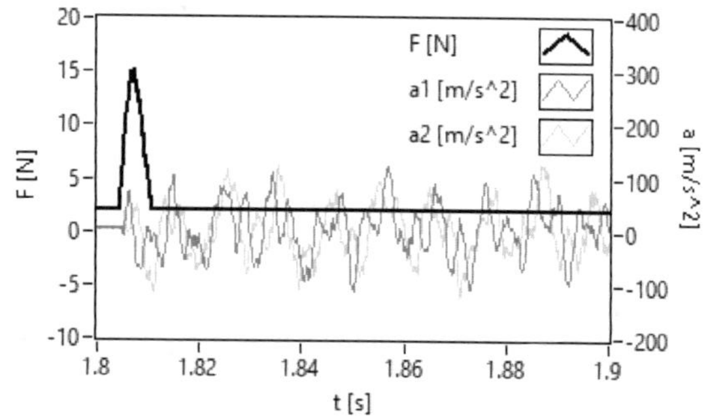


**Figure 6.** Example of measured impact force and accelerations of response vibrations

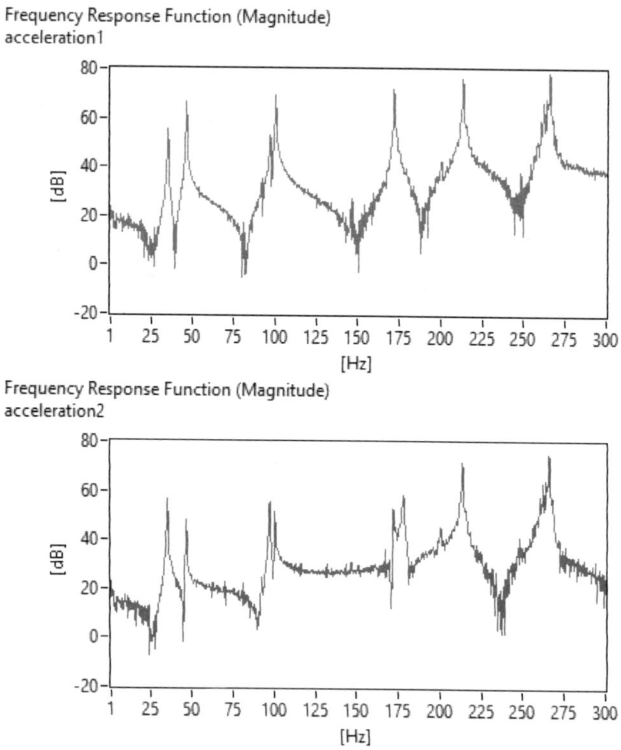


**Figure 7.** Example of frequency response functions obtained using measured signals

Experimental results have been processed from the two measuring points B and C, with impact force applied at different locations on the aluminium alloy sheet, and the natural frequencies obtained using frequency response functions are presented in table 1. The results show good repeatability between measurements with different impact locations, the differences are lower than 1 Hz with only one exception (at 200.3 Hz in the case of bottom right application of impact). The variations between results from the two accelerometers located at points B and C are very small (not exceeding 0.3 Hz).

**Table 1.** Natural frequencies obtained experimentally using frequency response functions

| | f [Hz] | | | | | |
|---|---|---|---|---|---|---|
| Impact location | bottom center | | center right | | bottom right | |
| Measuring point | B | C | B | C | B | C |
| | 35.1 | 35.1 | 35.0 | 35.0 | 34.9 | 34.9 |
| | – | 45.9 | – | 46.3 | – | 46.2 |
| | 97.0 | 97.0 | 97.1 | 97.4 | 97.2 | 96.9 |
| | – | 100.3 | – | 100.3 | – | 100.3 |
| | 171.9 | – | 171.8 | – | 171.8 | – |
| | – | 178.6 | – | 177.9 | – | 177.4 |
| | 198.1 | 198.4 | 198.4 | 198.4 | 200.3 | 200.3 |
| | 213.5 | 213.5 | 212.9 | 212.9 | 213.3 | 213.3 |

However it can be noticed that some natural frequencies are detected only in one measuring point, for instance 171.8÷171.9 Hz is obtained only in the frequency response function from measuring point B, while 45.9÷46.3 Hz, 100.3 Hz and 177.4÷178.6 Hz are identifiable only from measuring point C. This can be explained by the position of peaks and nodal points for each of the mode shapes, which can coincide with the position of measuring points. Therefore, if the accelerometer C is positioned in an area of nodal points for the mode shape that occurs at 171.8÷171.9 Hz, then the acceleration measured at corner does not include this value in the frequency response function. Similarly, the frequency response function of the acceleration measured at point B does not include several frequency values that correspond to mode shapes with nodal points on the edge midpoint.

Therefore some additional information about the mode shape can be obtained using two accelerometers, but many more measuring points are required for obtaining the entire mode shape. Since the number of transducers and measuring channels is limited not only by costs, but also by change of structure mechanical properties, and because sequential measurements with accelerometers moved in all locations across the surface would not provide precision or efficiency, the investigation of modal behaviour required the development of a finite element model of the aluminium alloy sheet.

## 3. Numerical analyses of the modal behaviour

Equations of natural vibrations performed in free conditions by the mechanical system displaced from equilibrium position are expressed depending on mass, damping and stiffness, according to [17], [18]:

$$[M][\ddot{x}] + [C][\dot{x}] + [K][x] = [0]$$

The solutions of differential equations system are represented by harmonic displacements of each component of mechanical system, which can be obtained through modal finite element analysis [19]. The natural frequencies characterize the motion of the mechanical system in each of vibration modes, while the displacements magnitudes depend on the initial conditions that generate the free vibrations.

Through modal finite element analysis, the natural frequencies of the aluminium alloy sheet and the corresponding mode shapes have been investigated. In view of elaborating the finite element model in Ansys Workbench, the CAD model of the aluminium alloy sheet has been developed in Catia V5 environment based on the coordinates measured on the real plate. Model geometry has been obtained directly from Catia V5 file using the import option available in Ansys Workbench [20], and the mesh has been generated automatically with hexahedral elements, including the region of small mounting hole for elastic support existing on the real plate. Linear elastic behaviour has been assumed for the aluminium alloy; no boundary conditions have been applied, taking into account the case of free-free conditions vibrations [21], [22]. Because the degrees of freedom of the aluminium alloy sheet model have not been restricted, the first six modes are obtained from the finite element modal analysis with 0 Hz frequency, without deformation, corresponding to rigid body displacements. The following eight modes are presented in figures 8÷15 as relative displacements, each plotted in two different moments throughout a period of the vibration, corresponding to the extreme values of displacements.

IOP Conf. Series: Materials Science and Engineering **1311** (2024) 012030 doi:10.1088/1757-899X/1311/1/012030

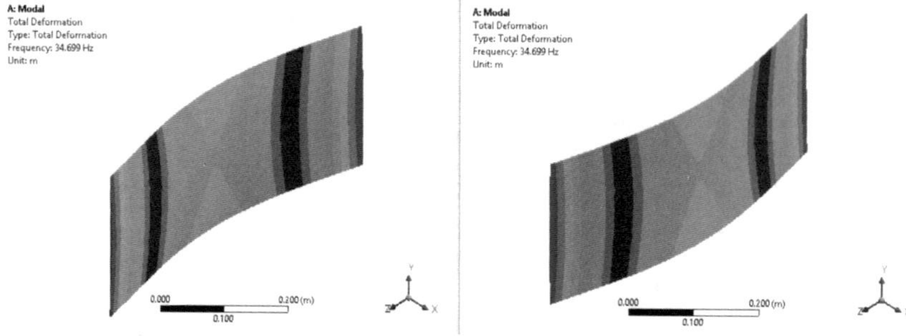


**Figure 8.** Mode shape with frequency 34.7 Hz

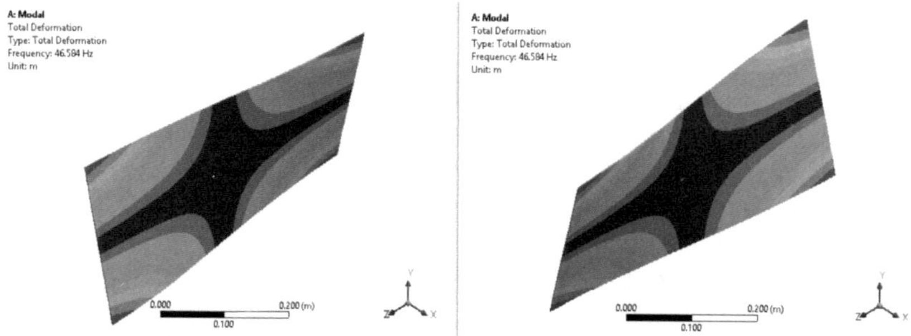


**Figure 9.** Mode shape with frequency 46.6 Hz

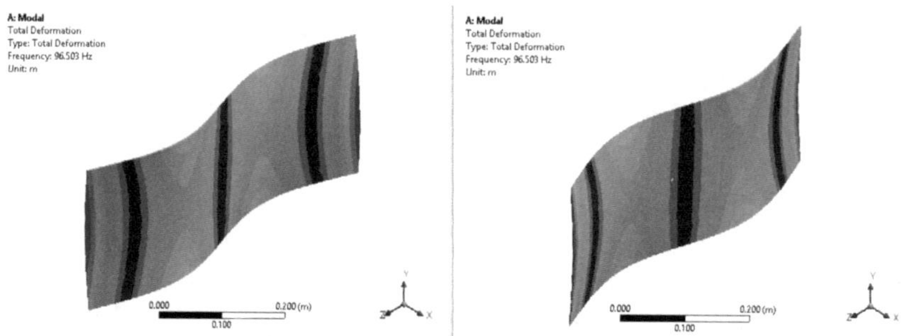


**Figure 10.** Mode shape with frequency 96.5 Hz

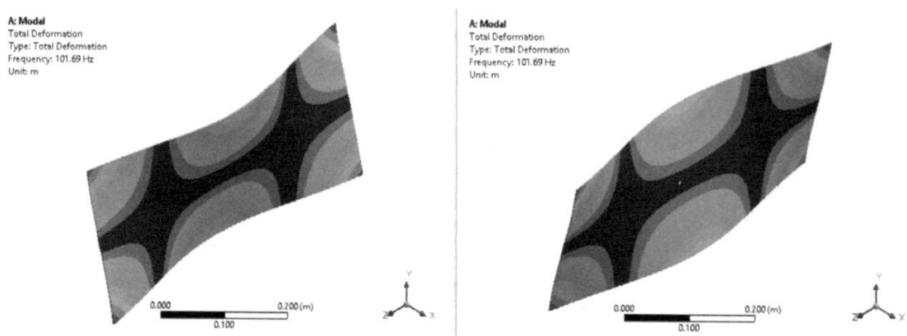


**Figure 11.** Mode shape with frequency 101.7 Hz

IOP Conf. Series: Materials Science and Engineering **1311** (2024) 012030 doi:10.1088/1757-899X/1311/1/012030

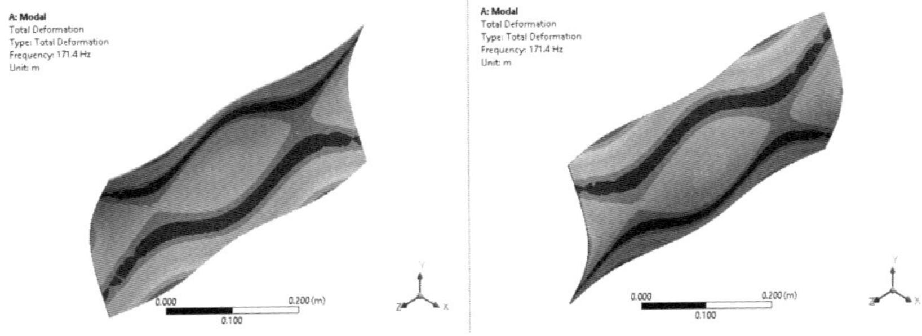


**Figure 12.** Mode shape with frequency 171.4 Hz

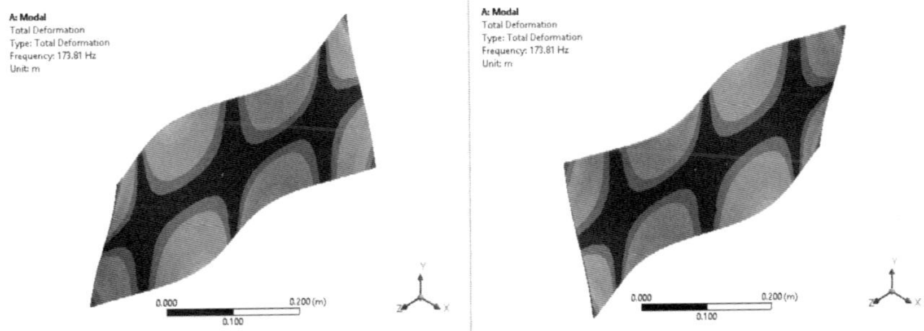


**Figure 13.** Mode shape with frequency 173.8 Hz

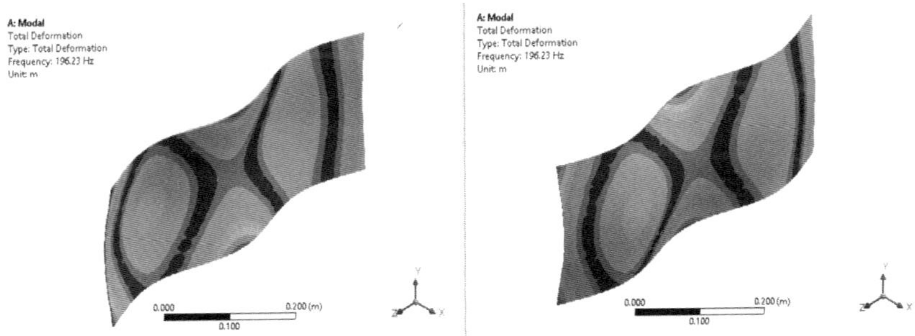


**Figure 14.** Mode shape with frequency 196.2 Hz

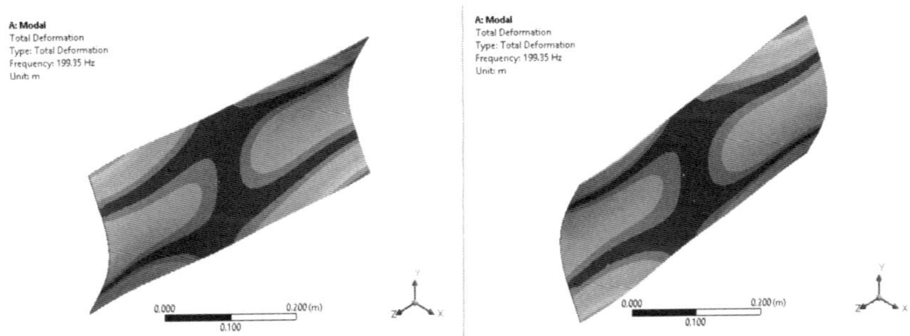


**Figure 15.** Mode shape with frequency 199.4 Hz

The mode shapes can be grouped from the point of view of nodal lines distribution in three main categories: transversal nodal lines (parallel to short edge) shown in figures 8, 10; intersecting longitudinal and transversal (crossed lines parallel to short edge and long edge) represented in figures 9, 11, 13, 15; bias nodal lines (diagonal or oblique with respect to edges) shown in figures 12, 14. The characteristics of mode shapes determined through finite element analysis are presented in table 2, where status of points B and C has been added (marked with ≈ symbol if located somewhat close to a peak) and the category of shapes has also been symbolized.

**Table 2.** The characteristics of mode shapes obtained from finite element analysis

| f [Hz] | Point status | | Type of mode shape | Type and number of nodal lines | | | Figure no. |
|---|---|---|---|---|---|---|---|
| | B | C | | longitudinal | transversal | oblique | |
| 34.7 | peak | peak | transversal | | 2 | | 8 |
| 46.6 | node | peak | intersecting | 1 | 1 | | 9 |
| 96.5 | peak | ≈peak | transversal | | 3 | | 10 |
| 101.7 | node | peak | intersecting | 1 | 2 | | 11 |
| 171.4 | peak | node | bias | | | 2 | 12 |
| 173.8 | node | peak | intersecting | 1 | 3 | | 13 |
| 196.2 | ≈peak | peak | bias | | | 4 | 14 |
| 199.4 | ≈peak | peak | intersecting | 2 | 1 | | 15 |

When frequency increases, the number of nodal lines and peaks also increases, thus resulting in higher order mode shapes. The mode shapes with longitudinal nodal lines only (without any transversal nodal lines) have not been obtained from finite element analysis in the frequency range studied (up to 800 Hz), because their frequencies are higher, as a consequence of smaller width than length and thus greater stiffness of the plate on lateral direction.

Table 3 presents the comparison between natural frequencies of the finite element model and experimental results from the frequency response functions. The natural frequencies are slightly higher in the case of experiments, with the only notable difference at 213.5 Hz which results in a 6.6 % difference from the simulation results at 199.4 Hz (explainable by the very close value of the previous mode at 196.2 Hz which could have influenced the results). The other differences are justifiable from material properties of the finite element model, but also from manufacturing tolerances of the real plate. Thus it can be stated that the finite element model is validated on basis of the very good similarity (with the abovementioned exception) between experimental and simulation results.

**Table 3.** Comparison between experimental results and finite element model results

| f [Hz] | | | | Status of measuring points | | Relative difference [%] | |
|---|---|---|---|---|---|---|---|
| experimental results | | finite element model | | | | | |
| B | C | B | C | B | C | B | C |
| 35.1 | 35.1 | 34.7 | 34.7 | peak | peak | 1.14 | 1.14 |
| – | 45.9 | – | 46.6 | node | peak | – | -1.53 |
| 97.0 | 97.0 | 96.5 | 96.5 | peak | ≈peak | 0.52 | 0.52 |
| – | 100.3 | – | 101.7 | node | peak | – | -1.40 |
| 171.9 | – | 171.4 | – | peak | node | 0.29 | – |
| – | 178.6 | – | 173.8 | node | peak | – | 2.69 |
| 198.1 | 198.4 | 196.2 | 196.2 | ≈peak | peak | 0.96 | 1.11 |
| 213.5 | 213.5 | 199.4 | 199.4 | ≈peak | peak | 6.60 | 6.60 |

It is also noticeable that the smallest relative differences are obtained for transversal and bias types of mode shapes, while in the case of mode shapes with intersecting longitudinal and transversal lines there are somewhat larger differences. This comparison confirms the assumption that some natural frequencies were not detected experimentally from frequency response functions because the position of measuring points was coincident with nodal points for some of the mode shapes.

## 4. Conclusions

The measuring system including impact hammer with force transducer and two piezoelectric accelerometers allows the experimental investigation of modal behaviour for the aluminium alloy sheet, which can be instrumented so that the mechanical properties are not influenced by transducers mass or dimensions. Data acquisition and processing applications developed in LabVIEW graphical programming language provide adequate support for the measuring system and represent valuable tools for computing the frequency response functions of the investigated system, as well as for interpretation and presentation of experimental results.

The natural frequencies determined by applying impact force at different locations on the aluminium alloy sheet show very good repeatability. Comparison between frequency response functions of acceleration signals measured by the two transducers allows obtaining additional information on the position of nodal points on the surface, but the entire mode shape is not achievable using only two accelerometers.

Through modal finite element analysis, considering the case of free-free conditions vibrations, the natural frequencies and corresponding mode shapes are obtained simultaneously. The natural frequencies characterize the motion of the mechanical system in each of the vibration modes, while the displacements magnitudes depend on the initial conditions that generate the free vibrations. When frequency increases, the number of nodal lines and peaks also increases, thus resulting in higher order mode shapes.

The natural frequencies obtained from the finite element analyses have been compared to the results of measurements and very good similarities have been ascertained. This comparison proved that some natural frequencies were not detected experimentally from frequency response functions because the position of measuring points was coincident with nodal points for some of the mode shapes. The smallest relative differences between measured results and simulation results are obtained for transversal and bias types of mode shapes, while in the case of mode shapes with intersecting longitudinal and transversal lines there are slightly larger differences. The experimental investigation of modal behaviour contributed to the validation of the finite element model, but further research is required to investigate the possibilities of improving the modal behaviour of the aluminium alloy sheet.

## 5. References

[1] Anghelache G et al 2007 Study on the aluminium alloy elements of bus upper-structures with detachable joints *AMCSIT Conference CeEx 2007 (Braşov Romania)* paper C131
[2] Pavaloiu I B et al 2020 Small-scale trials for autonomous vehicles study *INTED2020 Proceedings (Valencia Spain)* pp 8607–15 doi: 10.21125/inted.2020.2344
[3] Covaciu D, Florea D and Timar J 2018 Analysis of traffic noise from a main road crossing a city *Ro. J. Techn. Sci. Appl. Mechanics* vol **63** (Bucharest: Editura Academiei Romane) no 1 pp 9–27
[4] Constantinescu D M et al 2012 A study on the parameters that influence the performance of sandwich panels with chiral topology cores *The 4th International Conference Advanced Composite Materials Engineering COMAT 2012 (Braşov Romania)* vol **1** 56 pp 297–302
[5] Nicolae V et al 2011 The experimental determination of natural frequencies of a bio-composite material with applications in the car interior trim *15th International Conference of Modern Technologies, Quality and Innovation ModTech 2011 (Vadul lui Vodă Republic of Moldova)* pp 761–764
[6] Nicolae V et al 2011 Numerical model for the analysis of modal behaviour of a bio-composite

plate with applications to automotive parts manufacturing *15th International Conference of Modern Technologies, Quality and Innovation ModTech 2011 (Vadul lui Vodă Republic of Moldova)* pp 757–760

[7] Iliescu M et al 2009 Experimental and theoretical studies on mechanical characteristics of an important composite material *Materiale Plastice* vol 46 no 1 pp 62–66

[8] Mihon L, Lontis N and Deac S 2018 The behaviour of a vehicle's suspension system on dynamic testing conditions *IOP Conf. Ser.: Mater. Sci. Eng.* 294 012083

[9] Covaciu D, Florea D and Timar J 2019 Practical use of accelerometers in road traffic studies *The 30th SIAR International Congress of Automotive and Transport Engineering SMAT 2019* eds I Dumitru et al (Cham: Springer) p 46 doi: 10.1007/978-3-030-32564-0_46

[10] Negrus E, Anghelache G and Stanescu A 1997 Finite element analysis and experimental analysis of natural frequencies and mode shapes for a non-rotating tyre *Vehicle System Dynamics* **27(S1)** 221–224

[11] Negrus E, Anghelache G and Sorohan S 1998 Tire radial vibrations at high speed of rolling *SAE International Congress & Exposition (Detroit, Michigan, USA)* p 980260

[12] Anghelache G, Negrus E M and Sorohan S 2000 Radial vibrations of tire tread-band at different rolling speeds *SAE Technical Paper* no 2000-05-0254

[13] Tzou H S and Bergman L A 2007 *Dynamics and Control of Distributed Systems* (Cambridge University Press)

[14] Ewins D J 2009 *Modal Testing: Theory Practice and Application* (John Wiley & Sons)

[15] Greenspon J E 2003 Acoustics Linear *Encyclopedia of Physical Science and Technology (Third Edition)* ed R Meyers (Cambridge: Academic Press) pp 129-167

[16] Johnson G W and Jennings R 2006 *LabVIEW Graphical Programming* (New York: McGraw-Hill)

[17] Sorohan Şt 2011 *Modelare cu elemente finite în analiza dinamică a structurilor* (Bucharest: MatrixRom)

[18] Sorohan Şt et al 2016 Assessment of effective elastic properties of honeycomb cores by modal finite element analyses *Proceedings of the European Automotive Congress EAEC-ESFA 2015* eds Cr Andreescu, A Clenci (Cham: Springer) doi: 10.1007/978-3-319-27276-4_41

[19] Bathe K J 2014 *Finite Element Procedures* (New Jersey: Prentice Hall)

[20] Lee H H 2020 *Finite Element Simulations with ANSYS Workbench 2020* (Mission, Kansas: SDC Publications)

[21] Yang X, Olatunbosun O and Bolarinwa E 2010 Materials testing for finite element tire model *SAE Int. J. Mater. Manuf.* **3(1)** pp 211–220

[22] Moisescu R and Anghelache G 2010 Finite element analysis of truck tyre modal behaviour *Int. Automotive Congr. CONAT2010 (Brasov Romania)* paper CONAT(2010)4062

# Experimental Passivity-based Control for PMSM

**J J Vásquez-Sanjuan[1],S L Bocîi[2], T M Espinosa[1], M A Alonso[1], J A Arízaga Silva[1], E M Yescas[3], M C Portillo**

[1]Engineering of Automotive Systems Polytechnical University of Puebla, Puebla, México 72750
[2]University Aurel Vlaicu, Arad, Romania, 310130
[3]Technological University of Mixteca, Oaxaca, México 69000

jacob.vasquez@uppuebla.edu.mx, liviu.bocii@uav.ro

**Abstract**. This paper presents the design, simulation and experimental validation of a Permanent-Magnet Synchronous Motor (PMSM) speed control. The technique applied was passivity based-control. All the process of design, simulation and experimental test is described for each state. The experimental test is carried out with the system TMDSHVMTRPFCKIT of Texas Instruments, all processes for implementation are described and the final diagram for obtaining the code. The simulation and experimental results validating the control strategy are illustrated and analysed.

**Keywords:** passivity based-control, speed tracking, PMSM, Delfino, Texas Instruments

## 1. Introduction

Electric motors are used in a wide range of industrial, commercial, residential, agricultural and transportation applications. Typically, electric motors are part of a device that converts electrical power into mechanical energy [1].

The development in electrical machines and the improvement of power electronics have stimulated various applications:

- Computer hardware
- Home applications
- Land, sea and air transportation
- Renewable energy generation

Motors used in all sectors of society: industry, services, business, infrastructure, health, defense and domestic life.

PMSMs are replacing induction and direct current motors in various industrial, robotics and machine applications. On the other hand, in recent years the price of rare earth materials, with which magnets are made, has been significantly reduced [2], [3], [4], [5]. The reasons why its use has spread are: Greater power density, high performance in motion control and faster and better accuracy [6].

The PMSM is widely used in several applications principally by small size, high reliability, and excellent servo performance. In automotive systems, they are employed in hybrid and electric vehicles [8]; in drive by wire systems, when the motor is used substituting mechanical system in the steering [9]; in washing machines it uses a sensorless system for control speed [10]; the PMSM is used for applications of railway traction [11]; like motor or generator in energy systems or in wind energy conversion [12], [13]. The PMSM is preferred in applications where high-performance control is needed for efficiency applications. The costs of PMSM and Digital Signal Controllers (DSC) are decreasing continuously, making these types of motors attractive.

Passivity was used originally in classical mechanics problems before being extended to control problems [14]. Passivity–based control with energy molding for induction motors was carried out by Ortega [15]. Sira et al., showed the use of passivity for converter CD/CD [16]. Another's examples of passivity–based control was designed and implemented by Linares [17], [18], [19].

Various articles have been published related to control of PMSM. Burgos used a linear system for a controller PI with estimated position and speed of the rotor, the controller was implemented in a Digital Signal Processing (DSP) and a Field Programmable Gate Array (FPGA) for a PMSM of 3.5 kW, [6]. In the publication of Lu , some characteristics of a DSP for sensorless control are explained, [20]. The work of Kim , shows a sensorless control based on a sliding mode observer robust for variations of stator resistance; the use of low pass filter that can produce delay is avoided, replaced with a signum function; a motor of 1 kW and a DSP were used for experimental results, [21]. A sensorless controller PI–based on a FPGA was realized by Kung, co-simulation was done between Matlab/Simulink and ModelSim, the results were obtained with a motor of 750W, [3]. Urbanski used a Kalman filter for control, estimating the speed, the motor used was of 1.23 kW and the results were obtained only for low velocities, [22] . Quang used a Kalman filter to estimate the speed and a controller PI was implemented in a FPGA, [23]. A comparison between Model Reference Observer and Reduced Order Observer of PMSM Torque was done by Dong et. al. [24] .

There are other developments for several controllers, integral state feedback by Apte [7], which applies the state-space form; Martinez designed and simulate an IDA-PBC passivity-based control [8]; the proposal of Zaltni et al., who developed a controller that follows the reference speed from values close to zero to positive values utilizing a backstepping technique [20]; the Thau observer based sensorless control of Campos et al. presents good results [21]; Urbanski control for low speeds through the Luenberger observer [22] are examples controllers.

A conventional control requires the correct rotor position and speed measurement to perform effective control, usually from the sensor mounted on the shaft (resolver, encoder). In this study, the implementation of the PMSM passivity–based control drive is presented for velocity control. The controller requires knowledge of the signals i_d, i_q, ω and T_L (load applied torque).

The document is organized as follows: The model of the motor will be described; the controller is designed, and the stability analysis is performed; the experimental platform Is describes as the analysis of the circuit enveloped; the results obtained in the simulation and experimental are shown; finally, the conclusions obtained from the implemented system are discussed.

## 2. Mathematical model of PMSM

The typical model of a PMSM in a two–axis d–q synchronous rotating reference frame is represented by [25], [26].

$$L_d \frac{di_d}{dt} = \omega L_q i_q P - R i_d + v_d \tag{1}$$

$$L_q \frac{di_q}{dt} = -\omega (L_d i_d P + K_m) - R i_q + v_q \tag{2}$$

where $v_d$, $v_q$, $i_d$ and $i_q$ are the axis voltages and currents on $d$ and $q$; $R$ is the stator resistance; $L_d$ y $L_q$ are the inductances; $\omega$ is the electrical speed of the rotor, $K_m$ is the back-emf term, with $K_m = P\Psi_{PM}$, $P$ the pole–pair number and $\Psi_{PM}$ is the magnitude of the permanent magnet flux.

The mechanical model of the rotor is determined by:

$$\frac{2}{3}J\frac{d\omega}{dt} = K_m i_q - \frac{2}{3}(T_L + B\omega) \tag{3}$$

where T_L is the load torque, B the viscous friction and J rotor inertia constant.

### 2.1. *Passivity-based control*

This technique, also known as energy shaping and damping injection, allows for system regulation. Initially, the molding part modifies the potential energy of the system in such a way that it will now have a global and unique desired equilibrium point. On the other hand, the injection of damping modifies the dissipation function to ensure asymptotic stability.

In order to apply this strategy, the PMSM model must be converted to the representation where the state vector is given by $X = \begin{bmatrix} i_d & i_q & \omega \end{bmatrix}^T$. The representation is shown:

$$A\dot{X} = J(\omega)X - RX + Bu + \varepsilon \tag{4}$$

In this equation of a passive system, $A$ is a symmetric positive definite matrix; $u = \begin{bmatrix} v_d & v_q \end{bmatrix}^T$ is the input–matrix control, $J(\omega)$ is a skew symmetric matrix, $R$ is the losses matrix, $B$ is a matrix that dependents on the inputs and $\varepsilon$ is the disturbance vector. Rearranging terms it is possible to represent the system (4) as shown in (5):

$$\left[L\ 0\ 0\ 0\ L\ 0\ 0\ 0\ \frac{2J}{3}\right]\left[i_d\ i_q\ \dot{\omega}\ \right] = \left[0\ \omega PL_q\ 0\ -\omega PL\ 0\ -K_m\ 0\ K_m\ 0\ \right]\left[i_d\ i_q\ \omega\ \right] - \left[R\ 0\ 0\ 0\ R\ 0\ 0\ 0\ \frac{2B}{3}\right]\left[i_d\ i_q\ \omega\ \right] + \left[1\ 0\ 0\ 1\ 0\ 0\right]\left[v_d\ v_q\ \right] + \left[0\ 0\ -\frac{2}{3}\right]T_L \tag{5}$$

To apply the technique of passivity-–based control, a copy of the original system is created (4) where $X^*$ represents the desired state variables and $u^*$ the desired input, so:

$$A\dot{X}^* = J(\omega^*)X^* - RX^* + Bu^* + \varepsilon \tag{6}$$

Subtracting (4) from (6) and changing the variables so that $e = X - X^*$ · and $e_u = u - u^*$. Using the Taylor series for $J(\omega)$ at the point $\omega^*$, it is possible to express the following equation where $e_\omega = \omega - \omega^*$:

$$A\dot{e} = \frac{\partial J}{\partial \omega_r}\Big|_{\omega^*} X^* e_\omega - Re + Be_u + J(\omega)e \tag{7}$$

To guarantee the stability of system with the controller, the second Lyapunov principle is used; the candidate function proposed is the positive defined function:

$$V(e) = (1/2)e^T Ae \tag{8}$$

With $e = [i_d - i_{d*}\ \ i_d - i_{d*}\ \ \omega - \omega^*]^T$. Taking the derivative with respect to the time we get:

$$\dot{V}(e) = e^T A\dot{e} = e^T\left[\frac{\partial J}{\partial \omega_{\omega^*}} X^* e_\omega - Re + Be_u + J(\omega)e\right] \tag{9}$$

using

$$e_u = -\gamma B^T e - B^T \frac{\partial J}{\partial \omega_{\omega^*}} X^* e_\omega \tag{10}$$

where $\gamma = [\gamma_1\ 0\ 0\ \gamma_2\ ]$. Substituting (7) in (6), we get:

$$\dot{V}(e) = -e^T[R + B\gamma B^T]e \tag{11}$$

To guarantee the stability of the controller, $[R + B\gamma B^T] > 0$ needs to be fulfilled whit the matrix $\gamma$ positive, in other words with $\gamma_1 > 0$ and $\gamma_2 > 0$. Substituting the matrices in (10), we obtain:

$$e_u = -[\gamma_1\ 0\ 0\ ⍰_2\ ][1\ 0\ 0\ 0\ 1\ 0\ ][i_d - i_{d*}\ i_q - i_{q*}\ ⍰ - \omega^*\ ] \\ -[1\ 0\ 0\ 0\ 1\ 0\ ][0\ LP\ 0\ -LP\ 0\ 0\ 0\ 0\ 0\ ][i_{d*}\ i_{q*}\ ⍰^*\ ]e_\omega \tag{12}$$

When $e_u = u - u^*$ in (12), it is then possible to express:

$$u = [v_d\ v_q\ ] = [v_{d*} - \gamma_1(i_d - i_{d*}) - PL_q i_{q*}(\omega - \omega^*)\ v_{q*} - \gamma_2(i_q - i_q\ *) + PL_d i_{d*}(\omega - \omega^*)\ ] \tag{13}$$

The desired values for $v_{d*}$, $v_{q*}$, $i_{d*}$ and $i_{q*}$ can be determined from the system equilibrium point. Considering the fact that to get the maximum motor torque must $i_d = 0$ and solving:

$$v_{d*} = -PL_q\omega^* i_{q*}\ v_{q*} = K_m\omega^* + Ri_{q*}\ i_{q*} = \frac{2}{3K_m}(T_L + B\omega^*) \tag{14}$$

Such that the controller can be rewritten as follows:

$$u = [v_d\ v_q\ ] = [-\gamma_1 i_d - PLi_{q*}\ K_m - \gamma_2(i_q - i_q\ *) + Ri_{q*}\ ] \tag{15}$$

### *2.2. Experimental platform*

During the tests performed on the Texas platform, the incremental encoder was used. A block diagram of all the elements involved during this stage is shown in figure 1.

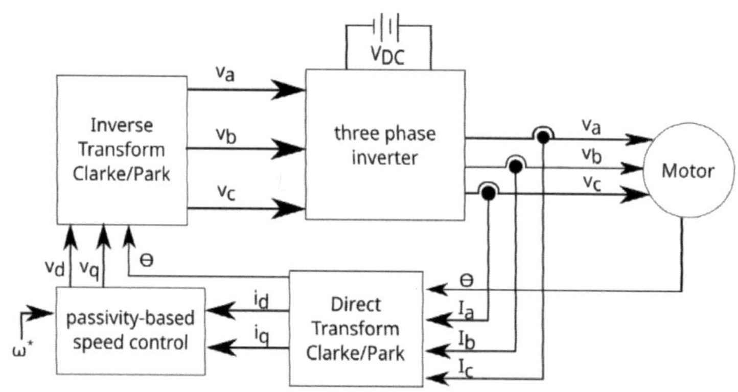


**Figure 1**. Diagram block of the system implemented.

The direct transformation has the function of converting the three-phase system to a two-phase system in order to reduce the system order. The voltage generated by the controller is in the $d - q$ frame so another inverse transformation is used. Finally, the inverter converts the voltage from direct current to three phases; since this system is supply by direct current (DC) voltage.

The TMDSHVMTRPFCKIT high voltage motor control platform is used to experiment with PMSM motor speed and position controllers, see figure 2. Its main features include the following:

- Maximum operating voltage of 350V DC
- Maximum power of 1.5 kW
- Inputs for quadrature incremental encoder or Hall effect sensors
- Connection for CAN (Controller Area Network) communications
- JTAG interface

**Figure 2**. Texas Instruments PMSM motor control kit TMDSHVMTRPFCKIT.

A block diagram of the kit is shown in figure 3. A rectifier is used at the AC voltage supply input to convert it to DC Bus, this voltage feeds the six transistors (IGBT) of the PS21765 inverter integrated circuit of Mitsubishi Company. The voltages of each of the phases can be measured, for measure speed with the controller can be connected to an encoder or a Hall Effect sensor.

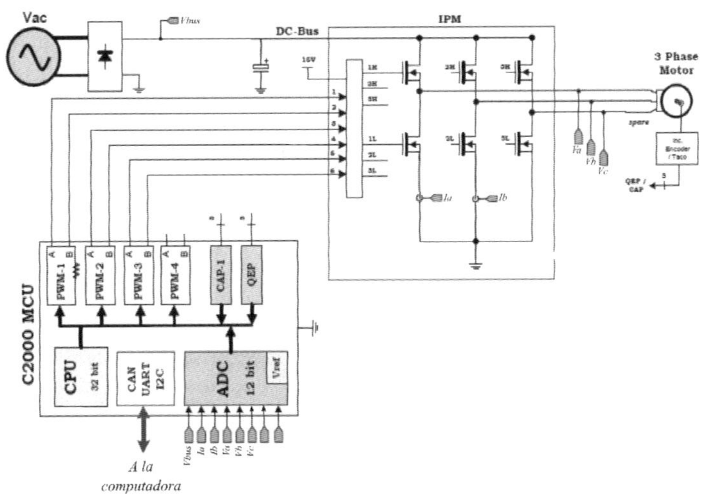


**Figure 3**. Block diagram of the Texas Instruments kit.

In each of the branches of the inverter a series resistor is used with which it is possible to measure a voltage V (Ia), V (Ib), which is proportional to the corresponding current and the operational amplifiers connected to these voltages conditionates the signals for the digital card. The voltages are measured in each of the phases Va , Vb , Vc and the direct current voltage Is obtained after the rectifier. The board generates the digital signals that activate the inverter transistors via the PWM terminals (1A, 1B, 2A, 2B, 1C, 2C), for more detail see [27].

The control card (C2000 MCU) for this kit was the Delfino card (28335) with a floating-point processor, the main characteristics are described below.

### *2.3. TMS320F28335 Digital Signal Controller TMS320F28335*

This microcontroller (MCU) belongs to the Delfino family. It has a floating-point unit, maximum speed of 150 MHz, 512 kB of flash memory, 68 kB of RAM, 18 PWM channels, 1 input for encoder or Hall Effect sensor, 16 timers, 16 analog/digital converters with conversion time of 80ns.

The programming of this microcontroller is done through PSIM, which generates code C compiled by the Texas Instruments Code Composer Studio development environment. Table 4.1 shows the terminals and signals used for motor control.

**Table 1**. Terminal and signals used for motor control

| Name | Description | Terminal | Port |
|---|---|---|---|
| Vfb-Bus | Bus voltage | 59 | ADC-A1 |
| Vfb-U | Phase voltage A | 13 | ADC-B3 |
| Vfb-V | Phase B voltage | 11 | ADC-B2 |
| Vfb-W | Phase C voltage | 9 | ADC-B1 |
| Ifb-U | Voltage proportional to the current of phase A | 15 | ADC-B4 |
| Ifb-V | Voltage proportional to the current of phase B | 19 | ADC-B6 |
| Ifb-Ret | Voltage proportional to total current | 65 | ADC-A4 |
| PWM-1H | It to the PWM (triggers the upper PWM transistor) phase A) | 23 | GPIO-00 |
| PWM-1L | PWM (triggers lower PWM transistor) phase A) | 73 | GPIO-01 |
| PWM-2H | It to the PWM (triggers the upper PWM transistor) phase B) | 24 | GPIO-02 |
| PWM-2L | It to the PWM (triggers the lower PWM transistor) phase B) | 74 | GPIO-03 |
| PWM-3H | It to the PWM (triggers the upper PWM transistor) phase C) | 25 | GPIO-04 |
| PWM-3L | It to the PWM (triggers the lower PWM transistor) phase C) | 75 | GPIO-05 |
| QEP1-A | Signal A of encoder | 40 | GPIO-20 |
| QEP1-B | Signal B of encoder | 90 | GPIO-21 |
| QEP1-I | Coding Index | 91 | GPIO-23 |
| DAC-1 PWM-6A | Analog output signal 1 | 28 | GPIO-08 |
| DAC-2 PWM-6B | Analog output signal 2 | 29 | GPIO-10 |
| DAC-3 PWM-7A | Analog output signal 3 (only 28F035) | 30 | GPIO-40 |
| DAC-4 PWM-7B | Analog output signal 4 (only 28F035) | 80 | GPIO-41 |

To control the speed of the motor, it is required to measure the position and angular velocity of the motor and the currents in each of the phases. The motor forms a balanced connection, that is:

$$I_a + I_b + I_c = 0 \tag{16}$$

it is possible to obtain the current Ic by clearing from equation (16). So, it is only necessary to measure two motor currents, as well as obtain the position and speed of the motor by means of the incremental encoder. The following analysis describes the circuit used on the platform to measure the current in each of the A and B branches of the motor, using a 0.002 Ω resistor.

### *2.4. Current sensors*

The partial diagram of the inverter provided by the manufacturer, is shown in figure 4, with this information it is possible to determine the method used by TI to measure the currents, in phases A and B of the kit.

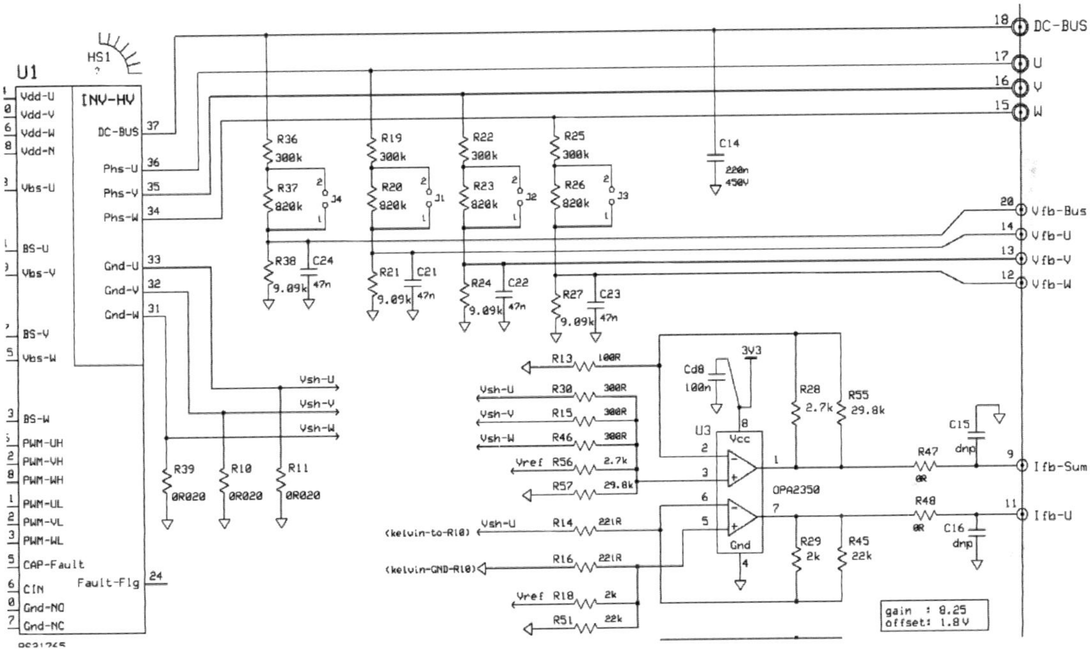


**Figure 4**. Partial diagram of TI kit inverter.

It is possible to observe in figure 4 that the ground terminals (31, 32 and 33) are connected to resistors of 0.02 Ω to measure the current in each of the inverter branches. On the other hand, it is possible to observe that the terminal Vsh - U , phase A, is connected to an operational amplifier to obtain a voltage proportional to the current, Ifb - U . The circuit used is the same for measuring the current of phase B, the case for phase A is shown below. Figure 5, shows the circuit used to obtain a voltage proportional to the current on phase A.

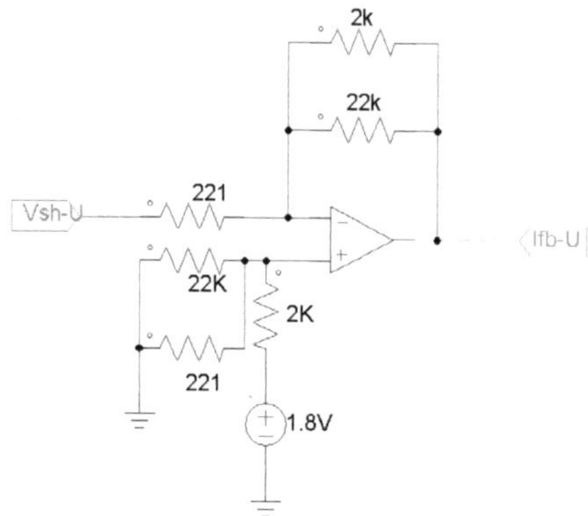


**Figure 5**. Circuit used to obtain a voltage proportional to the current.

The following is the analysis of the circuit in Figure 5, to determine the relationship between the measured current and the voltage obtained at the output of the operational amplifier. The resistors between the inverting and output terminals are connected in parallel, so that an equivalent resistance is obtained,

$$R_{eq1} = 22K\Omega||2k\Omega = 1833.33\ k\Omega \quad (17)$$

Resistors connected between ground and non-inverting input of 221Ω and 22kΩ are in parallel, so that a new equivalent resistance is obtained.

$$R_{eq2} = 221\Omega||\ 22K\ \Omega = \ 218.8\Omega \quad (18)$$

The equivalent circuit is shown in Figure 6.

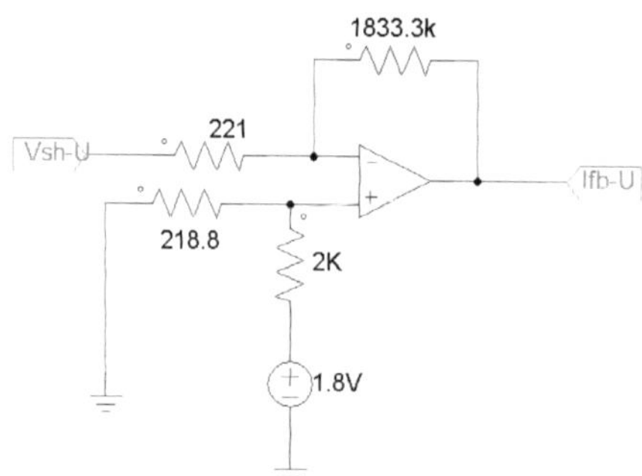


**Figure 6**. Simplified circuit to obtain the voltage proportional to the current.

The circuit is analyzed by superposition and the outputs are obtained for each of the cases, the final result will be the sum of the voltages obtained in each case. The first case, is obtained by making zero the voltage of the source connected to the positive terminal, so the resulting circuit is an inverter amplifier. Since the dc voltage connected to the 221 Ω resistor, (Vsh-U), is $V = -ia - 0.02\ \Omega$, equivalent to $-ia/50$. Thus, the output voltage will be:

$$V_1 = -i_a/50 * \left(-\frac{1833.33}{221}\right) = 0.165 * i_a \quad (19)$$

IOP Conf. Series: Materials Science and Engineering **1311** (2024) 012031 doi:10.1088/1757-899X/1311/1/012031

For the other case, the input voltage connected to the negative terminal resistor is made zero and the resulting circuit is a non-inverting amplifier, with a DC voltage at the positive terminal:

$$V_+ = 1.8V(218.8/2218.8) = 0.1775V \tag{20}$$

And at the output of the non-inverting amplifier, you have:

$$V_2 = 0.1775 * \left(\frac{1833.3}{221} + 1\right) = 1.65V \tag{21}$$

Therefore, the total output voltage of the circuit is the sum of the voltages individuals, V1 and V2:

$$V_O = V_1 + V_2 = 1.65V + 0.165 * i_a \tag{22}$$

This circuit allows to measure positive and negative current signals with the help of a DC voltage of 1.8V, the minimum and maximum currents will be ±10A, that is, if a current of -10A is measured the output voltage of the circuit will be 0V and if the current is 10A the output voltage will be 3.3V, which are the maximum and minimum values, respectively, of the A/D converter of the TMS28335 board. This circuit is a signal conditioner so that the voltages proportional to the currents can be connected to the analog-to-digital converter of the DSP.

To verify the results obtained from analysis, the circuit shown in figure 7, was simulated, from which it was possible to determine that the current measured in the 0.02 Ω resistor is equal in magnitude but of different sign to the current measured in phase A of the motor, i.e., $i_{-0.02} = -ia$ , which verifies the analysis described above.

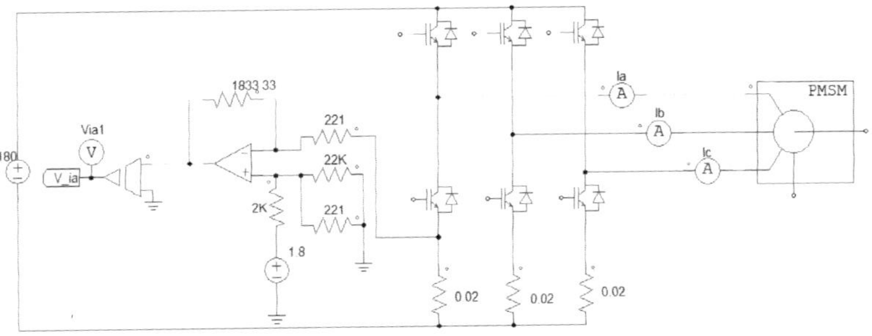


**Figure 7**. Circuit to simulate and compare the current measured in the resistance and in the phase of the motor.

*2.5. SimCoder*

Is possible obtain the simulation of the complete system using PSIM, however, another feature of this program is that it provides the SimCoder module. This module allows automatic code generation, which can be exported to a DSP for implementation. However, the block diagrams require certain adjustments.

*2.5.1. Incremental encoder.* In simulation a sensor is used to measure the motor speed as shown in Figure 7, but to implement it in SimCoder the block diagram shown at figure 8 b), was used. The input terminals of the incremental encoder are user specified and the output pulses are used to determine the speed and position. In this case encoder 1 i s used, which corresponds to GPIO terminals 20 and 21, see table 1, signals A and B, respectively.

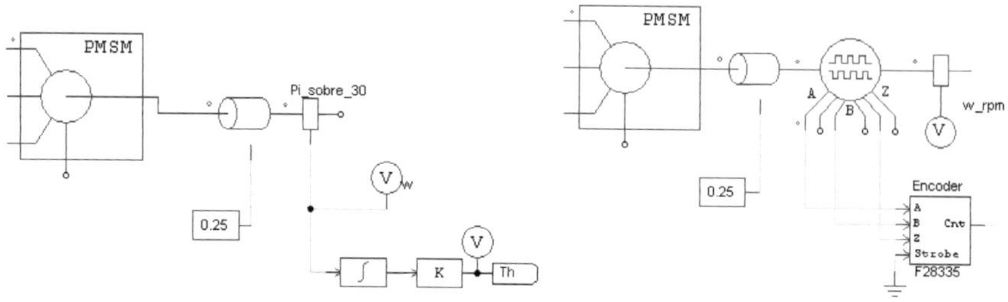


a) Simulation b) Simcoder

**Figure 8**. Speed measurement in simulation and implementation in simcoder.

The encoder used is of 2500 pulses/rev, the term Nl is used to indicate the number of lines per revolution. Figure 9, shows the block diagram.

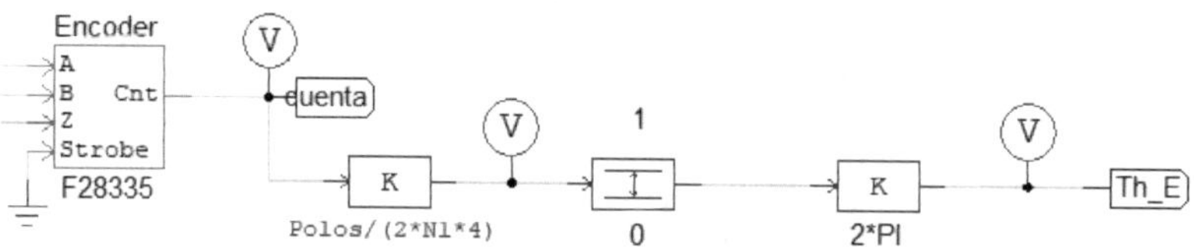


**Figure 9**. Measure of angular position.

The relationship between mechanical and electrical position is determined by the following equation:

$$\theta_e = \frac{P}{2}\theta_m \tag{23}$$

where $\theta e$ = electric angle [ rad], $P$ = number of poles and $\theta m$ = mechanical angle[ rad]. So to convert the mechanical position to electrical it is necessary to multiply the encoder signal by $P/2$.

Since the encoder sends signals in quadrature, you have 10,000 pulses for one turn. So, when multiplying by the term $\frac{P}{2\times Nl\times 4}$ the angle is obtained. The following block converts these values to a range between 0 and 1, which is then multiplied by $2\pi$. The next block converts these values to a range between 0 and 1, which is then multiplied by $2\pi$ and outputs a range of 0 to $2\pi$ radians.

*2.5.2. Calculation of angular velocity.* As is known the velocity can be obtained from the sampled position by its derivative:

$$v(k) = \frac{x(k)-x(k-1)}{T} = \frac{\Delta X}{T} \tag{24}$$

where:

v(k) = Velocity at instant k [rad/s]
x(k) = Position at instant k [rad]
x(k - 1) = Position at instant k - 1 [rad]
T = Sampling period [s]

The formula of equation (24) is implemented in the following block diagram, figure 10. To obtain the previous value of the position a unit delay is used, which represents the previous sample of the signal. The unit delay connected between the encoder value is used to avoid algebraic loops and sets a sampling frequency. The encoder changes from the maximum count to zero, so that in that case the derivative will increase abruptly, producing a peak value of -10000, to eliminate this a comparator with -5000 is included so that when this happens the value of 10000, the resolution of the incremental encoder, is added to the derivative. Finally, a gain block converts the signal to units of [rad/s]. To reduce the noise produced by this method, a gain block is added.

The output value is the electrical speed of the motor. The value obtained at the output is the electrical speed of the motor

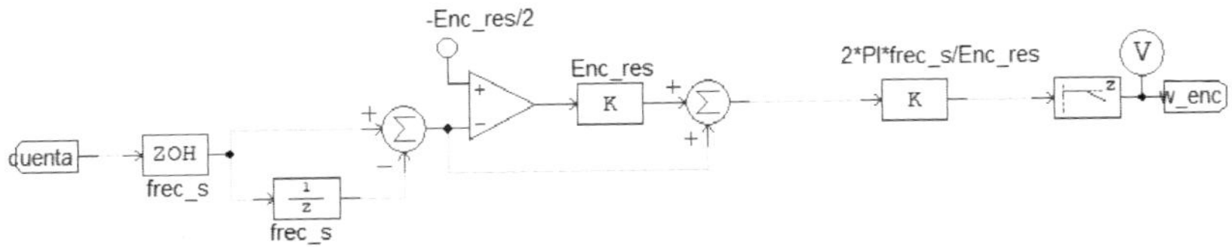


**Figure 10**. Speed calculation in SimCoder.

The added low-pass filter modifies the phase of the system, so that the cut-off frequency with the shortest delay time and stability is selected. For this case different frequencies were tested and the selected one was 300Hz, for more details it is recommended [53].

*2.5.3. Analog to digital converter.* Figure 11, shows how the current measurements are obtained for each phase ia and ib. The output voltage of the operational amplifiers, for the currents ia and ib, respectively is:

$$V_{ia} = 1.65V + 0.165 * i_a \quad (25)$$

$$V_{ib} = 1.65V + 0.165 * i_b \quad (26)$$

These voltages are connected to ADC terminals A1 and B1, respectively. At the output of the converter the following diagram is used to remove the DC voltage of 1.65V from both signals and normalized by multiplying it by 0.99*2/3. At the output of the diagram, the normalized currents of $ia$ and $ib$ are obtained. That is, the output voltages will be:

$$Vf_{ia} = 1.65V - 1.65V + (0.165i_a)0.99 * 2/3 = 1.01i_a \quad (27)$$

$$Vf_{ib} = 1.65V - 1.65V + (0.165i_b)0.99 * 2/3 = 1.01i_b \quad (28)$$

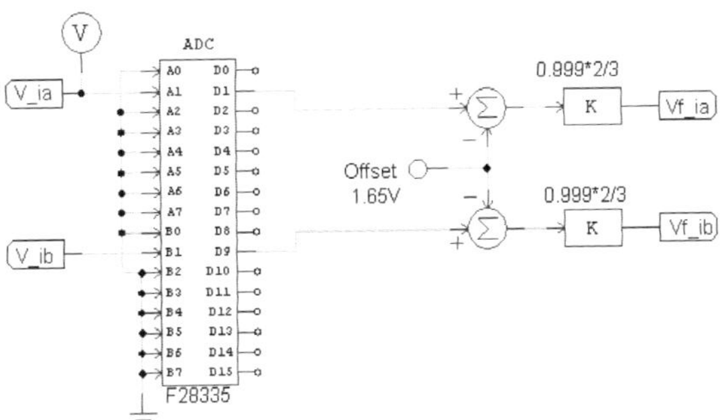


**Figure 11**. Measurement and conditioning of current signals.

*2.5.4. Pulse width modulation.* Figure 12, shows the block used to convert the analog three-phase voltages into digital signals to drive the power transistors, known as the as an inverter. This circuit was used during the simulation.

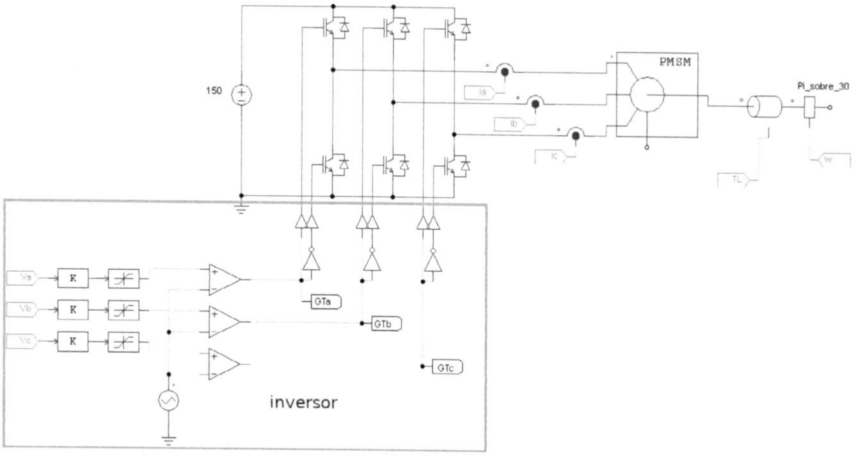


**Figure 12**. Analog three-phase voltage to digital signal convertor.

The block used in Simcoder for the inverter is shown in Figure 13. The PWM-3 module has the parameters used in the simulation. There are two options to implement this inverter, the first one replaces this block directly, for the second one it is necessary to normalize the voltage signals (compatible with version 10). The blocks used are shown below.

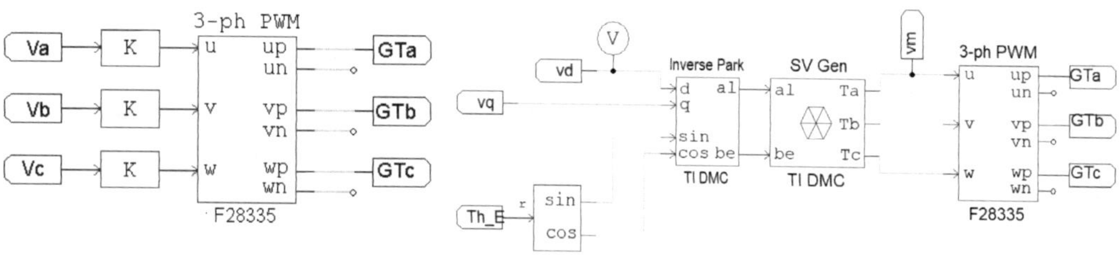


**Figure 13**. Two different implementations for simulation and SimCoder.

*2.5.5. Controler.* This block was developed using C language code, which corresponds to equations (14) and (15). The block has 5 inputs corresponding to the signals id ,iq ,ω,TL and wd ; the two outputs are vd and vq.

```
// x1=id, x2=iq, x3=w, x4=TL, x5=wd
#define L 0.00665
#define B 7.8e-5
#define P 4
#define Km 0.1744
#define R 2.35

float iqd;

iqd=2*(x4+B*x5)/(3*Km);

y1=-P*L*x3*iqd-x6*x1;
//y2=iqd;
y2=Km*x5+R*iqd-x7*(x2-iqd);
```

IOP Conf. Series: Materials Science and Engineering **1311** (2024) 012031 doi:10.1088/1757-899X/1311/1/012031

The diagram used to generate the code for the DSP, is shown in the Figure 14.

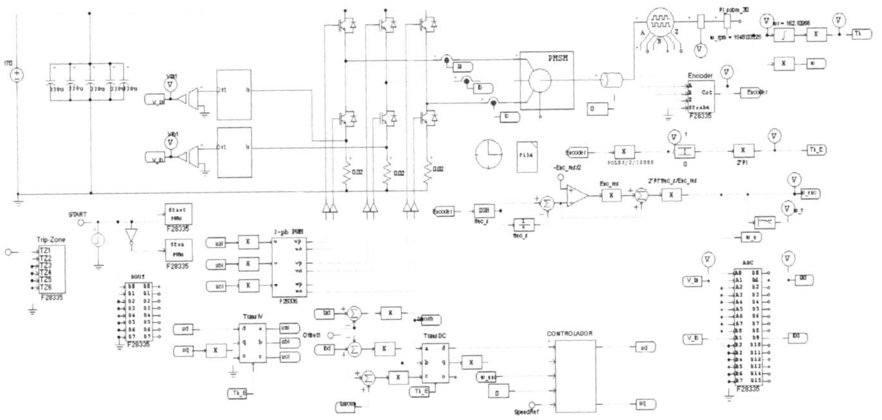


**Figure 14**. Diagram for simulation and implementation in PSIM.

## 3. Results

In this section is shown the numerical simulation obtained and the experimental results obtained of the system implemented. The parameters used are shown in the next table.

**Table 2**. Parameters of PMSM

| Parameter | Value |
|---|---|
| Voltage (VDC) | 150V |
| Poles pair | 4 |
| Magnetic flux $\Psi_{PM}$ | 0.0436 Wb |
| Inertia constant J | 0.000031 $kgm^2$ |
| Resistance R | 2.35 $\Omega$ |
| Inductance Ld | 0.007 H |
| Inductance Lq | 0.007 H |
| Viscous friction B | 0.0000775 Nm/rad/s |

In PSIM the system is simulated for a speed of 100 rad/s of reference, obtaining the results shown in figure 15.

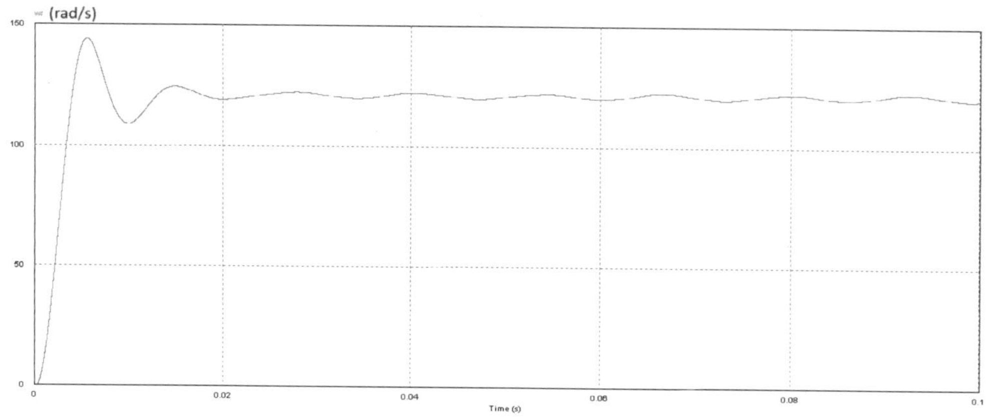


**Figure 15**. Result of simulation for a speed of 100 rad/s.

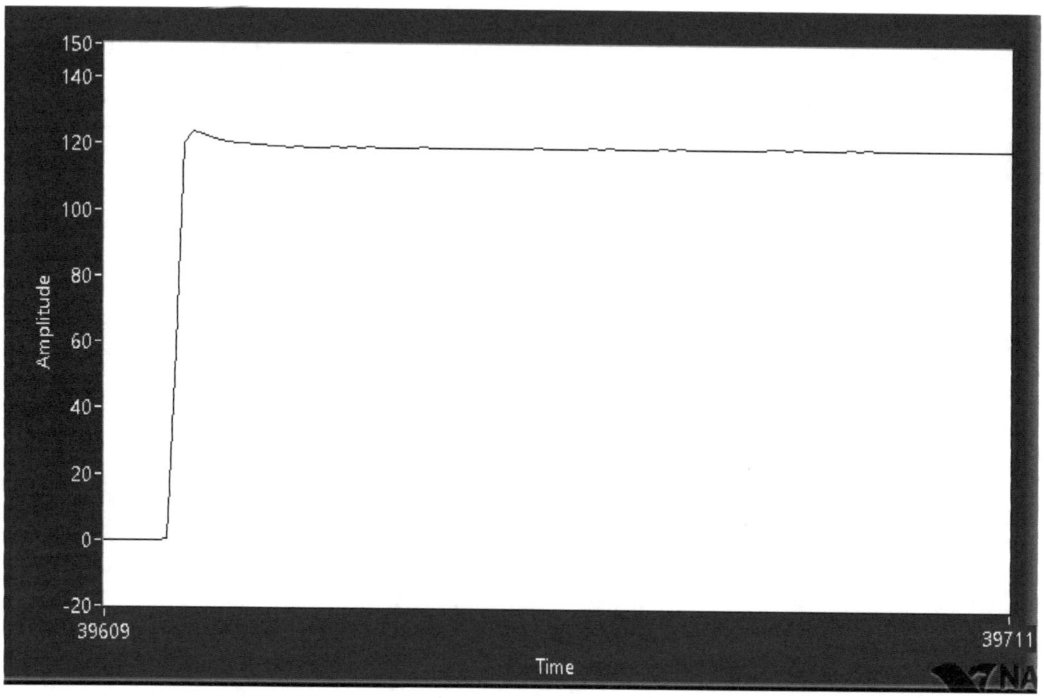


**Figure 16**. Experimental results for a speed of 100 rad/s

In PSIM is obtained the code for a reference speed of 100 rad/s, obtaining the results shown in figure 16.

## 4. Conclusions

This paper has presented a mathematical model and control design methodology for the passivity-based control of PMSM. The implementation of this controller is explained with detail for another implementation in DSP, the principal condition is that the sample time must be of 100 μs. The simulation and experimental results, shown differences between speed measure and desired, for the speed shown but for another reference the percentage of error is biggest. Is possible establish that this controller requires an integral term for reduce the stable state error. A future work will be adding an integral term to this controller or use another with this characteristic.

**References**

[1] P. Waide, Energy-Efficiency Policy Opportunities for Electric Motor-Driven Systems. OECD Publishing, 2011.

[2] F. Giri, *AC Electric Motors Control: Advanced Design Techniques and Applications*. John Wiley & Sons, 2013.

[3] Y.-S. Kung, R. Risfendra, Y.-D. Lin, y L.-C. Huang, «FPGA-realization of a sensorless speed controller for PMSM drives using novel sliding mode observer», *Microsyst. Technol.*, vol. 24, n.º 1, pp. 79-93, ene. 2018, doi: 10.1007/s00542-016-3179-6.

[4] E. W. Zurita-Bustamante, H. Sira-Ramirez, y J. Linares-Flores, «On the sensorless rotor position control of the permanent magnet synchronous motor: An active disturbance rejection approach», en *2016 13th International Conference on Power Electronics (CIEP)*, jun. 2016, pp. 12-17. doi: 10.1109/CIEP.2016.7530723.

[5] S. Hassaine, B. Sari, S. Moreau, y B. Mazari, «Rapid prototyping of a multivariable control with pole placement by state feedback of a PMSM: LMI approach», en *IECON 2012 - 38th Annual Conference on IEEE Industrial Electronics Society*, oct. 2012, pp. 1970-1975. doi: 10.1109/IECON.2012.6388900.

[6] R. P. Burgos, P. Kshirsagar, A. Lidozzi, F. Wang, y D. Boroyevich, «Mathematical Model and Control Design for Sensorless Vector Control of Permanent Magnet Synchronous Machines», en *2006 IEEE Workshops on Computers in Power Electronics*, jul. 2006, pp. 76-82. doi: 10.1109/COMPEL.2006.305655.

[7] J. F. Gieras, *Permanent Magnet Motor Technology: Design and Applications, Third Edition*, 3.ª ed. Boca Raton: CRC Press, 2009.

[8] A. Glumineau y J. de L. Morales, *Sensorless AC Electric Motor Control: Robust Advanced Design Techniques and Applications*. Springer, 2015.

[9] M. Arshad Khan, M. Faisal Aftab, E. Ahmed, y I. Youn, «ROBUST DIFFERENTIAL STEERING CONTROL SYSTEM FOR AN INDEPENDENT FOUR WHEEL DRIVE ELECTRIC VEHICLE», *Int. J. Automot. Technol.*, vol. 20, n.º 1, pp. 87-97, 2019, doi: 10.1007/s12239−019−0008−9.

[10] S. Chi, Z. Zhang, y L. Xu, «Sliding-Mode Sensorless Control of Direct-Drive PM Synchronous Motors for Washing Machine Applications», *IEEE Trans. Ind. Appl.*, vol. 45, n.º 2, pp. 582-590, mar. 2009, doi: 10.1109/TIA.2009.2013545.

[11] C. Calleja, A. López-de-Heredia, H. Gaztañaga, L. Aldasoro, y T. Nieva, «Validation of a Modified Direct-Self-Control Strategy for PMSM in Railway-Traction Applications», *IEEE Trans. Ind. Electron.*, vol. 63, n.º 8, pp. 5143-5155, ago. 2016, doi: 10.1109/TIE.2016.2572661.

[12] M. Zoghlami, R. Kharoui, N. Fnaeich, y F. Bacha, «Direct Power Control strategy for variable speed Wind Energy Conversion System based on PMSM Generator», en *2016 International Symposium on Power Electronics, Electrical Drives, Automation and Motion (SPEEDAM)*, jun. 2016, pp. 1348-1353. doi: 10.1109/SPEEDAM.2016.7525951.

[13] I. Ducar y C. Marinescu, «Comparative study for reversible pump at variable speed in PMSM applications», en *2015 9th International Symposium on Advanced Topics in Electrical Engineering (ATEE)*, may 2015, pp. 205-210. doi: 10.1109/ATEE.2015.7133765.

[14] R. Ortega, A. Loría, P. J. Nicklasson, y H. Sira-Ramírez, «Euler-Lagrange systems», en *Passivity-based Control of Euler-Lagrange Systems: Mechanical, Electrical and Electromechanical Applications*, R. Ortega, A. Loría, P. J. Nicklasson, y H. Sira-Ramírez, Eds., en Communications and Control Engineering. , London: Springer, 1998, pp. 15-37. doi: 10.1007/978-1-4471-3603-3_2.

[15] R. Ortega y G. Espinosa-Pérez, «PASSIVITY BASED CONTROL WITH SIMULTANEOUS

ENERGY SHAPING AND DAMPING INJECTION: THE INDUCTION MOTOR CASE STUDY», *IFAC Proc. Vol.*, vol. 38, n.º 1, pp. 477-482, ene. 2005, doi: 10.3182/20050703-6-CZ-1902.00734.

[16] H. Sira-Ramirez, R. A. Perez-Moreno, R. Ortega, y M. Garcia-Esteban, «Passivity-based controllers for the stabilization of DC-to-DC power converters», *automatica*, vol. 33, n.º 4, pp. 499-513, 1997.

[17] J. Linares-Flores, H. Sira-Ramirez, E. F. Cuevas-Lopez, y M. A. Contreras-Ordaz, «Sensorless passivity based control of a DC motor via a solar powered Sepic converter-full bridge combination», *J. Power Electron.*, vol. 11, n.º 5, pp. 743-750, 2011.

[18] J. Linares-Flores, J. Reger, y H. Sira-Ramírez, «Load torque estimation and passivity-based control of a boost-converter/DC-motor combination», *IEEE Trans. Control Syst. Technol.*, vol. 18, n.º 6, pp. 1398-1405, 2010.

[19] J. L. Flores, J. L. B. Avalos, y C. A. B. Espinosa, «Passivity-based controller and online algebraic estimation of the load parameter of the DC-to-DC power converter Cuk type», *IEEE Lat. Am. Trans.*, vol. 9, n.º 1, pp. 784-791, 2011.

[20] Q. X. Lu y X. R. Dong, «Application of DSP in sensorless PMSM control system», en *2009 International Conference on Measuring Technology and Mechatronics Automation, ICMTMA 2009*, 2009, pp. 357-360. doi: 10.1109/ICMTMA.2009.324.

[21] G. Xie, K. Lu, S. K. Dwivedi, J. R. Rosholm, y F. Blaabjerg, «Minimum-Voltage Vector Injection Method for Sensorless Control of PMSM for Low-Speed Operations», *IEEE Trans. Power Electron.*, vol. 31, n.º 2, pp. 1785-1794, feb. 2016, doi: 10.1109/TPEL.2015.2426200.

[22] K. Urbanski, «Sensorless control of PMSM at low speed range using reference model», en *2015 17th European Conference on Power Electronics and Applications (EPE'15 ECCE-Europe)*, sep. 2015, pp. 1-8. doi: 10.1109/EPE.2015.7309200.

[23] N. K. Quang, D. D. Tung, y Q. P. Ha, «FPGA-based sensorless PMSM speed control using adaptive extended Kalman filter», en *2015 IEEE International Conference on Automation Science and Engineering (CASE)*, ago. 2015, pp. 1650-1655. doi: 10.1109/CoASE.2015.7294338.

[24] X. Dong, W. Tianmiao, y W. Hongxing, «Comparison between model reference observer and reduced order observer of PMSM torque», en *2011 6th IEEE Conference on Industrial Electronics and Applications*, jun. 2011, pp. 663-667. doi: 10.1109/ICIEA.2011.5975669.

[25] H.-H. Chou, Y.-S. Kung, N. Vu Quynh, y S. Cheng, «Optimized FPGA design, verification and implementation of a neuro-fuzzy controller for PMSM drives», *Math. Comput. Simul.*, vol. 90, pp. 28-44, abr. 2013, doi: 10.1016/j.matcom.2012.07.012.

[26] B. K. Bose, *Modern Power Electronics and AC Drives*. Prentice Hall PTR, 2002.

[27] Y.-S. Kung, V. Q. Nguyen, C.-C. Huang, y L.-C. Huang, «Simulink/ModelSim co-simulation of sensorless PMSM speed controller», en *2011 IEEE Symposium on Industrial Electronics and Applications*, sep. 2011, pp. 24-29. doi: 10.1109/ISIEA.2011.6108709.

[28] E. W. Zurita-Bustamante, A. Luviano-Juárez, y H. Sira-Ramírez, «On the Robust Flat-Filtering Control of MIMO nonlinear systems: The PMSM Experimental Case Study», en *2018 Annual American Control Conference (ACC)*, jun. 2018, pp. 1-4. doi: 10.23919/ACC.2018.8431316.

[29] H.-H. Chou, Y.-S. Kung, N. Vu Quynh, y S. Cheng, «Optimized FPGA design, verification and implementation of a neuro-fuzzy controller for PMSM drives», *Math. Comput. Simul.*, vol. 90, pp. 28-44, abr. 2013, doi: 10.1016/j.matcom.2012.07.012.

IOP Conf. Series: Materials Science and Engineering **1311** (2024) 012032 doi:10.1088/1757-899X/1311/1/012032

# Heavy-duty vehicles with solar panels as a regenerative power source

**D.M.Popa, D.Popa, S.Tabacu**

University of Pitesti, Faculty of Mechanics and Technology, 1, Targu din Vale street, 110040, Pitesti, Arges, Romania

Email: dragos.popa@upit.ro

**Abstract.** Road transportation represents more than a third of all $CO_2$-producing sources on Earth, being the main producer of greenhouse gas (GHG) emissions. Nowadays, continuous development of technology has reduced fossil-fuel consumption by integrating clean energy sources as a range extender. Solar energy has the largest potential of all renewable energy sources worldwide. The photovoltaic and energy storage research fields have grown exponentially during the last decade and, forced by the current energy crisis, had a great contribution to the hybrid solar vehicles (HSV) industry. Vehicle Integrated Photovoltaics (VIPV) comply with requirements on vehicles aesthetics, and they convert solar energy into electric energy to supply the low-voltage and high-voltage battery packs onboard HSV. Heavy-load road transportation by commercial vehicles equipped with PV-integrated modules or retrofit installation will contribute to global decarbonization. A study is done on a configuration of a heavy-duty vehicle with its cuboid cargo body covered with solar panels and a regulated driving route during different weather conditions [3]. The energy management system is designed to maximize solar energy conversion in order to reduce fuel consumption and extend the vehicle's total driving distance.

## 1. Introduction

According to the European Environment Agency declaration, during the year 2016, up to 21% of total $CO_2$ emissions on this continent were produced by road transport, which represents more than 80% of total $CO_2$ of all types of transport.

Innovative technology has made solar power one of the most efficient regenerative power; the development of photovoltaic cells and energy storage systems has led to a very affordable cost for converting solar energy into electric energy and using on board light-weight and heavy-duty passenger vehicles. The photovoltaic cost per watt is opposite to solar module efficiency.

Hybrid Solar Vehicles (HSV) represent the ultimate technology of electric vehicles as BEV (battery electric vehicles), HEV (hybrid electric vehicles), PHEV (plug-in hybrid electric vehicles), and FCEV (fuel cell electric vehicles – the same as BEV but having a fuel cell instead of battery). BEV and PHEV require charging from an energy storage system which is the utility power grid or any other external power supply source using a charging plug. The main issues remain the cost of the energy supplying the batteries and the dependency on the power supply infrastructure.

HSV is an HEV which is also using solar power sources to supplement the energy necessary for traction and all the other consumers onboard the vehicle. The hybrid vehicle has the meaning of combining two or more types of energy and power sources that together or alternatively can drive the vehicle. The goal of adding an extra power source, preferably a regenerative one, is to significantly reduce fossil fuel consumption. This is equivalent to extended driven distance, longer refueling intervals, fewer pollutant emissions, and an extended lifetime of the internal combustion engine.

The global solar vehicles (full solar EV and HSV) industry is predicted to increase 18% annually for the next five years, reaching around 680 million dollars in 2026 and 4 billion in 2030. The estimated trend is shown in below figure:

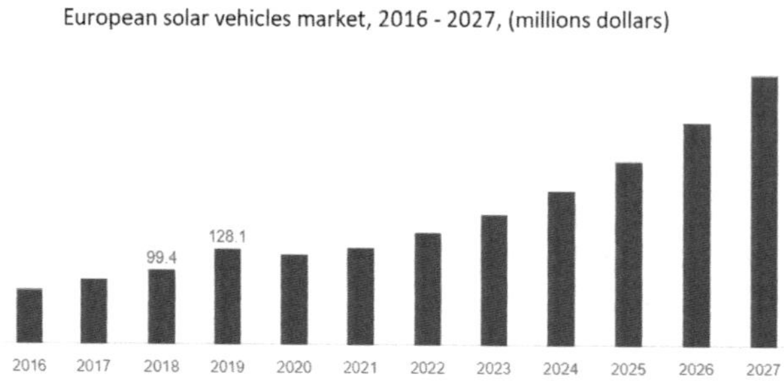


**Figure 1.** European solar vehicles market tendency for 2016 – 2027 interval (millions of dollars)

## 2. Driveline architecture of HSV

An optimized parallel driveline designed for a heavy-duty Hybrid Solar Vehicle is illustrated in figure 2 stated below:

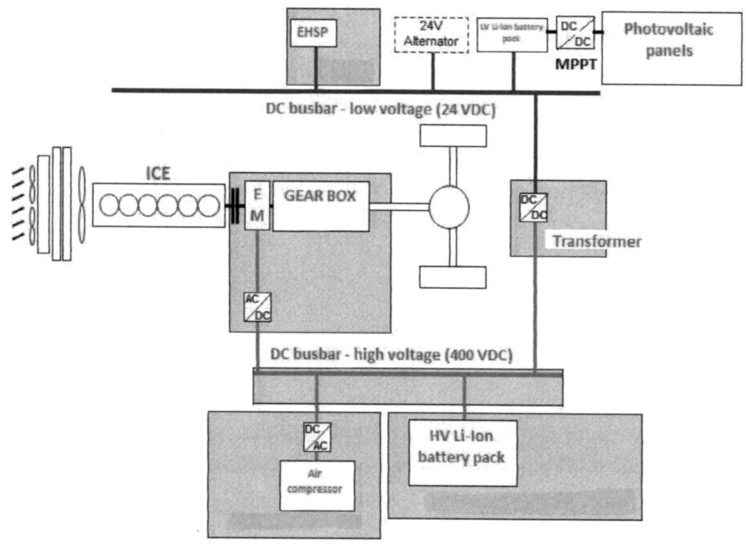


**Figure 2.** Heavy-duty HSV layout

Some of its components are listed here:

- ICE as the primary power source of the driveline, a Euro VI standard diesel engine
- Electric motor; it is installed between the clutch and the gearbox, and it can provide drive as well as function as a generator
- Automatic shift gearbox
- Low voltage lithium-ion battery pack (12V, 24V, 36V, or 60V), which receives electric power directly from the photovoltaic array system through the charging controller
- MPPT (maximum power point tracker) charging controller
- Solar panels array
- High voltage lithium-ion battery pack (400V or more), which is charged from the LV battery pack via the DC/DC converter or from the electric motor via AC/DC converter
- Electro-hydraulic servo pump, power supplied from LV busbar

- Air compressor to supply pressurized compressed air to operate heavy-duty vehicle brakes; it is power supplied from the HV busbar via a DC/DC converter
- DC/DC inverter, which connects LV and HV busbars
- AC/DC transformer between the electric motor and HV busbar, which converts the alternating current supplied by the generator to direct current for the HV batteries
- Regenerative braking system which supplies LV battery pack through LV busbar
- Cooling unit for the AC/DC current transformer
- Air conditioning compressor
- Infotainment system
- Other consumers inside the vehicle's cabin

An energy management system (EMS) is the most important component whose role is to coordinate the continuous energy flow from primary and alternative energy sources towards LV and HV consumers and to control both battery packs simultaneously according to their state-of-charge (SoC) settings [5], [6]. EMS has to improve energy efficiency and minimize conventional fuel consumption without jeopardizing safety and normal functioning parameters during the vehicle's acceleration, deceleration, stopping, starting, and parking.

## 3. Heavy-duty vehicles empowered with solar energy. Vehicle Integrated Photovoltaics (VIPV)

Heavy-duty vehicles such as trucks, trailers, cargo vans, and buses have much larger available external surfaces where VIPV can be installed. This is the reason why their VIPV's economic viability can be achieved earlier than for small passenger vehicles.
The below figures present all possible areas on the external vehicle's body that can easily be covered by solar cells as a supplementary power source:

**Figure 3.** Vehicles with VIPV

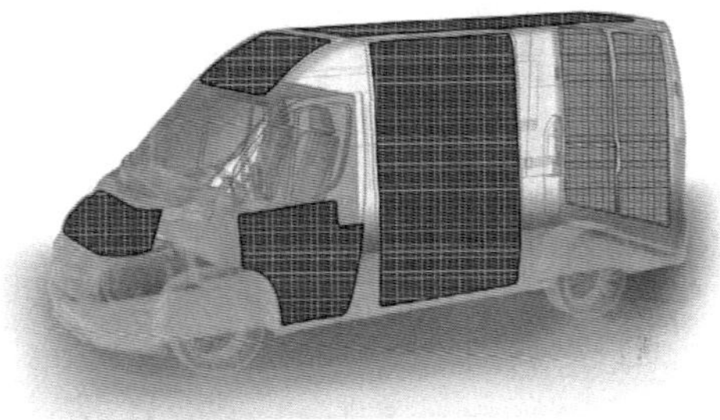

**Figure 4.** Cargo van: examples of outer body surfaces where VIPV can be installed

**Figure 5.** Truck: examples of outer body surfaces where VIPV can be installed

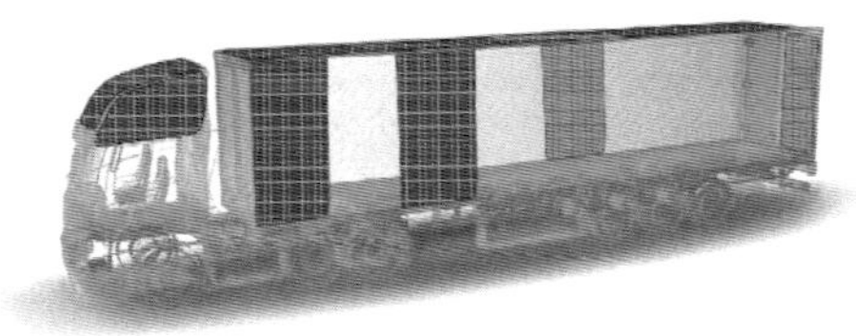

**Figure 6.** Trailer: examples of outer body surfaces where VIPV can be installed

With a large body surface covered by photovoltaic modules, heavy-duty HSVs receive a huge amount of energy compared to conventional passenger solar vehicles. They are becoming mobile solar parks. The solar irradiance side-to-roof ratio calculated for the same photovoltaic surface is more than 2, depending on environmental conditions such as season, weather, time of the day, and road environment, which creates a transient shading effect during driving [1].

## 4. Driving modes

Figures 7 to 11 illustrate the operating modes of a hybrid solar vehicle. Energy flows are "arrow" marked accordingly for each driving mode.

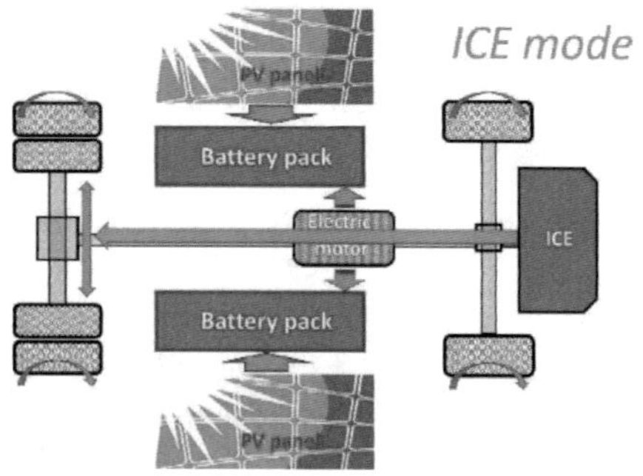


**Figure 7.** Traction is done by ICE only; battery packs are charged by PV panels and by electric motor functioning as a generator (producing energy into battery packs)

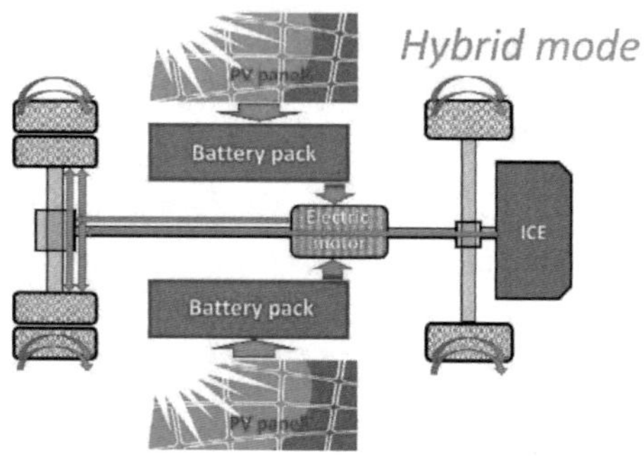


**Figure 8.** Traction is done by ICE together with an electric motor (consuming energy from battery packs); PV panels charge battery packs

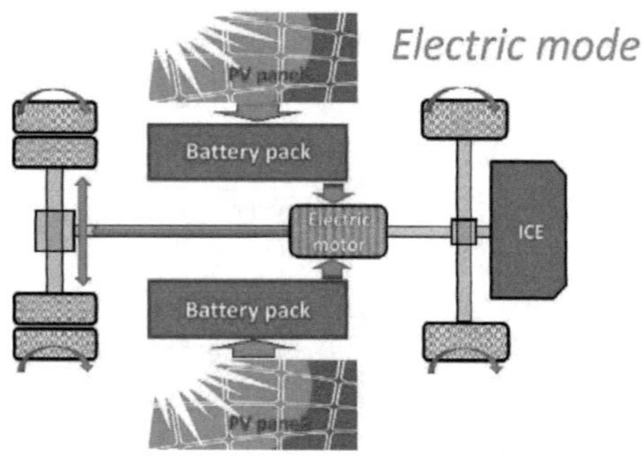


**Figure 9.** Traction is done by the electric motor only (consuming energy from battery packs); ICE is stopped; PV panels charge battery packs

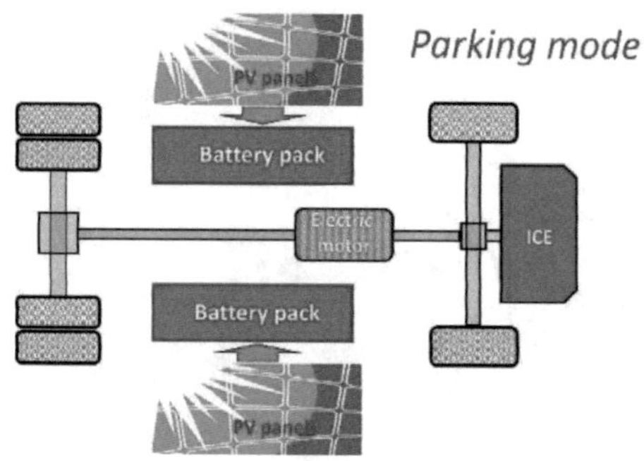


**Figure 10.** The vehicle is parked, no active consumer; both ICE and electric motor are off; PV panels still charge battery packs

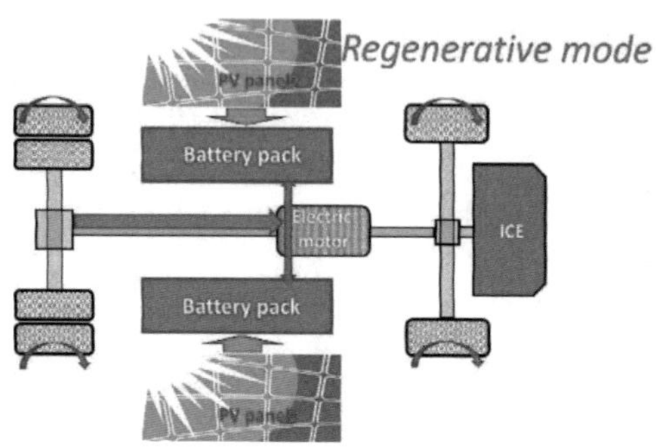


**Figure 11.** A regenerative braking system charges the battery packs during the vehicle's slowing-down or mechanical braking

PV-generated electricity is fed into the LV battery pack during all the above-presented operating modes, independent of the functioning status of the internal combustion engine and electric motor. During

IOP Conf. Series: Materials Science and Engineering **1311** (2024) 012032 doi:10.1088/1757-899X/1311/1/012032

outdoor parking daytime periods, the vehicle orientation shall be favorable for receiving maximum solar irradiance outside of any shading from nearby obstacles.

## 5. Energy management system (EMS)

The main target of EMS is to improve the overall fuel economy by increasing the energetic driveline efficiency [12]. The integration of photovoltaic panels into the bodywork of the HSV, therefore a continuous energy conversion and charging of the low voltage battery pack, represents an extra task for the energy management system to be dealt with. All of the parallel hybrid architecture components are going to be controlled for optimized co-operation.

From the PV array, the converted energy is transferred through the MPPT into the LV battery pack. The onboard low voltage consumers are fed from this battery pack, but it is also acting as a buffer battery for charging the HV energy storage batteries via the high efficiency (up to 90%) DC/DC converter. LV and HV battery packs are charged according to their state-of-charge set levels (LV max SoC is set to 90% and HV max SoC, as the main storage device, may be set to 98%) [4].

During driving, the state-of-charge level of both battery packs decreases despite continuous PV charging, just because the PVs can not supply the entire energy required for driving. That's why the EMS has to permanently analyze the power balance for starting/stopping the internal combustion engine [8].

The MPPT charge controller has to handle the variable output power from the PV system due to the vehicle's transitions through different shadow conditions (transitions from high to low irradiance and vice-versa), trying to smoothen the power fluctuations [10].

The electric motor should have enough strength and available power to fully drive the heavy-duty HSV during the electric drive mode, see Figure 9, especially needed during traveling through city's green zones. Depending on the HV battery pack SoC, mainly charged during the parking period, the electric motor can move the vehicle from a standstill position without support from the ICE if its mechanical peak power is high enough.

During the hybrid driving mode, the EMS shares the load between the ICE and the electric motor; a lower load on the ICE also lowers fuel consumption. During traffic standstill periods, when sufficient stored energy is available, running the ICE idle will be unnecessary to generate electricity.

## 6. Solar irradiance

Yearly solar irradiance varies according to the geographical position; its average hourly measurement ranges from 88 W/m$^2$ in Finland to 218 W/m$^2$ in Cyprus, as the Research Center's Photovoltaic Geophysical Information System (PVGIS) data logs are presented. In Europe, the maximum hourly solar irradiance for the above year was recorded in Malta, and its value was 1012 W/m$^2$ [2].

The standard testing conditions of solar panels have considered 1000 W/m$^2$ as the reference value of the solar irradiance with a panel surface temperature of 25°C and an air mass AM1.5 [2].

About 42% of total solar irradiance which may reach the photovoltaic panels' surface is lost due to the intermittent shading effect, especially when the HSV drives in an urban environment with various obstacles along the roads (buildings, trees, walls, bridges, etc.) [1].

The time frame of the day and environmental conditions like weather, season, and buildings density have different influences on roof-mounted solar panels compared to sides-mounted solar panels. The weather conditions have a greater impact on roof-mounted solar irradiance, while the irradiance received by sides-mounted panels is more dependent on the season.

## 7. Route planning

The solar-powered vehicles' road transportation increases its efficiency if the driving routes are planned to avoid irradiance sudden changes between full irradiance and full shading moments (light-dark and dark-light transitions), which create fluctuating photovoltaic power production [1]. This fluctuation is higher during sunny days and lowers during cloudy days. Also, the VIPVs are negatively affected by often passing through dynamic partial shading areas. The MPPT has difficulties in leveling the charging into the LV battery pack.

Urban roads have been categorized into 3 types of roads:

a) Narrow streets with low-speed limits and with along buildings and trees whose height is higher than the distance between the street and building/tree; most of these streets have only one line for each direction [1];
b) Wide streets with low-speed limits with more than one line for each direction; the distance from the street to both sides of buildings is higher than the building's height; these streets are directly connected to first-category streets [1];
c) Wide streets with a speed limit of 80 km/h or more and with few or no buildings close by; these city highways have no shadows except tunnels or underpasses [1].

Based on experimental irradiation measurements on the roof and sides surfaces of a heavy-duty HSV driving on these three types of streets, the results are shown in the below two figures:

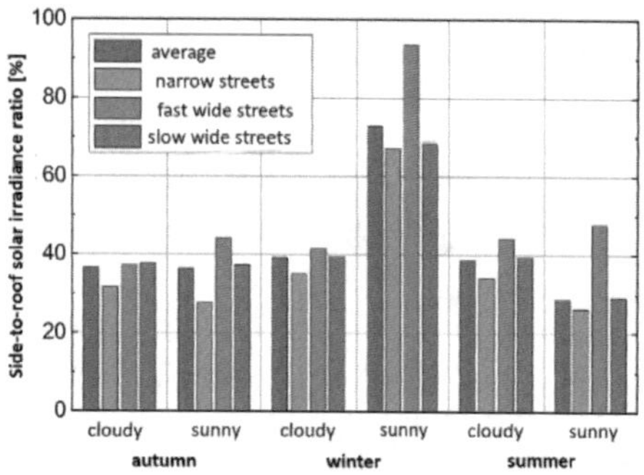


**Figure 12.** Season and weather dependant side-to-roof solar irradiance ratio for each type of streets

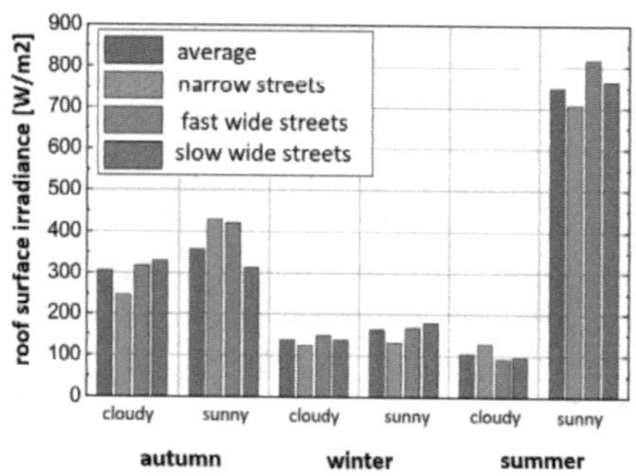


**Figure 13.** Season and weather-dependent solar irradiance for each type of streets

The solar irradiance measurements are done for entirely PV panels covered roofs and both sides of an HSV cargo trailer.

Figures 14 and 15 present daily average irradiance and average daily air temperature during the warmest month in Constanta for latitude/longitude coordinates of 44.1°/28.6°, based on PVGIS-5 geo-temporal irradiation database. Similar data is useful when planning an HSV urban-extra-urban combined route. The most favorable time frame for maximized solar energy production is between 11.00 and 13.00 hours for a two-way estimated journey of 80 km total distance with an average speed of 40 km/h.

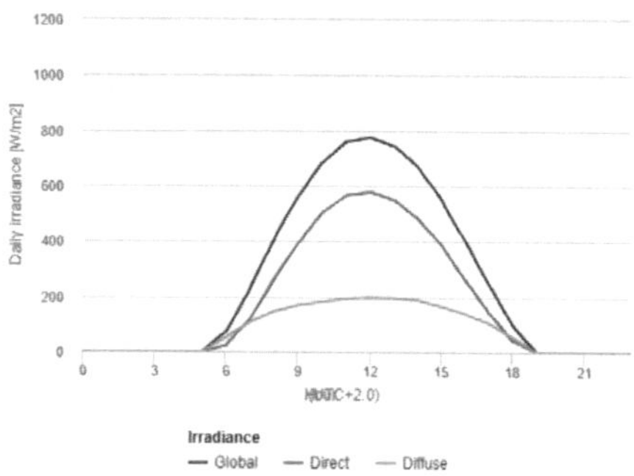


**Figure 14.** Daily average irradiance [W/m$^2$] on the fixed plane with slope 0° and azimuth 0° during the month of August in Constanta

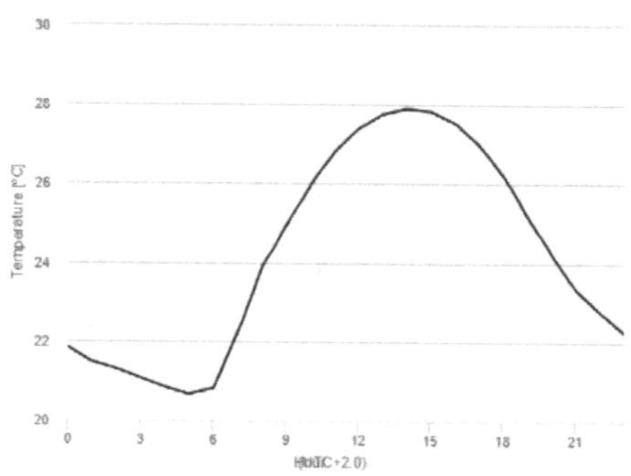


**Figure 15.** Daily average temperature [°C] during the month of August in Constanta

Another important piece of information in choosing the traffic route is the side roads building orientation. Experimentally observed that high obstacles built on a north-south orientation have less impact on shading effects than those built on an east-west orientation [1].
The side-to-roof ratio is very important for gaining solar irradiation as much as possible during driving hours.

## 8. Conclusions

Considering the actual energy crisis, the higher pollution restrictions, and the accelerated technology development regarding photovoltaics, road cargo transportation can easily receive an alternative clean power source to reduce its fossil-fuel consumption and extend the driving distance. The new generation of VIPV has a short break-even period, and it's worth the investment for long-run heavy-duty vehicle companies. The HSVs are independent of public electrical charging network and, by planning the route, can absorb maximum solar power.
Compared to an HEV, the internal combustion engine runs fewer hours in idling mode to produce electrical power, reducing wear and fuel consumption.
The heavy-duty HSV can transit the urban areas without disturbing noise and pollutant emissions for a certain driving distance.
The future of road transportation will be linked more and more to solar power.

## References

[1] Wetzel G, Salomon L, Krugener J, Bredemeier D, Peibst R 2022 High time resolution measurement of solar irradiance onto driving car body for vehicle integrated photovoltaics Prog. Photovolt Res Appl. 30(5):543-551.doi:10.1002/pip.3526Another reference

[2] Lodi C, Gil-Sayas S, Curro D, Serra S, Drossinos Y 2021 *Full battery effect during on-board solar charging of conventional vehicles* Transportation Research Part D **96** 102862.

[3] Peibst R, Fischer H, Brunner M, Schiefl A, Wohe S, Wecker R, Haase F, Schulte-Huxel H, Blankemeyer S 2021 *Demonstration of feeding vehicle-integrated photovoltaic-converted energy into the high-voltage on-board network of practical light commercial vehicles for range extension* Reasearch article published by Wiley-VCH GmbH, doi: 10.1002/solr.202100516, pp 1-8.

[4] Verma S, Mishra S, Gaur A, Chowdhury S, Mohapatra S, Dwivedi G, et al. 2021 *A comprehensive review on energy storage in hybrid electric vehicle* J Traffic Transp Eng English Ed **8**:621–37 https://doi.org/10.1016/j.jtte.2021.09.001.

[5] Thiel C, Gracia Amillo A, Tansini A, Tsakalidis A, Fontaras G, Dunlop E, et al. 2022 *Impact of climatic conditions on prospects for integrated photovoltaics in electric vehicles* Renew Sustain Energy Rev **158** pp1–17 https://doi.org/10.1016/j.rser.2022.11210.

[6] Hoai-An T, Hoai Vu A T, Tri C D, Manh H N, Van D P, Kyoung K A 2022 *Optimization-based energy management strategies for hybrid construction machinery: A review* Energy reports **8** pp 6035-6057.

[7] Boström T, Babar B, Hansen JB, Good C 2021 *The pure PV-EV energy system – A conceptual study of a nationwide energy system based solely on photovoltaics and electric vehicles* Smart Energy **1** https://doi.org/10.1016/j.segy.2021.100001.

[8] Arsie I, Di Martino R, Rizzo G, Sorrentino M. 2007 *Toward a supervisory control of a hybrid solar vehicle* IFAC Proc **5**:359–66. https://doi.org/10.3182/20070820-3-us-2918.00049.

[9] Frank C, Holsten L, Sahin T, Vietor T 2022 *How to manage vehicle platform variants? A method to assess platform variance through competitive analysis* Science Direct, 32nd CIRP Design Conference, Procedia CIRP **109** pp 598-603.

[10] Meskani A, Haddi A 2019 *Modeling and Simulation of an Intelligent Hybrid Energy Source based on Solar Energy and Battery* Special issue on emerging and renewable energy: Generation and automation ScienceDirect, Energy Procedia **162** pp 97-106.

[11] Grandone M, Naddeo M, Marra D, Rizzo G 2016 *Development of a regenerative braking control strategy for hybridized solar vehicle* ScienceDirect IFAC-PapersOnLine 49-11 pp 497-504.

[12] Yang Y, Xu Y, Zhang H, Yang F, Ren J, Wang X, et al. 2022 *Research on the energy management strategy of extended range electric vehicles based on a hybrid energy storage system* Energy Reports **8**:6602–23 https://doi.org/10.1016/j.egyr.2022.05.013.

[13] Song Z, Guan X, Cheng M 2022 *Multi-objective optimization strategy for home energy management system including PV and battery energy storage*. Energy Reports **8** pp 5396-5411.

[14] *Lade-PV project begins : vehicle-integrated PV for clectrical commercial vehicles*, Fraunhofer ISE, Press release, 03-april-2020.

[15] Landu J, Yu H, Chen M and Xue L 2017 *SunChase: energy-efficient route planning for solar-powered EVs* IEEE 37th International Conference on Distributed Computing Systems pp 1-11.

IOP Conf. Series: Materials Science and Engineering **1311** (2024) 012033 doi:10.1088/1757-899X/1311/1/012033

# Bathtub curve and Allan-Plaît chart: two problematic graphs in reliability engineering

**Boroiu Andrei-Alexandru[1], Vieru Ionel[1], Udma Tiberiu-Aurelian[1], Boroiu Alexandru[1]**

[1] University of Piteşti, Târgul din Vale 1, 110040 Piteşti, Romania

E-mail: alexandru.boroiu@upit.ro

**Abstract.** In the reliability literature, there are completely different representations for two of the most common graphs: the graph that expresses the failure rate over the life of a product (bathtub curve) and the one that graphically identifies the Weibull triparametric model (linear regression of the Weibull curve using the Allan-Plaît graph). Even if the identification of the Weibull model is no longer done graphically, but with the help of dedicated calculation programs, the mathematical bases of the linearization of the F(t) fault distribution function by logarithm remain equally important. The present paper aims to establish which of the graphic variants are the correct ones and to identify the causes that led to the incorrect graphic representations. Based on an extensive theoretical documentation (analysis of numerous works in the field of reliability, published over the last decades, in the country and abroad) it is concluded that in foreign literature these graphs are always correct and that they were taken correctly in Romanian works from the 70's to the 80's of the last century, but starting with the 90's there are fundamentally different representations, especially in the field of reliability engineering. Finally, the causes that led to these non-compliant graphical representations are identified. As these misrepresentations seem to persist, this paper is intended to be a signal of this issue.

## 1. Introduction

The first stage of a scientific study is a literature review. This represents an overview of the previously published works on a specific topic. The term can refer to a full scholarly paper or a section of a scholarly work such as a book, or an article [1].

In the field of mathematical modeling, things can be quite complex, because the choice of the most suitable model requires its validation, which can be achieved through theoretical tests of verosimility and must be confirmed by the results obtained in the real use of the products or in the experiments organized especially in this purpose.

Going through this stage in the field of reliability, it is found that there are completely different representations for two of the most common graphs: the graph that expresses the failure rate over the life of a product (bathtub curve) and the one that graphically identifies the Weibull triparametric model (linear regression of the Weibull curve using the Allan-Plaît graph).

Even if the identification of the Weibull model is no longer done graphically, but with the help of dedicated calculation programs, the mathematical bases of the linearization of the F(t) fault distribution function by logarithm remain equally important.

IOP Conf. Series: Materials Science and Engineering **1311** (2024) 012033 doi:10.1088/1757-899X/1311/1/012033

## 2. Problem formulation

Based on an extensive theoretical documentation (analysis of numerous works in the field of reliability, published over the last decades, in the country and abroad) it is concluded that in foreign literature these graphs are always correct [2], [3], [4], [5], [6], [7], [8] and that they were taken correctly in Romanian works from the 70's to the 80's of the last century [9], [10], but starting with the 90's there are fundamentally different representations, especially in the field of reliability engineering. Few of these agree with the "correct variant" presented below [11], [12], [13], [14].

Thus, the very well-known graphic representation for the failure rate, known as the "bathtub curve", in the various works consulted appears in two variants on the first area - the period of early failures (fig. 1):

- *the correct variant* (black color), starting from a high value and decreasing continuously to stabilize at a low and constant value at the end;
- *the incorrect variant* (red color), which starts from the origin following a less and less increasing trend and then follows a turn that approaches the maturity zone.

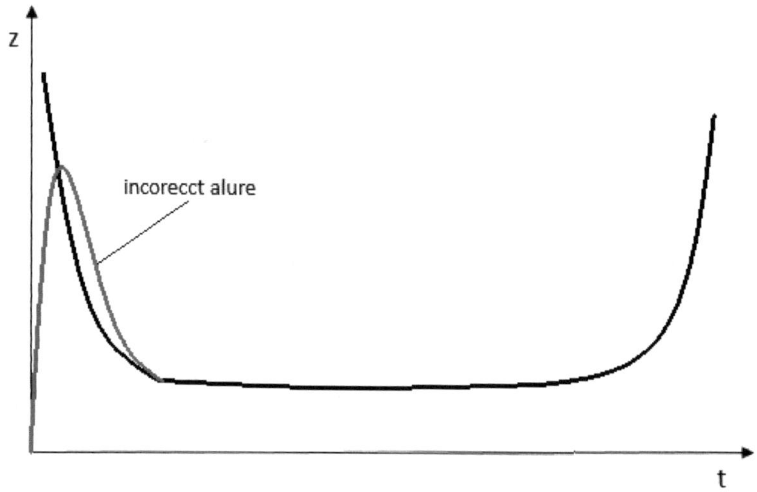


**Figure 1.** The two variants of representation for the early failure period on the bathcurve graph.

Obviously, the correct representation is the one in fig. 2 [13], [14], where the constancy in the period of maturity is explained by the fact that, along with the accidental defects that are constantly manifested, the infantile ones fade and the wear and tear ones start to appear.

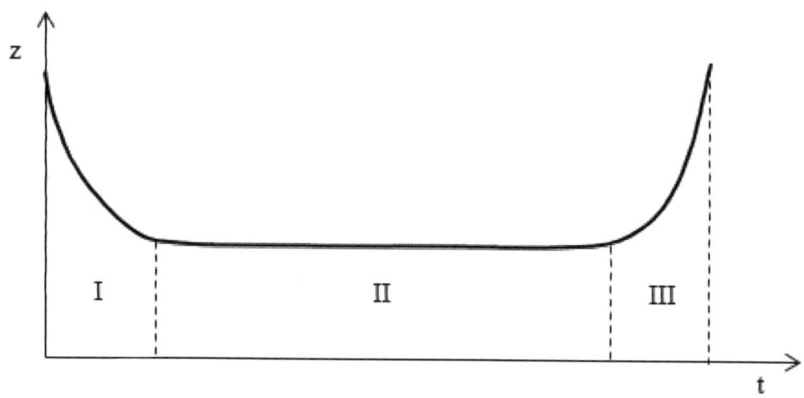


**Figure 2.** The three periods of the bathtub curve z(t).
I - Early defects; II – Useful life; III – Wear defects.

IOP Conf. Series: Materials Science and Engineering **1311** (2024) 012033 doi:10.1088/1757-899X/1311/1/012033

Regarding the graphic representations on special probabilistic networks, in order to identify the parameters of the three-parametric Weibull model, an incorrect representation of the Weibull line is frequently encountered (fig. 3).

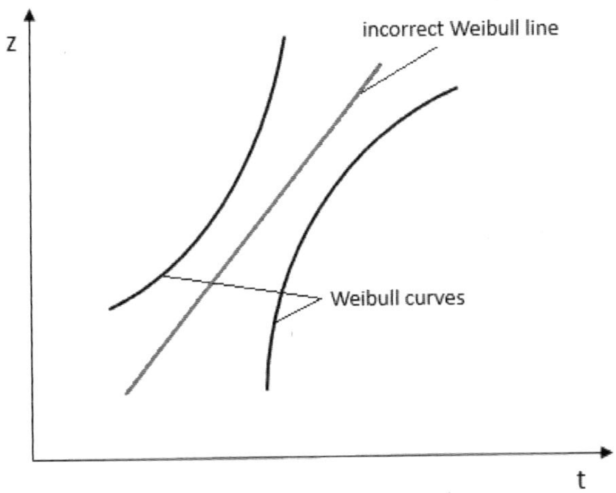


**Figure 3.** The incorrect Weibull line.

However, there are exceptions [8], [9], [11], [13], as in the representation in fig. 4 [13], where it is observed that the Weibull line represents one of the asymptotes for the Weibull curve (the one on the left or the right, depending on the value - negative or positive - adopted for the localization parameter $\gamma$), following the explanations that will be formulated below.

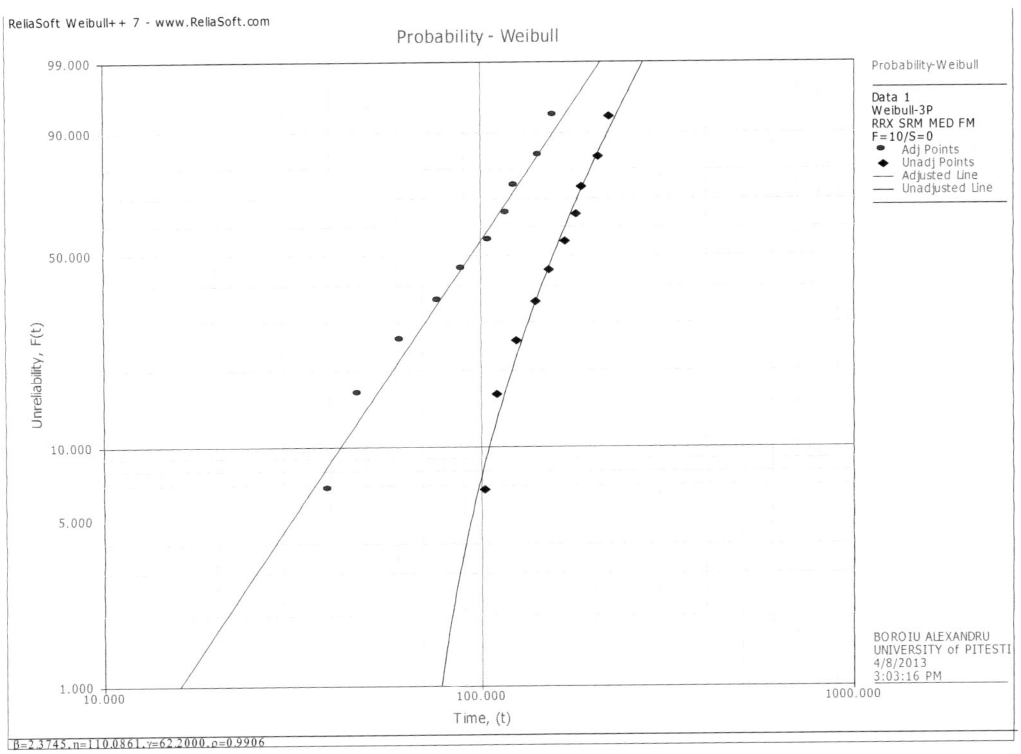


**Figure 4.** The correct Weibull line.

As a result of the above, the present paper aims to establish which of the graphic variants are the correct ones and to identify the causes that led to the incorrect graphic representations.

## 3. Research results

The researches carried out are based on the theoretical interpretation of the two problematic graphs, given that it is intuited that the errors come from the wrong application of the theoretical foundations in the construction of the graphic representations.

### *3.1. The bathcurve graph for failure rate z(t)*

The failure rate or the failure intensity, denoted z(t), represents the probability that an element that has functioned without defects until time t will fail within the range (t, t + Δt).

The failure rate expresses the number of failures that occur in the unit of time, at a given time, taking into account the number of elements still in operation at that time. Thus, the failure intensity is a local indicator of reliability, as well as the probability density of the operating time, but, unlike this, it refers not to the total number of tracked items, but to the number of items that are still in operation at the moment t.

In terms of reliability, the failure rate z(t) is the distribution density of the defects at time t, provided that the element has functioned without fail until that moment. By definition, the failure rate z(t) is the limit of the ratio between the probability of failure in the interval (t, t + Δt], conditioned by the good functioning in the interval (0, t], and the size of the interval Δt, when:

$$z(t) = \lim_{\Delta t \to o} \frac{P(t < T \leq \Delta t / T > t)}{\Delta t} \tag{1}$$

In a statistical definition, the failure rate z(t) represents the ratio between the number of failures that occur in the time unit, at time t, and the number of elements that were in working condition at time t:

$$z(t) = \frac{n(\Delta t)}{N(t) \cdot \Delta t} \tag{2}$$

Explaining the value N(t) as a function of R(t), it results:

$$z(t) = \frac{n(\Delta t)}{N(t) \cdot \Delta t} = \frac{n(\Delta t)}{N_0 R(t) \Delta t} = \frac{f(t)}{R(t)}, \tag{3}$$

relationship frequently used in calculations, because it expresses the direct link between the three reliability indicators.

The failure rate z(t) generally has a time evolution as shown in the previous figure 2, called *the bathtub curve*, due to its allure. On this curve there are 3 distinct zones:

1. *The period of early defects* or *the period of childhood*, characterized by high values of the intensity of defect z(t), but with a strongly decreasing allure.

2. *The period of useful life* or *the period of maturity*, characterized by a low and almost constant level of the failure rate, corresponds to the normal period of use, when the defects mainly have a random character, being of an accidental type, following normal degradation processes.

3. *The period of wear defects* or *the aging period*, characterized by an accelerated increase in the failure rate, caused by the processes of wear, fatigue, aging of the materials.

To understand the origin of the errors found in the representation of the first portion of the bathcurve graph it is very important the next observation.

So, through the statistical and analytical development of the calculations above, it is observed that z(t) were even defined by reference to an infinitesimal interval of time, the value calculated on an interval for the indicator being obligatorily assigned to the half of the interval. This means that these indicators are not defined at point $t = 0$, and the extrapolation (graphical or analytical) for the first half-time interval $(0, \Delta t/2)$ must respect the tendency of the function near the point corresponding to $t = \Delta t/2$.

Thus, only in this case (figure 5) is the representation correct (the curve asymptotically follows the z ordinate).

Most of the time, however, the origin ($t = 0$, $z = 0$) is also entered in the graphic (unjustified, however), which leads to obtaining a volte that cannot be explained, of course, by the failure processes that take place at the beginning of a product's life (fig. 6).

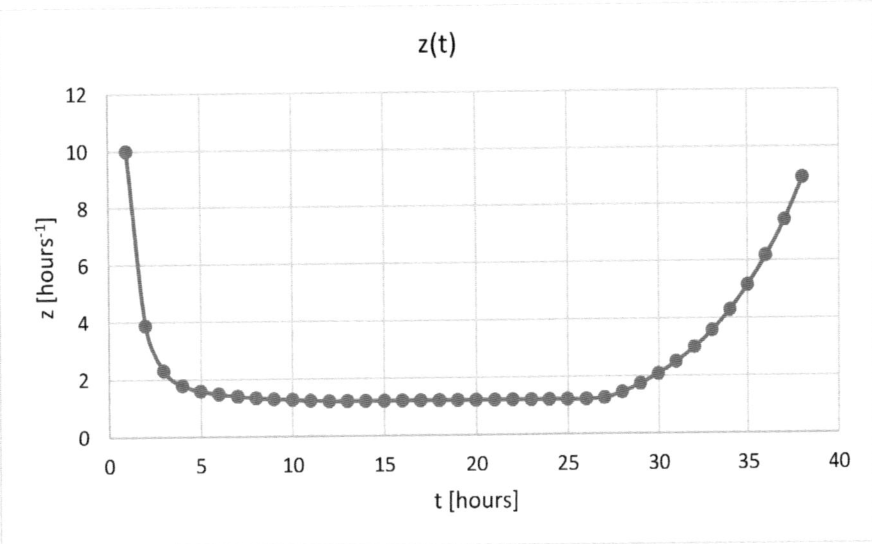


**Figure 5.** The correct bathcurve graph.

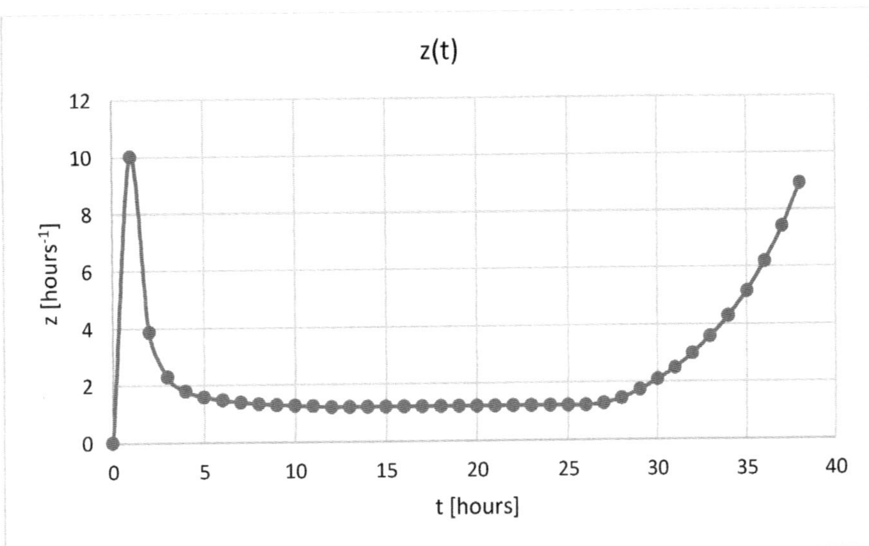


**Figure 6.** The incorrect bathcurve graph.

IOP Conf. Series: Materials Science and Engineering **1311** (2024) 012033 doi:10.1088/1757-899X/1311/1/012033

*3.2. The Allan-Plaît graph for three-parametric Weibull model*

The three-parametric Weibull model is defined by the following expression for the reliability function or the survival function:

$$R(t) = e^{-\left(\frac{t-\gamma}{\eta}\right)^{\beta}} \tag{4}$$

or by the function of unreliability, also called function of distribution of the time of good operation:

$$F(t) = 1 - R(t) = 1 - e^{-\left(\frac{t-\gamma}{\eta}\right)^{\beta}} \tag{5}$$

where:

- $\gamma$ - location parameter or position parameter;
- $\eta$ - scale parameter;
- $\beta$ - shape parameter.

The identification of the Weibull model can be achieved through analytical methods for which special calculation programs have been created, such as Reliasoft.com [13], but a long time ago, special probabilistic networks were used, such as the Allan-Plaît graph [12].

Even if these probabilistic graphs are no longer used today, they continue to be remembered for explaining modeling principles, and a special case is the representation of the significance of the localization parameter $\gamma$. The most well-known probability network used is the one proposed by Allan-Plaît in 1962.

The grapho-analytical modeling is based on the linearization of the F(t) curve through two successive logarithms of the reliability function R(t) [15] given by the relation (4), as presented below.

After the first logarithm results:

$$\ln R(t) = -\left(\frac{t-\gamma}{\eta}\right)^{\beta} \Rightarrow \ln\frac{1}{R(t)} = \left(\frac{t-\gamma}{\eta}\right)^{\beta} \tag{6}$$

and, after the second logarithm, the final shape is obtained,

$$\ln\ln\frac{1}{R(t)} = \ln\ln\frac{1}{1-F(t)} = \beta\ln(t-\gamma) - \beta\ln\eta \tag{7}$$

which represents a linear equation of the form:

$$Y = \beta X + c \tag{8}$$

where:

$$Y = \ln\ln\frac{1}{1-F(t)},\ X = \ln(t-\gamma) \text{ and } c = -\beta\ln\eta \tag{9}$$

It follows that in the probabilistic network having on the ordinate the Y scale (a double logarithmic scale) and on the abscissa the X scale (a logarithmic scale), a distribution of statistical data defined by the Weibull law of parameters $\gamma$, $\eta$ and $\beta$ will be aligned after a line, also called "Weibull line", having the inclination equal to the value of the parameter $\beta$ [16].

But in order to get to the Weibull right, it is necessary to identify (through various graphical or analytical methods) or adopt the location parameter $\gamma$.

The Weibull curve obtained in the first phase is for the case $\gamma = 0$ (the biparametric Weibull model) and its straightening in a straight line (the Weibull line) implies a change of variable (the variable t is changed to the variable t – γ). Thus, the conversion of the curve of the variable t into the line of the variable (t – γ) is realized as in figure 7.

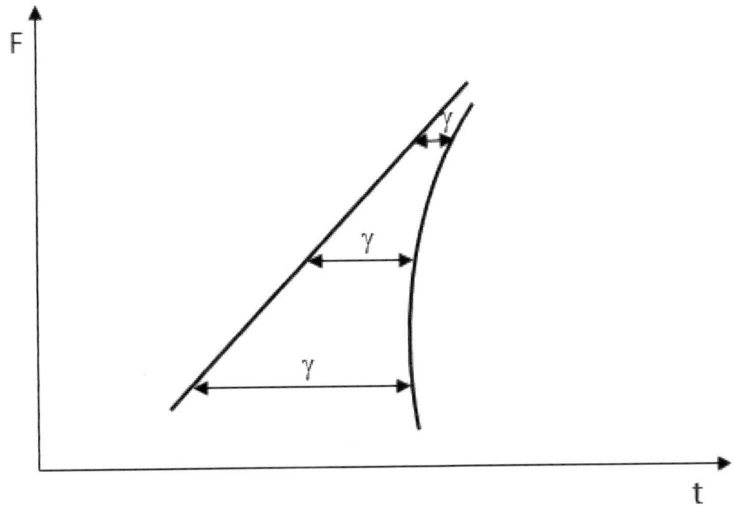


**Figure 7.** Different lengths of the same γ value on the logarithmic time scale.

Since on the abscissa t, which has a logarithmic scale, the same value γ will have different lengths depending on the position of the segment (see also figure 4, represented previously), it is obvious that the correct representation can only be like the one in the figure above, i.e. the difference γ = t - (t – γ) is represented graphically over a shorter and shorter length as t increases, so the Weibull line must necessarily approach the Weibull curve asymptotically.

## 4. Conclusions

The research carried out led to the establishment of the correct representations for the two graphs and to the understanding of the causes that led to obtaining the incorrect representations.

So, the causes that led to these non-compliant graphical representations are identified:

• for the bathcurve graph, that incorrect voltage in the first part of the graph (the period of early failures) appears in the construction of the graph because it is not taken into account that the function z(t) is not defined for t = 0 (as a result, incorrect explanations are forced for failure processes in that area).

• for probabilistic graphs such as the Allan-Plait graph, the Weibull line is not rigorously constructed, so that it faithfully follows the values obtained by changing the variable $t \rightarrow (t - \gamma)$ , that is, the correct conversion of the points on the Weibull curve of the variable t into the Weibull line of the variable (t-γ) is not followed, but simply a right is drawn arbitrarily in the area of the concave part of the Weibull curve.

As these misrepresentations seem to persist, this paper is intended to be a signal of this issue.

## References

[1] * * * *https://en.wikipedia.org/wiki/Literature_review*
[2] Smith C 1976 *Introduction to Reliability in Design* (Tokyo: McGraw Kogakusha Publishing House)
[3] * * * *https://www.sciencedirect.com/topics/engineering/weibull-probability-distribution*
[4] * * * *https://www.weibull.com/hotwire/issue14/relbasics14.htm*
[5] * * * *https://www.quora.com/What-does-a-beta-parameter-of-1-3-on-Weibull-distribution-mean-Alternatively-from-what-value-above-1-does-the-beta-parameter-signify-wear-out-*

*failure-on-the-bathtub-curve*
[6] * * * *https://reliawiki.org/index.php/The_Weibull_Distribution*
[7] Lyonnet P 1988 - *La maintenance. Mathématiques et méthodes* (Paris : Lavoisier Publishing House)
[8] Zwingelstein G 1996 *La maintenance basée sur fiabilité* (Paris : Hermes Publishing House)
[9] Oprean A, Dorin A, Drimer D, Paris A and Ionescu IA 1979 *Fiabilitatea maşinilor-unelte* (Bucureşti: Technical Publishing House)
[10] Panaite V and Munteanu R 1982 *Control statistic şi fiabilitate* (Bucureşti: Didactic and Pedagogical Publishing House)
[11] Andreescu C *https://gradua.ro/cursuri/transporturi/fiabilitatea-si-mentenanta-autovehiculelor*
[12] Burlacu G, Dăneţ N, Brandabur C, Duminică T 2005 *Fiabilitatea, mentenabilitatea şi disponibilitatea sistemelor tehnice* (Bucureşti: Matrix Rom Publishing House)
[13] Boroiu A, Boroiu A-A 2019 *Vehicle reliability* (Piteşti: University of Piteşti Publishing House).
[14] Boroiu A-A, Boroiu A. 2022 *Product reliability. Theory and practice* (Piteşti: University of Piteşti Publishing House)
[15] Sindhu, Tabassum & Atangana, Abdon. (2021). Reliability analysis incorporating exponentiated inverse Weibull distribution and inverse power law. Quality and Reliability Engineering International. 37. 10.1002/qre.2864.
[16] Pedrosa, Bruno & Correia, José & Rebelo, Carlos & Veljkovic, Milan. (2020). Reliability of Fatigue Strength Curves for Riveted Connections Using Normal and Weibull Distribution Functions. 6. 10.1061/AJRUA6.0001081.

IOP Conf. Series: Materials Science and Engineering **1311** (2024) 012034 doi:10.1088/1757-899X/1311/1/012034

# Optimization of the intake system for the Setra NS 913 bus

**C Birtok Baneasa[1], A Budiul Berghian[1], M Ardelean[1], R Bucevschi[1]**

[1]Politehnica University Timişoara, Faculty of Engineering Hunedoara, 5 Revolutiei Street, Hunedoara, Romania

corneliugroup@gmail.com

**Abstract**. The paper presents the implementation on buses of a dynamic air transfer system (SDTA) brand Air by Corneliu. That system is the subject of a patent entitled Dynamic Air Transfer Device RO 200900028. At the initiative of the Alba Iulia public Transport society, with a fleet of about 100 buses, we started a study on the efficiency of air intake in the internal combustion engine by implementing an SDTA. Since the air filter location on this bus model, Setra NS 913, is behind the rear axle wheels, the filter is exposed to dust, water and other particles thrown by the drive wheels, which leads to premature clogging and blockage of the filtration area, in the context in which the bus runs mainly on asphalt-free roads. The gas-dynamic resistance (filter permissiveness) would increase considerably, resulting in a severe reduction in the degree of filling of the engine cylinders. In order to eliminate the disadvantages mentioned above, respectively to increase the life of the air filter with implications on the performance of the engine, we have designed a dynamic system consisting of the following components: the air collector, located on the top of the bus; the grid housing, located on the suction grid; tubing, connecting element between the air collector and the grid housing. In order to realize the 3D model of SDTA, the specialized software VISI - CAD / CAM was used.

## 1. Introduction

At the initiative of the Alba Iulia public Transport society, with a fleet of about 100 buses, we started a study on the efficiency of air intake in the internal combustion engine by implementing a dynamic air transfer system (SDTA) [1].

**Figure 1.** Position of the air filter grille

Because the air filter location (figure 1) on this bus model, Setra NS 913, is behind the rear axle wheels, the filter is exposed to dust, water and other particles thrown by the drive wheels, which leads to premature clogging and blockage of the filtration area. The gas-dynamic resistance (filter

IOP Conf. Series: Materials Science and Engineering **1311** (2024) 012034 doi:10.1088/1757-899X/1311/1/012034

permissiveness) would increase considerably, resulting in a severe reduction in the degree of filling of the engine cylinders [2].

In order to eliminate the disadvantages mentioned above, respectively to increase the life of the air filter with implications for the performance of the motor, we have designed a dynamic system consisting of the following components:

- catcher - located the upsetting part of the bus;
- grid housing – located on the suction grid;
- piping – connecting element between the sensor and the grid housing.

The solution adopted facilitates the aspiration of relatively clean air flow (with minimum impurities) through the filter into the engine cylinders.

In order to achieve the dynamic air transfer system for buses, it was decided to use composite materials.

A very important problem is the rational use of raw materials, from this point of view it is necessary to reduce the mass of the car, in order to reduce the consumption of materials and increase the percentage of its recyclable mass. In 2005, around 85% of car's mass was recyclable and now 95% for new vehicles designed of the car's mass is expected to be recyclable. Composite materials prove to be competitive both in terms of price and the possibilities of successful replacement and/or completion of traditional materials (metal, ceramics, glass) [3], [4], [5].

The dynamic air transfer system components are made of fiberglass or carbon fiber, for this purpose five types of materials with different combinations of materials and thicknesses have been designed and tested. The tests performed aimed at establishing the hardness and breaking strength on the LabTest 6.50 Universal Mechanical Testing Machine within the Materials Laboratory of the Hunedoara Faculty of Engineering.

The adopted solution facilitates the aspiration of the relatively clean air flow, with a minimum of impurities, increasing the availability of the bus to run on unpaved roads without affecting the life of the filter element.

The dynamic air transfer system for buses is realized through the educational program "DEXTER" Laboratory with the support of the CorneliuGroup Research and Innovation Association [6].

## 2. Conditions and measurements

### *2.1. Determination of mechanical characteristics of composite materials*

Five samples were taken with the following characteristics:

• Sample 1 – a layer of 300g/sqm stratimatate + epoxy resin (thickness 1.5mm);
• Sample 2 – two layers of 300g/sqm stratimate + epoxy resin (thickness 2.2mm);
• Sample 3 – two layers of 150g/sqm carbon + epoxy resin (thickness 1.4mm);
• Sample 4 – a layer of stratimatate 300g/sqm + a carbon layer 150g/sqm + epoxy resin (thickness 1.8mm);
• Sample 5 – two layers of 150g/sqm carbon + two layers of 150g/sqm carbon + polyesteric resin (multilayer casting process) (4mm thickness).

Sample size 25x80mm

➢ Determination of hardness

For the determination of hardness, shore A type durometer was used (figure 2). This type of durometer is intended for the following categories of materials: Soft rubber, elastomers, natural rubber products, resin for casting, leather, neoprene, polyester, soft PVC [16].

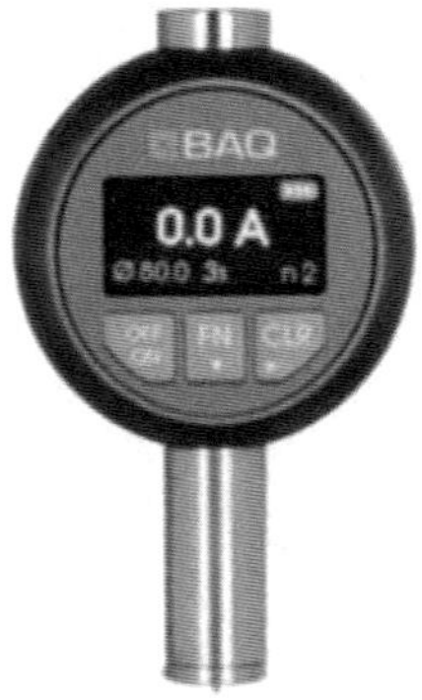


**Figure 2**. Durometer

**Table 1**. Test hardness values

| Nr. | Sample mass [g] | Hardness 1 Shore A | Hardness 2 Shore A |
|---|---|---|---|
| P1 | 2,3 | 92 | 65 |
| P2 | 4,7 | 96 | 82 |
| P3 | 2,2 | 96 | 96 |
| P4 | 3,2 | 97 | 90 |
| P5 | 10,8 | 98 | 98 |

➢ Testing of samples from composite materials

In order to establish fracture strength we used the universal mechanical testing machine LabTest 6,50 (figure 3) of the materials Laboratory of the Hunedoara Engineering Faculty.

Technical data Universal Mechanical testing machine LabTest 6,50:

- Force cell ASTGmbH, KAP-S 50kN;
- software testing;
- set of parallel plane compression plates;
- set of flexible compression plates with spherical joint;
- grips – jaws with pyramidal striations.

**Figure 3.** Universal mechanical testing machine LabTest 6,50

Figures 4 to 7 show the variation in tensile force over time, stating that the test was carried out until the samples were destroyed.

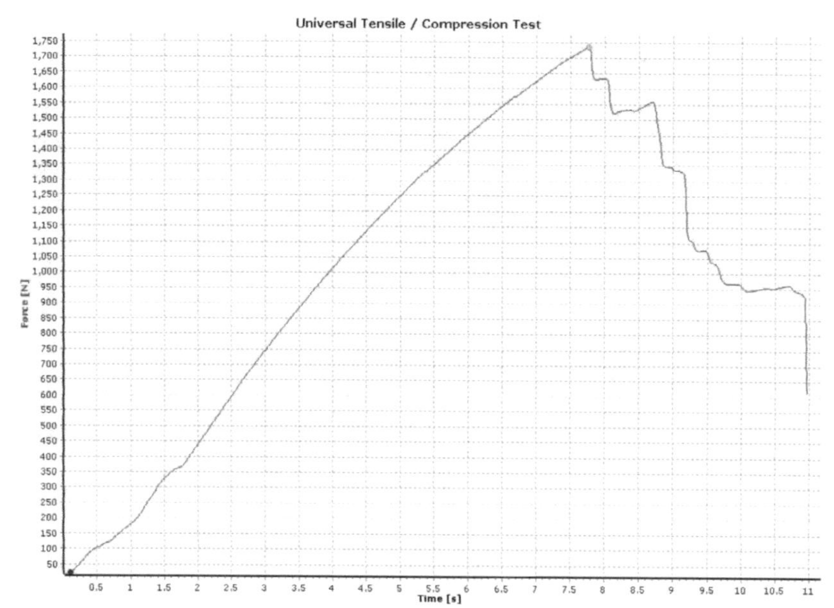


**Figure 4.** Sample 1

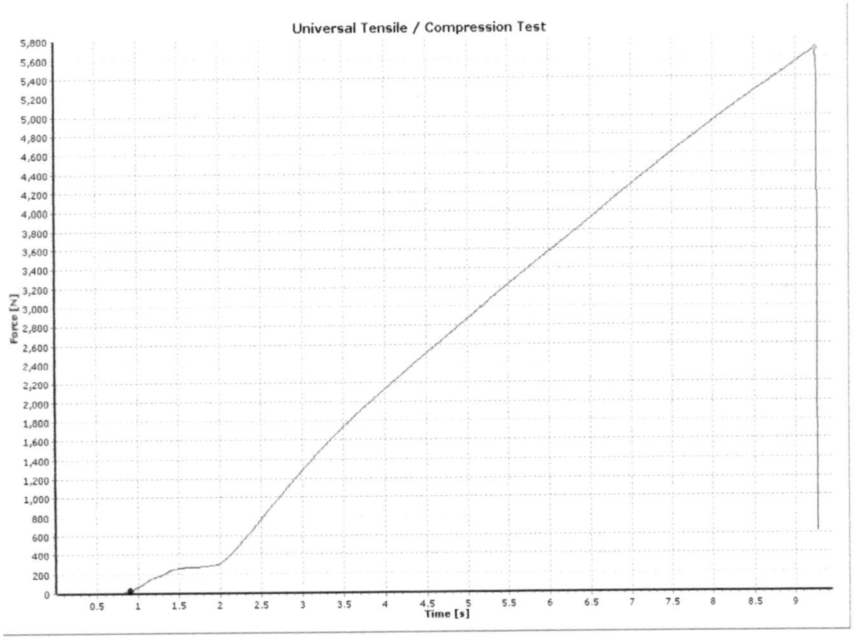


**Figure 5.** Sample 2

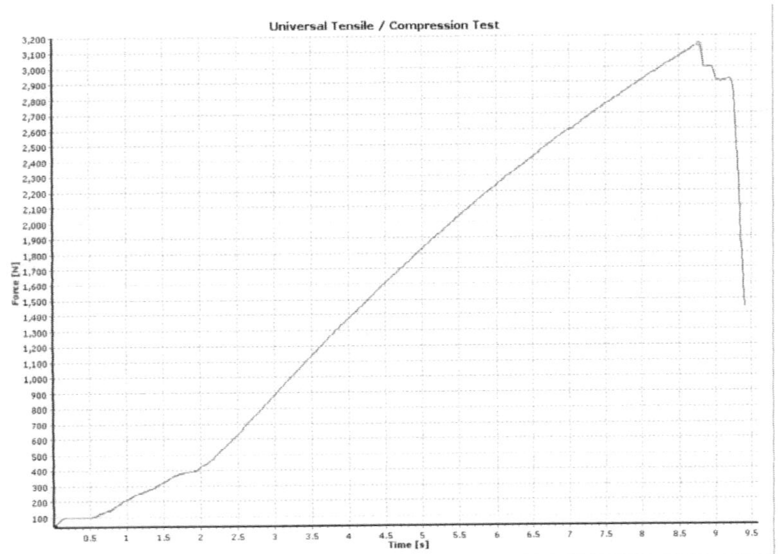


**Figure 6.** Sample 3

IOP Conf. Series: Materials Science and Engineering **1311** (2024) 012034 doi:10.1088/1757-899X/1311/1/012034

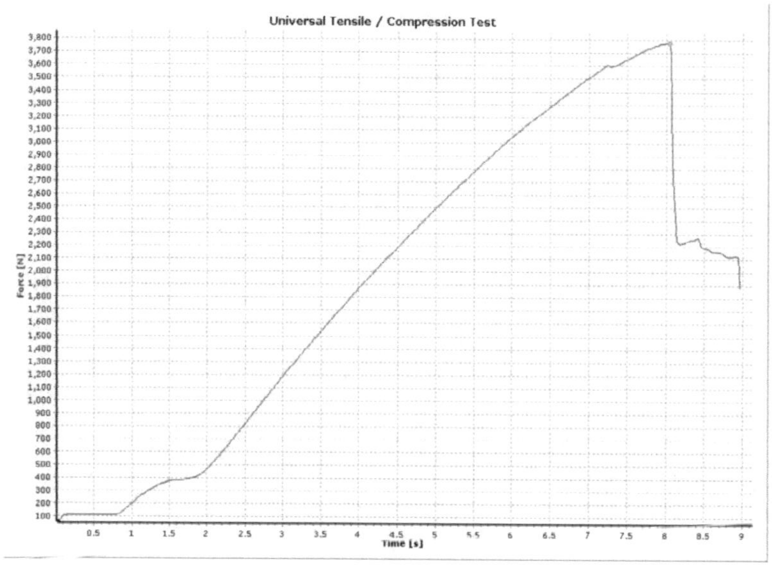


**Figure 7.** Sample 4

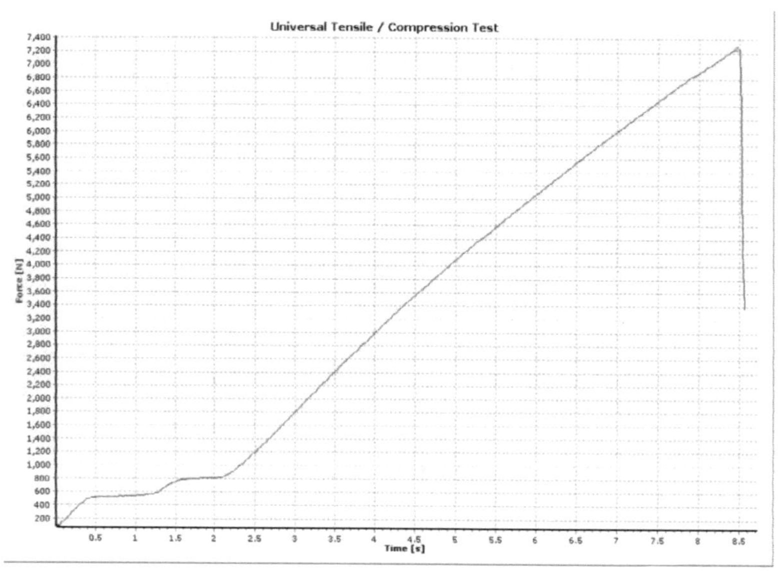


**Figure 8.** Sample 5

The interpretation of the results from the above figures shows that sample 1 was destroyed at a force value of 1700 [N] and sample 5 at a force value of 7300 [N].

The difference in values in terms of breakage force is explained by:
- number of layers;
- the type of material used for the layers;
- the type of binder used – resin;
- the recipe applied.

To study the type of crack, we used an existing stereomicroscope within the materials Laboratory of the Hunedoara Faculty of Engineering. The Stereomicroscope is a high-quality optical system to meet high-precision requirements on modern biology, medicine, scientific research, inspection lines in modern electronics industry and other technological industries.

IOP Conf. Series: Materials Science and Engineering **1311** (2024) 012034 doi:10.1088/1757-899X/1311/1/012034

Figure 9 shows the samples and the cracks recorded after the traction test.

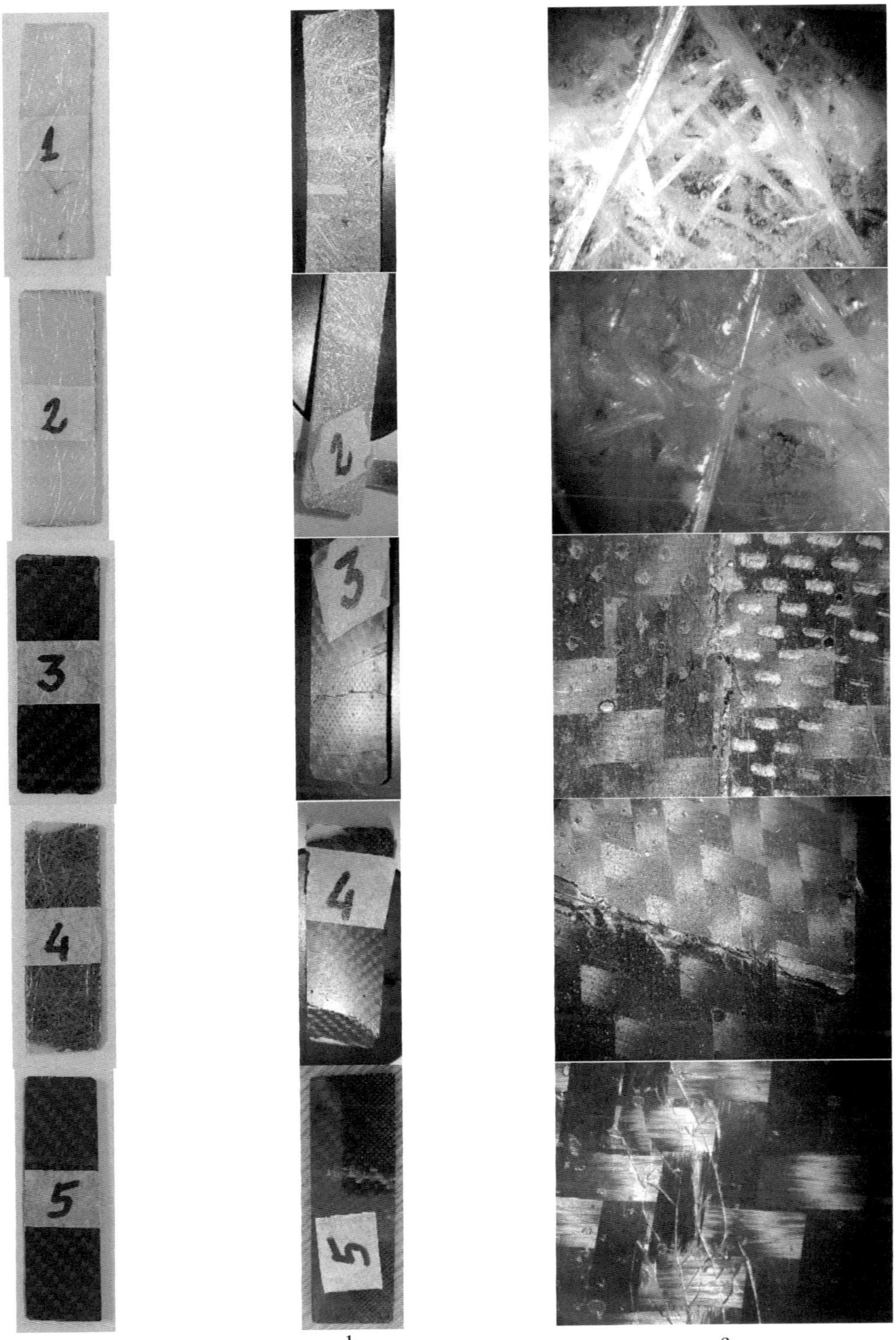

a b c

**Figure 9.** Composite samples
a-sample; b-crack; c-image microscope

IOP Conf. Series: Materials Science and Engineering **1311** (2024) 012034 doi:10.1088/1757-899X/1311/1/012034

*The stages of SDTA construction*
Drawing the template to determine the shape and dimensions (figure 10);

**Figure 10.** SDTA template

**Figure 11.** Obtaining SDTA quotas

Obtaining the template quotas for the 3D model generation (figure 11);
The specialized software VISI - CAD / CAM was used to create the 3D model

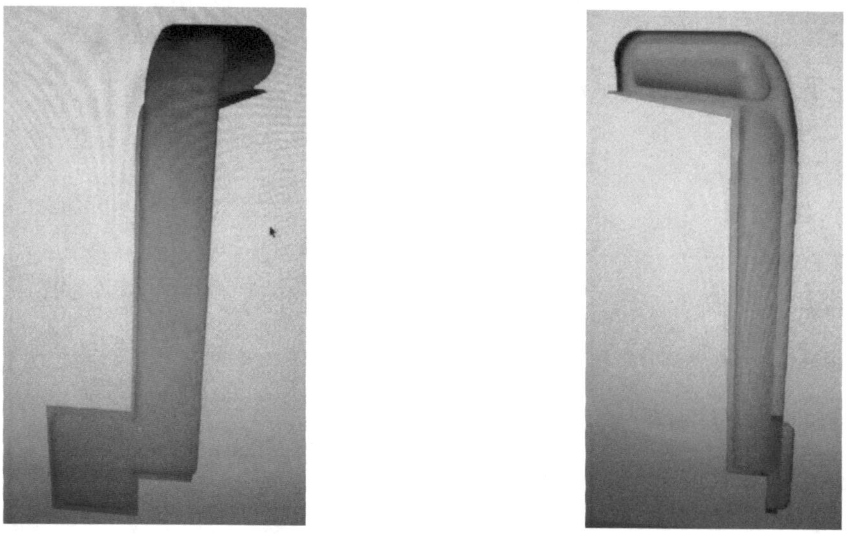

**a** **b**

**Figure 12.** SDTA 3D Model
a-exterior; b-interior

Based on the obtained 3D model (figure 12) the core of the polystirene SDTA was achieved by CNC machining Pfleghar JP 2000 S

Figures 13 and 14 show the sections of the SDTA core;

IOP Conf. Series: Materials Science and Engineering **1311** (2024) 012034 doi:10.1088/1757-899X/1311/1/012034

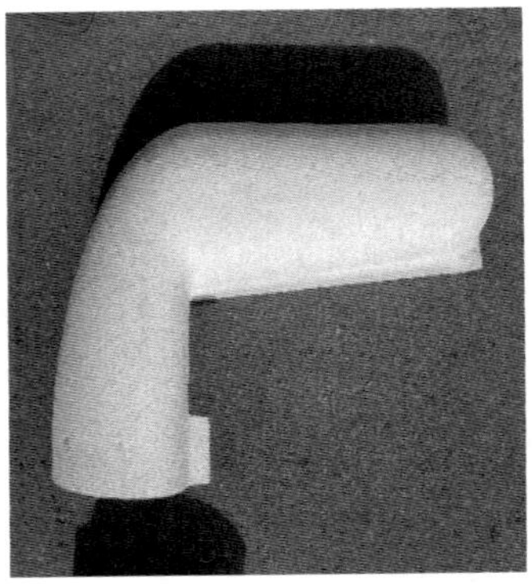

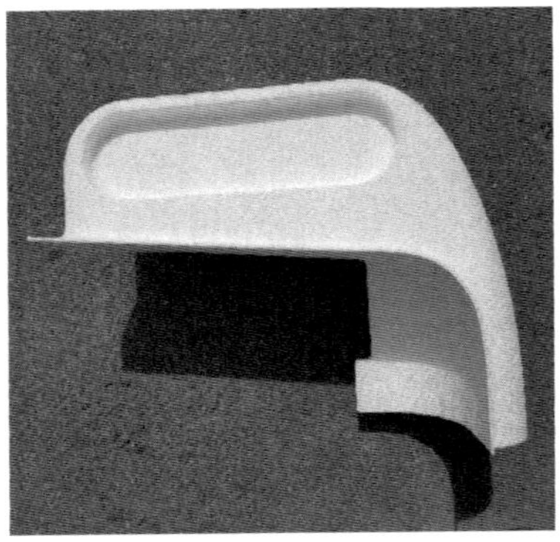

a b

**Figure 13**. Sensor core SDTA
a-outer surface; b-grid

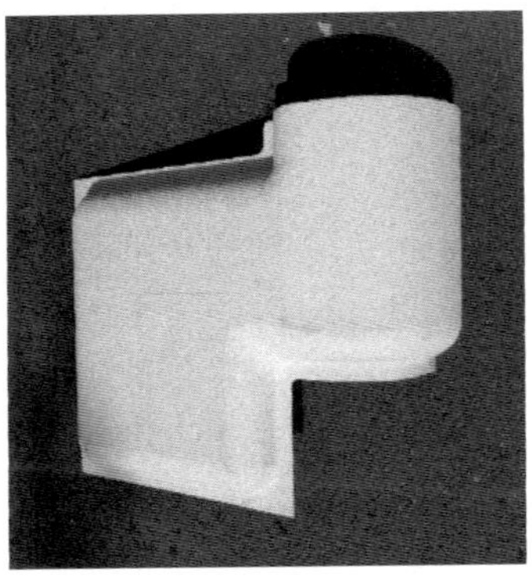

a b

**Figure 14.** Grid housing core
a-outer surface; b-inside

A series of molds shown in Figure 15 have been designed for the execution of the fiberglass SDTA. Mold construction solutions differ depending on the complexity of the components in this case permanent molds were made.

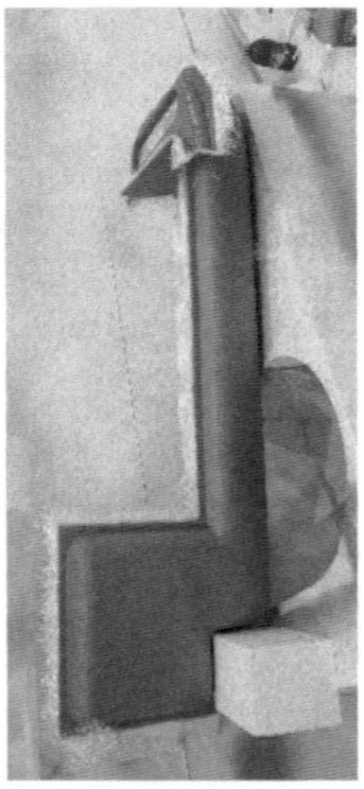

**a**

**b**

**Figure 15.** SDTA molds
a-making mold; b-final mold

The casting of SDTA in previously made molds is shown in figures 16; 17, the material used being glass fibers in tissue form. The stages of realization of components by casting are as follows:

-
- - application of anti-adhesion solution (decrypting);
- - application of the gelcoat;
- - the static application of glass fiber;
- - impregnation of the fiber with rasin;
- - molding material;
- - in the case of double molds, the joint;
- - vacuum to remove air between the casting layers;
- - maintaining at a temperature that allows the glass fiber to solidify;
- - mold removal
- - elimination of burrs and excess material;
- - additional finishing operations (soldering, drilling, cutting, etc.).

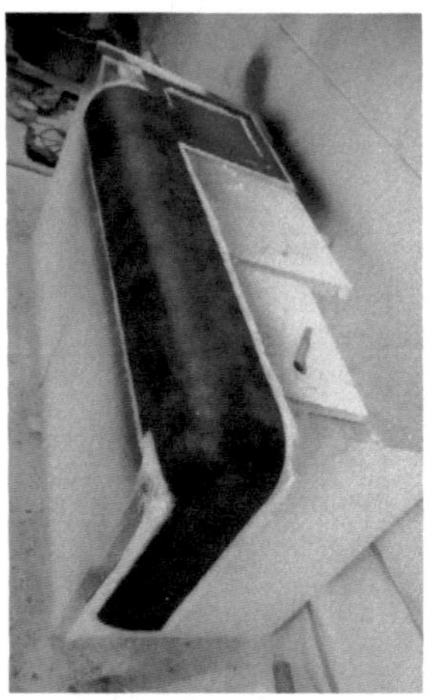

**Figure 16.** Casting SDTA

**Figure 17.** SDTA semi-finished

➢ The final form of the SDTA was obtained by bonding the two components

**Figure 18**. Dynamic air transfer system - mounting on the bus

Final stage: SDTA montage (figure 18) on the Setra NS 319 bus of the public Transport society Alba Iulia.

## 3. Result and Conclusions

Kevlar is a polyamide with superior mechanical properties, being five times more resistant than steel. The strength and modulus of the Kevlar are comparable to those of glass fibers, while the density is reduced by almost half.

The use of carbon fiber glass fibers to make the components of SDTA allows the shape of the constructive geometry to be efficentified while reducing their mass.

The interpretation of the results shows that sample 1 was destroyed at a force value of 1700 [N] and sample 5 at a force value of 7300 [N].

The difference in values in terms of breakage force is explained by:
- number of layers;
- the type of material used for the layers;
- the type of binder used – resin;
- the recipe applied.

The dynamic air transfer system implemented on the Setra NS 319 bus facilitates the aspiration of relatively clean air flow (with minimum impurities), which leads to the extension of the service life, eliminating the risk of premature destruction of the filter element.

**Acknowledgment:** This paper was financially supported by the Project "Network of excellence in applied research and innovation for doctoral and postdoctoral programs / InoHubDoc", project co-funded by the European Social Fund financing agreement no. POCU/993/6/13/153437

## References

[1] Backer, A, Dutton S and Donald Kelly – *Composite Materials for Aircraft Structures;*
[2] Lupescu, M B – *Fibre de Armare pentru Materialele Compozite;*
[3] Birtok Băneasă, C *Cercetări privind utilizarea materialelor avansate în optimizarea procesului de admisie a motoarelor cu ardere internă*, Teză de doctorat, Universitatea Politehnica Timișoara, 2019;
[4] Gheorghe V, *Structuri cu rigiditate ridicată din materiale composite utilizate în construcția*

*autovehiculelor*, Teză de doctorat, Universitatea Transilvania Braşov, 2013;
[5] www.imperial.ac.uk;
[6] Maurer A, *Cercetări privind optimizarea producţiei de componente pentru industria de automobile în condiţiile globalizării*, Teză de doctorat, Universitatea Transilvania Braşov,2015;
[7] Chiru A, Scutaru L, Vlase S, Cofaru C, *Materiale plastice şi composite în industria autovehiculelor*, Editura Universitaţii Transilvania din Braşov, 2010;
[8] http://www.acea.be/statistics/tag/category/key-figures;
[9] Chiru A, Benea B, *Tehnologii noi pentru fabricarea autovehiculelor*, Editura Universitaţii Transilvania din Braşov, 2014;
[10] ISO 14040: 2006, Environmental management - Life cycle assessment - Principles and framework, 2006;
[11] ISO 14044: 2006, Environmental management - Life cycle assessment - Requirements and guidelines, 2006;
[12] https://www.scribd.com/doc/244621227/Materiale-Composite-Folosite-in-Industria-Auto
[13] Ilie A.,: Suba A. F., Cataniciu A., *Drift*, Simpozion Științific Studențesc HD 48 STUD, Hunedoara, 2018
[14] www.corneliugroup.ro
[15] https://www.visicadcam.com/products
[16] https://www.multilab.ro/duritate/durometre_shore.html

IOP Conf. Series: Materials Science and Engineering **1311** (2024) 012035 doi:10.1088/1757-899X/1311/1/012035

# Study concerning methods for managing used engine oil. Analysis of a questionnaire-based survey

**D M Armioni[1], I Ionel[2] and S A Rațiu[1]**

[1] Politehnica University of Timişoara, Faculty of Engineering Hunedoara, Revoluției Street 5, 331128, Hunedoara, Romania
[2] Politehnica University of Timişoara, Faculty of Mechanical Engineering, Mihai Viteazu Street, 1, 300222, Timişoara, Romania

diana.armioni@student.upt.ro

**Abstract**. This paper focuses on methods of managing used engine oil by presenting an analysis of the results of a survey based on a questionnaire designed by the authors, which aims to illustrate the real, objective situation of the management of used engine oil generated by vehicles powered by internal combustion engines. The answers are analyzed for each question separately and both conclusions and implementable objectives that could improve the current situation are highlighted. In order to facilitate the understanding of the current framework, general aspects regarding the management of used engine oil are also presented, with an emphasis on the legislative perspective.

## 1. Introduction

The phrase "used engine oil" generally refers to any semi-solid or liquid used product composed primarily or in part of mineral oil or synthetic hydrocarbons (synthetic oils), whose initial properties have changed over time due to physico-chemical and thermal processes in the internal combustion engine, therefore rendering them unacceptable for future use for the original purpose [1].

The oxidation process of mineral oils results in oxygenated compounds (acids, aldehydes, ketones) as well as resin-type products. Also, under certain conditions of temperature, pressure and in the presence of certain catalysts, by the reaction of the hydrocarbons in the oil with oxygen, insoluble products are formed in the oil that lead to an increase in viscosity. In the same way, resins, under certain temperature conditions, form hard deposits (varnishes) on the surfaces of segments and pistons. These deposits immobilize the rings in the piston channels leading to loss of combustion chamber sealing. Resins, due to deposits, can also cause clogging of the channels and holes of the lubrication system [2]. Oxidizing substances in the oil have a detrimental effect on its characteristics. The resulting deposits reduce the thermal conductivity of the walls of the parts (piston, segments, cylinders) and favor the formation of sludge in the oil bath at low temperatures. Also, the acids resulting from the oil oxidation process have a corrosive effect on the metal parts as well as on the anti-friction layers on the pistons, etc. [3].

Oil degradation also occurs due to its contamination with other substances. Engine oil can be contaminated with fuels, water or coolant, particles or impurities [4]. Contamination with fuels (gasoline or diesel) has the effect of decreasing viscosity, which implies poor lubrication of moving parts. It also lowers the flammability temperature of the oil, which can cause its self-ignition [5]. Contamination with water or coolant leads to emulsification of the oil and reduced lubricity. Also, the viscosity of the oil increases and the additives can be separated from the oil, which leads to the drastic reduction of the lubricating capacity of the oil [6]. Particulate contamination occurs due to incomplete combustion of the fuel and leads to increased oil viscosity. Also, the oil can become contaminated with metal particles (from the parts) or impurities from the air. Oil particles, if they reach the metal surfaces in contact, lead to the intensification of the wear process [7].

In this regard, it is necessary to establish a clear and well-defined management plan, based on the principles of sustainable development, to ensure the introduction of used motor oil into the circular economy. The American Petroleum Institute (API) states that recycling used engine oil requires 50–85% less energy than refining crude oil to generate the same amount of product. Used engine oil is subsequently transformed from hazardous waste into a significant resource and a profitable petroleum base product [8].

Since the number of cars is constantly expanding, the amount of used engine oil generated by them also increases. The purpose of this article is to present the current situation of used motor oil, going through the general aspects related to the legislative framework and the factors involved in its management process and analyzing the answers obtained following a questionnaire-based survey on the management of used engine oil.

## 2. Aspects regarding the management of used engine oil

According to Emergency Ordinance no. 92 of August 19, 2021 regarding the waste regime, published in the Official Gazette no. 820 of August 26, 2021, used oils must be collected separately and treated, giving priority to regeneration or, alternatively, other recycling operations that have an overall equivalent or better environmental result than regeneration [9]. The analysis of the used motor oil management system in Romania requires, first of all, the identification of the main factors involved in the engine oil life cycle. They are illustrated in figure 1.

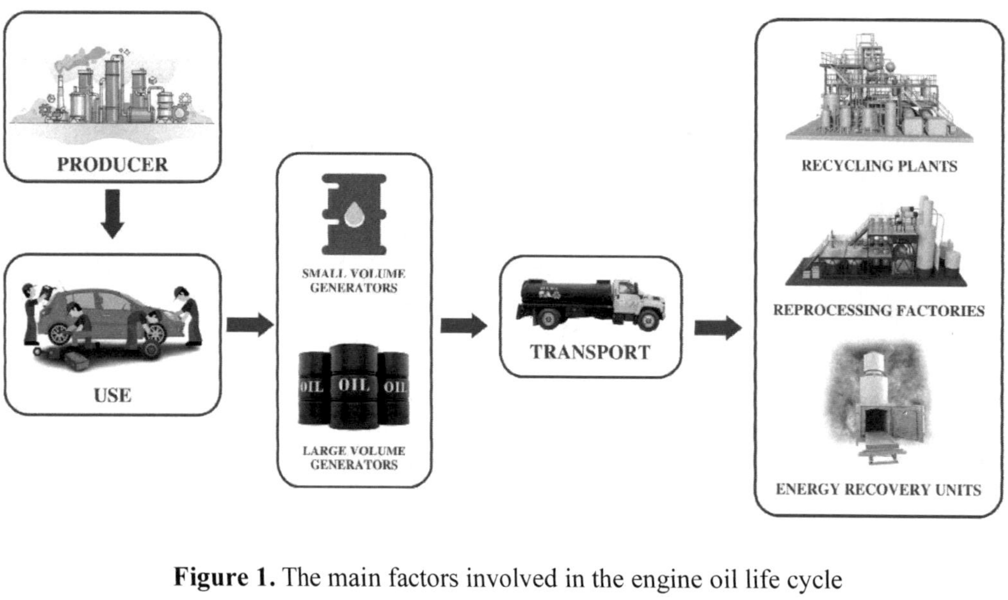


**Figure 1.** The main factors involved in the engine oil life cycle

The National Waste Management Plan (NWMP) mentions the fact that a significant part of the quantities of used oils collected are mainly used as fuel or as another source of energy (including recovery by co-incineration in cement kilns, recovery through other recovery operations, which has the right purpose of obtaining a fuel, valorization through combustion in authorized thermal power plants) [10]. The current scheme of the management of used motor oils in Romania, according to the NWMP, is presented in figure 2.

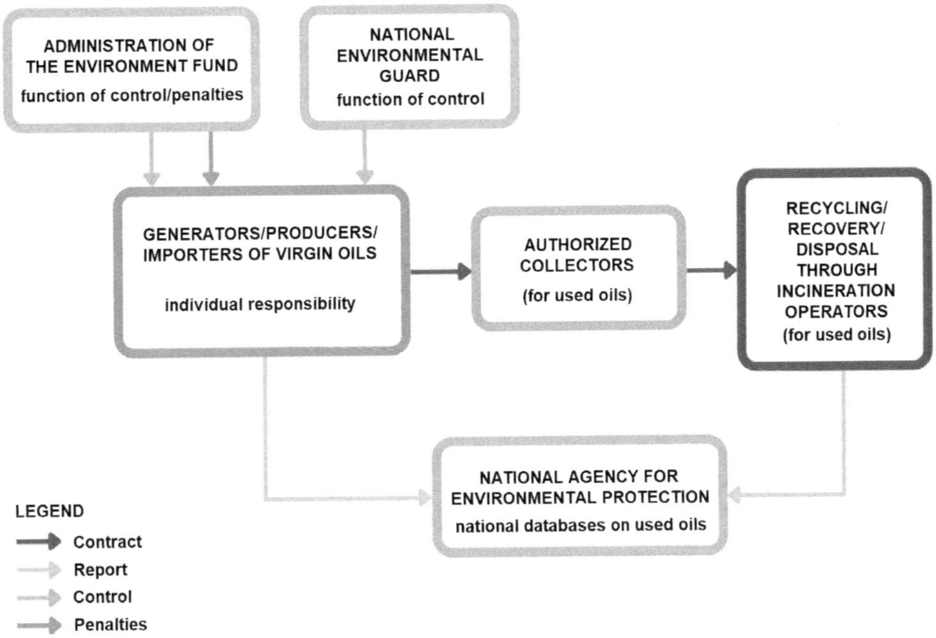


**Figure 2.** The current scheme of the management of used engine oils in Romania

According to Government Decision 856/2002, used motor oil is classified as hazardous waste, with waste code 13 02 05*. According to the Emergency Ordinance 92/2021 regarding the waste regime and the Government Decision 856/2002 regarding the waste management record, used engine oil is irritating (code: H4) and ecotoxic (code H14). If it is used, oil must be recycled or disposed of in an environmentally responsible and secure manner, efficient collection and transportation from the site of generation to recycling or end-use destinations is vital. Thus, for safe handling, adequate airing and ventilation of the workplace and the warehouse, including at ground level, will be ensured. It is mandatory to avoid contact with skin, eyes and clothing. Also, situations of product leakage will be avoided. It will not be mixed with other types of waste [11].

The minimum conditions imposed by the collection companies and the legislation in force, for the collection of used oils from customers are [12]:

I. the condition of admissibility/collection of used oil - water and sediment maximum [1% (collection of used motor oils is done in hermetically closed containers, resistant to mechanical and thermal shock, and storage in properly arranged, fenced and secured spaces, to prevent leaks uncontrolled);

II. it is forbidden to mix this used engine oil with solvents, brake fluid and coolant. It is forbidden to use it as a fuel mixed with diesel;

III. the collection of used engine oils will initially be carried out in the generator's containers, but in time, the collection companies will be responsible for providing containers in the custody of the used oil generator.

Unfortunately, it is not always possible to ensure a proper management of used engine oil, especially for the lubricant from oil changes carried out outside the organized framework (services or authorized representatives). Thus, this oil ends up being used in an inappropriate way and, consequently, it contaminates the environment and becomes a danger to the ecosystem and human health. Table 1 it is the result of an own study based on a series of observations and it shows the main ways of inappropriate applications of used engine oil identified in the Hunedoara area.

IOP Conf. Series: Materials Science and Engineering **1311** (2024) 012035 doi:10.1088/1757-899X/1311/1/012035

**Table 1.** Improper applications of used engine oil

| Applications | Effects on the environment |
|---|---|
| Protecting wood against degradation | Infiltration and soil pollution |
| Rust prevention of ferrous objects | Corrosion and generation of contaminated waste |
| Engine lubricant/transmission fluid | Corrosion, oil leaks and soil pollution |
| Combating weeds and agricultural pests (pesticide/herbicide) | Soil and air pollution |
| Lubricant for hydraulic systems | Oil leaks and soil pollution |
| Fuel in thermal power plants without appropriate filters | Toxic emissions, air pollution |

## 3. Methodology and Results

In order to gain a real perspective on the concrete situation regarding the management of used motor oil in the Hunedoara region, a questionnaire-based survey (figure 3) addressed to the owners of vehicles in this area was carried out. The survey included 6 questions that aimed to touch the essential elements regarding the problem of the used engine oil that is generated.

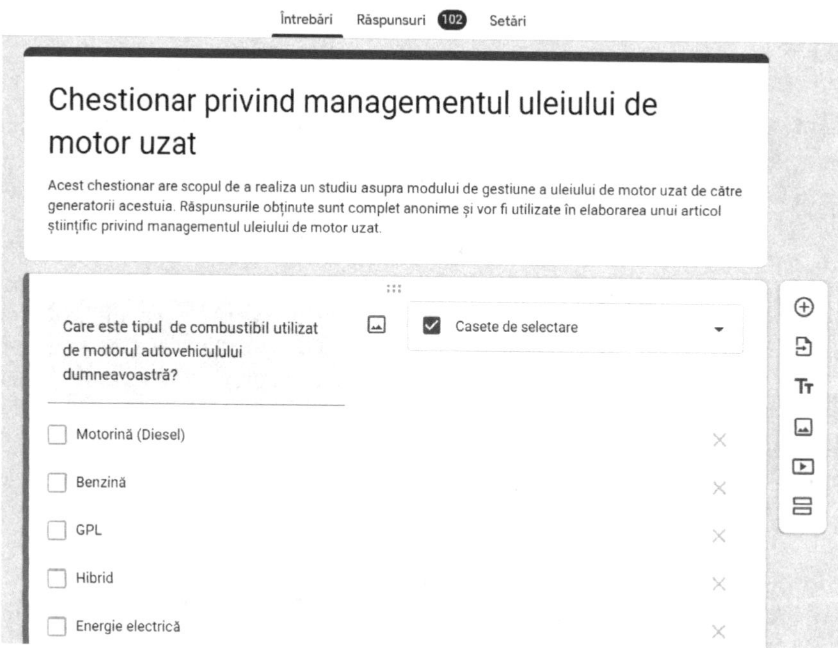


**Figure 3.** Used engine oil management questionnaire addressed to car owners in the Hunedoara area

The questionnaire was created in the Google Forms survey management software and the answers were verified and centralized in an Excel file. The survey in question obtained 102 responses, of which 5 respondents mentioned that they owned at least two vehicles, powered by different types of fuel.

The first question of the survey was "What is the type of fuel used by the engine of your vehicle?". Figure 4 shows the variation of the answers regarding the type of fuel used by the vehicles whose owners answered the questionnaire.

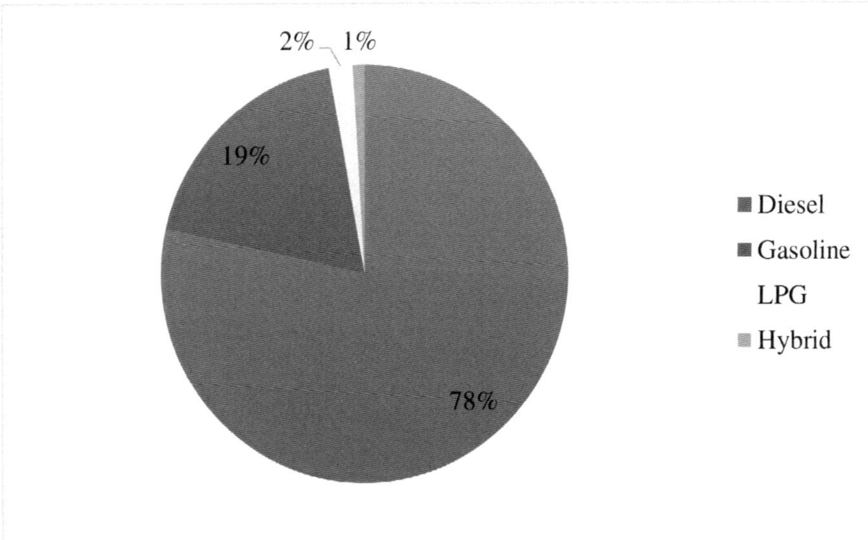


**Figure 4.** The type of fuel used by vehicles, in percentage, as resulted from the conducted survey

According to figure 4, a number of 84 vehicles (78%) use diesel fuel, 20 (19%) are powered by diesel engines, 2 (2%) run on liquefied petroleum gas (LPG) and one vehicle (1%) is hybrid. This is due to the fact that in the Hunedoara area, the car fleet is not yet updated according to the latest automotive trends and diesel vehicles still predominate.

The second question of the survey was "What is the presumed interval (expressed in km) at which one changes the engine oil?".

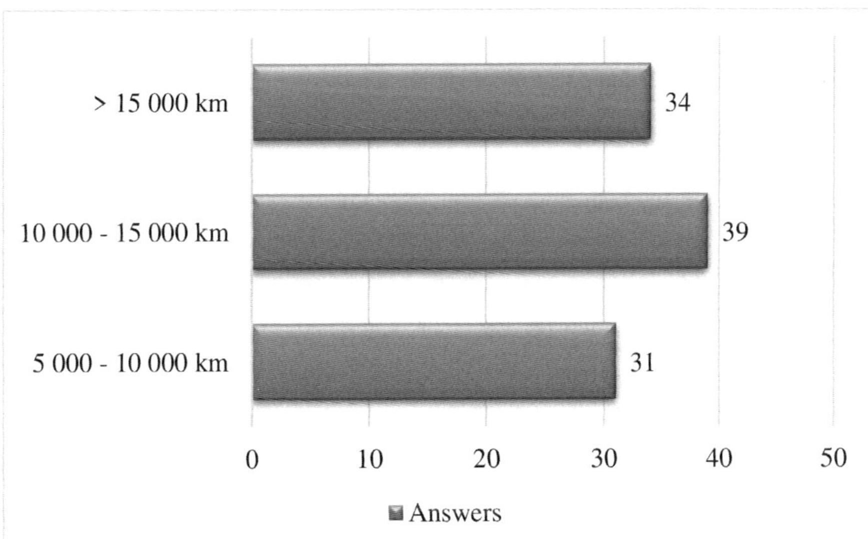


**Figure 5.** The interval for the oil change, in kilometers, as resulted from the conducted survey

As it can be seen in figure 5, a number of 31 respondents confirmed that they changed the oil between 5,000 and 10,000 kilometers traveled with the vehicle. The largest number of responses (39) were recorded for the 10,000 – 15,000 km interval, and 34 responses mentioned changing the oil after more than 15,000 km traveled.

The third question of the survey was "What is the time interval (months) at which one changes the engine oil?"

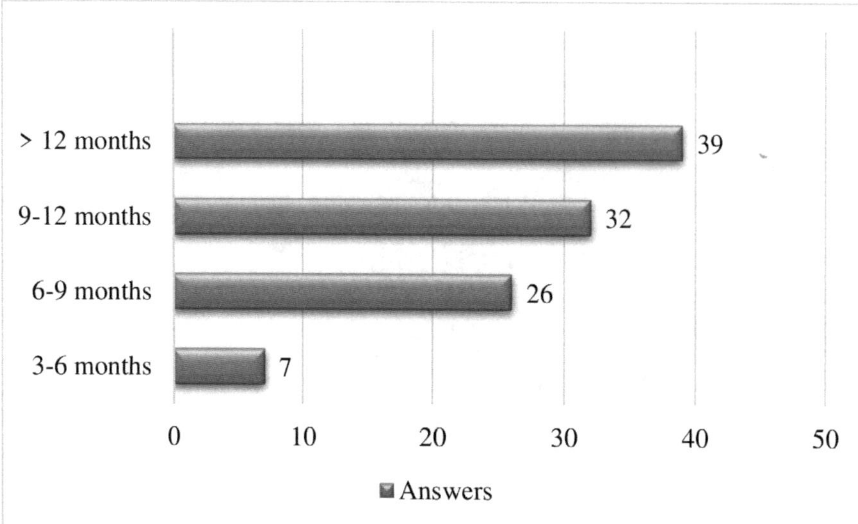


**Figure 6.** The interval for the oil change, in months, as resulted from the conducted survey

In figure 6 it can be seen that most answers (39) highlighted a period of more than 12 months between two consecutive engine oil changes. A number of 32 respondents confirmed that they change the oil at intervals between 9 and 12 months, while 26 answers fell within the interval of 6-9 months. The lowest number of responses (7) was for the time interval 3-6 months.

The 4th question of the survey required textual completion by the respondents (unlike the rest of the questions, which were multiple-choice) and was "What is the amount of engine oil (liters) needed for a change of oil in the vehicle?"

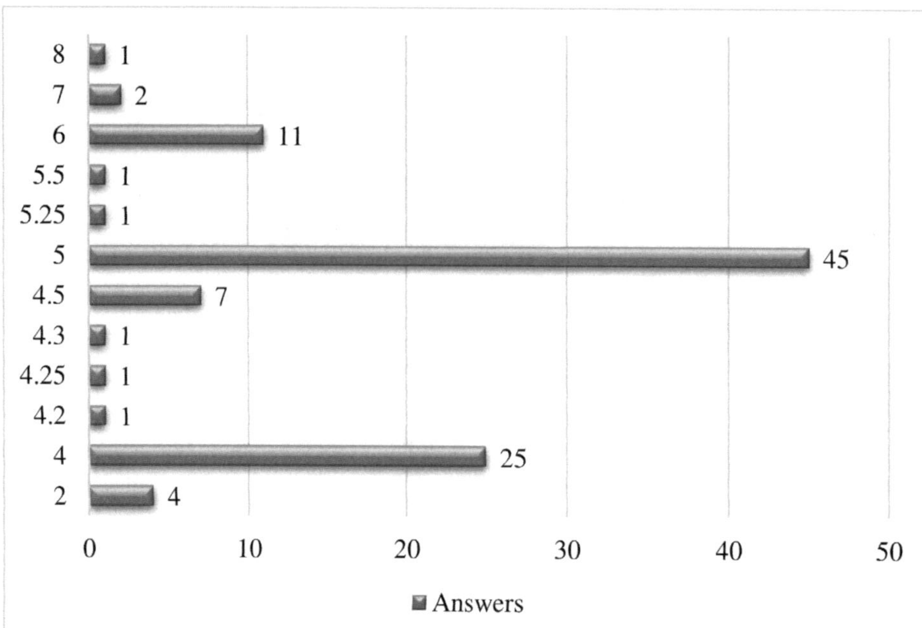


**Figure 7.** The amount of lubricant required for a change of oil, in liters, as resulted from the conducted survey

Figure 7 shows that the largest amount of engine oil used during a change by the vehicles whose owners participated in the survey is 8 liters, while the smallest is 2 liters. However, these extreme values do not occur frequently. The most frequently encountered quantities are 5 liters (45 answers) and 4 liters (25 answers).

The 5th question of the survey was " Where is the oil change for the vehicle performed?".

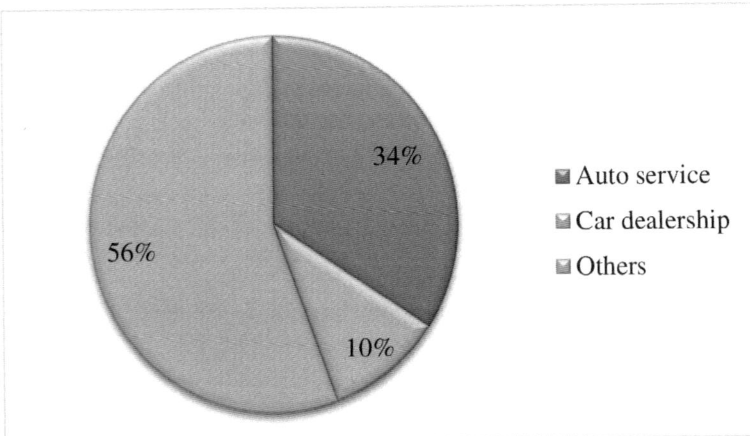


**Figure 8.** The place where the oil change is performed, in percentage, as resulted from the conducted survey

Surprisingly, figure 8 illustrates the fact that a remarkable number of oil changes (60 responses, 56% of the total) take place in places or spaces that are not authorized. Compared to these values, a number of 37 oil changes (34%) are performed in car services, respectively only 11 changes in car dealerships (10%). This denotes the fact that the quantities of lubricant resulting from exchanges carried out in places other than the authorized ones are not officially registered and it is difficult to follow the management of the entire volumes of used engine oil generated from these activities.

The last question of the questionnaire was not mandatory, as it was addressed mainly to the respondents who, in the previous question, confirmed that they change oil in places other than those authorized in this regard. The question was "If the oil change is not performed at a car service/dealership, what is the way of managing the used engine oil resulting from the oil change?"

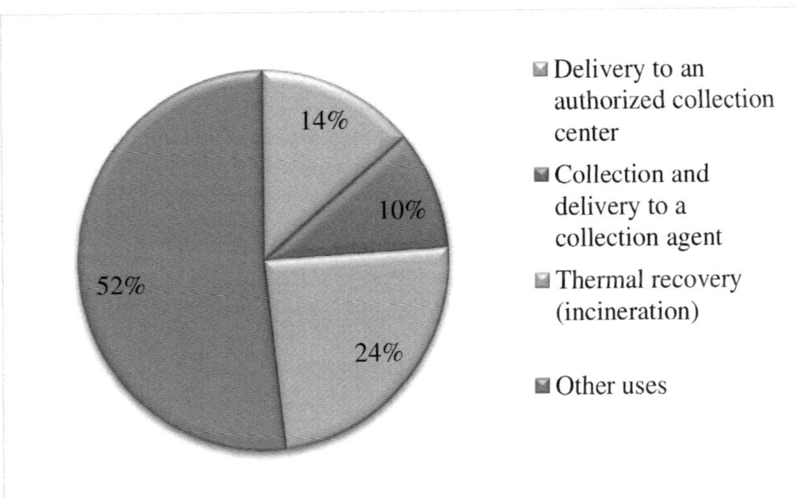


**Figure 9.** Management options for the resulting used oil, in percentage, as resulted from the conducted survey

IOP Conf. Series: Materials Science and Engineering **1311** (2024) 012035 doi:10.1088/1757-899X/1311/1/012035

As seen in figure 9, the percentages of used motor oil properly managed (by sending it to collection centers or handing it over to an authorized collection agent) are low (14% of respondents ticked "handover to a collection center" authorized collection" and 10% chose "collection and delivery to a collection agent"). A percentage of 24% confirms that the used engine oil is incinerated as fuel, ensuring energy recovery, while the remaining 52% of the answers selected "other uses", a worrying aspect regarding the quality of the environment and the degree of pollution at which it is subject to, in case the used engine oil ends up, by various means, in the soil, water or the atmosphere.

As a result of this analysis, it proves necessary to implement measures at the local level to improve the situation of used engine oil management. In this regard, a series of objectives can be stated for the regional environmental authorities to be considered. Thus one concludes that it is necessary to:

- provide direction and facts to all used oil generators regarding what is and is not an acceptable application for used oil;
- make available reliable and secure methods for used oil collection and transportation;
- encourage lube oil retailers to set up used oil collecting centers that are accessible to both their customers and the public;
- convince small-scale oil generators to bring their used oil to nearby collection points;
- avoid contaminating used oil with unwanted substances;
- ensure that recyclers and end users handle used oil properly.

## 4. Conclusions

Engine oil is a vital element for the proper functioning of vehicles. Following the degradation process during its operation within the engine's lubrication system, the oil loses its characteristic properties and becomes contaminated with various substances, most of which are dangerous pollutants for the environment. For this reason, it is necessary to integrate some eco-responsible management strategies, in accordance with legislative norms and the principles of sustainable development.

Following the analysis of the results obtained in the questionnaire-based survey on the management of used engine oil, it can be seen that most vehicles have an oil change between 10 000 and 15 000 km and more than 12 months in traffic. Regarding the amount of oil that a change requires, most answers indicated the values of 5 liters and 4 liters, with an average value of 4,76 liters/change. A considerable percentage (56%) of the respondents indicated that the place where the oil change is performed is not officially authorized in this sense (for instance, a car service or dealership), which includes the risk of a faulty management of the resulting used engine oil. With regard to the finality of the life cycle of the used engine oil resulting from an exchange, a total of only 24% of the respondents confirmed that they hand over the used engine oil to collecting companies or that they deliver it to an authorized collection center. A large number (52%) chose the option of other uses, which brings risks of environmental pollution and economic losses, as the used oil could be recycled and reintroduced into the economic circuit. The rest of the answers fell under the option of energy recovery through incineration. The questionnaire was filled in by car owners or users living in the region of Hunedoara. Similar methodology could be applied for different regions as well.

## References

[1] *** Technical Working Group of the Basel Convention 1995 *Basel Convention. Technical Guidelines on Used Oil Re-Refining Or Other Re-Uses of Previously Used Oil (R9)*, Châtelaine, Switzerland

[2] Hamawand I, Yusaf T and Rafat S 2013 Recycling of Waste Engine Oils Using a New Washing Agent, *Energies* **6** 1023-1049

[3] Raţiu S and Alexa V 2017 *Sisteme auxiliare ale autovehiculelor rutiere*, Editura Politehnica Timişoara, Romania

[4] Ali M K et al. 2017 Effect of Lubricant Contaminants on Tribological Characteristics During Boundary Lubrication Reciprocating Sliding, *Appl. Phys. Cornell Uni.* 1710 04448

[5] Addison J A and Needelman W M 1986 *Diesel Engine Lubricant Contamination and Wear*, Scientific and Laboratory Services Department Pall Corporation, New York, USA

[6] Raţiu S A, Mihon N L, Laza I and Armioni D M 2020 A review on the contamination of used engine oil, *Ingineria Automobilului* **57** 20-25

[7] Aljabiri N A 2018 A Comparative Study of Recycling used Lubricating Oils using Various Methods, *Journal of Scientific and Engineering Research* **5**(9) 168-177

[8] *** American Petroleum Institute 1997 *Study Of Used Motor Oil Recycling In Eleven Selected Countries*, Washington, D.C., USA

[9] *** Guvernul României 2021 *Ordonanţă de Urgenţă nr. 92 din 19 august 2021 privind regimul deşeurilor* https://legislatie.just.ro/Public/DetaliiDocumentAfis/245846, accessed on August 29th 2022

[10] *** Ministerul Mediului, Apelor şi Pădurilor 2018 *Planul Naţional de Gestionare a Deşeurilor* http://mmediu.ro/app/webroot/uploads/files/2018-01-10_MO_11_bis.pdf, accessed on August 29th 2022

[11] *** Guvernul României 2002 *Hotărâre nr. 856 din 16 august 2002 privind evidenţa gestiunii deşeurilor şi pentru aprobarea listei cuprinzând deşeurile, inclusiv deşeurile periculoase* https://legislatie.just.ro/Public/DetaliiDocument/38294, accessed on August 29th 2022

[12] *** Autonet Group 2022 *Condiţii de colectare ale uleiurilor şi filtrelor de ulei uzate* https://www.autonet-group.ro/Reciclare-la-Uleiuri-filtre-uzate-, accessed on August 29th 2022

IOP Conf. Series: Materials Science and Engineering **1311** (2024) 012036 doi:10.1088/1757-899X/1311/1/012036

# Study of Alternatives to Manufacturing of Long Pipes with Many Holes

**DI Poiana, I Ionel, M Vasilescu and I Vetreş**

Politehnica University Timisoara, Faculty of Mechanical Engineering, Piata Victoriei no. 2, postal code 300006, Timisoara, county Timis, Romania.

dinu.poiana@student.upt.ro, ioana.ionel@upt.ro

**Abstract:** In industrial manufacturing, certain long, thin-walled pipe-type parts that have many holes arranged along their longitudinal axis are required. In the automotive industry, such thin-walled long pipe parts are needed for body frames, for heavier component supports, for protection bars, etc. Frequent cases are encountered, for example, in the manufacture of modular shelves for storage. In these cases, long pipes, up to 6 meters long, with thin walls are used. Such long thin-walled pipes, which have many holes arranged along them, sometimes even equally spaced between them, are frequently made with a manufacturing technology that has several characteristic steps: 1) making holes in sheet metal strips, 2) bending the strips and welding them lengthwise, resulting in perforated pipes having a rectangular or square section. In this paper, two variants of obtaining pipes with holes arranged along the length of the pipe, but which are obtained from laminated pipes with thin walls of rectangular section, are proposed and analyzed. The characteristics of obtaining holes in these pipes are presented and the advantages and disadvantages of the proposed solutions are analyzed. The finite element method was also used for the analysis using SolidWorks software.

## 1. Introduction

The study aims to:

- finding alternative solutions to the manufacture of thin-walled long pipes with many holes,
- determining the characteristics of the proposed solutions,
- Finding the advantages and disadvantages of the proposed solutions compared to the usual current solutions.

Currently, a technology frequently used in the industry for the manufacture of long pipes with many holes consists of perforating some sheet metal strips, which are then bent and welded lengthwise (for example, for the manufacture of modular shelves). But also in the automotive industry, such thin-walled long pipe parts are needed, for body frames, for heavier component supports, for protection bars, etc. Such long thin-walled pipes, which have many holes arranged along them, sometimes even equally spaced between them, are frequently made with a manufacturing technology that has several characteristic steps:

1) making holes in sheet metal strips,

2) bend the strips and connect them lengthwise, resulting in perforated pipes having a rectangular or square section.

For cutting of materials, oxygas cutting, laser cutting, and plasma cutting are widely used today. These procedures are already known in the industry and are described in the specialized literature. The use of these procedures is explained in many works. The plasma and laser cutting is used industrially

IOP Conf. Series: Materials Science and Engineering **1311** (2024) 012036 doi:10.1088/1757-899X/1311/1/012036

and in various forms of automation [1]. Both laser cutting [2], [3] and plasma cutting [4], [5] are widely used in non-automated forms. In recent years, high-precision plasma cutting has been perfected [6].

The question arises whether these pipes could be made by perforating laminated pipes of different sections, annular, square, or rectangular, in which case the bending and welding of sheet metal strips would be eliminated.

The proposed solutions for obtaining holes in laminated pipes are as follows:

- making holes in long rolled pipes, by plasma or laser perforation with the help of robots,
- Obtaining holes in long rolled pipes, by plastic deformation, by perforating with a single edge, using punches with different cutting edges.

## 2. Solution 1. Perforation of laminated pipes with the help of industrial robots

In industry, the performance of industrial robots has already improved a lot, thanks to the acquisition of some components of these robots with clearly superior characteristics [1]. These technically advanced components are mainly: motors, actuators, transducers, and control systems. With the help of these technically advanced components, plasma cutting precision has become very high, of ± 0.2 mm [6], [7]. The technical conditions for making precise cuts are as follows:

- bearing capacity, at the level of 60-80 N.
- The control system of industrial robots, to ensure the movement along the cutting path by contouring and the possibility of starting and stopping cutting when needed.

The control part of industrial robots (RI) provides mainly the movement along the continuous cutting path [7] and the start/stop of the cutting.

Figure 1 shows a schematic diagram of the controls and supply of an oxygas cutting head in the case of robotic cutting/drilling [8].

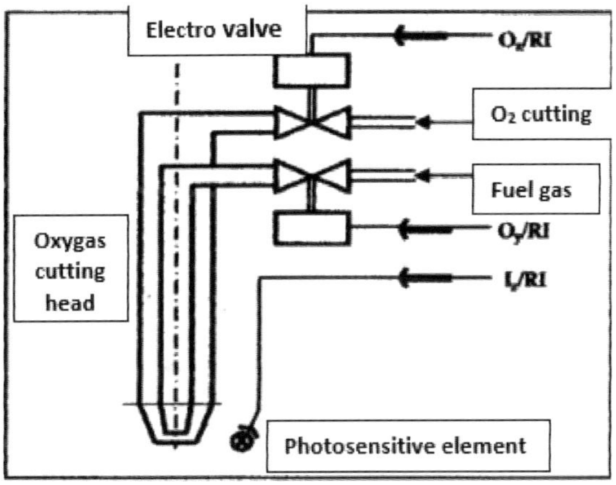


**Figure 1.** Schematic diagram of the controls and supply of an oxygas cutting head in robotic cutting/drilling [8].

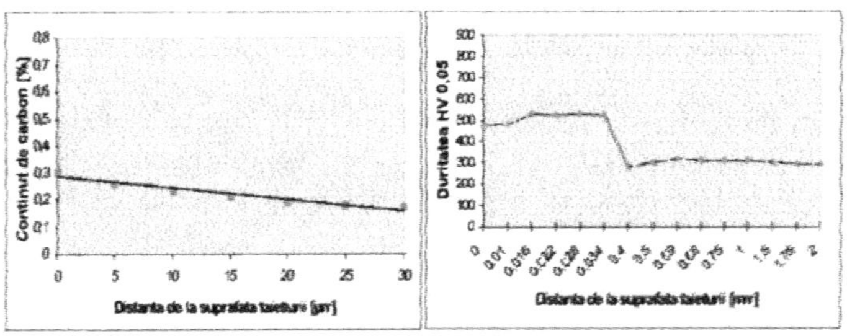

**Figure 2.** a) Changes at the cutting edge in the laser and oxygen

IOP Conf. Series: Materials Science and Engineering **1311** (2024) 012036 doi:10.1088/1757-899X/1311/1/012036

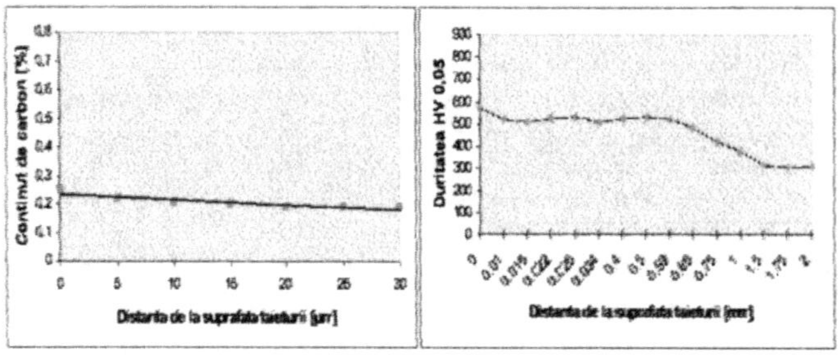


**b)** cutting process (a), respectively, in the plasma and oxygen cutting process (b) [8]

When using plasma, this means connecting/disconnecting the power source to the plasma arc. In the case of cutting with oxygas, the robot will control, according to a program, the electro valves that supply oxygen, acetylene, and methane, and in the case of cutting with a water jet, it will control the water valve under pressure.

Figure 2 shows changes at the cutting edge in the laser and oxygen cutting process (a), respectively, in the plasma and oxygen cutting process (b) [8].

## 3. Solution 2. Obtaining holes by punching with a single cutting edge

In this case, a pair of active cutting edges is not used, but only the cutting edges of a punch. It can be practically demonstrated that, in the case of using a single cutting edge, with a punch, in the case of pipes, a phenomenon of slight displacement/local deformation of the pipe wall subjected to perforation appears in the phase before cutting.

Long pipes perforating using a single cutting edge can be studied using the finite element method using virtual process simulation programs. One of these programs is SolidWorks. With the help of this program, the tension, displacement, and deformation limit situations can also be studied, as well as the different cases in which certain parameters of the drilling process can be modified [9].

Figure 3 shows a rectangular pipe made with holes, Figure 4 shows a type 1 punch to perforate the pipe, and Figure 5 shows a type 2 punch, which differs in the shape of the cutting edges.

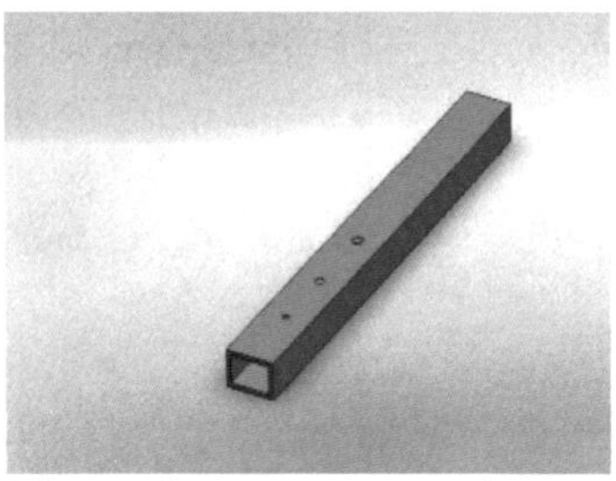
**Figure 3.** Rectangular pipe with holes.

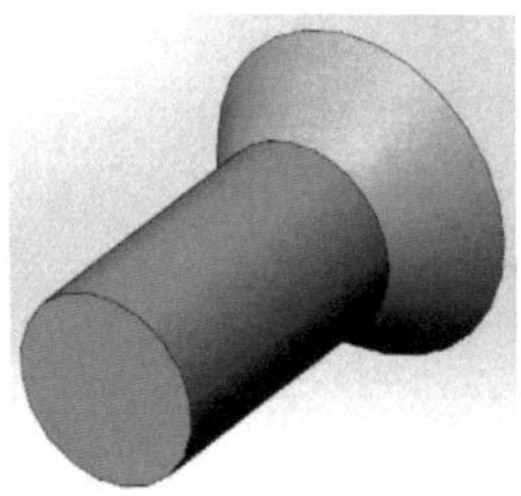
**Figure 4.** Punch **Type** 1.

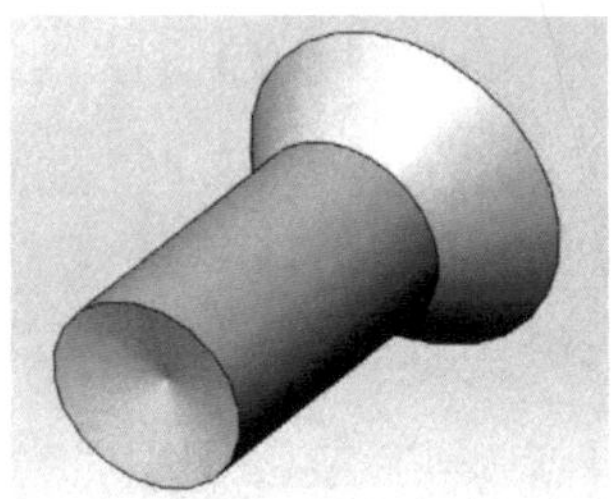
**Figure 5.** Punch Type 2.

## 4. Research and results obtained during the simulation of the perforation process of some pipes with a rectangular section

Research was carried out to simulate the process of perforating a rectangular pipe by making a hole with a diameter of 6 mm on the face of the piece with a size of 40 mm. Only the pipe length in the area of the hole to be perforated was considered, that is, 60 mm in length for the area of perforation. Considerations regarding the stability zone of perforated pipes presented in the articles [10], [11].

The results obtained during the perforation simulation with a single edge, for the case of rectangular pipes with a section of 40 x 30 x 3, made of different materials taken from the SolidWorks (SW) program database, are presented in Figures 6-14. In these figures, the images resulting from the simulation of the punching process using the SolidWorks program are presented, without considering the use of the different possible cutting edges of the punches.

IOP Conf. Series: Materials Science and Engineering **1311** (2024) 012036 doi:10.1088/1757-899X/1311/1/012036

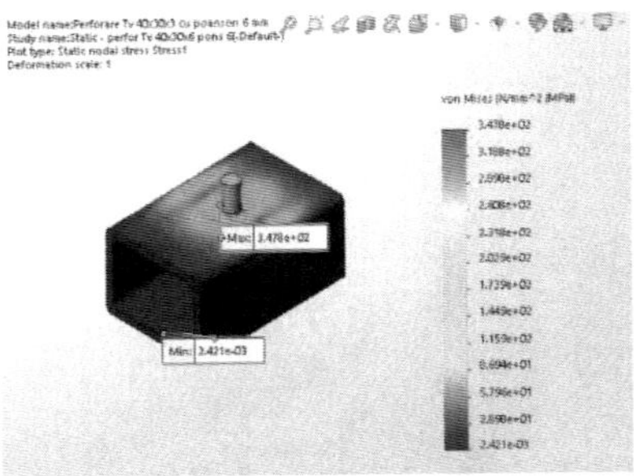

**Figure 6.** The result of the stress analysis (-von Mises-) on the aluminum alloy part.

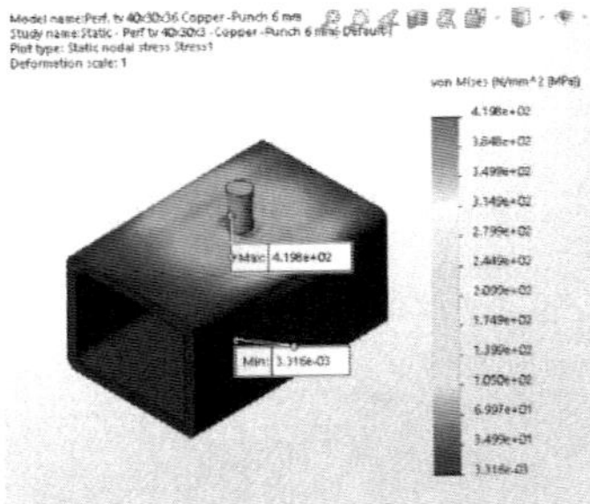

**Figure 7.** The result of the stress analysis (-von Mises-) on the copper part.

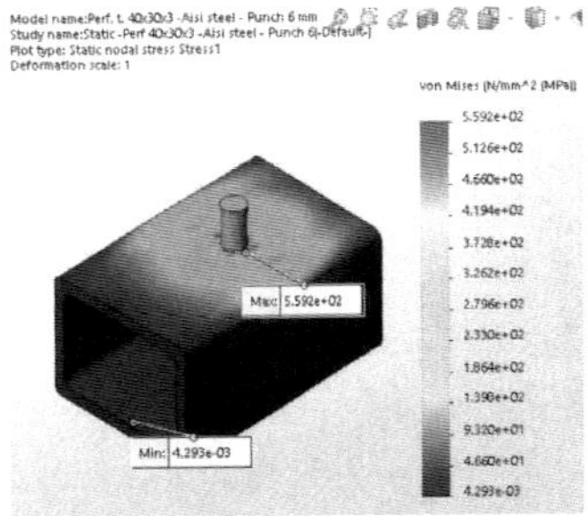

**Figure 8.** The result of the stress analysis (-von Mises-) on the steel part.

The data obtained from the analysis of the perforation process, with the help of the finite element method, demonstrate the possibility of using this technological variant to obtain holes in long thin-walled pipes with a maximum pipe wall thickness of 4 mm.

The results of the research made to determine the stresses in the piece are shown in figures 6, 7 and 8. Figure 6 shows the stress analysis on the aluminum alloy part. Figure 7 shows the stress analysis on the copper part. Figure 8 shows the stress analysis of the steel part.

The results of the research performed for the determination movements in the parts are shown in Figures 9, 10 and 11. Figure 9 shows an analysis of the displacements on the aluminum alloy part. Figure 10 shows an analysis of the displacements in the copper part. Figure 11 shows the analysis of displacements in the steel part.

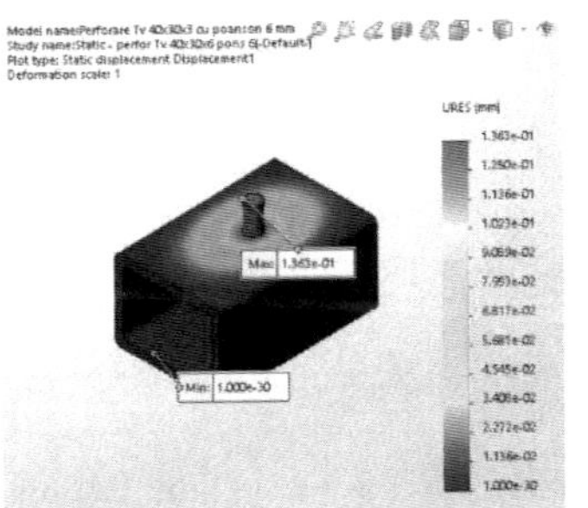

**Figure 9.** The result of the analysis of the displacements on the aluminum alloy part.

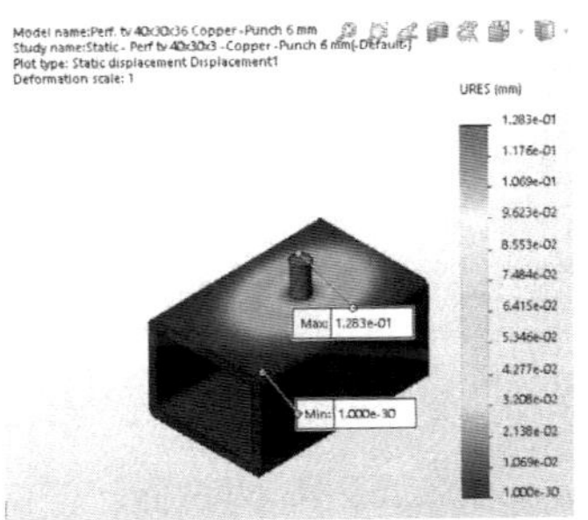

**Figure 10.** The result of the displacement analysis of the copper part.

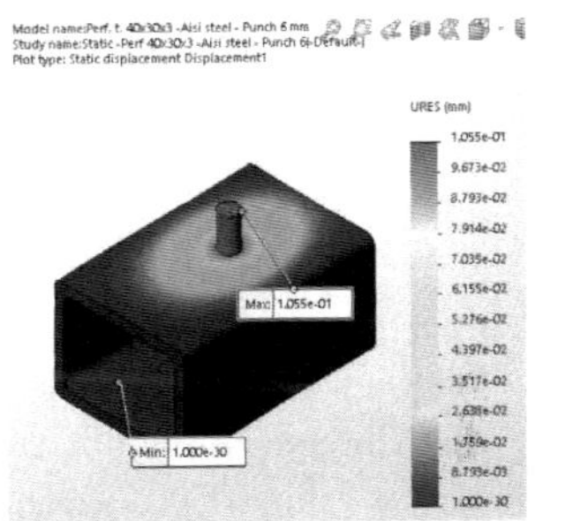

**Figure 11.** The result of the analysis of the displacements on the steel part.

The results of the stress strain (equivalent) are shown in Figures 12, 13 and 14. Figure 12 shows the analysis of the stress strain (equivalent) on the aluminum alloy part. Figure 13 shows the analysis of the stress strain (equivalent) in the copper part. Figure 14 shows the stress strain (equivalent) analysis in the steel part.

Data from the analysis of the perforation process, with the finite element method, demonstrate the possibility of using this technological variant to obtain holes in long thin-walled pipes with a maximum pipe wall thickness of 4 mm.

The research was done for a case of a rectangular pipe with dimensions 40x30 (mm) with a wall thickness of 3 mm, for 3 types of materials, aluminum alloy, copper, and steel. Research was carried out to study the stability behavior of the piece in the case of perforation with a punch with an active diameter of 6 mm.

For the punch action, pressures corresponding to the three types of materials investigated were used: aluminum alloy, copper, and steel. The characteristics of the materials were considered to be those found in the database of the SolidWorks program. The pressures used for the punch action were considered:

- 100 MPa for aluminum alloy material
- 150 MPa for copper

- 200 MPa for steel

Conclusions The action of the punch did not endanger the stability of the pipe.

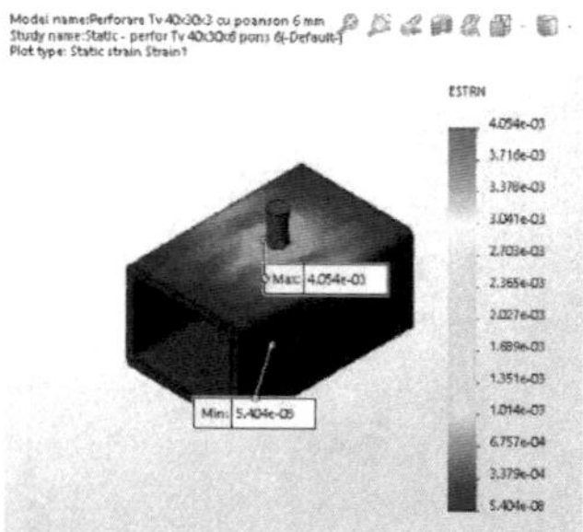

**Figure 12.** The result of (equivalent) strain analysis on the aluminum alloy part.

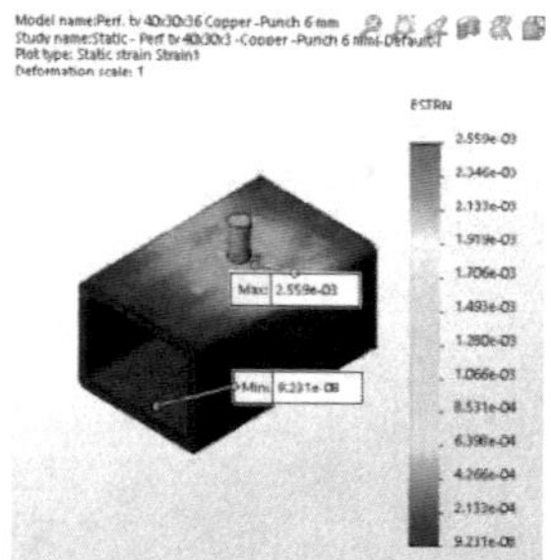

**Figure 13.** The result of the (equivalent) strain analysis on the copper part.

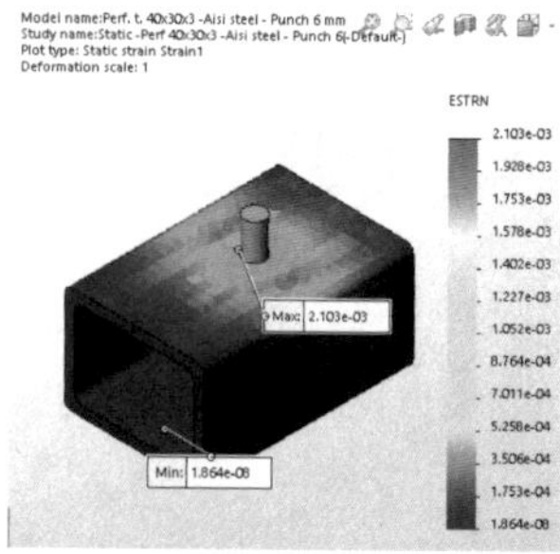

**Figure 14.** The result of the analysis of the strain (equivalent) in the steel part.

## 5 Conclusions

Following the results of the research carried out to determine the stresses, displacements, and strain - equivalent on the parts, the following conclusions can be highlighted:

1) In the paper are presented two technological options that can be used in industry, not only in specialized laboratories.
   - Perforation of laminated pips with the help of industrial robots and,
   - Obtaining holes by punching with a single cutting edge.
2) In Chapter 4, the results of the research concerning solution 2 mentioned in Chapter 3 are presented.
   The research was done for a case of a rectangular pipe with dimensions 40x30 (mm) with a wall thickness of 3 mm, for 3 types of materials, aluminum alloy, copper, and steel. Research was carried out to study the stability behavior of the piece stability in the case of perforation with a punch with an active diameter of 6 mm.
3) The presented technological variants will have to be verified in industrial conditions, depending on the needs of the interested companies.
4) When obtaining parts through the procedures specified in this document, it is important that these parts meet the quality requirements of the designer.
5) For single edge perforation/piercing, the solution can be used for pipes with a maximum pipe wall of 4mm.
6) When punching with plasma or laser, pipe wall thicknesses greater than 4 mm are also accepted.
7) The advantages of the two solutions proposed in the paper consist in the fact that they use laminated pipes, which are cheaper semi-finished products to obtain parts of the type of long pipes with many holes arranged along the pipe.
8) The disadvantage of using laser or plasma robots is the high costs of these robots, but for a very large series production or mass production, these costs are amortized over time.
9) The disadvantage of using single edge perforation is the limitation of the maximum diameter of the holes in order not to lose the stability of the piece around the hole.

**Acknowledgements**

The study is part of the Ph.D. program of the first author at the Politehnica University of Timisoara.

**References**

[1] Irmer W and Karpenko M 2001 *Brenn-Plasma und Laserstrahlschneiden mit dem Sauerstoff als Schneidgas* Der Praktiker 9/2001 (DVS Verlag Dusseldorf)

[2] Duta Majundar J and Manna I 2003 *Laser Processing of Materials Sadhana* **28** 3 and 4 pp 495-562

[3] Tiechuan Z and Jimin C 2000 3-Dimensional welding and cutting using high power CO2 or TAG *Laser Process Control and Inspection for Industry Proceedings of SPIE* **4222** p 1
[4] Vanschen W 2005 Plasma cutting Part I Principle, nozzle geometry, gases Welding no. 6 years XVI 58
[5] Ruset C and Grigore E 2005 Current trends in Plasma surface engineering of mechanical components *13th Conference on plasma physics and applications* Iasi 57
[6] Simler H, Krink V and Laurisch F 2006 Potential largit de apkicatie a procedeului de taiere cu jet de plasma de inalta precizie Partile 1-3 Sudura nr. 1.2.4 an XVI
[7] Kovacs F, Tusz F and Varga S 1999 *Factory of the future-introduction to production: Computer Integration of Design, Manufacturing, and Management* (Ed. Multimedia International, Arad)
[8] Joni N, Trif I N, and Joni A 2003*Industrial robots used in cutting and related processes*, Robotics & Management Vol 8 no 1(Ed. TIM Resita)
[9] Tichkiewitch S 2005 *Virtual Design and Automation* (Weiss Z) Publisher Poznan University of Technology Poznan Poland 7143-201-1 pp 29-46
[10] Poiana D I, Ionel I, and Draghici G 2022 Research on the influence of the clamping devices on the size of the perforated holes in long round pipes *IOP Conf. Ser.: Earth and Environ. Sci.* **960** 012023
[11] Poiana D I, Ionel I, and Draghici G 2022 Research concerning the influences related to the perforated hole diameters in long pipe *IOP Conf. Ser.: Mater. Sci. Eng.***1235** 012070

IOP Conf. Series: Materials Science and Engineering **1311** (2024) 012037 doi:10.1088/1757-899X/1311/1/012037

# Comparative analysis of the manufacturing processes for 316L material

**I A Sandu[1], A G Badea[1] and S Tabacu[1]**

University of Pitesti, Faculty of Mechanics and Technology, 1, Targu din Vale str., 110040, Pitesti, Arges, Romania.

ana.badea@upit.ro

**Abstract**. The present work focuses on the investigation of a comparative analysis of the manufacturing processes for 316L material. It aims to highlighting the advantages but also the disadvantages offered by conventional and additional manufacturing methods. This analysis was developed for the austenitic material, stainless steel 316L (1.4404). Due to its properties, this material is used in an ample range of applications spread across various industries (automotive, medicine and medical devices, chemical, biomedical) by using the additive manufacturing techniques (AM).

## 1. Introduction

Nowadays, additive manufacturing (AM) has been the main method of rapid prototyping since the 1980s. Part manufacturing consists of creating solid material progressively from powder utilizing additive methods from the previously designed model (CAD). The final parts, in AM, are obtained, starting with the CAD model. This can also be achieved by wire feedstock [1].

AM technology is also called rapid prototyping, solid free-form fabrication, layered manufacturing, or 3D printing. Usually, the process consists of a joining layer on top of a layer of material.

The most significant advantages of AM technology are time reduction and the removal of the traditional process tools and procedures; at the same time, samples with complex geometry are archived, geometry that would be hard or impossible to achieve using conventional technology [2].

AM is a rapid prototyping method as well as a fabrication technology allowing the manufacturing of complex geometrical shapes and giving access to a multitude of options regarding surface adjustments. The most important AM technologies for manufacturing are laser and electron-based techniques, including SLM (Selective Laser Melting, DMLS (Direct Metal Laser Sintering), DED (Directed Energy Deposition), and EBM (Electron Beam Melting) [1].

The technology that allows obtaining the sample directly from the CAD model using additive layer manufacturing techniques without the necessity of manufacturing process planning, tooling, or fixtures, is called Rapid Prototyping (RP). In the past, RP equipment was used to produce models and prototype parts, but meanwhile, they are widely used even for small production parts [2], [3].

By using SLS/ SLM techniques, one can process many types of materials, having a broad coverage in vast fields of use such as automotive, aerospace, tooling, biomedical, and architecture [2], [4].
A wide variety of materials can be manufactured by additive manufacturing technology, including metal alloys, ceramics, polymers, and composite materials [5], [6].

The present study review article summarises the mechanical properties, composition, and applications of the austenitic material, stainless steel SS316L. Also, this paper provides a comparison between conventionally manufactured methods and additive manufacturing methods (SLM – Selective Laser Melting).

The literature [7], [8], [9] presents an ample range of applications spread across various industries (automotive, medical devices, chemical, biomedical) by using additive manufacturing techniques.

Additive manufacturing (AM) is a group of technologies that is essential to meet the demands of Industry 4.0 [7]. The 4.0 Industry movement is a strategic and intentional approach to shaping the future. This strategy involves the trend towards automation, including cyber-physical systems (CPS), the internet of things (IoT), the industrial internet (IIoT), artificial intelligence, but also 3D printing. The main benefits of this industry are: improved efficiency, increased productivity, flexibility and agility, reduced costs, and obtaining a return on investment [10].

3D Printing technologies such as selective laser melting (SLM) are becoming more and more used nowadays and important, already being applied in wide industries. As happens with other manufacturing technologies, it is impossible to produce specimens without defects, so the SLM method is no exception to the rule [11].

Changes in the microstructure are achieved by fast solidification. Widening the alloy range of uses (e.g., non-miscible or supersaturated alloy compositions) is allowed by high solidification rates [12].

Mirzababaei and Pasebani affirm that the near-net-shape is manufactured through an incremental process that involves a laser melting the metal powder followed by the application of a new powder layer that is again melted in conformity with the geometry that was previously defined [13].

## 2. Additive manufacturing by SLM – 316L

The material studied in this work, the SLM-processed 1.4404 stainless steel, shows mechanical and corrosion performance enhancements compared to the more conventional manufacturing technologies and have applications in various industries. An anisotropic corrosion behavior is determined by way of the layer's topology of the SLM processing, resulting in the anisotropy of mechanical properties. In the study of Kaijie s.a., the authors investigated the effect of micro-structural differences of SLM-processed 1.4404 in line with the direction of building on corrosion behavior. According to them, corrosion resistance can be enhanced by increasing the distance from the bottom plane in line with the building direction. Also, a different corrosion behavior can be achieved by modifying the crystalographic orientation at the microstructural level and grain dimensions along the fabrication direction in the SLM process [14], [15].

Austenitic stainless steel (SS316L) is often utilized in a broad range of engineering fields due to its high strength and corrosion resistance [16], [17] oxidation resistance [18], high heat resistance, and good weldability [17]. This kind of material has a particularity, a negative Poisson's ratio. The elements from the composition of the material lead to its main properties. The SS316L contains, in addition to iron, chromium to improve corrosion resistance and nickel to stabilize the austenite phase. Austenitic stainless steel contains a small carbon percentage for minimized intergranular fracture. The chemical composition is indicated in Table 1. Table 2 and Table 3 present the physical and mechanical properties of 1.4404 [19].

**Table 1.** Chemical composition for SS316L (1.4404).

| Alloy | Elements | | | | | | | | |
|---|---|---|---|---|---|---|---|---|---|
| | C | Mn | Si | Cr | P | Ni | Mo | S | Fe |
| SS316 L | 0.011 | 0.14 | 0.75 | 17.1 | 0.007 | 12.9 | 2.33 | 0.006 | Balance |

**Table 2.** Physical properties for SS316L.

| Density | Melting point | Thermal Expansion | Modulus of elasticity | Thermal Conductivity | Electrical Resistivity |
|---|---|---|---|---|---|
| 8 [g/cm$^3$] | 1400[°C] | $15.9\cdot10^{-6}$/K | 193 [GPa] | 163 [W/m·K] | $0.74\cdot10^{-6}$[Ω·m] |

IOP Conf. Series: Materials Science and Engineering **1311** (2024) 012037 doi:10.1088/1757-899X/1311/1/012037

**Table 3.** Mechanical properties for SS316L.

| Proof Stress | Tensile Strength | Elongation A50mm | Hardness Brinell | Yield strength |
|---|---|---|---|---|
| 200 Min MPa | 500-700 MPa | 40 Min % | 215 Max HB | 205 MPa |

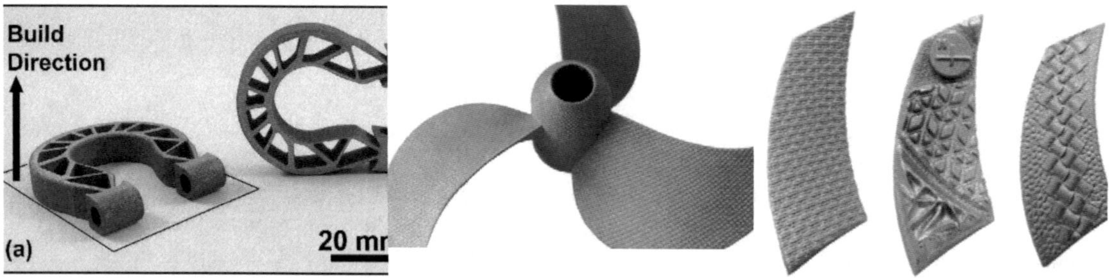


**Figure 1.** Parts manufactured via SLM [20], [21]

Usually, austenitic stainless steels obtained through AM have higher yield strength, ultimate tensile strength and hardness, and lower ductility than their traditionally processed counterparts [22], [23]. Components manufactured from SS316L obtained with selective laser melting (SLM) offer improved mechanical properties. Some authors say sintered stainless steel has lower corrosion resistance than cast or wrought stainless steel.

AM printers have the following main limitations:

- Relatively small commercial SLM metal printers: The larger dimension is about 1–2 m in length;
- Higher costs compared to other fabrication techniques (e.g., metal casting or machining);
- The limitation imposed by the reduced diversity of common commercial alloys. The common 3D-printed steels are AISI 316L/1.4404, maraging steel 18Ni-300/1.2709, and Inconel [24].

The following figure illustrates the stress-strain curves for 1.4404, manufactured with SLM on different conditions.

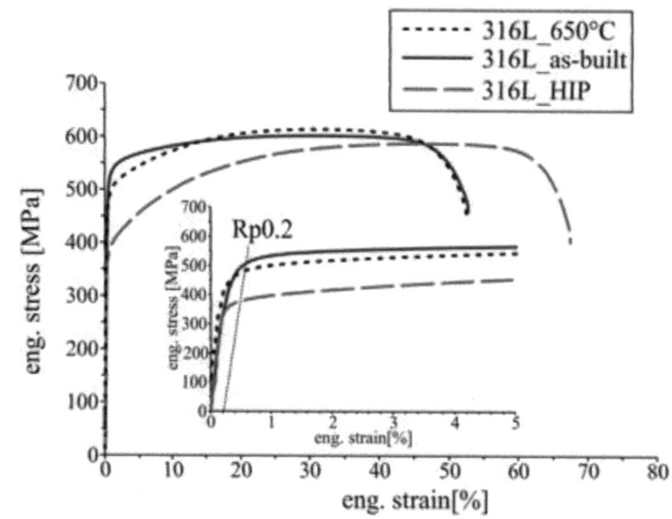


**Figure 2.** Stress-strain curves for SLM processed SS316L [26].

## 3. Parameters affecting the performance of SLM parts

The building direction and the building parameters influence the parts' property. Moreover, because of the manufacturing process, the parts generally show high surface roughness, isolated defects at overall low porosity, and significant macroscopic residual stresses in the raw state.

Due to the characteristics of the SLM process, a broad range of final material properties can be chosen [11].

## 4. Parameters – 3D printing process

The final properties of the sample are influenced by all variables and all parameters like layer thickness, required drying time, and post-manufacturing variables: habitat, time, cooling/ heating rate, and sintering condition [13]. According to previous studies, it is recommended as the layer thickness be greater than the king-size particle and no less than roughly three times that of the particle diameter. These recommendations are helpful in obtaining a polished surface finish and higher packing density [25].

Generally, the fatigue lives of SLM parts are considerably lower than traditionally manufactured ones, and the fatigue behavior depends on the material. Leuders et al. [26] make public the monotonic strength as a leading parameter for SS316L.

The work of YingChen s.a. shows nitriding's capability to improve the mechanical performance of open cellular structures obtained by AM [28].

One of the most important process parameters for AM techniques is layer thickness; it gives the littlest possible design characteristic dimension and limitation resolution. Layer thickness is given by the distance between the worktable from the initial moment and its position after printing a layer. Furthermore, the greater the thickness of the layer, the better the properties of the piece obtained; on the other hand, the production time increases. Because after printing each layer of the part, its construction surface will move under a heater in order to dry the binder, the heater power ratio needs to be set so that it doesn't cause any defects in the final parts.[13]

The major properties of the powder bed material that defines a good process for SLS and SLM technology are presented in the following table. These properties are important for an accurate simulation of the process.[29]

**Table 4.** Powder Bed Material Properties

| Optical Properties | Granulomorphometric characteristics of powder grains | Thermal properties | Rheological properties | Mechanical Properties | Metallurgical properties |
|---|---|---|---|---|---|
| Reflection/ Absorption | Particle size | Thermal conductivity | Viscosity | Elastic Modulus | Oxidation potential |
| Optical penetration | Particle size distribution | Specific heat Latent heat | Surface tension | Yield point | Alloy composition |
| | Particle shape | Melting Temperature | | Tensile strength | Diffusion coefficients |
| | Particle roughness | Thermal expansion | | | Liquid solid range |

## 5. 3D printing process

The SLM technique followed the process of SLM powders working with a laser device, which melts the powder layer after layer. Conventional operations of the SLM machine involve pre-build machine getting ready. The operations mainly include pathway heating, inert gas flow inside the build chamber, and covering the pathway with powder material. Afterward, a controlled laser is sent to the desired area to melt the powder. After the powder solidification, another layer of powder is covered, representing steps that are repeated several times until the product is completed. Maintaining constant inert gas flow inside the build chamber during the process is necessary to guarantee minimum presence of oxygen to prevent oxidation. Besides inert gas flow, the SLM machine's construct rate significantly impacts the electric power consumed by the machine. Inert gas can be argon or nitrogen [30].

It has a variable energy consumption throughout the printing process, its variation depending on intermediary factors or the geometry of products. The need for powder coating increases while the layer height reduces, thus increasing the use of the laser for melting. So, a proportional impact on power consumption is given by the rise in the scan speed of the laser. Supplementary, it is necessary to connect to the machine's cooling system to decrease the warming of internal equipment. Further, a PSM (Powder Sieving Machine) is required to be screened during the printing process for purity and performed by a PSM (Powder Sieving Machine). Following, in the tables are presented the machine specifications for SLM solutions.

**Table 5.** SLM 280$^{HL}$ machine specification

| Features | SLM Solutions |
|---|---|
| Machine type | Selective Laser Melting |
| Rated electrical power | 3.5-6 kW |
| Build chamber | (100 x 100 x 365) mm |
| Laser power | 400W |
| Build speed | 55 cm$^3$/h |
| Maximum scan speed | 10 m/s |
| Inert gas consumption in operation- Flooding | 3 l/min- 130 l/min |

**Table 6.** Powder Sieving and cooling machine specifications

| **Features** | **SLM Solutions PSM 100** | **WKS 51** |
|---|---|---|
| Machine type | Powder Sieving | Cooling |
| Rated electrical power | 0.5 kW | 2.5-3.5 Kw (20°C outlet, 32°C ambient) |
| Inert gas consumption in operation- Flooding | 100 l/min | |

According to Bharathwajanprabu Ravisankar [30], the specific operation of an SLM process put on a kind of pseudo-code as follows:

```
START:
        Heating: platform, chamber
        Feed: powder feed
        Flooding: inert argon gas
FOR:
    Layer
        Motor: platform motor movement
        Motor: coater motor movement
        Laser: laser operation
        Flow: argon inert gas flow
        Cooling: cooling operation inside the machine
END
```

There on the market are high-performance equipment with multiple applications, where printing speed is important, and printers that use SLM technology to create parts directly from a 3D-CAD model.

There are printers on the market with a large construction area and decreased printing time. They typically have two/four independent high-precision multi-lasers, which means parts will be printed up to four times faster. Such a system is equipped with two feeding chambers, which reduces printing time and increases productivity; having advanced software, a printer of this type is designed for industrial-level applications.[31]

IOP Conf. Series: Materials Science and Engineering **1311** (2024) 012037 doi:10.1088/1757-899X/1311/1/012037

The SLS (Selective Laser Sintering) technology appeared in 1989 and involved using a high-power laser that fuses reduced particles of raw material to obtain 3D parts. Figure 3 illustrates the schematic principle of the SLS and SLM process.

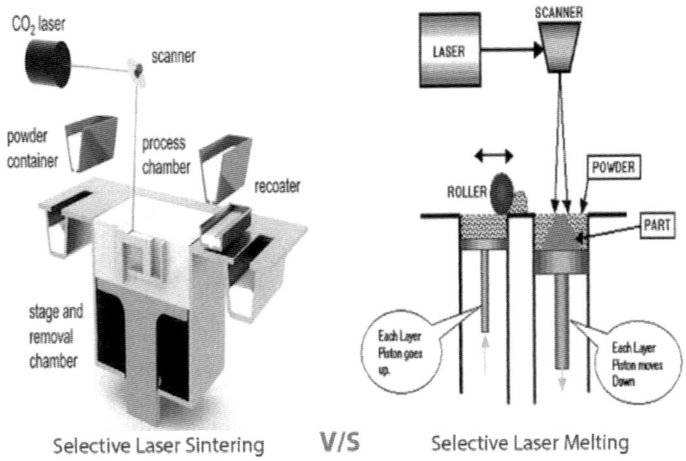


**Figure 3.** Schematic presentation of SLS/SLM process [32]

## 6. Conclusion

The study aimed to highlight AM's importance, advantages, and use. The research in this study was a review of AM for SS316L, a common alloy for studies.

This study discusses the general convenience of AM technologies. Layer-by-layer technologies is time-cost-efficiency from the modeling to the manufacturing phase without any investment in designing and producing the tools and fixtures.

The main advantages of AM technologies are listed in the following:

- economic advantages;
- better efficiency;
- improvements in the field of design, analysis, testing, and manufacturing;
- immediately begins production;
- avoiding delays due to tooling.

Also, an important advantage of additive manufacturing (AM) is given by the possibility of manufacturing components with complex geometry directly from CAD models, parts that conventional manufacturing processes can't produce (e.g., cooling channels). Therefore, AM processes also offer good geometrical freedom, giving the ability to create complex geometries impossible by conventional manufacturing techniques. That means a reduction in cost, energy, raw material, process steps, and human resources, advantages according to Industry 4.0.

Starting from the AM characteristics and advantages and the interest in new manufacturing methods, it can be seen that the need to use AM is justified.

The high strength of the austenitic stainless steel (SS316L) combined with hardness, high heat resistance, corrosion resistance, and good weldability makes it suitable to be used in a wide range of industrial fields (medical, construction, automotive, aerospace).

SLS and SLM allow obtaining metallic parts with complex geometry, so with CAD software, the topology of the structures can be optimized while keeping the same strength and load requirements, also reducing the component's mass which represents the main goal of the automotive and aerospace industry. An important machine parameter for this technique is layer thickness, which determines how small a possible design feature can be and constraint resolution.

**References**

[1] Nouri, A. Rohani Shirvan, Y. Li, and C. Wen, "Surface modification of additively manufactured metallic biomaterials with active antipathogenic properties," *Smart Mater. Manuf.*, vol. 1, p. 100001, Dec. 2023. https://doi.org/10.1016/j.smmf.2022.100001.

[2] G. Gong *et al.*, "Research status of laser additive manufacturing for metal : a review," *J. Mater. Res. Technol.*, vol. 15, pp. 855–884, 2021. https://doi.org/10.1016/j.jmrt.2021.08.050.

[3] K.-H. Chang, "Rapid Prototyping," *e-Design*, pp. 743–786, Jan. 2015. https://doi.org/10.1016/B978-0-12-382038-9.00014-4.

[4] S. Kumar, "Selective Laser Sintering/Melting," *Compr. Mater. process.*, vol. 10, pp. 93–134, Jan. 2014. https://doi.org/doi: 10.1016/B978-0-08-096532-1.01003-7.

[5] P. K. Gokuldoss, S. Kolla, and J. Eckert, "Additive Manufacturing Processes : Selective Laser Melting , Electron Beam Melting and Binder," 2017. https://doi.org/10.3390/ma10060672.

[6] D. Bourell *et al.*, "CIRP Annals - Manufacturing Technology Materials for additive manufacturing," *CIRP Ann. - Manuf. Technol.*, 2017.

[7] A. Haleem and M. Javaid, "Additive Manufacturing Applications in Industry 4.0: A Review," *J. Ind. Integr. Manag.*, vol. 04, no. 04, p. 1930001, 2019. https://doi.org/doi: 10.1142/s2424862219300011.

[8] M. Javaid and A. Haleem, "Additive manufacturing applications in medical cases : A literature based review," *Alexandria J. Med.*, vol. 54, no. 4, pp. 411–422, 2018. https://doi.org/doi: 10.1016/j.ajme.2017.09.003.

[9] A. Bandyopadhyay, K. D. Traxel, M. Lang, M. Juhasz, N. Eliaz, and S. Bose, "Alloy design via additive manufacturing: Advantages, challenges, applications and perspectives," *Mater. Today*, vol. 52, no. February, pp. 207–224, 2022. https://doi.org/doi: 10.1016/j.mattod.2021.11.026.

[10] "Industria 4.0 - Evoluția către a 4-a Revoluție industrială - GreenSoft." https://www.greensoft.ro/industria-4-0/ (accessed Oct. 08, 2022).

[11] M. Zeißig and F. Jablonski, "Comparison of different approaches to model fatigue for additively manufactured specimens considering production-related characteristics," *Procedia Struct. Integr.*, vol. 38, no. C, pp. 60–69, 2021. https://doi.org/doi: 10.1016/j.prostr.2022.03.007.

[12] C. Doñate-Buendía, D. Gu, M. Schmidt, S. Barcikowski, A. M. Korsunsky, and B. Gökce, "On the selection and design of powder materials for laser additive manufacturing," *Mater. Des.*, vol. 204, p. 109653, Jun. 2021. https://doi.org/doi: 10.1016/J.MATDES.2021.109653.

[13] S. Mirzababaei, "A Review on Binder Jet Additive Manufacturing of 316L Stainless Steel," pp. 8–12, 2019. https://doi.org/10.3390/jmmp3030082.

[14] K. Lin *et al.*, "Selective laser melting processing of 316L stainless steel: effect of microstructural differences along building direction on corrosion behavior," *Int. J. Adv. Manuf. Technol.*, vol. 104, no. 5–8, pp. 2669–2679, 2019. https://doi.org/doi: 10.1007/s00170-019-04136-9.

[15] Z. Sun, X. Tan, S. B. Tor, and W. Y. Yeong, "Selective laser melting of stainless steel 316L with low porosity and high build rates," *Mater. Des.*, vol. 104, pp. 197–204, Aug. 2016. https://doi.org/doi:10.1016/J.MATDES.2016.05.035.

[16] A. Simchi, A. Rota, and P. Imgrund, "An investigation on the sintering behavior of 316L and 17-4PH stainless steel powders for graded composites," *Mater. Sci. Eng. A*, vol. 424, no. 1–2, pp. 282–289, May 2006. https://doi.org/doi: 10.1016/J.MSEA.2006.03.032.

[17] J. P. Choi, G. Y. Lee, J. Il Song, W. S. Lee, and J. S. Lee, "Sintering behavior of 316L stainless steel micro–nanopowder compact fabricated by powder injection molding," *Powder Technol.*, vol. 279, pp. 196–202, Jul. 2015. https://doi.org/doi: 10.1016/J.POWTEC.2015.04.014.

[18] A. Bautista, F. Velasco, and J. Abenojar, "Oxidation resistance of sintered stainless steels: effect of yttria additions," *Corros. Sci.*, vol. 45, no. 6, pp. 1343–1354, Jun. 2003. https://doi.org/doi: 10.1016/S0010-938X(02)00217-2.

[19] "Stainless Steel – Austenitic - 1.4404 (316L) Bar and Section." https://www.aalco.co.uk/datasheets/Stainless-Steel-14404-316L-Bar-and-Section_39.ashx

(accessed Oct. 09, 2022).

[20] B. Rankouhi, K. M. Bertsch, G. Meric de Bellefon, M. Thevamaran, D. J. Thoma, and K. Suresh, "Experimental validation and microstructure characterization of topology optimized, additively manufactured SS316L components," *Mater. Sci. Eng. A*, vol. 776, 2020. https://doi.org/10.1016/j.msea.2020.139050.

[21] "for 3d printing – Taktilesdesign." https://www.taktilesdesign.com/textures/for-additive-manufacturing/ (accessed Oct. 10, 2022).

[22] T. DebRoy *et al.*, "Additive manufacturing of metallic components – Process, structure and properties," *Prog. Mater. Sci.*, vol. 92, pp. 112–224, Mar. 2018. https://doi.org/doi: 10.1016/J.PMATSCI.2017.10.001.

[23] M. Ghayoor, K. Lee, Y. He, C. Chang, B. K. Paul, and S. Pasebani, "Microstructural Analysis of Additively Manufactured 304L Stainless Steel Oxide Dispersion Strengthened Alloy," *Microsc. Microanal.*, vol. 25, no. S2, pp. 2594–2595, 2019. https://doi.org/doi: 10.1017/s1431927619013709.

[24] B. Utela, D. Storti, R. Anderson, and M. Ganter, "A review of process development steps for new material systems in three dimensional printing (3DP)," *J. Manuf. Process.*, vol. 10, no. 2, pp. 96–104, Jul. 2008. https://doi.org/doi: 10.1016/J.JMAPRO.2009.03.002.

[25] B. Utela, D. Storti, R. Anderson, and M. Ganter, "A review of process development steps for new material systems in three dimensional printing (3DP)," *J. Manuf. Process.*, vol. 10, no. 2, pp. 96–104, Jul. 2008. https://doi.org/doi: 10.1016/J.JMAPRO.2009.03.002.

[26] S. Leuders, T. Lieneke, S. Lammers, T. Tröster, and T. Niendorf, "On the fatigue properties of metals manufactured by selective laser melting - The role of ductility," *J. Mater. Res.*, vol. 29, no. 17, pp. 1911–1919, 2014. https://doi.org/doi: 10.1557/jmr.2014.157.

[27] N. H. J. S. D. D. Andreas Jaksch, "Selective laser sintering of polyamide 12 - Effect of flow-enhancing additives on processability and part properties," *HANSER eLibrary*, 2021. https://doi.org/10.3139/9783446471733.002.

[28] Y. Chen, C. Liu, H. Yan, Y. Fan, J. Wang, and Y. Cui, "Effect of gas nitriding on 316 L stainless steel lattice manufactured via selective laser melting," *Surf. Coatings Technol.*, vol. 441, p. 128559, Jul. 2022. https://doi.org/doi: 10.1016/J.SURFCOAT.2022.128559.

[29] E. L. Papazoglou, N. E. Karkalos, P. Karmiris-Obratański, and A. P. Markopoulos, *On the Modeling and Simulation of SLM and SLS for Metal and Polymer Powders: A Review*, vol. 29, no. 2. Springer Netherlands, 2022. https://doi.org/10.1007/s11831-021-09601-x.

[30] B. Ravisankar, E. Jaeger, M. Meißner, A. Wirtz, P. Wiederkehr, and C. Rehtanz, "Investigation of electrical power consumption of an additive process chain and empirical modelling as feature selection for machine learning algorithms," *Procedia CIRP*, vol. 99, pp. 358–363, Jan. 2021. https://doi.org/doi: 10.1016/J.PROCIR.2021.03.103.

[31] "SLM - Metal - Imprimante 3D - Nutechnologies." https://www.nutechnologies.ro/imprimante-3d/metale-slm (accessed Oct. 09, 2022).

[32] "Green Mechanic: Selective Laser Melting (SLM) and Selective Laser Sintering (SLS), a comprehensive comparison." https://www.green-mechanic.com/2016/12/comparison-between-selective-laser.html (accessed Oct. 10, 2022).

IOP Conf. Series: Materials Science and Engineering **1311** (2024) 012038 doi:10.1088/1757-899X/1311/1/012038

# Development of a coupled process - design numerical model for an automotive assembly

**M Stirosu[1], A Badea[1], A I Sandu[1] and S Tabacu[1]**

[1]University of Pitesti, Faculty of Mechanics and Technology, str. Targu din Vale, nr. 1, Pitesti, ROMANIA

stefan.tabacu@upit.ro

**Abstract**. The product development process is a very challenging one, especially in the case of the automotive industry. The actual constructions require performance and quality while focusing on mechanical performance, lightweight and cost-effective. Thus, the virtual development stage is critical. Once the product is validated according to its geometrical definition and features, it is necessary to evaluate its mechanical performance. This process is usually performed using numerical simulation methods. In this paper, the numerical process performed for the validation of a safety part is discussed. As a specific feature, the assembly is finished using a riveting process. The manufacturing process's influence on the assembly's mechanical performance is further investigated. A discussion of the methods available for the design validation summarizes the finding of this specific work.

## 1. Introduction

Product development is now very challenging due to the currently existing requirements [1][2][3][4]. Better structures [5], better materials [6] [7] [8], and cheaper solutions [9] are not easy to get. Therefore, experiments and design should be closely put together. This paper presents the procedure performed for developing a product designed for the automotive industry.

The procedure includes the design process, experimental analysis, and numerical simulation. These three data sets should be closely connected, and the experience should be valuable knowledge for further products [10].

The procedure discussed is related to the inclusion of the results of a numerical simulation of a specific manufacturing process [11], namely riveting, in the global model of a striker tested according to currently available regulations.

## 2. Introduction to the Technical Specifications

The assembly investigated in the paper is used in a safety mechanism. The assembly is mounted on the vehicle's body in white to ensure the locking function of the fifth door. The technical specification associated with the locking mechanism is prescribed in Regulation ECE 11 [12]. The assembly is tested under loads applied, considering the principal axes of the vehicle. The document indicates the nominal testing values, although the manufacturer can impose higher parameters in some cases.

The present work is focused on the investigation of the mechanical response under a load applied in a longitudinal direction or considering the actual position of assembly on the car along $X$ axis.

Prior to product and process validation, prototypes were manufactured and tested under laboratory conditions. Figure 1 presents the assembly further identified also as the "striker," the mounting fixture, and the loading part.

**Figure 1.** Striker mounted on the fixture ready to be tested.

The loading part is moving at a speed of 5 $[mm/min]$ according to the product validation program. The striker must be capable of supporting a load of 11.5 $[kN]$ (the prescribed values are amended by 20%).

## 3. Geometrical and numerical model

The life of every product starts with a geometrical model used to outline the main features of the assembly, validate within the environment, provide an estimate of the raw materials, and analyze the manufacturing process.

Figure 2 presents the striker's CAD model as it was integrated into the global layout of the vehicle.

Figure 3 presents an update of the CAD model being prepared for the numerical simulation. Some geometrical features (mainly small radii fillets) need to be removed from the model to improve the efficiency and accuracy of the numerical simulation process.

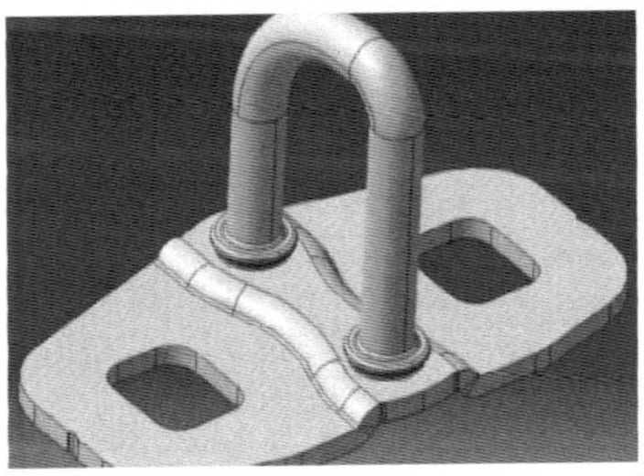

**Figure 2.** The CAD model of the striker.

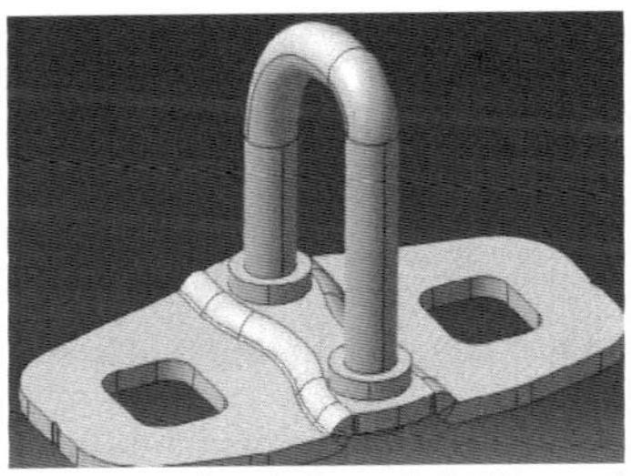

**Figure 3.** The defeated CAD model

The mechanical response of the assembly will be investigated using numerical simulation implemented by the Finite Element Method (FEM).

In this case, the foundation of the numerical simulation resides in developing an accurate numerical model capable of embedding interfaces and various material behaviors.

For the definition of the numerical model, it is rather usual to use some automatic tools, yet for the current analysis, a user-driven method was implemented.

Experiences from the past revealed that the assembly's critical section is the wire's riveted end.

The process required by the simulation strategy [13]:

- Scenario 1: the simulation will be performed on the assembly defined as it is;
- Scenario 2: the simulation will be performed with wires connected by rivets with a true shape;
- Scenario 3: the effects of the riveting process (residual stresses) will be embedded in the numerical model.

The presentation of the proposed simulation strategy reveals that multiple steps should be performed.

### 3.1. Material data

The assembly is manufactured using common-grade materials available for the automotive industry. The materials are selected considering the mechanical performances and the manufacturing process.

During the early design stages the validation process is performed under these circumstances, although it is recommended to perform material analysis using specific equipment. Table 1 presents the materials currently used to manufacture the assembly. The values listed in Table 1 were extracted from public sources.

**Table 1.** Physical and mechanical properties of the materials.

| | $\rho\ [kg/m^3]$ | $R_p\ [MPa]$ | $R_m\ [MPa]$ | $A\ [\%]$ |
|---|---|---|---|---|
| **S460MC** | 7800 | min. 460 | 520-560 | 14 |
| **42CrMo4** | 7800 | min. 750 | 1000-1200 | 11 |

The values reported in Table 1 are directly measured from a tensile test. These values are often referred to as "engineering" data. Thus, performing a conversion from "engineering" to "true" stress data is necessary.

The ultimate stress is assumed to be extracted from the tensile strength $(R_m)$ using the equation:

$$\sigma_u = R_m \cdot (1 + A) \tag{1}$$

A very convenient way to represent stress-strain data is the use of a power law function [14]. A good representation can be obtained using the formula:

$$\sigma_i = \left(\frac{\varepsilon_i}{\varepsilon_u}\right)^n \cdot \sigma_u \tag{2}$$

where:

- $\sigma_i$ is the current stress;
- $\varepsilon_i$ is the current strain;
- $\varepsilon_u$ is the strain at failure (is equal to the absolute value of $A$);
- $n$ is an exponent that can be determined by data fitting. The parameter that is monitored for the data fitting process is the yield stress $(\sigma_y)$ that is derived from $R_p$when the measured strain is of 0.2%.

The complete dataset is presented in figure 4.

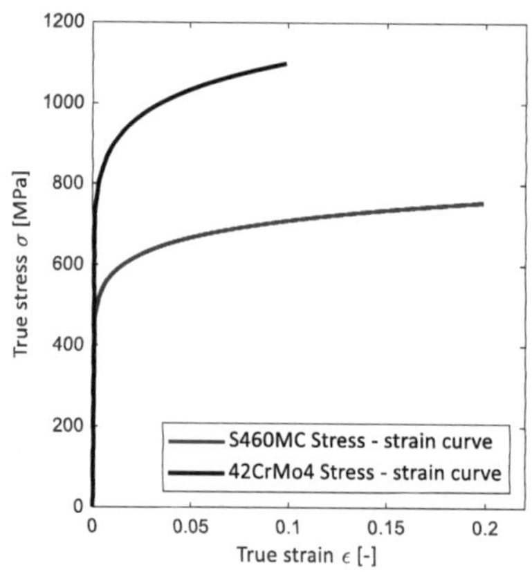


**Figure 4.** Stress-strain data (true data)

### 3.2. Material model

The paper's numerical models discussed and analyzed were developed for the Ls-Dyna solver. The material model assigned to the deformable structures is *MAT_PIECEWISE_LINEAR_ PLASTICITY. The Von Mises yield function (Φ) is implemented by:

IOP Conf. Series: Materials Science and Engineering **1311** (2024) 012038 doi:10.1088/1757-899X/1311/1/012038

$$\Phi = \frac{1}{2} \cdot s_{ij} \cdot s_{ij} - \frac{\sigma_y^2}{3} \leq 0 \tag{3}$$

where:

- $s_{ij}$ is the deviatoric stress
- $\sigma_y$ is the current radius of yield surface defined:

$$\sigma_y = \beta \cdot \left[\sigma_0 + f_h\left(\varepsilon_{eff}^p\right)\right] \tag{4}$$

In equation (4), $f_h\left(\varepsilon_{eff}^p\right)$ is the hardening function defined by tabulated data considering the effective plastic strain $\left(\varepsilon_{eff}^p\right)$.

### *3.3. Numerical models*

The most robust finite element is the one defined using eight nodes and identified as "hexahedral" or "brick" element. The definition of a numerical model using only this type of element is sometimes difficult to obtain. Thus, the focus was oriented on the most critical section, identified as the wire.

Figure 5 presents a step performed during the modeling of the wire. It can be observed the exclusive use of brick elements. Figure 6 presents the numerical model of the striker. The riveted sections are modeled as ideal using the geometrical definition existing in the CAD model.

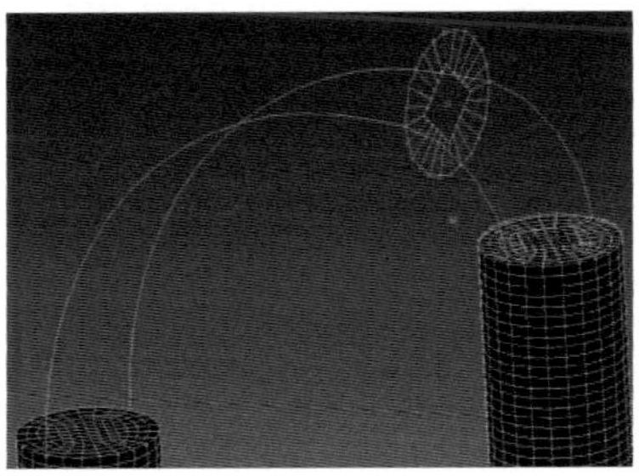

**Figure 5.** The technical process of the definition of the numerical model.

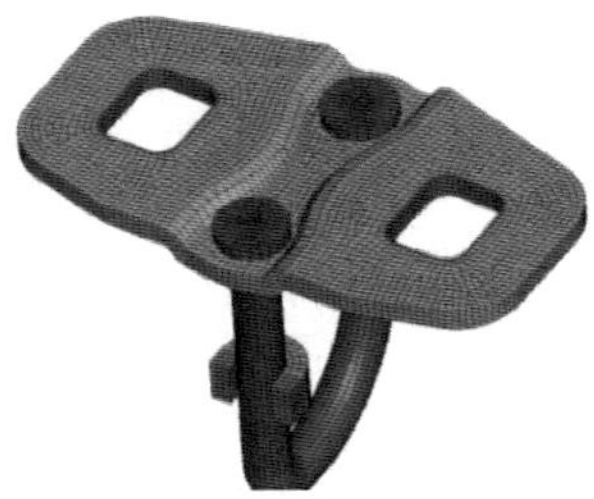

**Figure 6.** The complete definition of the numerical model.

A special configuration of the manufacturing process was defined to obtain the actual shape of the riveted section,. The riveting process is performed using a specialized industrial machine that shapes the workpiece (end of the wire) using a forming head. Thus, the main feature that is required in the numerical model of the riveting process is the workpiece, forming the head and the rotating head of the machine.

A series of elements were added to provide the complete functionality of the numerical model. These include kinematic joints, the definition of some sub-sections as rigid bodies, merged rigid bodies, tool paths defined as displacement vs. time curves, and interfaces in the form of contacts.

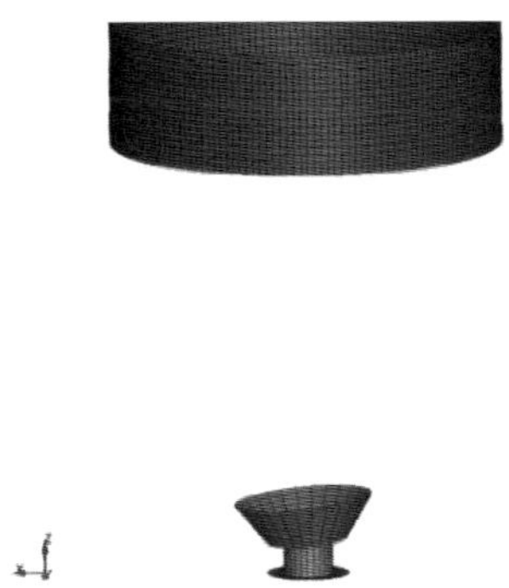

```
*CONTROL_CONTACT
*CONTROL_ENERGY
*CONTROL_TERMINATION
*CONTROL_TIMESTEP
*DAMPING_GLOBAL
*CONTROL_HOURGLASS
*DATABASE_BINARY_D3DUMP
*INTERFACE_SPRINGBACK_LSDYNA
*INCLUDE
*CONSTRAINED_JOINT_TRANSLATIONAL_ID
*DEFINE_COORDINATE_NODES
*CONSTRAINED_JOINT_STIFFNESS_TRANSLATIONAL
*CONSTRAINED_RIGID_BODIES
*CONSTRAINED_EXTRA_NODES_NODE
*PART
*MAT_RIGID_TITLE
*CONTACT_FORMING_SURFACE_TO_SURFACE_TITLE
```

**Figure 7.** The numerical model of the riveting process. Some features are listed on the right.

IOP Conf. Series: Materials Science and Engineering **1311** (2024) 012038 doi:10.1088/1757-899X/1311/1/012038

## 4. Results

The available data obtained from experiments and simulations will be investigated.

### *4.1. Experimental data*

The assembly was fitted by the interfacing fixture on the universal testing machine. A particular part with a thickness defined according to the testing specifications was used to load the assembly. The moving part was set to move with a speed of 5 $[mm/min]$. Figure 8 presents the assembly during the testing process.

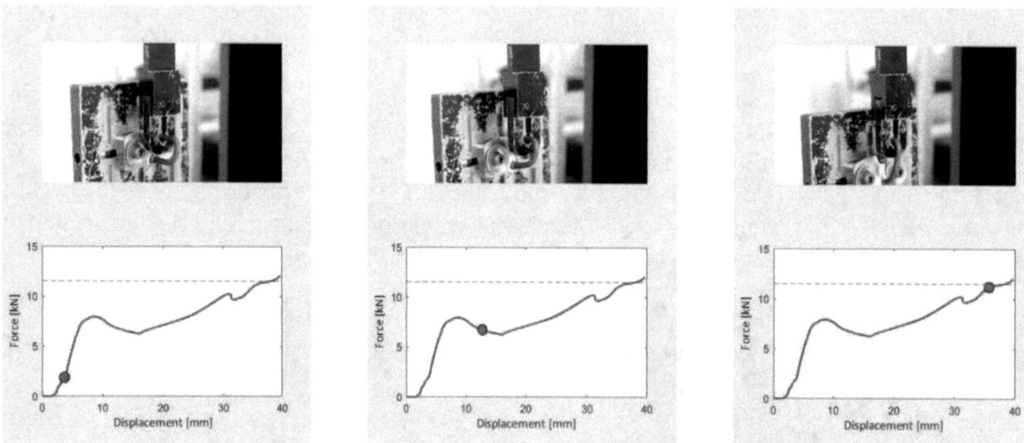


**Figure 8.** Experimental results: left to right – the loading part is located at the required distance; once the load increases, the elasticity of the fixture influences the loaded section of the striker, and the loading part slides; once the loading part was relocated, the testing process continues.

The details presented in figure 8 reveal that, due to the stiffness of the fixture, the loading part slid along the wire to a new loading position. This explains the force drop presented in figure 9.

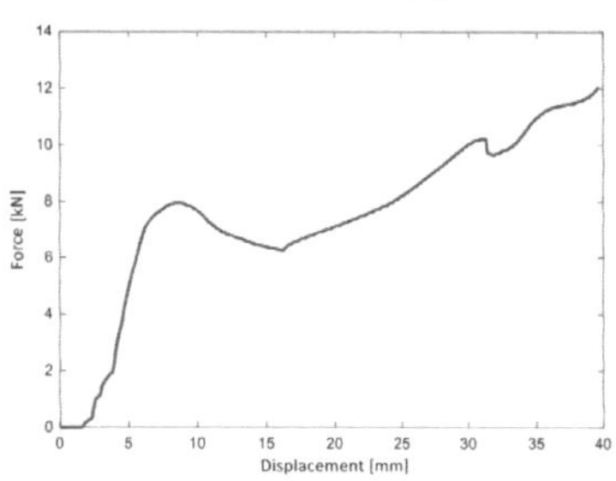


**Figure 9.** Experimental data.

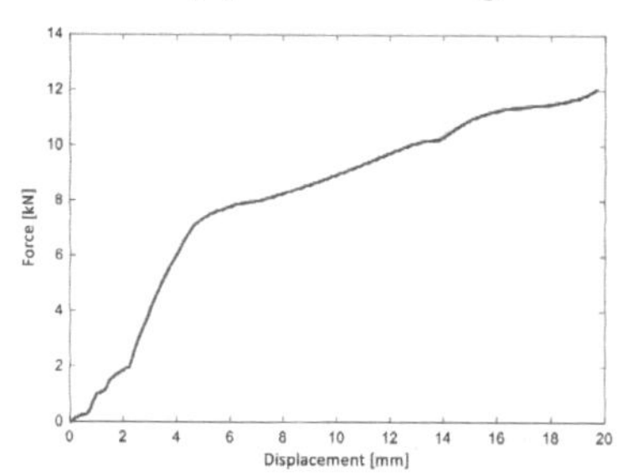


**Figure 10.** Updated experimental data.

This phenomenon requires updates of the experimental dataset. The acquired dataset was post-processed using a custom procedure developed using MATLAB. The procedure identifies the points where the force is dropping and constructs a continuous ascending dataset. The resulting dataset is presented in figure 10.

### *4.2. Riveting process*

As mentioned before, the riveting process was simulated using Ls-Dyna software [15][16][17]. Figure 11 presents a series of captions obtained during the processing process of the simulation data. The images points to the contact phase, followed by the forming stages, calibration of the head of the rivet and tool removal.

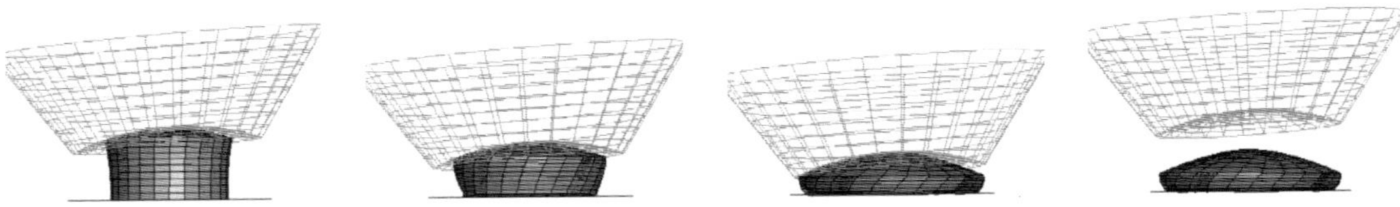

**Figure 11**. Riveting process: left to right – contact; forming, calibrating, and finishing.

The simulated section represents only the top of the wire. This procedure was adopted to fit the resulting shape to the initial numerical model of the assembly. For this purpose, a special numbering solution was adopted for the nodes and elements to have a quick model-fitting process.

### *4.3. Load cases analysis*
A first simulation set was performed with the loading part positioned according to the technical specification. As the model does not include the fixture and the machine, there is no stiffness; thus, no deformations associate with these structures. The loading part is not sliding along the wire, changing the wire section where the load is applied. As can be noticed, the deformations are rather small compared to the experimental data.

**Figure 12.** Numerical simulation of the assembly. The load is applied at the nominal position.

A second simulation was performed with the load part located in an updated position that corresponds to the section where the load was applied after sliding. The deformations of the structures are increasing, showing behavior similar to the one identified in the experiment.

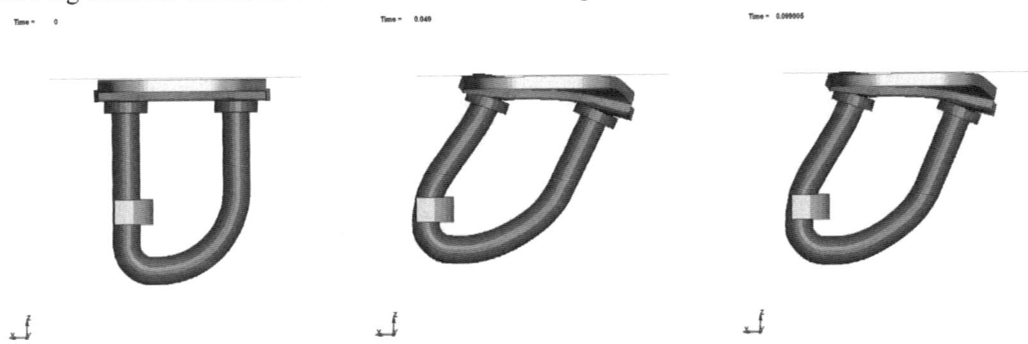

**Figure 13.** Numerical simulation of the assembly. The load is applied at the updated position.

## 5. Discussion
The data obtained from the experiments and simulations are presented in figure 14. Curve A points to the experimental dataset, while numerical data are represented by curves B to D.

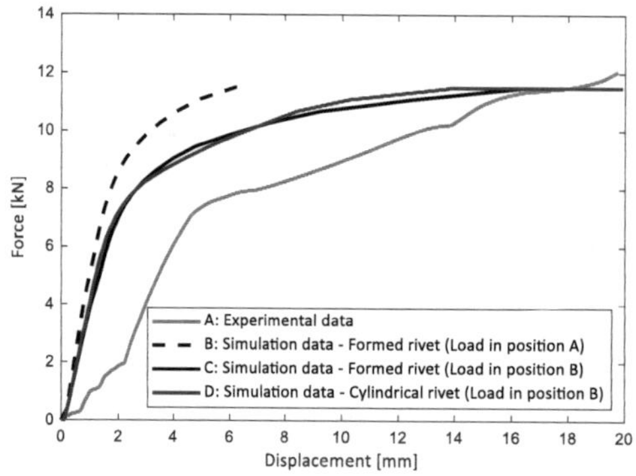


**Figure 14.** Cumulated experimental and simulation data.

Although there is not a perfect match to the experimental data, it can be noticed that the deformation paths are quite similar.

The simulation results that changing the loading section of the assembly gives a larger displacement, as indicated by curves B and C.

An interesting phenomenon was observed while comparing curves C and D. The formed rived assembly is stiffer than the cylindrical rivet assembly. This can be explained by the actual shape of the rivet. The section near the hole, where is wire is passing, is thicker in the case of forming rivet, thus giving an increase in the capable shear force. The actual deformations of the rivets can be observed in figure 15.

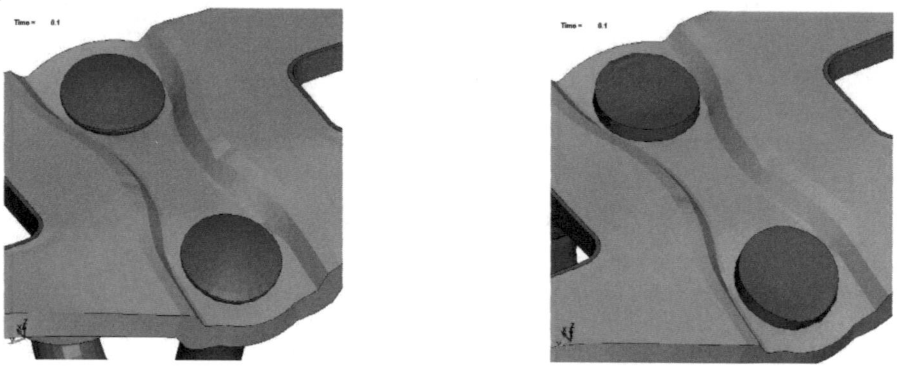

**Figure 15.** Deformed shapes of the rivet: with the formed rivet and with the cylindrical rivet.

## 6. Conclusions

The work presented in this paper lays its foundations on the procedure used during the product validation process. The product is validated once the experiments performed on prototypes show the expected performances. The use of prototypes required a design validation process. The process includes a numerical simulation stage.

The numerical simulation should be as realistic as possible and capable of capturing the most representative features of the model. This includes geometry and the manufacturing process. For this, a co-simulation model was developed. The steps and the strategy adopted during the numerical simulation give great flexibility to the model. The riveting process can be reused for other products while keeping the specifications of the development of the numerical model.

Results show that satisfactory results can be obtained by a proper combination of experiments and numerical simulations.

IOP Conf. Series: Materials Science and Engineering **1311** (2024) 012038 doi:10.1088/1757-899X/1311/1/012038

## References


[1] X. Li, K.M. Nam, Environmental regulations as industrial policy: Vehicle emission standards and automotive industry performance, Environ. Sci. Policy. 131 (2022) 68–83. https://doi.org/10.1016/j.envsci.2022.01.015

[2] J.M. Ikome, O.T. Laseinde, M.G. Kanakana Katumba, The Future of the Automotive Manufacturing Industry in Developing Nations: A Case Study of its Sustainability Based on South Africa's Paradigm, Procedia Comput. Sci. 200 (2022) 1165–1173. https://doi.org/10.1016/j.procs.2022.01.316.

[3] V. Palea, C. Santhià, The financial impact of carbon risk and mitigation strategies: Insights from the automotive industry, J. Clean. Prod. 344 (2022). https://doi.org/10.1016/j.jclepro.2022.131001.

[4] P. Munten, J. Vanhamme, F. Maon, V. Swaen, A. Lindgreen, Addressing tensions in coopetition for sustainable innovation: Insights from the automotive industry, J. Bus. Res. 136 (2021) 10–20. https://doi.org/10.1016/j.jbusres.2021.07.020.

[5] X. Yu, J. Zhou, H. Liang, Z. Jiang, L. Wu, Mechanical metamaterials associated with stiffness, rigidity and compressibility: A brief review, Prog. Mater. Sci. 94 (2018) 114–173. https://doi.org/10.1016/j.pmatsci.2017.12.003.

[6] M.S. Sarfraz, H. Hong, S.S. Kim, Recent developments in the manufacturing technologies of composite components and their cost-effectiveness in the automotive industry: A review study, Compos. Struct. 266 (2021) 113864. https://doi.org/10.1016/j.compstruct.2021.113864.

[7] B. Stojanovic, J. Glisovic, Application of Ceramic Matrix Composite in Automotive Industry, Elsevier Ltd., 2006. https://doi.org/10.1016/B978-0-12-819724-0.00018-5.

[8] G. Atxaga, A. Arroyo, B. Canflanca, Hot stamping of aerospace aluminium alloys: Automotive technologies for the aeronautics industry, J. Manuf. Process. 81 (2022) 817–827. https://doi.org/10.1016/j.jmapro.2022.07.032.

[9] J. Santos, R.M. Gouveia, F.J.G. Silva, Designing a new sustainable approach to the change for lightweight materials in structural components used in truck industry, J. Clean. Prod. (2017). https://doi.org/10.1016/j.jclepro.2017.06.174.

[10] S. Suhaib Kamran, A. Haleem, S. Bahl, M. Javaid, C. Prakash, D. Budhhi, Artificial intelligence and advanced materials in automotive industry: Potential applications and perspectives, Mater. Today Proc. 62 (2022) 4207–4214. https://doi.org/10.1016/j.matpr.2022.04.727.

[11] E. Nitu, S. Tabacu, M. Iordache, D. Iacomi, Finite element analysis and experimental validation of the wedge rolling process, Proc. Inst. Mech. Eng. Part B J. Eng. Manuf. 227 (2013) 1325–1339. https://doi.org/10.1177/0954405413485568.

[12] U. Nations, Addendum 10: Regulation No. 11: Revision 2, (2013). http://www.unece.org/trans/main/wp29/wp29regs1-20.html .

[13] S. Tabacu, S. Parlac, C. Tomus, V. Baba, Computational modelling of vehicle interior components for impact applications: moulding residual stress, Int. J. Crashworthiness. 17 (2012) 591–605. https://doi.org/10.1080/13588265.2012.703473.

[14] S. Tabacu, C. Ducu, Numerical investigations of 3D printed structures under compressive loads using damage and fracture criterion: Experiments, parameter identification, and validation, Extrem. Mech. Lett. 39 (2020) 100775. https://doi.org/10.1016/j.eml.2020.100775.

[15] A. Nasser, A. Yadav, P. Pathak, T. Altan, Determination of the flow stress of five AHSS sheet materials (DP 600, DP 780, DP 780-CR, DP 780-HY and TRIP 780) using the uniaxial tensile and the biaxial Viscous Pressure Bulge (VPB) tests, J. Mater. Process. Technol. 210 (2010) 429–436. https://doi.org/10.1016/j.jmatprotec.2009.10.003.

[16] A. Reyes, O.S. Hopperstad, O.G. Lademo, M. Langseth, Modeling of textured aluminum alloys used in a bumper system: Material tests and characterization, Comput. Mater. Sci. 37 (2006) 246–268. https://doi.org/10.1016/j.commatsci.2005.07.001.

[17] Y. Xiao, Y. Hu, An extended iterative identification method for the GISSMO model, Metals (Basel). 9 (2019). https://doi.org/10.3390/met9050568.

IOP Conf. Series: Materials Science and Engineering **1311** (2024) 012039 doi:10.1088/1757-899X/1311/1/012039

# Diagnosing the truck engine operation by evaluating engine oil wear and issuing maintenance recommendations

**C Zaharia[1*], R Niculescu[1], M Năstase[1], A Clenci[1]**

[1]University of Pitesti, Faculty of Mechanics and Technology, Department Automobiles and Transport, 1, Tg. din Vale Street, Pitești, 110040, Argeș, Romania

E-mail: catalin.zaharia@upit.ro

**Abstract.** The lubricating oil of an internal combustion engine (i.c.e.) performs several functions having a major contribution to the reliability of the system, implying proper operation during the period of exploitation. Replacing the engine oil ahead reaching the turnover / time corresponding to preventive maintenance it turns out to be sometime an imperative action, with negative implications on operating costs. Any user of today vehicles desire to use the engine performances at full percentage, and for that an important condition becomes the changing of the oil at defined intervals of time/rollover intervals. Thus, it is very important to be aware when to replace because a late change of oil can affect the engine operation and, surely, the performance of the vehicle. Premature oil change influences costs, the environment and is considered a harmful action in terms of the process of depletion of resources. As regards the operating conditions, it is intended that they are similar for all vehicles taken into consideration. It is imperative that the study be carried out by investigating the operation of a single type of engine, with a single type of engine oil, for several vehicles with the same technical characteristics. It is considered that the transport fleets are the concrete source that falls under the already mentioned conditions and a methodology based on the analysis of the 10W30 oil samples used in the MX - 11 330 H2, EURO 6 engines of the DAF vehicles within the fleet of one transport company from Argeș county will be used. The authors of this paper will analyze the following properties of the oil: viscosity, density, flash point, contamination with fuel and water content, these being the most important in illustrating the degree of wear and tear of it. The information about the evolution during the exploitation period of these characteristics is the basis for the elaboration of maintenance strategies considering the three desiderata: the environmental issues, the vehicle's reliability, and the operating costs of the vehicle.

## 1. Introduction

The development of the automobile it is based on the need for information and understanding of the relationship between the internal combustion engine (i.c.e.) and its lubricant. The truck fleet owner's expectations/desires of reliability grew with the automobile development. Considering this growth and accompanying warranty responsibility, engine manufacturers began specifying minimum performance stages using engine and bench tests and this specification rise with time and understanding of engine requirements, [1].

There are many conditions of engine oils that allows to the engine to run properly over time. The most important ones are lubrication, cooling, cleaning, protection against corrosion and sealing. The engine temperature may rise and that is the cause of the energy lost through combustion and the friction between mechanical parts. The characteristic of lubrication offered by the engine oil helps to partly address the heat through the lubrication circuit. The engine also has the coolant, which can only cool certain parts of the engine. While less known, the cleaning power of engine oil is fundamental. One of the significant factors it was demonstrated to be the wear resistance of friction pairs. This resistance depends on the metal-physical characteristics of the friction surfaces as well as the physical and chemical properties of engine oils, methods of their transfer to friction modules, and the design of the lubrication system, [5].

Engine oil also provides the sealing of pistons and cylinders. A protective film is maintained between the various parts and for that the oil level must be checked regularly, action that contribute to the change of oil at the right time.

Fuel combustion engender corrosive acid that can cause to the damaging of i.c.e. metal parts. With the existence of additives in the modern engine oils, corrosion is decelerated, but over time, and in the presence of oxygen, engine oil can oxidize and will no longer play its corrosion role. That's the reason engine oil must be replaced systematically. The outspread utilization hybridization and alternative fuels impose challenges to automotive lubricant improvement. With the innovative fuel contents and species of combustion by-products, which are newly to the engine operations nowadays, the oil degradation can be affected and may induce increased wear at the level of contact pairs and bearing surfaces of an i.c.e., [7]. At cold engine starting, it operates in a way that richens the fuel mixture, which makes excess fuel to flow down through the piston rings and into the crankcase. This dilutes the oil and breaks down its chemical properties, finally modifying the oil's ability to flow.

The change in the properties of the oil can rely on its contamination with impurity, metal residues due to wear, oil particles caused by the modifying in its composition, and dilution with cooling liquid, water and fuel, [1], [20]. The quantity of engine oil in the lubrication system varies during engine operation, increasing with the contamination level of the oil with a specific amount of water, coolant and / or fuel or decreases by oil oxidation or loss on the seals.

Various factors affect the recommended interval between oil changes. Sometimes drivers complete their oil change relying on the sticker in the engine compartment or on the board computer information. Driving style is one of the main factors that influence oil change intervals. The recommended oil change schedule must take in consideration the difference between normal and severe driving and for that, manufacturers will present two sets of maintenance planning: one for normal driving and one for severe driving.

Regarding the maintenance of the truck engine MX – 11 330 H2, the subject of this study, the manufacturer imposed the next maintenance intervals for the oil DAF Xtreme FE 10W30: for severe driving, oil and filter change at 48000 km (or 800 hours), for normal driving (> 20% idle time), oil change and filter at 80000 km (or 1500 hours) and for normal driving (< 20% idle time), oil change and filter at 120000 km (or 2250 hours), [22].

Nowadays most of vehicles have an engine oil life monitoring equipment. Their operations are based on monitoring of some parameters such as: the crankcase temperature, combustion chamber events, and the interval passed since the last oil replacement. Then, working with a computer method it will be informed the proximate percent of usable life left in the oil. These kinds of monitoring can sustain an oil replacement before the date/mileage presented in the service interval planned in the owner's manual or board computer. If the oil life monitoring equipment indicates to change the oil, then it is compulsory to do so whatever the mileage of the vehicle is. For that, in paper [2] the authors presented a predictive method for the estimation of the value of a quality factor of engine oil based on mission profile results. In the proposed method, it was analyzed the relation between indirect indicators and direct indicators with statistical analysis. For finding the relationship between an oil quality indicator and mission profile data, were utilized regression models and then, it was approximated the engine oil state, [2].

Thus, for a normal operation, the quality and quantity of engine oil are extremely important factors in the period between two planned actions of maintenance. The manufacturers of automotive equipment have come up with relatively simple technical devices, the oil condition sensors, which, at the beginning, do not identify any specific variable. These instruments cannot however take over the outcome of lab tests, they can indicate some information to decide preventive maintenance choice ahead fault appear, [8], [16], [20].

Vehicle manufacturers developed software that are guided by the injection calculator of the internal combustion engine. The strategy of activating of the warning light "service" is based on knowing the operating conditions and the value of rotations of the crankshaft in a specified interval of time.

Late change of engine oil can involve the damaging of engine components with extra costs of repairs actions, which are often significant. The premature replacement of engine oil influences the environment and can be treated like a weak procedure if the resource reduction activity is considered [9], [10]. So, when talking about about oil as a important resource, every force should be involved to recycle it on any occasion. Thus, it is a normal process and an obligation to collect waste oils separately and to organize and develop recycling or removal activities [11], [12].

SO, with the content of this paper, the authors intent to develop a procedure based on the analysis of the 10W30 oil samples used in the MX - 11330 H2 engines of the DAF vehicles within a company's fleet from Argeş county. The following parameters will be analyzed: viscosity, density, flashpoint, contamination with fuel and water content, these being the most important in illustrating the degree of wear of the oil.

## 2. Engine oil physico - chemical properties

For the developing of a methodology for characterizing the actual properties of the used engine oil, the authors will review parameters such as: kinematic viscosity, flash point, density, fuel dilution/content and water content [17].

The property of a fluid's resistance to flow at a given temperature is the viscosity, which in the case of oils, it is indicated as the thickness of the lubricant [13]. The oils producers are basing on viscosity tests as an important factor in finding the state of used oil, or details of new oil. Modification in viscosity of used oils can be an important sign which indicates lubricant degradation or oil dilution, [4], [20]. The viscosity is known as kinematic viscosity and is commonly noted kV. Kinematic viscosity is the absolute or dynamic viscosity in (Pa · s) divided by the fluid density. This property is determined by measuring at a well-known temperature the time a fluid demand to pass through a specific capillary, and it is presented as a kV@40 or kV@100 test result. The viscosity of oils is decided at 100°C corresponding to the SAE J300 Industry Classification System and is presented in $mm^2/s$ or centistokes (cSt). The viscosity is also determined at 40 °C and some used oil investigation labs announced the results of viscosity at 40 °C, 100 °C or both, [20].

Changes in lubricant viscosity appear because of oil molecules that breaks in engines that operates to high temperatures [18]. For that reason, lubricants can receive additives to improve their viscosity. But, at their turn, the additives may decompose while are used and break, and again the viscosity decreases, and it is well known that the decreasing is more dangerous than increasing for the mentioned parameter. If the viscosity decrease comparing with the reference, the thickness of the oil film will be damaged, and there can appear the risk of damaging the bearings by blocking them. The studies conducted on this area reported that the factors that involve the reductions in viscosity are firstly, the contamination of the oil with fuel, the addition of fresh oil with other diminished viscosity or its contamination by cleaning fluids, [18].

Flash point (FP) it is measured under controlled laboratory conditions and indicates the capacity of an oil sample to create a flammable mix combined with air. The results of flash point measuring are commonly utilized in shipping and safety settlements to define flammable and combustible materials. Also, flash point results offer information about the existence of highly volatile and flammable material in a nonvolatile or nonflammable material, such as the dilution of lubricating oils by some quantities of diesel or gasoline. Flash point could also change, and the process is altered by fuel dilution and to a specific extent by oil depreciation [14]. The industry of flash point test methods is including the "open cup" and "closed cup" type approach [3].

IOP Conf. Series: Materials Science and Engineering **1311** (2024) 012039 doi:10.1088/1757-899X/1311/1/012039

Density (d) is a part of product details/specifications and, is a significant chemical property of products based on petroleum.

## 3. Methodology proposal and experimental settings

In this paper the authors proposed a working approach which is developed by acquiring engine oil samples directly from the garage of the transport fleet company in the moment when the truck was coming to the planned maintenance process. The next step is the investigation in the lab and after at the accomplishment of the laboratory tests and analysis, established on the results gathered, will be delivered recommendations on the actions of maintenance to be made (figure 1).

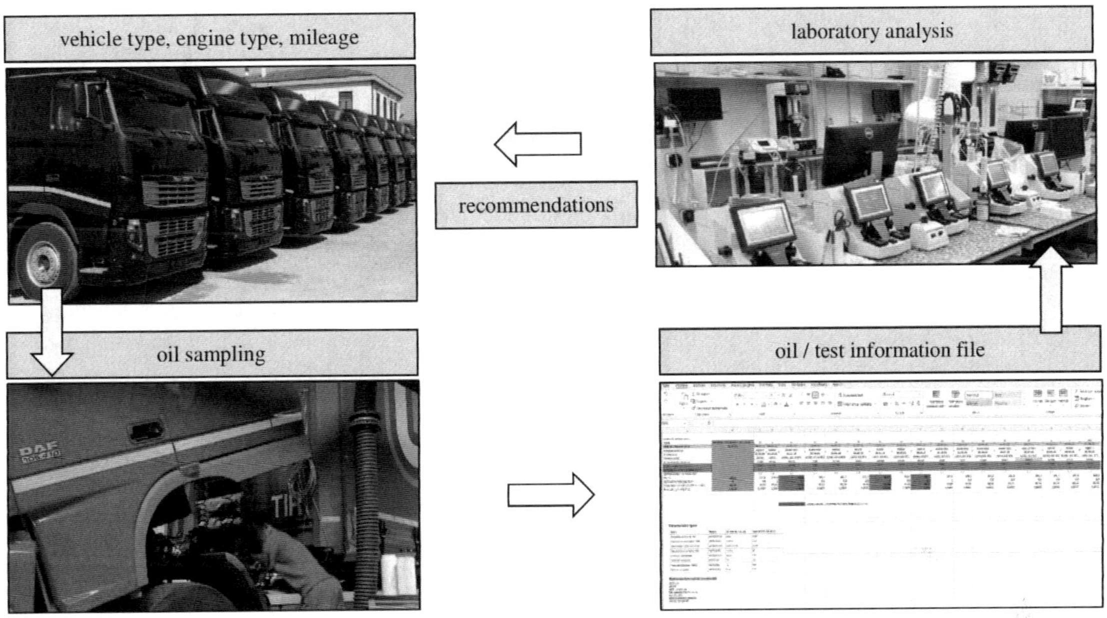


**Figure 1.** The four steps approach for oil sample analysis

The research was conducted on oil samples from DAF i.c.e. with manufacture year between 2016 and 2021, with the next specifications: type of the engine oil (DAF Xtreme FE 10W30, approved by the manufacturer), engine type/model (diesel, EGR, DPF, SCR / Paccar MX – 11330 H2), cylinder capacity (10800 $cm^3$), engine power (449 HP @ 1600 rpm), mileage (100000 - 800000 km) and emissions standards (EURO 6).

Viscosity - equipment description. U-VIsc is the appropriate apparatus for the study of viscosity, and it is grant as a specific tool for the study of oils, both mineral and synthetic and for any fluid within the viscosity range of the apparatus [15], [20]. U-VIsc is an automated equipment based on the glass capillarity to obtain details of kinematic viscosity of Newtonian fluids.

Method description. Standard test approach for kinematic viscosity of liquids both, transparent and opaque is indicated under the code D 445 and it is under the administration of ASTM Committee D-2 for Petroleum Products and Lubricants. The time must be known and for that a well-defined volume of liquid will pass under gravity through the capillary of a viscometer under a reproducible driving head. The hole process considers that the temperature is known and strictly controlled [15], [20].

Flash point and dilution - equipment description. To determine the oil flash point and dilution degree the authors used for the measurement the Eralytics apparatus (figure 4). It is well known that the measurement is precisely under the standards ASTM D6450 & D7094 and relies on the Continuous Closed Cup method for determining the Flash Point (CCCFP).

Method description. The approach used is ASTM D7094 known as an exceptionally adaptable flash point method that can be utilized for the correct flash point tests and, to find the impact to the flash point of extremely flammable fractions in samples (diesel in oil, gasoline in diesel, or gas in oil).
Density - equipment description. The apparatus is Density Meter DS7800 [15], [20].

Method description. The equipment works according to the frequency oscillator principle. The authors used it for the determination of the density or relative density of crude oils with high vapor pressures. The results are significant and help to the prevention of vapor loss during transfer of the sample to the density analyzer. The method relies on the addition of a volume of liquid sample into an oscillating sample tube. To determine the density of the sample the change in oscillating frequency caused by the change in the mass of the tube is utilized in association with calibration data [15].

Water can contaminate the engine oil causing its damage. Thus, it is recommended that its value be kept below 2000 ppm [3]. Water could cause corrosion of various metal components of the engine. It can also enter a chemical reaction with the additives leading to their exhaustion with profound consequences on the quality of the oil, it can lead to its acidification. Regrettably, with any i.c.e., water in its oil seem to be a reality. To find the causes will be needed to look after internal leaks and by defining the causes of the presence of condensation.

Water content - equipment description. To determine the water content, the authors used for the measurement the Aquamax KF Coulometric made by GRScientific –UNITED KINGDOM. It is mentioned that the measurement is precisely under the standards: ISO 10101-3, 10337&12937, ASTM D6304, D4928, D1533, IP 438, IP 386, API MPMS Chapter 10.9, IEC 60814.

Method description: Karl Fischer titration is simply a means to measure water content of samples. Modern instruments, such as the Aquamax KF, use the Coulometric principle, whereby the water present in the sample is coulometrically titrated to a predefined end point at which there is a minute excess of free iodine present. Combining the Coulometric technique with Karl Fischer titration, Aquamax KF titrators determine the water content of the sample by measuring the amount of electrolysis current necessary to produce the required iodine. This is an absolute technique which does not require calibration of the reagents.

## 4. Results and discussions

Engine oil samples were collected from 43 trucks engines. The collected oil was studied considering the physico-chemical properties of the oil, finding values for the next parameters: flash point, FP (°C), diesel content, DC (% vol.), kinematic viscosity at 40 °C, kV (mm²/s), density at 15° C, d, (g/cm³), and water content (ppm). The limits of use in terms of the values of the properties of the engine oil used are not provided in the standards, but according to [3] the specialists, from experience, make some recommendations, presented in the table below (table 1).

**Table 1.** The limits of oil samples evaluation

| Oil properties | Reference value | Recommended area of precaution |
|---|---|---|
| Flash point (° C) | 204.1 | min. 190 |
| Diesel content (%) | 0 | - |
| Kinematic viscosity (mm²/s) | 81.5 | -25% … +25% |
| Density (g/cm³) | 0.8669 | min. 0,7 … max. 0,95 [23] |
| Water content (ppm) | - | max. 2000 ppm |

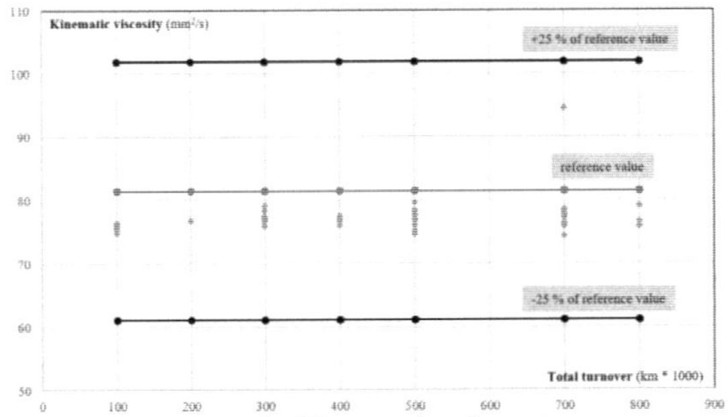


**Figure 2.** Kinematic viscosity

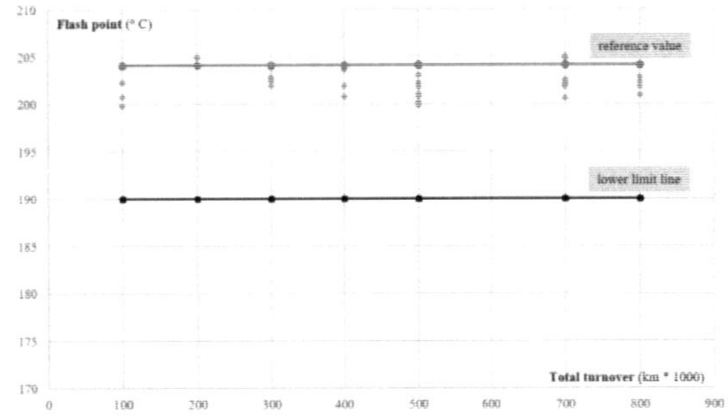


**Figure 3.** Flash point

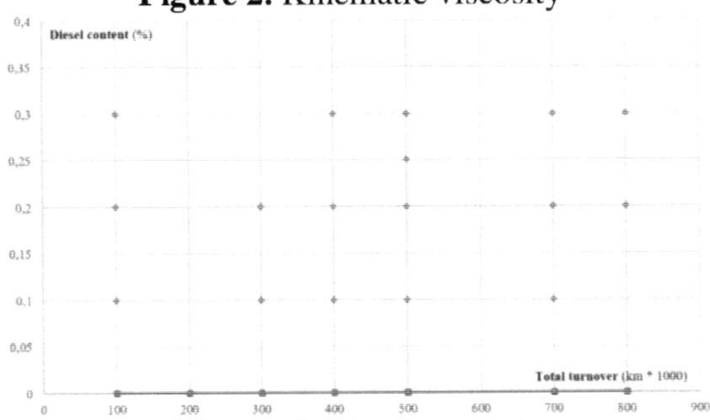


**Figure 4.** Diesel content

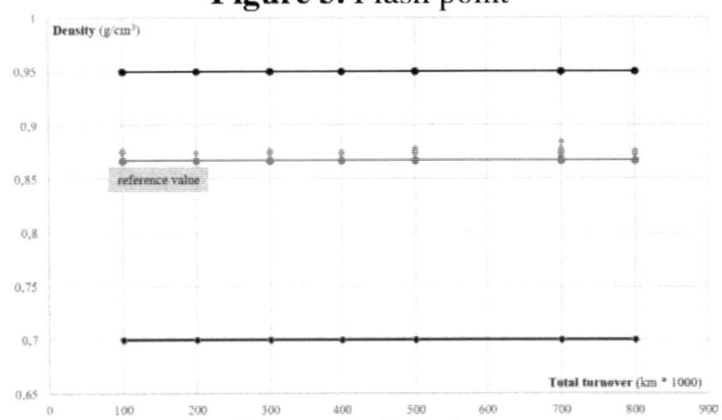


**Figure 5.** Density

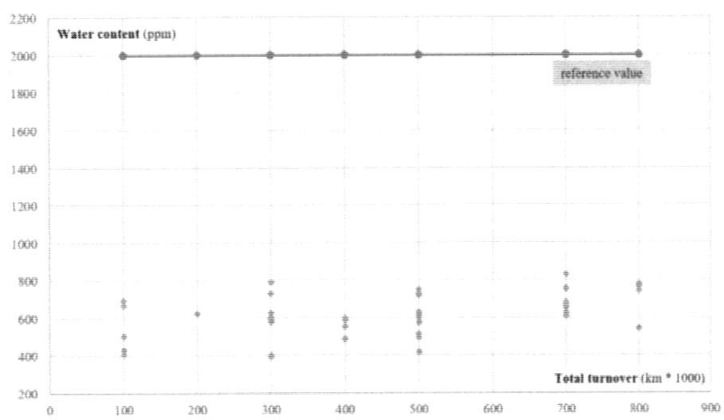


**Figure 6.** Water content

The value of 81.5 mm²/s is the reference data for kinematic viscosity (kV, table 1) for the new oil (10W30). From figure 2 all the samples fall within the recommended area of caution. A number of 42 samples out of 43 analyzed are near the reference value, their variation not being influenced by the total turnover of the trucks. A single sample has a higher viscosity value (94.53 mm²/s) in the case of a truck with a total turnover of 700000 km for which it is recommended to carry out the diagnostic process based on possible causes involving increased viscosity: low level of engine oil (not being completed), degree of oxidation, sediment level. It is recommended to check the fuel combustion process and the engine operating temperature level.

When it comes to analyze the flash point, the reference data for the oil 10W30, for specified engine, is 204.1 $^0$C. After the analysis of the obtained results, it is noticed that all the samples have values at least 10 $^0$C above the minimum precautionary value (figure 3).

The diesel content was determined with the help of the results obtained for flash point. In the figure 4 is shown the diesel dilution of the oil and it is noticed that the limits are normal, as seen from the results also obtained for the flashpoint.

The highest fuel content in oil is found in engines with a turnover of 800000 km, with a value of 0.225%, the result being obtained by the average of the analyzed samples in the turnover category. The increase in the diesel content with the increase in the engine turnover can be influenced by:
- excessive idling operation of the engine or faulty injectors, factors that are responsible to the passing of fuel nearby the piston rings straight into the oil pan.
- diesel engines have diesel particulate filter; their regeneration processes, make possible the additional introduction of diesel into the cylinders or by individual injection of fuel which will be used to eliminate by burning the ash and particles found in the filter [19]; knowing the fact that can exist tolerances between the engine cylinders and the pistons, a quantity of diesel fuel can form a mixture with engine oil.
- another source of contamination of the oil are the engine fuel pumps, also lubricated with oil.

So, for all these reasons the reduction of the oil viscosity appears and that could make possible the metal-to-metal friction which increases wear of components and also, the raise of corrosivity of the oil.

Analyzing the results obtained for the oil density, it is noticed that the values are significantly increased above the reference limit due to different amounts of water, solid residues, and oxidation level, but they are kept in the normal range 0.7 – 0.95 (figure 5). As can be observed in figure 6, the water content of all samples is below the recommended limit (2000 ppm). However, a higher value can be identified in the case of trucks with higher mileage, the statement is made considering the average of the samples resulted from the engines by turnover categories. A possible permeability (gaskets, flexible connections) to moisture of the cooling system may be the cause of the increase in the water content in the oil. To determine punctually the problem that led to the modification of the studied parameters, it is necessary to diagnose the elements mentioned above.

The characteristics of the oil engine revealed changes after analyzing the oil samples taken in the period of operation until the truck engine reaches 100000 km. Their evolutions are not important now and it can be said the fact that the oil samples were collected from truck engines with years of production between 2016 – 2021, which do not have important wear.

## 5. Conclusions

It is not anymore, a novelty that the structure of the maintenance process changes in a continuous way. The process of strict control of the state condition of the trucks has implications on the volume of maintenance action for the correction of those. Thus, the result of a present-day precautionary maintenance process based on real-time results and fast evaluation is highlighted, so that the reliability planners can generate fast maintenance choices.

This kind of approach, structured on the investigation of the engine oil used are very helpful in finding in advance the wear level and possible faults that can appear in the period of engine operation and resolving them before they become imminent and any kind of capital consuming.

Taking the example of truck fleets companies, one of their dominant concerning is the reduction of costs for maintenance, diagnostics, and repair activities.

For sector of road transportation, the authors of this study can come up with the following method by periodically studying engine oils before the truck reaches turnover limits, evidently knowing: the same oil used for the entire truck fleet, the turnover made with the analyzed engine oil and total turnover of the truck, year of manufacture of the truck, quantity and type of oil completed if there was the case and even driving conditions.

Taking in consideration the collected results, the truck company can have information about the level of engines wear, and there can be recommended preventive maintenance actions before the truck reaches the turnover maximum limit to efficiently administer their costs.

## References

[1] Theodore W. Selby Jonathan C. Evans Gordon C. Cox Tina P. Dasbach 2014, *A Global Perspective of Engine Oil Properties,* 19th International Colloquium Tribology – Lubricants, Materials and Lubrication Technische Akademie Esslingen, Stuttgart/Ostfildern, Germany.

[2] Dimitris Kiritsis Hong-bae Jun 2008 *A predictive algorithm for estimating the quality of vehicle engine oil,* International Journal of Industrial Engineering, 15(4), 386-396.
[3] CIMAC 2011 *The International Council on Combustion Engines, Used engine oil analysis - user interpretation guide*, No. 3.
[4] Bogdan Landowski Monika Baran 2019 *Analysis of selected results of engine oil tests*, 18$^{th}$ International Conference Diagnostics of Machines and Vehicles.
[5] Vladimir Baskov Anton Ignatov Vladislav Polotnyanschikov 2020, *Assesing the influence of operating factors on the properties of engine oil and the environmental safety of internal combustion engine*, XIV International Conference 2020 SPbGASU "Organization and safety of traffic in large cities.
[6] Nehad A. Ahmad Khaled M. Abdelbary Samy M. Younis 2018, *Chemical analysis of engine oils as an indicator to estimate the rate of wear*, Egypt. J. Chem. Vol. 61, No.4 pp.581 – 590.
[7] Adam Agocs Andras Lajos Nagy Zsolt Tabakov J´ozsef Perger Jan Rohde Brandenburger Michael Schandl Charlotte Besser Nicole D¨orr 2021, *Comprehensive assessment of oil degradation patterns in petrol and diesel engines observed in a field test with passenger cars – Conventional oil analysis and fuel dilution*, Tribology International 161 107079.
[8] Perić Sreten, Bučko Mihael Nedić Bogdan Radovanović Radovan 2020, *Condition monitoring through engine oil analysis tests*, International Congress Motor Vehicles & Motors 2020 Kragujevac, Serbia.
[9] Paul Farless 2021 *The DOs and DONT's of Used Oil* (Poseidon Systems, Machinery Lubrication).
[10] Michael C. Brown 2015 *Why You Should Regenerate, Reclaim or Recycle Used Oil* (Poseidon Systems, Machinery Lubrication).
[11] DIRECTIVE 2008/98/EC OF THE EUROPEAN PARLIAMENT AND OF THE COUNCIL 2008 *on waste and repealing certain Directives*.
[12] HOTARARE nr. 235 2007 privind gestionarea uleiurilor uzate (MONITORUL OFICIAL nr. 199 / 2007.
[13] Isuru Premaratne A J M Dayawantha 2017 *Vehicle service date prediction system using lubricant viscosity degradation pattern* (Jaipur, India: International Conference on Computer, Communications and Electronics).
[14] Dana Kareem Hameed Kameran Ali 2020 *Analysis of 0W-20 Totachi Brand Oil to Determine the Rate of Oil Deterioration* (UKH Journal of Science and Engineering, Volume 4, Number 2).
[15] Zaharia C Niculescu R Iorga Ducu C Clenci A Aron B 2016 *Diagnosing the operation of a locomotive diesel engine based on the analysis of used oil in the period between two technical revisions* (Brașov, Romania: International Congress of Automotive and Transport Engineering, 319-327, Editor Springer).
[16] Hugo Raposo José Torres Farinha Inácio Fonseca L. Andrade Ferreira 2019, *Condition monitoring with prediction based on diesel engine oil analysis: a case study for urban buses*, MDPI, Actuators.
[17] Noria Corporation 2021 *Density's Role in Lubricant Performance* (Poseidon Systems, Machinery Lubrication).
[18] Story R 2019 *Why an Oil's Viscosity Drops* (Poseidon Systems, Machinery Lubrication).
[19] Tingjun H Teng H Xuwei L Bin C 2015 *Impact of Fuel Injection on Dilution of Engine Crankcase Oil for Turbocharged Gasoline Direct-Injection* (SAE 2015 World Congress & Exhibition).
[20] Zaharia C Niculescu R M Năstase Clenci A Iorga S 2021 *Engine oil analysis in order to evaluate the degree of its wear during the period of operation of the vehicle,* AITS 2021.
[21] www.totalparco.com
[22] www.paccarpowertrain.com
[23] https://www.neste.com/products/all-products/fossil-products/lubricants/basic-conceptslubricants

**Acknowledgments:**
**Author Contributions:** The authors have equal contribution to the work presented in this article.

IOP Conf. Series: Materials Science and Engineering **1311** (2024) 012040 doi:10.1088/1757-899X/1311/1/012040

# The influence of particle filter regeneration on oil consumption in euro 5 Diesel Engine

**D. Ostoia[1*], A.S. Negoițescu [1],**

[1] Politehnica University of Timişoara, Timişoara, Romania

*E-mail: daniel.ostoia@upt.ro

**Abstract**. The transport sector is a significant contributor to air pollution through the concentrations of emissions contained in the road vehicles flue gases. Although the implementation of EURO standards and the use of modern de-pollution technologies have greatly contributed to the reduction of these emissions, the concentrations of pollutants in the atmosphere are still high. In this context, the paper analyses the cases in which the stability of the de-pollution system indirectly influences the oil consumption during the operation of a EURO 5 diesel engine fitted to an M1 car, which is caused by its improper operation. Thus, oil consumption is analyzed according to the degree of particulate filter blockage and its influence on engine performance. The results of the study showed that new technical solutions are needed to improve the stability of the engine operation. These solutions also lead to a reduction in the pollution generated by the car operation.

## 1. Introduction

The constructive-functional stability of an engine's parameters influences its performance both from an energy point of view and from the point of view of the parameters that characterize its pollution degree [1], [2].

Manufacturers of engines for modern motor vehicles are subject to legislative restrictions on pollution standards, which are characterized by the limitation of pollutants produced. As such, the technical condition of the engine and its auxiliary systems must be taken as a diagnostic parameter. The noxious emissions produced by an engine are directly dependent on functional design factors, which are linked to the technical solutions adopted by design [2], [3].

On the other hand, these damages are also produced indirectly by some factors that are dependent on the technical condition of the engine that has changed, either as a result of wear and tear, but also as a result of the way the vehicle is operated [4].

A parameter of the technical condition of an engine is the oil consumption. If the oil enters the combustion chamber in excess, the oil consumption may increase above the value established by the manufacturer, which leads to an increase in pollutant emissions.

IOP Conf. Series: Materials Science and Engineering **1311** (2024) 012040 doi:10.1088/1757-899X/1311/1/012040

## 2. Engine oil consumption analysis

### *2.1. Technical data of the vehicle engine*

In this paper there were analyzed the technical conditions from the point of view of the excessive oil consumption of a Diesel engine that equips a car. The car technical configuration is presented in Table1.

**Table 1.** The car technical configuration

| | |
|---|---|
| Engine type | 4 in-line cylinders |
| Cubic capacity | 1998cmc |
| Fuel type | diesel |
| Standard of emissions | Euro 5 |
| Fuel consumption - urban | 7.5 l/100 km |
| Fuel consumption - extra-urban | 5.1 l/100 km |
| Fuel consumption - mixt | 5.9 l/100 km |
| Particle filter | yes |
| Adblue system. | no |
| Quantity of $CO_2$ | 156 g/km |
| Engine power | 150 HP at 3600 rot/min |
| Engine power/liter | 75.1 HP/l |
| Torque | 350 Nm at 1800-2400 rot/min |
| Diameter of the cylinders | 86 mm |
| Cylinder stroke | 86 mm |
| Compression ratio | 16.3 |
| Number of valves per cylinder | 4 |
| Fuel System | Common-rail Diesel |
| Engine intake | Turbine, Intercooler |
| Distribution system | DOHC |
| Oil tanker capacity | 5.9 l |

### *2.2. Experimental testing of the technical condition of the engine*

While the vehicle is running, the on-board computer display indicates a low oil level via the MIL (Malfunction Indicator Lamp) after a run of 2500km from the maximum reference level.

As the lack of oil during engine operation can lead to an accelerated wear process inside the engine [4], as well as thermal stresses, determining the causes of this condition is important both in terms of engine functionality and in terms of the pollutants emitted because of the engine poor operation.

As the oil consumption was high for a low number of kilometers driven (20000 km), oil consumption per 1000 km driven was further monitored (Table 1) to see if the consumption exceeded the maximum value of 1l oil/1000km allowed by the manufacturer. This monitoring aimed to determine the evolution of the engine wear degree [4]. As similar problems of excessive wear have been observed in these types of engines under warranty, it was assumed in a first analysis that this would be a possible cause of the excessive oil consumption.

The method of determining oil consumption was carried out by monitoring the oil amount which must be added to restore the maximum reference level retained on the dipstick.

The moment of intervention regarding the oil filling was appreciated through the oil level sensor which displays a minimum permissible oil level on board the vehicle.

Initially, the oil consumption monitoring was carried out without analyzing the oil quality and other diagnostic parameters which were recorded in parallel by the engine computer.

IOP Conf. Series: Materials Science and Engineering **1311** (2024) 012040 doi:10.1088/1757-899X/1311/1/012040

**Table 2**. The oil consumption ml/1000 km

| ***No.*** | ***Date of intervention*** | ***No. of km recorded on board/no. of km travelled between inspections*** | ***Works requested/Needed work additional found*** |
|---|---|---|---|
| 1. | **21.05.2012** | **22144 km** | Oil indicator light on/*1 liter of 5W30 oil is added* |
| 2. | **02.08.2012** | **26742km**<br>*4598 km travelled since last recording* | Oil indicator light on/*1 liter of 5W30 oil is added* |
| 3. | **13.09.2012** | **29960 km**<br>*3218 km travelled since last recording* | Service 30000km, very high oil consumption, when changing gears sometimes does not accelerate/ *replace air filter, oil filter, pollen filter, oil* |
| 4. | **02.10.2012** | **30150km** | Particulate filter indicator light on/engine regeneration |
| 5. | **08.10.2012** | **31292 km**<br>*1332 km travelled since last recording* | Particulate filter indicator light on, lack of power especially when accelerating, does not accelerate immediately/*90% particulate filter regeneration, fuel replacement recommended.* |
| 6. | **12.10.2012** | **31478 km**<br>*186 km travelled since last recording* | Particulate filter indicator light on, lack of power especially when accelerating/ *soft reset, particulate filter regeneration* |
| 7. | **23.10.2012** | **32403km**<br>925km *travelled since last recording* | Check/*300 ml engine oil is added*, normal driving behavior, return to the corresponding to 2000km |
| 8. | **10.12.2012** | **34536 km**<br>2133km *travelled since last recording* | Oil level check */150 ml engine oil is added* |
| 9. | - | **36948 km**<br>2412 km *travelled since last recording* | Oil level check */900 ml engine oil is added* |
| 10. | **26.02.2013** | **39303 km**<br>2355 km *travelled since last recording* | Oil level check, occasionally lack of acceleration when changing gear */oil consumption of 300 ml* |
| 11. | **22.05.2013** | **44985 km**<br>2412 km *travelled since last recording* | Service 45000km, check steering, brake system/front and rear pads 50% wear, need wheel geometry/ *replaced: air filter, oil filter, pollen filter, oil* |
| 12. | **23.07.2013** | **48923 km**<br>2412 km *travelled since last recording* | Oil level indicator on */1l oil is added* |
| 13. | **13.08.2013** | **49396 km**<br>473 km *travelled since last recording* | Particulate filter indicator light flashes / *regeneration 59% clogged* |
| 14. | **12.11.2013** | **51409 km**<br>2013 km *travelled since last recording* | Check/the turbocharger is checked/ *400 ml of oil filled, oil check necessary from 2000km* |
| 15. | **14.01.2014** | **53476 km**<br>2067 km *travelled since last recording* | Oil level check/ *900 ml oil is added* |
| 16. | **11.02.2014** | **54365 km**<br>889 km *travelled since last recording* | Particulate filter indicator light flashes / *Particulate filter regeneration* |

As a result of periodically monitoring, there was a fluctuating variation in oil consumption (Figure 1), which led to the conclusion that it was not the engine wear that was causing the excessive increase in oil consumption.

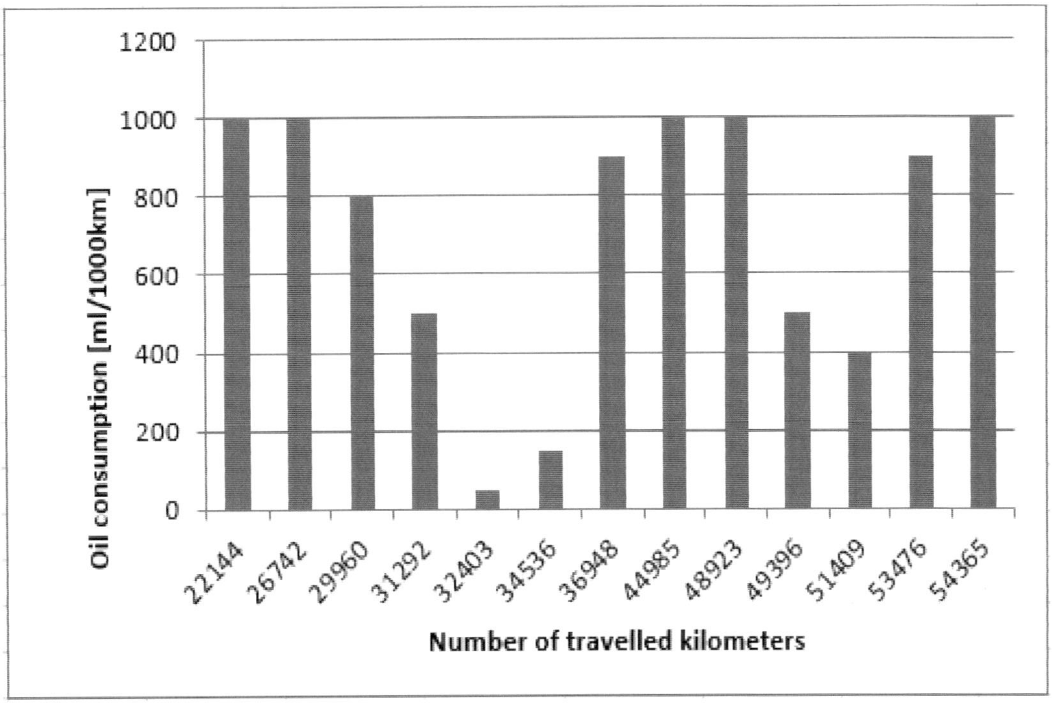


**Figure 1**. Variation in oil consumption

To determine the real cause of the excessive oil consumption, the vehicle was subjected to a computerized diagnosis which resulted in the following error code [5]:

| | |
|---|---|
| Fault P1469 retained in particulate filter (DPF) | DTC P1469 Limp Home normally indicates a blocked "DPF" Diesel Particulate filter, in this case an in-depth diagnosis is required to determine whether the filter requires replacing or another fault exists.<br>Items such as faulty Injectors, poor operating EGR and multiple aborted regenerations can cause a blocked filter [5] |

Following the analysis of the error code significance, the engine functional parameters values were also checked (Table 3).

**Table 3**. Values of engine operating parameters [6]

| Item | Value | Unit | Maximum | Minimum |
|---|---|---|---|---|
| Cylinder #4 quantity correction value | 0 | ms | - | - |
| Cumulative ash ratio | 2 | % | 2 | |
| Estimated catalyst temperature | -20 | °C | -15 | -20 |
| Estimated DPF temperature | -25 | °C | -15 | -20 |
| Soot accumulation ratio | 163 | % | 163 | 163 |
| Oil dilution ratio | 10 | % | 10 | 10 |
| DPF regeneration count | 181 | Time | - | |
| Running distance after last regeration | 4141 | km | 4141 | 4141 |

*2.3. Analysis and interpretation of the engine operating parameter values*

An analysis of the values shown in Table 3 reveals the following**Error! Reference source not found.**:

The particulate filter clogging of 163% indicates that the filter is well above the permissible limit of 30%, which causes the engine computer to initiate the active regeneration procedure. The fact that the filter clogging of 163% was reached in combination with the high number of regenerations (181 regenerations) shows that the regeneration process was carried out too frequently and not properly.

The high number of regenerations attempts of the particulate filter was a consequence of the sudden interruption of the regeneration process due to engine shutdown, as the vehicle under analysis was driven short distances in urban traffic.

The consequence of this sudden interruption of the regeneration process was the accumulation of unburned post-injected fuel in the combustion chamber and its mixing with the oil. This is evidenced by the 10 % oil dilution value.

The significant dilution of the oil led to a decrease in the viscosity of the oil, which favored the penetration of more oil into the combustion chamber, resulting in an excessive increase in oil consumption.

## 3. Conclusions

The analysis carried out on the excessive increase of oil consumption in the engine operation led to the following conclusions:

Failure to correctly identify the real causes of the increased oil consumption led to poor engine operation due to high mileage.

The high number of attempts to regenerate the particulate filter due to the sudden interruption of the regeneration process also led to increased fuel consumption and thus increased pollutant emission concentrations.

IOP Conf. Series: Materials Science and Engineering **1311** (2024) 012040 doi:10.1088/1757-899X/1311/1/012040

The cause of the fault was generated by operating the vehicle in a different thermal mode than the optimal one. Since the real cause of the excessive increase in oil consumption was identified with difficulty, it is necessary to equip the engines with additional oil quality sensors of the OLM (Oil Life Monitor) type, in conjunction with a calculation algorithm that would consider the number of starts/stops, the thermal regime of the engines, but above all the number of failed regenerations. The presence of these sensors and the calculation algorithm would lead to an improvement in the rapid and correct diagnosis of faults that increase oil consumption and reduce engine reliability.

## References

[1] İbrahim Aslan Reşitoğlu, Kemal Altinişik & Ali Keskin. *The pollutant emissions from diesel-engine vehicles and exhaust aftertreatment systems*. Clean Technologies and Environmental Policy volume 17, pages15–27 (2015).

[2] Băldean, D. L., *Cercetarea aspectelor disfuncționale legate de filtrul de particule diesel (dpf) în cazul motorului dacia logan e5 1.5dci*, A XVII-A Conferința Internatională-multidisciplinară Profesorul Dorin Pavel-Fondatorul hidrotehnicii romanesti, Sebeş, 2017.

[3] *REGULAMENTUL (CE) NR. 715/2007 AL PARLAMENTULUI EUROPEAN ŞI AL CONSILIULUI din 20 iunie 2007 privind omologarea de tip a autovehiculelor în ceea ce priveşte emisiile provenind de la vehiculele uşoare pentru pasageri şi de la vehiculele uşoare comerciale (Euro 5 şi Euro 6) şi privind accesul la informațiile referitoare la repararea şi întreținerea vehiculelor.* Jurnalul Oficial al Uniunii Europene., 29.06.2007, https://eur-lex.europa.eu.

[4] Dimaratos, A., Giechaskiel, B., Clairotte, M., Fontaras G. *Impact of Active Diesel Particulate Filter Regeneration on Carbon Dioxide, Nitrogen Oxides and Particle Number Emissions from Euro 5 and 6 Vehicles under Laboratory Testing and Real-World Driving*. Energies 2022, 15, 5070, https://doi.org/10.3390/en15145070.

[5] WONG, W. V., TUNG, C.S., *Overview of automotive engine friction and reduction trends–Effects of surface, material, and lubricant-additive technologies*, Massachusetts Institute of Technology, Cambridge, MA 02139, USA2 Vanderbilt Chemicals LLC, Norwalk, CT 06856, USA 06 January 2016

[6] Subaru Select Monitor III SSM3 Diagnostic Scan Tool.

# The influence of data aggregation levels on accident severity researches

**A Drosu[1*], C Cofaru[1], M V Popescu[1] and C Slaff[2]**

[1]Transilvania University of Brasov, Brasov, Romania

[2]Misericordia University, Dallas PA, USA

*E-mail: drosu.alin.gabriel@unitbv.ro

**Abstract**. This paper aims at assessing the influence of data aggregation on the estimation of the accidents' severity. For this purpose, 27,400 accidents are investigated on three aggregation levels: individually (idl) which is highly disaggregated hourly (hrl), and daily (dil). A Severity Score (SS) is defined based on the injuries' severity and relative cost weights. Multiple Linear Regressions are developed by assessing the SS's variation as a collective explanatory effect of 83 predictors related to road environment, weather and lighting conditions, adherence conditions, and number of the involved vehicles. F and t tests are employed in order to assess the developed models and to investigate which variable is of statistical significance. This research shows that the higher the level of aggregation is, the higher the coefficient of determination. SS_dil model has the higher $R^2$ value and direct relationships between SS and the number of accidents occurring in "U" curves, on roads with no shoulders, where the roadway was covered with slime or black ice and where the roadway was wet was found. One of the interesting findings of this paper shows that in the models with low $R^2$ values (based on individual and hourly aggregated data), snow and fog significantly explain the variation of SS. As these are seasonally and short time phenomena, a further research should be focused on how the estimated models based on individual or hourly aggregated data could be improved by adding other significant predictors related to driver's and vehicle's characteristics and road environment' features.

## 1. Introduction

Road safety represents an interesting research topic since the United Nations (UN) launched, in 2019, a worldwide strategy aimed at reducing the number of road crashes. This strategy represents the "first multi-sectorial and holistic approach to road safety management" implemented by the UN in line with the Global Decade of Action for Road Safety (2011-2020) and SDG 3.6 [1]. The next year in 2020 the UN General Assembly adopted resolution A/RES/74/299 with the same aim to improve the global road safety and proclaimed the "Decade of Action Road Safety 2021-2030" [2] with a highly ambitious goal of reducing road traffic deaths and injuries by 50% by 2030 in partnership with other organizations who have such goals like the WHO, FIA, etc.

The road accident severity is therefore, an interesting and important research topic often discussed and analyzed in the literature. This research looks at accident data via statistical methods widely used by researchers in this field in order to assess the road accident severity. These are principally based on aggregated data, the outputs which at a large extent depend on the way the raw data are prepared to be

IOP Conf. Series: Materials Science and Engineering **1311** (2024) 012041 doi:10.1088/1757-899X/1311/1/012041

analyzed. This paper aimed at assessing the influence of data aggregation levels on modeling the accident severity by using the linear regression techniques.

Road accidents are a negative consequence of the rapid development of the transport industry and are a burden, especially for the economies of low income or lower-middle income countries lacking the infrastructure or with underdeveloped infrastructures. Among the states of the European Union, Romania records the highest mortality rate in terms of deaths caused by road accidents, 85 deaths / $10^6$ inhabitants, while the European average is 42.3 deaths/$10^6$ inhabitants [3]. For this reason, the investigation of road accidents should be a priority for the Romanian authorities.

The level of data aggregation is a topic of interest in road safety research [4] since this influences the result of road safety researches [5]. Depending on the level of aggregation, information is lost that leads to certain biases in the results obtained, and the variables lose their significance. The models investigated on the basis of a high-level aggregate data (monthly, annual, etc.) start from the hypothesis that there are correlations between the data studied at the individual level [6]. The regression coefficients have an identical behavior, but their trend is not constant in different types of aggregation [7]. The literature shows that there is no recommended or unique level of aggregation; it depends on the objective of the research and its type [8], [9]. However, the choice of the level of aggregation depends on the type of data collected because the responsible authorities use different methods of aggregating them [8]. The approach to data aggregation significantly influences the results of road accident modeling, especially when different regression techniques are used [5].

The road accident severity is an interesting research topic often analyzed in the literature. As the statistical methods used by the researchers in order to assess the road accident severity are principally based on aggregated data, the outputs, at a large extent, depends on the way the raw data are prepared to be analyzed. This paper aims at assessing the influence of data aggregation levels on modeling the accident severity by using the linear regression techniques.

## 2. Data preparation

To carry out this research a complex database, provided by the Ministry of Internal Affairs, which contains the road accidents produced in Romania during 2013 – 2018 (6 years) was used, because only these years were available through the database. In order to be able to correctly assess the influence of the road environment on traffic accidents, including the influence of geometrical characteristics of the road, weather conditions and speed, the records containing data of traffic accidents related to pedestrian hit and those that occurred inside the built-up areas (such as cities, villages etc.) were eliminated. In this way, a dataset containing 27,400 individual accident records (27,400 rows) resulted, this being set as the individual level of aggregation (idl). After this operation, the data were aggregated on hourly intervals of one hour (hrl) and daily (dil).

## 3. Methodology

The severity of road accidents depends on the number of victims and the degree of their injury and can be assessed using a Severity Score (SS) which takes into account the number of victims and the degree of their injury (death, serious injury or light injury) and relative cost weights for each type of injury (fatality, serious and slight) [4].

The severity of road accidents depends on the number of victims and the degree of their injury and can be assessed using a Severity Score (SS) which takes into account the number of victims and the degree of their injury (death, serious injury or light injury) and relative cost weights for each type of injury (fatality, serious and slight) [4].

The $SS_i$ Severity Score for an accident is defined by the following relationship:

$$SS_i = V_{1i}*p_1 + V_{2i}*p_2 + V_{3i}*p_3 \tag{1}$$

where,

$V_{1i}$ = the number of deaths generated by accident i.

$V_{2i}$ = number of people seriously injured in accident i.

$V_{3i}$ = number of people slightly injured in accident i.

$p_1$ = relative cost weight relating to road deaths.
$p_2$ = relative cost weight relating to seriously injured victims.
$p_3$ = relative cost weight relating to slightly injured victims.

The cost weights $p_1$, $p_2$, $p_3$ are calculated relative to the social costs of seriously injured victims [10]. The calculated values of the severity score for road accidents investigated in this paper and whose social costs are estimated in [11] are the following [2]:

$p_1 = 7.19$, $p_2 = 1.00$ and $p_3 = 0.08$

Once the $SS_i$ are calculated, they will be estimated by using the Multiple Linear Regression (MLR) technique applied for explanatory purposes. The linear model of MLR is the following:

$$SS_i = b_0 + b_1 * x_{1i} + b_2 * x_{2i} + \ldots + b_k * X_{ki} + \ldots b_n * X_{ni} \tag{2}$$

where,
$SS_i$ = the estimated severity score value corresponding to accident i.
$b_0, b_1, \ldots .b_k, \ldots, b_n$ = regression's coefficients
$X_{1i}, X_{2i}, \ldots\ldots, X_{k1}, \ldots\ldots, X_{ni}$ = the independent variables taken into account in this research.

A number of 83 independent variables $X_{1i}, X_{2i}, \ldots\ldots, X_{k1}, \ldots\ldots, X_{ni}$ that refer to the characteristics of the road environment, weather and lighting conditions and the conditions of road adherence are taken into account. Also, the number of vehicles involved in the analyzed accidents is taken into account.

The explanatory power of the model presented in the relationship (2) is shown by the value of the F test (Fisher test). The model is statistically significant if the value $p<0.05$ and then all independent variables lead to a better explanation of the variation $SS_i$ and the coefficient of multiple determination $R^2$ will show in what proportion the dispersion of the dependent variable is explained by the common action of all the variables.

However, even if F is significant statistically, it does not mean that all variables included in the analysis will significantly add their contribution to explaining the variation of $SS_i$. For this reason, each variable will be tested individually using the t-test, and the insignificant variables will be removed from the model, following that a new multiple regression will be generated, this process will be repeated until all the remaining independent variables are statistically significant.

## 4. Results and discussions

The SSi values calculated using the relation (1) are shown in the frequency distributions' curves in Figures 1, 2 and 3.

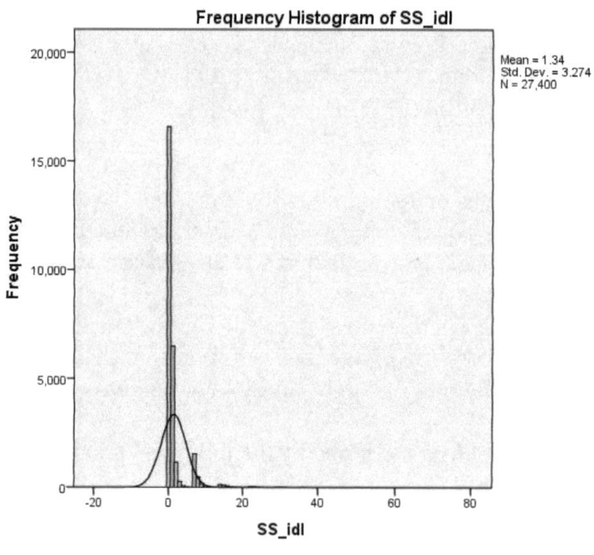


**Figure 1.** Frequency histogram of $SS_i$ in case of highly disaggregation level of data (individual).

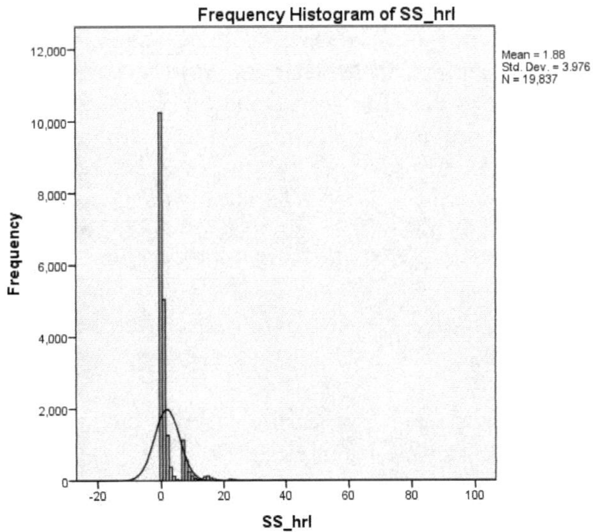


**Figure 2.** Frequency histogram of $SS_i$ in case of hourly aggregation level of data.

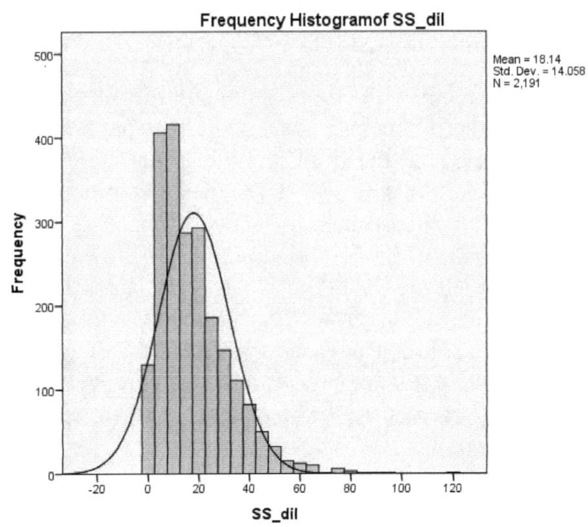


**Figure 3.** Frequency histogram of $SS_i$ in case of daily aggregation level of data.

In Table 1 are presented the main descriptive parameters of the curves from Figures 1, 2 and 3.

**Table 1.** Number of the analyzed samples, mean and the values of Skewness and Kurtosis coefficients for each frequency curve.

| Descriptive parameter | SS_idl | SS_hrl | SS_dil |
|---|---|---|---|
| **Number of samples N** | 27,400 | 19,837 | 2,191 |
| **Mean** | 1.34 | 1.88 | 18.14 |
| **Skewness** | 4.886 | 4.111 | 1.421 |
| **Kurtosis** | 40.865 | 27.992 | 3.346 |

Decreasing the number of units as the aggregation level increases is logical, since a superior level of aggregation involves the compression of the data sets. In all three cases, the Skewness parameter has positive values, which shows an asymmetrical character of all distribution curves, where lower values are less numerous than those higher values. The decrease in the Skewness value shows that with the increase in the aggregation, the number of lower SS values decreases, leaving the place of the high ones. The Kurtosis parameter also has a downward trend, its values decreasing as the level of aggregation increases. With positive values, the Kurtosis parameter shows the leptokurtic character of the curves, where the values are homogeneous and most of them grouped around the mean. Therefore, the increase in the level of aggregation leads to a decrease in the number of SS values that are grouped around the mean.

The results of the testing of the models at global significance after the application of the multiple linear regression techniques, according to the methodology mentioned under pct. 3, are presented in Table 2.

**Table 2.** The coefficient of determination $R^2$, values of the Fisher test and Significance test of the tested models.

| Aggregation level of data | R Square | F | Sig. |
|---|---|---|---|
| **Individual (idl)** | 0.011 | 18.695 | 0.000 |
| **Hourly (hrl)** | 0.072 | 80.730 | 0.000 |
| **Daily (dil)** | 0.365 | 89.296 | 0.000 |

As can be seen from tab. 2, all three models are globally significant, since the Sig. values are less than 0.05 (they have values close to zero which shows a good explanatory power of the models). The coefficient of determination $R^2$ increases as the level of aggregation increases. At the lowest level of aggregation (individual) it can be observed that only 1.1% of the common action of the independent variables explains the SS_idl variation, because $R^2 = 0.011$. In the case of hourly aggregation, the independent variables explain only 7.2% of the SS_hrl variation, while in the case of daily aggregation $R^2 = 0.365$, which shows that 36.50% of the SS_dil variation is explained by the common action of all the variables included in the analysis. Although at first glance the value of 0.365 may lead to the conclusion of a model with a low explaining power, in reality the value $R^2$ depends on the type of research carried out [12], where in the exploratory researches a value $R^2 = 0.10$ is a common one. In fact, in the researches in which the involvement of the human factor is studied, as it is in those concerning road safety, $R^2$ has values between 0.10 – 0.50 [13].

The results obtained from the testing at individual significance of the independent variables for each model are presented in Tables 3, 4 and 5.

**Table 3.** Unstandardized regression coefficients B and the t and Sig. values in case of the highly disaggregation level of data (individual).

| Model | Variables related to: | Variables | Unstandardized Coefficients | | | |
|---|---|---|---|---|---|---|
| | | | B | Std. Error | t | Sig. |
| **Idl** | - | (Constant) | 1.286 | .040 | 32.489 | 0.000 |
| | Road | Street | -0.623 | 0.183 | -3.407 | 0.001 |
| | Road | On the bridge | 0.538 | 0.218 | 2.467 | 0.014 |
| | Road | Railway crossing | 1.736 | 0.349 | 4.969 | 0.000 |
| | Road | Railway | 1.349 | 0.500 | 2.700 | 0.007 |
| | Road | Roundabout | -0.709 | 0.194 | -3.647 | 0.000 |
| | Road | Simple T intersection | -0.474 | 0.097 | -4.891 | 0.000 |

| | | | | | |
|---|---|---|---|---|---|
| Road | Lane width 3.25 m and 3.75 m | 0.140 | 0.040 | 3.494 | 0.000 |
| Road | Median road markings | -0.124 | 0.040 | -3.104 | 0.002 |
| Road | No hard shoulder | -0.203 | 0.057 | -3.584 | 0.000 |
| Adherence conditions | Black ice on the road | -0.682 | 0.182 | -3.745 | 0.000 |
| Adherence conditions | Road covered by snow | -0.515 | 0.134 | -3.834 | 0.000 |
| Visibility | No lighting systems | 0.348 | 0.049 | 7.046 | 0.000 |
| Visibility | Lighting systems not functional | 0.574 | 0.182 | 3.150 | 0.002 |
| Visibility | At down lighting conditions | 0.497 | 0.115 | 4.320 | 0.000 |
| Weather | Snow | 0.449 | 0.149 | 3.017 | 0.003 |
| Weather | Rain | 0.337 | 0.068 | 4.979 | 0.000 |

**Table 4.** Unstandardized regression coefficients B and the t and Sig. values in case of the hourly aggregation level of data.

| **Model** | **Variables related to:** | **Variables** | **Unstandardized Coefficients** | | | |
|---|---|---|---|---|---|---|
| | | | **B** | **Std. Error** | **t** | **Sig.** |
| **hrl** | - | (Constant) | 0.070 | 0.059 | 1.197 | 0.232 |
| | Vehicles | No. of vehicles | 0.419 | 0.025 | 17.095 | 0.000 |
| | Road | Street | -0.571 | 0.216 | -2.650 | 0.008 |
| | Road | Curves | 0.610 | 0.065 | 9.336 | 0.000 |
| | Road | No road feature | 0.424 | 0.060 | 7.020 | 0.000 |
| | Road | On the bridge | 1.133 | 0.257 | 4.400 | 0.000 |
| | Road | Railway crossing | 2.405 | 0.348 | 6.906 | 0.000 |
| | Road | U curves | -0.417 | 0.197 | -2.119 | 0.034 |
| | Road | Roundabout | -0.466 | 0.232 | -2.010 | 0.044 |
| | Road | Lane width 3.25 m and 3.75 m | 0.100 | 0.046 | 2.176 | 0.030 |
| | Road | Median road markings | -0.107 | 0.045 | -2.366 | 0.018 |
| | Road | Roadway with rare bumps | 0.337 | 0.129 | 2.610 | 0.009 |
| | Adherence conditions | Ice on the road | -0.892 | 0.352 | -2.533 | 0.011 |
| | Adherence conditions | Black ice on the road | -0.699 | 0.201 | -3.481 | 0.001 |
| | Adherence conditions | Road covered by snow | -0.445 | 0.136 | -3.275 | 0.001 |
| | Visibility | No lighting systems | 0.408 | 0.052 | 7.874 | 0.000 |

| | | | | | |
|---|---|---|---|---|---|
| Visibility | Lighting systems not functional | 0.566 | 0.211 | 2.680 | 0.007 |
| Visibility | At dawn lighting conditions | 0.809 | 0.127 | 6.351 | 0.000 |
| Weather | The Fog | 0.519 | 0.167 | 3.101 | 0.002 |
| Weather | Rain | 0.510 | 0.074 | 6.932 | 0.000 |

**Table 5.** Unstandardized regression coefficients B and the t and Sig. values in case of the daily aggregation level of data.

| **Model** | **Variables related to:** | **Variables** | **Unstandardized Coefficients** | | | |
|---|---|---|---|---|---|---|
| | | | **B** | **Std. Error** | **t** | **Sig.** |
| **dil** | - | (Constant) | 0.698 | 0.582 | 1.199 | 0.231 |
| | Vehicles | No. of vehicles | 0.553 | 0.035 | 15.924 | 0.000 |
| | Road | Curves | 0.877 | 0.139 | 6.295 | 0.000 |
| | Road | Railway crossing | 2.099 | 1.001 | 2.096 | 0.036 |
| | Road | U curves | -1.175 | 0.558 | -2.104 | 0.035 |
| | Road | No hard shoulder | -.447 | 0.179 | -2.502 | 0.012 |
| | Road | Median safety barriers | 4.138 | 1.227 | 3.373 | 0.001 |
| | Road | Roadway with multiple bumps | 1.374 | 0.518 | 2.651 | 0.008 |
| | Adherence conditions | Slime on the road | -1.652 | 0.662 | -2.495 | 0.013 |
| | Adherence conditions | Black ice on the road | -1.336 | 0.453 | -2.948 | 0.003 |
| | Adherence conditions | Wet roadway | -0.290 | 0.147 | -1.975 | 0.048 |
| | Visibility | No lighting systems | 0.938 | 0.145 | 6.456 | 0.000 |
| | Visibility | Functional lighting systems | 1.118 | 0.399 | 2.804 | 0.005 |
| | Visibility | At dawn lighting conditions | 0.787 | 0.391 | 2.011 | 0.044 |
| | Weather | Rain | 0.793 | 0.193 | 4.115 | 0.000 |

Analyzing Tables 3, 4 and 5, we observe, in the three models, the existence of a single common variable characteristic of the road ("Railway crossing"), of one characteristic of the grip conditions ("Black ice on the road"), of two variables characteristic of the visibility conditions ("No lighting systems" and "At dawn lighting conditions"), as well as of a single one related to the weather ("Rain").

In all three levels of aggregation studied, the values of the standardized B coefficients are positive for the variables "Railway crossing", "No lighting systems", "At dawn lighting conditions" and "Rain" which shows a direct relationship between them and the SS values, this being in line with other similar researches [14]. In other words, if the number of accidents occurred at railway crossings, where no public lighting systems are installed, in the lighting conditions at dawn or during rain increases, then the value of the severity score increases accordingly with the value of the coefficient B of the respective variable, given that the rest of the variables remain constant (they are statistically controlled). If the values B are negative, they show that between the variables and the value of SS there is an indirect relationship, which leads to the decrease of the SS value. For example, as can be seen from Tables 3, 4 and 5, of all the common variables in all three models, the highest value of the regression coefficient B is the one corresponding to the variable "Railway crossing", this demonstrating the severity of accidents occurring in such characteristics of the road, because the variable that makes the greatest contribution to explaining the variation of the SS is the one that has the highest value of the coefficient B. Therefore, in the case of the dil model, the increase by one unit of the number of road accidents occurred at railway crossings leads to an increase in the severity score by 2.099, given that the other variables remain constant. In the case of the individual aggregation level, the variable "Snow", characteristic of the meteorological conditions, significantly explains the variation in the value of the severity score (Snow, $t=3.017$, $p<0.05$), and in the case of hourly aggregation, the variable "Fog" has a significant influence (Fog, $t=3.101$, $p<0.05$). Fog and snow are meteorological phenomena characterized by a high degree of spatial and temporal volatility, their presence being seasonal [15]. Therefore, with the increase in the aggregation of data, the influence of meteorological factors of interest is lost. Also, the aggregation level of data is a very important topic in the statistical research, since the significance of variable cand be affected by the aggregation level [16].

## 5. Conclusions

The increase in the level of aggregation of the data leads to the decrease of the number of SS values that are grouped around the mean, increasing in this way the standard deviation of the values of the dependent variables.

The values of the multiple determination coefficients $R^2$ improve as the degree of aggregation of the data increases. The highest value of $R^2$ is the one corresponding to the daily aggregation level $R^2 = 0.365$, which shows that 36.50% of the Severity Score variation is explained by the common action of the analyzed variables, the remaining 63.50% being represented by other factors that were not taken into account. In order to improve the values of the coefficients of determination, other dependent variables must be included for the drivers, the vehicles involved or those related to the geographical and environmental positioning of the places of occurrence of road accidents.

The increase of the aggregation level leads to the increase of the regression coefficient values and, implicitly, to the increase of the estimated values of the severity scores. This fact should be viewed as a research disadvantage, since the severity score has an individual character. The degree of aggregation of the research data influences the obtaining of conclusions regarding the influence on traffic accidents of some meteorological phenomena with volatile and seasonal character, such as fog or snow. The same cannot be said about rain, a meteorological phenomenon encountered in most time frames.

The research is limited to analyzing the variables registered by the traffic police when the accidents are investigated. More road environment related variables should be included in order to investigate the data aggregation level's effect on the accidents' severity. The applicability of this study should be seen as a theoretical approach and base for future researches based on accident severity as related to road conditions, weather and other traffic related factors.

## References

[1] ***. https://www.un.org/en/safety-and-security/road-safety (accessed 14.01.2022).
[2] ***. https://www.who.int/teams/social-determinants-of-health/safety-and-mobility/decade-of-action-for-road-safety-2021-2030 (accessed 14.01.2022).

IOP Conf. Series: Materials Science and Engineering **1311** (2024) 012041 doi:10.1088/1757-899X/1311/1/012041

[3] Romanian Road Safety Report 2020, 2021. https://www.politiaromana.ro/files/pages_files/Buletinul_sigurantei_rutiere_raport_anul_2020.pdf (accessed 14.01.2022).

[4] Drosu A., Cofaru C., 2019. Explanatory Models of the Weather Factors' Influence on Road Traffic Accidents Produced in Romania in Different Road's Features. *In: Burnete N., Varga B. (eds) Proceedings of the 4th International Congress of Automotive and Transport Engineering (AMMA 2018). AMMA2018 2018. Proceedings in Automotive Engineering. Springer, Cham.* https://doi.org/10.1007/978-3-319-94409-8_92

[5] Usman, Taimur & Fu, Liping & Miranda-Moreno, Luis., 2012. Accident Prediction Models for Winter Road Safety. Does Temporal Aggregation of Data Matter?. *Transportation Research Record Journal of the Transportation Research Board*. 2237. 144-151. 10.3141/2237-16.

[6] McCullagh, P., and J. A. Nelder. Generalized Linear Models, 2nd ed. Chapman and Hall, London, 1989.

[7] Clark, W.A.V. and Avery, K.L., 1976. *The Effects of Data Aggregation in Statistical Analysis. Geographical Analysis, 8*: 428-438. https://doi.org/10.1111/j.1538-4632.1976.tb00549.x

[8] Thomas I., 1996. Spatial data aggregation: exploratory analysis of road accidents. *Accid Anal Prev*. Mar;28(2):251-64. doi: 10.1016/0001-4575(95)00067-4. PMID: 8703283.

[9] Thomas, I. Les difficult&s de la cartographic des accidents de la route. In Cartographie des accidents de la route: modes d'analyse et de representation. Berthoz, E.; Brossard, T.; Lassare, S. (Eds). Paris: INRETS; 1993: 41-49.

[10] Miranda-Moreno, L. F., Fu, L., Ukkusuri, V. S., Lord, D., 2009. *How to Incorporate Accident Severity and Vehicle Occupancy into the Hot Spot Identification Process?*. Transportation Research Record, 2102(1), pp. 53–60.

[11] Huib van, E., Lisanne van, W., Arno, S., Sander de, B., 2019. Handbook on the External Costs of Transport, Report for European Commission. Directorate General for Mobility and Transport. ISBN 978-92-79-96917-1, doi: 10.2832/27212.

[12] Mooi, E., Sarstedt, Marko, 2011. A Concise Guide to Market Research: The Process, Data, and Methods Using IBM SPSS Statistics (p.179). New York: Springer. 10.1007/978-3-642-12541-6.

[13] Sava, F., 2011. Analiza datelor în cercetarea psihologică [Analysis of data in psychological research], p. 219. Cluj Napoca: ASCR Publishing House.

[14] Drosu, A., Cofaru, C. & Popescu, M.V. Fatal Injury Risk Model (FIRM) of the Road Accidents that Occurred in Rainy Conditions — A Probabilistic Approach. *Int.J Automot. Technol.* **22**, 1415–1426 (2021). https://doi.org/10.1007/s12239-021-0123-2.

[15] Tobin, D. M., H. D. Reeves, M. N. Gibson, and A. A. Rosenow, 2022: Weather Conditions and Messaging Associated with Fatal Winter-Weather-Related Motor-Vehicle Crashes. Wea. Climate Soc., 14, 835–848, https://doi.org/10.1175/WCAS-D-21-0112.1.

[16] Usman, Taimur & Fu, Liping & Miranda-Moreno, Luis. (2012). Accident Prediction Models for Winter Road Safety. Does Temporal Aggregation of Data Matter?. Transportation Research Record Journal of the Transportation Research Board. 2237. 144-151. 10.3141/2237-16.

# Weather and road environment's influence on multi-vehicle crashes in Romania

**A Drosu[1*], C Cofaru[1] and C Slaff[2]**

[1]Transilvania University of Brasov, Brasov, Romania

[2]Misericordia University, Dallas PA, USA

drosu.alin.gabriel@unitbv.ro

**Abstract**. The aim of this paper is to assess the weather and road environment's influence on Multi Vehicle Crashes (MVC) in Romania. In order to capture the real effect of these factors, the road accidents recorded between the months of November and March of each year from 2013 to 2018 are analysed, as this is the period of each year when most of the meteorological phenomena relevant to our research are present. The analysed data set contains complex records of a number of 8,538 accidents which are assessed at the highly disaggregation level (individual), a type of analysis which is rarely found in the road safety studies. A Binary Multiple Logistic Regression (BMLR) is employed for explanatory purposes in order to estimate the probability of MVC. Road accidents that involve at least 3 vehicles are declared as MVC and the probability of their occurrence is labelled as "1". The rest of the crashes in the data set are assigned the value "0" of the probability. The results of this research show that the highways have the highest risk of a MCV occurring as the odds values show and the sleet is the weather phenomena which highly increase the probability of having a MCV. These risk factors should be further researched in connection with other variables that relate to the driver and weather characteristics.

## 1. Introduction

At the European level, Romania has the highest mortality rate among all European Union states, 93 deaths/$10^6$ inhabitants [1]. The road accidents that occur on the Romanian road network generate huge costs that put pressure not only on national budgets, but also affect local communities.

The Multiple-Vehicles Crashes (MVC) that occur in Romania need to be further investigated, since this type of collisions generate victims and high volumes of damages. Although they do not have a high degree of injury severity, these types of collisions are very common all over the world [2]. The material damages resulting from these types of collisions are much higher than other types of accidents.

Multiple vehicle crashes are also referred to as "chain reaction crashes". The chain reaction is produced by the first collision, which is considered the primary collision, and after it the secondary collisions occur that give the character of a chain collision [3], being caused by the "tail effects" [4]. The causes of MVC are different from other collisions. Thus, single accidents are mainly caused by a loss of control, while MVC are caused by faulty interactions with other road traffic participants. These types of collisions affect a large part of the road surface and cause road traffic congestions and delays. Also, collisions with multiple vehicles delay the medical and extrication teams.

Accidents with multiple vehicles are of interest to researchers due to the fact that they have a low frequency, therefore having a seasonal character, being influenced by fog and snow [5]. Furthermore, snow increases the risk of MVC occurrence on highways [6]. The weather conditions, the lighting conditions, the condition of the roadway, the width of the shoulders and the traffic lanes are factors that contribute to the occurrence of multiple accidents involving heavy vehicles [7]. Moreover, the weather conditions, intersections and curves are considered risk factors in regards to the occurrence of accidents in which multiple vehicles are involved [8]. These collisions occur easily in foggy conditions [9] due to not properly assessing the following distance of the vehicle in front, this type of accident being a common one in such conditions [10]. Most accidents with multiple vehicles occur during phenomena that affect visibility (fog, smoke etc.) [3].

## 2. Data preparation

For this research a complex database, made available by the Ministry of Internal Affairs, was used. The database contains records of the traffic accidents that occurred between 01.01.2013 and 31.12.2018 on the entire Romanian road network. The traffic accidents which occurred in the months of November and December, as well as those recorded in the months of January, February and March of each year, were selected. The accidents that occurred outside built-up areas have been selected and only the vehicle to vehicle collisions were kept. In this way, a data set containing 8,009 lines resulted. The monthly distribution of MVC and non-MVC is presented in figure 1.

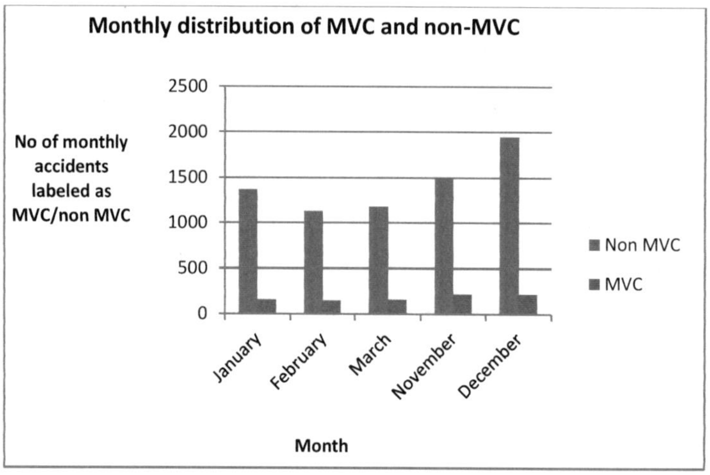


**Figure 1.** Monthly distribution of the multiple vehicles accidents (MVC) and other types of accidents labelled as non-MVC.

The probability of a multi-vehicle accident (MVC) was labelled as 1, and for the rest of the accidents the probability was labelled as 0.

The categorical variables that have more than two response variables were transformed into Dummy variables. A categorical variable that has k response levels are transformed into k-1 Dummy variables ($DV_{k-1}$), and each Dummy variable will have two response values: "0" and "1". Applying this procedure, a number of $D_{k-1}$ newly created dummy variables will result, and the variable with the highest frequency of occurrence will be considered as the reference level, following which all other dummy variables will be compared with it. Such a comparative approach will show the influence of some factors that do not have a high frequency of occurrence, compared to those that have the highest frequency. Therefore, the predictors with the highest frequency of occurrence will be considered as the reference level.

## 3. Methodology

The Multiple Binary Logistic Regression (MLR) method will be applied to estimate the probability of occurrence of an MVC. The mathematical model of probability is as follows:

$$P=\frac{e^y}{1+e^y} \tag{1}$$

where,

P= the probability of occurring a MVC

y = chance of occurring a MVC. In other words, y is the ratio between the probability of occurrence of an MVC and the probability that such an accident does not occur. Mathematically, y is calculated with the following relationship:

$$y= e^{b0+b1*X1+b2*X2+\cdots+bn*Xn} \tag{2}$$

where,

$b_0$, $b_1$, $b_2$, ..., $b_n$ are the logit coefficients, with $b_0$ associated with the intercept.

$X_1$, $X_2$, ..., $X_n$ are the predictors included in this analysis.

The odds y can take values between {0, 1, ... ∞}, and the probabilities only in the interval {0; 1}.

The predictors used in the regression will be interpreted using the odds and not the probabilities. Thus, using the logit coefficient of the predictors, the odds ratio (OR) is calculated, which shows how much the odd of a road accident being a MVC changes, when a predictor increases by one unit, in the conditions where all the others predictors remain constant. The odds ratio is calculated with the formula:

$$\text{OR (odds ratio)} = e^{b_k} \tag{3}$$

The statistical significance of an MLR model is assessed using the $G^2$ test of likelihood - "chi-square likelihood statistic" and the "Hosmer and Lemeshow's Goodness of Fit Test".

Mathematically, the $G^2$ test is written according to the relation:

$$G^2 = (-2LLnull) - (-2LLmodel) \tag{4}$$

The $G^2$ test check if all logit coefficients of the predictors, without intercept, have zero values.

If the "Hosmer and Lemeshow's Goodness of Fit Test" test is statistically insignificant ($p>0.05$), then there are no differences between the estimated and observed data and, therefore, the model is well chosen.

To test the hypothesis that the chosen predictors significantly influence the variation of the dependent variable, two parameters will be used, namely Cox & Snell R Square and Nagelkerke R Square, which establish the explanatory efficiency of the chosen model. In fact, these two indicators are similar to the multiple correlation coefficients $R^2$ from linear regression.

The testing of the logit coefficients will be done by sing the Wald statistical test, which tests the null hypothesis when a coefficient has a zero value. The interpretation of the logit coefficients is carried out through factors, their negative values showing a decrease in the odds of occurring a MVC (indirect relationship between the predictor and the odds). At the same time, a positive logit coefficient shows an increase in the odds of occurring a MVC.

## 4. Results and discussions

After running the Multiple Binary Logistic Regression (MLR), the primary data show that the generated model is significant because $G^2(83) = 1467.223$, $p<0.001$ is statistically significant, as shown in table 1 .

**Table 1**. Multiple binary logistic regression's primary output data

| Type of output data | Output's data coefficients | Output data coefficients' values |
|---|---|---|
| Variables in the Equation | B | -2,065 |
| | it | 0.034 |
| | Wald | 3636,318 |
| | df | 1 |
| | Sig. | 0,000 |
| | Ex(B) | 0.127 |

| | | |
|---|---|---|
| **Omnibus Tests of Model Coefficients** | Chi-square | 1467,223 |
| | df | 83 |
| | Sig. | 0,000 |
| **Model Summary** | -2 Log likelihood | 5218,500 |
| | Cox & Snell R Square | 0.158 |
| | Nagelkerke R Square | 0.313 |
| **Hosmer and Lemeshow Test** | df | 8 |
| | Sig. | 0.135 |

Since $G^2$ is statistically of significance, the null hypothesis is rejected and therefore there is at least one predictor that significantly influences the logarithm of the odds that the dependent variable changes towards 0 or towards 1. The Hosmer and Lemeshow test is not statistically significant (Sig. = 0.135, $p>0.001$ ), this showing that the model is well chosen.

As the tested model is well chosen, the values of the Cox & Snell $R^2$ and Nagelkerke $R^2$ indicators (tab. 1), show that the predictors included in the model explain between 15.8% and 31.3% of the variation in the probability of occurring a MVC.

Using the Wald test, the individual testing of the logit coefficients is performed. After performing the tests and eliminating those predictors for which the Wald test is not significant, i.e. those predictors that do not significantly influence the probability of producing a MVC, the results obtained are presented in table 2.

**Table 2.** The predictors that significantly explain the variation of the probability, the logit coefficients (B), the value of the Wald test, the significance threshold, the exponents of the logit coefficients (Exp( B)) and the confidence intervals.

| | | | | | | | **95% CIfor EXP(B)** | |
|---|---|---|---|---|---|---|---|---|
| **Variables** | **B** | **it** | **Wald** | **df** | **Sig.** | **Ex(B)** | **Lower** | **Upper** |
| **Road category** | - | - | 38,275 | 5 | 0.000 | - | - | - |
| **Motorway (HI)** | 0.702 | 0.248 | 7,990 | 1 | 0.005 | 2.017 | 1.240 | 3.281 |
| **Inter villages roads (IVR)** | -1.807 | 0.610 | 8787 | 1 | 0.003 | 0.164 | 0.050 | 0.542 |
| **Inter county roads (ICR)** | -0.516 | 0.132 | 15.357 | 1 | 0.000 | 0.597 | 0.461 | 0.773 |
| **Street (STR)** | -2.185 | 0.725 | 9.088 | 1 | 0.003 | 0.112 | 0.027 | 0.466 |
| **Road Feature (RF)** | - | - | 18.192 | 6 | .006 | - | - | - |
| **Curves (CRV)** | -0.478 | 0.210 | 5.175 | 1 | 0.023 | 0.620 | 0.411 | 0.936 |
| **Intersection (INT)** | -0.988 | 0.329 | 9.008 | 1 | 0.003 | 0.372 | 0.195 | 0.710 |
| **Road Configuration (RC)** | - | - | 27.602 | 12 | 0.006 | - | - | - |

| | | | | | | | | |
|---|---|---|---|---|---|---|---|---|
| **Other types of configurations (OCF)** | 0.317 | 0.127 | 6.204 | 1 | 0.013 | 1.373 | 1070 | 1.761 |
| **Lane width (LW)** | - | - | 20.520 | 3 | 0.000 | - | - | - |
| **LW between 2.75-3.25 (LW1)** | -0.271 | 0.100 | 7.387 | 1 | 0.007 | 0.762 | .627 | 0.927 |
| **LW > 3.75 m (LW2)** | 0.225 | 0.110 | 4.165 | 1 | 0.041 | 1.252 | 1.009 | 1.554 |
| **LW < 2.75 m (LW3)** | -1.960 | 0.727 | 7.267 | 1 | 0.007 | 0.141 | 0.034 | 0.586 |
| **Shoulders (SHD)** | - | - | 4.673 | 2 | 0.097 | - | - | - |
| **Not consolidated shoulders (NCS).** | -0.203 | 0.101 | 4.053 | 1 | 0.044 | 0.816 | 0.670 | 0.995 |
| **Road Safety Measures (RSM)** | - | - | 16.665 | 8 | 0.034 | - | - | - |
| **Safety barriers (RSB)** | -1.621 | 0.794 | 4.167 | 1 | 0.041 | 0.198 | 0.042 | 0.937 |
| **Median separation space (RMS)** | -0.756 | 0.322 | 5.497 | 1 | 0.019 | 0.470 | 0.250 | 0.883 |
| **Roadway surface (RS)** | - | - | 4.243 | 3 | 0.236 | - | - | - |
| **Roadway with bumps (RBS)** | -0.642 | 0.317 | 4.104 | 1 | 0.043 | .526 | 0.283 | 0.979 |
| **Weather conditions (WC)** | - | - | 12.711 | 6 | 0.048 | - | - | - |
| **Fog (FOG)** | 0.470 | .182 | 6.662 | 1 | 0.010 | 1.600 | 1.120 | 2.287 |
| **Sleet (SLT)** | 0.709 | 0.303 | 5.465 | 1 | 0.019 | 2.032 | 1.121 | 3.683 |
| **Constant** | -5.790 | 0.490 | 139.389 | 1 | 0.000 | 0.003 | - | - |

As can be seen from table 2, for all the selected predictors the value is $p<0.05$, which shows that all these predictors significantly explain the variation of the probability of occurring a MVC. It should be noted that none of the variables that refer to the inclination of the road surface, signposting, the number of lanes, the road markings, the adhesion of the roadway and the degree of brightness, do not significantly explain the variation of the probability of producing an MVC.

Interpreting the value of the logit coefficients Exp(B) for the "road category" variable, we can see that on the highways there are 2,017 times more odds of occurring a MVC comparing with the national roads, the variable that is considered the reference (HI: Exp(B ) = 2.017, Wald(1) = 7.990, $p<0.05$). On inter villages roads there are 0.164 times more odds (IVR: Exp( B) = 0.164, Wald(1) = 8.787, $p<0.05$), and on inter county roads there are 0.597 times more odds (ICR: Exp(B) = 0.597, Wald(1) = 15.357, $p<0.05$). On the streets there are the fewest odds of an MVC occurring, a logical thing if we consider that they are a category of road on which very low volumes of road traffic are registered (STR: Exp(B) = 0.112, Wald(1) = 9.088, $p<0.05$).

Regarding the characteristics of the road (RF), tab. 2 shows that in curves there are 0.620 more ods for an accident to involve at least three vehicles (CRV: Exp(B) = 0.620, Wald(1) = 5.175, $p<0.05$) compared to those road sections without any characteristic (alignments), and in intersections they are 0.372 higher than the same characteristic (INT: Exp(B) = 0.372, Wald(1) = 9.008, $p<0.05$).

Other types of road configurations (OCF) generate odds of producing a MVC of 1.373 higher than on road segments without configuration (OCF: Exp( B) = 1.373, Wald(1) = 6.204, $p<0.05$).

Regarding the width of the road lane (LW), where it is > 3.75 m, the odds of producing an MVC are 1.252 times higher than on roads where the traffic lane has a width between 3.25 - 3.75 m and which is considered as a reference value (LW2: Exp( B) = 1.252, Wald(1) = 4.165, $p<0.05$). Where the width of the traffic lane is between 2.75 - 3.25 m, the odds are 0.762 times higher (LW1: Exp( B) = 0.762, Wald(1) = 7.387, $p<0.05$), and for the lane width below 2.75 m they are 0.141 times higher (LW3: Exp(B) = 0.141, Wald(1) = 7.267, $p<0.05$). A traffic lane width >3.75 m is specific to highways, which shows the fact that highways create high odds of accidents with multiple vehicles.

On roads with unconsolidated shoulders (UNS), which are specific to roads of lower categories, there are 0.816 times more odds of producing an MVC than on those roads where the shoulders are consolidated (UNS: Exp( B) = 0.816, Wald(1) = 4.053, $p<0.05$), which is in contradiction with the results of other researches that show that these elements of the road have no influence on MVC [11].

From tab. 2, we observe that road safety measures (RSM) decrease the odds of producing a MVC, especially safety barriers (RSB: Exp( B) = 0.198, Wald(1) = 4.167, $p<0.05$). On roads where there is a median separation space, specific to highways and expressways, the odds are 0.470 times higher than where there are no such road devices installed (RMS: Exp( B) = 0.470, Wald(1) = 5.497, $p< 0.05$).

On the roads with road surfaces that have bumps (RS), the odds of occurring a MVC are 0.526 times higher than those where the road surface is normal and smooth (RBS: Exp( B) = 0.526, Wald(1) = 4.104 , $p<0.05$). This confirms previous research by which the deterioration of the road surface leads to the increase of the probability of occurrence of MVC [5].

Fog and sleet are the only weather characteristics that influence the probability of a MVC occurring. During fog, the odds of producing a MVC are 1.600 times higher than those produced in normal weather (FOG: Exp( B) = 1,600, Wald(1) = 6,662, $p<0.05$), a fact that confirms the results other researches [5]. If sleet is present, the ods of producing a MVC are 2.032 times higher than normal weather conditions (SLT: Exp( B) = 2.032, Wald(1) = 5.465, $p<0.05$). Although fog is the meteorological phenomenon that drastically affects the visibility of drivers and influences them on a psychological level, here is that the odds of producing a MVC during fog are lower than those during sleet. The sleet affects the visibility of drivers less than fog, but the efficiency of the brakes drops considerably during such a meteorological phenomenon.

## 5. Conclusions

This research shows that the highways are the most dangerous in terms of the occurrence of accidents with multiple vehicles, since their odds are 2.017 times higher than streets and have the highest values among all categories of roads considered as significant variables that explain the probability of occurrence of a MVC. Moreover, this is also demonstrated by the high value of the odds on roads where the lane width is >3.75 m, this being a characteristic of highways.

Fog and sleet are the only meteorological phenomena that significantly influence the probability of an accident with multiple vehicles, the odds values being above unity. The sleet also has the greatest explanatory power between the two variables, contrary to previous researches that demonstrated that fog has a major influence on the production of MVC. There is a direct relationship between fog, sleet and the odds of producing a MVC. As the number of road accidents caused by fog or sleet increases, so do the odds that they involve more than 3 vehicles.

Curves, intersections, unconsolidated approaches, safety barriers, median spaces for traffic flows and the presence of bumps in the roadway are considered variables that significantly explain the variation in the probability of producing a MVC, but being in an indirect connection with them.

The conclusions of this research will help the road authorities and the traffic police to ensure measures needed to prevent road accidents and to ensure a better maintenance of the road network in Romania. Also, further researches should be focused on studying the driver behaviour during a MVC and how the fog or sleet influences it.

IOP Conf. Series: Materials Science and Engineering **1311** (2024) 012042 doi:10.1088/1757-899X/1311/1/012042

**References**


[1] ***. https://transport.ec.europa.eu/2021-road-safety-statistics-what-behind-figures_en (accessed 01.03.2022).

[2] Mathis, Annabelle Marie, "Statistical Analysis of Fatalities Due to Vehicle Accidents in Las Vegas, NV" (2011). UNLV Theses, Dissertations, Professional Papers, and Capstones. 1084. http://dx.doi.org/10.34917/2465740.

[3] Call, DA, Wilson, CS and Shourd, KN (2018), Hazardous weather conditions and multiple-vehicle chain-reaction crashes in the United States. Meth. Apps, 25: 466-471. https://doi.org/10.1002/met.1714 .

[4] Meng F, Xu P, Song C, Gao K, Zhou Z, Yang L. Influential Factors Associated with Consecutive Crash Severity: A Two-Level Logistic Modeling Approach. Int J Environ Res Public Health. 2020 Aug 4 ;17 (15):5623. two : 10.3390/ijerph17155623. PMID: 32759863; PMCID: PMC7570167.

[5] Hong, J.; Tamakloe, R.; Park, D. A Comprehensive Analysis of Multi-Vehicle Crashes on Expressways: A Double Hurdle Approach. *Sustainability* **2019** , *11* , 2782. https://doi.org/10.3390/su11102782 .

[6] Becker, N., Rust, HW . & Ulbrich, U. Weather impacts on various types of road crashes: a quantitative analysis using generalized additive models. Eur. Transp. Res. rev. 14 , 37 (2022). https://doi.org/10.1186/s12544-022-00561-2.

[7] M. Moomen, MM Rezapour Mashhadi, and K. Ksaibati, "An Investigation of Influential Factors of Truck Crashes on Two-Lane Downgrades in Wyoming: A Logistic Regression Approach", The 97th Transportation Research Board Annual Meeting, vol. 18, 2018. 05285.

[8] K.Vachal, "Analysis of Risk Factors in Severity of Rural Truck Crashes", Upper Great Plains Transportation Institute North Dakota State University, Fargo, 2016, https://www.ugpti.org/resources/ reports/downloads/mpc16-308.pdf (accessed 01.09.2022).

[9] Jinhua Tan *et al* 2019 *IOP Conf. Ser.: Earth Environ. Sci.* **304** 032077.

[10] Tan, Chunqing & Qin, Gong & Qin ,. (2019). Effect of Imitation Phenomenon on Two-Lane Traffic Safety in Fog Weather. International Journal of Environmental Research and Public Health. 16. 3709. 10.3390/ijerph16193709.

[11] Yasmin, S., Eluru, N., Lee, J., & Abdel-Aty, M. (2016). Ordered fractional split approach for aggregate injury severity modeling. Transportation Research Record: Journal of the Transportation Research Board, (2583), 119–126.

IOP Conf. Series: Materials Science and Engineering **1311** (2024) 012043 doi:10.1088/1757-899X/1311/1/012043

# The integrated methodology for assessing the safety of the road network infrastructure in Romania.

**C I Dimitriu[1,2] and C R Bujor[1,2]**

[1]PhD. Stud. Eng., Department of Road Vehicles, Faculty of Transport, POLITEHNICA University of Bucharest, RO-060042, Romania
[2] Road Infrastructure Safety Department, Romanian Road Authority – A.R.R., Bucharest, RO-010873, Romania

corneliu.dimitriu@stud.trans.upb.ro, cosmin_robert.bujor@stud.trans.upb.ro

**Abstract**. Romania has the highest road fatality rate in the EU and the lowest perception of road quality. This study presents the implementation of a preliminary procedure and tools provided by the European Commission for the safety assessment of the DN2-E85 road infrastructure, which covers 472.23 km. This road was selected due to the high number of road deaths between 2017 and 2022. The research focuses on the application of both proactive and reactive methodologies as required by Directive (EU) 2019/1936, which requires full road network safety assessments by 2024, with subsequent assessments every five years. The proactive methodology assesses the built-in road safety of the road infrastructure based on design parameters, while the reactive methodology is based on historical data on accidents and traffic flows. The integration of these methodologies allows the classification of road segments into five risk levels. Preliminary results indicate that approximately 59% of the length of the DN2-E85 road is classified as “very high priority”, highlighting significant safety deficiencies, which require actions to immediately improve traffic safety. This study highlights the need for systematic road infrastructure safety assessments and prioritisation of high-risk segments to reduce road fatalities and improve overall road safety. The findings serve as crucial information for road safety auditors and the Romanian Road Authority - ARR, supporting the identification of high-risk areas and the allocation of resources to improve safety. In addition, this research provides a basic basis for future adaptations of road infrastructure safety assessment guidelines, considering the specific characteristics of national and county roads. By aligning with EU directives and national standards, the study aims to increase road safety and reduce the incidence of serious accidents in Romania. The insights gained from this research are essential for developing effective road safety strategies and implementing remedial measures that address the most dangerous segments of the road network.

## 1. Introduction

Directive (EU) 2019/1936 amending Directive 2008/96/EC on road infrastructure safety management introduces a new method for classifying and ranking road segments according to road risk. This approach moves from “road network classification” to “road network assessment” to identify the most dangerous segments. The goal is to reduce road risks by implementing improvement measures, prioritizing the most dangerous segments.

Similar research on road network safety assessment, which considers built-in road safety, has also been carried out in New Zealand. For example, the New Zealand Transport Agency has developed a process called the Road Infrastructure Safety Assessment (RISA), which monitors the performance of local road authorities over time and identifies the benefits to road users by improving road safety infrastructure [1].

The European Commission provides a standardized assessment methodology to ensure that all Member States operate under the same system. The main goal is to reduce road risks by prioritizing and implementing measures to improve safety on the most dangerous sections of the network.

In addition to the road infrastructure safety assessment methodologies proposed by the European Commission, there are other recognized international programmes, such as the International Road Assessment Programme (iRAP). iRAP is a road assessment tool that helps reduce road accidents by assessing high-risk routes and developing star ratings and safer roads investment plans (SRIPs). Recent research on the application of iRAP worldwide has highlighted the state of implementation, advantages and disadvantages of this approach. The studies were carried out by reviewing the literature and analysing the implementation of iRAP in countries such as Thailand, Bangladesh, Malaysia, India, Brazil and China, highlighting the effectiveness of star ratings and investment plans to improve road safety [2].

Recent research highlights the importance of Integrated Road Safety Assessment (IRSA) methodologies in identifying and mitigating infrastructure-related safety deficiencies. According to a comprehensive analysis, IRSA methodologies are essential for supporting road authorities in preventing and reducing future road accidents by assigning risk scores to road segments, making it easier to identify high-risk areas that need improvement. This approach is in line with the objectives of the EU Directive, stressing the importance of systematic and consistent safety assessments in national road networks [3].

According to the provisions of Directive (EU) 2019/1036 [4], Member States shall carry out the first road safety assessment at road network level by 2024. Subsequently, the evaluation will be carried out once every 5 years. Directive (EU) 2019/1936 was transposed into Romanian legislation by the OG no. 3/2022 [5], which amended and supplemented Law no. 265/2008 on traffic safety management on road infrastructure [6].

According to data published in the National Road Safety Strategy – SNSR, 2022-2030, based on the 2019 World Economic Forum Report, Romania ranks 119th out of 141 countries analysed and occupies the last position in the European Union in terms of the "perceived quality of roads" indicator. This position is reflected by the coefficient of 2.7 out of a total of 7, awarded for the best roads in the world, p. 9 of the SNSR [7].

According to data from the National Institute of Statistics – INS, at the end of 2022, the rural road network in Romania had a length of 86,336 km [8]. The analysis of the development of the road network in the last 10 years shows a modest increase in the number of kilometres for national, county and communal roads. In the same period, the length of highways increased by 42%. However, it is noted that the percentage of modernization of county and communal roads is low, highlighting the need for investments in modernizing the infrastructure of these categories of roads.

The research highlights that serious accidents are concentrated on national roads. The analysis of the results of the distribution of serious accidents shows a worrying situation regarding the higher number of accidents recorded on national/county roads that have the function of street during the transit of localities.

The study presents how Romania implemented the preliminary procedure using the tools provided by the European Commission for the safety assessment of the European and main road network. The Commission has developed a Methodological Guide to guide all Member States in applying similar procedures for road network assessment. The final form of the unitary evaluation procedure at EU level will be determined based on Member States' comments.

The main objective of this research is to assess the safety of the road infrastructure on the national road DN2 – E85 (Bucharest – Siret Customs), on a length of 472.23 km. According to the statistics provided by the Romanian Police and the analysis of the road accident database in the period 2017-

2022, DN2-E85 recorded the highest number of road deaths. The results of the research allow the observation of the distribution by risk categories in relation to the total length of the evaluated road.

## 2. Road network infrastructure in Romania

A large part of these roads pass through the localities. The road segments inside the localities are also considered, which, although they have the function of a street, keep the name of the transit road. From an administrative point of view, rural roads are classified into categories: highways, express roads, national and communal roads. Based on the statistical data extracted from the table „TEMPO_TRN139A" [8], the graph in Figure 1 shows the evolution of the development of the rural road network by category in the last 10 years.

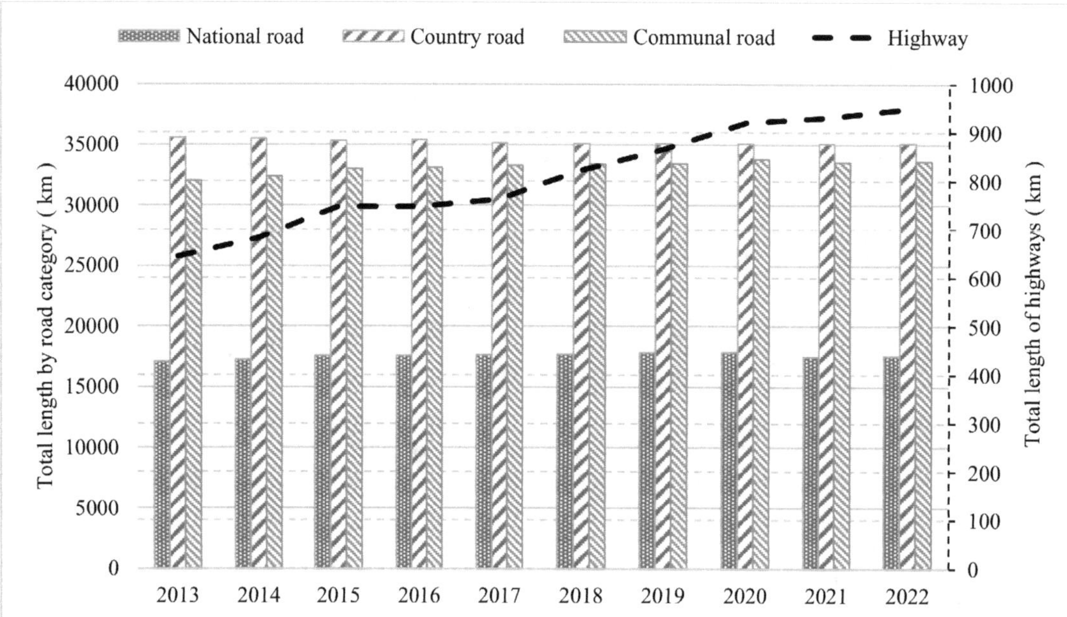


**Figure 1**. Development of the road network by road category

The analysis of the development of the road network in recent years shows a stagnation - in the last decade, the growth was 1.92%. The highway network where there was an increase of approx. 47% which stands for approx. 300 km.

In Figure 2 is presented the percentage of modernization of the road network at the level of 2022. While national roads are modernized to the extent of approximately 96%, the percentage drops to approximately 50% for county roads and below 25% for communal roads. The composition of the national road network by road category at the end of 2022 is graphically presented in Figure 3. The national rural road network spans 86,336 km, of which approximately 48.35% (41,653 km) are modernized. The percentage of 48.35% is represented by 949 km of highways, 16,813 km of national roads, 17,268 km of county roads, and 7,572 km of communal roads. Road modernization has steadily increased from 29,166 km in 2013 to 41,653 km in 2022. Source: Table „TEMPO_TRN139A" [8].

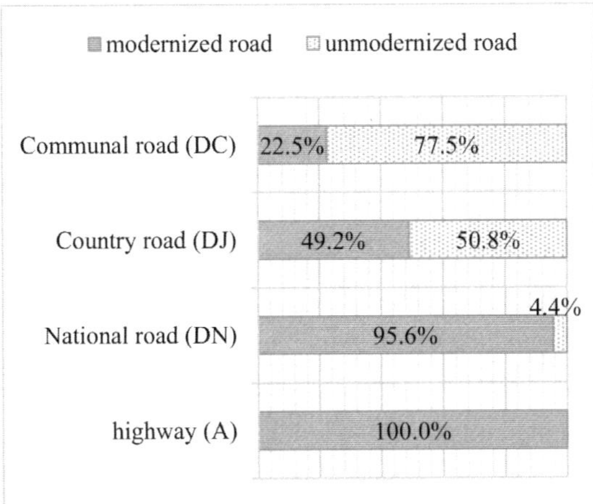


**Figure 2**. Percentage of modernized and non-modernized roads by category (2022)

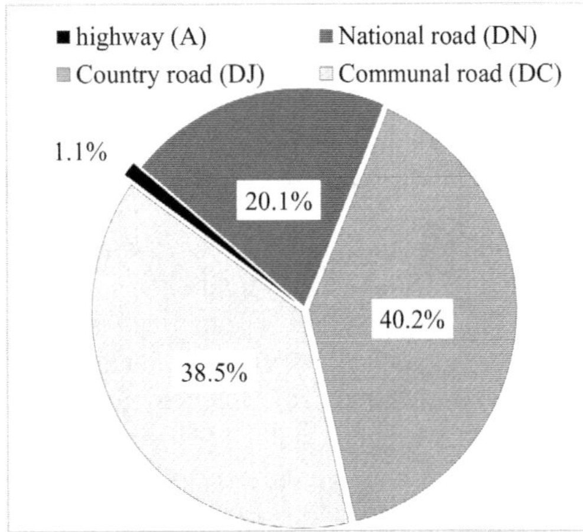


**Figure 3.** Composition of national road network by road categories (2022)

To determine the category of roads with the most serious accidents, the research focused only on the data of road accidents from the period 2017-2022, recorded on the roads represented in Figure 3.

The graph in Figure 4 shows the percentage distribution of serious accidents on each road category. The analysis of data from the period 2017-2022 says that on national roads, which represent 20.1% of the entire national road network, 56.5% of all serious accidents resulting in deaths and serious injuries were recorded. In comparison, 30.4% of serious accidents were recorded on county roads (40.2%), while 11.2% of these serious accidents were recorded on communal roads (38.5%) [9].

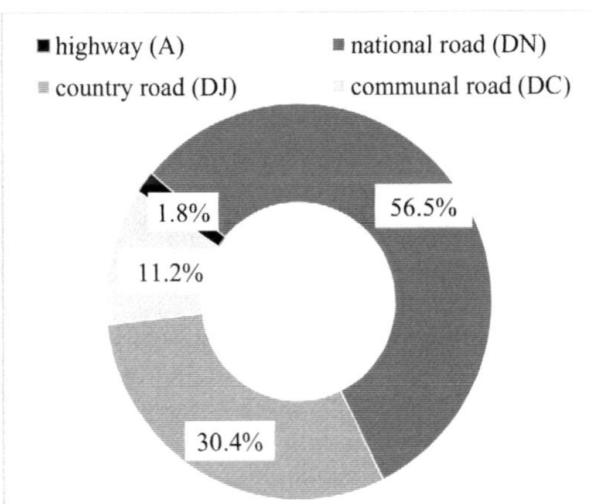


**Figure 4.** Distribution serious accidents by roads categories (2017-2022)

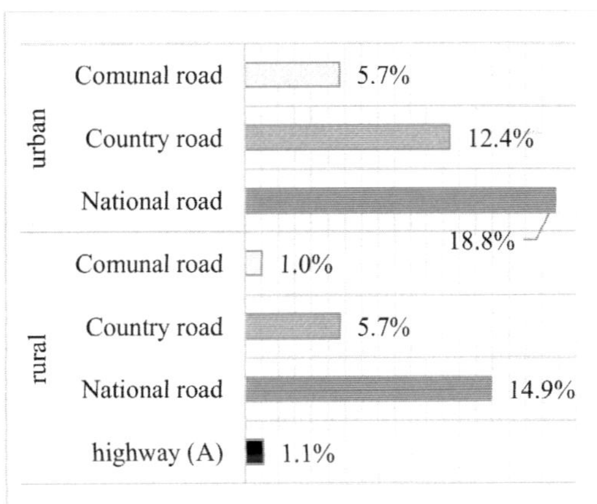


**Figure 5.** Distribution of serious accidents by categories of urban and rural roads (2017-2022)

As a result of filtering the database according to the road environment (urban/rural) in which the serious accidents occurred, Figure 5 shows the results of the research on the percentage distribution of these accidents by urban and rural road categories. For national roads, we note that the sum of the percentages for accidents recorded in urban and rural road environments (18.8% + 14.9% = 33.7%) is lower than the total percentage of 56.6% ("Figure 4") which, represents the percentage of serious accidents registered on national roads after filtering the database according to the road category on which they occurred. The difference of 22.9 percent is explained by the fact that, in most cases, the roads that cross localities are named after the names of the streets on which they overlap, and thus accidents are registered on the "street" road category.

One of the common shortcomings of the national road infrastructure is the absence of bypass roads for localities. This leads to the mixing of transit traffic with local traffic and with that of vulnerable users within localities. This situation increases the likelihood of road conflicts and, implicitly, serious accidents. Also, the road infrastructure faces the problem of the lack of specific facilities for disciplining drivers in the sense of reducing the speed of vehicles at the entrances to localities [10].

## 3. Applied methodology and work tools.

According to Directive (EU) 2019/1936, the European Commission sent member countries a „Study on the evaluation methodology of the road network" [11], which includes three parts:

- Proactive methodology (called NWA-proactive) - analyses road safety based on design parameters, classifying them as: "High risk", "Intermediate risk" or "Low risk".
- Reactive methodology (called NWA-reactive) - is applied based on available accident data for at least three years, classifying sections into: „High risk", „Inconclusive risk" or „Low risk".
- Integrated methodology (called NWA-integrated) - combines proactive and reactive approaches, classifying sections into five classes, according to the needs of road safety investments: "Very High Priority", "High Priority", "Intermediate Priority", "Low Priority" and "Very Low Priority".

The proactive methodology consists in the segmentation of the road network, the field collection of road parameter data, the processing of the data using the Excel tool and the ranking of the road segments as follows:

- Segmentation and the safety parameters: Assessing the built-in safety of roads involves detailed analysis of homogeneous segments with predefined values, including road type, topography, traffic lanes, roadside (RHR) and the annual average daily vehicles (AADT).
- Proactive Excel file: Collected data is entered into a dedicated Excel file, which calculates the estimated safety score for each road type (urban/rural road and divided/undivided road), Excel-proactive file [11].
- Accident Modification Factors (CMF): These factors indicate expected changes in accidents after implementation of countermeasures [12]. The CMFs in the proactive Excel file are used to calculate the estimative scores.
- Safety ranking: Estimated scores determine the classification of segments into risk categories for highways (low, intermediate, high) and other roads (low, intermediate, high). „High risk" segments are checked according to AADT volume, maintaining, or changing to the „Intermediate risk" class.

The reactive methodology is based on historical road accident data and the use of statistical methods to identify high-risk road sections. It requires data on serious accidents (road deaths and serious injuries) for at least 3 years as well as data on traffic flows of the road population of the evaluated roads.

- Road segmentation is based on criteria other than the proactive methodology, depending on the type of road and the structure of the data available in Romania.
- Reference populations differ by road type and performance parameters include crash rate and crash density.
- Thresholds for defining the degree of risk are set for each section, and the method involves using the Poisson distribution and cumulative probability to classify sections as „low risk", „unsure risk" or „high risk".
- The data and calculation are performed in an Excel-reactive file [11] and the sections are classified according to the highest risk identified between accident rate and accident density.

By applying the integrated methodology, the results of the proactive and reactive methodologies are combined to achieve a final ranking of 5 risk classes. The scheme of this process is presented in Table 1 which is made by processing the information from *„Figure 4.1: Integration of NWA-proactive and NWA-reactive results."* from the methodological implementation manual [11]. It is shown how the results of the proactive methodology (Excel file NWA-proactive) and those of the reactive methodology (Excel file NWA-reactive) are integrated.

In the methodological manual [11], in „*Figure 4.3: Example of combining results with different segmentation in each method.*" an example is provided of how the different segmentation results for each method are combined according to the scheme in Table 1.

**Table 1.** Schema for integrating NWA-proactive with NWA-reactive results

| ***PROACTIV ASSESSMENT RESULTS*** | ***REACTIVE ASSESSMENT RESULTS*** | | | |
|---|---|---|---|---|
| | **High Risk (Class r.3)** | **Risk Unsure (Class r.2)** | **No Data** | **Low Risk (Class r.1)** |
| **High risk (class p.3)** | Very High Priority (Class 5) | High Priority (Class 4) | High Priority (Class 4) | Low Priority (Class 2) |
| **Intermediate risk (Class p.2)** | Very High Priority (Class 5) | Intermediate Priority (Class 3) | Intermediate Priority (Class 3) | Low Priority (Class 2) |
| **Low risk (class p.1)** | Very High Priority (Class 5) | Low Priority (Class 2) | Very Low Priority (Class 1) | Very Low Priority (Class 1) |

In the methodological manual [11], in „*Figure 4.3: Example of combining results with different segmentation in each method.*" an example is provided of how segments, the results of applying proactive and reactive evaluation methodologies, they are combined and they are coloured according to the scheme in Table 1.

To arrive at a final ranking, we need to organize the results of these two methodologies in a way that provides a clear and meaningful view. This is done by using an Excel file - integrated, which automatically colours the cells corresponding to road segments according to the specifics of each methodology.

## 4. Road safety assessment of the infrastructure of the national road DN 2 - E85

To emphasize the results obtained by using the tools provided by the European Commission within the pilot program for road infrastructure safety assessment in the Member States, the results of the implementation of the evaluation guides for the analysis of the national road DN 2-E85 on a length of 472.23 km out of the total length of 482.23 km are shown. The road segment between km 0 and km 10 is under the administration of the municipality of Bucharest and has been excluded from the current assessment.

The road infrastructure assessment for the 10 km segment of DN2-E85, which also includes category I and II streets, presents several significant challenges. The complexity of this segment derives from the functional and structural diversity of the roads included, which requires a detailed and nuanced approach to ensure an accurate road safety assessment. Category I and II streets in Bucharest are characterized by high traffic density, multiple intersections, and a variety of users, including pedestrians, cyclists, and motor vehicles. This diversity creates a dynamic and often unpredictable road environment, which can negatively influence road safety. After the end of the pilot programme, this section of road will be subject to an evaluation using the evaluation tools calibrated by the European Commission.

### *4.1. The proactive methodology*

After applying the proactive assessment guidelines, the road was divided into two categories: undivided segments and divided segments. Each segment was evaluated according to 10 parameters. According to the data presented in Table 2, there were 243 undivided road segments (94% of the road length), with an average length of 1.83 km, and 24 divided road segments (6% of the road length), with an average length of 1.21 km.

**Table 2.** DN2-E85 road segmentation results

| Risk classes | No. sections | Total length (km) | % | No. sections | Total length (km) | % |
|---|---|---|---|---|---|---|
| | **Undivided** | | | **Divided** | | |
| **High Risk** Class p.3 | 135 | 228.44 | 48.4% | 13 | 2 x 21.9 | 4.6% |
| **Intermediate Risk** Class p.2 | 71 | 136.25 | 28.9% | 11 | 2 x 6.7 | 1.4% |
| **Low Risk** Class p.1 | 37 | 78.94 | 16.7% | 0 | 0 | 0.0% |
| **Total** | **243** | **443.63** | **93.9%** | **2 x 24** | **2 x 28.6** | **6.1%** |

The preliminary results indicate for the undivided road, approx. 50% fall into the "High Risk" class and approx. 30% fall into the "Intermediate Risk" class. For the divided road, which represents 6.1% of the total length of DN2-E85, approx. 4.6% fall into the "High Risk" class and approx. 1.4% fall into the "Intermediate Risk" class.

The results reveal that more than half of the evaluated section (53%) has low road safety. Undivided segments, which are more frequent and have a longer average length, have increased complexity in terms of road infrastructure safety assessment, compared to divided segments, which, although shorter on average, have specific requirements that have imposed an assessment on both directions of road traffic.

In Table 3 you can find the final results of the evaluation, which were obtained after checking the criterion AADT > (15% AADT), according to the technical indications presented in subchapter 3.9, p.66, [11]. As a result, of the 135 segments initially classified as having a "High Risk", 98 segments have been reclassified as having an "Intermediate Risk".

**Table 3.** Results of the proactive assessment of the DN2-E85 road (NWA – proactive)

| Risk classes-p | No. Sections-p | Total length (km) | % | No. Sections-p | Total length (km) | % |
|---|---|---|---|---|---|---|
| | **Undivided** | | | **Divided** | | |
| **High Risk** (Class p.3) | 37 | 60.8 | 12.9% | 4 | 2 x 4.3 | 0.9% |
| **Intermediate Risk** (Class p.2) | 169 | 303.9 | 64.4% | 20 | 2 x 24.3 | 5.1% |
| **Low Risk** (Class p.1) | 37 | 78.94 | 16.7% | 0 | 0 | 0.0% |
| **Total** | 243 | 443.64 | 93.9% | 24 | 2 x 28.6 | 6.1% |

As a result, for the undivided road, 12.9% of the total length of the DN2-E85 road falls into the "High Risk" class, 64.4% falls into the "Intermediate Risk" class and 16.7% falls into the "Low Risk" class. For the divided road, 0.9% of the total length of the DN2-E85 road is in the "High Risk" class, 5.1% is in the "Intermediate Risk" class and 0% is in the "Low Risk" class.

### *4.2. Reactive methodology*

Two reactive evaluation methods are available. The first method can be applied when data on serious accidents as well as intersection traffic flows (AADT) are known. Reactive evaluation can be done separately for intersections and separately for road segments between intersections. The research is carried out by applying the second method, that by which the intersections are evaluated together with the road alignments.

The segmentation of the DN2-E85 road for the application of the reactive evaluation method is done according to several criteria. These include separating divided from undivided road sections, dividing the road into segments when the average daily traffic level (AADT) changes, creating road segments homogeneous in terms of traffic speed and road environment, and including curve length and of the functional area of the intersection within a road segment.

In order to calculate the specific safety indicators, data relating to the population of national roads are entered in the „*Info Preliminary*" spreadsheet, such as „*Time period of accident data*" (4 years), „*Total no. accidents*" on national road DN2-E85 (673), „*Total length of all road sections*", DN2-E85 (472.23 km), „*Total km of national roads*" (17.913 km), „*Total no. accidents on national roads*" (10,316), „*Average* $AADT_{2020}$" (8,000 vehicles/ 24 h/ year). This information is fundamental for the correct assessment of road safety on this road network. The recommended lengths for the divided road segments are 10±5 km and for the undivided road 5±2 km.

The results of the reactive evaluation of the DN2-E85 road were obtained by applying the segmentation criteria, AADT allocation and accidents for each road segment, both in the downstream and upstream directions. This reactive - segmentation was performed so that the locations of the sections corresponded to those of the proactive assessment, thus simplifying the analysis process and avoiding artificially increasing the number of sections to be assessed. Table 4 shows the results of the reactive assessment, including the number of sections, the total length, the percentage of the total and the number of accidents for each risk category, for both undivided and divided road segments.

**Table 4.** DN2-E85 road segmentation results (reactive NWA)

| Risk classes-r | No. Sections-r | Total length (km) | % | No. Accidents | No. Sections-r | Total length (km) | % | No. Accidents |
|---|---|---|---|---|---|---|---|---|
| | Undivided | | | | Divided (a-d)* | | | |
| **High Risk** | 59 | 263.27 | 55.8% | 514 | 11 - 8 | 9.7-7.9 | 1.9% | 18 - 12 |
| **Unsure Risk** | 30 | 143.89 | 30.5% | 116 | 3 - 3 | 10.1-10.1 | 2.1% | 5 - 7 |
| **Low Risk** | 12 | 36.47 | 7.7% | 0 | 10 - 13 | 8.8-10.6 | 2.1% | 1 - 0 |
| **Total** | 101 | 443.63 | 93.9% | 630 | 24 - 24 | 28.6-28.5 | 6.1% | 24 - 19 |

**(a = upward direction of traffic; d = downward direction of traffic)*

### 4.3. *The integrated method*

The evaluation of the DN2-E85 road and the determination of the risk classes are carried out through the integrated method, which involves combining the results obtained from the application of both the proactive and the reactive assessment methodology. Using the combined scheme presented in Table 1. *Schema for integrating NWA-proactive with NWA-reactive results,"* road segments are classified into five road risk classes. In ascending order of risk, Table 5 presents the results of the DN2-E85 road infrastructure safety assessment.

**Table 5.** The result of the road safety assessment of the infrastructure of the national road DN2-E85.

| Class risk | Class colour | Undivided | % | Divided (a) | % | Divided(d) | % |
|---|---|---|---|---|---|---|---|
| **Length-Class 1** | green dark | 5.80 | 1.23% | 0.00 | 0.00% | 0.00 | 0.00% |
| **Length-Class 2** | green | 50.76 | 10.75% | 8.80 | 1.86% | 10.60 | 2.24% |
| **Length-Class 3** | yellow | 115.70 | 24.50% | 10.10 | 2.14% | 10.10 | 2.14% |
| **Length-Class 4** | orange | 8.10 | 1.72% | 0.00 | 0.00% | 0.00 | 0.00% |
| **Length-Class 5** | red | 263.27 | 55.75% | 9.70 | 2.05% | 7.90 | 1.67% |
| **Total** | | 443.63 | 93.94% | 28.60 | 6.06% | 28.60 | 6.06% |

More than half of the total length of the road, representing 58.8%, falls into the category of "Very High Priority" (class 5), indicating road segments with low road safety. The assessment highlights sectors with a significant level of risk, highlighting the need to implement measures to improve road safety on these sections.

The complexity of Bucharest's category I and II streets, these streets require specific strategies for managing the safety of the street infrastructure and adapted remedial measures. Thus, the 10 km segment in Bucharest, which includes category I and II streets, represents an additional challenge in the assessment due to its infrastructural complexity and diverse traffic flows.

## 5. Implementation of preliminary measures for completing the pilot project

The Romanian Road Authority - ARR recently hired many specialized inspectors (2021-2022), who lack practical experience and specific knowledge about road infrastructure. Intensive continuous professional training courses are necessary to address this gap.

Modern planning tools, such as Gantt diagrams (Figure 6), are essential for organizing and managing activities efficiently. These tools help schedule and track tasks, ensuring projects meet deadlines. Continuous professional development of ARR's specialized staff is crucial for enhancing their expertise.

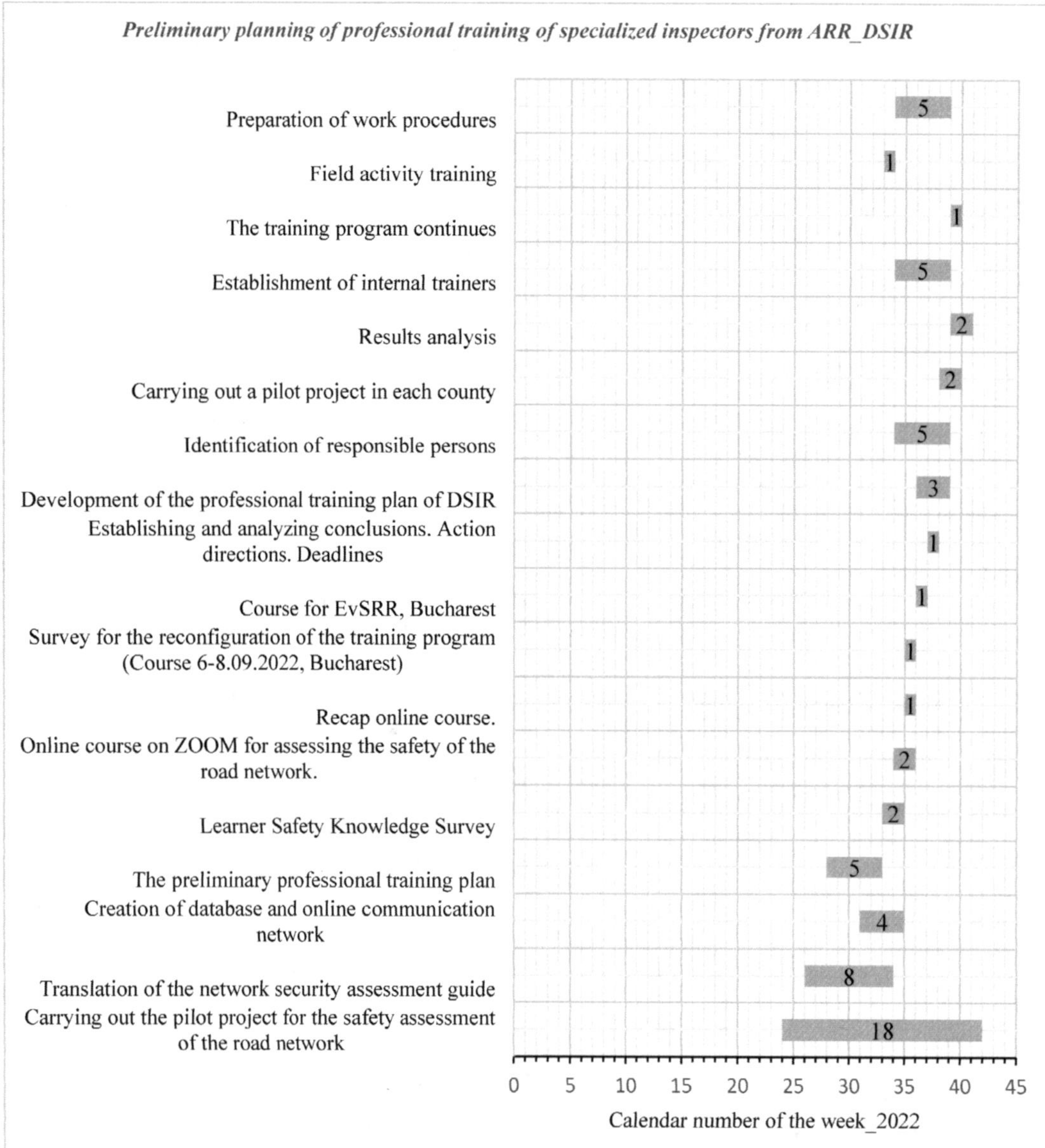


**Figure 6.** Gantt diagram of the preliminary measures for the completion of the pilot project

The objective of completing the road network safety assessment, as mandated by the updated Law 265/2008, can only be achieved through the meticulous application of a well-structured preliminary measures plan. The Gantt diagram in Figure 6 illustrates the timeline and sequence of necessary steps to accomplish this goal. This diagram includes key activities such as preparation of work procedures, field activity training, establishment of internal trainers, and the execution of pilot projects in each county.

By adhering to this comprehensive plan, ARR aims to enhance the safety of the national road network, ultimately reducing road accidents and improving overall traffic safety. The structured training and planning approach will empower the new inspectors with the knowledge and skills required to effectively contribute to the road safety assessment process.

## 6. Conclusions

The study is particularly valuable for road safety auditors and specialized personnel of the Romanian Road Authority (ARR). By providing detailed information and data, the study improves their ability to assess and improve the safety of the infrastructure of the national road network. This is a significant contribution as it equips these professionals with the tools and insights needed to identify and mitigate road safety risks.

The analysis of serious accidents by road categories, with a specific focus on differentiating between rural and urban roads, is crucial. It helps to identify areas with high road risk, thus facilitating a better understanding of the causes of infrastructure-related accidents. This targeted approach allows for more efficient interventions and resource allocations.

A clear indication of the need to allocate financial resources to improve road safety in high-risk segments is a key conclusion. Prioritising investments in these areas can significantly reduce road accidents and improve overall safety. This is an important recommendation for policymakers and funding bodies.

The study lays a solid foundation for future research and adjustments to road infrastructure safety assessment guidelines. By highlighting the need to consider the specific characteristics of county roads and streets in categories I and II, it paves the way for more adapted and effective safety measures. This forward-looking perspective ensures that the impact of the study will outweigh its immediate findings.

Highlighting the importance of improving road safety across the country, especially on county roads and urban streets where many serious accidents are recorded, underlines the wider implications of the study. This concentration is in line with national safety objectives and can lead to significant improvements in road safety.

The study highlights the importance of training specialised staff and raising awareness among decision-makers about the need to allocate additional resources for road safety. This is essential to sustain long-term improvements and to ensure that safety measures are implemented and maintained.

The conclusions are comprehensive and actionable, providing a clear plan for improving road safety in Romania. They address both immediate needs and long-term objectives, making them highly relevant to various stakeholders, including road safety auditors, policymakers and researchers. By focusing on high-risk areas and supporting targeted financial investments, the study provides practical solutions that can lead to significant reductions in road accidents and fatalities. Overall, the conclusions are well justified and present a compelling case for continued efforts and investment in road safety.

The implementation of the preliminary measures for the continuous professional training of specialized inspectors, as presented in the Gantt chart, emphasizes the importance of using modern planning tools and continuous staff development to achieve road safety objectives.

These measures are essential to ensure that inspectors are well prepared to assess and improve the safety of road infrastructure, in accordance with legal requirements and European standards. This creates a solid framework for the effective application of the integrated safety assessment methodology on national roads.

## References

[1] Appleton I, Hannah J, Noone M, Wilkie S. Road infrastructure safety assessment. New Zealand Transport Agency. 2009 September.

[2] Md Murozi AF, Ishak SZ, Nusa F, Hoong APW, Sulistyono S. The Application of International Road Assessment Programme (iRAP) as a Road Infrastructure Risk Assessment Tool. IEEE Xplore 13th Control and System Graduate Research Colocvium (ICSGRC). 2022 August 09;:

p. 237-242.

[3] Persia L, Usami SD, De Simone F, Feypell De La Beaumelle V, Yannis G, Laiou A, et al. Management of road infrastructure safety. Transport Research Arena TRA2016. 2016 April 18-21; 14: p. 3436-3445.

[4] European Parliament, Council of the European Union. https://eur-lex.europa.eu. [Online].; 2019 [cited 2024 February 10. Available from: http://data.europa.eu/eli/dir/2019/1936/oj.

[5] The Government of Romania. Ordinance no. 3/2022 for the amendment and completion of Law no. 265/2008 regarding the management of traffic safety on the road infrastructure. The Official Monitor of Romania. 2022 January 28: p. 1-8.

[6] The Romanian Parliament. Law no. 265/2008, republished and updated with subsequent amendments, regarding the management of traffic safety on the road infrastructure. The Official Monitor of Romania. 2012 August 23: p. 6-18.

[7] The Government of Romania. Annex to Government Decision no. 682/2022 for the approval of the National Road Safety Strategy for the period 2022-2030. The Official Monitor of Romania. 2022 May 31: p. 3-74.

[8] National Institute of Statistics - INS, Romania. https://insse.ro/cms/ro. [Online].; 2023 [cited 2024 May 12. Available from: http://statistici.insse.ro:8077/tempo-online/#/pages/tables/insse-table.

[9] Romanian Police. www.politiaromana.ro. [Online].; 2022 [cited 2023 July 07. Available from: https://www.politiaromana.ro/ro/utile/statistici-evaluari/statistici.

[10] Burlacu FA, Burlacu A, Răcănel C. The auditor's manual road safety. 1st ed. Adrian B, editor. Bucharest: CONSPRESS; 2019.

[11] The European Commission_Mobility & Transport - Road Safety_. https://road-safety.transport.ec.europa.eu/. [Online].; 2023 [cited 29 06 29. Available from: https://road-safety.transport.ec.europa.eu/eu-road-safety-policy/priorities/infrastructure/road-infrastructure-guidelines_en.

[12] U.S. Department of Transportation - Federal Highway Administration. Crash Modification Factors Clearinghouse. [Online].; 2023 [cited 2023 06 30. Available from: https://www.cmfclearinghouse.org/.

IOP Conf. Series: Materials Science and Engineering **1311** (2024) 012044 doi:10.1088/1757-899X/1311/1/012044

# Geometrical optimization and evaluation of the bridge safety guardrail

**M I Ghereş, N Cordoş and A Todoruţ**

Technical University of Cluj-Napoca, Department of Automotive Engineering and Transports, Romania

marius.gheres@auto.utcluj.ro

**Abstract**. Lately, the rapid growth of traffic led to the construction of a new freeway bridge to ease the road congestion. This paper studied the crash behaviour of steel guardrails for semi-heavy traffic roads used on bridges. The guardrail structure is tested according to the SR EN 1317 standards at two velocity cases, corresponding to the TB 11 and TB 51 impact test description. To perform this evaluation study, the guardrail structure is first modelled in CAD environment software, respecting the commercial dimensions. The boundary conditions, the material properties assignment, and the load case creation are done by using dedicated FEA software. The obtained results are analysed and are proposed some design improvements to the base guardrail structure. The improved guardrail structure is tested under the same work conditions as the base structure. The conclusions of this research are presented in the last part of the paper.

## 1. Introduction

Increasing the safety of vehicle's passengers is a continuing goal of the public road safety device designers. The guardrails systems placed on public roads are vital to increase the passenger's safety and the transportation systems. The guardrails systems are defined as a passive device placed on the lateral roadside to prevent run off the road of the vehicles. Nowadays, the most common type of the guardrail used to increase the roadside protection is the W-beam guardrail, which has a great ability to reduce and redirect the vehicle on the roadside. In the case of the guardrail system placed on bridge, in addition to reduce the crash impact, the focus is also to the vehicle retention on the roadside. In the event of the crash, breaking of the guardrails system, it may cause the vehicle to fall of the bridge, resulting severe injury of the car passenger's and significant damage of the vehicles [4]. To improve the guardrails performance, a series of evaluation studies from the domain are analysed [8], [11].

A full-scale crash and the simulation of a bridge guardrail was studied by Atahan to analysis and establish the high containment level guardrail to be used on bridges around city, placed on bridges with on high-risk location. The crash simulations were performed in or using Ls-Dyna environment, and the real crash test were run at CSI-SPA located in Italy. The simulation was performed according to EN 1317 standard, TB 11, and TB 81 requirements test. The results of this study establish the performance of the guardrail proving the adequacy for use on bridge [1]. A safety and economical optimization of two guardrails system are studied by Teng et all. The optimization criteria used as the objective and constrained function are the guardrail system working width and exit angle of the vehicle. Following the optimization process the specific crash tests of H1W4 and H2W4 barrier system used on Turkey roadside are performed. The vehicles models used to perform the crash test are chosen to TB 11 (car with mass of 900 kg), TB42 (HGV with mass of 10000 kg) and TB51 (BUS with mass of 13000 kg) test descriptions. The optimized results of the real guardrails were being compared to results obtained from the crash test performed in Ls-Dyna environment, resulting that the optimized models fulfil the safety requirements from EN 1317 standard [6]. An important study performed to find the optimal W-beam guardrail (AG04-2.0 type) height and post spacing for small car protection during the crash impact. To performing the crash test according to the TB11 test descriptions it is chosen a Geo Metro vehicle model (900 kg) available in the National Crash Analysis Center database. The virtual crash test was solved by using the Ls-Dyna environment. The both end of the guardrail continuation are connected by the elastic springs according to the cross-sectional area of the W-beam, the steel Young's modulus and the length of the unmodeled guardrail and the nonlinear spring elements was applied to describe the post-soil interaction. The crash test analysis was run for three distance cases between guardrail post: 1333, 2000, and 4000 mm. To verify the numerical stability of the simulation the global energy balance curves are represented. After W-beam evaluations the barrier with 200 mm spacing show the better performance comparing to remaining system [10]. The safety assessment of the W-beam guardrails system based on TB32 test description are presented by Wide et all. The crash study is based on the safety evaluation of the damaged guardrail system hit again by a vehicle. The W-beam guardrails are analysed in three configurations according to TB32 test description: in the first case the impact angle of the vehicles is changed from 20° to 7°, the other conditions remain unchanged; in the second test the undamaged guardrails system is tested according TB32 test description; and in third crash test simulation the initially damaged guardrail system is tested. The results show that the damaged system keeps the initial functionality, however, due to the impact severity and geometric deformation during the crash, the damaged parts should be replaced or repaired [12].

The goal of this study is to analysis and compare two W-beam guardrails device H2 W2 A, designed in two variants: first model with the W-beam placed on the C-post and the second model that have added a shock-absorber between the W-beam and C-post and a reinforcement element between base plate and supporting C-post. The crash study fulfils the requirements described in EN1317 standards.

## 2. Bridge safety barriers

The guardrails are defined as a roadside safety system that improve the vehicles safety on the road sections considered dangerous. This assembled device constitutes a safety barrier that allows to mitigate de impact of the vehicles and redirect it on the road, in the event of the crash. In this study a semi-heavy bridge traffic guardrail, type H2 W2 A is tested regarding EN 1317 standard requirements. Figure 1 describes the symbol meanings of the proposed guardrail for impact test and improvement solution.

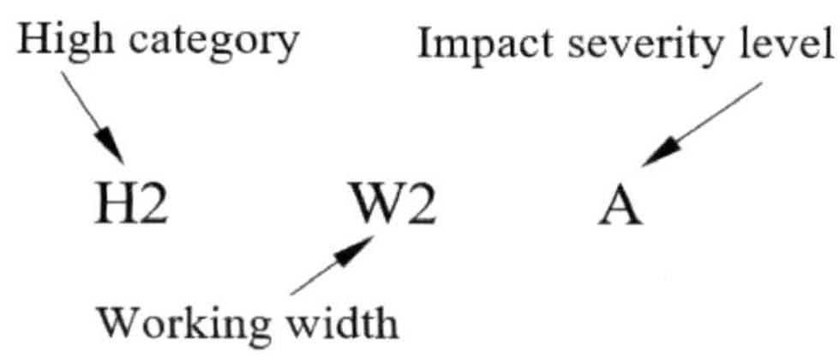


**Figure 1.** Symbol description of H2 W2 A bridge guardrail system.

In this study the CAD model of the bridge guardrails are designed in two configurations: first basic model made up of three main components and auxiliary connections parts presented in figure 2a; the second improved model, that have added reinforcement part, the gusset welded by the supporting post and base plate and spacer mounted between the W-beam and supporting post, as can be seen in figure 2b. Both guardrails' models are designed by using the SolidWorks CAD environment, modelling each part, and then assembling the guardrails parts [5].

The distance between the C-supporting post is 1500 mm and the total length of the guardrail system studied in this paper is 18000 m. The lateral width of the base model is 211 mm and of the improved model is 408 mm. In figure 3 is presented a detail where can be observed the main dimensions of the improved guardrail model.

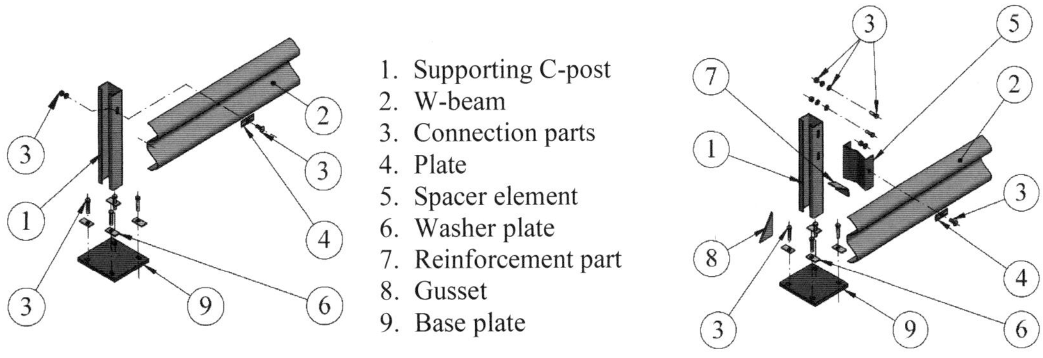


a) Base model - parts descriptions

b) Improved model - parts descriptions

**Figure 2.** Technical details of bridge W-beam guardrail.

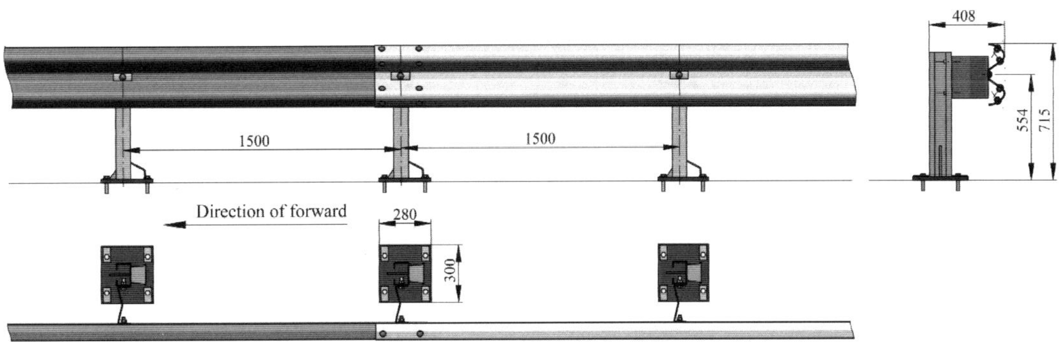


**Figure 3.** Dimension of the improved model.

### *2.1. European Standards (EN) 1317 requirements*

The safety guardrails are evaluated by the performance impact classes established by three main characteristics regarding the passenger's protection, vehicles, and protected areas: level of protection, acceleration severity index, and guardrail deformation expressed by the levels of working width.

**Table 1.** Vehicle impact crash test criteria.

| Test | TB 11 | TB 51 |
|---|---|---|
| Impact speed (km/h) | 100 | 70 |
| Impact angle (°) | 20 | 20 |
| Total mass (kg) | 900 | 13000 |
| Type of vehicle | Car | Bus |

**Table 2.** Approved levels.

| Impact severity level | Index values |
|---|---|
| A | ASI ≤ 1.0 |
| B | ASI ≤ 1.4 |
| C | ASI ≤ 1.9 |
| | THIV ≤ 33 km/h |

Some requirements from EN 1317 standard are presented in table 1, that represent the crash criteria applied in this study. It can be observed, that for the crash test simulation are used two different crash test criteria for both guardrails systems, as recommended in standards, table 1. The presented guardrails systems are classified according to the performance classes established after the crash tests results according to the EN 1317 acceptance criteria. The impact severity level represents a theoretical parameter used for evaluation of the injury's severity or other risk factors pf the vehicles passengers. In table 2 are presented the impact severity levels, highlighting the level of the tested guardrail.

**Table 3.** Levels of working width – according EN 1317 standard [2].

| Working width classes | W1 | W2 | W3 | W4 | W5 | W6 | W7 | W8 |
|---|---|---|---|---|---|---|---|---|
| Working width $W_m$ (m) | ≤ 0.6 | ≤ 0.8 | ≤ 1.0 | ≤ 1.3 | ≤ 1.7 | ≤ 2.1 | ≤ 2.5 | ≤ 3.5 |

The working width of the guardrail system represent the maximum lateral distance between the any undeformed face of the barrier exposed to the roadside and the maximum dynamic position of the any deformed parts of the barrier after crash. In table 3 are presented the working width levels depending on the working width classes. The containment levels of the guardrails system must be in accordance with EN 1317-2 standard requirements where the guardrails are tested respecting the vehicle impact criteria. In table 4 are presented the containment levels according to standard descriptions. It can be observed that the guardrail system chosen for testing has a high containment level.

**Table 4.** Guardrails system - containment levels [2].

| | Containment levels | Acceptance test |
|---|---|---|
| Low angle containment | T1 | TB21 |
| | T2 | TB22 |
| | T3 | TB41, TB21 |
| Normal containment | N1 | TB31 |
| | N2 | TB32, TB11 |
| Higher containment | H1 | TB42, TB11 |
| | L1 | TB42, TB32, TB11 |
| | H2 | TB51, TB11 |
| | L2 | TB51, TB32, TB11 |
| | H3 | TB61, TB11 |
| | L3 | TB61, TB32, TB11 |
| Very high containment | H4a | TB71, TB11 |
| | H4b | TB81, TB11 |
| | L4a | TB71, TB32, TB11 |
| | L4b | TB81, TB32, TB11 |

In addition to the guardrail deformation, the intrusion level in the guardrails system and vehicles should be as low as possible. Table 5 shows the nineteen intrusion classes that provide a range of 0.6 to 3.5 m. The generic sketches of the vehicles and guardrail impact were also presented.

## 3. Finite element analysis

### *3.1. Finite element model of the generic vehicles*

The current work involves the modelling of two generic vehicles to mimic the real crash behaviour between the guardrail system and vehicle, according to both TB 11 and TB 51 test crash criteria. The pre-processing and processing of the models are done in Altair Hypermesh and simulations are runed in Radioss solver. The proposed model of the impactor corresponding to TB 11, is organized into five vehicle parts zone with different thickness, as shown in table 6.

**Table 5.** Levels of vehicles intrusion, corresponding EN 1317 [2].

| Classes of vehicle intrusion levels | Levels of vehicle intrusion (m) |
|---|---|
| VI1 | $VI_N \leq 0.6$ |
| VI2 | $VI_N \leq 0.8$ |
| VI3 | $VI_N \leq 1.0$ |
| VI4 | $VI_N \leq 1.3$ |
| VI5 | $VI_N \leq 1.7$ |
| VI6 | $VI_N \leq 2.1$ |
| VI7 | $VI_N \leq 2.5$ |
| VI8 | $VI_N \leq 3.5$ |
| VI9 | $VI_N \leq 3.5$ |

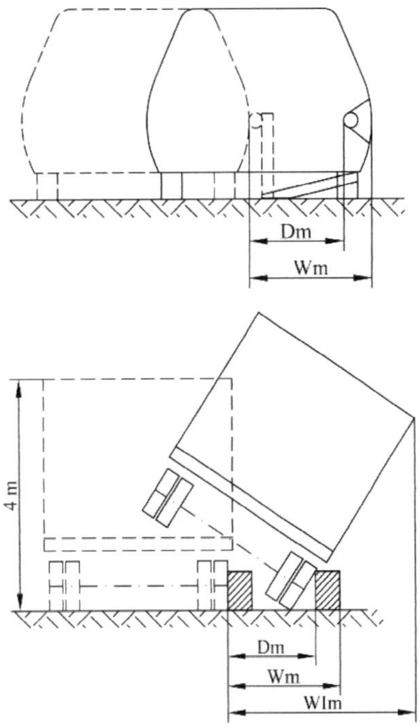


**Table 6.** Impactor vehicle model for TB 11 crash test criteria.

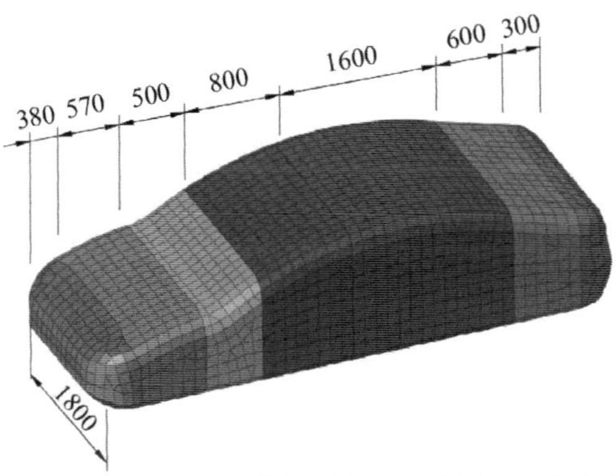


| Vehicle parts | Thickness (mm) | Target length (mm) | Number of shell elements | Mass (kg) |
|---|---|---|---|---|
| Part 1 | 2.5 | | | |
| Part 2 | 3.0 | | | |
| Part 3 | 4.0 | 100 | 2669 | 900 |
| Part 4 | 6.0 | | | |
| Part 5 | 7.2 | | | |

The numbering of the parts begins from the frontal side of the vehicle, and the last two parts from the rear side are composed from the first two parts. The first part of the vehicle is made of aluminium and all other parts are made of steel. The length of the vehicle is about 4750 mm. For TB 51 crash test criteria, the proposed impactor model mimics the shape and dimension of a bus. This model is also divided into five parts of different thickness, and the front and rear part are made of aluminium. To reduce the element number, the original length of the bus model is reduced about 8050 mm.

A difference from reality is that the vehicles models do not have wheels, and during the crash event the vehicle dynamics may differ from the real crash.

### *3.2. Finite element model of the guardrail*

The parts of the improved guardrail system are presented in table 8, where can be seen the front and back view detail in the connection zone between two W-beam. All guardrail parts are meshed into mixed shell elements.

**Table 7.** Impactor vehicle model for TB 51 crash test criteria.

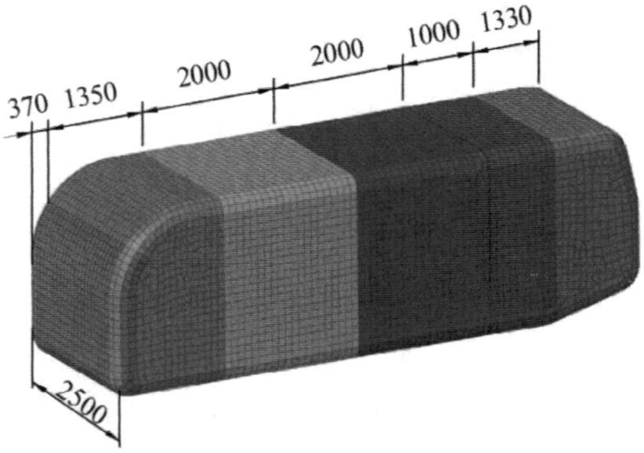


| Vehicle parts | Thickness (mm) | Target length (mm) | Number of shell elements | Mass (kg) |
|---|---|---|---|---|
| Part 1 | 3.0 | | | |
| Part 2 | 5.0 | | | |
| Part 3 | 10.0 | 100 | 8628 | 13000 |
| Part 4 | 45.0 | | | |
| Part 5 | 48.0 | | | |

**Table 8.** Finite element model of W-beam improved guardrail.

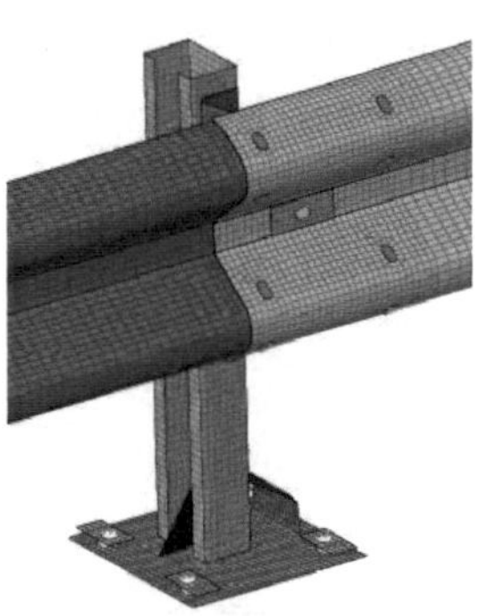
a) Front view

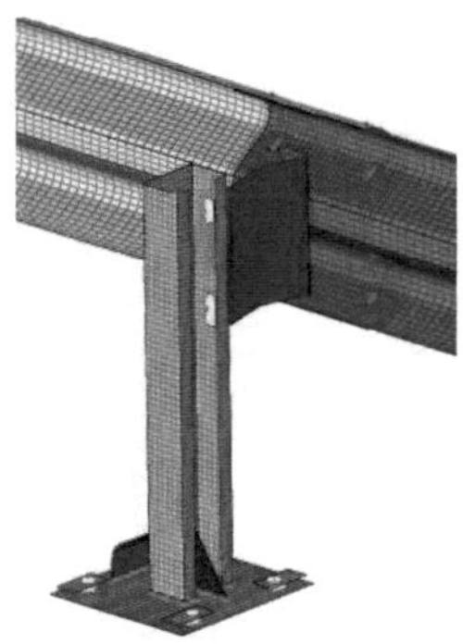
b) Back view

| **W-beam guardrail parts** | **Thick (mm)** |
|---|---|
| Supporting C-post | 4.5 |
| W-beam | 3 |
| Base plate | 20 |
| W-beam connection plate | 5 |
| Washer plate | 5 |
| Spacer element | 5 |
| Reinforcement part | 6 |
| Gusset | 10 |

Because the total guardrail length is about 18000 mm, both ends of this device are rigidly constrained, considering that on this length the guardrail can take the impact loading. Also, the bolted connections are represented by using RB2 rigid elements. The improved guardrail model representation was chosen because it contains the most complete form of the device. The base model of the guardrail device does not contain the spacer elements, gussets, reinforcements parts and specific connecting elements of this parts.

In this study are used two materials models, the steel material assigned to the guardrail device and impactor models at parts 2÷5 and aluminium assigned to the frontal and rear parts of the generic vehicles [7]. The mechanical properties of the used materials are presented below in tables 9 and 10. The material models assigned in this study use the Zerilli-Amstrong plasticity model and are isotropic elasto-plastic properties, [3].

### *3.3. Crash test setup*

The crash simulation presented in this study are performed according to the TB 11 and TB 51 crash test criteria described in EN 1317 standard. In first stage, the base model of the W-beam guardrail is tested according to the TB 11 crash test criteria at an initial velocity of 100 km/h and 20° impact angle of the generic car model, presented in table 6. In the second crash simulation the base guardrail is tested according to TB 51 crash test criteria. The generic bus model hit the guardrail devis at 70 km/h and 20° impact angle. The improved model of the guardrail is tested at the same test criteria as the base guardrail model.

**Table 9.** Mechanical properties of the assigned steel [9].

| | |
|---|---|
| Initial density | $7.85e^{-6}$ kg/mm$^3$ |
| Poisson ratio | 0.3 |
| Hardening exponent | 0.5 |
| Hardening parameter | 0.450 |
| Maximum Stress | 0.425 GPa |
| Failure plastic strain | 0.3 |
| Young's modulus | 210 GPa |

**Table 10.** Mechanical properties of the assigned aluminium alloy.

| | |
|---|---|
| Initial density | $2.7e^{-6}$ kg/mm$^3$ |
| Poisson ratio | 0.33 |
| Ultimate tensile stress | 0.175 |
| Yield stress | 0.09026 GPa |
| Failure plastic strain | 0.75 |
| Young's modulus | 60.04 GPa |

## 4. Results

The simulations results are presented and compared for each case separately. Following the energy balance curves are plotted, the obtained working width of the guardrail and the vehicles intrusion are presented. To a better view the picture that represents the impact between the vehicle and guardrail are cropped to highlighting the deformed region.

### *4.1. Crash results of the base guardrail model*

The guardrail and generic car model behaviour are presented in figure 4, where can be observed the maximum value of the von Mises is 6.786E-01 GPa.

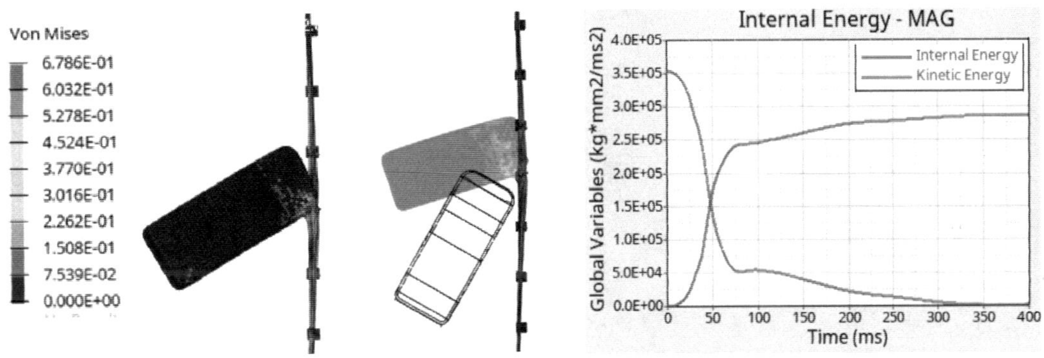


200 ms 400 ms Energy balance during impact

**Figure 4.** Simulation results according to TB 11 crash criteria for base guardrail model.

Guardrail impact and vehicle position is presented in two positions in figure: first position after impact, at 200 ms and second position simulation at the final at 400 ms. In the second position are presented the initial shape of the guardrail and deformed shape after impact. Also, is presented the initial position of the generic car and the position after impact, determined the vehicles dynamics during the impact. It can be observed that the kinetic energy curve starts from the initial value of 3.533E+05 J and drops continuously intersecting the internal energy curve at 50 ms, after decrease until 80 ms. After this time, the curve has a smoothly decrease until 350 ms, representing the stopped of the vehicle by the guardrail. In figure 5 are presented results according to TB 51 crash test criteria. The maximum value of the von Mises is 7.396E-1 GPa. The kinetic energy curve starts from initial energy value of 2.472, after drops continuously until 300 ms. At the 177 ms the kinetic energy curve intersects the internal energy curve.

After 300 ms the kinetic energy curve decreases smoothly until the end of simulation. After simulation it can be observed that the generic bus impactor remains on the roadside.

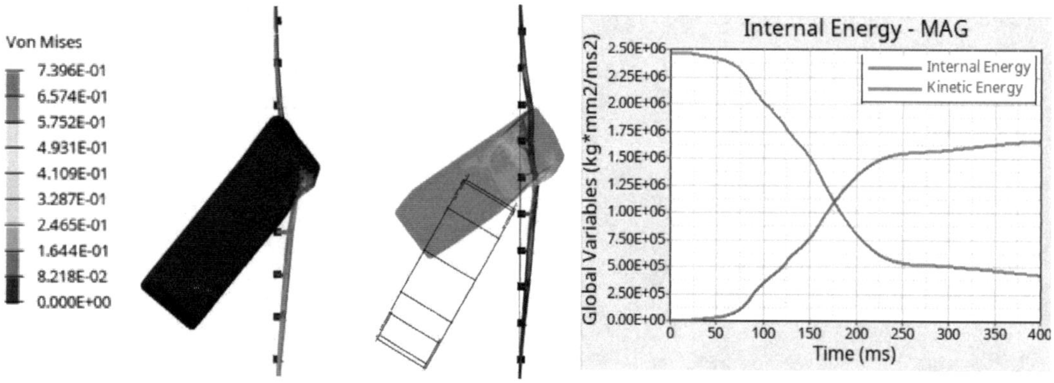


200 ms 400 ms Energy balance during impact

**Figure 5.** Simulation results according to TB 51 crash criteria for base guardrail model.

### *4.2. Crash results of the improved guardrail model*

The results according to TB 11 crash criteria for improved guardrail model are presented in figure 6. Maximum von Mises stress value is 6.872E-01 GPa. The kinetic energy curve starts from initial value of 3.533+05 J and drops continuously intersecting the internal energy curve at 100 ms, after decrease until 140 ms. After this time the kinetic energy curve decrease smoothly until the end of the simulation.

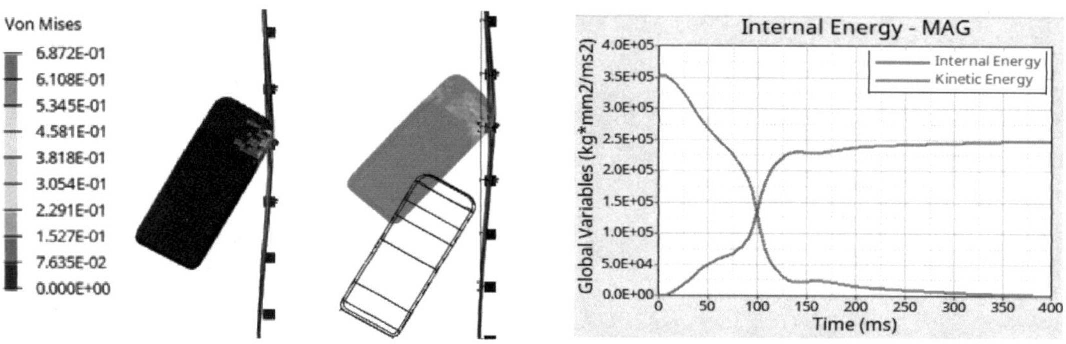


200 ms 400 ms Energy balance during impact

**Figure 6.** Simulation results according to TB 11 crash criteria for improved guardrail model

The results according to TB51 crash criteria are presented in figure 7, where can be observed the maximum value 7.972E-01 of the von Mises stress. The kinetic energy curve starts from initial value of 2.472E+06 J and remains constant to the first 4 ms, after drops continuously until 260 ms. After this time, the curve decreases smoothly until de end of simulation.

**Table 11.** Guardrails deformations and vehicles intrusions.

| **W-beam guardrail deformation** | | | | **Vehicles intrusions** | | | |
|---|---|---|---|---|---|---|---|
| Base model | | Improved model | | Base model | | Improved model | |
| Car | Bus | Car | Bus | Car | Bus | Car | Bus |
| 179 | 615 | 206 | 1166 | 413 | 1180 | 271 | 762 |

IOP Conf. Series: Materials Science and Engineering **1311** (2024) 012044 doi:10.1088/1757-899X/1311/1/012044

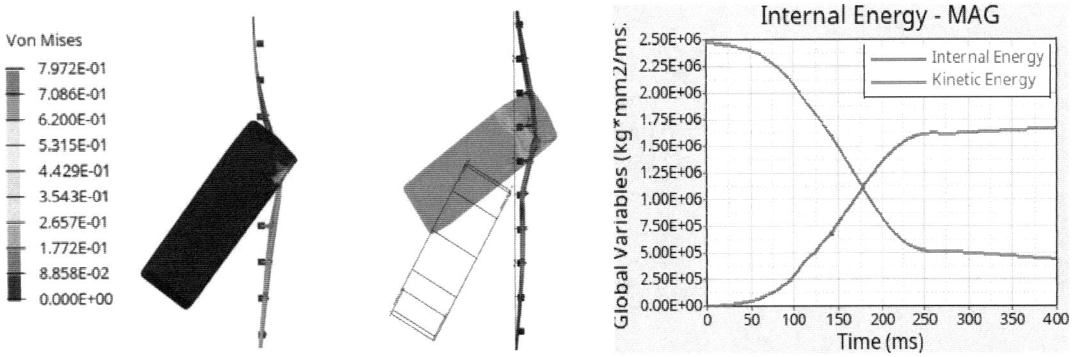


200 ms 400 ms Energy balance during impact

**Figure 7.** Simulation results according to TB 51 crash criteria for improved guardrail model.

In all simulation cases the vehicles remain on the roadside after the crash simulation, which shows the guardrail device has a good retention performance. In table 11 are presented the approximate values of the W-beam guardrail deformation and the car intrusions.

The comparative results of the vehicles intrusions are presented in figure 8. It can be observed that the values of the intrusions are lower at the impact from the improved guardrail model.

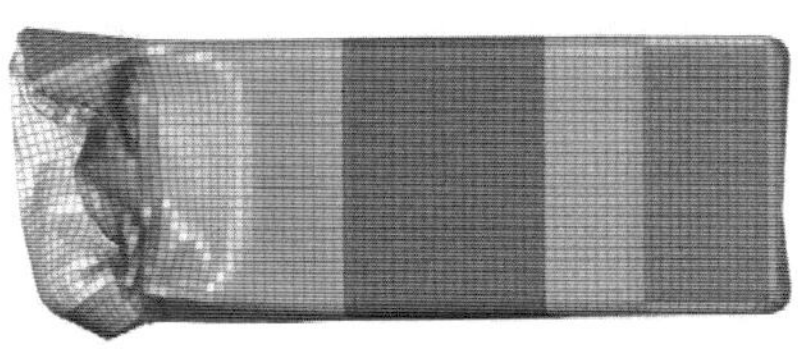
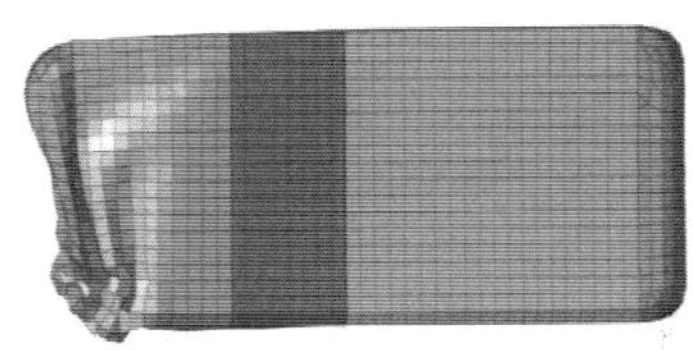

a) Bus intrusion-impact from base guardrail model

b) Car intrusion-impact from base guardrail model

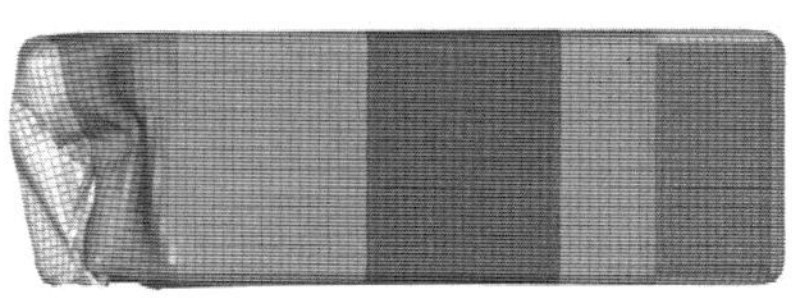
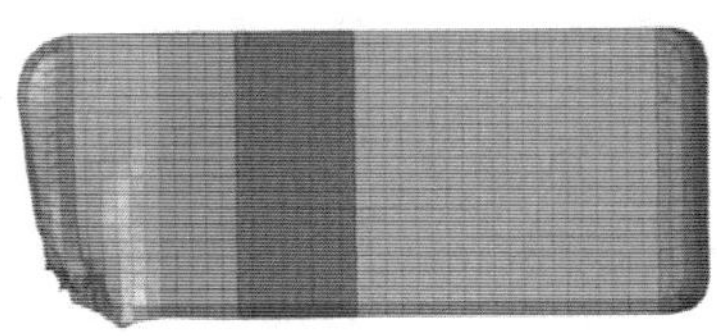

b) Bus intrusion-impact from improved guardrail model

c) Car intrusion-impact from improved guardrail model

**Figure 8.** Comparison of simulation results – deformation of vehicles.

## 5. Conclusion

This paper presents a comparative study between two guardrails models: the first base guardrail model represents the commercial model of a guardrail designed to use on the bridge roadside and the second guardrail model that represents the improved model of the base guardrail model. The evaluation between two guardrails models was performed in four simulations cases, were it established the von Mises stress distribution values during the de crash, the energy balance and the deformation of the guardrail and generic vehicles. Analysing the measured deformation from table 11, it can be observed that the improved model of the guardrail presents the higher deformation than the base model. This is because the improved guardrail model has added the spacer elements which reduce the vehicles impact. Analysing the obtained values that shows the vehicles intrusion it can be observed that the car deformation and damages is lower than base model. Following the results of this study, the improved guardrail model reduces the damage of the vehicles during a real crash about 65%.

## References


[1] Atahan, A. O. 2018. *Development of a heavy containment level bridge rail for Istanbul. Latin American Journal of Solids and Structures, 15* https://doi.org/10.1590/1679-78254684

[2] European Committee for Standardization, European *Standard EN 1317-1, EN 1317-2, Road Restraint Systems*, 2010

[3] Gheres, M. I., Scurtu, I. L. 2022. *Crash testing and evaluation of W-beam guardrail using finite elements method.* In IOP Conference Series: Materials Science and Engineering (Vol. 1220, No. 1, p. 012049). IOP Publishing. DOI:10.1088/1757-899X/1220/1/012049

[4] Gutkowski, R. M., Shigidi, A. M., Peterson, M. L. 2007 *Dynamic impact load tests of a bridge guardrail system* (No. MPC Report No. 07-188). Mountain-Plains Consortium.

[5] Kiraly,A., 2002 *Grafica Inginereasca (Engineering Graphics)*, Editura U.T.PRESS Publishing, ISBN 973-8335-35-3, Cluj-Napoca

[6] Ozcanan, S., Atahan, A. O. 2019 RBF surrogate model and EN1317 collision safety-based optimization of two guardrails. *Structural and Multidisciplinary Optimization*, 60(1), 343-362.

[7] Radioss Theory Manual, *Large Displacement Finite Element Analysis,* 2021

[8] Sassi, S., Sassi, A., & Ghrib, F. 2017 Effect of crushable blockouts on a full-scale guardrail System *International Journal of Crashworthiness* 22(1) 63-82 DOI:10.1080/13588265.2016.1223525

[9] Scurtu I.L. and Lupea I. 2014 *Frontal Crash Simulation of a Chassis Frame,* vol 57-III, Acta Technica Napocensis-Series: Spplied Mathematics, Mechanics, and Engineering, pp 411-414

[10] Teng, T. L., Liang, C. C., & Tran, T. T. 2015 *Effect of various W-beam guardrail post spacings and rail heights on safety performance* Advances in mechanical engineering 7(11) DOI: 10.1177/1687814015615544

[11] Tso-Liang Teng, Cho-Chung Liang, Ching-Yu Hsu, Chien-Jong Shih, Thanh-Tung Tran. 2016 *Impact Performance of W-beam Guardrail Supported by Different Shaped Posts* International Journal of Mechanical Engineering and Applications. Vol. 4, No. 2, 2016, pp. 59-64. DOI: 10.11648/j.ijmea.20160402.1

[12] Wilde, K., Bruski, D., Burzyński, S., Chróścielewski, J., Pachocki, Ł., & Witkowski, W. 2021. *On analysis of double-impact test of 1500-kg vehicle into w-beam guardrail system.* Archives of Civil Engineering, 67, 101-115.

IOP Conf. Series: Materials Science and Engineering **1311** (2024) 012045 doi:10.1088/1757-899X/1311/1/012045

# Study of the forces in the pretensioned seat belt and their effects on the human body

**A Soica[1], A Budala[1]**

[1]Automotive and Transport Engineering Department, Transilvania University of Brașov, RO

E-mail: a.soica@unitbv.ro

**Abstract.**
The paper aims to analyze the potential injuries induced by the seat belt in case of triggering the pretensioning mechanism. Two mannequins, with different geometries, have been installed on the front seat of the vehicle and were subjected to the efforts generated by the pretensioning of the seat belt, in the absence of an impact of the vehicle. The force due to the impact has been analyzed in various bibliographic resources, in the proposed model it can be taken into account. Efforts were measured in the belt strap for adult passenger and adult pregnant female passenger, or adult passenger with high BMI body mass index. The efforts that act on the human body, with emphasis on the abdominal loop, have been determined. It has been found that stress on the abdomen can cause internal damage, even just by simply triggering the pretensioning system, without taking into account the force on the body due to the impact.

## 1. Generalities

The seatbelt continues to be the main retaining and protection of cars' passengers, in case of collisions. Many authors analyzed the systems and sub-systems of seat-belt in-deep, from different point of views and under a lot of circumstances. Few of the experimental or statistically studies of different researchers are presented in next paragraphs.

The goal of the study conducted in [1] was to investigate the influence of the occupant characteristics on seat belt force vs. payout behavior based on experiment data from different configurations in frontal impacts. The seat belt behavior was characterized in terms of the shoulder belt force vs. belt payout behavior. A univariate linear regression was used to assess the factor significance of the occupant body mass or stature on the peak tension force and gross belt payout.

In [2] Yulong et al says that higher body mass index (BMI) people was associated with greater head and hip mouvement and larger normalized chest deflection. Higher belt routing increased the hip excursion and torso angle, which indicates a higher submarining risk, whereas the anchor pretensioner reduced hip excursion and torso angle. Lower load limits decreased the normalized chest deflection but increased the head excursion.

In [3] load limiters at seat track, seat, seat back, and lap belt were investigated to determine their effect on pelvis and lumbar spine loading, using the SAFER human body model and models of a semi-rigid seat and a triple-pretensioned, load-limited belt system. Seat track and seat load limiting were most effective at reducing lumbar spine loading. Seat track and lap belt load limiting were most effective at reducing pelvis loading, while seat back load limiting affected pelvis and lumbar spine loading only marginally.

In [4] Upper torso restraint and tie-down straps also help prevent a situation known as "submarining" from occurring (Figure 1). This is where the lap belt rides over the pelvic brim and compresses the soft tissues of the abdomen resulting in serious abdominal and spinal injuries. Submarining occurs due to the pelvis rotating under the lap belt, usually due to inappropriate location of the lap belt anchors or due to poor design of the seat bottom or a combination of both. Lap belt only restraints so commonly inflicted serious injuries on users in automobile crashes that the medical community coined a new term, "the seat belt syndrome".

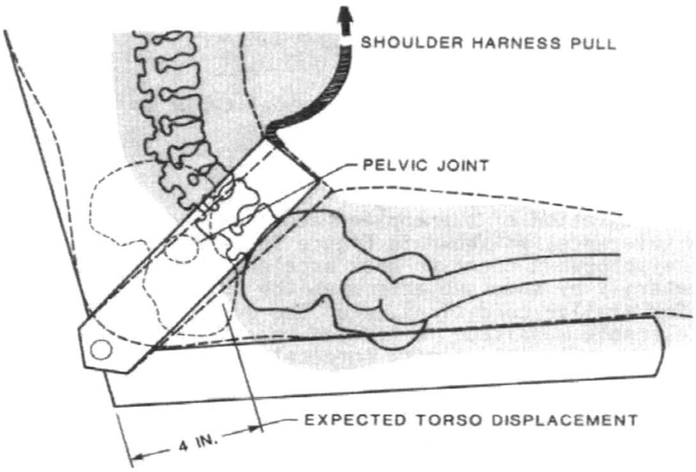


**Figure 1.** Lumbar spine injuries mechanism [4]

In [4] the author analyze the tolerance limits of human body to different loads. In order to maximize survivability in a crash, one must have an understanding of the tolerance of humans to abrupt acceleration.

Based on literature studies it was found that the pretensioners cause the belt tension at about 15-20 ms from the moment the car hit the barrier. The pretensioner tightens the belt with the force of 100-300 daN. When the speed during the impact amounts 56 km/h, the deceleration time of the dummy on the driver's seat is about 120 ms. Without a pretensioner, the effective interaction of the seat belt with the pretensioner starts about 15-20 ms earlier. Earlier affecting of the seat belt on the dummy body effectively limits the forward movement of the dummy against the seat. [5]

Pregnant women in the second trimester have a higher risk of fetal injuries when are involved in a serious motor vehicle crash compared to that of pregnant women in the first or third trimester or non-pregnant women. Serious accidents during pregnancy can result in fetal and neonatal deaths. [8]

In [9] the authors examined the seatbelt use by pregnant women drivers, in order to improve appropriate seatbelt wearing.

By increasing information on the correct method of wearing seat belts during pregnancy may increase the number of pregnant women who wear seat belts, it is say in [10].

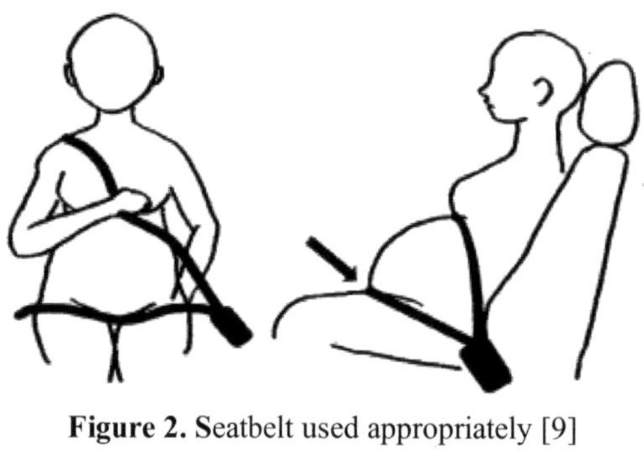

**Figure 2.** Seatbelt used appropriately [9]

In this paper, the forces on the human torso and pelvis, measured by the triggering of the seat belt pretensioning system, were analyzed. The mathematical and geometric model used to determine the forces acting on the body, is based on the FMVSS210 norms for representing the human body. Most of the performed studies imply a collision and thus the level of developed forces is much higher. In present paper this aspect was not taken into account, considering a stationary vehicle and triggering the pretensioning system. The attention has been focused on the efforts in the belt strap, especially the abdominal one, comparing types of passengers with different BMI, "vulnerable" passengers (pregnant women, or obese people) on one hand and occupants with normal body mass on the other hand. No standard dummies have been used, considering the wide variety of mass and dimensions of car passengers. After triggering pretensioning system, the forces that occurred in the abdominal strap were compared, for high and normal BMI people.

## 2. Mathematical model

According to the Figure 3, during a collision, on the thorax and pelvis of the occupants of a vehicle, act the forces given by the pretensioning system of a three-point seat belt, as well as the inertial force of the body given by the intensity of the impact.

It is considered that the force $F_{diag}$ from the diagonal loop of the belt generates the $N_{sbd}$ reaction on the body and the force $F_{lap}$ from the abominable loop generates the reaction $N_{sbl}$. If the impact velocity is different from zero then in the mass center of the upper part of the body (consisting of the head, thorax and pelvis), acts its inertial force $F_{imp}$. With $N_{sb}$ was noted the sum of the reactions given by the efforts in the belt strap, the relation (1).

$$N_{sb} = N_{sbd} + N_{sbl} \tag{1}$$

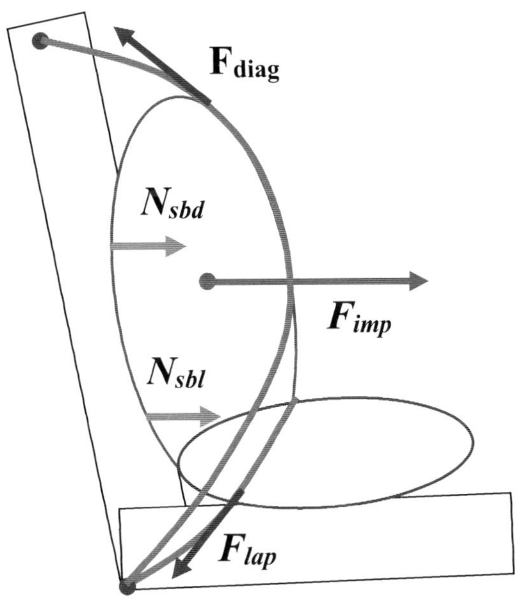


**Figure 3.** Forces that acting on human body - RRP pretensioner mechanism

The pretensioning system pulls the belt strap around the body, through the abominable loop and the diagonal loop, exerting a force of pre-tightening of the body in the seat. Two tensioning systems were used, one of PBP type, mounted on the seat reinforcement, in the area of the central tunnel and one of

RRP type, mounted in the area of the B-pillar. These two types of pretensioners produced different forces in the two loops of the seat belt strap, due to the different working ways and point of action.

The values of the reactions $N_{sbd}$ and $N_{sbl}$ can be determined, in the absence of the impact velocity $F_{imp}$ = 0, with the relations:

$$N_{sbd} = \sqrt{F_{diag}^{2} + F_{diag}^{2} + 2 \cdot F_{diag}^{2} \cdot \cos(\alpha)} = \sqrt{2 \cdot F_{diag}^{2} \cdot (1+\cos(\alpha))} \quad (2)$$

$$N_{sbl} = \sqrt{F_{lap}^{2} + F_{lap}^{2} + 2 \cdot F_{lap}^{2} \cdot \cos(\beta)} = \sqrt{2 \cdot F_{lap}^{2} \cdot (1+\cos(\beta))} \quad (3)$$

If the impact occurs, the force of inertia of the upper part of the human body amplifies the effect of the $N_{sbl}$ and $N_{sbd}$ reactions on the body. Thus, the tension in the seat belt strap will increase dramatically, and in the absence of an additional load-limiter system, high values of forces on the rib cage will be obtained, as well as its large deflections. The value of the resulting force acting on the body can be expressed as:

$$R = F_{imp} + N_{sbd} + N_{sbl} \quad (4)$$

This will be transmitted to the body through the two loops of the seat belt. In order to perform some calculations, the geometric model from FMVSS 210 [7] can be adopted.

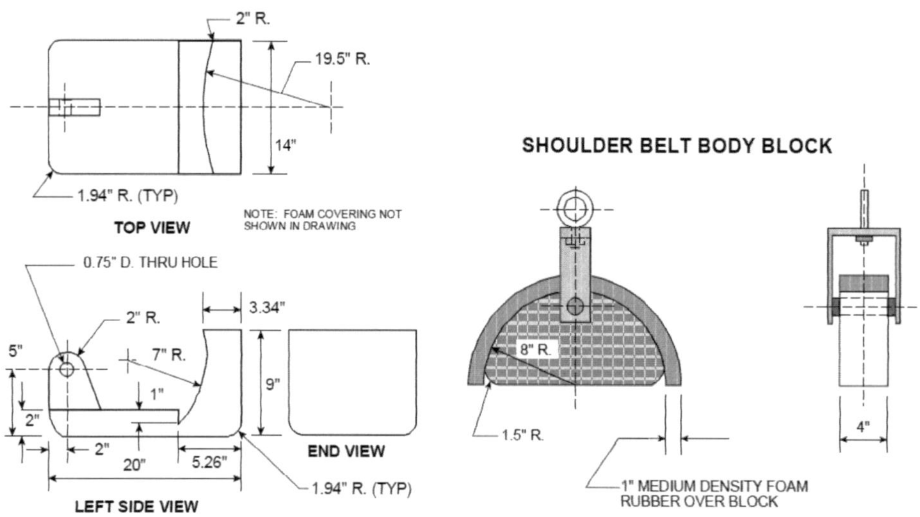


**Figure 4.** FMVSS210 model for seatbelt anchorage [6]

In [6] is presented a model for car seats anchorage improvement; the same model is also used in [7]. In [7] the author marks what was said by Patil et al. and Hu et al. They explained the effect of changing the mounting location of each seat belt anchorage point on occupant injuries, and performed an optimization of the seat belt anchorage mount locations for occupant injury reduction in frontal crashes. Rear seat restraint design optimizations were carried out to find the ranges of optimum restraint system configurations that offer the best safety guides for children.

Starting from the geometry proposed in FMVSS210, a simplified geometric model is adopted for the occupant's body. There are two types of passengers, one with high BMI, represented in cross section of the torso (abdomen) by a circle and other with normal BMI, represented by an ellipse, figure 5. Points A and B represent the seat belt anchorages on the B-pillar base and between the seats. It is considered a seat belt assimilated to some wires. The directions that start from the anchor points and are tangents to the two body typologies make the angles $\alpha$ and $\beta$; it can be easily observed that the angle for the occupant with high BMI (circle) is little than the one for normal BMI (ellipse) : $\alpha < \beta$. From the tangent points with the circle, respectively the ellipse, towards to the anchoring points A and B, the forces in the abdominal strap of the $F_c$ and $F_e$ seat belt were drawn.

As can be seen, the resultant value for the two pairs of forces $F_c$- $F_c$ and $F_e$- $F_e$ is higher form people with higher BMI.

These resultant forces are actually the forces with which the belt acts on the abdomen, when triggering the pretension system.

In the case of an impact, an inertial force (not shown here) of the body (thorax-pelvis) occur and its action cumulates with the resultant force occurred by tightening the body in the chair, due to pretensioning at the beginning of the impact.

It can be concluded that people with high BMI or pregnant women, are subjected, in the abdomen, with higher forces resulting from pretension than people with normal BMI.

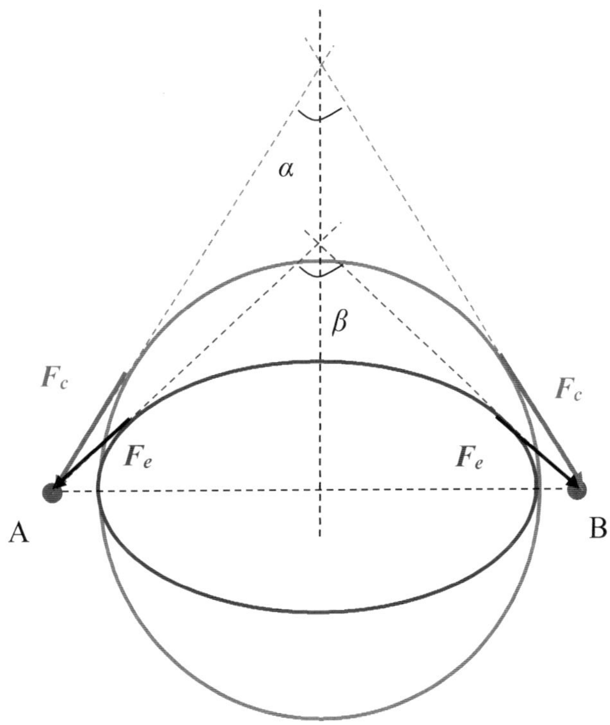


**Figure 5.** Forces on abdomen for normal and high BMI passenger

IOP Conf. Series: Materials Science and Engineering **1311** (2024) 012045 doi:10.1088/1757-899X/1311/1/012045

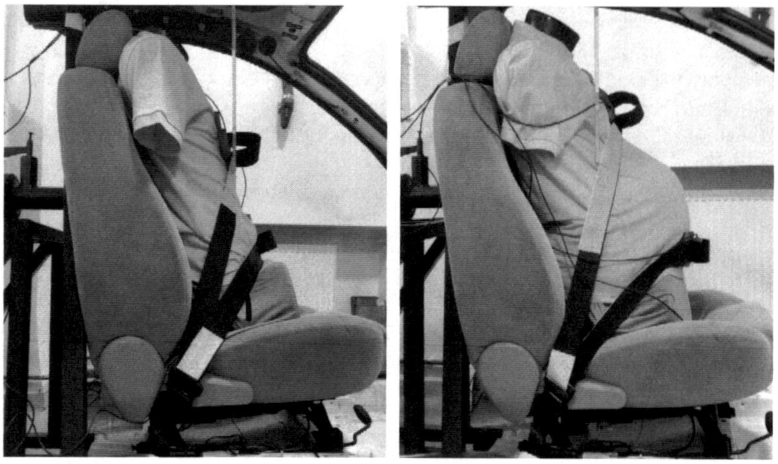

**Figure 6.** Test facilities with two types of dummies

**3. Results and discussion**

In case of zero impact speed of the vehicle (no impact occurs), the tension in the two loops of seat belt strap has been determined, by triggering the pretensioner. The measuring system comprises two belt-mounted tensile force measuring sensors. A sensor was also mounted on the seat in order to determine the body pressure ("pressing force") on the seat.

Special marks were made on the seat belt strap, in order to visualize the pre-tightening belt payout given by its pretensioning system. It should be noted that a different posture of the body in the chair, at various tests showed different values of pre-tithing and in-belt efforts.

The targeted passengers were adult-type, male, respectively pregnant female. The present study was focused only on the pelvic and thoracic parts, on which act the forces given by the seat-belt, during collisions. These body-parts were isolated from the rest of the body segments and mounted on the seat of the vehicle. The mass of the dummy used was 36 kg. Results are presented in table 1.

**Table 1.** Test results

| No. | Pretensioner type - passenger type | Time of delay between thorax and hip gain force [ms] | Time of pretensioning at thorax loop [ms] | Thorax max force [N] | Hip max force [N] | Seat max force [N] |
|---|---|---|---|---|---|---|
| 1 | RRP | 3 | 17 | 711 | 120 | 65 |
| 2 | RRP | 3 | 18 | 798 | 73 | 65 |
| 3 | RRP | 4 | 18 | 757 | 118 | 63 |
| 4 | RRP-normal BMI | 4 | 19 | 812 | 50 | 122 |
| 5 | PBP | <1 | 18 | 745 | 518 | 215 |
| 6 | PBP-normal BMI | <1 | 20 | 464 | 415 | 74 |
| 7 | PBP | <1 | 15 | 878 | 478 | 209 |
| 8 | PBP-normal BMI | <1 | 21 | 1035 | 452 | 166 |
| 9 | PBP-normal BMI | <1 | 20 | 688 | 448 | 153 |

It can be observed that, for RRP type pretensioner, due to the degree of initial tightening of the strap on the body of the dummy, there is an increase of the effort on the hip loop with a delay of 3-4 ms compared to the thorax loop.

The maximum force developed on the thorax loop is between 700...800 N and between 50...120 N on the hip. These values can also be influenced by the initial body position on the seat and the initial degree of pre-tightening of the strap.

For PBP type pretensioner, due to the simultaneous pulling of the two belt loops, the above-mentioned time gap does not appear.

The forces that arise in the thorax loop are in the range of 450...1000 N, while in the abdominal loop there are efforts 3-4 times higher than in case of RRP pretensioner. This effort, in the case of pregnant women, can cause damage to the fetus. To limit this effect, it is necessary to use devices to divert to the bony areas the forces from the loop designed for the abdominal/pubic area of pelvis.

There is also an increase in loading force on the seat for this type of pretensioning.

For both types of systems, the pretensioning time is approximately 20 ms.

For both types of systems, RRP and PBP, the difference in effort between the hip and the thorax strap, depending on the measuring points, is lost by friction in the ear of the clasp and friction of the strap on the dummy's body, as well as by deforming the human body and seat (foam structure).

## 4. Conclusions

The pretensioner generates different tighten forces of the human body in the chair depending on its type. The effort on the chest is approximately identical for PBP and RRP, respectively, while PBP generates greater efforts on the pelvis.

In the event of an impact of a certain intensity, the inertial force of the body is added to those given by the seat belt pre-tightening.

The correct adjustment of the seat belt and the compliance of a correct position of the occupant in the seat can limit the injuries in the thoracic and pelvis area during the accidents.

People with high BMI or pregnant women are more likely to have more severe injuries in the pelvic area due to the forces developed while belt pretensioning and the impact.

The use of approved exertion devices by pregnant women can be a solution in limiting abdominal and fetal injuries.

## References

[1] Bingbing Niea, David Poularda, Damien Subitb, John-Paul Donlona, Jason L. Formana, and Richard W. Kent, Experimental investigation of the effect of occupant characteristics on contemporary seat belt payout behavior in frontal impacts., Traffic Injury Prevention, Taylor & Francis, 2016, 17 (4), pp.374-380. ff10.1080/15389588.2015.1088944ff. ffhal-02473262f.

[2] YULONG WANG, ZHONGHAO BAI, LIBO CAO, MATTHEW P. REED, KURT FISCHER, ANGELO ADLER, and JINGWEN HU, A Simulation Study on the Efficacy of Advanced Belt Restraints to Mitigate the Effects of Obesity for Rear-Seat Occupant Protection in Frontal Crashes, Traffic Injury Prevention (2015) 16, S75–S83 Published with license by Taylor & Francis ISSN: 1538-9588 print / 1538-957X online DOI: 10.1080/15389588.2015.1010722.

[3] Krystoffer Mroz, Martin Östling, Rachel Richardson, Jason Kerrigan, Jason Forman, Bronislaw Gepner, Nils Lubbe, Bengt Pipkorn, Effect of Seat and Seat Belt characteristics on the Lumbar Spine and Pelvis Loading of the SAFER Human Body Model in reclined Postures, IRC-20-58, IRCOBI 2020.

[4] Dennis F. Shanahan, Human Tolerance and Crash Survivability , M.D., M.P.H. Injury Analysis, LLC 2839 Via Conquistador Carlsbad, CA 92009-3020 USA.

[5] Andrzej Zuchowski, Jerzy Jackowski, ANALYSIS OF PROPERTIES OF OPERATION OF THE SUPPORTING EQUIPMENT FOR THE SEAT BELTS, Journal of KONES Powertrain and Transport, Vol. 18, No. 1 2011.

[6] LABORATORY TEST PROCEDURE FOR FMVSS 210 Seat Belt Assembly Anchorages.

[7] Chang Yong Song, Enhanced Robust Design Optimization in Seat Belt Anchorage Strength for Front Crash Safety of Multi-Purpose Vehicle, MDPI, Appl. Sci. 2021, 11, 1023. https://doi.org/10.3390/app11031023.

[8] Shizuka Yamada, Koji Nishijima, Jin Takahashi, Nozomu Takahashi, Chiyo Tamamura, Yoshio Yoshida, Intrauterine fetal death caused by seatbelt injury, Taiwanese Journal of Obstetrics and Gynecology, Volume 56, Issue 4, August 2017, Pages 558-560.

[9] Kyoko Hanahara , Masahito Hitosugi & YumikoTateoka, Education for appropriate seatbelt use required for early-phase pregnant women drivers, http://www.nature.com/scientificreports, Scientifc Reports (2020) 10:17609 https://doi.org/10.1038/s41598-020-74730-5.

[10] Masamichi Hoshino, Hiromitsu Shinozaki, Shota Ogawa, Kunihiko Hayashi, Masahiro Itoh, and Akira Iwase, Exposure to Information and Seat Belt Use among Pregnant Women, The Kitakanto Medical Journal , KMJ2020 ; 70 : Pages 207-213.

IOP Conf. Series: Materials Science and Engineering **1311** (2024) 012046 doi:10.1088/1757-899X/1311/1/012046

# Possibilities to evaluate the vehicle pre-collision travel velocity in the case of a frontal impact with a fixed object, through numerical modelling

**I Duma, N Burnete, A Todoruț, I Barabás and M D Dragoste**

Technical University of Cluj-Napoca
Department of Automotive Engineering and Transports

irina.duma@auto.utcluj.ro

**Abstract**. The present paper evaluates, from a physical-mathematical perspective, the pre-collision travel speed of a vehicle, in the case of a frontal impact with a rigid, fixed pole-type obstacle. The evaluation considers both energy methods and deformation coefficients, taking into account the parameters resulting from the primary investigation of the accident site. The reconstruction of such road accidents aims to determine the kinematic parameters that influence the behavior of the involved vehicle, and to allow the analysis of avoidance possibilities of such events. The developed working algorithm can be applied to solve a large number of road accident cases by changing the input data and to obtain the necessary kinematic and dynamic parameters for the accident reconstruction process. Thus, it is possible to establish the dynamics of such road traffic events, which facilitates the assessment and comparison of different conditions under study. The validation of the fore-mentioned reconstruction algorithm is performed by comparing the results obtained through numerical modelling with those determined experimentally. The relative error between the two is, as well, assessed.

## 1. Introduction

Since accidentology represents a vast field of research considering the individuality of each collision case, the present paper considers a frontal collision between a road vehicle and a fixed rigid object (pole) under test conditions.

The study of the vehicle's behavior in collisions with various objects or with other vehicles has the primary purpose of improving traffic safety. By determining the causes and circumstances, it is possible to identify safety measures to be implemented in order to avoid or reduce the effects of a road vehicle crash [11]. Thus, over time, this inexhaustible topic was approached with the help of real-scale experimental tests, the results of which were then compared with the predictions obtained from numerical modeling and computer simulations, by each individual car manufacturer [7], [8]. The behavior of the vehicle upon impact is also studied from the perspective of its deformations [16], also known as the ability to absorb the impact energies in order to protect both the vehicle's occupants and other traffic participants [6].

The aim is to determine the pre-collision vehicle velocity through numerical modeling, by using the energetic method and the deformation coefficients, respectively. Based on the obtained results and the specificity of the considered case, an analysis of the viability of these two methods can be conducted, in order to choose the most suitable option when dealing with car accidents reconstructions.

The validation of the proposed numerical modeling methods is made through the comparison of the results obtained from the calculations with the vehicle velocity value registered during the experimental test [3].

## 2. Numerical modelling of the vehicle-pole frontal collision

Mathematical modelling of a road event represents a rather quick procedure for determining the interest parameters of a car crash, such as pre-collision velocity of the vehicle. The considered case represents a frontal impact between a vehicle of mass m and a pole, as the fixed rigid object, following the trajectory depicted in figure 1. The obtained results are then compared and analyzed based on the used variable parameters.

### *2.1. Numerical modelling using the energetic method*

For the determination of the pre-collision vehicle velocity, the use of energies method is proposed. This method [10] involves the identifying of the energies consumed during the impact, thus determining the pre-collision velocity based on the energy balance.

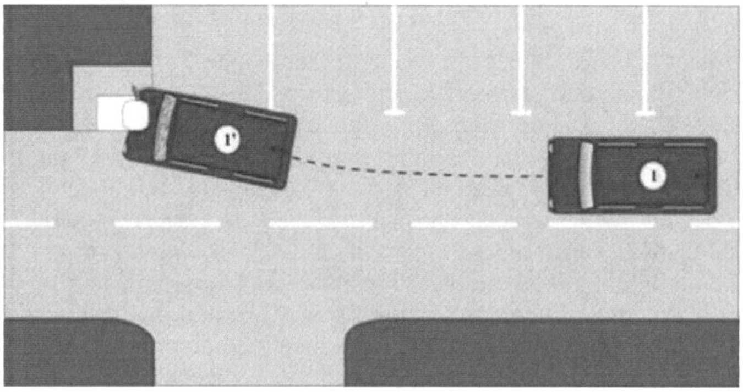


**Figure 1.** Vehicle-pole frontal collision scheme used for the determination of pre-collision velocity using energetic methods
(1 – initial position of the vehicle; 1' – final position of the vehicle).

The considered algorithm involves the successive determination of the following quantities, starting from the energy balance [9], [12]:

$$E_c + E_p = \sum E, \qquad (1)$$

where $\sum E$ represents the sum of all energies consumed by the vehicle during the collision; $E_c$ – kinetic energy; $E_p$ – potential energy. Since the collision considered took place on a road without any longitudinal or transversal inclination, it can be considered that the potential energy is null ($E_p = 0$).

Thus:

$$E_c = \sum E. \qquad (2)$$

The kinetic energy of the moving vehicle, prior to the collision is determined by the following equation:

$$E_c = \frac{m_{tot} \cdot v^2}{2}, \text{ in J}, \qquad (3)$$

where $m_{tot}$ is the total mass of the vehicle, together with its two occupants, in [kg]; v – initial vehicle speed to be determined, in [m/s].

Hence, a 2$^{nd}$ degree equation is obtained, having the pre-collision vehicle velocity as the only unknown parameter:

$$\frac{m_{tot} \cdot v^2}{2} = \sum E. \quad (4)$$

The energy consumed for the plastic deformation of a part of the vehicle's metallic mass:

$$E_d = \frac{\sigma^2{}_c}{E} \cdot \frac{m'}{\rho}, \text{ in J,} \quad (5)$$

where: $\sigma_c$ is the yield strength of the deformed material, in [N/m$^2$]; E – longitudinal modulus of elasticity (Young's modulus), in [N/m$^2$]; m′ - plastically deformed mass of the vehicle, in [kg]; ρ – density of the deformed material, in [kg/m$^3$].

The energy consumed to rotate the vehicle on the road surface after the impact:

$$E_{rot} = \frac{1}{2} \cdot m_{tot} \cdot g \cdot \varphi_y \cdot L \cdot \Delta\alpha, \text{ in J,} \quad (6)$$

in which g is the gravitational acceleration, in [m/s$^2$]; $\varphi_y$ – lateral adhesion coefficient; L – total length of the vehicle, in [m]; Δα – vehicle's angle of rotation on the roadway, in [°].

The energy consumed for bending some elements of the load-bearing structure of the vehicle:

$$E_i = p_s \cdot h_c \cdot l_c \cdot \beta, \text{ in J,} \quad (7)$$

where $p_s$ is the specific contact pressure between the vehicle and the pole, in [N/m]; $h_c$ – height of the contact surface, in [m]; $l_c$ – length of the contact surface, in [m]; β – deviation angle, in [°].

By replacing the energies identified as consumed during the impact from equations (5), (6) and (7) into equation (4), the pre-collision velocity is determined, in [m/s]:

$$\frac{m_{tot} \cdot v^2}{2} = E_d + E_{rot} + E_i, \quad (8)$$

thus obtaining:

$$v = \sqrt{\frac{2 \cdot (E_d + E_{rot} + E_i)}{m_{tot}}} \quad (9)$$

*2.2. Numerical modelling using deformation coefficients*

The method of deformation coefficients [4], [5] consists in determining the vehicle's impact velocity by using the deformations resulting from the collision (figure 2). Having the deformations ζ of the frontal part of the vehicle's body measured at the crash site, the deformed surface was divided into equidistant widths, noted with *l*.

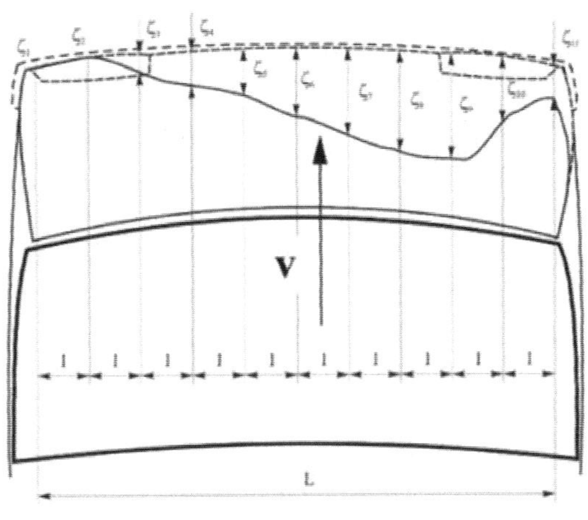


**Figure 2.** Vehicle's measured deformations.

Thus, the energy consumed to deform the vehicle is calculated by [4], [5]:

$$E_d = \frac{A^2}{B} \cdot K_1 + A \cdot K_2 + B \cdot K_3, \tag{10}$$

where A represents the ratio between the highest impact force that does not cause deformations and the width of the contact surface, in [N/m]; B – the ratio between the impact force distributed over the width of the contact surface between the vehicle and the pole, and the average amplitude of the deformation along the width, in [N/m$^2$]; K – coefficients expressed with relations (11), (12) and (13), respectively [4, 5]:

$$K_1 = \frac{L}{2}, \tag{11}$$

$$K_2 = \frac{L}{2 \cdot (n-1)} \cdot \left(\zeta_1 + \zeta_n + 2 \cdot \sum_{i=2}^{n-1} \zeta_1\right), \tag{12}$$

$$K_3 = \frac{L}{6 \cdot (n-1)} \cdot \left(\zeta_1^2 + \zeta_n^2 + 2 \cdot \sum_{i=2}^{n-1} \zeta_i^2 + \sum_{i=1}^{n} \zeta_i \cdot \zeta_{i+1}\right), \tag{13}$$

in which n is the number of points where the deformations are measured; L – the width with respect to which the deformations are measured, in [m].

The vehicle velocity equivalent to the energy consumption, in [m/s], is determined according to the deformation energy of the vehicle, in a head-on collision with the rigid assembly:

$$v = \sqrt{\frac{2 \cdot E_d}{M_i}}, \tag{14}$$

where $M_i$ is the mass moment of inertia, in [N·m].

## 3. Results and discussions

Considering the input data described in Table 1 for the energetic method and in Table 2 when using the deformation coefficients, respectively, the pre-collision travel velocity was obtained. In the first case, through the variation of the yield strength of the deformed material, which was considered to be steel, $\sigma_c$ between 180...340 N/m$^2$, a vehicle velocity variation of 10.46...10.60 m/s is obtained (figure 3).

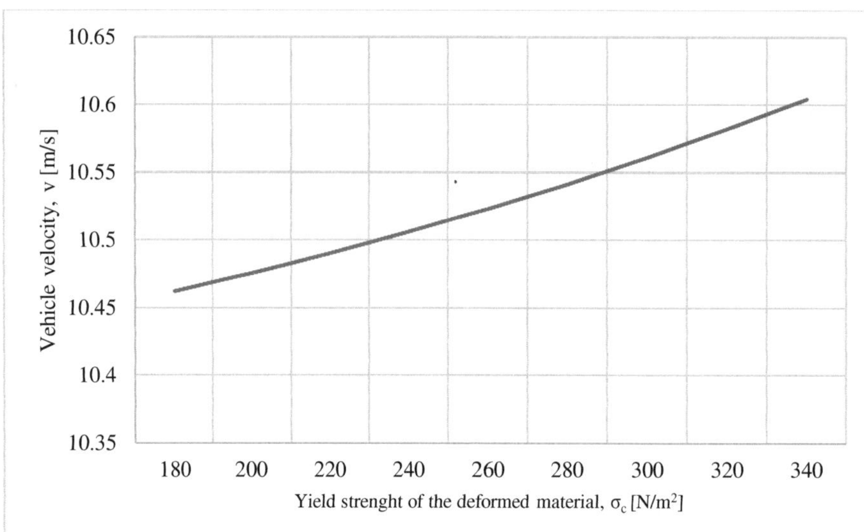


**Figure 3.** Variation of the obtained vehicle velocity through the energetic method, with respect to the yield strength of the deformed material.

**Table 1.** Input data and obtained results by using the energetic method.

| Parameter | Annotation | Value | Measurement unit |
|---|---|---|---|
| **Input data** | | | |
| *Vehicle mass* | $m_a$ | *1530* | *kg* |
| *Vehicle occupant's mass* | $m_0$ | *2·80* | *kg* |
| *Yield strength of the deformed material (steel)* [1], [2] | $\sigma_c$ | *180...340·10⁶* | $N/m^2$ |
| *Longitudinal elasticity modulus (Young's modulus)* [1], [2] | $E$ | $2.1\cdot10^5\cdot10^6$ | $N/m^2$ |
| *Plastically deformed mass* | $m'$ | *50* | *kg* |
| *Density of the deformed material (steel)* [1], [2] | $\rho$ | $7.8\cdot10^3$ | $kg/m^3$ |
| *Gravitational acceleration* | $g$ | *9.81* | $m/s^2$ |
| *Longitudinal adhesion coefficient (for dry concrete-asphalt road)* [4], [5], [10], [12], [13], [14], [15] | $\varphi$ | *0.75* | - |
| *Transverse adhesion coefficient* | $\varphi_y$ | $0.8\cdot\varphi$ | - |
| *Vehicle's total length* | $L$ | *4.707* | *m* |
| *Vehicle's angle of rotation on the road surface* | $\Delta\alpha$ | *25* | $^o$ |
| *Height of the contact surface between the vehicle and the pole* | $h_c$ | *1.50* | *m* |
| *Length of the contact surface between the vehicle and the pole* | $l_c$ | *0.50* | *m* |
| *Deviation angle* | $\beta$ | *20* | $^o$ |
| **Obtained results** | | | |
| *Energy consumed for the plastic deformation of a part of vehicle's mass* | $E_d$ | $1.908\cdot10^3$ | *J* |
| *Energy consumed through the rotation of the vehicle on the road surface* | $E_{rot}$ | $1.029\cdot10^4$ | *J* |
| *Energy consumed for the bending of a part of vehicle's load-bearing structure* | $E_i$ | $8.188\cdot10^4$ | *J* |
| *Vehicle's impact velocity* | $v$ | *10.46...10.60* | *m/s* |
| *Mean vehicle's impact velocity* | $v_m$ | *10.53* | *m/s* |

As for the approach using deformation coefficients based on the measured deformations of the vehicle's body (table 2), a vehicle travel velocity of 10.28 m/s is obtained.

**Table 2.** Input data and obtained results by using the deformation coefficients.

| Parameter | Annotation | Value | Measurement unit |
|---|---|---|---|
| **Input data** | | | |
| *Number of points in which deformations are measured* | $n$ | *11* | - |
| *Width of the measured deformations* | $L$ | *1.84* | *m* |
| *Vehicle total mass (with occupants included)* | $m_1$ | $1.702 \cdot 10^3$ | *kg* |
| *Deformations measured for the frontal part of the vehicle* | $\zeta_i$ | - | *m* |
| *Mass of the fixed obstacle (the pole)* | $m_2$ | *1300* | *kg* |
| **Obtained results** | | | |
| *Mass moment of inertia* | $M_i$ | $5.762 \cdot 10^3$ | $kg \cdot m^2$ |
| *Width of the deformed parts of the vehicle's body* | $l$ | *0.167* | *m* |
| *Mean amplitude of the deformations* | $\zeta_{med}$ | *0.735* | *m* |
| *Ratio between the highest impact force that does not cause deformations and the width of the contact surface* | $A$ | $7.479 \cdot 10^4$ | *N/m* |
| *Ratio between the impact force distributed over the width of the contact surface between the vehicle and the pole, and the average amplitude of the deformation along the width* | $B$ | $1.018 \cdot 10^5$ | $N/m^2$ |
| | $K_1$ | *0.92* | - |
| *Deformation coefficients* | $K_2$ | *1.34* | - |
| | $K_3$ | *0.604* | - |
| *Energy consumed for the deformation of the vehicle body* | $E_{d1}$ | *212296.559* | *J* |
| *Equivalent moment* | $M_e$ | *722.261* | *N·m* |
| *Vehicle's impact velocity* | $v$ | *10.28* | *m/s* |

A comparison between the two used methods is made by calculating the relative error $\varepsilon_r$ of the two obtained vehicle velocities, by using the following relation:

$$\varepsilon_r = \frac{v_m - v}{v} \cdot 100, \text{ in } \%. \quad (15)$$

Considering $v_m$, the mean vehicle velocity obtained through the energy method to be 10.53 m/s and v, the vehicle velocity obtained by using deformation coefficients of 10.28 m/s, the relative error between the two methods is 2.37%.

On the other hand, when comparing each of the methods with the registered mean vehicle velocity during impact of [3], 10.5 m/s, a relative error of 0.28% is obtained for the energy method and 2.09% for the deformation coefficients method, respectively.

## 4. Conclusions

The frontal impact considering a vehicle-to-pole case occurs with major energy losses, thus the numerical model based on energy methods was considered to be a suitable one. Following the identification of the energy types consumed during the collision, a vehicle impact velocity was obtained, the result being only 2.37% different from the one registered during the experimental test for the considered vehicle. This distortion may be considered as acceptable, being influenced by the choice of parameters or coefficients in the calculations, such as the yield point of the material from which the deformed body parts are manufactured, the longitudinal modulus of elasticity or the deformed mass of the vehicle.

When applying the deformation coefficients method based on the dimensions of the vehicle body's part measured at the crash site, a relative error of only 0.28% is obtained compared to the real-time data provided by the GPS during the experimental test.

From here, it follows that both considered analytical methods are viable in the case of a frontal vehicle-pole collision, by the relatively low deviations of 2.37% and 0.28%, respectively compared to the reference data.

Finally, the numerical modeling of such events is an important tool for road vehicle accidents reconstructionists in the determination of the event's dynamic parameters. The vehicle impact velocity can be used further for establishing the causes of the collision and, consequently, proposing the means of avoidance.

**References**

[1] Bejan, M., *Rezistența materialelor,* Vol. 1, 5th Edition, Bucharest, AGIR Publishing House; Mega Publishing House, Cluj-Napoca, 2009;
[2] Bejan, M., *Rezistența materialelor,* Vol. 2, 4th Edition, Bucharest, AGIR Publishing House; Mega Publishing House, Cluj-Napoca, 2009;
[3] Duma, Irina, Burnete, N., Todoruț, A., Trușcă, D., *Determination of pre-collision travel speed in the event of a frontal collision between a vehicle and a fixed obstacle, using video recordings,* in: XXIX International Scientific Conference on Transport, Road-Building, Agricultural, Hoisting & Hauling and Military Techniques and Technologies 'trans&MOTAUTO2021', ISSN: 2535-0307, Year IV, Issue 1 (8), Sofia, Bulgaria, 2021;
[4] Gaiginschi, R., *Reconstrucția și expertiza accidentelor rutiere,* Tehnical Publishing House, Bucharest, 2009;
[5] Gaginschi, R., *Siguranța circulației rutiere,* Vol. I, Technical Publishing House, Bucharest, 2004;
[6] Limpert, R., *Motor vehicle accident reconstruction and cause analysis,* Transportation Research Board, 1984;
[7] McHenry, R.R., DeLeyes, N.J., *Vehicle dynamics in single vehicle accidents – validations and extension of a computer simulation,* Transportation Research Board, 1994;
[8] Pawlus, W., Nielsen, J.E., Karimi, H.R., Robbersmyr, K.G., *Comparative analysis of vehicle to pole collision models established using analytical methods and neural networks,* in: 5th IET International Conference on System Safety, Manchester, UK, doi: 10.1049/cp.2010.0817, 2010;
[9] Tabacu, I., Ivan, F., Cristea, D., Neagu, Elena, *Metode de calcul pentru parametrii cinematici ai autovehiculului în momente anterioare producerii accidentului,* in: Expertiza Tehnică, no. 22, Vol. 2, pg. 1-5, Symposium: Actualitatea în expertiza tehnică, Sinaia, 1995;
[10] Todoruț, A., Cordoș, N., *Modele fizico-matematice în dinamica accidentelor de circulație rutieră,* U.T. Press, Cluj-Napoca, 2017;
[11] Todoruț, A., Barabás, I., Burnete, N., *Siguranța autovehiculelor și securitatea în transporturi rutiere,* UT Press Publishing House, Cluj-Napoca, 2012;
[12] Todoruț, A., *Dinamica accidentelor de circulație,* UT Press Publishing House, Cluj-Napoca, 2008;
[13] Todoruț, A., *Bazele dinamicii autovehiculelor: Algoritmi de calcul, teste aplicații,* Sincron Publishing House, Cluj-Napoca, 2005;
[14] Untaru, M., Poțîncu, Gh., Stoicescu, A., Pereș, Gh., Tabacu, I., *Dinamica autovehiculelor pe roți,* Didactical and Pedagogical Publishing House, Bucharest, 1981;

[15] Untaru, M., Câmpian, V., Ionescu, E., Pereş, Gh., Ciolan, Gh., Todor, I., Filip, Natalia, Câmpian, O., *Dinamica autovehiculelor,* Transilvania University of Braşov, 1988;
[16] Wood, D.P., Doody, M., Mooney, S., *Application of a generalised frontal crash model of a car population to pole and narrow object impacts,* SAE Technical Paper 930894, https://doi.org/10.4271/930894; 1993.

# Road Safety in The European Road Transport System and the Correlation with Sustainable Development

**Gheorghe NEAMȚU[1], Aurel Mihail ȚÎȚU[2,*], Alina Bianca POP[3], Adrian BOGORIN – PREDESCU[4]**
[1]University Politehnica of Bucharest, Faculty of Industrial Engineering and Robotics, 313 Splaiul Independenţei, 6th District, Bucharest, Romania, geluneamtu@yahoo.com
[2]Lucian Blaga University of Sibiu, 10 Victoriei Street, Sibiu, România, mihail.titu@ulbsibiu.ro
[3] Technical University of Cluj-Napoca, Northern University Centre of Baia Mare, Faculty of Engineering - Department of Engineering and Technology Management, 62A, Victor Babes Street, 430083, Baia Mare, Maramures, Romania; bianca.bontiu@gmail.com
[4]University Politehnica of Bucharest, Faculty of Industrial Engineering and Robotics, 313 Splaiul Independenţei, 6th District, Bucharest, Romania adrian_bogorin@yahoo.com

E-mail: mihail.titu@ulbsibiu.ro

**Abstract.** In May 2018, the European Commission unveiled a medium-term strategic action plan as part of the Europe on the Move package, as well as a novel approach to road safety policy. In the long run, the European Union's Vision Zero initiative has maintained an ambitious target of nearly zero fatalities by 2050. This mindset must grow stronger than it has been in the past, both among decision-makers and in society. The European Union Council, facilitating the Valletta Declaration on Road Safety of March 2017, set, for the first time, an important road safety target of halving the number of serious injuries across all EU countries by 2030, using 2020 as their benchmark. Road accidents are those quiet criminals since they often go unreported in our culture. Most Western countries, including EU members, boast some of the world's safest roadways. In this connection, they aid other less developed European nations by emphasizing the knowledge acquired in this sector for their growth and advocating the highest possible standards on a worldwide scale. When compared to other continents, Europe is performing well in this regard, owing to decisive action at all levels, including national, regional, and municipal. At the European level, cumulative road accidents kill around 500 people every week. Deaths in other modes of transportation are no longer tolerated, and deaths on the road are no longer acceptable, and this concept must be integrated into the road safety decision-making process. It is absolutely crucial, and also required, to adopt the Safe System at the EU level as soon as possible. The scientific study includes specific research undertaken by the authors to apply various technological and managerial principles, in which correct and up-to-date facts and numbers concerning road safety and its management at the European level are provided. Those interested can thus become acquainted with the current state of play in the field approached about the new safety systems on vehicles, which have become mandatory for all car manufacturers in Europe as of January 1, 2022, but also about how these systems intervene to ensure the safety of the occupants of a vehicle. Finally, the results and future research and development directions are discussed.

## 1. Introduction

Road safety is an important concern in European Union policies. All road users, whether traveling for personal, professional, or recreational reasons, are entitled to the greatest level of safety and care. In April 2019, the European Parliament approved new rules on the equipping of road vehicles with advanced active and passive safety equipment and systems as part of the technical and economic policy pursued by the European Parliament on the road safety of car users in EU countries and to eliminate human error in traffic. They will help to reduce the number of fatalities and injuries. The direct visibility systems for buses and trucks, as well as the extension of the head impact zone to automobiles and vans, are exceptions to this rule. Due of structural design adjustments that must be made, they will be put on the vehicles later. Advanced safety technologies will prevent traffic accidents while also ensuring the automated mobility and connection of road vehicles. They will boost innovation and the European car industry's worldwide competitiveness. The safety measures that will be installed on the cars are expected to save more than 25,000 lives and prevent more than 140,000 major injuries by 2038. They will help the EU meet its long-term zero vision aim of zero fatalities and major injuries by 2050.

## 2. Road safety and road safety management at European level

Between 2010 and 2020, the number of road deaths in the EU fell by 37%, from 29691 to 18844, falling short of the 50% reduction intended for this decade. In 2020, 18844 persons died on European Union roads because of a car accident, a 17% drop from the previous year (3809 instances). [1] Based on this data, we strongly state that in the long term, the European Union has reaffirmed an ambitious objective of almost zeroing deaths by 2050 through the Vision Zero program. This mentality will have to become stronger than it has been so far, both among decision-makers but also among society at large. The European Union Council, facilitating the Valletta Declaration on Road Safety in March 2017, set, for the first time, an important road safety target of halving the number of serious injuries across all EU countries by 2030, using 2020 as their benchmark. Road accidents are those silent criminals because they sometimes go unnoticed in the society in which we live. Most Western countries, members of the EU, have some of the safest roads in the world. In this context, they support the other less developed European countries, highlighting the experience gained in this field for their development and promoting the highest possible standards on the global stage.

Road safety in the European Union has improved significantly in recent decades, and when compared to the rest of the globe, the European Union has some of the safest roads in the world, with forty-two fatalities per million people, compared to 180 deaths per million inhabitants overall. [1] As part of the Europe on the Move package, the European Commission revealed a medium-term strategic action plan in May 2018, as well as a new approach to road safety legislation. Deaths in other forms of transportation are no longer accepted, and road deaths are no longer acceptable; this notion must be integrated into road safety decision-making. It is vital that the Safe System be implemented at the EU level as soon as feasible. This system includes the provision of safe vehicles, safe infrastructure, safe road use (speed, lucid driving, wearing seat belts and helmets), and improved post-collision care, all of which are well-established and crucial factors in the approach to the safe system [2]. Furthermore, urgent action is required at the EU level to improve road safety management in the following areas: institutional road safety organization; road infrastructure; vehicles and safety devices; education and road education campaigns; driver training; traffic law application; rehabilitation and diagnosis; post-accident care; road safety data and their collection.

## 3. Romanian road safety and road safety management

When compared to other countries in Eastern and Central Europe, our country's road safety falls significantly short of the projected aim of lowering the number of victims from road accidents by half by 2020. In 2020, 6,273 significant traffic accidents occurred on Romanian public roads, resulting in the deaths of 1,646 people, serious injuries to 5,484 people, and minor injuries to another 2,155 people. Every day in 2020, on average, seventeen significant traffic accidents occurred, with five people killed and fifteen seriously injured. [1] However, between 2010 and 2020, Romania saw a 31% decrease in the number of individuals died in road accidents (from 2,377 to 1,646 people), close to the

IOP Conf. Series: Materials Science and Engineering **1311** (2024) 012047 doi:10.1088/1757-899X/1311/1/012047

European average of 37%. The number of major traffic accidents fell by 32.2%, while the number of those seriously injured fell by 35.6%. In 2020, there were an average of 26.2 deaths per 100 significant road accidents, an increase from 2019, indicating that the severity of road accidents increased in 2020. Despite a decrease in the frequency of catastrophic traffic accidents, the proportion of those ending in fatalities has climbed. [1] Figure 1 depicts the national road accident situation in 2019 and 2020, while Figure 2 depicts the national traffic accident situation in 2019 and 2020, and Figure 3.2 depicts the national fatality index for the same years.

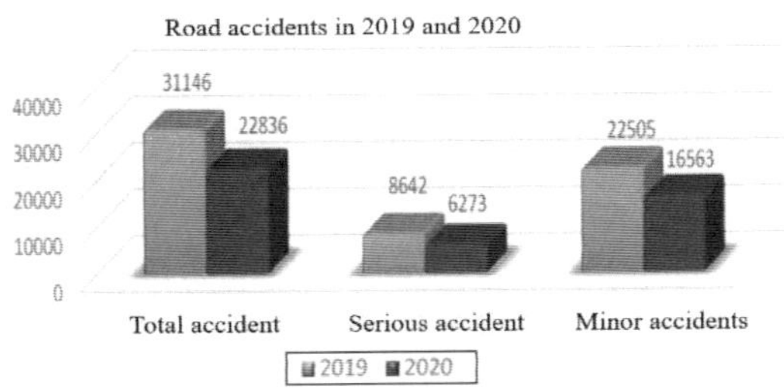


**Figure 1.** Situation of traffic accidents at national level in the years 2019-2020 [1].

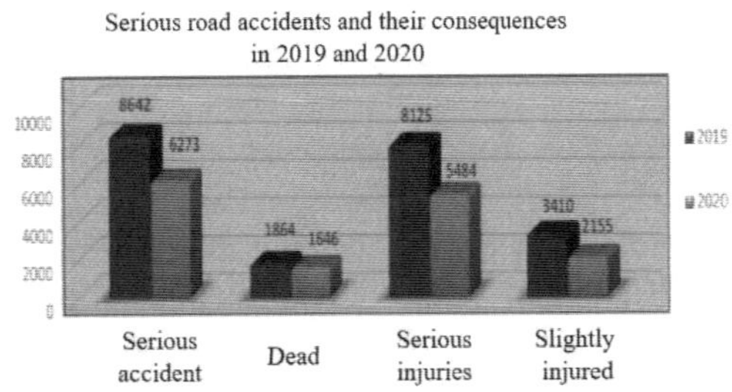


**Figure 2.** National mortality index in the years 2019-2020 [1].

According to these stats, Romania's performance in road safety has not improved in the last year. The mortality index, which measures the proportion of people killed in major traffic accidents, increased from 21.6% in 2019 to 26.2% in 2020 [1]. When we compare the number of fatalities to the number of road accidents (figure 3), we can infer that the mortality index was high on all seven days of the week in 2020. In rural regions, the most vulnerable road users were walkers, cyclists, motorcyclists, and the elderly; nevertheless, cyclists are represented less statistically as deceased individuals, but they are more gravely wounded (figure 3).

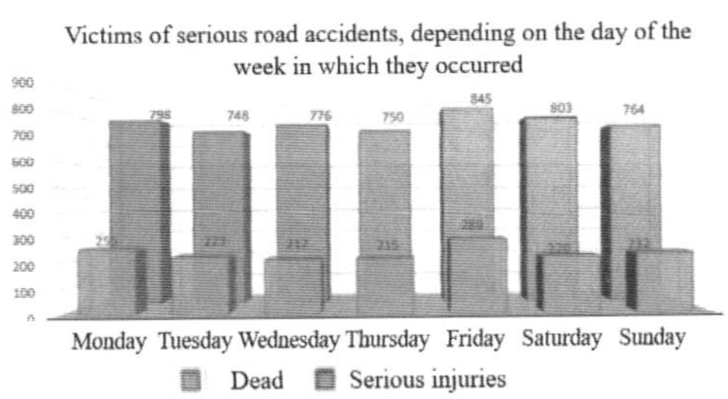


**Figure 3.** National mortality index in 2020 according to the days of the week. [1].

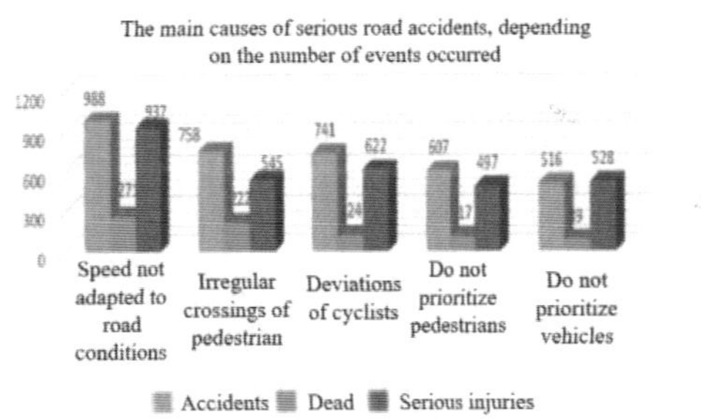


**Figure 4.** The primary causes of significant traffic accidents in Romania, based on the number of incidents in 2020. [1].

Pedestrians were gravely wounded in around one-third (30.4%) of major traffic accidents in 2020 (1,909 accidents). There is a considerable decline from the previous year, with 915 fewer such incidents occurring in 2020 (-32.4%). [1] According to the reports provided by the irregular crossing of pedestrians caused, in 2020, 758 serious accidents that resulted in the death of 222 people and the severe injury of 545 people, down from 2019 (- 418 events, -35.5%). Compared to 2019, cyclists' deviations produced 110 fewer serious road events in 2020 (-12.9%), thus remaining in the third place in the ranking of the main causes generating serious road accidents. Failure to give priority to pedestrians produced 210 fewer accidents in 2020 (-25.7%), while failure to give priority to vehicles produced 258 fewer serious road accidents than in 2019 (-12.9%).[1] Figure 4 depicts the primary causes of significant road accidents in Romania based on the number of occurrences created in 2020. As regards the number of victims represented by the third category of vulnerable persons participating in road traffic, drivers of powered two-wheelers, according to the data presented in figure 5 Results that the largest share is held by young people, ranging in the age range of 25-36 years, identical as in previous years.

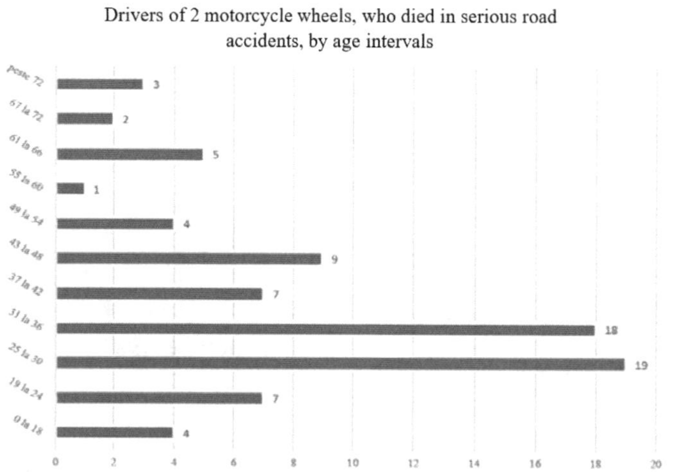


**Figure 5.** The age structure of deaths among those who drove powered two-wheelers. [1].

It follows that an impressive number of vulnerable road users were seriously injured on the streets, in urban areas, namely: 216 pedestrians, sixteen motorcyclists, forty-eight cyclists. A large category of victims is represented by pedestrians injured on the streets in urban areas. In the graph in the figure 6 the situation of persons killed in serious road accidents is presented, depending on the category of participant and the category of road.

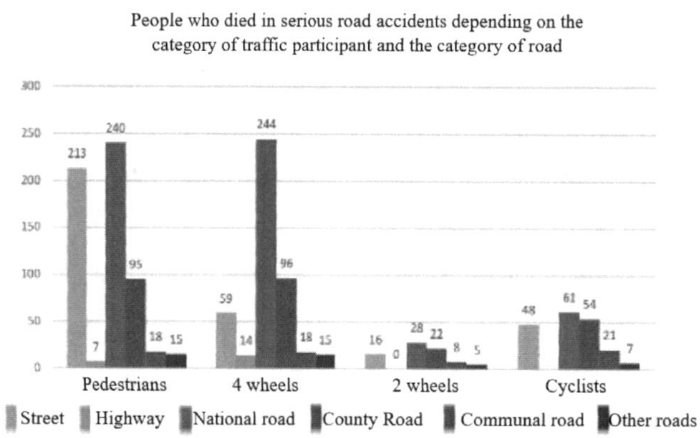


**Figure 6.** Situation of injured pedestrians by road categories. [1].

For a rigorous management in the field of road safety, since 1995, by Romanian Government Decision nr. 437 of 20 June 1995, amended by Government Decision no. 901 of August 20, 2008, in Romania was established Inter-ministerial Council for Road Safety. It is the Romanian Government's advisory body, with the goal of ensuring the overall concepts and coordination at the national level, based on the national road safety strategy and the national program of priority actions for the strategy's implementation, of activities aimed at improving road safety. In this direction, the activities conducted by the specialized bodies of the public administration and other institutions and organizations with attributions in these fields are concerned, and the evaluation of public policies on road safety. According to the main elements that endanger the safety of the road users, based on the conclusions drawn from the risk analyses and the observations on the evolution of the road phenomenon, in 2020, the structures of the Traffic Police organized 46,657 actions (figure 7). [1]

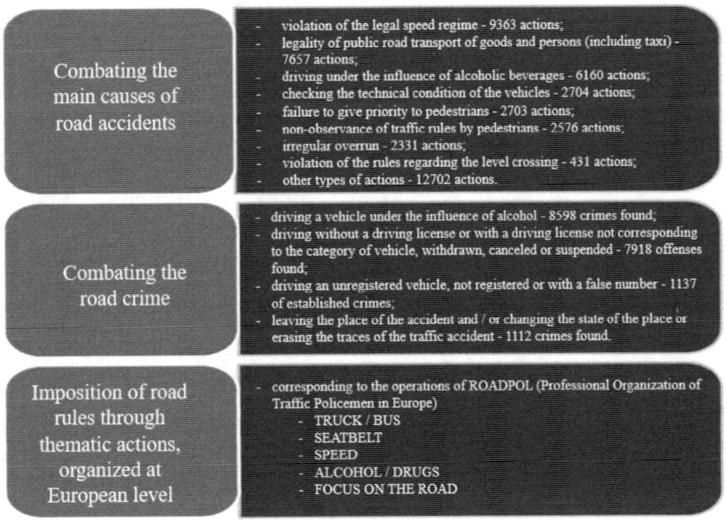


**Figure 7.** National actions on road safety management in 2020. [1].

In the area of our investigation, infractions of traffic laws were finished with the application of 1,666,413 contravention punishments, of which we highlight the most important, according to data provided by: [1]

- non-observance of traffic rules by pedestrians - 76,628 contravention sanctions.
- failure to give priority to pedestrian crossing - 18,763 contravention sanctions.
- non-use of seat belts and retention devices by road vehicle users – 192,697 contravention sanctions.
- exceeding the legal speed regime – 667,699 contravention sanctions.

## 4. Technical and economic guidelines and policies on road safety at European level, in the context approached.

Based on the European Parliament and Council Regulation (PE-CONS 82 of 18.10.2019) amending Regulation (EU) 858 of 30 May 2018, MEPs propose new proposals on the endowment of motor vehicles with safety systems for the protection of passengers, pedestrians, and cyclists as part of the European Parliament's technical and economic policy to improve the quality and safety of vehicles on European roads. They go into effect in 2022 and include the following: [3]

a) System of lane-keeping assistance Upgraded emergency braking systems and seat belts with enhanced impact testing, as well as enlarged head protection zones for pedestrians and cyclists in the event of an accident.

b) For trucks and buses, there are systems to improve drivers' direct sight, systems to reduce blind spots, and front and side systems that detect and indicate the presence of vulnerable road users such as pedestrians and cyclists. These are mandatory and aid big truck drivers when turning.

c) Driver sleep and inattention warning systems (use of smartphones or other gadgets while driving), intelligent speed control assistance (ISA), rear-speed warning and visibility systems; on-board recorders (black box) in the event of an accident; alcohol-testing anti-slip systems; driver distraction warning systems; emergency braking signals; TPW tire pressure monitoring systems (some manufacturers have a system). Figure 8 depicts new vehicle, passenger, and pedestrian safety technologies that will be required beginning in 2022, except for direct visibility for buses and trucks and the extension of the head impact zone to cars and vans, which will be required later owing to structural design modifications. [3].

IOP Conf. Series: Materials Science and Engineering **1311** (2024) 012047 doi:10.1088/1757-899X/1311/1/012047

**Figure 8.** Modern technologies for the safety of cars, passengers and pedestrians that become mandatory from 2022 [3].

All such tests, approvals and active safety systems shall be fitted to all types of motor vehicle, including hybrid or electric motor vehicles. In hybrid or electric vehicles, all passive and active safety systems work, being coordinated by an electronic control unit. Thus, they can act individually or in tandem with other safety systems (e.g., anti-slip system (ASR), anti-slip system (ESP) and traction control system (TCS). An electronic control unit (UCE) controls the management and, if required, the way they collaborate. Human mistakes are responsible for 90% of vehicle accidents on European highways, according to research. Since 2022, modern technology for the safety of automobiles, passengers, and pedestrians have been required. New regulations for the emergency introduction of Intelligent Transport Systems, cooperatives on European roads, have also been enacted. They will make it possible for automobiles to interact easily with one another, with road infrastructure, and with other road users. Some of the mounted or to be mounted safety equipment will connect with road traffic coordination and control centers (e.g., cellphones or black boxes). We think that active and passive safety systems and elements can significantly reduce the number of traffic accidents. Their presence on a motor vehicle supports drivers and protects passengers, reduces their severity, increases the chances that in the event of an impact the victims will survive, increase the quality, comfort, and safety of environmentally friendly road transport vehicles. Other active and passive safety systems that will be mounted on motor vehicles are also: the anti-temperature system with ethyl-test; driver distraction warning system; the rear-facing inspection system of the vehicle; the device for recording traffic accident data in the black box. All European countries will have to amend and at the same time tighten their legislation against drivers who do not use hands-free devices while driving, being distracted by using mobile devices. Sustained efforts on the part of the European authorities and beyond are being made to find new forms of personal mobility.

## 5. The role of advanced driver assistance systems in road safety

All automotive accidents are caused by human error, which may be avoided by using Advanced Driver Assistance Systems (ADAS). The role of ADAS is to lessen the incidence and severity of automotive accidents that cannot be avoided to prevent fatalities and injuries. Key ADAS applications include pedestrian detection/avoidance, lane departure warning/correction, traffic sign recognition, automatic emergency braking, and blind spot detection. These life-saving systems are important to the success of ADAS applications because they contain the most recent interface standards and execute multiple vision-based algorithms to enable real-time multimedia subsystems, sensor co-processing, and fusion [4], [5]. Figure 9 depicts the relationship between the ADAS system's functionality, and the electronic devices utilized to gather information from the surroundings. Sensors such as image cameras, medium and long-range radar cameras, ultrasound, and LIDAR acquire data from the surroundings. The data acquired by these sensors is consolidated and processed in such a way that the ADAS computer's software program can execute the essential activities to activate the driver assistance system and make the correct judgments. These selections concern pedestrian recognition, automated emergency braking, parking assistance, lane change, 360-degree vision, and, eventually, driver sleepiness. [6]

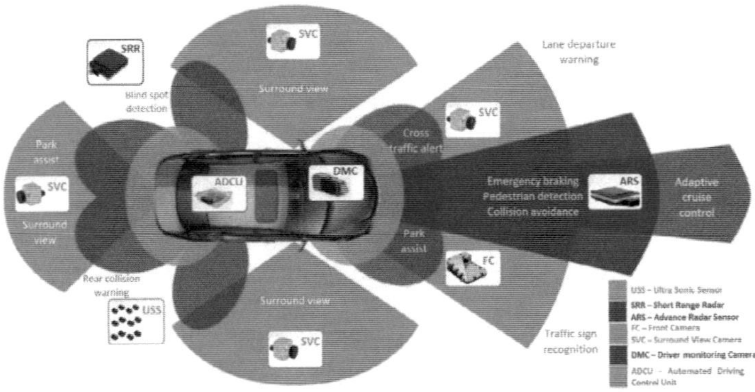

**Figure 9.** Components and functionalities of the ADAS system [4].

Figure 10 shows the link between the diverse types of sensors and the ADAS architecture.

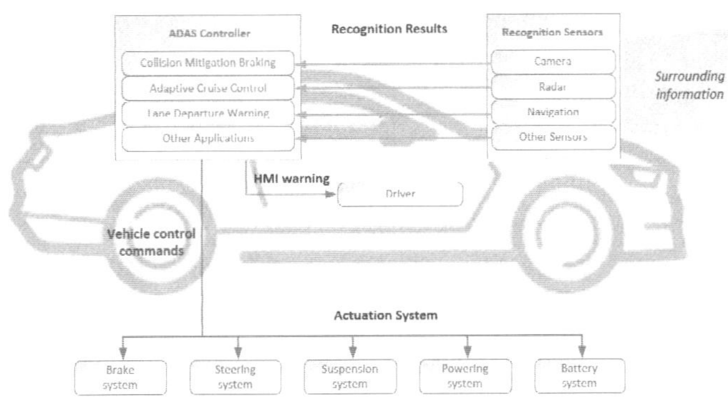

**Figure 10.** Drive system in the architecture of the ADAS vehicle [7].

The actuation of the output elements such as: the braking, steering, suspension, and propulsion system are performed on the data bus of the vehicle by ADAS according to the event occurring in traffic [8], [9], [10].

## 6. Conclusions

In recent years, great progress has been made at the European level in lowering the fatality rate from vehicle accidents. It has been demonstrated that they have stalled in recent years, but it is almost guaranteed that a current medium-term aim of halving the number of persons killed in road accidents between 2010 and 2020 will be met at the EU level. In rural regions, the most vulnerable road users were walkers, cyclists, motorcyclists, and the elderly; nevertheless, cyclists are less represented statistically as deceased individuals than pedestrians, but they are more gravely wounded. The new safety measures will safeguard pedestrians and cyclists, as well as other road users, and will assist drivers while driving their vehicles. They will have to gradually become accustomed to this newly installed equipment on road cars to do this. Advanced road vehicle automation will minimize driver mistakes in traffic and assist the elderly and disabled, boosting their comfort and mobility. These advantages will make autonomous cars more appealing and will encourage independent driving. A message on linked and autonomous mobility has been released by the European Commission. This statement aims to position Europe as a global leader in safe autonomous transportation systems. This mobility must assume efficient, ecologically beneficial, and population-safe forms to decongest urban traffic. There are parallels in guaranteeing safe and affordable mobility for all community members, particularly the elderly, but also those with impairments. This will be a significant accomplishment, paving the way for a new goal of lowering serious injuries from accidents by 50% between 2020 and 2030. The framework will also address novel issues, such as the growing number of vulnerable road users, the risk of road degradation, and future vehicle automation. Any recommended action should be based on data on road safety obtained from the European Union's 28 member states and beyond. The data gathered will allow the transportation strategy to be adjusted through the ideas and recommendations offered. To improve road safety management at the EU level, immediate action is required in areas such as: institutional road safety organization; road infrastructure; vehicles and safety devices; education and road education campaigns; driver training; traffic law application; rehabilitation and diagnosis; post-accident care; and road safety data collection. In Romania, the most vulnerable road users were walkers, cyclists, motorcyclists, and the elderly in rural regions, yet cyclists are less represented statistically than pedestrians. In terms of law, Romania is currently completely aligned with European road transport legislation.

**References**

[1] Ministry of Internal Affairs – IGPR 2021 Road safety bulletin. Annual Report 2020, Bucharest. Retrieved on 02 03, 2022, from https://www.politiaromana.ro/ro/prevenire/buletinul-sigurantei-rutiere/buletinul-sigurantei-rutiere-raport-anul-2020.

[2] Commission, E 2018 Europe on the move. Sustainable mobility for Europe: safe, connected and clean. Communication from the Committee to the European Parliament, the Council, the European Economic and Social Committee and the Committee of the Regions. (P. European, Ed.) Brussels: European Parliament COM (2018) 293 final. Retrieved on March 23, 2022, from https://eur-lex.europa.eu/legal-content/RO/TXT/HTML/?uri=CELEX:52018DC0293&from=FI.

[3] Commission, E 2019, March 26 Road safety: Commission welcomes agreement on new EU rules that help save lives. Retrieved March 25, 2021 from Press Release: https://ec.europa.eu/commission/presscorner/detail/ro/IP_19_1793.

[4] What is ADAS? 2022 Retrieved from www.synopsys.com: https://www.synopsys.com/automotive/what-is-adas.html.

[5] Aguilar, Adiac. (2020). Safety Analysis and Risk Assessment of Automated Driving (SAE Level 2-4) by Adiac Aguilar.

[6] Pradip, K S 2020 Driver Assistance Technologies. Transportation Systems for Smart, Sustainable, Inclusive and Secure Cities.

[7] Hitachi Astemo Americas, I 2022 Advanced Driver Assistance System. Retrieved from Hitachi Inspire the NExt: https://www.hitachi-automotive.us/Products/oem/DCS/ADAS/index.htm.

[8] CHAABAN, K., SHAWKY, M., & CRUBILLE, P. (2009). A distributed embedded architecture for the evaluation of ADAS systems. In IFAC Proceedings Volumes (Vol. 42, Issue 15, pp. 237–244). Elsevier BV. https://doi.org/10.3182/20090902-3-us-2007.0095.

[9] Payá-Vayá, G., & Blume, H. (2022). Towards a Common Software/Hardware Methodology for Future Advanced Driver Assistance Systems. River Publishers. https://doi.org/10.1201/9781003339847.

[10] Ravi, Aravind. (2024). Optimizing ADAS and Autonomous Driving Systems with Advanced Ethernet Protocols and Machine Learning. International Journal of Science and Research (IJSR). 12. 2147-2155. 10.21275/ES24611084428.

IOP Conf. Series: Materials Science and Engineering **1311** (2024) 012048 doi:10.1088/1757-899X/1311/1/012048

# Theoretical and experimental considerations regarding the airbag system operation

**O V Oţăt, I Dumitru, D Tutunea, L Matei and G Marinescu**

Faculty of Mechanical Engineering, University of Craiova, Craiova, Romania

otatoana@yahoo.com

**Abstract.** The concept of Passive Safety aims to reduce the harmful effects on vehicle occupants and pedestrians in the event of an accident. Intelligent car bodies and restraint systems have been designed. Thus, side, front, knee and head airbag systems have been introduced, as well as pre-strained seat belts and force limiters. The objectives of car manufacturers are not only to reduce injuries that can cause death, but also a significant reduction of those that can result in permanent damage such as a whiplash. Advanced airbag systems are currently introduced, which offer protection to the occupants at the knees, torso, hips and abdomen. The topic proposed in the paper concerns both theoretical research on the operation of passive security systems, respectively the airbag system, and experimental research. Regarding the theoretical research, this paper proposes a virtual prototyping method by means of the LS Dyna software package of the airbag system functioning in order to protect the driver in the event of a frontal collision. For the analysis of the airbag system functioning, a complex assembly which contains of a dummy, seat, steering wheel, passive safety systems was used. For the experimental research a stand was designed for the analysis and testing of the airbag system, which has the role to protect the driver in the event of a frontal collision of the vehicle. In order to carry out the experimental tests, it was proposed to create a sledge type stand, which simulates frontal collisions.

## 1. Introduction

Road Safety [1] is a complex phrase that implies optimal collaboration and coordination between several factors directly or indirectly involved in order to ensure safe conditions for all public road users within a national transport system.

All factors of the road system are involved in improving driving conditions. The concern of specialists for the study and improvement of security measures is also seen through the numerous published studies. In [2], [3] are treated both research regarding the Active Safety Systems from a theoretical point of view, but the most common road safety engineering deficiencies are also highlighted.

Over time, some manufacturers have used different terminology for airbags. In the 1970s, the term Air Cushion Restraint System was used. In North America, Supplemental Restraint System (SRS) or Supplemental Inflatable Restraint (SIR) were used as general terms.

### *1.1. Proper use of airbags*
From the very beginning, the use of airbags was recommended to be done in conjunction with the use of seat belts. Only used in conjunction with seat belts, the airbag ensures that the risk of injury is significantly reduced.

Airbags can generate accidents, although their main effect is to provide additional protection for passengers restrained in seatbelts. If the passenger is close to the airbag at the time of inflating it, serious injuries occur by a strong blow to the neck and chest, the inflating being very rapid and can be similar to an explosion in the area of contact with the human body. According to [4], the injuries can be severe in specific passenger positions called out of position or when the airbag receives the inflation command with delay.

The danger zone is 5-8 cm from the inflating airbag cover. The safety zone is about 25 cm from the airbag. It is recommended that the airbag in the steering wheel to be directed towards the driver's chest area and not towards his head or neck. The safety position is obtained either by changing the position of the steering wheel (when the car has such a steering wheel adjustment system) or by changing the position of the driver's seat, in height or in the forward-backward direction. Starting from this principle, in [5], is presented a research study regarding the occupant kinematics in the vehicle at the front and rear impact.

### *1.2. Airbags reaction time*
The signal sent by the impact sensors is analyzed by the microprocessor of the airbag control system and it commands their deployment according to the received data. There are several constructive variants of impact sensors: some feel deceleration, other pressure changes. However they work, they transmit to the control unit an electrical signal indicating the intensity of the deceleration, or rather, the crash pulse produced after the accident. Any delay in this moment can cause serious injury if the passenger hits an expanding airbag. Likewise, a premature opening greatly decreases the efficiency of the airbags. Airbags are designed to deflate very shortly after inflating and taking over the passenger's shock to allow the rapid evacuation of the vehicle. A preoccupation in this direction is presented in [6], where the study was directed at two types of airbag miss-deployments: late deployment and non-deployment.

Constructively, an airbag is composed of 3 distinct parts:

1. The air bag itself is constructed from a thin nylon cloth that is folded to enter the recess inside the steering wheel, instrument panel or seat/doors.

2. The impact sensor is a device that gives the command to inflate the airbag. It is usually positioned in the front bumper and gives the command to inflate the airbag under certain impact conditions. The sensor is designed to sense car collisions with objects at speeds of at least 20 km/h.

3. Airbag inflator system. Inflating the airbag actually occurs as a result of a chemical reaction between 2 compounds. These react and result in nitrogen in the gaseous form which being warm and having a large volume, rapidly inflates the air bag. The Airbag practically jumps out of its seat at a speed of almost 300 km/h.

## 2. Virtual Prototyping and testing of the airbag system
LS-Dyna software is a specialized modeling software which contains: the analysis and reconstruction of traffic accidents, the analysis of the driver's behavior for determining the degree of his injury. It has available a virtual prototyping module of passive safety systems, respectively airbag and seat belt. This program package uses the finite element method.

For the analysis of the operating mode of the airbag system, using the LS Dyna software package, a complex assembly consisting of dummy, seat, steering wheel, passive safety systems, respectively seat belt and front airbag [7] was used -figure 1.

IOP Conf. Series: Materials Science and Engineering **1311** (2024) 012048 doi:10.1088/1757-899X/1311/1/012048

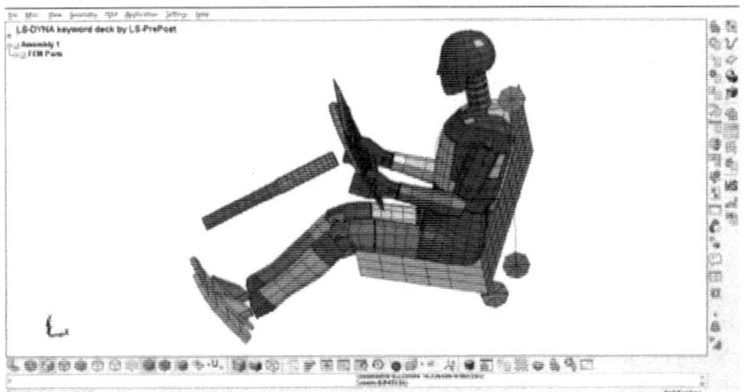

**Figure 1**. Complete model for airbag testing [7]

For the proposed analysis, a model of the front airbag was used, which was located on the steering wheel of the vehicle. This model consists of two parts, respectively: airbag lower and airbag upper-figure 2. The next step in the analysis was to discretize the model into finite elements, using the AutoMesher method, available in the LS-Dyna software package.

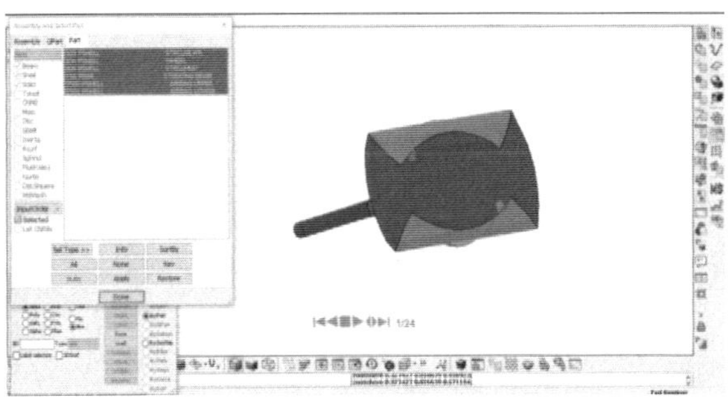

**Figure 2.** Lower and upper parts of the airbag model

### *2.1. Initial data*

To study the behavior of the airbag it is necessary that the entire assembly has a law of motion, which will simulate the frontal impact. In Figure 3 is presented the motion law that was defined for the considered model. It is noted that the analyzed time interval is 150 ms, and the maximum speed is 17.45 mm/ms.

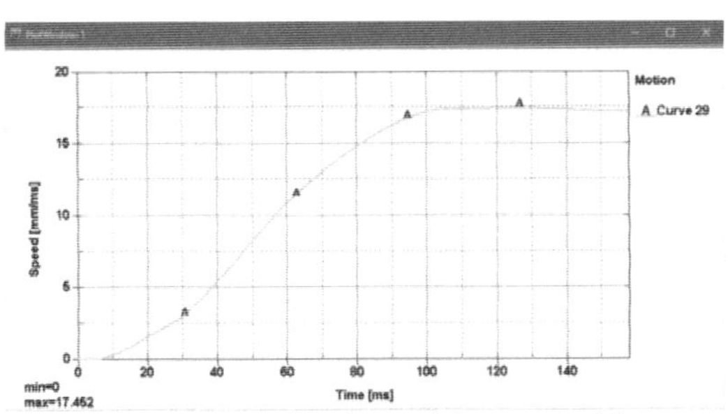


**Figure 3.** The motion law of the ensemble

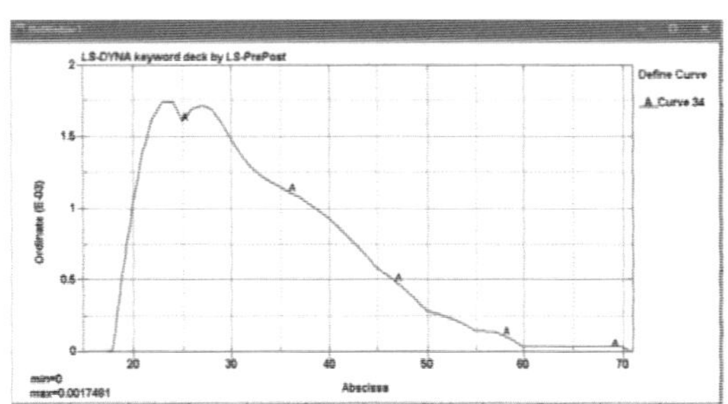


**Figure 4.** Airbag load curve

For airbag modeling, the LS-Dyna software has a module for wrapping and defining the system properties. From figure 4 it can be observed the interface with this module, and also the definition method, respectively Wang_Nefske. The load curve specifies the input mass flow rate or tank pressure as a function of time. If the tank volume, TVOL, is nonzero, the ID curve is considered to be the tank pressure as a function of time. During the dynamic relaxation phase the airbag is ignored if the curve is not marked to act during this phase.

*2.2. Presentation of successive phases of the airbag action*

In order to observe the operating mode of the airbag, successive sequences were extracted from the simulation performed, at different time intervals:

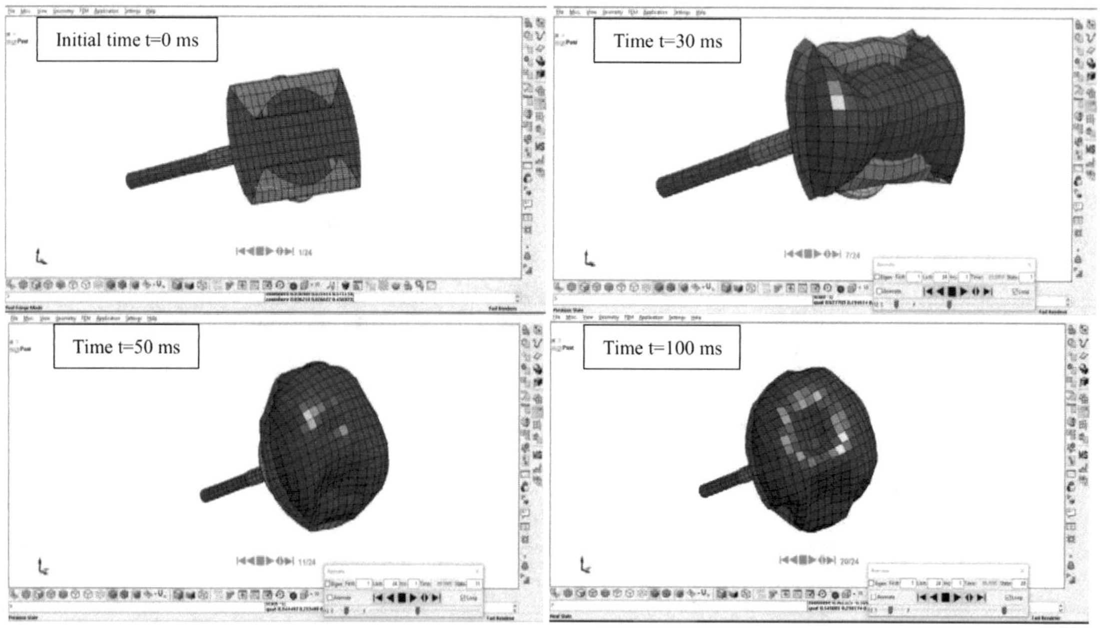


**Figure 5.** Airbag-steering wheel assembly

*2.3. Simulation processing and presentation of results*

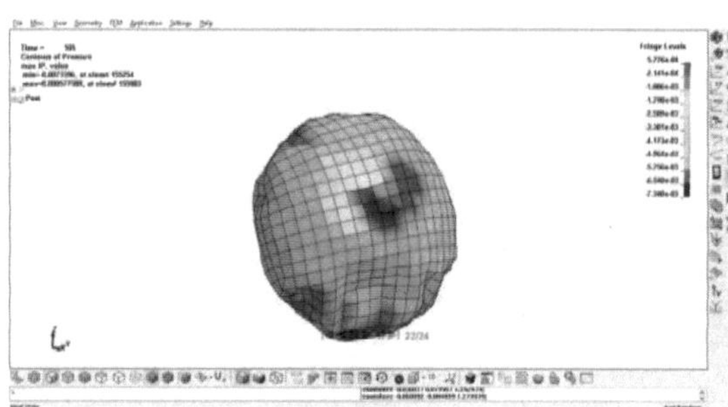

**Figure 6.** Pressure [kN/mm$^2$] at t=105 ms

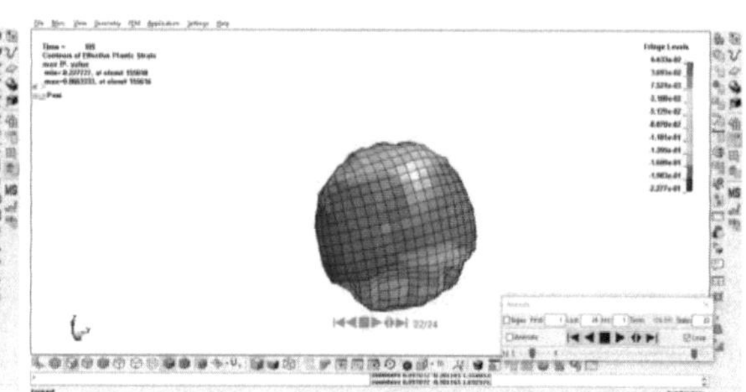

**Figure 7.** Plastic deformations at t=105 ms

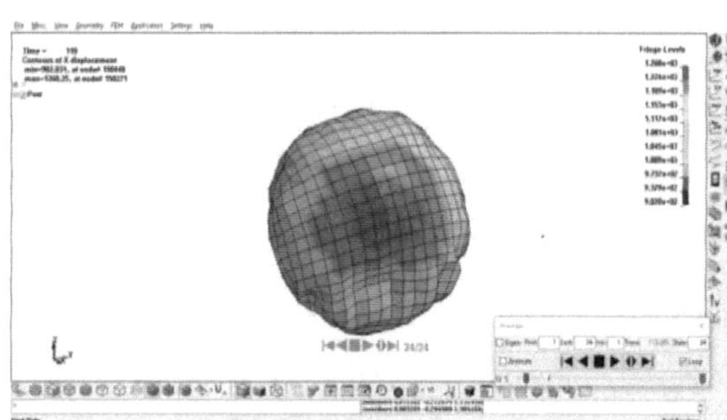

**Figure 8.** Displacement on longitudinal axis X at t=110 ms

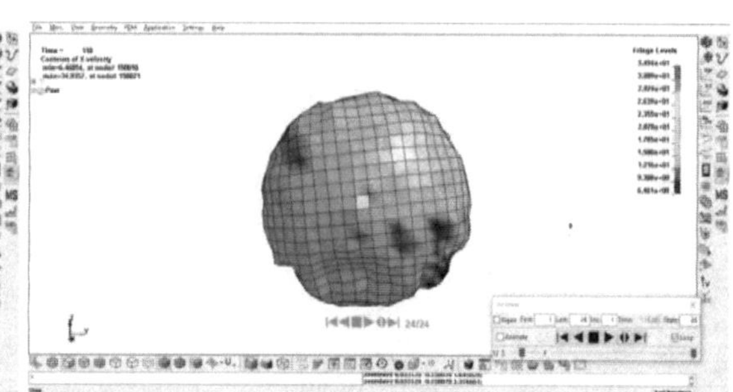

**Figure 9.** Speed [mm/ms] on longitudinal axis X to t = 110 ms

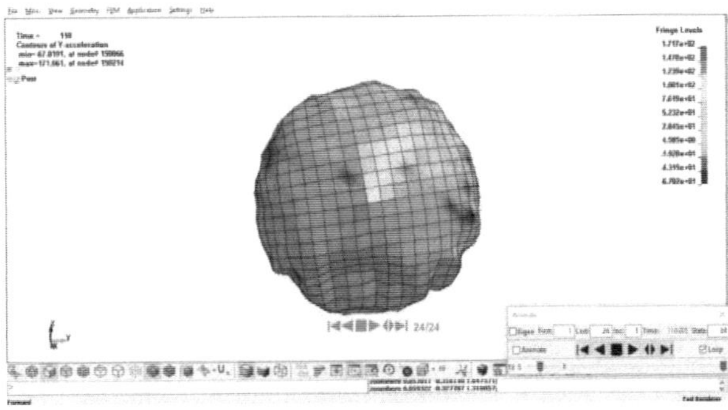

**Figure 10.** Acceleration [mm/ms$^2$] on the transverse axis Y to t=110 ms

### *2.4. Presentation of the results obtained in graphic form*

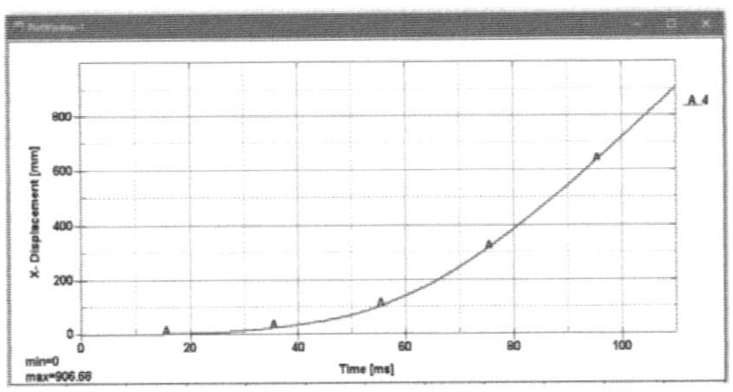


**Figure 11.** Displacement of the assembly on longitudinal X-axis as a function of time

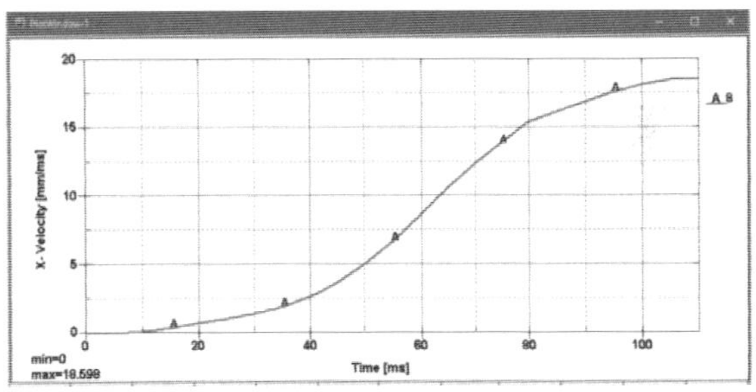


**Figure 12.** Assembly speed [mm/ms] on longitudinal X-axis as a function of time

From figure 11 it is observed that the maximum value of 906 mm of the assembly displacement after the longitudinal X axis is recorded at the time moment of 110 ms, which coincides with the final moment of the simulation.

Making a comparison between the values recorded for the speed of the assembly on the longitudinal X axis as a function of time (figure 12) and the values of the resultant of the speeds of the assembly as a function of time it is found that the maximum speed is recorded at the time moment of 110 ms, and has the value of 18.6 [mm/ms], this being oriented according to the longitudinal direction.

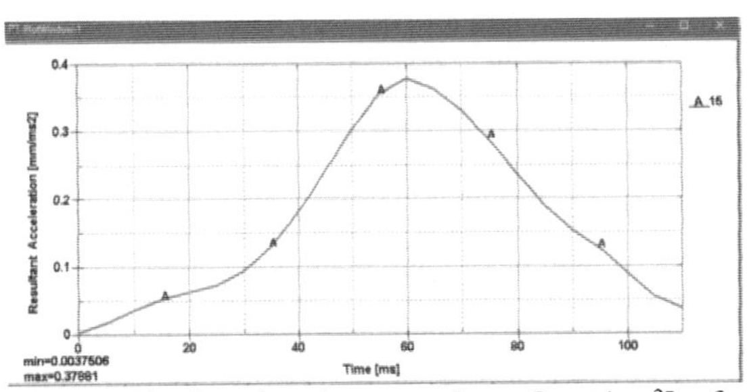


**Figure 13.** Accelerations resultant [mm/ms$^2$] of the assembly as a function of time

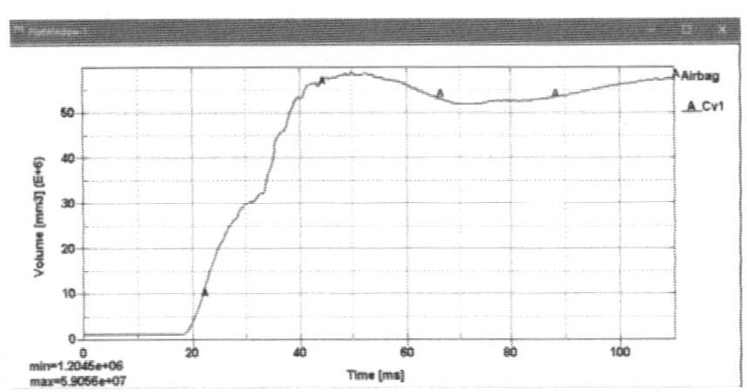


**Figure 14.** Airbag volume variation [mm$^3$] in time

Maximum value of 0.37 [mm/ms$^2$] acceleration is achieved at the time of 60 ms, at which time it coincides with the impact between the airbag and the upper part of the dummy. As in the case of displacements, speed, and for acceleration is observed the maximum values are obtained after the

IOP Conf. Series: Materials Science and Engineering **1311** (2024) 012048 doi:10.1088/1757-899X/1311/1/012048

longitudinal direction of the assembly, relative to the global coordinate system. The maximum volume value of the airbag is 5.9 * 107 $mm^3$ at t= 50 ms. The maximum airbag temperature is 810 K at t=35 ms.

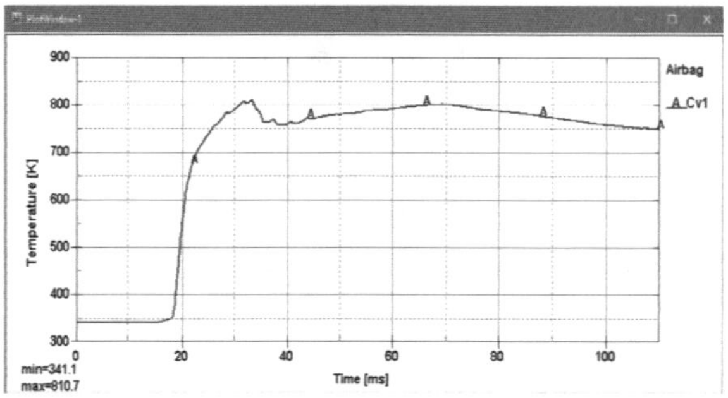


**Figure 15.**Airbag temperature variation [K] as a function of time

## 3. Experimental development of the airbag test stand

The topic proposed in the paper concerns both theoretical research on the functioning of passive security systems, namely the airbag system, as well as experimental research. In order to carry out experimental research, it was necessary to design a stand for the analysis and testing of the airbag system, intended for the protection of the driver in the event of a frontal collision of the car.

For the experimental tests, a sledge stand was designed, which simulates frontal collisions. This stand includes the frontal part of a vehicle, with the driver's and passenger's right seat, as well as the frame on which the vehicle is running, and a rigid, fixed barrier with which the vehicle will come into contact.

### *3.1. Conducting experimental tests*

After the design and construction of the experimental stand, tests on the functioning of passive security systems were carried out. The first step consisted of placing the dummy on the driver's seat and retaining it with the seat belt system.

**Figure 16.** Positioning the dummy in the driver's seat

Through experimental tests, in addition to observing the operating mode of the airbag, the kinematic parameters of the vehicle both before and at the time of impact were also determined.

For the determination and recording of kinematic parameters, several applications were used, which were downloaded and run on the Android operating system. In order to determine the maximum speed, average speed, time and distance traveled, the application *GPS speedometer* was used, this is installed on the phone which was fixed on the right passenger seat.

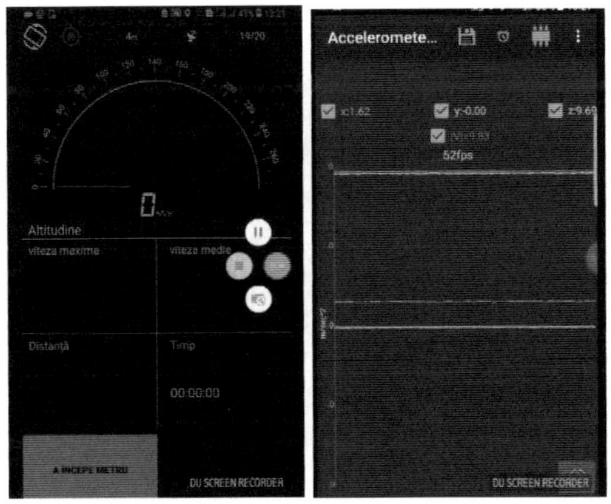


**Figure 17.** Application interface Accelerometer Analyzer and GPS Speedometer

The determination of the accelerations after the three directions and the speed of the assembly was possible by installing on the device located on the left lower limb of the dummy two applications, respectively: *Accelerometer Analyzer* and *GPS Speedometer*.

**Figure 18.** Collision moment

## 4. Conclusions

The present study proposes a virtual prototype and testing method of the airbag system functioning in order to protect the driver in the event of a frontal collision. For the proposed analysis, the front airbag, located in the steering wheel of the vehicle, was taken into account. For the analysis of the airbag system functioning, by means of the LS Dyna software package, a complex assembly consisting of a dummy, seat, steering wheel, passive safety systems was used. At the same time, the seat belt with three-point fastening was modeled, through the LS-Dyna software package, specialized software for the analysis and reconstruction of traffic accidents and for the analysis of the driver's behavior. The analyzed time interval is 150 ms and the maximum speed is 17.45 mm/ms. After processing the simulations, the results obtained for the following parameters were presented: Pressure, Plastic Deformations, Displacement, Speed, Acceleration.

The topic proposed in the paper concerns both theoretical research on the operation of passive security systems, respectively the airbag system, and experimental research. In order to carry out the experimental research, a stand was made for the analysis and testing of the airbag system, designed to protect the driver in the event of a frontal collision of the vehicle. In order to carry out the experimental tests, it was designed a sledge type stand, which simulates frontal collisions. This stand contains the front part of a vehicle, with the driver's and passenger's seat, as well as the frame on which the vehicle is running, and a rigid, fixed barrier with which the vehicle will come into contact.

**References**

[1] Nistor, N., Expertiza tehnica a accidentelor de circulaţie, Publishing House Militara , Bucharest, 1987

[2] Beles H, Preda I, Dragomir G, Blaga V and Ardelean F, Theoretical Research Regarding the Active Safety Systems, Annals of DAAAM for 2008 & Proceedings of the 19th International DAAAM Symposium, Pages: 87-88, Published: 2008, ISBN: 978-3-901509-68-1, ISSN: 1726-9679

[3] Jovanov D, Vollpracht H J, Beles H, Popa V, Tolea B A, Most common road safety engineering deficiencies in South Eastern Europe as a part of safe system approach, International Congress of Automotive and Transport Engineering - Mobility Engineering and Environment (CAR2017), IOP Conference Series-Materials Science and Engineering, Volume: 252, Article Number: UNSP 012021, DOI: 10.1088/1757-899X/252/1/012021, Pages: 12-21, Published: 2017, Publisher: IOP Publishing LTD, England, ISSN: 1757-8981

[4] Soica A., Aspects of the passenger airbag E.C.U. location, IOP Conference Series: Materials Science and Engineering, Volume 147, 7th International Conference on Advanced Concepts in Mechanical Engineering 9-10 June 2016, Iasi, Romania

[5] Popa V, Beles H, Tolea B, Dragomir G, Mitran T, Studies Regarding Occupant Kinematics in the Vehicle at the front and Rear Impact, Applied Mechanics and Materials, Vol. 822, Pages: 512-522, Jan. 2016, Publisher: Trans Tech Publications, ISSN print: 1660-9336, ISSN web: 1662-7482, ISBN: 978-3-03835-703-2

[6] Toganel G. R., Soica A., A late and failure of airbag deployment case study for drivers of passenger cars in rear-end collisions, IOP Conference Series: Materials Science and Engineering, Volume 252, CAR2017 International Congress of Automotive and Transport Engineering - Mobility Engineering and Environment 8-10 November 2017, Pitesti, Romania

[7] http://www.lstc.com/models

IOP Conf. Series: Materials Science and Engineering **1311** (2024) 012049 doi:10.1088/1757-899X/1311/1/012049

# Top causes of road traffic accidents in Romania

**C I Dimitriu[1] and G Frățilă[2]**

[1] PhD. Stud., auditor of road safety, Romanian Road Authority -ARR, Bucharest
[2] Prof. Ph.D. Eng., Faculty of Transport, POLITEHNICA University of Bucharest, Splaiul Independenței, no.313, RO-060042, Bucharest, Romania

corneliu.dimitriu@stud.trans.upb.ro, ghe_fratila@yahoo.com

**Abstract**. Road accidents worldwide impose a significant burden on national budgets, particularly affecting the least developed countries. According to the latest WHO "Global Status Report on Road Safety 2023," road fatalities are the leading cause of death among young people aged 5 to 29. Among the European Union countries, Romania has the highest mortality rate. Between 2011 and 2021, approximately 2.8% to 1.1% of Romania's GDP represented the social costs caused by road accidents. Statistical analysis of serious road accidents in Romania during this period reveals five main causes responsible for at least 50% of the total deaths recorded on the national public road network. The most significant cause is "speed not adapted to road conditions," accounting for approximately 19% of the total. A road accident occurs due to the action or interaction of three determining factors: human, vehicle, and road (road environment). In most cases, the blame for road accidents is attributed to human error or behaviour. This conclusion is often influenced by data collection methods that do not consider all the influences of road infrastructure on the accident's dynamics. A closer analysis shows that road infrastructure deficiencies often influence or cause road accidents, highlighting the importance of continuous evaluation and improvement of infrastructure to reduce accident risk. The implementation of Directive 2008/96/EC on road infrastructure safety management has prompted authorities to pay particular attention to infrastructure deficiencies in fatal accidents. During 2018-2021, following road safety inspections, auditors provided police and road administrators with detailed reports on deficiencies that caused or favored fatal accidents. Processing and analysing these reports can form the basis for developing a statistical model that provides a unified approach to road infrastructure deficiencies. This model can set up best practices for road safety, contributing to the uniform treatment of infrastructure problems found by auditors. With the transposition of Directive (EU) 2019/1936, which amends Directive 2008/96/EC, through Romanian Government Ordinance no. 3/2022, an innovative approach to managing road infrastructure deficiencies has been introduced. At the EU level, emphasis is placed on finding main road sections with low safety levels. Specific inspections are directed towards road sections with multiple risks, and remedial measures will focus on cost-effective solutions to enhance safety. Continuous evaluation and improvement of road infrastructure are crucial for reducing accident risk. By addressing deficiencies contributing to serious accidents, authorities can create safer road environments. This proactive approach involves detailed inspections and thorough data analysis to find and rectify potential hazards. The goal is to set up a system that not only responds to accidents but also prevents them by mitigating infrastructure risks. In conclusion, road safety management needs a multi-faceted approach that integrates human, vehicle, and road factors. By implementing directives and conducting rigorous inspections, authorities can significantly reduce road accidents. The development of a statistical model from inspection data will systematically address deficiencies, leading to safer roads and reduced fatalities, highlighting road safety as a public health issue.

## 1. Consequences of serious accidents

Road accidents, in addition to the suffering and pain caused to the families of injured or deceased persons, constitute a significant public health problem and generate social costs that place a burden on national budgets and affect economically mainly the least developed countries. The leading cause of death worldwide for the 15-29 age group is the death of people in road accidents. Globally, about 1.35 million people die in road accidents each year and about 50 million are injured. It is estimated that by 2030, road deaths will become the seventh leading cause of death globally by 2030 [1].

### *1.1. Romania's place in the EU ranking*

According to the data published in The National Road Safety Strategy 2022-2030, Romania stays among the top countries with the highest rate of road traffic deaths. Thus, in 2019, a rate of 96 deaths per million inhabitants recorded by Romania was the highest in the EU, whilst the average was at 51 deaths. The analysis of the death rate per 10,000 registered vehicles shows that Romania is also recording negative performance among Member States, with the highest death rate in 2019, namely 2.26 compared to the average rate in the EU of 0.88 [2].

### *1.2. Costs of serious road traffic accidents in Romania*

The Romanian Road Authority – ARR calculated for 2016 the average unit social costs of road accidents resulting in deaths and seriously injured persons [3]. These values were updated in 2017 and are presented in Table 1.

**Table 1.** Social cost of an accident unit

| The year | Monetary unit | Accident with serious injuries | Fatal accidents |
|---|---|---|---|
| 2016 | RON | 760,158 | 4,883,603 |
| | EURO | 169,270 | 1,087,468 |
| 2017 | RON | 770,040 | 4,947,089 |
| | EURO | 168,569 | 1,082,964 |

The analysis on the evolution of the consequences of serious traffic accidents resulting in deaths and serious injuries over the last 10 years shows in Figure 1 a trend of a higher decrease in the number of seriously injured people compared to actual deaths. The data on the Romanian Police website shows that in 2020 and 2021 the volume of road traffic registered a decline due to reduced mobility during the COVID19 pandemic [4]. The main consequence was the reduction in both mortality and serious injuries caused by road accidents, which were also seen in other countries.

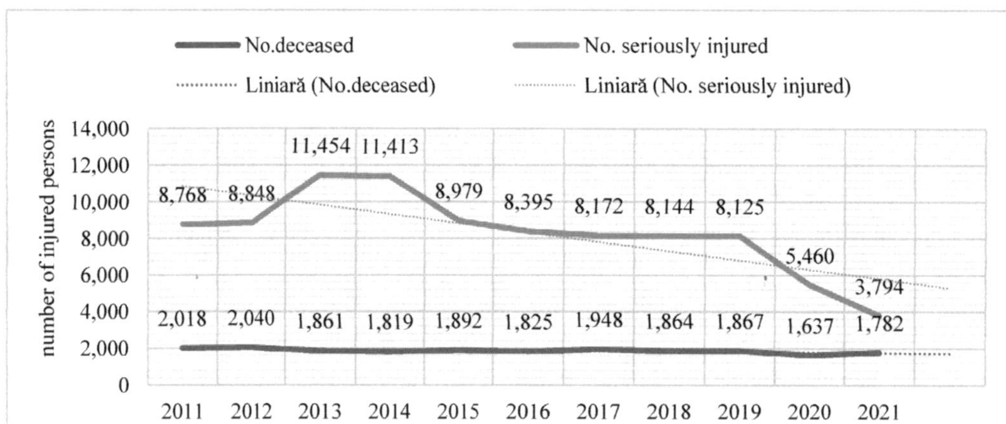


**Figure 1.** Evolution of the consequences of serious traffic accidents

IOP Conf. Series: Materials Science and Engineering **1311** (2024) 012049 doi:10.1088/1757-899X/1311/1/012049

According to a study in Sweden, the restriction measures have led to a significant decrease in the number of road accidents, but their severity has increased, reflecting a trend like that observed in Romania [5].

There is research that has looked at differences in slopes, trends in 'road deaths' and 'seriously injured'. Although the total number of accidents decreased, the accidents that occurred were often more severe, with a higher likelihood of being fatal due to higher speeds and other aggravating factors [6].

Figure 2 shows an overview on social costs impact generated by accidents on the national budget. For the period 2011-2021, the dynamics of serious traffic accidents' influence on gross domestic product - GDP is presented. Data is taken from national Institute of Statistics'– INS website [7].

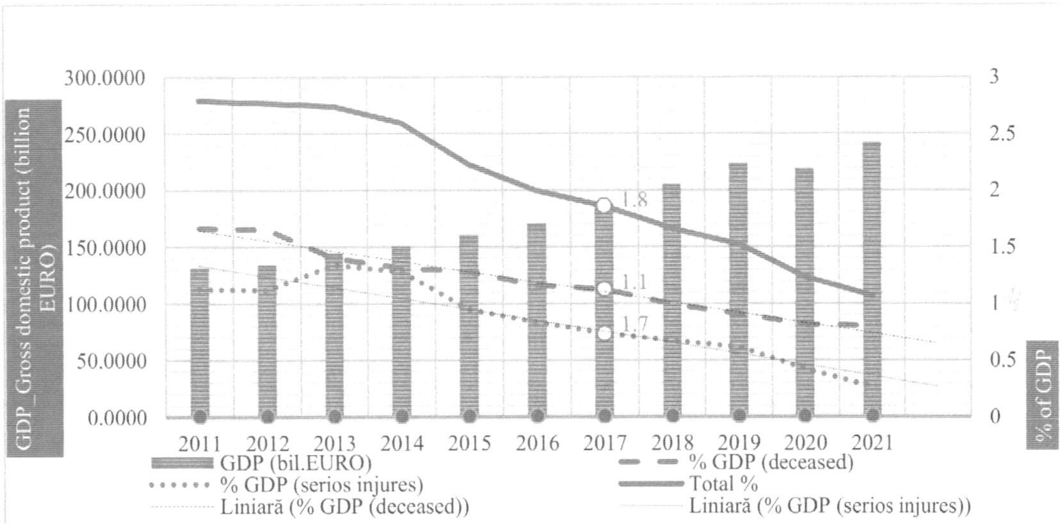


**Figure 2.** Costs dynamics for serious traffic accidents' influence on GDP

In order to correctly understand the relationship between the number of deaths, the number of seriously injured and their economic impact on GDP, an analysis of the slopes of the two graphs, in Figure 1 and Figure 2, is performed.

Although GDP has grown gradually, the share of social costs due to road accidents in GDP has decreased, but not at a rate as fast as the decrease in the number of serious injuries. This underlines the need for continuous analysis and the implementation of new measures to further reduce the economic and social impact of road accidents.

## 2. Statistics on traffic accidents causes

WHO statistics show that almost 54% of people killed in traffic accidents are vulnerable road users: pedestrians, cyclists, and motorcyclists [1]. A similar statistic showed that at EU level, 47% of the deaths are happening for vulnerable road users. This statistic presented in Directive (EU) 2019/1939 amending Directive 2008/96/EC is related with the result of the analysis on Member States that have applied road infrastructure safety management procedures for the entire road network and for all categories of vulnerable users. It is highlighted that this approach has led to significant progress in improving road safety and that notable results have been achieved in reducing the number of fatalities and serious injuries from road traffic accidents [8].

One of the methods for analysing traffic accidents resulting in injuries is to filter specific data according to the main cause of their occurrence. The Traffic Police attribute to each accident a primary, deciding cause. There are 47 main causes as seen in Figure 3. Starting with 2013, the Romanian Police has excluded from its statistics five main causes of road accidents, these being marked in the graph with the symbol "(!)".

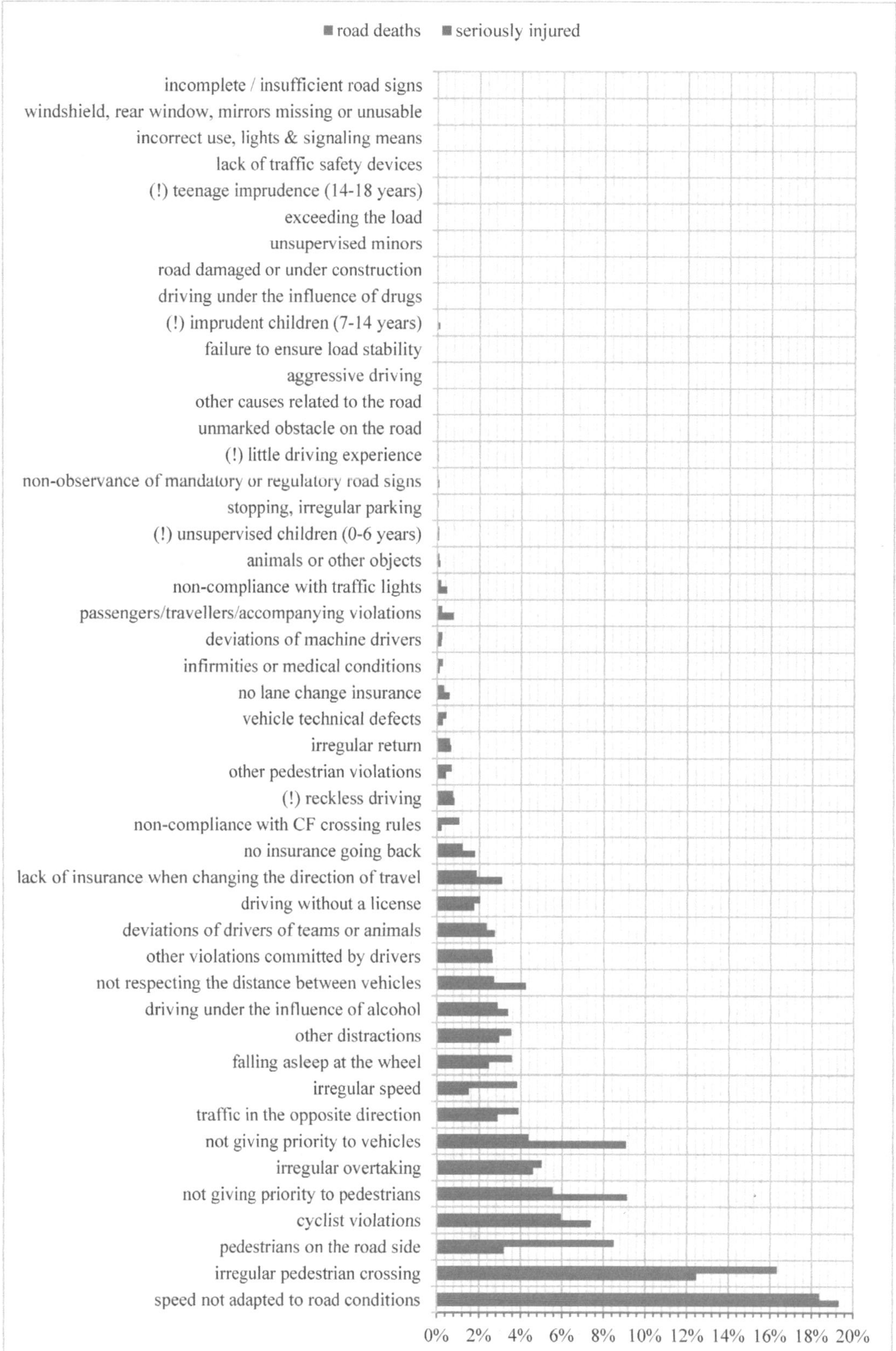


**Figure 3.** Top causes of serious road traffic accidents

Table 2 presents a classification of the main causes (47) of serious road accidents in the period 2011-2021, highlighting both the proportion of deaths and serious injuries associated with each cause.

The combined total of these causes covers 59.68% of deaths and 51.48% of serious injuries, highlighting that a significant part of serious road accidents are the result of risky behaviours and a lack of compliance with traffic rules. Interventions to reduce these behaviours can have a substantial positive impact on road safety [9].

**Table 2.** Ranking of the main causes of serious road accidents (2011-2021)

| **Top causes** | **Deaths** | **Serious injuries** |
|---|---|---|
| Speed not adapted to road conditions | 18.38% | 19.32% |
| Pedestrian crossing violation | 16.33% | 12.45% |
| Pedestrians on the roadside | 8.49% | 3.18% |
| Cyclist violations | 5.94% | 7.39% |
| Not giving priority to pedestrians | 5.53% | 9.13% |
| **Total** | **59.68%** | **51.48%** |

The study of data from 2011-2021 shows the need to implement multidimensional measures to address the main causes of serious road accidents. These measures should include improvements in infrastructure, education and awareness campaigns, as well as strict enforcement of traffic laws. Such an integrated approach can significantly contribute to reducing the number of deaths and serious injuries on public roads.

The graphical representation of the data collected by the Romanian Police for the period 2017-2020 and which exclusively reflects the 42 main cases used since 2013, is illustrated in Figure 4.

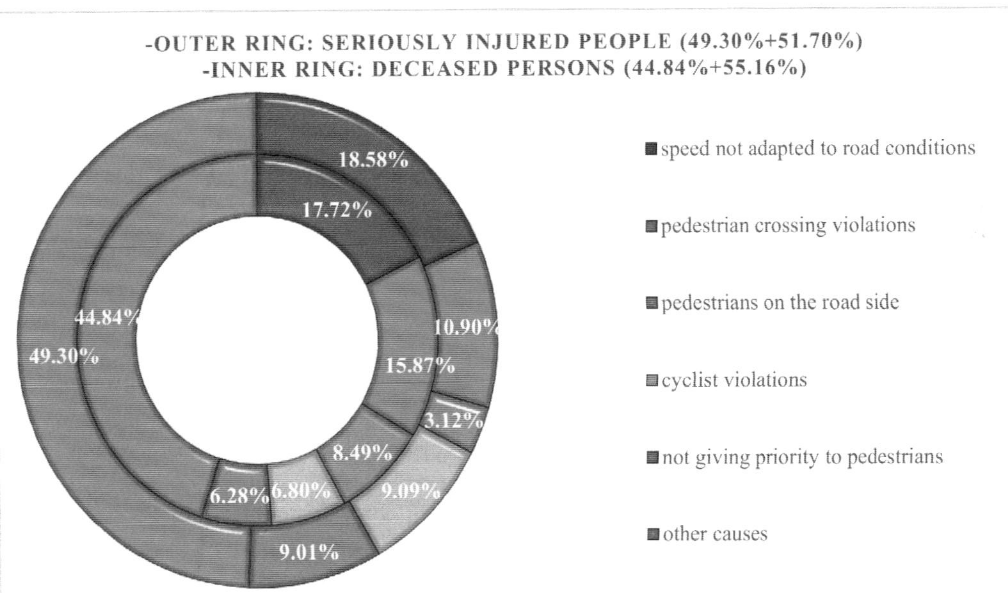


**Figure 4.** Top causes affect over the occurrence of serious road traffic accidents

The statistical analysis of the evolution of serious accidents in Romania highlights an alarming trend of increasing severe incidents, highlighting the need to implement effective strategies to reverse this negative trend and improve road safety [10].

The Executive Director of the European Transport Safety Council (ETSC), Antonio Avenosso, in an open letter published by the Financial Times, criticizes the perpetuation of the "myth" related to the main cause of road accidents, namely human errors [11]. Mr. Antonio Avenosso states that this myth: "Human error causes 90% of road accidents", is "misleading and dangerous" especially because it is repeated and perpetuated. If we want to look at the matter from this point of view, to grasp a correct approach of what the honourable gentleman has suggested, we need statistics to substantiate claims. The information in the statistics is important because it helps us understand what the determining factors in the occurrence of individual road accidents are.

The details obtained can help to set up the necessary road safety policies and actionable directions, including showing authorities the scope of the applicable road safety initiatives, preventing accidents, injuries and assessing the impact of the measures taken.

## 3. Addressing traffic accidents that result in casualties

The Traffic Police plays a crucial role in investigating serious accidents, measuring and documenting details for reporting. Along with this, the entities responsible for providing first aid, transporting victims and controlling the legality of commercial transport contribute to a rapid and effective intervention to resume traffic and prevent further incidents.

The integrated management of RISM (Road Infrastructure Safety Management) procedures in Romania involves close collaboration between various entities, such as the Romanian Police, the Romanian Road Authority - ARR, the National Company for Road Infrastructure Management - CNAIR and other road administrators. The collection and management of road accident-specific data is essential for understanding the causes and improving road safety [12].

### *3.1. Data on serious road traffic accidents*

The collection and management of data specific to road accidents is conducted by the Romanian Police – an entity empowered to manage and control compliance with the legislation on road traffic on public roads. The Romanian Police notifies the Romanian Road Authority – ARR on the cases of accidents with victims by art. 18 para. (5) of Law no. 265/2008 [13]. By amending Law no. 265/2008 by OG no. 3/2022, the data on fatal road accidents are transmitted every six months on paper to the Romanian Road Authority (ARR). This method of transmission prevents the generation of an efficient database for road safety statistics, requiring electronic transmission of data [14].

In the field of road infrastructure safety, the Ministry of Transport and Infrastructure, through the Romanian Road Authority – ARR, is the entity responsible for transposing Directive 2008/96/EC on road infrastructure safety management, as amended by Directive (EU) 2019/1936.

According to Article 7 of the Directive, the Romanian Police is the entity responsible for drawing up a report for each road accident resulting in deaths, including the elements listed in Annex IV of the Directive. Submitting these reports in electronic format to the ARR would ease a detailed statistical analysis of the data, offering a deeper insight into the causes and circumstances of road accidents. This approach would support road safety auditors in conducting field inspections, thereby improving the effectiveness of road safety assessments and interventions.

To improve the management and analysis of road accidents resulting in fatalities and to increase the efficiency of road safety inspections, measures are needed such as: digitization of accident reporting, improvement of data collection, statistical analysis and evaluation, preventive measures and intervention, better coordination and institutional collaboration through the Interministerial Council for Road Safety - CISR. The implementation of these measures can ensure more efficient management of road accident data and a significant improvement in road safety at national level.

### *3.2. Proactive and reactive approach to road traffic accidents*

To carry out periodic road safety inspections, in the period 2018-2022, the Romanian Road Authority – ARR concluded service contracts with the National Company for Road Infrastructure Administration – CNAIR and with the Bucharest Municipality Streets Administration – ASB [15].

Table 3 shows the statistics of the performance of the contracts over a period of 4 years, showing that the 89 sections included in the contract with the National Company for Road Infrastructure Management (CNAIR) represent 17.75% of the national roads and highways network. Road safety inspection is essential, constituting a proactive approach to find and remedy potential risks before accidents occur.

**Table 3.** Statistics of the execution of periodic road safety inspection contracts between ARR and CNAIR/ ASMB

| | | Periodic road safety inspections | | | | Year of execution |
|---|---|---|---|---|---|---|
| | | Planning | | Execution | | |
| | | No. of sections (National Road) | Length (km) | No. of sections (National Road) | Length (km) | |
| Contract no. 1 2018/2020 | CNAIR / ARR | 38 | 1486.2 | 14 | 590.5 | 2018 |
| Contract no. 2 2019/2021 | CNAIR / ARR | | | 23 | 857.6 | 2019 |
| | | 35 | 1025.6 | 3 | 87.3 | |
| Contract no. 4 2020/2022 | ASMB / ARR | | | 1 | 38.1 | 2020 |
| | | | | 26 | 762.3 | |
| | | 92 | 211.4 | 9 | 9.9 | |
| Contract no. 3 2021/2023 | CNAIR / ARR | | | 12 | 17.1 | 2021 |
| | | | | 6 | 176 | |
| | | 16 | 486.3 | 12 | 380.5 | |
| | | | | 4 | 105.8 | 2022 |

The current inspection rate is insufficient to cover the entire network in a best time. It is necessary to speed up the inspection process by giving added resources and implementing modern technologies to ensure efficient and complete monitoring of road infrastructure.

The reactive approach involved conducting inspections following accidents resulting in victims, started at the notification of the Romanian Police. Following the legal competences, the Romanian Road Authority - ARR conducted added road safety checks in approximately 97% of the places where fatal accidents occurred. The accident reports were drawn up exclusively for accidents resulting in deaths on the road network within the scope of Law no. 265/2008.

Table 4 presents a numerical analysis of the reactive approach through added road safety inspections for accidents in the period 2019-2021.

**Table 4.** Statistics of road safety inspections by ARR upon notification of the Romanian Police

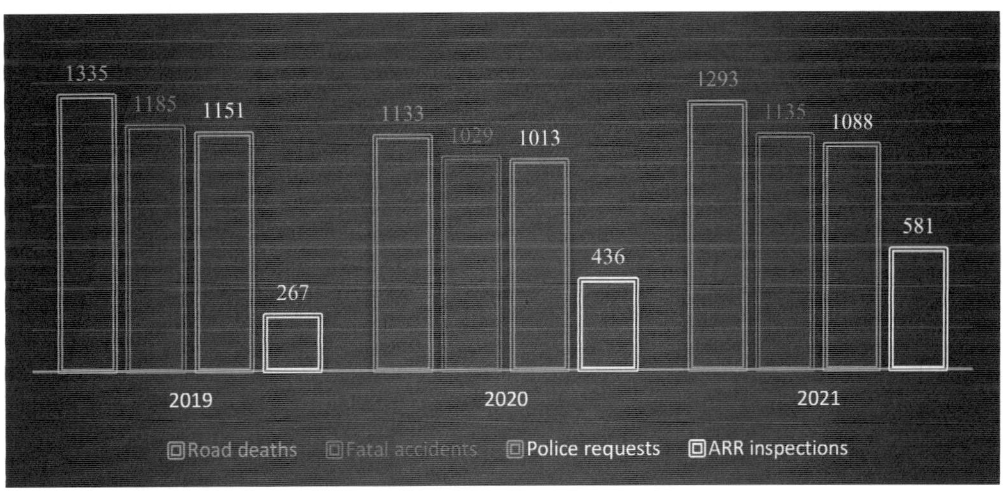

The road safety inspection reports (periodic or accident reports) had valuable information on deficiencies in road infrastructure. However, with no provision in the normative acts for centralizing in databases, they were limited only to the establishment of mandatory provisions for the road manager and aimed at cutting deficiencies found by the road safety auditors on a punctual basis. To achieve full transposition, in the sense and spirit of Directive 2008/96/EC, express obligations were necessary for the competent authority to build databases useful for road safety statistics. This found those components of road infrastructure that handled, or conducive to, serious accidents.

The statistical analysis of the data obtained, together with the effects of improving road safety generated by some works – because of the provisions of the reports – would have led to the establishment of directions of action on the road infrastructure, aiming to reduce the risks on the most dangerous sections of the national network. At the same time, these data could lead to different conclusions than the first ones, proving that the fault for accidents is not exclusively the driver's behaviours.

After a road accident resulting in serious injuries or deaths, a laborious process follows to decide the culprits and their responsibilities. In general, this process consumes precious resources and time. Modern technology supports road accident research, easing the entire investigation process and providing a greater degree of accuracy for specific data, helping to reach pertinent conclusions and getting closer to reality than ever before.

### *3.3. Factors deciding the occurrence of road traffic accidents*

It is estimated that the occurrence of road traffic accidents is the faulty interaction between the factors, man – road – car or in a general and all-encompassing way, road users – road environment – vehicles. Table 5 provides a description of what each of these factors could stand for.

**Table 5.** Faulty interaction between the factors

| No. Crt. | Road users (man) | Road environment (road) | Vehicles (car) |
|---|---|---|---|
| 1 | The driver | Public road | Vehicles |
| 2 | Vulnerable motorized/non-motorized users | Human factors | 2-wheel vehicles |
| 3 | Other drivers | Weather and light conditions | Vehicle assemblies |
| 4 | Maintenance/ repair/ accident clearance personnel | Isolated animals on the road herds of animals | Special priority vehicles |

The influence of each factor varies from case to case. Every serious accident has its own peculiarities, and it is quite difficult to decide the correct weight of each factor. If we look at the priorities of road safety investments, we have a problem with the length of time in which the expected results occur. Investments and measures to correct human behaviour or to develop automated systems in motor vehicles have the disadvantage of measurable and efficient results over a longer period. Immediate results come from investments in road infrastructure modernization or measures to remedy/correct the deficiencies identified following proactive/reactive approaches.

## 4. Road infrastructure safety legislation

Until 2009, Romania did not have a legislative framework dedicated to road infra-structure safety activities. 19.12.2010 was the deadline for transposing the provisions of Directive 2008/96/EC on road infrastructure safety management into national legislation. This date marked the beginning of the pioneering activity in the field and at the same time gave the title of the first normative act that considered the requirement of the moment: the need to conduct the road safety audit for new road infrastructure projects. Law no. 265/2008 on road safety audit was amended by Government Ordinance no.6/2010, which changed the original title, "traffic safety management on road infrastructure", was after approved by Law no.125/2012.

Directive 2008/96/EC has now been amended by Directive (EU) 2019/1936, following the Commission communication and the Council conclusions which, analysing the stagnation of progress in reducing

road fatalities, approved a new intermediate aim, namely, halve the number of serious injuries by 2030 compared to 2020. In order to achieve the new objective, new concepts have been introduced by Directive (EU) 2019/1936 such as the road infrastructure safety approach based on the concept of a "safe system" which states that fatalities and serious injuries in road accidents can be prevented to a large extent, as well as the RISM concept – Road infrastructure Safety Management, which opens the pathway for Member States to use, where appropriate, the constantly developing technologies when inspecting road sections, documenting road safety conditions and collecting other road safety data.
In relation to the presented aspects, it is necessary now, after 10 years of road safety management, years in which the process of improving road safety has stagnated in our country, to modernize the primary legislation starting with the change of title, namely: law on road infrastructure safety management.

## 5. Conclusions

Initially, the provisions of the law no.265/2008 were mainly a methodological norm, because this approach was essential to draw some necessary milestones for the elaboration of follow-up acts that would respect both the directions of action and the spirit of Directive 96/2008/EC, but now:

- It is necessary to assuage the body of the law of the provisions that, over time, have been copied, transposed, and developed in subsequent acts: these are those provisions that have entered the current methodological and contractual practice and no longer have the nuances at the level required by a modern law.
- It is necessary to introduce in the body of the law new provisions that have the role of improving the level of institutional management, management of the implementation of road safety procedures, and to comply with the new directions, guidelines and accents required by the amendments and additions of Directive (EU) 2019/1936.
- It is necessary to correct and/or correlate some of the provisions that unjustifiably change the terminology in the field by introducing new notions, such as the deletion of the term "recommendation" or the partial replacement of the term "provision" with "measure", and by changing the structure of the reports prepared by road safety auditors, so that, these changes are likely to cause confusion, while valuable time is wasted to readjust specialist staff to unwarranted changes.
- Provisions need to be introduced to initiate close collaboration between institutions to organize existing databases and adapt them to allow statistical analyses to guide, motivate and justify the managerial decision to align us with European objectives.
- To increase the quality and number of the body of road safety auditors available, a more careful concern of the competent authority is needed for material stimulation, intensification of continuous professional training and attracting new people available from those currently carrying out other activities, although they have the appropriate training to carry out road safety procedures.

It is vital to strengthen interinstitutional cooperation:

- On the background of extending the scope of the law to other categories of roads, such as county roads, and streets of category I and II, provisions are necessary to determine close cooperation between ministries responsible for coordinating the road managers included in the road network covered by this law.
- Proper attention will be paid to the realization of close cooperation with county road administrators and managers of the first and second category streets, so that the remedial and action plans for improving road safety must be sufficiently flexible and adaptable to the real needs but consider the available resources.
- First, the efforts of inspection teams should focus on the awareness by local governments of the importance of conducting agreed plans.
- It is necessary to strengthen the composition of the Standing Delegation of the interministerial Council for Road Safety and to attract representative persons who have the necessary availability, ability, determination, knowledge, and support to achieve notable results because of the completion of the plans in this field.

- It is necessary that the members of the inspection teams be supported materially, logistically, and professionally by the institutions with direct attributions in the field and by the actors involved in the joint effort.

- It is necessary to update Law no. 265/2008 to cut redundant provisions and introduce new provisions that improve institutional management and compliance with EU Directives. It is vital to strengthen interinstitutional cooperation, improve the professional training of road safety auditors and create effective databases for statistical analysis to support managerial decisions and achieve European aims.

**References**

[1] World Health Organization. www.who.int. [Online].; 2023 [cited 2024 April 10. Available from: https://www.who.int/publications/i/item/9789240086517.

[2] The Government of Romania. Annex to Government Decision no. 682/2022 for the approval of the National Road Safety Strategy for the period 2022-2030. The Official Monitor of Romania. 2022 May 31: p. 3-74.

[3] Romanian Road Authority - A.R.R. www.arr.ro. [Online].; 2017 [cited 2022 July 07. Available from: https://www.arr.ro/s-i-r_doc_512_costuri-medii-ale-accidentelor-cu-persoane-decedate-si-ale-accidentelor-rutiere-grave_pg_0.htm.

[4] Romanian Police. www.politiaromana.ro. [Online].; 2017-2023 [cited 2024 July 07. Available from: https://www.politiaromana.ro/ro/utile/statistici-evaluari/statistici.

[5] Zheng Q, Sharmeen F, Xu C, Liu P. Assessing regional road traffic safety in Sweden through dynamic panel data analysis: Influence of the planned innovative policies and the unplanned COVID-19 pandemic. Transportation Research Part A: Policy and Practice. 2024 January; 179(103918).

[6] Yasin J. Y, Grivna M, Friki M. AZ. Global impact of COVID-19 pandemic on road traffic collisions. World Journal of Emergency Surgery. 2021 September 28; 16(51).

[7] National Institute of Statistics - INS, Romania. https://insse.ro/cms/ro. [Online].; 2022 [cited 2022 July 07. Available from: https://insse.ro/cms/ro/tags/comunicat-pib-anual.

[8] European Parliament, Council of the European Union. https://eur-lex.europa.eu. [Online].; 2019 [cited 2024 February 10. Available from: http://data.europa.eu/eli/dir/2019/1936/oj.

[9] Moslem S, Farooq D, Ghorbanzadeh O, Blaschke T. Application of the AHP-BWM Model for Evaluating Driver Behaviour Factors Related to Road Safety: A Case Study for Budapest. Symmetry. 2020 January; 12(2) (243).

[10] Grzebieta R, Czapski R, Burlacu A, Sakashita C, Soames Job R, Burlacu A. Adoption Recommendations of Road Safety Management Capacity Review: addressing a tragic decade of road safety in Romania. Journal of Road Safety. 2020; 31(3): p. 85-97.

[11] Avenoso A. https://e-sigurantarutiera.ro. [Online].; 2022 [cited 2023 July 07. Available from: https://e-sigurantarutiera.ro/avenoso-etsc-90-din-accidentele-rutiere-provocate-de-erori-umane-este-un-mit-periculos-si-inselator.

[12] Burlacu A, Burlacu AF, Răcănel C. Road safety management in Romania. In Fifth International Conference on Road and Rail Infrastructure; 2018; Zadar, Croatia.

[13] The Romanian Parliament. Law no. 265/2008, republished and updated with next amendments, regarding the management of traffic safety on the road infrastructure. The Official Monitor of Romania. 2012 August 23: p. 6-18.

[14] The Government of Romania. Ordinance no. 3/2022 for the amendment and completion of Law no. 265/2008 about the management of traffic safety on the road infrastructure. The Official Monitor of Romania. 2022 January 28: p. 1-8.

[15] Romanian Road Authority - A.R.R. www.arr.ro. [Online].; 2022 [cited 2024 February 10. Available from: https://www.arr.ro/arr_doc_230_rapoarte-si-studii_pg_0.htm.

IOP Conf. Series: Materials Science and Engineering **1311** (2024) 012050 doi:10.1088/1757-899X/1311/1/012050

# Case study of optimization of junction for the access to Iulius Mall Suceava

**A Vilcan[1], V Nicolae[2], I Martin[3], E Ionescu[4], M Istrate[5]**
[1]AV TRANSPORT PLANNING SRL, Ionel Perlea 11, Apt. 15, S1, Bucharest, Romania, adrian@vilcan.com
[2]University of Pitești, Faculty of Mechanics and Technology, Târgu din Vale Street, no.1, Pitești, Romania, viorel.nicolae@upit.ro
[3]AV TRANSPORT PLANNING SRL, Ionel Perlea 11, Apt. 15, S1, Bucharest, Romania, ivanamartin.mi@gmail.com
[4]Urban Trafic SRL, Bucharest, Romania, urbantrafic@yahoo.com
[5]University of Pitești, Faculty of Mechanics and Technology, Târgu din Vale Street, no.1, Pitești, Romania, mihaela.istrate@upit.ro

E-mail: viorel.nicolae@upit.ro ORCID ID : 0000-0001-7598-7434

**Abstract**. The impact of new commercial developments in urban areas is twofold, first one is given by the increased economic activity and increasing number of working places, and secondly increasing traffic in the area of development might lead to congestion. Therefore, it is of high importance the optimisation of access and junctions and street network in the area, to balance the two aspects mentioned before.

## 1. Introduction

This paper is based on the development, application and implementation of a transport model for testing and optimizing the traffic performance for the up-grade of a commercial development in Suceava, Romania.

The paper shows how the traffic performance can be enhanced through the optimisation of the junction configuration and control. The paper is focusing on the PM peak hour of a normal working day.

## 2. Model development and calibration

A local transport model has been developed in VISUM for assessing the current conditions and for optimizing the traffic performance in the area. The study area and local transport model structure is shown below.

IOP Conf. Series: Materials Science and Engineering **1311** (2024) 012050 doi:10.1088/1757-899X/1311/1/012050

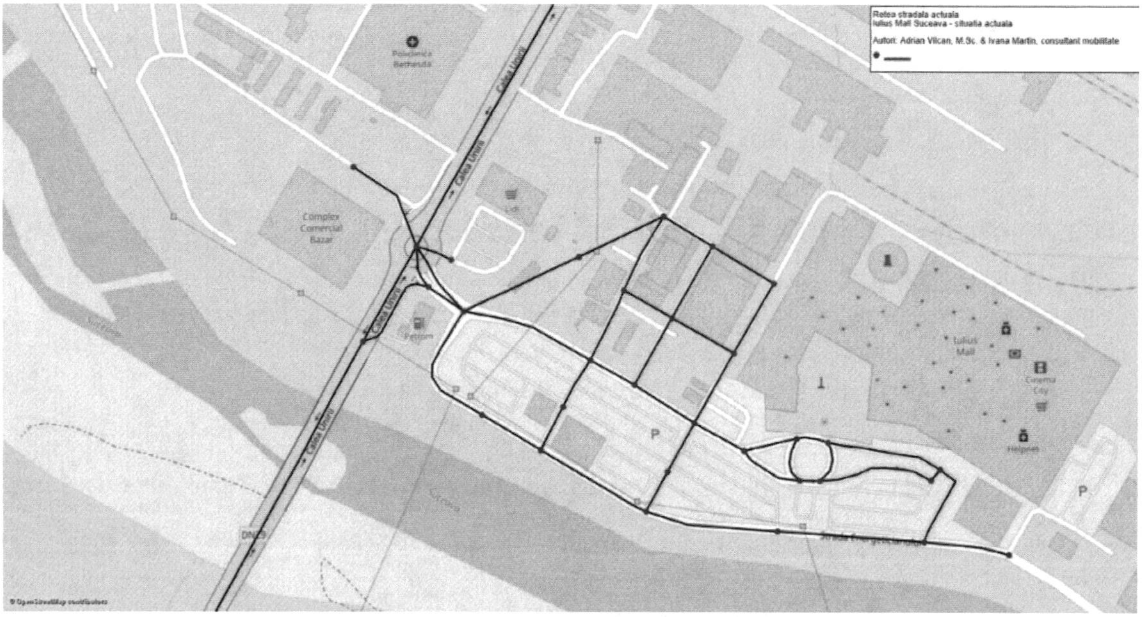


**Figure 1.** Study area and structure of the local transport model

In the following are shown the traffic flow and traffic performance – Volume/Capacity ratio and LOS in junctions for the PM peak hour.

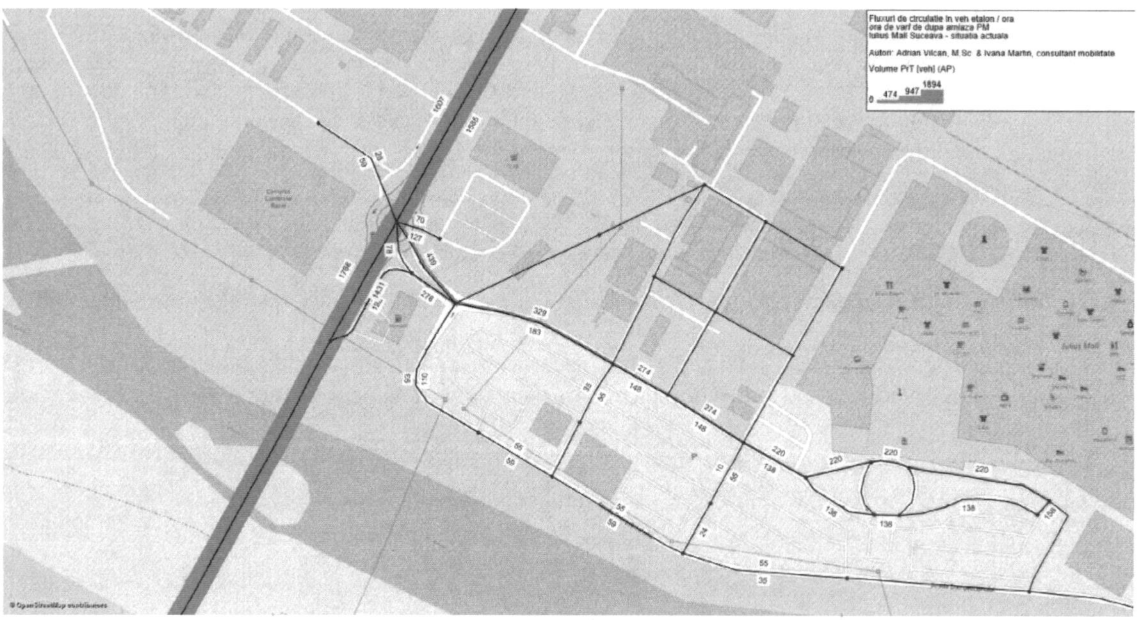


**Figure 2.** Traffic flows PM, PCU/hour, PM peak hour, current situation - calibration

It is observed that the traffic flows reach about 1,431 - 1,788 PCU per hour per direction on Calea Unirii. The incoming traffic flows to the Mall are 276 PCU/hour, and the outflows from the Mall are 439 PCU/hour.

IOP Conf. Series: Materials Science and Engineering **1311** (2024) 012050 doi:10.1088/1757-899X/1311/1/012050

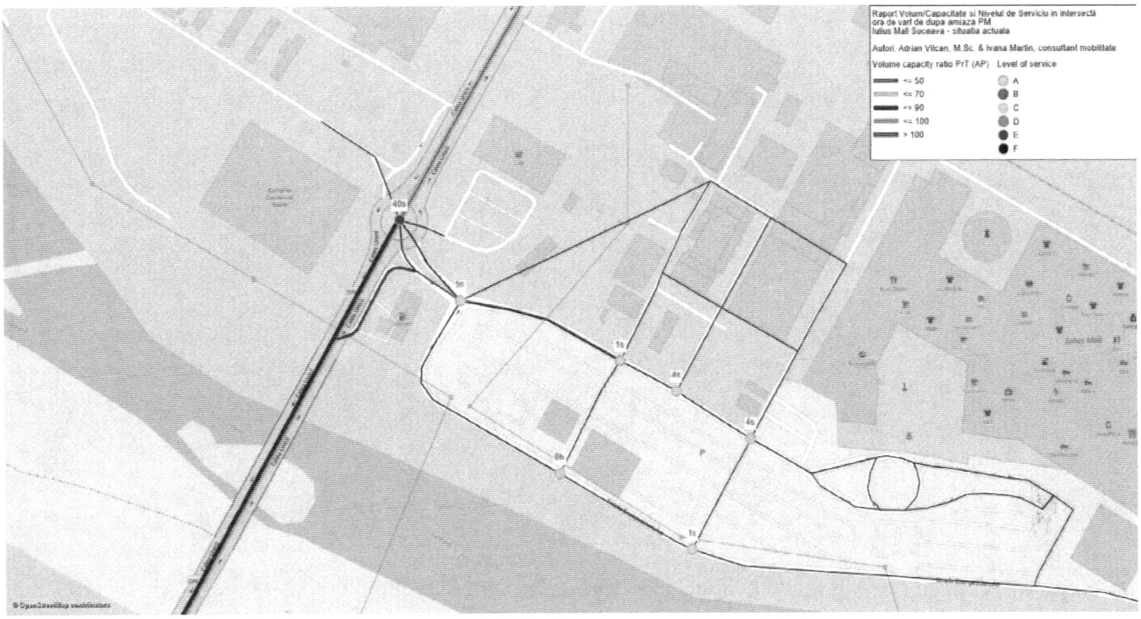


**Figure 3.** Volume / Capacity ratio and LOS in junctions, PM peak hour, current situation - calibration

It is observed that the capacity reserve is at least 10% on all road segments. The LOS is " E" at the access from Calea Unirii - the roundabout, with an average delay of 40 sec / PCU.

## 3. Scenario with development

First of all, the scenario with the up-graded development and without traffic performance optimization is presented.

The new generated and attracted traffic is as follows:

**PM peak hour**

Generated traffic: 566 PCU / hour

Attracted traffic: 878 PCU / hour

In the following are shown the traffic flow and traffic performance – Volume/Capacity ratio and LOS in junctions for the PM peak hour, with development, without optimisation of the traffic performance.

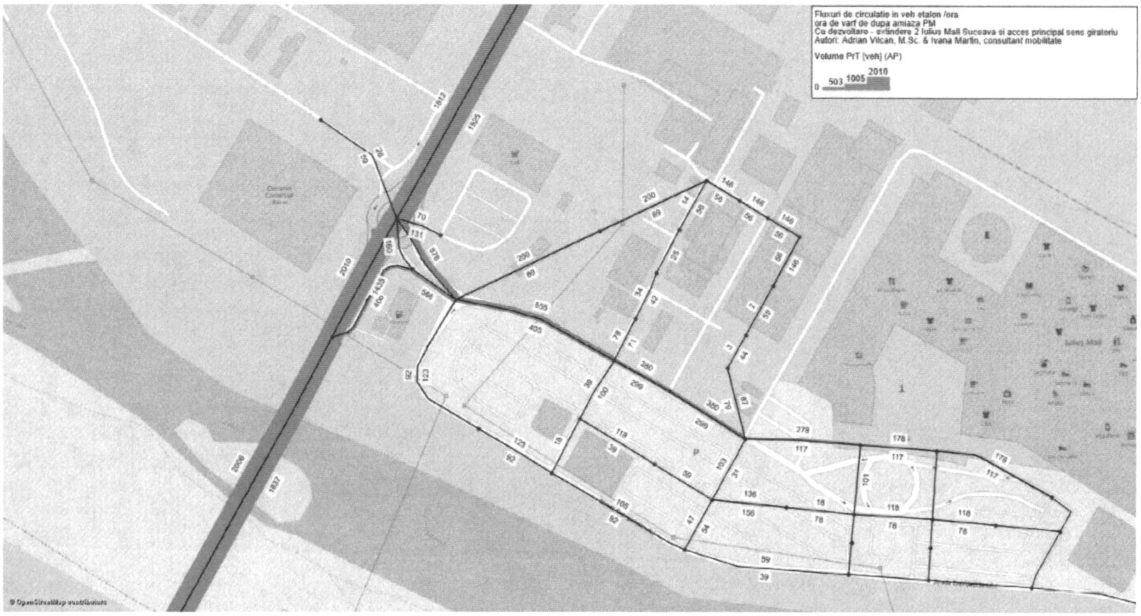


**Figure 4.** Traffic flows PM, PCU/hour, PM peak hour, with development up-grade, no optimisation of traffic performance

It is observed that the traffic flows reach about 1,812 – 2,006 PCU per hour per direction on Calea Unirii. The incoming traffic flows to the Mall are 566 PCU/hour, and the outflows from the Mall are 878 PCU/hour.

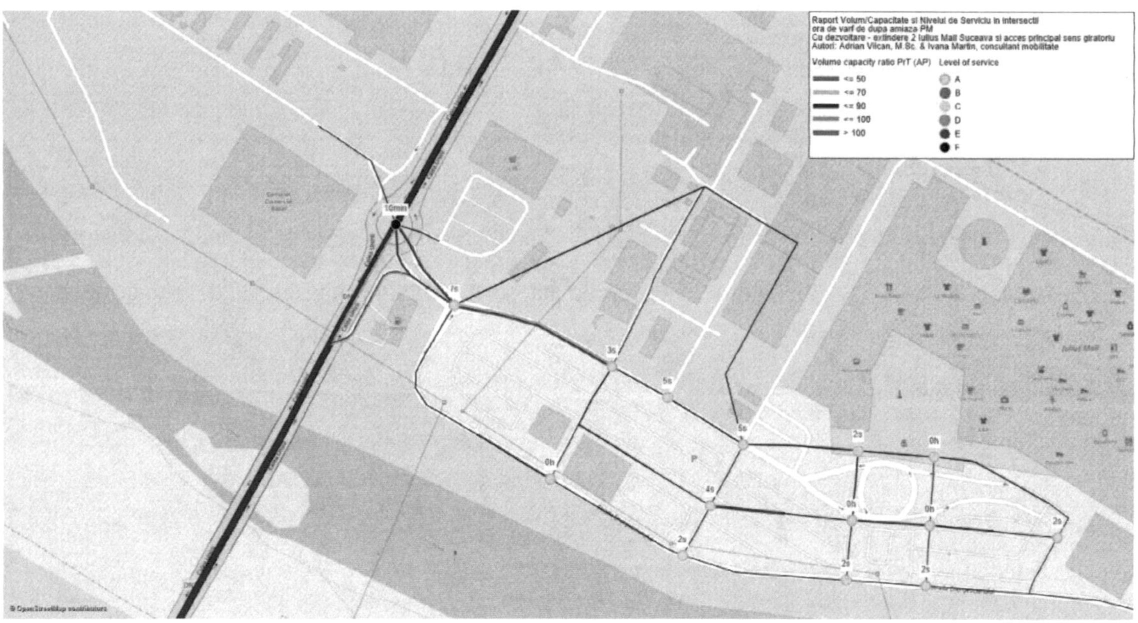


**Figure 5.** Volume / Capacity ratio and LOS in junctions, PM peak hour, with development up-grade, no optimisation of traffic performance

IOP Conf. Series: Materials Science and Engineering **1311** (2024) 012050 doi:10.1088/1757-899X/1311/1/012050

It is observed that the capacity reserve is at least 10% on all road segments. The LOS is " F" at the access from Calea Unirii - the roundabout, with an average delay of 10 min / PCU.
In order to optimize the traffic performance and reduce the delays in the roundabout from the access to the Mall in Calea Unirii, it is proposed a traffic light intersection and a lane dedicated to the left turn towards the Mall in Calea Unirii.

The proposed traffic light cycle for PM peak hour is shown below.

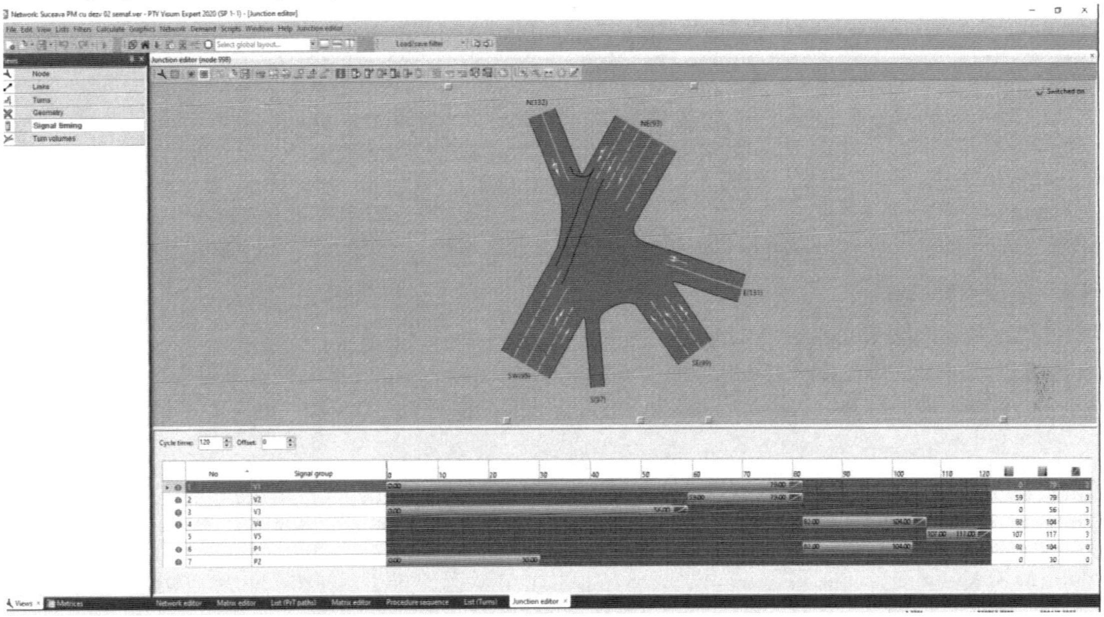

**Figure 6.** Optimised traffic light phases, PM peak

In the following are shown the traffic flow and traffic performance – Volume/Capacity ratio and LOS in junctions for the PM peak hour, with development, with the optimisation of the traffic performance.

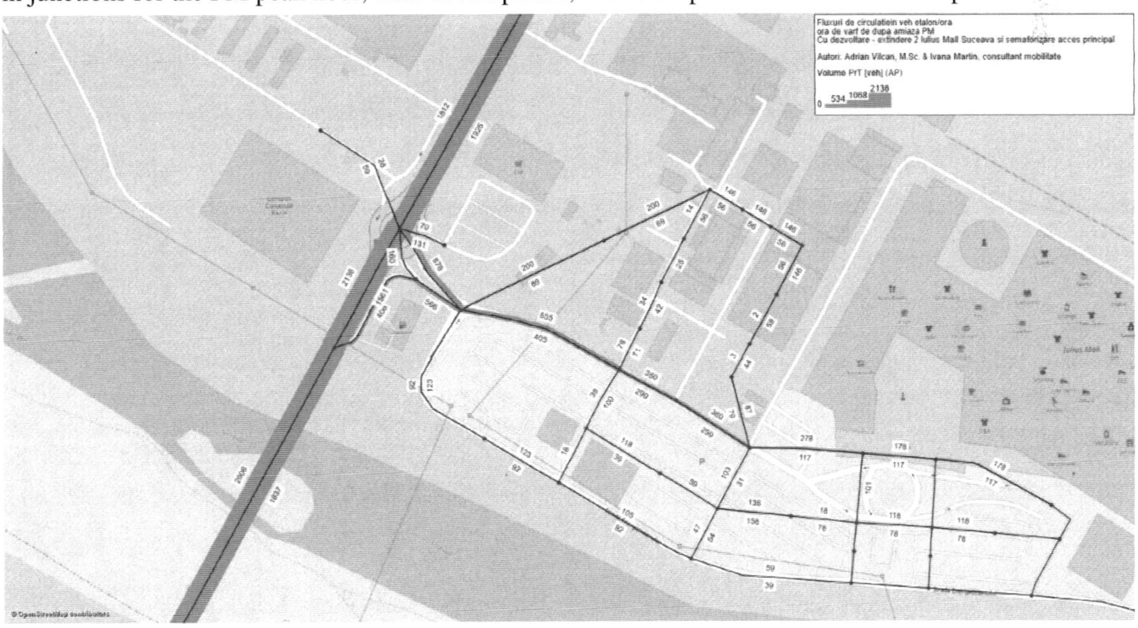


**Figure 7.** Traffic flows PM, PCU/hour, PM peak hour, with development up-grade, with optimisation of traffic performance

It is observed that the traffic flows reach about 1,812 – 2,006 PCU per hour per direction on Calea Unirii. The incoming traffic flows to the Mall are 566 PCU/hour, and the outflows from the Mall are 878 PCU/hour.

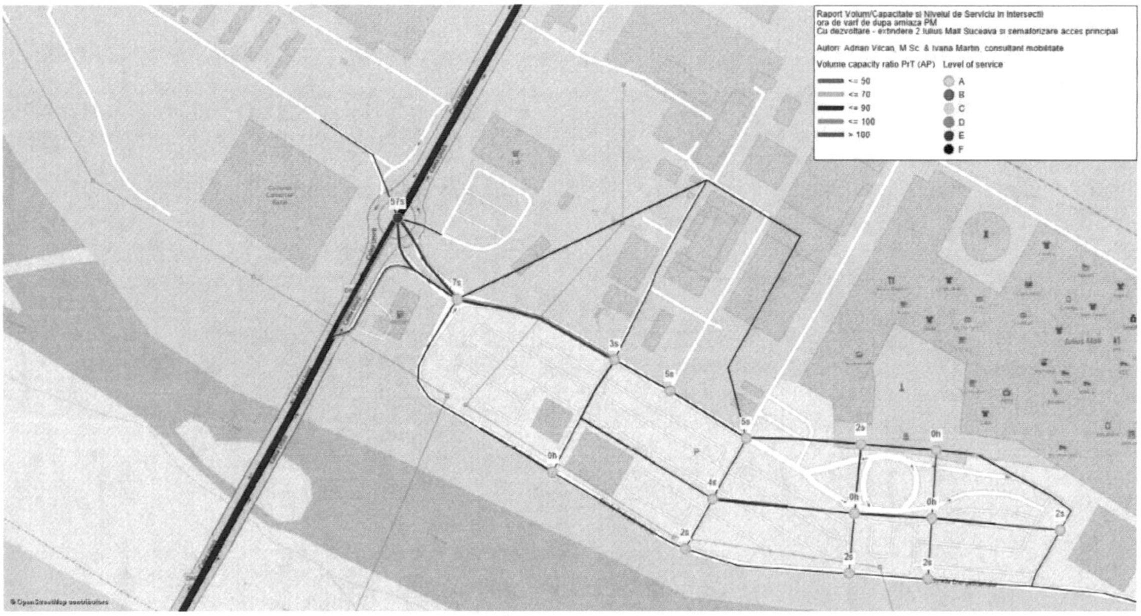


**Figure 8.** Volume / Capacity ratio and LOS in junctions, PM peak hour, with development up-grade, with optimisation of traffic performance

It is observed that the capacity reserve is at least 10% on all road segments. The LOS is " E" at the access from Calea Unirii - the roundabout, with an average delay of 57 sec/ PCU, thus a strong improvement compared to the scenario without optimisation, when the average delay is estimated at10 min / PCU.

In the following is presented the detailed LOS in the optimized junction.

IOP Conf. Series: Materials Science and Engineering **1311** (2024) 012050 doi:10.1088/1757-899X/1311/1/012050

Node 998:

| | |
|---|---|
| Control Type | Signalized |
| Method | HCM 2000 |
| Average Delay | 57.2 |
| Average LOS | E |
| V/C | 0.94 |
| Loss Time | 12 |

**Volume and Adjustments by Movement**

| Approach | N | | | | NE | | | | E | | | | SE | | | | S | SW | |
|---|---|---|---|---|---|---|---|---|---|---|---|---|---|---|---|---|---|---|---|
| Movement | L2 | L1 | T | R1 | L2 | L1 | T | R1 | L2 | L1 | R1 | R2 | L1 | T | R1 | R2 | | L1 | T |
| Base Volume | 27 | 0 | 0 | 32 | 123 | 160 | 1501 | 28 | 0 | 37 | 0 | 33 | 566 | 0 | 304 | 8 | | 0 | 1561 |
| PHF, Peak-hour factor | 1.000 | 1.000 | 1.000 | 1.000 | 1.000 | 1.000 | 1.000 | 1.000 | 1.000 | 1.000 | 1.000 | 1.000 | 1.000 | 1.000 | 1.000 | 1.000 | | 1.000 | 1.000 |
| Peak 15 Volume | 7 | 0 | 0 | 8 | 31 | 40 | 375 | 7 | 0 | 9 | 0 | 8 | 142 | 0 | 76 | 2 | | 0 | 390 |
| Adjusted Volume | 27 | 0 | 0 | 32 | 123 | 160 | 1501 | 28 | 0 | 37 | 0 | 33 | 566 | 0 | 304 | 8 | | 0 | 1561 |

**Volume and Adjustments by Lane Group**

| Approach | N | NE | | E | SE | | S | SW |
|---|---|---|---|---|---|---|---|---|
| Lane Group | C | L | C | C | L | C | | C |
| ID | 1709 | 1706 | 1707 | 1705 | 1702 | 1704 | | 1710 |
| Lanes | LTR | L | T, RT | LTR | L, L | RT | | T, T |
| Control Type | Perm | Prot | No Left | Perm | Split | No Left | | No Left |
| V, Volume | 59 | 283 | 1529 | 70 | 566 | 312 | | 1561 |
| PLT, Proportion Left Turns | 0.46 | 1.00 | 0.00 | 0.53 | 1.00 | 0.00 | | 0.00 |
| PRT, Proportion Right Turns | 0.54 | 0.00 | 0.02 | 0.47 | 0.00 | 1.00 | | 0.00 |

**Saturation Flow Rate**

| Approach | N | NE | | E | SE | | S | SW |
|---|---|---|---|---|---|---|---|---|
| Lane Group | 1709 | 1706 | 1707 | 1705 | 1702 | 1704 | | 1710 |
| Control Type | Perm | Prot | No Left | Perm | Split | No Left | | No Left |
| so, Base Saturation Flow Rate | 1900 | 1900 | 1900 | 1900 | 1900 | 1900 | | 1900 |
| N, Number of Lanes | 1 | 1 | 2 | 1 | 2 | 1 | | 2 |
| fw, Lane Width Adjustment | 0.983 | 0.983 | 0.983 | 0.983 | 0.983 | 0.983 | | 0.983 |
| Phv, % Heavy Vehicles | 0.000 | 0.000 | 0.000 | 0.000 | 0.000 | 0.000 | | 0.000 |
| fHV, HV Adjustment | 1.000 | 1.000 | 1.000 | 1.000 | 1.000 | 1.000 | | 1.000 |
| fg, Grade Adjustment | 1.000 | 1.000 | 1.000 | 1.000 | 1.000 | 1.000 | | 1.000 |
| fp, Parking Adjustment | 1.000 | 1.000 | 1.000 | 1.000 | 1.000 | 1.000 | | 1.000 |
| fbb, Bus Blocking Adjustment | 1.000 | 1.000 | 1.000 | 1.000 | 1.000 | 1.000 | | 1.000 |
| fa, Area Type Adjustment | 1.000 | 1.000 | 1.000 | 1.000 | 1.000 | 1.000 | | 1.000 |
| fLU, Lane Utilization Adjustment | 1.000 | 1.000 | 0.952 | 1.000 | 0.971 | 1.000 | | 0.952 |
| fLT, Left Turn Adjustment | 0.801 | 0.950 | 1.000 | 0.746 | 0.950 | 1.000 | | 1.000 |
| fRT, Right Turn Adjustment | 0.927 | 1.000 | 0.997 | 0.936 | 1.000 | 0.850 | | 1.000 |
| fLpb, Left Turn Ped. Adjustment | 1.000 | 1.000 | 1.000 | 1.000 | 1.000 | 1.000 | | 1.000 |
| fRpb, Right Turn Ped. Adjustment | 1.000 | 1.000 | 1.000 | 1.000 | 1.000 | 1.000 | | 1.000 |
| Has Short Lane | | | | | | | | |
| s, Saturation Flow Rate | 1386 | 1774 | 3545 | 1305 | 3445 | 1587 | | 3555 |

**Capacity, Control Delay, and Level of Service Determination**

| Approach | N | NE | | E | SE | | S | SW |
|---|---|---|---|---|---|---|---|---|
| Lane Group | 1709 | 1706 | 1707 | 1705 | 1702 | 1704 | | 1710 |
| Control Type | Perm | Prot | No Left | Perm | Split | No Left | | No Left |
| V, Volume | 59 | 282.9 | 1529 | 70 | 566.2 | 311.8 | | 1561.2 |
| s, Saturation Flow Rate | 1386 | 1774 | 3545 | 1305 | 3445 | 1587 | | 3555 |
| c, Capacity | 115 | 296 | 2334 | 109 | 632 | 291 | | 1659 |
| g/C, Green / Cycle | 0.08 | 0.17 | 0.66 | 0.08 | 0.18 | 0.18 | | 0.47 |
| X, Volume / Capacity | 0.51 | 0.96 | 0.66 | 0.64 | 0.9 | 1.07 | | 0.94 |
| d1, Uniform Delay | 52.66 | 49.57 | 12.32 | 53.28 | 47.89 | 49.00 | | 30.43 |
| k, Delay Calibration | 0.50 | 0.50 | 0.50 | 0.50 | 0.50 | 0.50 | | 0.50 |
| d2, Incremental Delay, | 15.99 | 70.70 | 1.46 | 28.65 | 22.07 | 191.20 | | 15.14 |
| d3, Initial Queue Delay | 0.00 | 0.00 | 0.00 | 0.00 | 0.00 | 0.00 | | 0.00 |
| Rp, Platoon Ratio | 1.00 | 1.00 | 1.00 | 1.00 | 1.00 | 1.00 | | 1.00 |
| P, Proportion Arriving on Green | 0.08 | 0.17 | 0.66 | 0.08 | 0.18 | 0.18 | | 0.47 |
| PF, Progression Factor | 1.00 | 1.00 | 1.00 | 1.00 | 1.00 | 1.00 | | 1.00 |
| d, Delay | 68.64 | 120.27 | 13.78 | 81.93 | 69.96 | 240.20 | | 45.57 |
| LOS | E | F | B | F | E | F | | D |
| dA, Approach Delay | 68.64 | 30.40 | | 81.93 | 130.41 | | 0.00 | 45.57 |
| Approach LOS | E | C | | F | F | | A | D |
| dI, Intersection Delay | 57.2 | | | | | | | |
| Intersection LOS | E | | | | | | | |

**Back of Queue**

| Approach | N | NE | | E | SE | | S | SW |
|---|---|---|---|---|---|---|---|---|
| Arrival Type | 3 | 3 | | 3 | 3 | | 3 | 3 |
| Lane Group | 1709 | 1706 | 1707 | 1705 | 1702 | 1704 | | 1710 |
| g/C, Green / Cycle | 0.08 | 0.17 | 0.66 | 0.08 | 0.18 | 0.18 | | 0.47 |
| PF2, Progression Factor | 1.00 | 1.00 | 1.00 | 1.00 | 1.00 | 1.00 | | 1.00 |
| Q1, First-term Queued | 1.88 | 9.35 | 16.08 | 2.26 | 9.50 | 10.39 | | 25.99 |
| Q2, Second-term Queued | 0.32 | 6.53 | 3.02 | 0.52 | 4.38 | 16.11 | | 13.32 |
| XU, Upstream V/C | 0.00 | 0.00 | 0.00 | 0.00 | 0.00 | 0.00 | | 0.00 |
| I, Upstream Adj. Factor | 1.00 | 1.00 | 1.00 | 1.00 | 1.00 | 1.00 | | 1.00 |
| kB, Early Arrival Adj. Factor | 0.31 | 0.60 | 1.61 | 0.30 | 0.64 | 0.59 | | 1.27 |
| Q, Average Back of Queue | 2.20 | 15.88 | 19.10 | 2.78 | 13.87 | 26.50 | | 39.31 |
| fB%, % Factor | 1.82 | 1.52 | 1.51 | 1.79 | 1.53 | 1.50 | | 1.50 |
| Q%, Back of Queue Percentile | 90 | 90 | 90 | 90 | 90 | 90 | | 90 |
| % Back of Queue | 4.01 | 24.15 | 28.86 | 4.97 | 21.24 | 39.82 | | 58.98 |

**Figure 9.** Detailed LOS, PM peak hour, with development up-grade, with optimisation of traffic performance

## 4. Conclusions

In the optimization scenario, it is considered a signalised junction at the main access to the Mall as previously presented.

Thus, it is observed that:

For PM peak hour:

- It is observed that the traffic flows reach about 1,812-2,006 standard vehicles per hour per direction on Calea Unirii. The incoming traffic flows to the Mall are 566 standard vehicles / hour, and the outflows from the Mall are 878 standard vehicles / hour;
- It is observed that the capacity reserve is at least 10% on all road segments;
- The Service Level is E at the access from Calea Unirii - the traffic light roundabout, with an average delay of 57 sec / standard vehicle.

From the above it is observed that in the scenario with the traffic lights of the access to Iulius Mall, the traffic performance is substantially improved, the average delays at each peak hour being reduced by about 8-9 minutes / standard vehicle. Thus, it is recommended to configure and traffic light access as presented above.

## References

[1] Vilcan A et al. 2012 *Update of feasibility study & preparation of traffic model for a new bridge over Danube* (TCS 30376) AV Transport Planning SRL / Komobile AT / ISEPP .
[2] NEA Transport Research and Training bv for Senter Agency of the Dutch Ministry for Economic Affairs, 2004 *Development of the National Road Transport Model for Romania*.
[3] * * * 2000 *Highway Capacity Manual* Institute of Transportation Engineers ISBN 0-309-06681-6.
[4] * * * 1999 Transportation Planning Handbook Inst. of Transp. Engineers ISBN: 0-935403-33-7.
[5] Traffic Impact Study Iulius Mall Suceava, Urban Traffic SRL – Adrian Vilcan, Ivana Martin & Eugen Ionescu, 2022

IOP Conf. Series: Materials Science and Engineering **1311** (2024) 012051 doi:10.1088/1757-899X/1311/1/012051

# The Hydrogen Vehicle in the Context of the Sustainable Development of the Green Road Transport System

**A M ȚÎȚU[1,*], G NEAMȚU[2], A B POP[3], C BULGARIU [2]**

[1]Lucian Blaga University of Sibiu, 10 Victoriei Street, Sibiu, România
[2]University Politehnica of Bucharest, Faculty of Industrial Engineering and Robotics, 313 Splaiul Independenței, 6th District, Bucharest, Romania
[3] Technical University of Cluj-Napoca, Northern University Centre of Baia Mare, Faculty of Engineering - Department of Engineering and Technology Management, 62A, Victor Babes Street, 430083, Baia Mare, Maramures, Romania

mihail.titu@ulbsibiu.ro

**Abstract.** The scientific work presents concrete research carried out by the authors to implement some concepts of technical and managerial nature, so that those interested in this field can get acquainted with a certain way of exposing the problems concerning the hydrogen-powered vehicle, in the context of the sustainable development of the ecological road transport system. At the moment, the car whose energy for propulsion is generated by means of hydrogen or hydrogen fuel cells does not enjoy a high popularity among users, due to the very high purchase price (where we have high technology, we also have a price to match - a fuel cell is more expensive than an electric battery), the reluctance of the population to use this type of fuel, defined by the danger of explosions and fires generated by hydrogen, and the lack of hydrogen refueling stations in Romania and in some European countries. The use of alternative fuels in internal combustion piston engines has the advantage of significantly reducing pollution due to combustion products. With regard to the direct use of hydrogen as a fuel in internal combustion piston engines, we can say that it has proved its inefficiency, because the hydrogen injected or sucked directly into the engine cylinders is unstable, produces uncontrolled burns and detonations, and in the liquid state it is very difficult to store, it requires special transport methods, hydrogen tanks must be armored with Kevlar because hydrogen in the event of a car accident can explode. For example, if we use it as a fuel in a diesel engine, hydrogen may not ignite itself because the self-ignition temperature in the cylinders of compression-ignition engines is high (600°C). However, at a normal stoichiometric ratio, hydrogen has a high flame rate, much higher than gasoline. This creates a thermodynamic cycle of the thermal engine almost perfectly, but ideally, let us not forget that in poor mixtures of air-hydrogen the flame propagation speed is considerably reduced. The compression ratio of the engine is of utmost importance because it is finally defining the thermal efficiency of the engine. In the following, we present a technical and managerial study with engineering nuances, through which readers can look and understand the stage of development, place and role of the vehicle propelled by hydrogen fuel cells. Finally, the conclusions and further directions of research and development in the field addressed are presented.

## 1. Introduction

Despite efforts by the authorities to reduce air pollution and greenhouse effect, they are far from being resolved. International bodies and institutions are constantly calling on transport organizations and car manufacturers to take all appropriate measures to reduce pollutant emissions and to switch to zero-emission vehicles [1]. In other respects, carbon-based fuels are becoming more limited each year, with liquid fuels being concentrated mainly in the Middle East, which is currently a politically unstable area [1]. Experts estimate that these resources are still available to mankind for about 35-40 years old. That is why the environmentally friendly, green car using green energy for propulsion is high on the agenda of engineers and researchers in the automotive sector. Among the viable solutions are increasingly being circulated ideas related to the use of hydrogen, a gas that is very widespread in nature and which by burning does not produce polluting emissions, being practically inexhaustible. Recent studies give us indication that if we move to the widespread use of hydrogen, we will have a much cleaner and less costly environment [1], [2]. Some experts believe that stronger public policies and educational efforts are needed to accelerate the process [1], [2]. Found in the water composition of the Earth's oceans and seas, experts have concluded that one km3 of water contains 113.108 tons of hydrogen [1]. To be used as a fuel in high-powered transport engines, it must meet the following conditions [1]: it must offer an equivalent level of maneuverability and a range and safety like that of the conventional cars (petrol, diesel, LPG, methanol, etc.). If we were to analyze the origin of hydrogen as a fuel, we agree with specialists who state that the disadvantages of hydrogen consumption are high carbon intensity emissions when produced from natural gas, oil or coal, where the following must be taken into account: the burden of capital cost, low energy content per unit volume under environmental conditions, production and compression of hydrogen, and the investment required in refueling stations for its distribution [3], [4], [5]. To power transport vehicles, hydrogen can be used directly in the piston engine as a fuel, in a mixture with air, or as an electricity generator in fuel cells. Introduced directly as fuel into the cylinders of a spark-ignition thematic engine, research has shown some deficiencies in its operation. We, the authors, we will try to explain and demonstrate the factors that cause the malfunctions in the operation of car engines that use hydrogen directly as a fuel.

## 2. References, studies, and research on the use of hydrogen as fuel in piston thermal engines

Hydrogen is a colorless, odorless, and tasteless gas, has a density to air = 0,06984 [1], is the lightest gas, has the letter H as its symbol, and can easily pass (diffuse) through extremely small orifices or spaces [1]. The first work and attempts at using hydrogen in piston engines date back to 1820, when Reverend Wiliam Cecil, a professor at the University of Cambridge, presented to the Cambridge philosophical society the use of hydrogen to produce power for mechanical installations. The engine presented on the vacuum principle, where power was produced by atmospheric pressure, pushing a piston in the direction of the vacuum. A partial vacuum is created by burning a hydrogen/air mixture, which is released and then cooled. Although the engine worked satisfactorily, vacuum engines did not find their usefulness and efficiency in practice [1]. There were then further attempts by Nikolaus August Otto, whose research was later continued by the German engineer Wilhelm Maybach. Otto found that the use of this fuel in combination with gasoline improved thermal efficiency at part loads [1] but abandoned the combination as dangerous. Of course, there were other attempts, such as those of German engineer Rudolf Erren, but to no avail. After all these experiments, given the energy-density ratio, hydrogen was used mainly as a fuel for space programs. Given the possible depletion of carbon-based fuels, it was not until the 1970's that hydrogen was given importance as a fuel for transport propulsion. Both spark ignition and Diesel engines behave differently when hydrogen is used as fuel directly in their cylinders, generating additional mechanical stress on the engine mechanism. This is determined by the pressure gradient, but also by the maximum pressure force applied to the piston after combustion. Experts estimate that these engines have a much higher efficiency than those that burn gasoline in their cylinders. Their efficiency is 35% higher and is obtained by making poor mixtures, compared to the air-gasoline mixture which has the value of ~ 1,1 [1], [6], the value $\lambda$ being contained in the so-called lambda window - the range 0,99-1,05, where the ideal value is $\lambda = 1$). We know that, to ensure complete combustion of the fuel introduced into the engine (gasoline, diesel, liquefied petroleum gas, biofuels, etc.), a certain amount of oxygen is needed. In this context, to completely burn one part (e.g., 1 kg of petrol, diesel fuel, LPG), 14,7 parts of air (e.g., 14,7 kg of air) are required. If this ratio is kept at the exact value of 14,7:1, we can say that the mixture in the engine cylinders is stoichiometric. Lambda is the ratio of the ideal dosage to the actual dosage: $\lambda$= 1/enrichment=stoichiometric dosage/actual dosage. Thus, according to the relationship presented, we can have the following situations: when $\lambda < 1$, (0,8-0,9), we have a rich mixture, defined

IOP Conf. Series: Materials Science and Engineering **1311** (2024) 012051 doi:10.1088/1757-899X/1311/1/012051

by the amount of fuel that is in excess, the amount of oxygen being insufficient for a complete combustion; when $\lambda = 1$, we have an ideal stoichiometric ratio in which combustion is complete; when $\lambda > 1$ (1,02-1,03), we have a poor mixture, defined by the amount of oxygen that is in excess, the amount of fuel being insufficient for a complete combustion. Figure 1 shows the curve of the evolution of the power dosage of a gasoline-powered thermal engine. Regarding the use of hydrogen as a fuel in MAS cylinders, we agree with the specialists who show that there is an increase in the efficiency with mixture depletion which is explained, on the one hand by the reduction of the specific heat of the gas mixture, and on the other hand, the molar coefficient of variation m=(0,5+2.381): (1+2.381) when burning hydrogen in air increases with mixture depletion [18]. We appreciate that filling the cylinders of a conventional engine with hydrogen leads to the following problem: the hydrogen reduces the power, and the efficiency of the conventional engine decreases by up to 75-80%. According to some information, the average power and pressure in the hydrogen engine are 20-25% lower than those found in gasoline operation, but the decrease is probably slightly lower (18-20%) considering the improvement in combustion [1], [6]. This is due to hydrogen which, once in the engine cylinders in gaseous form (even if it is liquefied at high pressures and low temperatures, at atmospheric pressure it instantly returns to the gaseous state), occupies the low density it has, large spaces in the volume of the intake galleries and engine cylinders, thus reducing the volume, and the fuel, that is, the air that must combine with it to maintain the combustion, no longer has the space required for the mixture.

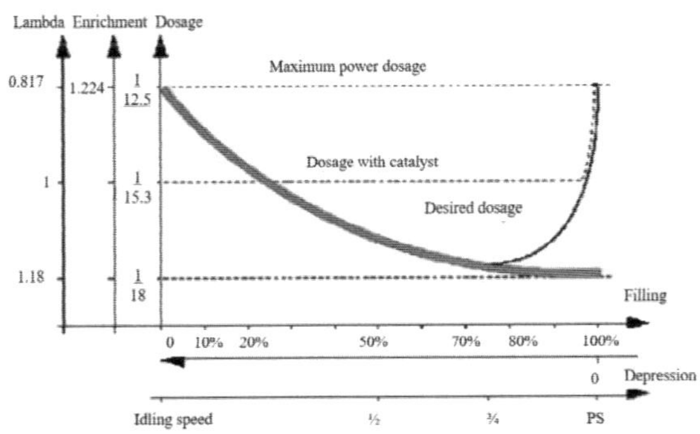


**Figure 1.** The curve of the evolution of the power dosage of a petrol-powered thermal engine [7].

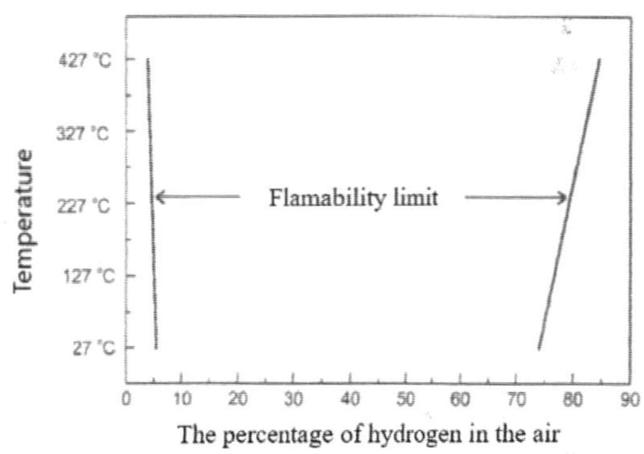


**Figure 2.** Variation of hydrogen flammability limit with temperature [1], [8].

Consequently, by adding hydrogen to the air admitted to the cylinder of an engine (when mixing is formed outside), a sensible reduction in the mass of the intake air is produced; at full load, for example, compared to the petrol-air mixture, the permissible air mass is reduced by about 30 % [1]. So, with a small amount of air being sucked in and a major volume expansion of hydrogen in the engine cylinder, a rich mixture of fuel results in a poor burn due to lack of oxygen. This results in reduced mechanical work, therefore, low power and low efficiency of the thermal engine. This impediment is supplemented by the self-ignition of excess fuel with a very high flammability index, which produces uncontrolled combustion. In the following, we will explain in detail each of these properties of hydrogen, but also how these properties make their mark on the proper functioning of the thermal engines powered directly by hydrogen. The wide flammability range of hydrogen is set between the lower (LII) and upper flammability (LIS) limits and varies with the temperature of the environment in which the gas evolves (figure 2). Looking at figure 2 we see that below the lower flammability limit the fuel-fuel mixture is lean, so there is not enough gaseous fuel to sustain combustion. Hydrogen has a wide flammability range compared to all other fuels. Consequently, hydrogen can burn in engines in several air-fuel mixtures. A significant advantage is that it can burn in a lean mixture. A lean mixture is one in which the amount of fuel is less than the theoretical, stoichiometric, or chemical ideal required for combustion (14, 7:1) with a certain amount of air [1], [9]. This leads to very light starting of spark ignition engines. Experts estimate that the combustion reaction is substantially improved when the engine is operated with a leaner air-hydrogen mixture, resulting in much better fuel economy due to the combustion reaction. It also results in a lower combustion temperature and a reduced amount of nitrogen oxides

IOP Conf. Series: Materials Science and Engineering **1311** (2024) 012051 doi:10.1088/1757-899X/1311/1/012051

(NOx) emitted into the atmosphere through the combustion gases. However, there is a limit to the lean mixture with which the engine can operate, which significantly decreases the engine power resulting in a reduction of the volumetric heating value of the air-fuel mixture [1]. The specific manifestations of hydrogen in engine cylinders are determined by the characteristics that define it. So, the characteristic of hydrogen combustion in air [1], are: the theoretical air requires for combustion, = 2.38 $Nm^3$ air/$Nm^3$ hydrogen; Lower calorific value = 10.760 KJ/$Nm^3$ and = 119.617 KJ/kg; Lower calorific value of the stoichiometric mixture ($p_0$ = 1,013x105 N/$m^2$, $T_0$ = 293k) = 3.183 KJ/Nm3 mixture; Minim ignition energy in air = 0.02mJ; Ignition temperature = 845K; Maximum velocity of the laminar flame in air relative to propane ($p_0$ = 1,013x105 N/m2, $T_0$ = 288k) = 6.79; Flamability limits ($p_0$ = 1,013x105 N/$m^2$, $T_0$ = 293k) = 2.96; Poor mixtures: % (vol) λls = 10; Rich mixtures: 75 % (vol) and λlb = 0.394. From the analysis of the data presented above, we can see that hydrogen has up to 8 times higher penetration speed compared to other fuels (e.g., gasoline or methanol). The characteristic confirms what the authors have previously stated. Hydrogen combines very easily with oxygen in the presence of a cold catalyst, the calorific value of its mixture with air being close to that of air and petrol. The low ignition energy of hydrogen is much lower than that of petrol. An important aspect to consider here is the detonation resistance of the spark ignition engine, which burns hydrogen in its cylinders. The low ignition energy produces hot gases from the combustion, which in turn create hot spots on the surface of the engine cylinders. They produce auto-ignition and detonation. Due to the wide flammability range of hydrogen any hot spot can cause an auto-ignition of almost any air fuel mixture [1]. We consider that the elimination of these auto-ignitions, is one of the major challenges for the correct operation of the thermal engine, using hydrogen as fuel. Another major shortcoming of hydrogen is the short burn time in piston engines. In this respect we agree with the experts that the burn time of hydrogen is shorter than that of petrol. As a result, hydrogen flames reach the cylinder walls closer than other fuels before they are extinguished [1]. This makes the combustion of hydrogen in the cylinder special compared to the combustion of gasoline or methanol, which creates great problems for the correct operation of the thermal engine. In this case, the air-hydrogen mixture in the engine cylinder burns near the valves and becomes dangerous due to the leakage created by the rapid diffusion of hydrogen. Flame from the cylinder can enter the intake manifold near the inlet valve, causing the fuel mixture in the manifold to self-ignite and create explosions (misfires in the intake manifold). We know that the auto-ignition of an air-fuel mixture is determined by the minimum temperature required to ignite a fuel mixture when the ignition source is inactive. This phenomenon is in fact caused by the high temperature of the fuel mixture in the engine cylinder which is self-igniting (self-igniting flame). Each fuel has a unique ignition temperature and hydrogen has a high auto-ignition temperature (585°C) [1]. The temperature aspect of the air-fuel mixture is of particular significance when time 2 of an internal combustion engine is spent. In this case, we specify that the temperature created during compression is created by the temperature of the auto-ignition phenomenon itself, which determines the compression ratio specific to each thermal engine. Below, we present the auto-ignition temperatures specific to some fuels used in the thermic engines of motor vehicles: Hydrogen = 585°C; Methane = 54°C; Propane = 490°C; Methanol = 385°C; Benzine = 230-480°C. But the increase in self-ignition temperatures is given by the relationship [1], [8]:

$$T_2 = T_1(V_1/V_2)^{g-1} \tag{1}$$

were: $V_1/V_2$ – compression ratio; $T_1$ – absolute initial temperature; $T_2$ – absolute final temperature; Γ – specific heat. If the auto-ignition temperature does not reach the value specified above, the auto-ignition of the air-hydrogen mixture does not take place, thus the absolute final temperature limits the compression ratio. The high hydrogen auto-ignition temperature allows for a higher compression ratio than in fossil fuel engines [1]. Another characteristic that distinguishes hydrogen from other fuels (e.g., gasoline, methane, or propane) is high diffusibility. This creates an advantage over the fuels presented because it favors and evens the air-hydrogen mixture, and in uncontrolled leakage its dispersion is fulminant. Specialists estimate that due to its low density (light weight compared to other fuels), hydrogen creates some problems when used as fuel in internal combustion engines [1]: a) a very large volume of storage is required to give the vehicle adequate range; b) the energy density of the hydrogen-

IOP Conf. Series: Materials Science and Engineering **1311** (2024) 012051 doi:10.1088/1757-899X/1311/1/012051

air mixture and the engine power are reduced. To see the efficiency of hydrogen in internal combustion engines, we determine the stoichiometric ratio of this fuel (oxygen-hydrogen ratio), which is given by the formula [1]:

$$2H_2 + O_2 = 2H_2O \qquad (2)$$

We know that $H_2$ mol for a complete reaction = 2 molecules, and $O_2$ mol for a complete reaction = 1 molecule. Because air is used as an oxidizer instead of oxygen, nitrogen must also be considered. Therefore, $N_2$ mol/air = $O_2$ mol x (79% N2 in air/ 21% $O_2$ in air) = 1 mol $O_2$ x (79% $N_2$ in air/ 21% $O_2$ in air) = 3.762 mol $N_2$. Air mole number = $O_2$ mole + $N_2$ mole = 1 + 3.762 = 4.762 mol of air. Weight $O_2$ = 1 mol $O_2$ x 32g/mol = 32g. Weight $N_2$ = 3.762 mol $N_2$ x 28g/mol = 105.33 g; Weight $H_2$ = 2 mol $H_2$ x 2g = 4g. Air weight = 32g + 105.33g = 137.33g. In contrast, the stoichiometric air/fuel ratio (A/C) for hydrogen and air is: A/C mass = 137.33g/4g = 34.33: 1, i.e. 34:1, compared to gasoline which is 14,7:1; A/C by volume = 4.762 / 2 = 2,4 : 1 Therefore, the percentage (%) in the combustion chamber of a piston engine occupied by hydrogen for an ideal stoichiometric mixture is: 2/(4.762+2) = 29.6%. As a result, hydrogen being gaseous under normal temperature and pressure, due to its dispersive properties, occupies a higher volume than other fuels (e.g., gasoline) in the combustion chamber of the combustion engine. Under stoichiometric conditions, hydrogen occupies about 30% of the combustion chamber volume compared to gasoline which can occupy only 1-2% [1].

To justify this, in Figure 3, we present the volumes occupied by hydrogen in the cylinder of the combustion engine, namely gasoline, and the energy input of these fuels. From the analysis of the data shown in Figure 3, the following results: the power of the hydrogen-powered thermal engine compared to that of liquid fuels (e.g., gasoline) is different, being in the range [85% (indirect injection) and 120% (indirect injection)] so this is directly determined by the method by which the hydrogen intake is made; hydrogen-powered thermal engines can operate with stoichiometric (air-fuel) ratios in the range [34:1-180:1]. Given the shortcomings demonstrated using hydrogen as fuel in piston engines, we believe that it finds its perfect efficiency and effectiveness in hydrogen fuel cells. To demonstrate this, we present the construction and the principle of their operation.

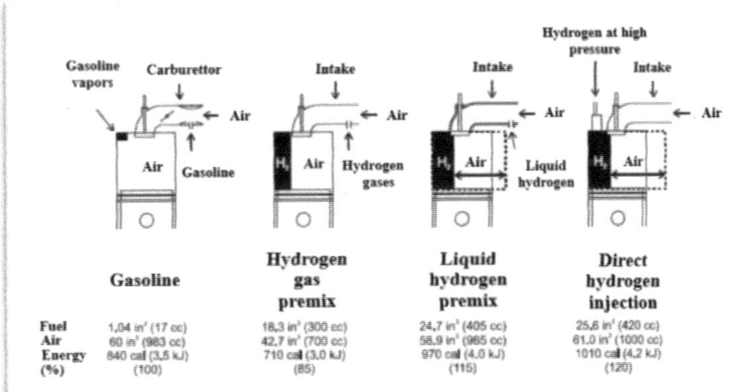


**Figure 3.** Volumes occupied in cylinders of hydrogen and petrol thermal engines and the energy input of these fuels, respectively [1].

## 3. Hydrogen fuel cells. Principle of operation

The operation of a fuel cell will be explained further, taking as an example the simplest fuel cell running on hydrogen and oxygen FUEL CELL HYDROGEN – OXYGEN. The irreversible kinetic processes associated with a fuel cell consist of a series of oxidation-reduction reactions.

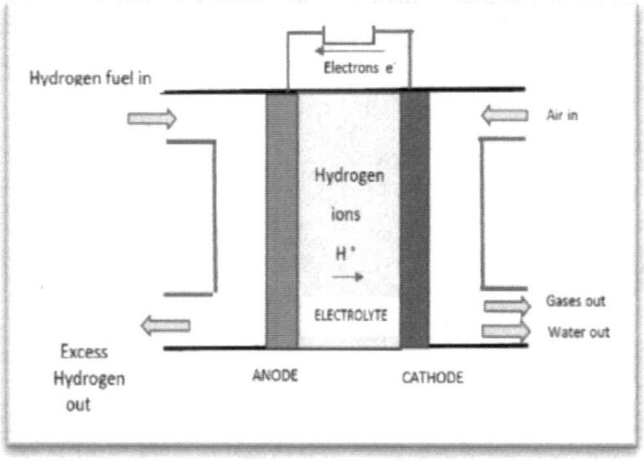


**Figure 4.** Schematic diagram of hydrogen fuel cell [12].

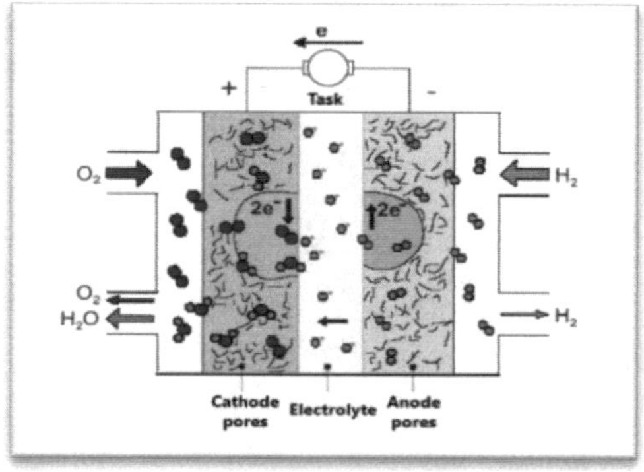


**Figure 5.** Principle of operation of a fuel cell [11].

A fuel A is transported to the porous anode where it is absorbed on its surface, then dissociated into ions and electrons in an oxidation process. After that, the migration of electrons from the anode occurs and the release of ionic gas to the anode surface. In the electrolyte, the transport of AZ+ ions from the anode to the cathode against the resulting electric field on account of the electrochemical imprinted field must be ensured. At the cathode, ions (arrived through the electrolyte), electrons (arrived through the outer circuit) and oxidizer B are found [10], [11]. The reduction reaction occurs, resulting in the reaction product to be eliminated. The fuel cell consists of three elements: The electrolyte, the electrodes, and the reactants (a fuel and an oxidizer). In figures 4 and 5, we schematically present the principle of operation of a hydrogen fuel cell. A fuel cell uses hydrogen as fuel, and as an oxidizer oxygen, alkaline electrolyte and electrodes that also play the role of catalysts. During operation, the electrodes do not undergo any structural changes, but they serve as support for the reaction. Catalytic oxidation of atomic hydrogen occurs at the anode, and catalytic reduction of atomic oxygen occurs at the cathode. Catalytic oxide return occurs in three phases (gas—liquid—solid) according to the overall reaction H2 + ½O2 =H2O and occurs at the surface of the catalyst. Individual fuel cells produce relatively small electrical potentials (about 0,7 V) and cells are usually connected in series to provide higher voltages or in parallel for high current with low voltage applications [12]. In general, the output voltage of a fuel cell decreases with the current load increases due to several factors including loss of activation (a fundamental element electrochemical phenomenon), internal resistance (analogue of the internal resistance of the battery) and mass transport losses due to depletion of anode and cathode reactants under high current conditions [12]. In addition to electricity, fuel cells produce water, heat, and very small amounts of other emissions [12]. The energy efficiency of a fuel cell is generally between 40-60% but if the waste heat can be applied usefully, efficiencies of up to 85% can be achieved [12]. Hydrogen fuel cells only demonstrate their efficiency if the hydrogen used for combustion is obtained from renewable sources (e.g., water electrolysis) [13], [14], [15].

## 4. Conclusions

Spark-ignition thermal engines fueled by hydrogen at air-hydrogen mixtures close to those of stoichiometric dosing have a brutal, harsh operation. If the fuelling is made by the pulsating introduction of hydrogen at the intake, the thermodynamic processes that take place in the engine are invariable, even in a variable range of the values obtained by $\lambda$ dosage coefficientWe believe, however, that specialists are right about the low power of the spark-ignition engine that uses hydrogen. It drops to 80%. It can only be increased by injecting hydrogen into the cylinder. In this case the power increase condition can only be achieved if the fuel mixture is formed inside the cylinders. The ideal value of the dosage coefficient l, it is obtained in the interval [2], [5], [8], analyzed in terms of specific consumption, indicated yield and nitrogen oxide emissions. At idling and low loads of the thermal engine using hydrogen fuel, the fuel consumption is lower than for petrol. The specialists also state that the use of hydrogen as a fuel in Diesel engines is not indicated because a substantial readjustment of the fuel system is necessary (a goal that can be achieved with high financial costs), and hydrogen used in vapor state is difficult to self-ignite and does not allow internal air-fuel mixture formations. If the air-hydrogen mixture is formed externally, ignition takes place much earlier. This is due to the high compression ratio values of these engines, defined by the ignition principle, but also to the thermal regime, which is much higher than in MAS. Direct injection of liquid hydrogen into diesel engines has been tried but the results have been disastrous. This was due to the very low auto-ignition temperature in the engine cylinder. (The cylinder temperature of a Diesel engine is 600 °C at ignition, and hydrogen self-ignites at 585 °C). As far as using hydrogen directly as a fuel in piston engines is concerned, the reality is much harsher than we would like because the biggest impediment, the biggest problem with this type of propulsion is still hydrogen as a fuel. Here are the arguments. Although it is a very common chemical element, it is not found in the free state, only in compounds. It can be extracted from these substances in three ways: thermochemically from natural gas; thermolysed from biogas obtained from the rotting of waste; and electrolysed from water. All three methods, we believe, are very expensive. The whole system of generating electricity for a hydrogen car is extremely expensive, and the installation involves certain risks. Pure hydrogen is known to be particularly dangerous as it is highly flammable, and the methods of separation/extraction, transport and storage require impressive logistics and a very large amount of fossil fuels to be produced in industrial quantities. Indeed the outcome of the processes described remains the same in the use of hydrogen for combustion in internal combustion engines, namely water. Taking into account the industrial manufacturing process, hydrogen worldwide is obtained 48% from natural gas by catalytic reforming [16], 30% from oil fractions [16], 18% from coal by gasification [9], [17], but the main source of hydrogen remains fossil fuels [17]. The reaction to obtain hydrogen results in carbon dioxide. This leads to the same situation as with electric cars: we do not pollute when we use them, but we pollute when we manufacture them. However, there is still 4% of the manufacturing process of green hydrogen left: the industrial process of manufacturing by electrolysis from water [18]. Fortunately, water is enough, but electricity is needed for the electrolysis extraction process. If the electricity is not from renewable sources, we find that we are back to the problem of environmental pollution. If hydrogen is produced from biomass or water using renewable energies, it is called *green hydrogen* [19]. Hydrogen in liquid form is very difficult to store, requires special transport methods, is stored at high pressures, and the tanks of hydrogen-powered vehicles are of special construction and must be shielded with a dowel. This is necessary because hydrogen can explode in the event of a car accident. The lines and connections of hydrogen fuel systems in cars must be very tight and away from all sources of heat because hydrogen ignites when it comes into contact with hot engine parts/parts. These are some reasons why people do not want such cars. Fuel cells can be included in *soft* energy systems because of the following characteristics: they produce direct current at low voltages and medium currents that can be used directly by end-users; they do not pollute the environment; they operate quietly, without vibration or noise, no moving parts, etc.

Considering the anticipated, we find that the vehicle powered by hydrogen fuel cells is far away, at least here on the European continent. Considering the large number of automobiles powered by the energy from electric batteries, which currently circulate on public roads, it seems that the hydrogen automobile has lost the start, but also the battle.

We are of the opinion that he will never make up for lost time. In addition to those previously shown, the major, ongoing problems that humanity will face worldwide with regard to the hydrogen car are economics, infrastructure and servicing. From what was shown previously, we believe that electric vehicles, compared to other types of vehicles, still represent, at this time, the most viable form of ecological transport in The World.

**References**

[1] https://pdfcoffee.com/utilizarea-hidrogenului-la-motoarele-cu-ardere-interna-1-pdf-free.html
[2] Dunn S 2002 Hydrogen futures: towards a sustainable energy system. Int. J. Hydrogen Energy, 27:235–264. January 2003 DOI: 10.1016/S0360-3199(01)00131-8.
[3] Cox, J 2014 Time To Come Clean About Hydrogen Fuel Cell Vehicles, CleanTechnica.com, June 4, 2014. Accessed on june, 29 2022.
[4] Davies, A 2013 Honda Is Working On Hydrogen Technology That Will Generate Power Inside Your Car. The Business Insider. November 22. Accessed on june, 29 2022.
[5] Berman, B 2013 Fuel Cells at Center Stage". The New York Times. Accessed on june, 29 2022.
[6] https://dantalmaciu.wordpress.com/2015/01/07/utilizarea-hidrogenului-in-transportul-rutier/.
[7] https://muhaz.org/instalaia-de-alimentare-cu-injecie-de-benzin.html?page=11.
[8] https://auto.howstuffworks.com/fuel-efficiency/fuel-consumption/fuel-processor.htm.
[9] Rapier, R 2020 Life cycle emissions of hydrogen. 4thgeneration.energy. May 26. 2020. Accessed at July 05.2022.
[10] Hacker, V and Mitsushima, S 2018 Fuel Cells and Hydrogen: From Fundamentals to Applied Research, Amsterdam, Elsevier.
[11] Hart, D, Howes, J, Madden, B. et al. 2016 Hydrogen and fuel cells: opportunities for growth, Report commissioned by Innovate UK, Department of Business, Energy and Industrial Strategy, Transport Scotland, Scottish Government, Scottish Enterprise, KTN, UKHFCA and SHFCA. E4tech and Element Energy. Accessed at July 05.2022 through: http://www.e4tech.com/reports/hydrogen-and-fuel-cells-opportunities-for-growth-a-roadmapfor-the-uk/.
[12] Murray-Smith, D 2019 Project: Rail transport applications of system modelling and simulation. Tehnical Report. A Review of Developments in Electrical Battery, Fuel Cell and Energy Recovery Systems for Railway Applications. A Report for the Scottish Association for Public Transport. November 30. 2019, University of Glasgow, DOI: 10.13140/RG.2.2.16555.67362. Accessed at July 06.2022. through: https://www.researchgate.net/publication/337991572.
[13] Carmo, M, Fritz, D, Mergel, J, Stolten, D, 2013 A comprehensive review on PEM water electrolysis. Journal of Hydrogen Energy. 38 (12):4901-4934. DOI:10.1016/j.ijhydene.2013.01.151.
[14] The European Commission 2010, A European strategy on green and energy-efficient vehicles. Communication from the Commission to the European Parliament, the Council and The European Economic and Social Committee, Publisher: Journal of the European Parliament, Pages 1-15, Brussels, COM(2010)186 final/28.04.2010, Acessed at July 06.2022, through: https://eur-lex.europa.eu/legal-content/RO/TXT/PDF/?uri=CELEX:52010DC0186&from=EN.
[15] The European Commission 2007 Establisment of the Joint Undertaking Fuel cells and hydrogen, P6_TA(2008)0203, European Parliament resolution of May 20, 2008 on the proposal for a Council regulation setting up the "Fuel Cells and Hydrogen" Joint Undertaking (COM(2007)0571 - C6-0446/2007 - 2007/0211(CNS)) (2009/C 279 E/29), Official Journal of the European Union, 19.11.2009, C 279 E/121-144, accessed an July 13, 2021 by https://eur-lex.europa.eu/LexUriServ/LexUriServ.do?uri=OJ:C:2009:279E:0121:0144:RO:PDF.
[16] Häussinger, P, Lohmüller, R, Watson, AM 2011 Hydrogen, 1. Properties and Occurrence. Ullmann's Encyclopedia of Industrial Chemistry. DOI: 10.1002/14356007.a13. P. 297. pub2. ISBN 978-3-527-20673-2.
[17] Liu, K, Song, C, Subramani, V 2009 Hydrogen and Syngas Production and Purification Technologies. DOI 10.1002/9780470561256. ISBN 9780470561256. November. 30.2009.
[18] Roman J, Santhanam, KSV, Miri, Massoud, J, & all 2008 Introduction to hydrogen Technology. John Wiley & Sons. p. 249. ISBN 978-0-471-77985-8.
[19] Purtil, J 2021 What is green hydrigen, how is it made and will it be the fuel of the future? ABC News. Australian Broadcasting Corporation. Accessed at July 06. 2022, through: https://www.abc.net.au/news/science/2021-01-23/green-hydrogen-renewable-energy-climate-emissions-explainer/13081872.

IOP Conf. Series: Materials Science and Engineering **1311** (2024) 012052 doi:10.1088/1757-899X/1311/1/012052

# Overview on potential, opportunities and trends for internal combustion engines

**S A Rațiu[1], N L Mihon[2] and D M Armioni[1]**

[1] Politehnica University of Timişoara, Faculty of Engineering Hunedoara, Revoluției Street 5, 331128, Hunedoara, Romania
[2] Politehnica University of Timişoara, Faculty of Mechanical Engineering, Mihai Viteazu Street, 1, 300222, Timişoara, Romania

sorin.ratiu@fih.upt.ro

**Abstract**. As one of the most influential markets on the planet, the automotive industry is expanding and, despite the increase in the number of electric cars, its internal combustion engine sector is no exception. Thus, it needs to regularly adapt to the latest trends and changes imposed by the technological innovations that are constantly appearing. Also, in addition to the technical aspect, environmental trends must be taken into account, because they directly influence the evolution of the automotive field, leading it towards a greener future, with an emphasis on ecological and sustainable solutions for current problems. This article aims to offer an overview on the most relevant trends at European level, manifested in the field of sustainability, urbanization, individualization and digitalization, which make possible the restructuring process of the automotive industry, successfully integrating it in the sustainable sectors of the industrial economy. Also, the current situation regarding the European and Romanian automotive sector is presented and there are exemplified a number of ways in which internal combustion engines can be optimized for sustainability, in order to improve the performance of vehicles and their integration into the ecological trends of the moment.

## 1. Introduction
Until now, diesel engine technology has been the mainstay of the automotive industry in Europe. But recent years have brought harsh criticism for this technology in European urban areas. In other words, the automotive industry is currently undergoing a process of transformation by experimenting with other technologies, which means that the usual procedures and conventions have become obsolete.

In this sense, the imminent restructuring of the automotive industry must be understood in the sense of being part of a transformation process of the transport sector. And this process, in turn, is part of a fundamental process of transformation related to the international approach to sustainability.

## 2. Megatrends in the field of mobility
In order to understand the dynamics of the transformation process that the automotive industry is currently facing, it is necessary to develop an analysis of the global trends that shape the present.

A trend is a concept that describes the changes manifested in all social areas, a concept that allows us to make statements about possible developments in the future. Trends are observable, but difficult to measure accurately. We can appreciate their evolution, but we can only partially influence them. According to futurist John Naisbitt, the notion of megatrends describes trends characterized by a specific intensity, persistence and amplitude. In the case of the automotive industry, the interaction between megatrends that is currently relevant generates, in turn, concrete trends from a mobility perspective [1].

The four megatrends that facilitate, but also require large-scale innovations in the field of automotive technology and mobility, are those related to sustainability, urbanization, individualization and digitization, illustrated in figure 1.

**Figure 1.** The four megatrends that transform the mobility sector

### *2.1. Sustainability*

In general terms, sustainability is defined as the ability to meet the needs of the present without compromising the resources of the future. More specifically, sustainability can be considered a collective notion that describes all attempts to reduce the external effects generated by industrial production, fossil fuels and consumption methods. The main purpose of this term, in harmony with socio-cultural and economic interests, is to control these effects in a sustainable manner [2].

According to the permanent message sent by the World Transport Outlook publication, the volume of traffic worldwide will increase at least three times by 2050 [3]. Given this situation, experts have yet to find an answer to the question of how the mobility requirements of up to 12 billion people can be met in a truly sustainable manner in the near future.

The only consensus reached is that if development remains unregulated, there will be a sharp increase in fossil fuel consumption, greenhouse gas emissions, air pollutants and particulate matter, noise pollution, costs related to accidents, and above all the material and space requirements related to mobility. Regarding modern automobiles, a specific problem is represented by vehicle emissions. The diesel engine is currently facing a specific critical current, and this technology has given rise to a dilemma between climate protection and public health protection, a dilemma that is unlikely to be resolved in the long term. The internationally agreed goals for climate protection - an extensive decarbonisation process by 2050 - can only be met in the mobility sector if, for example, combustion engines are banned from 2035 [4]. In this sense, the European Union proposed in June 2022 a ban on the sale of new cars with internal combustion engines. The measure would enter into force in 2035, as part of an extensive climate program that aims to reduce carbon dioxide emissions [6].

At the same time, at the EU level, clear objectives are established regarding the reduction of emissions, aimed at local efficiency from the point of view of vehicle emissions, these objectives representing legal regulations that must be respected.

*2.2. Urbanization*
In the 21st century, densely populated urban areas will represent the typical form of existence for the majority of the world's population. As the number of people in the situation of having to take care of their various needs and vital functions of life in ever smaller spaces increases, the functional space available for the form of motoring that we know today becomes more and more insufficient. In the world's rapidly developing metropolitan regions, there simply isn't enough space for expanding car parks and the externalities they produce [2].

*2.3. Individualization*
The term "individualization" refers to the transition from the determination of people through the prism of their external social environment to internal self-determination. The basic rule in the sociology of transport is that the more developed a society is, the greater the degree of individualization and the more specific, flexible and spontaneous the behavior in the transport sector, which makes the demand less and less receptive to grouping practices.

The use of automobiles has become increasingly pronounced in all locations where individualization is maintained. The car has become a means optimally adapted to this megatrend, both functionally and symbolically, i.e. in the sense of social differentiation and establishing one's own style through demonstrative consumption.

As it is possible that the most important economic misallocations related to ownership are manifested, to date, in the mobility sector, the dynamics underlying the development of the trend of sharing products for the purpose of efficient use are strongest at the time of face, also in this sector [3]. For example, in Germany a car is used on average one or two hours a day, so it remains unused for 22-23 hours.

*2.4. Digitization*
Digitization offers a diverse range of approaches and opportunities to address development challenges in the transport sector, challenges that are generated as a result of other megatrends.

It is expected that these opportunities will manifest themselves due to three possible effects of digitization: first, the enormous ability to increase efficiency in the use of transport infrastructures and vehicle fleets; second, automating and, as a result, optimizing command functions previously performed by humans; thirdly, highly efficient supply and demand regulation through networking technology, smart end-user devices, software applications and new service concepts in a platform system [2].

*2.5. The reflection of megatrends on the automotive sector*
Trends in the mobility sector may be more clearly differentiated spatially and temporally compared to global megatrends, but studies have found that they follow the same direction worldwide. Their emergence can be considered a result of the interaction between megatrends and the dynamics of change resulting from them.
It is still unclear how exactly these mobility patterns will combine and take shape in real-world settings around the globe. There is general agreement that the transportation conditions in dense, mixed, and complex metropolitan areas will result in specific market dynamics and notions of future mobility than those in rural and suburban residential and commercial locations. This represents a very visible difference from prior theories of a uniform worldwide car construction with vehicles that comply to universal standards [2]. The four main trends in the automotive sector are shown in figure 2.

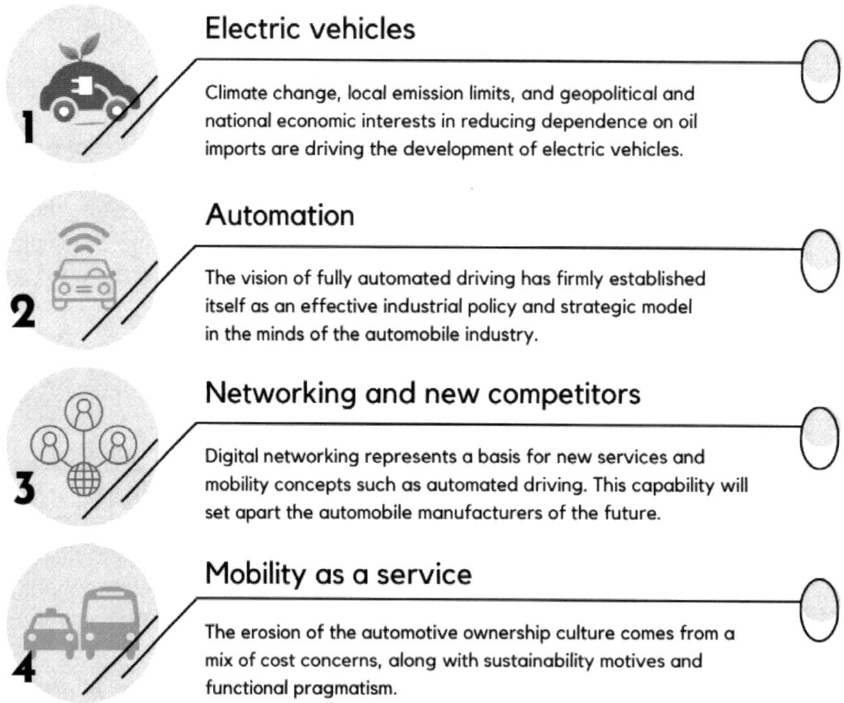


**Figure 2.** The four major trends in the automotive sector

## 3. Perspectives: cooperation, organization

The automotive industry is slowly moving towards a strategy based on cooperation, both with partners in other sectors and with those in sectors it is familiar with. The first category includes not only cooperation operations with partners in the IT sector, but also with municipalities. The second category refers to political factors and political entities at national and international level (for example, the national government, the European Union, China and so on).

Currently, three ideal types of corporate organization can be identified, in the future, of companies in the respective automotive sector [6]:

- B2C (business to consumer) car manufacturers: this type corresponds to the established, but moderately expanded, organizational concept of today's manufacturers. The key elements of the business will continue to focus on the development, production and sales of motor vehicles in the private and commercial fleet markets, together with a range of services. At customer level, the key benefit will be that the brand experience and the mobility package are offered from a single source in areas related to services related to the automotive industry (such as parking, refueling, health insurance, personal assistant systems).
- B2B (business to business) car manufacturers: this field involves the production of cars for mobility service providers. A B2B producer has no direct contact with the end customer.
- Mobility service providers: this ideal type focuses on the provision of mobility services. For customers, the main benefits are a wide range of integrated, intermodal mobility products with an increased comfort factor and increased flexibility in terms of availability, processing and payment.

## 4. Classic internal combustion engine propulsion in the current context of the automotive market

In the conditions of the radical transformation of the European automotive industry in recent years, the car market is increasingly polarized. The electrification of engines, connectivity and autonomous technology, as well as digitization along the production chain will lead to fundamental transformations of the way the entire automotive ecosystem works. That is why it is necessary to analyze the current state of the internal combustion engine in the car market and identify future perspectives for it.

According to the data provided by ACEA (Association des Constructeurs Européens d'Automobiles), the evolution of the number of vehicles produced in the European Union, respectively in Romania in 2020 and 2021, according to the type of fuel, is presented in table 1 and related graphs. Also, the difference between the values of the two years, in percentages, is illustrated.

**Table 1.** Number of vehicles produced in the EU and Romania by fuel type in 2020 and 2021 [7]

| Type of fuel | Number of units produced in the EU 2020 | 2021 | % of change 21/20 | Number of units produced in Romania 2020 | 2021 | % of change 21/20 |
|---|---|---|---|---|---|---|
| BEV | 538,734 | 878,432 | +63.1 | 2,845 | 6,342 | +122.9 |
| PHEV | 507,917 | 867,092 | +70.7 | 1,036 | 2,630 | +153.9 |
| HEV | 1,184,526 | 1,901,239 | +60.5 | 14,271 | 30,706 | +115.2 |
| NGV | 54,979 | 43,235 | -21.4 | 0 | 0 | - |
| LPG + Other | 153,549 | 226,702 | +47.6 | 9,278 | 15,952 | +71.9 |
| Petrol | 4,724,417 | 3,885,432 | -17.8 | 66,121 | 47,309 | -28.5 |
| Diesel | 2,776,665 | 1,901,191 | -31.5 | 33,834 | 19,206 | -43.2 |
| **Total** | **9,940,787** | **9,703,323** | | **127,385** | **122,145** | |

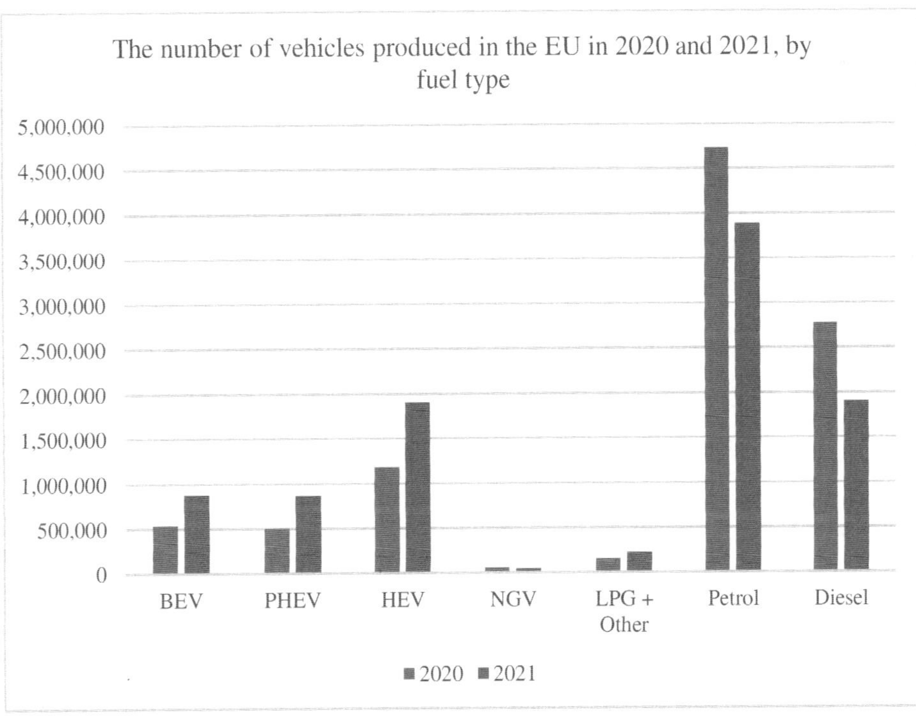


**Figure 3.** The number of vehicles produced in the EU in 2020 and 2021, by fuel type

From figure 3, it can be seen that the numbers of battery electric vehicles (BEV), plug-in hybrid electric vehicles (PHEV), hybrid electric vehicles (HEV) and other alternatively-powered vehicles (LPG + Other) are increasing in 2021 compared to the values of 2022, while the numbers of natural gas vehicles (NGV), petrol vehicles and diesel vehicles show a downward trend.

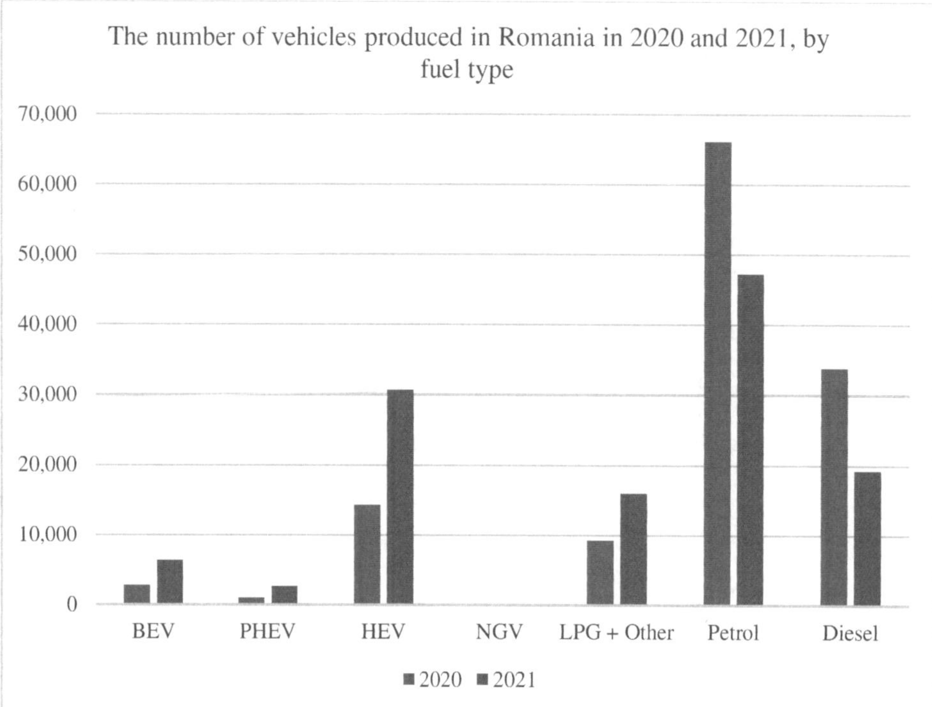

**Figure 4.** The number of vehicles produced in Romania in 2020 and 2021, by fuel type

From figure 4, it is noticeable that the values for 2020 and 2021 in Romania are in line with the trends at the EU level, with the exception of natural gas vehicles, which are not produced in our country. Thus, it can be seen that the numbers of battery electric vehicles (BEV), plug-in hybrid electric vehicles (PHEV), hybrid electric vehicles (HEV) and other alternatively-powered vehicles (LPG + Other) are increasing, while the numbers for petrol vehicles and diesel vehicles are in a steep decline.

## 5. Challenges for the internal combustion engine

### *5.1. Assessment of the current state of the internal combustion engine*

The internal combustion engine (ICE), which dominates traditional automobile powertrains, is facing more and more difficulties as the automotive industry comes under increasing pressure from energy consumption, environmental pollution, and climate change. In the meantime, internal combustion engine-powered vehicles face a serious challenge from battery electric vehicles (BEVs), which have developed significantly and have gained worldwide attention [8]. Is the use of internal combustion engines in transportation actually coming to an end? What are the prospects for transportation powertrain designs and internal combustion engines?

The technology potential of ICEs has been thoroughly researched from the angles of thermal efficiency and alternate fuels. Numerous eminent researchers and institutions worked hard to create cutting-edge low-emission and fuel-efficient engine technologies, and their findings showed that internal combustion engines' thermal efficiency is still far above their physical limit and that there is still plenty of room for improvement in terms of fuel savings [9] , [10], [11], [12].

Various powertrains have been compared in some research from different angles. These studies reached the conclusion that fossil fuels would continue to be the preferred energy source for the transportation industry. While fuel cell systems show promise for heavy-duty applications, battery electric powertrains are ideal for use in light-duty cars [13]. Hybrid cars also provide hope for reducing environmental impact without compromising the overall performance of the vehicle. Hybrid electric cars (HEVs) are suitable for the technological shift from conventional automobiles to new energy vehicles, despite the fact that new energy vehicles are not yet ready to compete with ICEVs. Electric motor coupling will make ICEs simpler in the future, but battery electric cars (BEVs) and fuel cell vehicles (FCVs) still have many obstacles to overcome [14].

Beyond the problems related to technical performance, the ecological aspect in the case of electric vehicles is not resolved. The construction of batteries and the source of electricity have a significant impact on the carbon emissions of electric vehicles. Additionally, the potential for major increases in human toxicity, freshwater eco-toxicity, metal depletion impacts, etc., is present for electric vehicles [15].

The „death theory" of ICEs was also refuted by a number of specialists and academics, who presented various arguments in support of ICEs' significance and prospective advancements. The editorial of the International Journal of Engine Research (2019), written by 37 specialists from the US, Japan, and Europe, is the most representative. In it, they emphasize the continued significance of ICEs in supplying the world's power and mobility demands [8]. The only reliable and economical energy source, particularly for developing nations, is fossil fuels, and since ICEs make up a substantial portion of the transportation infrastructure, replacing them would be extremely expensive and take many years. Additionally, potential alternatives—even those fueled by renewable energy sources like wind and solar energy—continue to have issues with pricing, electric range, security, and other topics currently. A number of professors also argued that the fight against climate change would suffer from a ban on ICEs. Combustion engines can be enhanced, but not majorly replaced as the primary source of power for transportation of people and cargo as well as mobile machinery. There will always be a demand for alternative synthetic and biogenic fuels for combustion engines (which are perfectly suited for these types of fuels) [16].

### *5.2. The three major challenges for the internal combustion engine*

- Higher environmental standards: to fulfill the new standard, ICEVs need to be equipped with more sophisticated postprocessing equipment, which significantly raises the cost [17].
- Stricter fuel consumption regulation: with more and more people owning cars nowadays, the amount of petroleum used in transportation is increasing constantly, and governments have kept up with updating their guidelines and restrictions to limit fuel consumption [18].
- New technologies: internal combustion engines are no longer the only choice for automobile powertrains because of the swift emergence of alternative energy and hybrid vehicle systems. Therefore, from the standpoint of the technology economy, using increasingly sophisticated internal combustion engines is no longer the best business option, regardless of whether ICEVs can meet future fuel consumption and environmental laws [19].

### *5.3. Potential of the internal combustion engine*

The promise of internal combustion engines should not be undervalued despite significant obstacles. It can be examined from the four directions outlined below:

- Ability to reach near-zero emissions: currently, ICEV technology can generate particle and nitrogen oxide emissions which are almost zero at an ever-increasingly cheap cost. ICEVs are anticipated to achieve zero pollution in the future by further enhancement of internal combustion engines through the electrification of the power system, the implementation of cutting-edge postprocessing technologies, and sustainable decarbonization strategies (such as $CO_2$ capturing and recycling technologies) [20]. In other words, it is possible to reduce pollution emissions to a level that does not harm the ecosystem or impair air quality.

- Carbon emissions not always higher than battery electric vehicles: in order to compare the carbon emissions of various cars, the entire lifecycle must be taken into account rather than just the usage phase. A supplementary energy source is electricity. Although electric vehicles emit no carbon during operation, the electricity comes from diverse sources, which results in varying degrees of carbon emissions. On the one hand, considering a vehicle's entire lifecycle, carbon emissions from electric vehicles with extended ranges are comparable to those of gasoline vehicles. Additionally, because batteries are manufactured with far higher carbon emissions than internal combustion engines, the carbon emissions of electric vehicles are much higher during the production stage than those of gasoline vehicles [21].
- Potential for improving thermal efficiency: engines' peak thermal efficiency is a key factor in lowering fuel use and carbon emissions. The highest peak thermal efficiency of gasoline engines in mass production has reportedly reached 41%, and in laboratory tests, it is higher than 50%, according to publicly available data from major automakers. Diesel engines' peak thermal efficiency in large scale production has reached 50%, and in lab tests, it is higher than 55%. According to research by the Japan Science and Technology Agency, a combination of exhaust gas recirculation (EGR), lean burn, lean boost, long stroke, low friction, high T/C efficiency, tumble + H2 addition, and heat insulation can increase a gasoline engine's peak effective thermal efficiency to 50.35%. By using multiple premixed compression ignition (MPCI), boost + EGR, combustion system optimization, and Miller cycle engines with low friction, low cooling heat loss, and high compression ratios (CR = 18), the peak effective thermal efficiency for compression ignition gasoline engines can be increased to 51.05 % [10]. As a result, there is still a lot of space for development in terms of fuel savings and internal combustion engines' thermal efficiency.
- Changes determined by the imminent electrification: the internal combustion engine had to operate in a broader range in the past to handle more challenging working circumstances. The optimal working area required to be increased in order to boost thermal efficiency, which also resulted in a large cost rise. Additionally, the ultimate thermal efficiency of internal combustion engines only takes into account the average value of each working mode rather than the highest thermal efficiency [12]. However, electrification can help the internal combustion engine achieve its full potential. Future powertrains will effectively merge the internal combustion engine and the electric motor. Because the ICE, battery, and motor will be connected, battery charging and discharging can be used to adjust for various working circumstances [22]. Consequently, internal combustion engines can always work in narrow and fixed thermal efficiency ranges, and they can even work only at the maximum thermal efficiency point [23].

## 6. Conclusions

In the conditions of the increasingly accentuated dynamism that the automotive industry has shown lately, it is necessary to analyze the current megatrends and their integration in the field of mobility. In this constantly updating landscape, the internal combustion engine is at a turning point. But despite the changes suffered by the trends in the automotive field, ICE has the potential of continuity on the market and, more than that, of improving the properties to ensure compliance with the legislative framework and a competitive advantage over other propulsion systems. Thus, as conclusions, a series of future opportunities for the internal combustion engine can be outlined.

First, the price of batteries and driving range are the biggest obstacles for electric vehicles. In the long run, the cost of batteries will be impacted by the link between the supply and demand of raw materials, which cannot be lowered in the short term. Second, there are still major impediments on the use of electric vehicles. These impediments are caused by the characteristics of the batteries and have proven challenging to fully solve. The performance of electric vehicles typically suffers when the temperature falls below -15°. The internal combustion engine, on the contrary, has strong flexibility and can function normally under harsh situations like extreme heat and cold. Lastly, the internal combustion engine already has a comprehensive industrial chain, a distinct division of labor, and significant accumulation after more than 100 years of development. Because of the internal combustion engine's strong industrial base and high level of dependability, it is challenging to replace it. Numerous studies and projections on the market shares of various powertrains have been made, and the majority of them have shown that plug-in electric vehicles (PEVs) will have a roughly 40% global market share in 2035 and a 50%–60% market share in 2050, while vehicles with internal combustion engines (ICE) as a partial or entire powertrain will still make up about 80% of the global passenger vehicle market in 2035 [22].

**References**

[1] Naisbitt J 1982 *Megatrends: Ten New Directions Transforming Our Lives*, Warner Books, New York, USA
[2] Bormann R, et al. 2018 *The future of the german automotive industry. Transformation by disaster or by design?*, Division for Economic and Social Policy, Bonn, Germany
[3] *** ITF/OECD 2017 *Transition to Shared Mobility: How Large Cities Can Deliver Inclusive Transport Services*, Studie des International Transport Forums der OECD, Paris, France
[4] *** Öko-Institut e.V. (Hrsg.) 2016 *Renewbility III: Optionen einer Dekarbonisierung des Verkehrssektors*, Berlin, Germany
[5] *** European Parliament 2022 *Amendments adopted by the European Parliament on 8 June 2022 on the proposal for a regulation of the European Parliament and of the Council amending Regulation (EU) 2019/631 as regards strengthening the CO2 emission performance standards for new passenger cars and new light commercial vehicles in line with the Union's increased climate ambition*, Bruxelles, Belgium
[6] Ramsauer H et al. 2017 *Transformation von Automobilunternehmen*, Norderstedt Books, Germany
[7] *** ACEA 2022 *Fuel types of new cars: battery electric 9.1%, hybrid 19.6% and petrol 40.0% market share full-year 2021* https://www.acea.auto/fuel-pc/fuel-types-of-new-cars-battery-electric-9-1-hybrid-19-6-and-petrol-40-0-market-share-full-year-2021/
[8] Reitz R D, Ogawa H, Payri R, et al. 2020 IJER editorial: The future of the internal combustion engine, *International Journal of Engine Research* **21**(1) 3-10
[9] Daw C S, Graves R L, Caton J A and Wagner R M 2010 *Summary Report on the Transportation Combustion Engine Effciency Colloquium Held at USCAR (No. ORNL/TM-2010/265)*, Oak Ridge National Laboratory, Oak Ridge, TN, USA
[10] *** Japan Science and Technology Agency 2019 *Achieving Thermal Efficiency of over 50% in Passenger Car Engines* https://www.jst.go.jp/EN/achievements/research/bt2019-04.html
[11] Farrell J T, Wagner R, Gaspar D and Moen C 2019 *Co-Optimization of Fuels & Engines: FY18 Year in Review (No. NREL/BR-5400-73282; DOE/GO-102019-5150)*, National Renewable Energy Lab (NREL), Golden, CO, USA
[12] Guzzella L and Onder C, 2009 *Introduction to Modeling and Control of Internal Combustion Engine Systems*, Springer Science & Business Media, Berlin/Heidelberg, Germany
[13] Yu X, Sandhu N S, Yang Z and Zheng M 2020 Suitability of energy sources for automotive application – A review, *Appl. Energy* **271** 115169
[14] Liu Z, Hao H, Cheng X and Zhao F 2018 Critical issues of energy effcient and new energy vehicles development in China, *Energy Policy* **115**, 92–97
[15] Hawkins T R, Singh B, Majeau-Bettez G and Strømman A H 2013 Comparative environmental life cycle assessment of conventional and electric vehicles, *J. Ind. Ecol.* **17** 53–64
[16] *** WKM 2017 *The Future of the Combustion Engine/Assessment of the Diesel Engine Situation* https://www.wkm-ev.de/images/20170708-englisch-Die-Zukunft-des-erbrennungsmotors.pdf

IOP Conf. Series: Materials Science and Engineering **1311** (2024) 012052 doi:10.1088/1757-899X/1311/1/012052

[17] Daisho Y 2010 Developing Advanced Low-Emission and Fuel-Effcient Vehicle Technologies Eyeing 2020 and beyond, *J. Automot. Energy Saf.* **1** 6–13
[18] Lipman T E 2017 Emerging technologies for higher fuel economy automobile standards, *Annu. Rev. Environ. Resour.* **42** 267–288
[19] Chen K, Zhao F, Hao H and Liu Z 2018 Synergistic Impacts of China's Subsidy Policy and New Energy Vehicle Credit Regulation on the Technological Development of Battery Electric Vehicles, *Energies* **11** 3193
[20] Zhao F 2006 Technologies for Near-Zero-Emission Gasoline-Powered Vehicles *(SAE Technical Paper vol 359)*, SAE International, Warrendale, PA, USA
[21] Qiao Q, Zhao F, Liu Z, Jiang S and Hao H 2017 Cradle-to-gate greenhouse gas emissions of battery electric and internal combustion engine vehicles in China, *Appl. Energy* **204** 1399–1411
[22] Zhao F, Chen K, Hao H and Liu Z 2020 Challenges, Potential and Opportunities for Internal Combustion Engines in China, *Sustainability* **12** 4955
[23] Moore T S 2001 *Dual Engine Hybrid Electric Vehicle*, U.S. Patent No 6306056

# Traffic analysis of different geometries of a junction for optimization of vehicle capacity and traffic safety increase

**M Lucian[1], D Ilie[2], L Racila[3], O Otat[4], A Oprica[5] and N Gencarau[6]**

[1, 2, 4, 5, 6] University of Craiova, Faculty of Mechanics, Department of Vehicle, Transportation and Industrial Engineering, Craiova, Romania
[3] University of Craiova, Faculty of Mechanics, Department of Applied Mechanics and Civil Constructions, Craiova, Romania

racila_laurentiu@yahoo.com

**Abstract**. Identifying various vehicle movement characteristics inside a network requires a careful assessment and different measurements technics, in regards to road traffic studies. These measures are influenced by a variety of variables, including the direction of traffic, changes in daily transportation demand, junction geometry and location in areal of study, style of road and also the size of the study area. Examining various geometries and arrangements of lanes with different distributions of vehicle platoons of the junction based on traffic-related elements is the goal of this essay. The analysis of the junction in regards to the speed, volume, capacity, pollution and traffic safety can give alternative solution and provide a foundation for the implementation of various political strategies in the area.

## 1. Introduction

In microscopic studies, the intensity and density of traffic are two factors that are looked at together with the behavior of an average group of automobiles [1], [2]. As a result, movement caused by vehicles and people concentrated on specific public road area will be characterized as road traffic and also traffic participants.

The following categories can be used to group the transportation challenges faced by drivers in merging European nations and also in their large cities:

- Requirement for more parking spaces
- Lack of transportation options;
- Unable to adapt the current street tram;
- Improper planning of the prior road transportation infrastructure without accounting for traffic patterns [3];
- Nonexistent data on these movements.

The purpose of this research study paper is to evaluate and compare the numerous sorts of junction settings and also rules and public configurations that might advantage from specific outputs, inclusive of speed, pollution [4], [5], [6]. Also, within this article the authors examine various transport policies and solutions that cannot be put to the field tests in the real world. The output of the study aims at mobility improvement also at increasing safety within the traffic participants (drivers and pedestrians).

## 2. Virtual network creation

The suggestions within the Romanian normative AND 557/2015, "Instructions for making the records of the road traffic on public roads" which was authorized by the Order of the Minister of Transportation No. 481/233.03.2015, were taken into consideration for the road traffic measurements [7], [8], [9].

In order to have as little as possible errors in the gathered data several techniques of evaluating the road traffic were used for movement of people and products, including. These techniques consist in intersections inquiries, O/D surveys and also of placement of special traffic metering video equipment at intersections [10], [11].

### *2.1. Virtual network design*

The virtual network that does the study of these paper was elaborated in Aimsun platform. Because the studied area was part of a larger project (a city traffic study), it was necessary to create a new virtual subnetwork. In this regard, the traffic network created for the entire project was used, figure 1, and then with the help of special tools from the traffic software we were able to create the new subnetwork that will be evaluated next in this paper.

**Figure 1.** Entire virtual network in Aimsun platform

The resulted virtual network after the utilization of the subnetwork command is the junction that we needed to analyze in order to give solutions for optimization of the traffic flow and safety of drivers and pedestrians. The initial subnetwork can be seen in the figure 2.

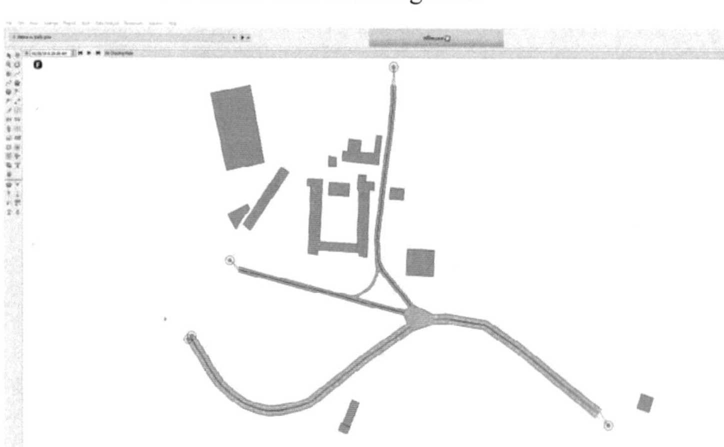

**Figure 2.** Subnetwork – initial junction

The subnetwork consists in 2 adjacent junctions that need to be analyzed as a whole. The virtual subnetwork system consists in sections of around 2 km, composed by lanes measuring almost 4 km. the whole subsystem is made by more than 12 sections and 3 intersections and accesses. It is important to mention that these characteristics will change accordingly to the modification of the junction in order to evaluate different geometries and structure.

### *2.2. Origin/Destination matrix creation*

The O/D matrix, is formed from the centroids of the whole network and concatenated into 5 demand data centroids that generating traffic into the system [12]. Complex iterations of original data and model are required to generate the complex origin-destination matrices used in the subnetwork. The result ca be seen in the figure 3 bellow.

A traffic simulation model will provide a more thorough analysis and, in the case of a microsimulation model, will also provide a vital visual depiction of traffic operation despite the fact that such models need a significant amount of time, work, and data to develop.

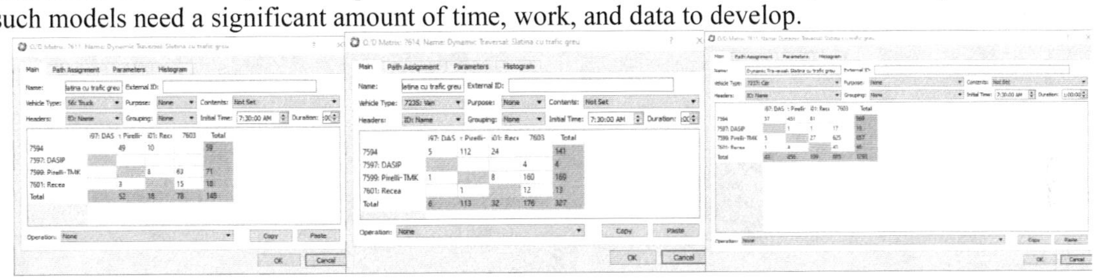

**Figure 3.** O/D matrix of the subnetwork

### *2.3. Traffic evaluation of different junction geometries*

The intersection area between Draganesti street, Ecaterina Teodoroiu Street and Recea Street presents certain problematic structures in geometry. These elements translate into a faulty dynamic of flows entering or leaving the 3 arteries, as well as a very problematic situation from the point of view of the safety of all traffic participants.

In order to increase safety at the level of the studied area, the simulation of 5 different geometries of modifying the street frame was carried out, namely:

1. The current situation of the intersection between the streets, figure 4a.
2. Roundabout 1 between Draganesti Street and Recea Street and a roundabout between the common flow coming out of roundabout 1 and Ecaterina Teodoroiu Street. At the same time, there will be a direct crossing in the west direction – it is, on Ecaterina Teodoroiu Street so that the flow is not dependent on roundabouts, figure 4b.
3. The intersection between Draganesti Street and Recea Street with the grouping of exit flows and entry in two lanes in one direction, following a roundabout at the meeting with Ecaterina Teodoroiu Street. At the same time, there will be a direct crossing in the west direction – it is, on Ecaterina Teodoroiu Street so that the flow is not dependent on roundabouts, figure 4c.
4. Roundabout between Draganesti Street and Recea Street and the intersection between the group of 2 lanes in the direction that exits and enters the roundabout with Ecaterina Teodoroiu Street. Ecaterina Teodoroiu Street having priority. At the same time, a brace will be made for the movement to the right from the roundabout to Slatina with the cession of the passage from the brace to the main artery, figure 5a.
5. Roundabout between Draganesti Street and Recea Street and the intersection between the group of 2 lanes in the direction that exits and enters the roundabout with Ecaterina Teodoroiu Street. Ecaterina Teodoroiu Street having priority. At the same time, a brace will be made for the movement to the right from the roundabout to Slatina with the cession of the passage from the brace to the main artery. In order to increase the safety of the road users, it will be done on the left movement, from Ecaterina Teodoroiu Street to Recea and from Recea to Ecaterina

IOP Conf. Series: Materials Science and Engineering **1311** (2024) 012053 doi:10.1088/1757-899X/1311/1/012053

Teodoroiu, the assembly braces of the road flows on the distance of at least 15 meters, figure 4b.

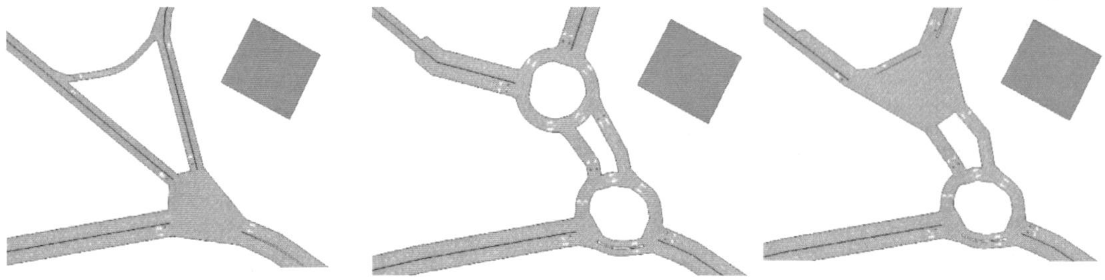

a b c

**Figure 4.** First set of geometries for the intersections

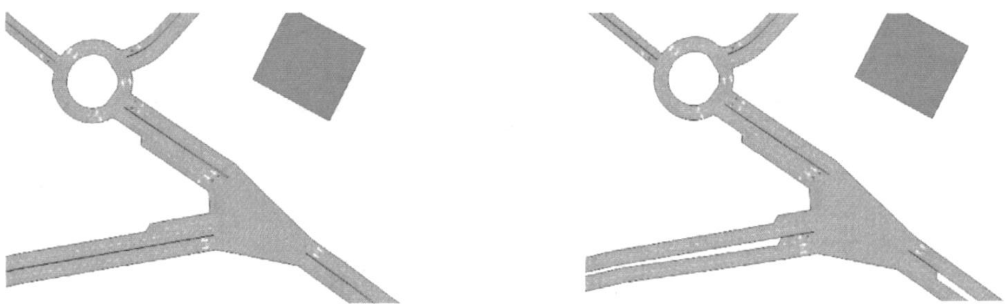

a b

**Figure 5.** Second set of geometries for the intersections

## 3. Simulation and results

Multiples types of strategies should be taken into account, regardless of the simulated method, when analyzing different types of geometries of the same junction. Therefore, a review of the essential components and outcomes of the Aimsun platform are required. Consequently, the simulation results were evaluated based on the following 4 criteria:

- Delay time
- Fuel consumption
- Chemical pollution, $CO_2$
- Chemical pollution, PM
- Driving Speed
- Stop time

In order to distinguish the differences between the simulations (the simulation parameters were kept identical in all 5 models), each geometry has assigned a distinct color so that each evaluation criterion can be evaluated properly.

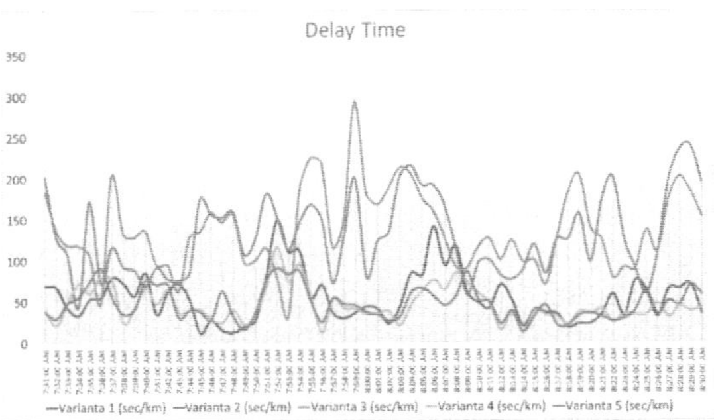


**Figure 6.** Delay time

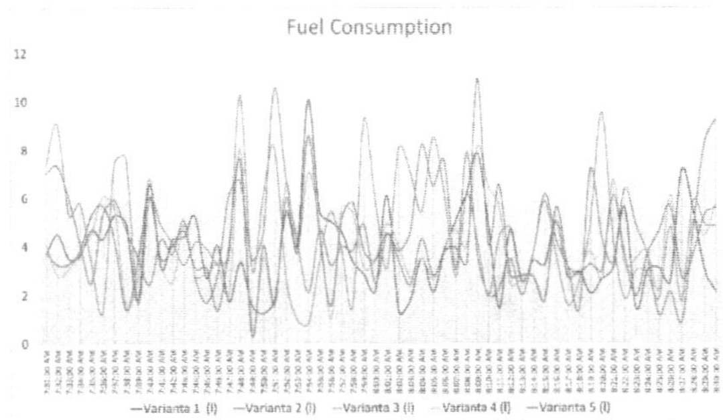


**Figure 7.** Fuel consumption

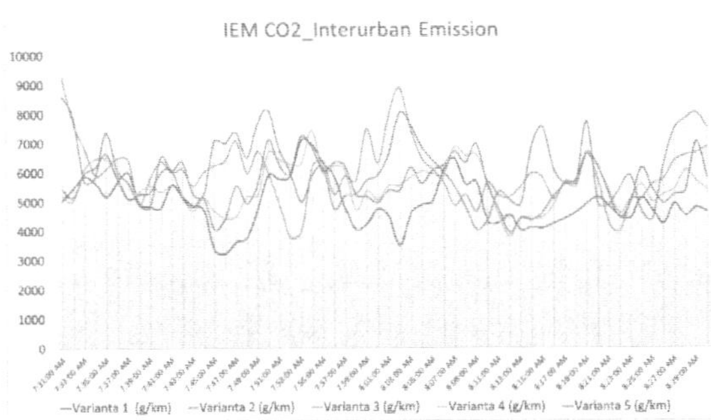


**Figure 8.** Chemical pollution, $CO_2$

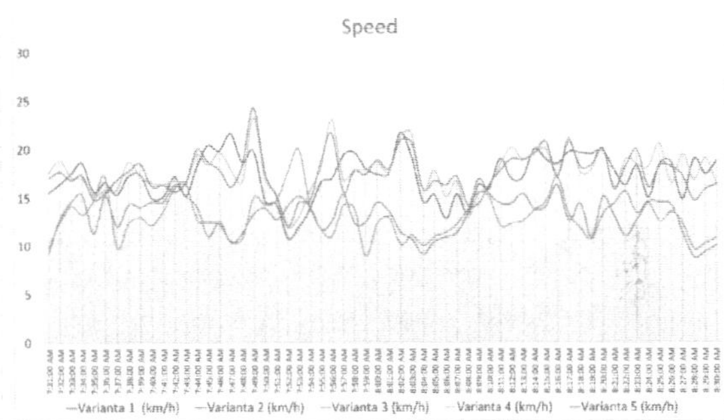


**Figure 9.** Driving Speed

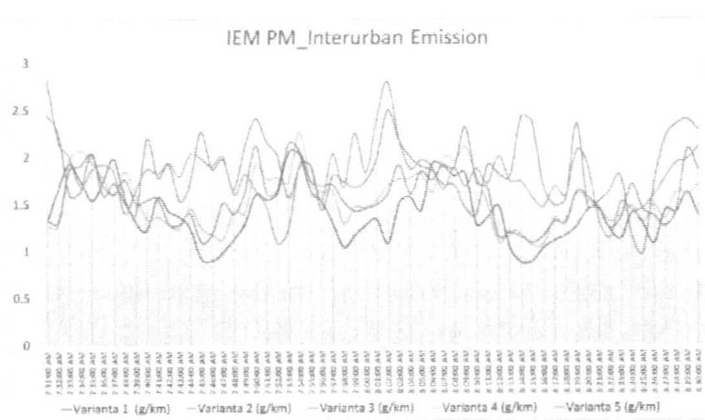


**Figure 10.** Chemical pollution, PM

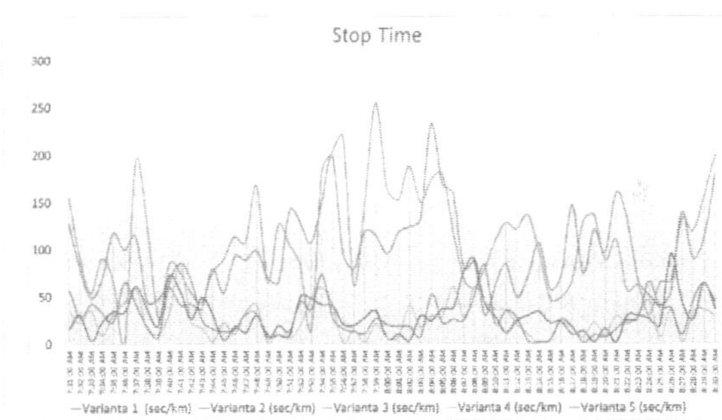


**Figure 11.** Stop time

## 4. Conclusion

It can be seen that the delay time is very high, on average 125 sec/km for variants 2 and 3 of geometric changes, figure 6. At the same time, it can be observed that the other variants fall within the area of 45 sec/km global delays at the level of the simulated network.

Fuel consumption and chemical pollution are maintained in about the same average values for all 5 variants of street tram geometries in the area of interest.

The speed in the area of interest can be seen to have an increasing trend, on average by 5 km/h for the 1, 4 and 5 variants. At the same time, you can see the much lower speed in the other variants, due to the roundabout on Ecaterina Teodoroiu Street, as well as to the phenomenon of calming the traffic and restoring the platoons in relation to the assessed geometry.

The stopping time increases considerably in the variants with a roundabout in the section of Ecaterina Teodoroiu Street. It can be seen that the downtime decreases by about 50% in variants 1,4 and 5, in relation to the simulated area.

IOP Conf. Series: Materials Science and Engineering **1311** (2024) 012053 doi:10.1088/1757-899X/1311/1/012053

**References**

[1] Moridpour, S., Majid, S., Geoff, R.: Lane changing models: A critical review. Transportation Letters: The International Journal of Transportation Research. 2., 10.3328/TL.2010.02.03.157-173, pp. 157-173, 2010.
[2] Virginia Tech Transportation Institute Cambridge Systematics, Inc.: Microscopic Analysis of Traffic Flow in Inclement Weather. Publication No. FHWA- JPO-09-066, 2009.
[3] Ilie Dumitru, Marian Valentin Zorilă, Răzvan Ştefan Ţolescu, Laurenţiu Racilă, Cristina Ileana Pascu, Alexandru Constantin Oprica, Daniela Vasilica Burghilă, Lucian Matei, Elena Janina Vîlcea, Cristina Popescu, Oana Badea-Voiculescu, Laurenţiu Mogoantă, Experimental model for the study of traumatic brain injury, Romanian journal of morphology and embryology, vol. 61, 2020
[4] Khan, Z. H., Aaron, T.: A macroscopic traffic model for traffic flow harmonization. European Transport Research Review. 10. 10.1186/s12544-018-0291-y, 2018.
[5] Corina Radu-Frent, Isabela Todirite, Dragos Capatina, Mihaiela Iliescu, Mechatronic system for increasing mobility of people with locomotor disability, The Romanian Journal of Technical Sciences. Applied Mechanics, vol. 66, 2021
[6] Gipps, P. G.: A model for the structure of lane-changing decisions. Transportation Research B 20(5), pp 403-414, 1986.
[7] Beles Horia, Tolea Bogdan Adrian, Crişan George Florin, Dogar Claudiu Alexandru, Ciotachiev Vlad Vasile. The assessment of pedestrian's head injury risk at the contact with the vehicle's hood. Annual Session of Scientific Papers "IMT ORADEA 2019" 30–31 May 2019, Oradea, , Romania. IOP Conf. Series: Materials Science and Engineering 568 (2019).
[8] Nicoleta GENCARAU, Theodor OPRICA, Oana OTĂT, Laurentiu RACILA, Lucian MATEI, Mihaela RACILA, Ilie DUMITRU, Alexandru OPRICA, ROAD TRAFFIC ANALISYS IN THE CONTEXT OF HEAVY TRAFIC TRANSPORT REMOVAL POLICY, Acta Techica Napocensis, Applied Mathematics, Mechanics, and Engineering, 2022
[9] Alexandru Cotfas, Ileana Dugaesescu, Cristina Nitu, Glencora-Maria Benec Mincu, Mihaiela Iliescu, Mechatronic system for increasing mobility of people with locomotor disability, The Romanian Journal of Technical Sciences. Applied Mechanics, vol. 66, 2021
[10] Dumitru I., Racilă L., Tutunea D., Matei, L., Dima A. and Oprica A., Experimental investigation of a vehicle behaviour using different complex data acquisition systems, CONAT 2016, 2016.
[11] Stewart, J. A., Aerde, M.V.: An Assessment of Adaptive Co-ordination of Traffic Signal offsets within integration, Traffic Engineering and Control, Volume 39, 1998.
[12] Dumitru I., Matei, L., Vinatoru M., Racila L., Oprica Th.: Aspects Regarding Priority Settings in Unsignalized Intersections and the Influence on the Level of Service, Proceedings of the European Automotive, Congress EAEC-ESFA, 2015.

# AUTHOR INDEX